Intermediate Dynamics

McGraw-Hill Series in Mechanical and Aerospace Engineering

Alciatore and Histand
Introduction to Mechatronics
and Measurement Systems

Anderson
Aircraft Performance and Design

Anderson
Computational Fluid Dynamics:
The Basics with Applications

Anderson
Fundamentals of Aerodynamics

Anderson
Introduction to Flight

Anderson
Modern Compressible Flow

Barber
Intermediate Mechanics of Materials

Baruh
Analytical Dynamics

Beer/Johnston
Vector Mechanics for Engineers

Beer/Johnston/DeWolf
Mechanics of Materials

Borman
Combustion Engineering

Borman and Ragland
Combustion Engineering

Budynas
Advanced Strength
and Applied Stress Analysis

Cengel and Boles
Thermodynamics:
An Engineering Approach

Cengel and Turner
Fundamentals of Thermal-Fluid
Sciences

Cengel
Heat Transfer: A Practical Approach

Cengel
Introduction to Thermodynamics
& Heat Transfer

Condoor
Mechanical Design Modeling
with ProENGINEER

Courtney
Mechanical Behavior of Materials

Curtis
Fundamentals of Aircraft
Structural Analysis

Dieter
Engineering Design: A Materials
& Processing Approach

Dieter
Mechanical Metallurgy

Doebelin
Measurement Systems: Application
& Design

Donaldson
Anaylsis of Aircraft Structures

Hamrock/Jacobson/Schmid
Fundamentals of Machine Elements

Henkel and Pense
Structure and Properties of Engineering
Material

Heywood
Internal Combustion Engine Fundamentals

Holman
Experimental Methods for Engineers

Holman
Heat Transfer

Hsu
MEMS & Microsystems: Manufacture
& Design

Hutton
Fundamentals of Finite Element Analysis

Hyer
Stress Analysis of Fiber-Reinforced
Composite Materials

Kays and Crawford
Convective Heat and Mass Transfer

Kelly
Fundamentals of Mechanical Vibrations

Kreider/Rabl/Curtiss
The Heating and Cooling of Buildings

Mattingly
Elements of Gas Turbine Propulsion

Meirovitch
Fundamentals of Vibrations

Nelson
Flight Stability and Automatic Control

Newman
Interactive Aerospace Engineering

Norton
Design of Machinery

Oosthuizen and Carscallen
Compressible Fluid Flow

Oosthuizen and Naylor
Introduction to Convective Heat Transfer
Analysis

Reddy
An Introduction to Finite Element Method

Ribando
Heat Transfer Tools

Schaffer et al.
The Science and Design of Engineering
Materials

Schey
Introduction to Manufacturing Processes

Schlichting
Boundary-Layer Theory

SDRC
I-DEAS Student Guide

Shames
Mechanics of Fluids

Shigley/Mischke/Budynas
Mechanical Engineering Design

Smith
Foundations of Materials Science
and Engineering

Stoecker
Design of Thermal Systems

Suryanarayana and Arici
Design and Simulation of Thermal Systems

Turns
An Introduction to Combustion: Concepts
and Applications

Ugural
Stresses in Plates and Shells

Ugural
Mechanical Design: An Integrated
Approach

Ullman
The Mechanical Design Process

Wark and Richards
Thermodynamics

Weisal
Spaceflight Dynamics

White
Fluid Mechanics

White
Viscous Fluid Flow

Zeid
CAD/CAM Theory and Practice

Intermediate Dynamics
Complemented with Simulations and Animations

M. R. M. Crespo da Silva
*Professor
Department of Mechanical, Aerospace,
and Nuclear Engineering
Rensselaer Polytechnic Institute*

Boston   Burr Ridge, IL   Dubuque, IA   Madison, WI   New York   San Francisco   St. Louis
Bangkok   Bogotá   Caracas   Kuala Lumpur   Lisbon   London   Madrid   Mexico City
Milan   Montreal   New Delhi   Santiago   Seoul   Singapore   Sydney   Taipei   Toronto

Higher Education

INTERMEDIATE DYNAMICS: COMPLEMENTED WITH SIMULATIONS AND ANIMATIONS

Published by McGraw-Hill, a business unit of The McGraw-Hill Companies, Inc., 1221 Avenue of the Americas, New York, NY 10020. Copyright © 2004 by The McGraw-Hill Companies, Inc. All rights reserved. No part of this publication may be reproduced or distributed in any form or by any means, or stored in a database or retrieval system, without the prior written consent of The McGraw-Hill Companies, Inc., including, but not limited to, in any network or other electronic storage or transmission, or broadcast for distance learning.

Some ancillaries, including electronic and print components, may not be available to customers outside the United States.

This book is printed on acid-free paper.

1 2 3 4 5 6 7 8 9 0 QPF/QPF 0 9 8 7 6 5 4 3

ISBN 0–07–285315–8

Publisher: *Elizabeth A. Jones*
Sponsoring editor: *Jonathan Plant*
Administrative assistant: *Rory Stein*
Marketing manager: *Sarah Martin*
Lead project manager: *Jill R. Peter*
Production supervisor: *Kara Kudronowicz*
Senior media project manager: *Stacy A. Patch*
Executive media technology producer: *Linda Meehan Avenarius*
Designer: *Rick D. Noel*
Cover designer: *John Rokusek Studios*
Cover illustrations by: *Marcelo R. M. Crespo da Silva*
Compositor: *Interactive Composition Corporation*
Typeface: *10.5/12 Times Roman*
Printer: *Quebecor World Fairfield, PA*

Library of Congress Cataloging-in-Publication Data

Crespo, da Silva, M. R. M.
    Intermediate dynamics: complemented with simulations and animations/Marcelo R. M. Crespo da Silva. — 1st ed.
        p.   cm.
    Includes bibliographical references and index.
    ISBN 0–07–285315–8
    1. Dynamics. I. Title.

QA845.D25   2004
531'.11—dc21
                                                    2003051273
                                                    CIP

www.mhhe.com

ABOUT THE AUTHOR

Marcelo R. M. Crespo da Silva Professor Crespo da Silva received his undergraduate engineering degree in 1963 from the *Universidade do Brasil* in Rio de Janeiro, Brazil (now, the *Universidade Federal do Rio de Janeiro*). After working in the electronics industry in Brazil during 1964 and 1965, he went to Stanford University, where he received his M.Sc. and Ph.D. degrees. His doctoral thesis dealt with satellite attitude dynamics. During 1969 and 1970 he was a National Research Council postdoctoral research associate at NASA Ames Research Center in California working on satellite dynamics. He joined the Rensselaer faculty in 1986 after rising through the ranks from assistant professor to professor at the University of Cincinnati, which he joined in 1971 as a faculty member in the aerospace engineering and engineering mechanics department. On a sabbatical leave during the first half of 1983 he continued research work on helicopter rotor blade dynamics at the Aeromechanics Laboratory at Ames Research Center. Professor Crespo da Silva's main interests are dynamics (including structural dynamics, satellite dynamics, and helicopter dynamics) and nonlinear oscillations. He has a number of publications in these areas and has received research grants from the National Aeronautics and Space Administration (NASA), the U.S. Air Force, the U.S. Army Research Office, and the National Science Foundation. He is a member of the American Academy of Mechanics, senior member of the American Institute of Aeronautics and Astronautics, and a fellow of the American Society of Mechanical Engineers.

CONTENTS

A Note to the Reader   x

Acknowledgments and Dedication   xvi

System of Units Used in the Book   xvi

Website for This Book   xvi

How to Use the Software on the Enclosed CD   xvi

Chapter 1
Essential Material for Dynamics   1

1.1 Vector and Scalar Quantities: A Brief Review   1
1.2 Transformation of Triads   4
1.3 Basic Operations with Vectors   7
1.4 Moment of a Vector about a Point   12
1.5 Newton's Laws of Motion   13
1.6 Kepler's Laws and Newton's Law of Universal Gravitation   15
1.7 Force of Attraction due to a Spherical Body   16
1.8 Gravitational Force, Weight, and the Meaning of Weightlessness   19
1.9 Equivalent Force Systems   20
1.10 Center of Gravity and Center of Mass   21
1.11 Free-Body Diagrams and Forces   27
1.12 Chain Rule of Differentiation and Exact Differentials   32
1.13 Some Basics on Differential Equations   35
1.14 Taylor Series Expansion and Linearization   48
1.15 Stability Analysis and Linearization   53
1.16 An Interesting Experiment   62
1.17 Problems   63

Chapter 2
Kinematics and Dynamics of Point Masses   68

2.1 How to Formulate a Dynamics Problem   69
2.2 A Note on the Reaction Force due to Springs   72
2.3 A Moment Equation Obtained from Newton's Second Law   73
2.4 Absolute Position, Velocity, and Acceleration of a Point   75
2.5 Solved Example Problems   77
2.6 Velocity and Acceleration Again and Angular Velocity   108
2.7 More Solved Example Problems   112
2.8 A Work-Energy Approach Obtained from Newton's Second Law   151
2.9 Problems   169

Chapter 3
Kinematic Analysis of Planar Mechanisms   179

3.1 Instantaneous Center of Rotation   179
3.2 The Four-Bar Mechanism   181
   3.2.1 General Analysis of the Four-Bar Mechanism   183
   3.2.2 Solution of Eqs. (3.2.1) and (3.2.2)   186
   3.2.3 Automated Solution for the Four-Bar Mechanism Using Simulink   193
3.3 The Geneva Wheel Mechanism   198
   3.3.1 Automated Solution for the Geneva Wheel Mechanism using Simulink   205
3.4 The Slider Crank Mechanism   208
   3.4.1 Automated Solution for the Slider Crank Mechanism Using Simulink   214

3.5 The Scotch Yoke Mechanism   218
3.6 Problems   220

Chapter 4
System of Point Masses: Special Dynamical Properties   225

4.1 Motion of the Center of Mass of a System of Point Masses   226
4.2 Time Rate of Change of Angular Momentum and Moment   227
4.3 An Integrated Form of Newton's Second Law and of the Moment Equation   229
4.4 Detailed Analysis of the Two-Body Problem   235
  4.4.1 *The Meaning of the Quantity E: Kinetic and Potential Energies of the Motion*   236
  4.4.2 *Determination of the Maximum and Minimum Values of $r(t)$*   237
  4.4.3 *Solution of the Orbital Differential Equation of Motion for r in Terms of θ*   239
  4.4.4 *Solved Example Problems*   243
4.5 Automated Solution for the Two-Body Problem   246
4.6 The Restricted Three-Body Problem of Space Mechanics   249
  4.6.1 *Differential Equations of Motion*   250
  4.6.2 *Equilibrium Solutions*   253
  4.6.3 *Stability of the Equilibrium Solutions*   256
4.7 Problems   260

Chapter 5
Dynamics of Rigid Bodies in "Simpler" Planar Motion   264

5.1 Degrees of Freedom of a Rigid Body   265
5.2 Types of Rigid Body Displacements   266
5.3 Dynamics of Simpler Planar Motion of a Rigid Body   268
5.4 Rolling of a Rigid Body   299

5.5 A Work-Energy Approach for Rigid Body Dynamics   311
5.6 Forces on Mechanisms   313
5.7 Problems   321

Chapter 6
Introduction to Dynamics of Rigid Bodies in General Motion   333

6.1 Juxtaposition of Two Vectors: Dyadics   334
6.2 The Two Basic Vector Equations of Rigid Body Dynamics   335
6.3 Rotational Kinematics of Rigid Bodies in General Motion   337
  6.3.1 *Absolute Angular Momentum*   337
  6.3.2 *The Parallel Axis Theorem*   342
  6.3.3 *Examples Involving Moments and Products of Inertia*   344
  6.3.4 *Principal Axes and Principal Moments of Inertia*   351
  6.3.5 *Examples of the Determination of Principal Axes and Principal Inertias*   353
6.4 Euler's Differential Equations of Rotational Motion   358
6.5 Kinetic Energy and the Work-Energy Approach   360
6.6 Classical Problems   362
  6.6.1 *Stability of a Body Spun about One of Its Principal Axes*   362
  6.6.2 *Dynamic Balancing of a Rotating Part of a Machine*   365
  6.6.3 *Motion of a Symmetric Top (or Gyroscope)*   371
  6.6.4 *On the Nutational Motion of a Symmetric Top*   380
6.7 Solved Example Problems   382
6.8 Angular Orientation of a Rigid Body: Orientation Angles   393
6.9 Angular Velocity and the Orientation Angles: Notes   408
6.10 Problems   410

Chapter 7
Introduction to Analytical Dynamics 417

- 7.1 Virtual Variations and Virtual Displacements 418
- 7.2 Virtual Work, Generalized Forces, and Constraint Forces 421
- 7.3 The Principles of d'Alembert and Virtual Work 432
- 7.4 Hamilton's Principle and Lagrange's Equation 433
  - 7.4.1 Hamilton's Principle 433
  - 7.4.2 Lagrange's Equation (Discrete Systems) Obtained from Hamilton's Principle 437
  - 7.4.3 General Form of the Kinetic Energy T 440
- 7.5 Conservation Principles: Classical Integrals of Lagrange's Equation 441
  - 7.5.1 What Is an Integral of Motion? 441
  - 7.5.2 Integrals of Motion Using Lagrange's Equation 441
  - 7.5.3 General Form of the Hamiltonian for Any Problem 442
- 7.6 A Brief Summary 445
- 7.7 Solved Example Problems 446
- 7.8 Constraints and Lagrange's Equation in the Presence of Constraints 463
  - 7.8.1 Constraints 463
  - 7.8.2 Lagrange's Equation in the Presence of Constraints 468
- 7.9 Solved Example Problems with Constraints 470
- 7.10 Problems 482

Chapter 8
Vibrations/Oscillations of Dynamical Systems 486

- 8.1 What Was Already Presented on Vibrations/Oscillations in Previous Chapters 486
- 8.2 Response of a Linear System to Sinusoidal Forcing Functions 487
- 8.3 Multiple-Degree-of-Freedom Linear Systems 491
- 8.4 Analysis of a Vibration Absorber Device 492
  - 8.4.1 Homogeneous Solution and Stability of the Equilibrium 493
  - 8.4.2 Particular Solution: Response to a Sinusoidal Force 494
- 8.5 Analysis of the Foucault Pendulum 495
  - 8.5.1 Problem Description 495
  - 8.5.2 Formulation Using Rectangular Coordinates 496

Appendix A
Simulink Tutorial and a Brief Introduction to MATLAB 502

- A.1 Introduction 502
- A.2 How to Use the Enclosed Software 507
- A.3 Description of Selected Simulink Blocks 508
- A.4 Special Simulink Blocks Prepared by the Author 513
- A.5 How to Construct a Simulink Model 514
- A.6 Invalid Loops in Simulink 527
- A.7 Brief Introduction to MATLAB 529

Appendix B
Notes on Sequential Rotations, Angular Velocity, and Acceleration 535

- B.1 Sequential Rotations and Angular Velocity 535
- B.2 A Note on the General Form of Acceleration Expressions 537

Appendix C
Properties of the Inertia Matrix of a Body 540

Appendix D
Suggested Computer Lab Assignments   545

Appendix E
Simulink and MATLAB Files on the CD   554

E.1 Simulink Models   555
E.2 MATLAB Files   558

Appendix F
Answers to Selected Problems   559

Appendix G
Some References for Advanced Studies   562

Index   563

A NOTE TO THE READER

Dynamics is a discipline that consists of the prediction and study of the motion of bodies under the influence of forces. It began to evolve as a rigorous scientific discipline based on solid mathematical principles during the seventeenth century when Sir Isaac Newton (1642–1727) published in July 1686 his famous book *Philosophiæ Naturalis Principia Mathematica*.[1] In that classical work he laid the foundations for dynamics by postulating his three famous laws of motion and his universal law of gravitation. The original work of Newton was directed to the motion of objects that were modeled as "point masses." A number of great scientists, including Leonhard Euler (1707–1783), Joseph Louis Lagrange (1736–1813), and William Rowan Hamilton (1805–1865), carried dynamics much further with their time-lasting contributions.

This book is aimed at an intermediate-level course in dynamics for college/university engineering students. It is also suitable for self-study or as a reference book. It is mostly a book on Newtonian dynamics, but, for completeness, a more advanced chapter (Chapter 7) covering the basics of *Lagrangian dynamics* is included for those readers who want to extend their capabilities in dynamics. As in the other chapters, many examples of application of the theory are also included in Chapter 7.

It is assumed that the reader has had courses in statics, calculus, and basic ordinary differential equations. A working knowledge of these subjects is a prerequisite for a solid understanding of dynamics. Hopefully, beginning-level engineering dynamics courses (which are generally taught in the second year of college) will evolve to be taught in a way to permit the usage, at that level, of this book or, at the very least, of the methodology presented here. The approach used is based on what an engineer actually needs to do to investigate dynamics problems.

This book stresses the use of *fundamentals* for setting up and solving each problem in dynamics, rather than the indiscriminate use of relatively elaborate "formulas" (which most humans will forget), the mere use of which does not contribute to the understanding of the subject matter.

A variety of solved problems is included in each chapter, and, for each problem, an organized approach to the formulation and analysis of the differential equations of motion of dynamical systems is presented. The methodology adopted here may be applied to a wide range of problems (from the very simple to the complex) encountered in engineering and in basic research. It is hoped that the repetitive approach used in this book will help eliminate the *mystery* that many students see in dynamics and the *fear* that many seem to have for the subject.

---

[1] *The Mathematical Principles of Natural Philosophy*. "Natural philosophy" is what is called "physical science" today.

Dynamics is a fascinating and powerful subject, and the author has been teaching it to university students, and working with it in his own research, for many years.

The differential equations that govern the motion of dynamical systems are generally nonlinear and cannot be solved analytically. In actual engineering practice, computers are used for integrating them numerically. The same is done in this book, especially for problems for which either the motion cannot be investigated analytically or advanced analytical methods that are beyond the scope of this book are needed for such investigations (such as "perturbation methods" for dealing with nonlinear oscillation problems, that are taught at the masters and doctoral levels). The implementation of the numerical integration of differential equations of motion is presented in the form of *block diagrams,* which are constructed simply by connecting appropriate blocks that perform some obvious operation to its input to generate an output. Examples are a time integration, which is done by an *integrator block;* the multiplication of the input by a desired constant, which is done by a *gain block,* etc. Inputs and outputs of blocks are represented by arrows into and out of the block. There is a software called Simulink that is actually based on fully operational block diagrams. It is the block diagram implementation of MATLAB,[2] and both software are widely used at many universities and companies worldwide. Simulink block diagram *models* are portable between different computers. Because of its simplicity, and because it does not involve the writing of computer programs, Simulink is the software used with this book for numerically integrating differential equations and for displaying solutions. Displays, such as plots, are simply obtained by including in a block diagram a block called a *scope* (which stands for "oscilloscope"). A tutorial is presented in Appendix A.

Although Simulink is widely used in this book because of its flexibility and ease of use, and mostly because preparing block diagrams is much less involved than writing computer programs and/or calling functions from a mathematical package for performing the same task, it must be emphasized again that this is a book on dynamics. As such, its main contents, which is dynamics and analysis of motion, should not be viewed as tied to any particular software. Readers and instructors using this book can, of course, use any other appropriate software of their choosing for performing the numerical tasks presented in this book, writing programs using their preferred language.

Several *animation MATLAB* files that can be simply added as another block to a number of Simulink block diagram models accompany this book. They were prepared by the author. Such blocks display, on the computer screen, drawings that resemble certain systems in order to impart to them a motion *governed by the differential equations that were formulated for the system* using the material presented in this book. Two special examples of such blocks are the `animate_nbars` and `animate3d_nbars` blocks. Both can be used in a variety of models (including point mass problems) that consist of an arbitrary number of interconnected straight bars. The `animate_nbars` block is for two-dimensional motion, and the `animate3d_nbars` block animates motion in

---

[2] Simulink and MATLAB are registered trademark of The MathWorks, Inc.

three-dimensional space. These blocks are described in Appendix A. Such a visualization capability is very helpful for better understanding the resulting motion governed by the differential equations of motion that are formulated for an object to model its dynamic behavior and to perform a number of numerical and visual experiments using such equations. By using this capability you will be able to change parameters and initial conditions in the differential equations of motion that you formulate and see the effect such changes have on the object's behavior. In engineering practice, such "numerical experimentations" are used in design involving dynamical systems and in research work.

A formidable combination for engineers, physicists, and others, for analyzing dynamical systems should have the following three ingredients, all three of which are part of this book:

1. Knowledge of dynamics for formulating the differential equations of motion of a system
2. Knowledge of approximate analytical methods for yielding essential information about the response of the system
3. Use of numerical software as a tool for solving the differential equations that govern the motion, and even for animating the motion

> The CD that comes with this book contains a number of Simulink models, animation blocks, and other files. They are listed and described in Appendix E. To use them, follow the instructions in Section A.2.

Chapter 1 presents essential material that is of fundamental importance for the study of subsequent chapters. It also introduces the notation used throughout the book. It is assumed that you have already learned most of that material from other courses and that you are familiar with it. However, you should read it carefully and study it thoroughly, even if some of the material may appear, at the beginning, to be too basic and simple to you at that stage of your reading. Not doing so may cause you difficulties in appreciating and understanding dynamics and the rest of the material in this book when you proceed to more advanced chapters. The same comment applies to the basic material presented at the beginning of the second chapter. Some of the material in Sections 1.12 to 1.15 of Chapter 1 may be new to you, though. If it is, you should study it more carefully since that material is essential for understanding the behavior of dynamical systems and for analyzing their motion.

> As you feel the material presented becomes progressively more complex, you should read it several times as your need dictates. As you read it again, make notes as you go along, put it aside for a little while and think about it, and then come back and study it again until things start to make more sense to you. This is actually unspoken commonsense. As you progress through the subsequent chapters, your patience and perseverance should start paying dividends to you. Hopefully, some of you will also become dynamicists and will start contributing to the dynamics literature yourselves.

The material presented progresses in a natural manner from kinematics and dynamics of a point mass (Chapter 2) to a system of point masses (Chapter 4), and then to continuum rigid bodies, which are bodies with an infinite number of point masses that are constrained so that the distance between any two of their points is constant.

Unlike in other books, kinematics and dynamics are not separated into different chapters, except in Chapter 3, which deals only with kinematic analysis of mechanisms whose motion takes place either in a plane or in parallel planes. In that chapter, the motion of one of the parts of the mechanism is specified without any concern for determining the external forces that cause the motion. The material in Chapter 3 is a natural extension of the application of the analysis involving velocity and acceleration of a point, and also the use of rotating unit vectors, that first appears in Chapter 2.

The first two sections of Chapter 4 deal with important properties of the motion of a system of point masses. Such properties involve the motion of the center of mass of the system, and the relation involving the moment of the forces applied to the system and the time rate of change of angular momentum of the system. The concept of angular momentum for a system of particles is introduced in that chapter. Chapter 4 constitutes a transition to the study of the motion of rigid bodies. In addition, two important classical problems in celestial mechanics and spacecraft trajectory dynamics are treated in reasonable detail in Chapter 4. Thus, by the time you have studied the material up to that chapter, you will have analyzed important problems in mechanical engineering and in aerospace engineering and been introduced to applications of the fundamentals of vibration analysis that were first presented in Chapter 1.

Chapter 5 deals with dynamics of rigid bodies in "simpler" planar motion so that the analysis of rotational motion involves only one moment of inertia of the body. Such motions are defined on p. 266 in that chapter. Dynamics of rigid bodies that are able to move in any manner in three-dimensional space is presented in Chapter 6, all at an intermediate level of sophistication. Examples 6.8.3 and 6.8.4 in Section 6.8 deal with important problems in spacecraft dynamics and further introduce you to that field of study. They are slightly more advanced but are still appropriate for an intermediate dynamics course.

Chapter 7 is a more advanced chapter that covers the basics of Lagrangian dynamics and is included in this book for those readers who want to extend their capabilities in dynamics. At least some of that material is covered in intermediate dynamics courses at several universities and colleges, especially when it comes to using the Lagrangian formulation for obtaining the differential equations of motion of dynamical systems. In contrast with the Newtonian methodology, the Lagrangian methodology does not involve the formulation of acceleration. Its fundamental quantities are the kinetic energy of the motion and the *virtual work* (which is defined in that chapter) done by the forces applied to the system, and both are scalar quantities. The Lagrangian methodology is actually simpler *to apply* to problems than the Newtonian methodology, and *classical integrals of the motion* (or *constants of the motion*) are easily obtained (when they exist, of course!) using that methodology. Free-body diagrams are rarely needed for working with the Lagrangian methodology. As in other chapters, many examples of the application of the methodology are also included in Chapter 7.

The subject of vibrations/oscillations is introduced throughout the book, starting in Chapter 1, as a natural application of "investigation of motion." Equilibrium solutions, linearization for small motions about any equilibrium, and the concept of stability are also presented in that chapter. Several examples throughout the book deal with the motion of dynamical systems, both linear and nonlinear, frequently accompanied by numerical integration of the differential equations that govern the motion.

Chapter 8 is short and complements the rest of the book with additional material on vibrations/oscillations, including the analysis of the response of a linear system to a sinusoidal forcing function, and a very brief introduction to the homogeneous response of multiple-degree-of-freedom linear systems. Two classical problems are presented in Chapter 8, namely, the analysis of the *Foucault pendulum* and the analysis of a *vibration absorber*. The Foucault pendulum demonstrates the effect of the rotation of the earth on the motion of a simple pendulum. It also brings into evidence when one can reasonably neglect such an effect and when one must take it into consideration when motion relative to a rotating planet as viewed by an observer on that planet is observed for a longer time. Vibration absorbers are practical devices used in engineering, and the one considered in Chapter 8 is one particular type of such device.

The book has several appendices. The first one, Appendix A, is a Simulink tutorial and a brief introduction to MATLAB. The best time to read that material is when you encounter the use of Simulink in Chapter 1. The use of the animation blocks mentioned earlier is also described in Appendix A. A third animation block, for use with the two-body problem of space mechanics, is presented in that appendix and described in Chapter 4. Appendices B and C consist of important material, but that material is somewhat more advanced and is optional. A suggested sample of computer problems, using numerical integration with Simulink and that can be used in formal courses, are included in Appendix D. Appendix E consists of the listing of the files included on the CD that accompanies the book, with a brief explanation about them. Answers to some selected problems are given in Appendix F. Several references for more advanced studies are listed in Appendix G.

The author has been extensively using the material in this book in two intermediate dynamics one-semester courses at Rensselaer taken by mechanical engineering seniors and by aerospace engineering seniors. Graduate students have also frequently taken them. The material in Chapter 1, parts of Chapter 2, the first two sections of Chapter 4, and Appendices A and B is common to both courses. The material in Appendix A is given in two or three hands-on tutorial classes (for a total of 3 hours or so) with the students having access to a computer. The material in Chapter 3 and parts of Chapters 5, 6, and 8 is covered in the mechanical engineering course. The material in Chapter 4 and parts of Chapters 5, 6, and 8 is covered in the aerospace engineering course. The material in Chapter 7 has been optional for some more advanced students and has been included as an additional assignment to a number of graduate students. Home assignments involving numerical integration of differential equations of motion formulated by the students has been an integral part of the learning experience in such courses. Lagrange's equation is also used in other senior-level courses. This book allows instructors to design their own course contents using selected parts of it.

Marcelo R. M. Crespo da Silva

ACKNOWLEDGMENTS AND DEDICATION

I am grateful to several anonymous reviewers for their useful suggestions and to Jonathan Plant, senior sponsoring editor at McGraw-Hill, for his interest in this book. Several students at Rensselaer provided me with very useful comments. In particular, I express my sincere appreciation to Mark Jensenius for checking most of my solutions to the proposed problems at the end of the chapters. A special thank you goes to Mrs. Harriet Borton, at Rensselaer, for her expert advice every time I could not solve a LaTeX problem on my own while typesetting the original manuscript, and to Ms. Elizabeth Doocey, from my department at Rensselaer, for her cheerful and expert answers to several of my editorial questions. I am also grateful to Ms. Courtney Esposito at The Mathworks, Inc., for her assistance and pivotal role in securing permission for me to include on the enclosed CD the *Auto-Scale Graph* scope block mentioned in Appendix A and two of the files mentioned in Section E.2. The entire book team did an excellent job in the production of this book and I sincerely appreciate their efforts.

I made many personal sacrifices while writing this book, and I am glad my cheerful wife Linda d'Escoffier Crespo da Silva has been at my side and that I have her support. I dedicate this book to her.

Marcelo R. M. Crespo da Silva

SYSTEM OF UNITS USED IN THE BOOK

This book uses the MKS (meter-kilogram-second) International System of units. As per standard convention, these units are abbreviated as m, kg, and s, respectively. The unit of force in this system, which is the newton, is abbreviated, in accordance with international convention, as N, where $1\text{ N} = 1\text{ kg} \cdot \text{m/s}^2$.

In some of the problems, lengths are also quoted either in centimeters (cm) or in millimeters (mm), and mass in grams (g).

In some examples, especially those involving space applications in Chapter 4, the equivalent distance in miles is also given after the distance is quoted in kilometers (km). In such cases, velocity may be quoted in both kilometers per hour (km/h) and in miles per hour (mi/h).

WEBSITE FOR THIS BOOK

A website for the book is available at *www.mhhe.com/crespodasilva*. This site contains a complete instructor's guide with solutions for all chapter problems, and further information on using the book and software for typical intermediate-level dynamics courses. Instructors using the text need to receive password information from their McGraw-Hill sales representative to access this protected information.

HOW TO USE THE SOFTWARE ON THE ENCLOSED CD

Follow the instructions in Section A.2 on p. 507. A list of all the files, and their descriptions, is given in Appendix E.

CHAPTER 1

Essential Material for Dynamics

The fundamental quantities that appear in Newtonian dynamics for analyzing the motion of a point mass (or mass particle[1]) are the *absolute acceleration* of the particle and the *resultant force acting on the particle*, and both are vector quantities. The absolute acceleration is, by definition, a vector that is equal to the time derivative of the *absolute velocity* of the particle (which is, in turn, the time derivative of the *absolute position vector* for the particle). The resultant force is determined directly from the free-body diagram for the particle, which is a diagram showing all the individual forces acting on the particle. These fundamental quantities, and the meaning of the adjective *absolute*, are introduced in Chapter 2.

This chapter presents the basic material that is essential in the study of dynamics. It includes a basic review of vectors and differential equations. Reviews of free-body diagrams, Newton's laws of motion, and Newton's law of universal gravitation are also included in this chapter. Details of the calculation of the gravitational force acting on a body are presented in Section 1.8, together with a discussion of the meaning of *weightlessness*.

1.1 Vector and Scalar Quantities: A Brief Review

A *scalar* quantity is one that can be represented by a number, a variable, or a function of one or more variables. Examples of scalar quantities are the mass, temperature, energy of a body and the speed of a car. Such quantities may be functions of a scalar variable (or variables), such as time. In such a case, one has a *scalar function*.

A *vector* is a quantity that has a magnitude (i.e., a "length") as well as a direction. In addition, to qualify as vectors, such quantities must obey the *parallelogram rule of addition* and be *commutative* (i.e., the result of the sum

---

[1] A *mass particle,* which is simply referred to as a *particle,* is an idealized point having a constant mass.

of two vectors must be independent of the order of summation). For example, to locate a point B relative to a point A in space one needs to specify not only the distance from A to B but also the direction of the line segment from A to B. The vector (which specifies distance and direction) that indicates the location of a point B relative to a point A is called the *position vector* of B relative to A. Additional examples of vectors are the velocity and acceleration of a point and the force acting on a body. Notice that, contrary to everyday speech, there is a distinction in dynamics between the velocity and speed of a point. *Speed* is a scalar quantity, always positive, that is equal to the magnitude of the velocity vector. The *velocity* of a point is a vector and, as such, contains information about the direction of the motion of the point.

In books and technical journals, vectors are commonly represented by boldface letters (e.g., **A**, **v**), but in this book they are represented by a letter (or by two letters written side by side) with an arrow on top (e.g., $\vec{r}$, $\vec{v}$, $\overrightarrow{AB}$). The representation adopted here is the one everyone uses either on their own or in classrooms because of the obvious inconvenience of using boldface symbols in those situations.

A *unit vector* is a vector of unit magnitude and no dimensions (i.e., no units). A unit vector is represented in this book by a symbol with a caret on top (e.g., $\hat{r}$, $\hat{e}_1$). Clearly, for any vector $\vec{r}$, whose magnitude is $|\vec{r}|$, the vector $\vec{r}/|\vec{r}|$ is the same as the unit vector $\hat{r}$ in the direction of $\vec{r}$.

Any vector $\vec{r}$ in three-dimensional space may be expressed in terms of any three independent basis vectors. However, it is most convenient to choose the basis vectors to be three orthogonal unit vectors, each one of which is chosen simply to indicate a reference direction in space. Such sets of unit vectors are called *unit vector triads,* or simply *triads*. For example, the vector $\vec{r} = \overrightarrow{AB}$ shown in Fig. 1.1 may be represented in the $\{\hat{x}, \hat{y}, \hat{z}\}$ triad (also shown in Fig. 1.1) as $\vec{r} = x\hat{x} + y\hat{y} + z\hat{z}$, where x, y, and z are distances measured from point A, along the three orthogonal axes that are parallel to the $\hat{x}$, $\hat{y}$, and $\hat{z}$ unit vectors shown in the figure.

Figure 1.1 An arbitrary vector $\vec{r}$.

SECTION 1.1 Vector and Scalar Quantities: A Brief Review

The unit vector triad $\{\hat{x}, \hat{y}, \hat{z}\}$ shown in Fig. 1.1 may be rotating in space, as would be the case if it were fixed to a maneuvering aircraft or to a rotating part in a machine, for example. The quantities x, y, and z are called the *scalar components* of the vector $\vec{r}$ along the directions defined by $\hat{x}$, $\hat{y}$, and $\hat{z}$, respectively, while the quantities $x\hat{x}$, $y\hat{y}$, and $z\hat{z}$ are called the *vector components* of $\vec{r}$. Notice that the magnitude of any vector $\vec{r}$ written in terms of three orthogonal unit vectors $\hat{x}$, $\hat{y}$, and $\hat{z}$ as $\vec{r} = x\hat{x} + y\hat{y} + z\hat{z}$ is, by the Pythagorean theorem, equal to the square root of the sum of the squares of its scalar components, i.e., $|\vec{r}| = \sqrt{x^2 + y^2 + z^2}$.

The scalar components of a vector may be either positive or negative and are often also referred to as the *measure numbers* of $\vec{r}$ in the triad that is used for expressing $\vec{r}$. The same vector $\vec{r}$ may also always be expressed in terms of a single component as $\vec{r} = r\hat{r}$, where $\hat{r}$ is a unit vector in the direction of the vector $\vec{r}$ itself. If the scalar component r is positive, it is the same as the magnitude $|\vec{r}|$ of the vector $\vec{r}$. If the unit vector $\hat{r}$ were chosen in the direction opposite to the vector $\vec{r} = r\hat{r}$, the quantity r would be negative (and, of course, would be the negative of the magnitude of $\vec{r}$).

You should keep in mind that the choice of unit vectors used for writing down an expression for a vector is simply that, an individual *choice*. Different persons will, in general, choose different unit vector triads to use, and the directions of the three unit vectors of the chosen triad may be either fixed in space or rotating. Different choices of unit vectors simply lead to different expressions for the same vector $\vec{r}$. Some judicious choices of unit vectors lead to simpler and more convenient expressions for a desired vector $\vec{r}$, and, as will be seen later, to simpler expressions for the derivatives of $\vec{r}$ when one needs to work with them.

In summary, to write down an expression for a vector, you need to choose a set of unit vectors to work with. But, that is not enough. You also need to choose a set of scalar variables to represent the components of the vector. In the previous example, where the expression written to represent a vector $\vec{r}$ was $\vec{r} = x\hat{x} + y\hat{y} + z\hat{z}$, the variables chosen for such representation were the three scalar quantities x, y, and z, which were distances, from the chosen reference point A shown in Fig. 1.1, measured along three perpendicular axes. Such variables are called *rectangular coordinates* (or *Cartesian coordinates*).

In dynamics, the variables that are chosen to describe a motion are frequently called the *motion coordinates*. In the expression $\vec{r} = x\hat{x} + y\hat{y} + z\hat{z}$, for the position vector of point B relative to point A, the motion coordinates are the distances x, y, and z shown in Fig. 1.1.

There are a number of coordinates one can use for expressing a vector. They include Cartesian, polar, cylindrical, and spherical coordinates, or any combination of them. *Spherical coordinates* consist of the magnitude $|\vec{r}|$ of the vector $\vec{r}$ (which we will denote here as r) and the angles θ_1 and θ_2 shown in Fig. 1.1. In terms of such variables, the coordinates x, y, and z shown in that figure are simply obtained by projecting the vector $\vec{r}$ along the $\hat{x}$, $\hat{y}$, and $\hat{z}$ directions, respectively. By looking at Fig. 1.1, the following expressions can then be written by inspection of the figure: $x = (r \cos \theta_2) \cos \theta_1$, $y = r \sin \theta_2$, and $z = (r \cos \theta_2) \sin \theta_1$. If we still use the unit vector triad $\{\hat{x}, \hat{y}, \hat{z}\}$, the expression for the vector $\vec{r}$ shown in

Fig. 1.1 is then

$$\vec{r} = r(\cos\theta_2 \cos\theta_1)\hat{x} + r(\cos\theta_2 \sin\theta_1)\hat{z} + r(\sin\theta_2)\hat{y} \qquad (1.1.1)$$

Cylindrical coordinates consist of the two distances $|\overrightarrow{AD}|$ and $y = |\overrightarrow{DB}|$ and the angle θ_1 shown in Fig. 1.1. Using the unit vectors $\{\hat{x}, \hat{y}, \hat{z}\}$, the expression for $\vec{r}$ in cylindrical coordinates is (see Fig. 1.1)

$$\vec{r} = |\overrightarrow{AD}|(\cos\theta_1)\hat{x} + |\overrightarrow{AD}|(\sin\theta_1)\hat{z} + y\hat{y} \qquad (1.1.2)$$

> The choice of the coordinates (such as rectangular, cylindrical, or spherical) to formulate any dynamics problem and of the unit vectors used for expressing the vectors that appear in the formulation is based on convenience for each problem under consideration.

Polar coordinates involve a distance and an angle and, therefore, appear in two-dimensional problems. In the illustration shown in Fig. 1.1, one would have a two-dimensional problem when either θ_1 or θ_2 is constant for all times. When θ_1 is constant, the polar coordinates that specify the position of point B relative to point A (which is simply the vector $\vec{r} = \overrightarrow{AB}$) in Fig. 1.1 are the distance from A to B and the angle θ_2. Conversely, when θ_2 is constant, the polar coordinates that specify the position of point D relative to point A (which is the vector $\overrightarrow{AD}$) in the same figure are the distance from A to D and the angle θ_1.

As mentioned earlier, different choices of unit vectors may certainly be made for writing down an expression for a vector. Another (among many others) perfectly valid expression for the vector $\vec{r}$ shown in Fig. 1.1, in which the unit vector triad $\{\hat{u}_1, \hat{u}_2, \hat{y} = \hat{u}_2 \times \hat{u}_1\}$ is used, is (see Fig. 1.1)

$$\vec{r} = \underbrace{r(\cos\theta_2)\hat{u}_1}_{\text{This is } \overrightarrow{AD}} + \underbrace{r(\sin\theta_2)\hat{y}}_{\text{This is } \overrightarrow{DB}} \qquad (1.1.3)$$

In Fig. 1.1, the unit vector $\hat{u}_1$ was chosen to be parallel to the projection $\overrightarrow{AD}$ of $\vec{r}$ on the x-z plane, and directed from A to D. The angle θ_1 in that figure specifies the direction of $\hat{u}_1$ relative to the $\hat{x}$ direction. As will be seen in later chapters where more advanced material and a number of solved example problems are presented, the considerations just discussed will play an important role in problems involving the formulation of the equations that govern the motion of any dynamical system.

1.2 Transformation of Triads

As seen in Section 1.1, the representation of any vector $\vec{r}$ in terms of a set of orthogonal unit vectors is not unique since any set of three orthogonal unit vectors (or a *triad*) may be chosen to represent $\vec{r}$. The complexity (or simplicity) of the expression for a vector $\vec{r}$ in terms of any chosen triad (whose unit vectors you name, say, $\hat{x}, \hat{y}, \hat{z}$) will depend on the particular spatial orientation of the chosen triad. The simplest representation of a vector $\vec{r}$ is obtained by writing $\vec{r} = r\hat{r}$. In such a case, the unit vector $\hat{r}$, itself, may be thought of as being one of the three unit vectors of some orthogonal triad. When necessary, to avoid confusion and to convey to the reader which triad some vector $\vec{r}$ is being represented in, each scalar component of $\vec{r}$ will be identified with a subscript indicating the unit vector that component is associated with. In such cases, the same symbol that is used for identifying the vector in question will be used in this book for identifying all its components. For example, consider a vector $\vec{r}$ and a triad $\{\hat{x}_1, \hat{x}_2, \hat{x}_3\}$ that was chosen (at this point, for no special reason at all) to represent $\vec{r}$. The expression for $\vec{r}$ in this triad might then be written as $\vec{r} = r_{x_1}\hat{x}_1 + r_{x_2}\hat{x}_2 + r_{x_3}\hat{x}_3$ (or even, $\vec{r} = r_1\hat{x}_1 + r_2\hat{x}_2 + r_3\hat{x}_3$ if there were no other triads for which the subscripts 1, 2, and 3 were used). In terms of another triad $\{\hat{n}, \hat{t}, \hat{z}\}$, for example, one might write $\vec{r} = r_n\hat{n} + r_t\hat{t} + r_z\hat{z}$ for the same vector $\vec{r}$.

The elements of a triad can be expressed in terms of the elements of any other triad by means of a transformation of triads. Such a transformation is nothing more than the expression for the projection of a unit vector in the direction of another unit vector, and such projections were illustrated in Section 1.1. To obtain such relationships between the unit vectors of two triads, start by drawing a figure showing the orientation of each triad of interest. This is illustrated in Fig. 1.2 where two unit vector triads $\{x\} = \{\hat{x}_1, \hat{x}_2, \hat{x}_3\}$ and $\{y\} = \{\hat{y}_1, \hat{y}_2, \hat{y}_3 = \hat{x}_3\}$ are shown.[2]

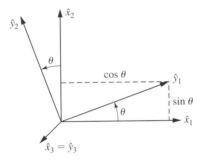

Figure 1.2 A simple transformation of triads.

Figure 1.2 shows how to obtain the expression for the unit vector $\hat{y}_1$ in terms of the unit vectors $\hat{x}_1$ and $\hat{x}_2$. For this, one simply draws the projections of the unit vector $\hat{y}_1$ along the directions parallel to the unit vectors $\hat{x}_1$ and $\hat{x}_2$. These projections are the dashed lines shown in Fig. 1.2. In terms of $\hat{x}_1$ and $\hat{x}_2$,

---

[2] Notice that there is no specific location or "origin" for unit vectors; such locations are irrelevant since unit vectors are only needed to define specific *directions* in space. Thus, any of the unit vectors shown in Fig. 1.2 could have been drawn anywhere on the page. However, for clarity, they are generally drawn intersecting at a common point. Such a practice also makes it easier to visualize the projections of a unit vector of one of the triads along the direction of a unit vector of the other triad.

the vector components of the unit vector $\hat{y}_1$ in the directions of $\hat{x}_1$ and $\hat{x}_2$ are then $|\hat{y}_1|(\cos\theta)\hat{x}_1 = (\cos\theta)\hat{x}_1$ and $|\hat{y}_1|(\sin\theta)\hat{x}_2 = (\sin\theta)\hat{x}_2$. The unit vector $\hat{y}_1$ is then equal to the vector sum of those components. A similar construction may also be made to obtain the vector components of $\hat{y}_2$ in terms of $\hat{x}_1$ and $\hat{x}_2$. The following expressions relating the unit vectors $\hat{y}_1$ and $\hat{y}_2$ to the unit vectors $\hat{x}_1$ and $\hat{x}_2$ may then be written by inspection of Fig. 1.2 (such relations are called *transformations of triads*).

$$\hat{y}_1 = (|\hat{y}_1|\cos\theta)\hat{x}_1 + (|\hat{y}_1|\sin\theta)\hat{x}_2 = \hat{x}_1\cos\theta + \hat{x}_2\sin\theta \quad (1.2.1)$$

$$\hat{y}_2 = -(|\hat{y}_2|\sin\theta)\hat{x}_1 + (|\hat{y}_2|\cos\theta)\hat{x}_2 = -\hat{x}_1\sin\theta + \hat{x}_2\cos\theta \quad (1.2.2)$$

Notice that if Eqs. (1.2.1) and (1.2.2) are written in matrix form as[3]

$$\begin{bmatrix}\hat{y}_1\\\hat{y}_2\end{bmatrix} = \begin{bmatrix}\cos\theta & \sin\theta\\-\sin\theta & \cos\theta\end{bmatrix}\begin{bmatrix}\hat{x}_1\\\hat{x}_2\end{bmatrix} \triangleq [A]\begin{bmatrix}\hat{x}_1\\\hat{x}_2\end{bmatrix}$$

the inverse transformation from $\{\hat{y}_1, \hat{y}_2\}$ to $\{\hat{x}_1, \hat{x}_2\}$ (i.e., the equations expressing $\{\hat{x}_1, \hat{x}_2\}$ in terms of $\{\hat{y}_1, \hat{y}_2\}$) is readily obtained by performing the matrix multiplication involving the inverse $[A]^{-1}$ of the matrix $[A]$, as shown in the following equation. The result is the same as that given by Eqs. (1.2.3) and (1.2.4), which were written by inspecting Fig. 1.2 a second time.

$$\begin{bmatrix}\hat{x}_1\\\hat{x}_2\end{bmatrix} = [A]^{-1}\begin{bmatrix}\hat{y}_1\\\hat{y}_2\end{bmatrix} = \begin{bmatrix}\cos\theta & -\sin\theta\\\sin\theta & \cos\theta\end{bmatrix}\begin{bmatrix}\hat{y}_1\\\hat{y}_2\end{bmatrix} = [A]^T\begin{bmatrix}\hat{y}_1\\\hat{y}_2\end{bmatrix}$$

As seen in this equation, the inverse $[A]^{-1}$ of a transformation matrix $[A]$ involving two unit vector triads is always equal to the transpose $[A]^T$ of $[A]$. Such a result is valid for any transformation matrix involving the unit vectors of two triads.

$$\hat{x}_1 = \hat{y}_1\cos\theta - \hat{y}_2\sin\theta \quad (1.2.3)$$

$$\hat{x}_2 = \hat{y}_1\sin\theta + \hat{y}_2\cos\theta \quad (1.2.4)$$

For a more elaborate example of transformations of triads, consider the triads $\{\hat{x}_1, \hat{x}_2, \hat{x}_3\}$ and $\{\hat{z}_1, \hat{z}_2, \hat{z}_3\}$ shown in Fig. 1.3. As shown in Fig. 1.3a, the triad $\{\hat{z}_1, \hat{z}_2, \hat{z}_3\}$ is obtained by first rotating $\{\hat{x}_1, \hat{x}_2, \hat{x}_3\}$ counterclockwise about the $\hat{x}_3$ direction by an angle θ_1 to obtain the triad $\{\hat{y}_1, \hat{y}_2, \hat{y}_3 = \hat{x}_3\}$, and then by rotating the triad $\{\hat{y}_1, \hat{y}_2, \hat{y}_3\}$ counterclockwise about the $\hat{y}_1$ direction by an angle θ_2 to obtain the triad $\{\hat{z}_1 = \hat{y}_1, \hat{z}_2, \hat{z}_3\}$. The two individual rotations are shown in Fig. 1.3b. From these, it follows that

$$\hat{y}_1 = \hat{x}_1\cos\theta_1 + \hat{x}_2\sin\theta_1 \quad (1.2.5)$$

$$\hat{y}_2 = -\hat{x}_1\sin\theta_1 + \hat{x}_2\cos\theta_1 \quad (1.2.6)$$

$$\hat{y}_3 = \hat{x}_3 \quad (1.2.7)$$

---

[3] The symbol $\triangleq$, which appears throughout the book, denotes "equal to by definition." Here, it is defining the matrix $[A]$.

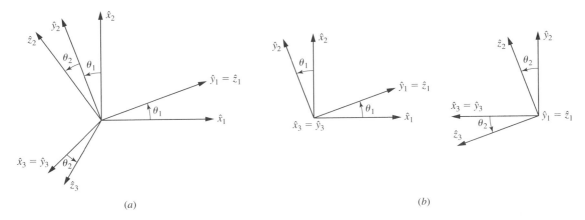

Figure 1.3 A more elaborate transformation of triads.

$$\hat{z}_1 = \hat{y}_1 \tag{1.2.8}$$

$$\hat{z}_2 = \hat{y}_2 \cos\theta_2 + \hat{y}_3 \sin\theta_2 \tag{1.2.9}$$

$$\hat{z}_3 = -\hat{y}_2 \sin\theta_2 + \hat{y}_3 \cos\theta_2 \tag{1.2.10}$$

Equations (1.2.5) to (1.2.10) may then be combined to yield the following representation of the transformation between the two unit vector triads $\{\hat{x}_1, \hat{x}_2, \hat{x}_3\}$ and $\{\hat{z}_1, \hat{z}_2, \hat{z}_3\}$:

$$\hat{z}_1 = \hat{x}_1 \cos\theta_1 + \hat{x}_2 \sin\theta_1 \tag{1.2.11}$$

$$\begin{aligned}\hat{z}_2 &= \hat{y}_2 \cos\theta_2 + \hat{y}_3 \sin\theta_2 \\ &= -\hat{x}_1 \sin\theta_1 \cos\theta_2 + \hat{x}_2 \cos\theta_1 \cos\theta_2 + \hat{x}_3 \sin\theta_2\end{aligned} \tag{1.2.12}$$

$$\begin{aligned}\hat{z}_3 &= \hat{y}_3 \cos\theta_2 - \hat{y}_2 \sin\theta_2 \\ &= \hat{x}_1 \sin\theta_1 \sin\theta_2 - \hat{x}_2 \cos\theta_1 \sin\theta_2 + \hat{x}_3 \cos\theta_2\end{aligned} \tag{1.2.13}$$

The need for such transformations appears frequently in dynamics when one has to convert the expression of a vector written in terms of a particular unit vector triad into another unit vector triad. These operations will show up in a number of solved examples and proposed problems throughout the remainder of the book.

1.3 Basic Operations with Vectors

In addition to the sum of two vectors (where the resulting vector is called the *resultant* of the two vectors) and the multiplication of a vector by a scalar, other basic operations that appear in dynamics are the dot product, cross product, and the time derivative of vectors.

Dot Product

The *dot product* of two vectors $\vec{a}$ and $\vec{b}$ (written as $\vec{a} \bullet \vec{b}$ and read as "a dot b") is, by definition, a scalar quantity given as

$$\vec{a} \bullet \vec{b} = |\vec{a}||\vec{b}| \cos \theta \qquad (1.3.1)$$

where $|\vec{a}|$ and $|\vec{b}|$ are the magnitudes of $\vec{a}$ and $\vec{b}$, respectively, and θ (with $0 \leq \theta \leq 2\pi$) is the angle between the two vectors, as shown in Fig. 1.4. This figure also shows the cross product of two vectors, which is presented after Example 1.3.1.

The dot product of two vectors $\vec{a}$ and $\vec{b}$ is a scalar that is proportional to the projection of either one of the vectors in the direction of the other. This is seen by observing Fig. 1.4 and noticing that the projection of vector $\vec{b}$, for example, in the direction of $\vec{a}$ is equal to $(|\vec{b}| \cos \theta) \hat{a}$, where $\hat{a} = \vec{a}/|\vec{a}|$. The quantity $|\vec{b}| \cos \theta$ is the scalar component of $\vec{b}$ along $\vec{a}$. Using the definition of the dot product, such a projection may be simply obtained as $(\vec{b} \bullet \hat{a})\hat{a}$. Similarly, the projection of $\vec{a}$ in the direction of $\vec{b}$ may be obtained as $(\vec{a} \bullet \hat{b})\hat{b}$.

If the vectors $\vec{a}$ and $\vec{b}$ shown in Fig. 1.4 are expressed in terms of an orthogonal triad $\{e\} = \{\hat{e}_1, \hat{e}_2, \hat{e}_3\}$ as $\vec{a} = a_{e_1}\hat{e}_1 + a_{e_2}\hat{e}_2 + a_{e_3}\hat{e}_3$ and $\vec{b} = b_{e_1}\hat{e}_1 + b_{e_2}\hat{e}_2 + b_{e_3}\hat{e}_3$, one obtains

$$\begin{aligned}\vec{a} \bullet \vec{b} = \vec{b} \bullet \vec{a} &= a_{e_1}\hat{e}_1 \bullet (b_{e_1}\hat{e}_1) + a_{e_1}\hat{e}_1 \bullet (b_{e_2}\hat{e}_2) + \cdots + a_{e_3}\hat{e}_3 \bullet (b_{e_3}\hat{e}_3) \\ &= a_{e_1}b_{e_1} + a_{e_2}b_{e_2} + a_{e_3}b_{e_3}\end{aligned} \qquad (1.3.2)$$

since $\hat{e}_1 \bullet \hat{e}_1 = \hat{e}_2 \bullet \hat{e}_2 = \hat{e}_3 \bullet \hat{e}_3 = 1$ and $\hat{e}_1 \bullet \hat{e}_2 = \hat{e}_1 \bullet \hat{e}_3 = \hat{e}_2 \bullet \hat{e}_3 = 0$. By using the dot product of $\vec{a}$ with $\vec{b}$, the cosine of the angle θ between two known vectors $\vec{a}$ and $\vec{b}$ can be readily obtained.

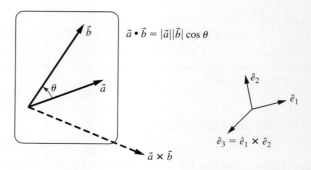

Figure 1.4 Two vectors $\vec{a}$ and $\vec{b}$ and a triad for illustrating the dot and cross products.

EXAMPLE 1.3.1

Determine the projection of the vector $\vec{a} = \hat{x}_1 - 5\hat{x}_2 + 2\hat{x}_3$ (where $\hat{x}_1$, $\hat{x}_2$, and $\hat{x}_3$ are orthogonal unit vectors) along the direction $\overrightarrow{AB}$ of the line that passes through the points A and B whose rectangular coordinates are $A(x_1 = 1, x_2 = 2, x_3 = 3)$ and $B(x_1 = 6, x_2 = 10, x_3 = 7)$. The axes x_1, x_2, and x_3 are parallel to the unit vectors $\hat{x}_1$, $\hat{x}_2$, and $\hat{x}_3$, respectively.

a. What is the scalar component of the projection of $\vec{a}$ along $\overrightarrow{AB}$?
b. Does the projection of $\vec{a}$ along $\overrightarrow{AB}$ point in the direction from A to B or from B to A?

■ Solution

a. Letting $\vec{c}$ denote the projection of $\vec{a}$ in the direction of $\vec{b} = \overrightarrow{AB}$, the vector $\vec{c}$ is obtained with the following calculations.[4]

$$\vec{b} = (6-1)\hat{x}_1 + (10-2)\hat{x}_2 + (7-3)\vec{x}_3 = 5\hat{x}_1 + 8\hat{x}_2 + 4\hat{x}_3$$

$$\hat{b} = \frac{\vec{b}}{|\vec{b}|} = \frac{5\hat{x}_1 + 8\hat{x}_2 + 4\hat{x}_3}{\sqrt{5^2 + 8^2 + 4^2}}$$

$$\vec{c} = [\vec{a} \bullet \hat{b}]\hat{b} = \left[(\hat{x}_1 - 5\hat{x}_2 + 2\hat{x}_3) \bullet \left(\frac{5\hat{x}_1 + 8\hat{x}_2 + 4\hat{x}_3}{\sqrt{105}}\right)\right] \frac{5\hat{x}_1 + 8\hat{x}_2 + 4\hat{x}_3}{\sqrt{105}}$$

$$= \frac{5 - 40 + 8}{\sqrt{105}} \frac{5\hat{x}_1 + 8\hat{x}_2 + 4\hat{x}_3}{\sqrt{105}} \approx -1.29\hat{x}_1 - 2.06\hat{x}_2 - 1.03\hat{x}_3$$

The scalar component of the projection of $\vec{a}$ along $\overrightarrow{AB}$ is the quantity $\vec{a} \bullet \hat{b} = -27/\sqrt{105}$.

b. Since the result obtained in this example is negative, the projection vector $\vec{c}$ points in the $\overrightarrow{BA}$ direction (i.e., in the direction from B to A, instead of A to B).

Cross Product

The *cross product* of two vectors $\vec{a}$ and $\vec{b}$ (written as $\vec{a} \times \vec{b}$ and read as "a cross b") making an angle θ with each other (see Fig. 1.4) is, by definition, a vector whose magnitude is equal to $|\vec{a}||\vec{b}||\sin\theta|$ and whose direction is perpendicular to the two vectors and is given by the *right-hand rule*.

The Right-Hand Rule

Open your right hand and keep your right thumb extended. Now, keeping the right thumb extended, curl your fingers in the direction indicated by the curved arrow (shown in Fig. 1.4) that defines the angle θ measured from $\vec{a}$ (i.e., the first vector in the expression $\vec{a} \times \vec{b}$) to $\vec{b}$ (i.e., the second vector in the expression

---

[4] The symbol $\approx$, which appears throughout the book, denotes "approximately equal to."

> $\vec{a} \times \vec{b}$). The rule defines the direction of the cross product $\vec{a} \times \vec{b}$ as being the direction indicated by your right thumb. For this, always use $\theta \leq \pi$; when doing so, $\sin\theta \geq 0$, and the magnitude of the cross product is equal to $|\vec{a}||\vec{b}|\sin\theta$.
>
> The right-hand rule is also known as the *screwdriver rule* since the direction of $\vec{a} \times \vec{b}$ is that which a regular screw, turned by a screwdriver, would travel if the screwdriver were turned by an angle θ (always using $\theta \leq \pi$) to align $\vec{a}$ with $\vec{b}$.
>
> Simply by using the right-hand rule, you can immediately conclude that $\vec{b} \times \vec{a} = -\vec{a} \times \vec{b}$.

If the vectors $\vec{a}$ and $\vec{b}$ are expressed in terms of their components in a triad $\{\hat{e}_1, \hat{e}_2, \hat{e}_3\}$ with, say, $\hat{e}_3 \overset{\Delta}{=} \hat{e}_1 \times \hat{e}_2$ (as shown in Fig. 1.4), the cross product $\vec{a} \times \vec{b}$ is then obtained as

$$\begin{aligned}
\vec{a} \times \vec{b} &= a_{e_1}\hat{e}_1 \times \left(b_{e_1}\hat{e}_1 + b_{e_2}\hat{e}_2 + b_{e_3}\hat{e}_3\right) \\
&\quad + a_{e_2}\hat{e}_2 \times \left(b_{e_1}\hat{e}_1 + b_{e_2}\hat{e}_2 + b_{e_3}\hat{e}_3\right) \\
&\quad + a_{e_3}\hat{e}_3 \times \left(b_{e_1}\hat{e}_1 + b_{e_2}\hat{e}_2 + b_{e_3}\hat{e}_3\right) \\
&= \left(a_{e_2}b_{e_3} - a_{e_3}b_{e_2}\right)\hat{e}_1 + \left(a_{e_3}b_{e_1} - a_{e_1}b_{e_3}\right)\hat{e}_2 + \left(a_{e_1}b_{e_2} - a_{e_2}b_{e_1}\right)\hat{e}_3
\end{aligned}$$

(1.3.3)

To determine the cross products $\hat{e}_1 \times \hat{e}_2$, $\hat{e}_2 \times \hat{e}_3$, etc., that show up in Eq. (1.3.3), see the unit vector triad shown in Fig. 1.4 and use the right-hand rule; as disclosed by that figure, $\hat{e}_1 \times \hat{e}_2 = \hat{e}_3$, $\hat{e}_1 \times \hat{e}_3 = -\hat{e}_2$, etc.

The cross product may also be determined in terms of the following determinant, if one prefers to use determinants to do so:

$$\vec{a} \times \vec{b} = \begin{vmatrix} \hat{e}_1 & \hat{e}_2 & \hat{e}_1 \times \hat{e}_2 = \hat{e}_3 \\ a_{e_1} & a_{e_2} & a_{e_3} \\ b_{e_1} & b_{e_2} & b_{e_3} \end{vmatrix}$$

(1.3.4)

In order for the calculation of a cross product using determinants to give the correct result given by Eq. (1.3.3), the following rule must be followed:

- The first two unit vectors written in positions 1-1 and 1-2 in the first row of the determinant are any two of the three unit vectors of triad $\{e\}$, but the third unit vector in the first row of that determinant must be equal to the cross product of the unit vector written in position 1-1 with the unit vector written in position 1-2.

- To form the cross product $\vec{a} \times \vec{b}$, the second and third rows of the determinant should contain the components of $\vec{a}$ and $\vec{b}$, respectively, along the unit vectors that appear in the first row. Thus, if the unit vectors $\hat{e}_3$ and $\hat{e}_2$ shown in Fig. 1.4 are listed first, the cross product $\vec{a} \times \vec{b}$ may also be correctly obtained as

$$\vec{a} \times \vec{b} = \begin{vmatrix} \hat{e}_3 & \hat{e}_2 & \hat{e}_3 \times \hat{e}_2 = -\hat{e}_1 \\ a_{e_3} & a_{e_2} & -a_{e_1} \\ b_{e_3} & b_{e_2} & -b_{e_1} \end{vmatrix}$$

(1.3.5)

Since the cross product is a vector, the three components of the cross product of two vectors $\vec{a}$ and $\vec{b}$ depend on the specific triad $\{e\}$ that one chooses to work with (the vector obtained as the result of the cross product *does not* depend on any triad, of course; to see this, look back at the definitions of the magnitude and direction of the cross product of two vectors).

EXAMPLE 1.3.2

Determine the unit vector $\hat{n}$ that is perpendicular to the plane defined by the vectors $\vec{a} = \hat{x}_1 - 5\hat{x}_2 + 2\hat{x}_3$ and $\vec{b} = \overrightarrow{AB} = 5\hat{x}_1 + 8\hat{x}_2 + 4\hat{x}_3$ of Example 1.3.1. The unit vector $\hat{x}_3$ is defined as $\hat{x}_3 = \hat{x}_1 \times \hat{x}_2$.

■ **Solution**

The perpendicular to the vectors $\vec{a}$ and $\vec{b}$ is parallel to the cross product $\vec{a} \times \vec{b}$ and also to the cross product $\vec{b} \times \vec{a} = -\vec{a} \times \vec{b}$. Therefore, either one of the unit vectors in the direction of $\vec{a} \times \vec{b}$ or $-\vec{a} \times \vec{b}$ is a solution. We then have

$$\vec{a} \times \vec{b} = (\hat{x}_1 - 5\hat{x}_2 + 2\hat{x}_3) \times (5\hat{x}_1 + 8\hat{x}_2 + 4\hat{x}_3)$$

$$= \underbrace{\hat{x}_1 \times (5\hat{x}_1)}_{=0} + \underbrace{\hat{x}_1 \times (8\hat{x}_2)}_{=8\hat{x}_3} + \underbrace{\hat{x}_1 \times (4\hat{x}_3)}_{=-4\hat{x}_2} - 5\underbrace{\hat{x}_2 \times (5\hat{x}_1)}_{=-5\hat{x}_3} - 5\underbrace{\hat{x}_2 \times (8\hat{x}_2)}_{=0}$$

$$- 5\underbrace{\hat{x}_2 \times (4\hat{x}_3)}_{=4\hat{x}_1} + 2\underbrace{\hat{x}_3 \times (5\hat{x}_1)}_{=5\hat{x}_2} + 2\underbrace{\hat{x}_3 \times (8\hat{x}_2)}_{=-8\hat{x}_1} + 2\underbrace{\hat{x}_3 \times (4\hat{x}_3)}_{=0}$$

$$= -36\hat{x}_1 + 6\hat{x}_2 + 33\hat{x}_3$$

$$\frac{\vec{a} \times \vec{b}}{|\vec{a} \times \vec{b}|} = \frac{-36\hat{x}_1 + 6\hat{x}_2 + 33\hat{x}_3}{\sqrt{36^2 + 6^2 + 33^2}} \approx -0.732\hat{x}_1 + 0.122\hat{x}_2 + 0.671\hat{x}_3$$

The unit vector $\hat{n}$ that is perpendicular to the plane defined by the vectors $\vec{a}$ and $\vec{b}$ is then either $\hat{n} = -0.732\hat{x}_1 + 0.122\hat{x}_2 + 0.671\hat{x}_3$ or $\hat{n} = 0.732\hat{x}_1 - 0.122\hat{x}_2 - 0.671\hat{x}_3$.

Time Derivatives of Vectors

The derivatives of vectors obey the *product rule*, so for any two vectors $\vec{A}$ and $\vec{B}$ and any scalar function of time $f(t)$

$$\frac{d}{dt}[f(t)\vec{A}] = \left[\frac{d}{dt}f(t)\right]\vec{A} + f(t)\frac{d}{dt}\vec{A}$$

$$\frac{d}{dt}(\vec{A} \bullet \vec{B}) = \left(\frac{d}{dt}\vec{A}\right) \bullet \vec{B} + \vec{A} \bullet \left(\frac{d}{dt}\vec{B}\right)$$

$$\frac{d}{dt}(\vec{A} \times \vec{B}) = \left(\frac{d}{dt}\vec{A}\right) \times \vec{B} + \vec{A} \times \left(\frac{d}{dt}\vec{B}\right)$$

Presentation of the time derivative $d\vec{A}/dt$ of any vector $\vec{A}$ is postponed until Chapter 2.

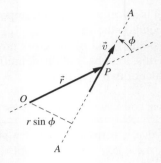

Figure 1.5 Moment of a vector about a point.

1.4 Moment of a Vector about a Point

The moment of a vector $\vec{v}$ about a point O in space, denoted by $\vec{M}_O$ (with the subscript O denoting the point about which the moment is taken), is, by definition, a vector that is equal to the cross product of the vector from O to *any* point P on the *line of action* of $\vec{v}$ (line AA in Fig. 1.5), and the vector $\vec{v}$, i.e.,

$$\vec{M}_O \triangleq \overrightarrow{OP} \times \vec{v} \tag{1.4.1}$$

Since (see Fig. 1.5)

$$|\vec{M}_O| = |\overrightarrow{OP} \times \vec{v}| = |\overrightarrow{OP}||\vec{v}||\sin\phi| \tag{1.4.2}$$

and, as disclosed by Fig. 1.5, the quantity $|\overrightarrow{OP}||\sin\phi|$ is the perpendicular distance from point O to the line of action of $\vec{v}$ (line AA), *the moment $\vec{M}_O$ is independent of the choice of point P on line AA.*

EXAMPLE 1.4.1

The rectangular coordinates (in meters) of three points A, B, and C are $A(x_1 = 1, x_2 = 2, x_3 = 3)$, $B(x_1 = 6, x_2 = 10, x_3 = 7)$, and $C(x_1 = 6, x_2 = 13, x_3 = 20)$. Determine the perpendicular distance between point C and line AB.

■ **Solution**

By letting $\hat{x}_1$, $\hat{x}_2$, and $\hat{x}_3$ denote three orthogonal unit vectors parallel to the x_1, x_2, and x_3 axes, respectively, the perpendicular distance d from point A to line AB is then determined, as presented in this section, as follows.

$$\overrightarrow{AB} = (6-1)\hat{x}_1 + (10-2)\hat{x}_2 + (7-3)\hat{x}_3 = 5\hat{x}_1 + 8\hat{x}_2 + 4\hat{x}_3$$

$$\overrightarrow{CB} = (6-6)\hat{x}_1 + (10-13)\hat{x}_2 + (7-20)\hat{x}_3 = -3\hat{x}_2 - 13\hat{x}_3$$

$$d = \left|\overrightarrow{CB} \times \frac{\overrightarrow{AB}}{|\overrightarrow{AB}|}\right| = \left|\frac{-15\hat{x}_2 \times \hat{x}_1 + 92\hat{x}_2 \times \hat{x}_3 - 65\hat{x}_3 \times \hat{x}_1}{\sqrt{5^2 + 8^2 + 4^2}}\right| = \frac{\sqrt{15^2 + 92^2 + 65^2}}{\sqrt{5^2 + 8^2 + 4^2}}$$

$$\approx 11.09 \text{ meters (m)}$$

Notice that the vector $\vec{AB}/|\vec{AB}|$ that appears in the expression for the distance d is the unit vector in the direction of $\vec{AB}$. The distance d may also be determined as

$$d = \left| \vec{CA} \times \frac{\vec{AB}}{|\vec{AB}|} \right|$$

and, as discussed in this section, both calculations give the same result.

1.5 Newton's Laws of Motion

Newton's three laws of motion, which were formulated for a single particle of constant mass, are the fundamental tools for the mathematical formulation of dynamics problems. They are presented here.

> **Newton's First Law**
> If the resultant $\vec{F}$ of all the forces acting on a particle is zero, the particle will move in space with constant velocity $\vec{v}$.

A vector is constant if its magnitude is constant *and* its direction does not change with time. Therefore, for a particle to move in space with constant velocity $\vec{v}$, its *speed* (i.e., the magnitude of its velocity) must be constant and its trajectory must be a straight line.

> **Newton's Second Law**
> A particle of mass m subjected to a system of forces whose resultant is $\vec{F}$ moves in such a way that the time rate of change of its absolute linear momentum vector, which is defined as the product $m\vec{v}$ (where $\vec{v}$ is the absolute velocity of the particle), is equal to the resultant force $\vec{F}$.

Mathematically, Newton's second law for a particle is expressed as follows:

$$\boxed{\vec{F} = \frac{d}{dt}(m\vec{v})} = m\frac{d}{dt}\vec{v} \triangleq m\vec{a} \tag{1.5.1}$$

where $\vec{a}$ is the *absolute acceleration* of the particle.

Newton's laws presume the existence of a reference point that is not accelerating (such a point is called an *inertial point*), and of an *inertial reference frame*. This requires the following explanations.

> A *reference frame* is a collection of points forming a three-dimensional space where the distance between all pairs of such points is constant. A reference frame is said to be *inertial* if all its points are inertial (which also implies that no line connecting two of its points changes direction in space).

The velocity of a point P, relative to another point O, is always associated with a reference frame; otherwise the concept of velocity is meaningless. The absolute velocity $\vec{v}$ of a particle P is, by definition, the time rate of change of the *absolute position vector*[5] $\vec{r} = \overrightarrow{OP}$ of P, as measured by an observer fixed to an inertial reference frame. Calculations involving the time derivative of vectors to determine velocity and acceleration (which is the time derivative of velocity) for analyzing dynamics problems are dealt with in detail in Chapter 2.

As mentioned, an inertial point is one that is not accelerating. In an attempt to find an example of an inertial point in the universe, one may loosely say that the "average position of the faraway stars" (!) is inertial. Other points in the universe may be approximated as inertial, but they are not convenient or particularly useful for engineering applications. Although the earth is not inertial because of the translation of its center (on an elliptic orbit) around the sun and of its spin about its own axis, many dynamics problems that call for the determination of the motion of a particle relative to the earth during a relatively small time interval may be reasonably approximated by neglecting the motion of the earth, thus treating it as inertial.

Newton's second law does not apply to particles in the atomic scale where quantum effects are dominant. Such problems fall in the realm of *quantum mechanics*. Also, the second law, as originally formulated by Newton, is only valid for particles whose speed is small when compared to the speed of light. Problems where speeds approach the speed of light belong to the realm of *relativistic mechanics* and are not treated in this book.

The vector equation (1.5.1) yields a set of scalar equations that involve the variables that were chosen for describing the motion of the particle and may also involve unknown forces. After such equations are combined to eliminate the unknown forces that appear in them, such as internal reaction forces, a set of *scalar differential equations of motion* for the particle are then obtained, and the problem one now has consists of extracting, from such equations, as much information as possible about the unknown motion of the particle. This may include numerical integration of such equations in order to simulate the motion of the system for desired initial conditions and/or analytical investigations in order to generate information that allows one to understand as much as possible about the dynamic behavior of the system.

In engineering practice, the analysis of the motion governed by the differential equations obtained as just discussed includes the investigation of small motions about constant solutions (which are called *equilibrium solutions*) by linearizing the differential equations about those solutions. The linearization process, which involves a Taylor series expansion of all the nonlinear terms that appear in the differential equations, is presented in Section 1.14.

---

[5] The *position vector* of a point P relative to a reference point O is simply the vector $\overrightarrow{OP}$ from O to P. A position vector $\overrightarrow{OP}$ is said to be an *absolute position vector* when the reference point O is inertial. Thus, the absolute position vector of a point P is a vector from an inertial point O to point P.

The engineer should be aware that dynamical systems may exhibit a variety of phenomena that are not disclosed at all by linearized differential equations, and that such phenomena may cause a physical system to exhibit undesirable behavior, or even physical damage in the system. The study of these phenomena belongs to the field of *nonlinear oscillations* and is presented in books that deal with such subject matter. An experiment describing one such phenomena is presented in Section 1.16.

When analyzing the motion of a system of particles interacting with each other, Newton's second law is still applicable to each individual particle provided that we isolate each particle from all others and include the force of interaction due to the other particles in the free-body diagram for the particle under observation. In such cases, Newton's third law plays a significant role in the analysis.

> **Newton's Third Law**
> The mutual forces of action and reaction between two mass particles i and j are collinear with the particles, equal in magnitude, and opposite in direction.

Expressed mathematically, Newton's third law states that

$$\vec{F}_{ij} = -\vec{F}_{ji} \tag{1.5.2}$$

where $\vec{F}_{ij}$ is the force on particle i due to particle j.

1.6 Kepler's Laws and Newton's Law of Universal Gravitation

Johannes Kepler (1571–1630) was an assistant to the Dutch astronomer Tycho Brahe (1546–1601) at the time of Brahe's death. He was entrusted in 1601 with the vast and precise amount of observation data on the motion of the planets that Brahe had gathered and collected. Using mathematical tools of the time, he analyzed the data for many years. He published his first two laws of planetary motion in 1609, and the third law in 1619.

> **Kepler's First Law**
> Each planet moves in an ellipse with the sun at one of its foci.

> **Kepler's Second Law**
> The motion of each planet is such that the radius vector from the sun to the planet sweeps out equal areas in equal times.

> **Kepler's Third Law**
> The square of the period of a planet in its motion around the sun is proportional to the cube of the semi-major axis of its elliptical orbit around the sun.

An important question whose answer was found by Newton is the following: Given two bodies that are subjected only to the forces of attraction between them, how does the force of attraction depend on the distance between the bodies so that the motion of one body relative to the other obeys Kepler's laws? The answer to this question is now known as Newton's law of universal gravitation.

> **Newton's Law of Universal Gravitation**
> Any two particles of masses m_1 and m_2, respectively, attract each other with a force whose magnitude is proportional to the product of their masses and inversely proportional to the square of the distance r between them.

Mathematically, Newton's law of universal gravitation is expressed as

$$|\vec{F}| = \frac{Gm_1 m_2}{r^2} \tag{1.6.1}$$

where G is the universal gravitational constant, whose value is $G \approx 6.67 \times 10^{-11}$ cubic meters per kilogram per second squared [m^3/(kg·s^2)].

The magnitude of the gravitational force acting on a particle of mass m located on the surface of a spherical homogeneous planet of radius R and mass M is equal to $|\vec{F}| = (GM/R^2)m \triangleq mg$ (see Sections 1.7 and 1.8). The constant g is called the *acceleration of gravity* at the surface of the attracting planet. For the earth, $g \approx 9.81$ m/s$^2 \approx 32.2$ feet (ft)/s^2.

1.7 Force of Attraction due to a Spherical Body

Newton's laws were formulated for particles, which are defined as material points that have mass but no dimension. Mass particles are, of course, idealizations. Several other researchers extended the work of Newton to both discrete and continuum systems of particles.

Using Newton's law of universal gravitation, which was presented in Section 1.6, let us determine the mutual force of attraction between a spherical body of radius R and mass density ρ kg/m^3, and a particle P of mass m located at a distance $D \geq R$ from the center O of the spherical body. For simplicity, it will also be assumed that the mass density ρ of the spherical body is constant, and thus, equal to $M/(4\pi R^3/3)$, where M is the mass of the body. Such a body could be a model for a planet. Of course, planets are neither perfectly spherical nor homogeneous, but such idealizations turn out to be good first approximations for calculating the forces of attraction due to them. More refined approximations are, of course, used in more precise calculations involving the motion of the planets and the motion of artificial satellites over a long time.

A point Q on a spherical portion of the body with radius $r \leq R$ and the mass particle P are shown in Fig. 1.6. To calculate the mutual force of attraction between the body and P, the body is divided into an infinite number of mass elements, each having the infinitesimal volume $dV = (r \sin\theta \, d\phi)(r \, d\theta) \, dr$ [since as seen

SECTION 1.7 Force of Attraction due to a Spherical Body

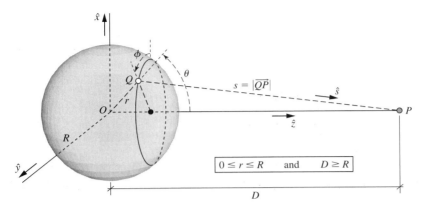

Figure 1.6 A portion of a spherical body of mass M, and a material particle P of mass m.

from Fig. 1.6, the lengths of the sides of the infinitesimal volume element are dr, $r\, d\theta$, and $(r \sin\theta)\, d\phi$]. The mass dM of the element located at point Q in Fig. 1.6 is then $dM = \rho\, dV = [M/(4\pi R^3/3)]\, dV$. When ϕ is varied from 0 to 2π, the mass element dM covers a strip on a cone of angle θ. After that, varying θ from 0 to π makes that cone generate a thin shell of radius r. Then, by varying r from 0 to R causes that shell to generate the solid sphere. This is what the triple integral in Eq. (1.7.3) will do.

Why did we divide the sphere into infinitesimal volume elements? Simply because we know how to calculate the force of attraction between two point masses, which is given by Newton's law of universal gravitation. Letting $s = |\overrightarrow{QP}|$ denote the distance between Q and P and $\hat{s} = \overrightarrow{QP}/s$ be the unit vector from Q to P (see Fig. 1.6), the infinitesimal force acting on P due to the attraction of the mass element at Q is then

$$d\vec{F} = -\frac{Gm\, dM}{s^2}\hat{s} = -\frac{Gm\, dM}{s^3}\overrightarrow{QP} = -\frac{GmM}{4\pi R^3/3}\frac{r^2(\sin\theta)\, dr\, d\theta\, d\phi}{s^3}\overrightarrow{QP}$$

(1.7.1)

The resultant force $\vec{F}$ acting on P due to attraction of the entire spherical body is simply equal to the integral, over the spherical body, of $d\vec{F}$.

By expressing the vector $\overrightarrow{QP}$ in terms of the quantities r, θ, ϕ, and D and of the unit vector triad $\{\hat{x}, \hat{y}, \hat{z}\}$ (see Fig. 1.6) as

$$\overrightarrow{QP} = \overrightarrow{QO} + \overrightarrow{OP} = -[(r\sin\theta)(\hat{x}\cos\phi + \hat{y}\sin\phi) + (r\cos\theta)\hat{z}] + D\hat{z}$$

(1.7.2)

and by noticing that

$$s = |\overrightarrow{QP}| = \sqrt{(r\sin\theta\cos\phi)^2 + (r\sin\theta\sin\phi)^2 + (D - r\cos\theta)^2},$$

which gives $s = \sqrt{D^2 + r^2 - 2Dr\cos\theta}$, the expression for the force $\vec{F}$ then becomes

$$\vec{F} = -\frac{GMm}{4\pi R^3/3} \int_{r=0}^{R} \int_{\theta=0}^{\pi} \int_{\phi=0}^{2\pi} \frac{(D - r\cos\theta)\hat{z} - (\hat{x}\cos\phi + \hat{y}\sin\phi)r\sin\theta}{(D^2 + r^2 - 2Dr\cos\theta)^{3/2}}$$

$$\times r^2 (\sin\theta) \, d\phi \, d\theta \, dr \qquad (1.7.3)$$

The integrals involving the $\hat{x}$ and the $\hat{y}$ components in Eq. (1.7.3) are zero because $\int_0^{2\pi} \cos\phi \, d\phi = \int_0^{2\pi} \sin\phi \, d\phi = 0$. Therefore, the resultant force is directed along the line joining the particle P and the center O of the spherical body. Physically, this occurs because the sum of the components of the force $d\vec{F}$ along the perpendicular to line OP is zero due to the spherical symmetry of the body centered at O. In other words, since the attracting sphere has constant density, for each point Q where a mass element $dM = \rho \, dV$ is located, there is another point Q^* on the body (because the body is spherical), located at $\phi = 180°$ from the position shown in Fig. 1.6, along the plane perpendicular to line OP, and, thus, the resultant of these two forces is along line OP.

The integrations associated with the $\hat{z}$ component in Eq. (1.7.3) may be readily performed by recognizing that $\sin\theta \, d\theta = -d(\cos\theta)$, and then letting $u = \cos\theta$. The resulting integrals can be found in a table of integrals. They are evaluated below using integration by parts. Since

$$v \triangleq \int \frac{du}{(D^2 + r^2 - 2Dru)^{3/2}} = \frac{1}{Dr\sqrt{D^2 + r^2 - 2Dru}}$$

and

$$\int \frac{u \, du}{(D^2 + R^2 - 2Dru)^{3/2}} \triangleq \int u \, dv = uv - \int v \, du$$

$$= \frac{u}{Dr\sqrt{D^2 + r^2 - 2Dru}} + \frac{\sqrt{D^2 + r^2 - 2Dru}}{D^2 r^2}$$

Eq. (1.7.3) then yields, after the integration with respect to $u = \cos\theta$ and ϕ,

$$\vec{F} = \hat{z} \frac{6\pi GMm}{4\pi R^3} \int_{r=0}^{R} \left[\frac{1}{\sqrt{D^2 + r^2 - 2Dr\cos\theta}} \left(\frac{1}{r} - \frac{\cos\theta}{D} \right) \right.$$

$$\left. - \frac{\sqrt{D^2 + r^2 - 2Dr\cos\theta}}{D^2 r} \right]_{\theta=0}^{\pi} r^2 \, dr$$

This yields

$$\vec{F} = -\frac{GMm}{D^2} \hat{z} \qquad (1.7.4)$$

By comparing Eq. (1.7.4) with Eq. (1.6.1), it is seen that an attracting spherical body of constant mass density (or, as mentioned earlier, one with mass distribution with ϕ symmetry, with ϕ as shown in Fig. 1.6) behaves as if it were a point with mass M concentrated at the center of the sphere. This conclusion is then applicable for two of such spheres attracting each other; i.e., their mutual force of attraction is directed along the line that joins their centers.

Let us now use these results to make some important observations about the force of gravity, weight, and what is commonly referred to in practice as *weightlessness*. These are important physical concepts that can be (and often are) easily misunderstood. They are discussed in Section 1.8.

1.8 Gravitational Force, Weight, and the Meaning of Weightlessness

A gravitational force is a force due to the mutual attraction between two masses. For most applications, the main *gravitational force* acting on a body B is the force acting on that body due to the gravitational attraction of the planet the body is near to. A body of mass m located on the surface of the earth, for example, is always subjected to the mutual force of attraction between the earth and the body. If the earth is approximated as a homogeneous sphere of mass M_E and radius R_E, the magnitude of that mutual force of attraction is equal to $GM_E m/R_E^2 = mg$, where $g \triangleq GM_E/R_E^2 \approx 9.81$ m/s² is called the *acceleration of gravity* at the earth's surface.[6] Such a mutual force of attraction is *always* acting on a body subjected to the earth's attraction, even if the body is in free fall or in orbit around the earth. If a body B is in contact with the earth, the body is also subjected to a reaction force $\vec{N}$, which is normal to the earth's surface at the point of contact. If the effects of the spin of the earth about its own axis and of the motion of the earth in space are neglected, the acceleration of B is zero and the magnitude $|\vec{N}|$ of the normal force is equal to the gravitational force mg. The magnitude $|\vec{N}|$ is the force that would be measured by a scale if body B were on the scale, and the scale were fastened to the earth.

If you imagine yourself being the body B described, the normal reaction $\vec{N}$ is the force you feel on your body when you are in contact with the earth, whether you stand, sit, or lie down. Now, imagine yourself inside an elevator, standing on a scale nailed to the floor of the elevator. The scale would read the value of the magnitude $|\vec{N}|$ of the normal force $\vec{N}$, which would be $|\vec{N}| = mg$ if the elevator were not accelerating. If the elevator were in free fall toward the earth's surface, with an acceleration equal to g m/s², you would also be in free fall with the same acceleration. In such a case, the scale underneath you would read $|\vec{N}| = 0$ (because both you and the elevator are falling toward the ground with the same acceleration and the same speed), and, therefore, you would feel

---

[6] The inhomogeneity of the earth and its departure from a perfect sphere cause the acceleration of gravity to vary slightly from the value given, depending on where the mass is located on the earth's surface.

as if you were weightless even though the gravitational force of attraction of the earth, with a magnitude equal to $GM_E m/(R_E + h)^2 \approx GM_E m/R_E^2 \stackrel{\Delta}{=} mg$, where h is the height from the earth's surface, would still be acting on your body (but you would not feel that force "pressing" your body).

From this discussion, it is seen that, although a body resting on the earth's surface and one in free fall are subjected to the same gravitational force mg, the sensation of "weightlessness" (which is characterized by the lack of the pressing sensation felt on the body) can only be explained if *weight* is associated with the normal reaction force applied by the earth to the body (instead of the gravitational force mg, which is always nonzero).

1.9 Equivalent Force Systems

The concept of equivalent force systems is introduced in statics courses. Two sets of forces are said to be equivalent to each other if they have the same resultant $\vec{F}$, and also the same moment $\vec{M}_P$ about the same point P, which is any point chosen as desired. A given set of forces can be replaced by its equivalent system, which consists of the force $\vec{F}$ whose line of action passes through P, and the moment $\vec{M}_P$. The resultant $\vec{F}$ of a system of forces $\vec{F}_1, \vec{F}_2, \ldots, \vec{F}_n$ is, by definition, the vector sum $\vec{F} = \vec{F}_1 + \vec{F}_2 + \cdots + \vec{F}_n$.

Two equivalent force systems have the same effect on a body they are acting on; i.e., they produce the same translational and rotational motions on that body. Figure 1.7b and c illustrates two force systems applied to a body that are each equivalent to the force system in Fig. 1.7a, where $F_1 = 8$ newtons (N), $F_2 = 20$ N, $F_3 = 10\sqrt{3}$ N, $L = 2$ m, and $h = 1$ m.

Inspecting Fig. 1.7a, one can see that the resultant force $\vec{F}$ of the system is

$$\vec{F} = \vec{F}_1 + \vec{F}_2 + \vec{F}_3 = \left[\left(20\cos\frac{\pi}{6} - 10\sqrt{3}\right)\hat{x} + \left(20\sin\frac{\pi}{6} - 8\right)\hat{y}\right] = 2\hat{y} \text{ N}$$

and this is the force that will appear in any other system that is equivalent to the one in Fig. 1.7a.

The sum of the moments about point A of each force of the given system is equal to zero (in fact, the moment about A of each force is zero). Therefore, the given system is equivalent to a single force $\vec{F} = 2\hat{y}$ whose line of action passes

(a) (b) (c)

Figure 1.7 An example of equivalent force systems.

through point A. This is the single-force system shown in Fig. 1.7b. You will also obtain a single-force equivalent system if you choose to take moments about any point that is in the line of action of the resultant $\vec{F}$ shown in Fig. 1.7b.

If one chooses point C to take moments about, one obtains

$$\vec{M}_C = \overrightarrow{CB} \times \vec{F}_1 + \overrightarrow{CA} \times \vec{F}_2 + \overrightarrow{CD} \times \vec{F}_3$$

$$= \underbrace{(-2\hat{x}) \times (-8\hat{y})}_{=16\hat{z}} + \underbrace{(-2\hat{x} - \hat{y}) \times \left[20\left(\hat{x}\cos\frac{\pi}{6} + \hat{y}\sin\frac{\pi}{6}\right)\right]}_{=-20\hat{z}+10\sqrt{3}\,\hat{z}}$$

$$+ \underbrace{(-\hat{y}) \times \left(-10\sqrt{3}\,\hat{x}\right)}_{=-10\sqrt{3}\,\hat{z}}$$

$$= -4\hat{z}\ \text{N}\cdot\text{m}$$

Thus, the force-moment system in Fig. 1.7c is also equivalent to the ones in Fig. 1.7a and b.

1.10 Center of Gravity and Center of Mass

The concept of *center of gravity of a body* of mass m is associated with the concept of equivalent force systems that was reviewed in Section 1.9. In contrast, the center of mass of a body has nothing to do with any forces acting on the body. These concepts will be briefly discussed here using Fig. 1.8 where two bodies are subjected to their mutual forces of attraction. The body centered at O is spherical, for simplicity, with its mass M being much greater than that of any other body in its vicinity. This spherical body might approximate the earth, for example. Its force of attraction exerted on other bodies, such as on body B shown in Fig. 1.8,

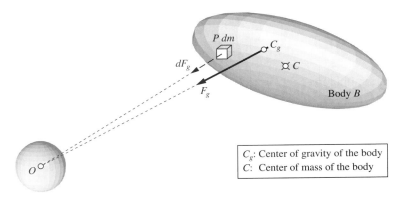

Figure 1.8 Distinction between the *center of gravity* and the *center of mass* of a body.

is what is called a *gravitational force*. The figure is not drawn to scale, of course, but that is not relevant to this presentation.

The *center of mass* of a body is a point C whose position is defined using a weighted average of the position of each mass element of the body. Using Fig. 1.8, the position vector $\overrightarrow{OC}$ (i.e., the vector from O to C) of the center of mass C of body B is defined as

$$\overrightarrow{OC} = \frac{\int_{\text{body}} \overrightarrow{OP}\, dm}{m} \qquad (1.10.1)$$

The center of mass of any body is a property of the body. It does not depend on any forces acting on the body, and it does not depend on how the body is oriented in space. The center of mass of a body plays a significant role in the formulation of the differential equations that govern the motion of a body. This is first addressed in detail in Chapter 4.

In contrast, the location of the *center of gravity* of a body depends on the gravitational force acting on each part of the body and, thus, on the orientation of the body. Points on body B in Fig. 1.8 that are closer to the attracting center O are subjected to a greater gravitational force than points that are farther away. In that figure, each infinitesimal element of body B, of mass dm, that is at a distance $|\overrightarrow{OP}|$ from the center O of the attracting body at O is subjected to the gravitational force $d\vec{F}_g$ shown in the figure. The resultant $\vec{F}_g$ is calculated according to Newton's law of universal gravitation and is also shown in Fig. 1.8.

$$\vec{F}_g = \int_{\text{body}} d\vec{F}_g = -\int_{\text{body}} \frac{GM\, dm}{|\overrightarrow{OP}|^2} \frac{\overrightarrow{OP}}{|\overrightarrow{OP}|} \qquad (1.10.2)$$

> The center of gravity of a body of mass m is, by definition, the point C_g in Fig. 1.8 that would be subjected to the same resultant $\vec{F}_g$ of the gravitational forces acting on the body if the entire mass of the body were concentrated at C_g.

The determination of the center of gravity of a body involves two steps:

Step 1. Replace the resultant of the gravitational forces acting on each part of the body by a *single equivalent force*. Such a force is the vector $\vec{F}_g$ given by Eq. (1.10.2). When this step is completed, the line of action of that force is then known. Such a calculation is illustrated in Example 1.10.1 at the end of this section.

Step 2. Choose C_g to be the point on the line of action determined in step 1 that would be subjected to the same resultant $\vec{F}_g$ of the gravitational forces acting on the body if the entire mass of the body were concentrated at that point. That is, C_g is the point located by the distance $|\overrightarrow{OC_g}|$, along the line of action of $\vec{F}_g$, defined as

$$|\vec{F}_g| = \frac{GMm}{|\overrightarrow{OC_g}|^2} \quad \text{thus} \quad |\overrightarrow{OC_g}| = \sqrt{\frac{GMm}{|\vec{F}_g|}} \quad (1.10.3)$$

In many applications, including most of those considered in this book, the gravitational field of the earth is approximated as that of an infinite plane (an exception to this is presented in Chapter 4). In such an ideal case, the force on a body due to gravity does not depend on the body's position, and its magnitude near the surface of the earth is simply approximated as mg where $g = GM_E/R_E^2$, as discussed in Section 1.8 (p. 19). In practice, such an approximation is justified when a body does not move too far from the line that is perpendicular to the local vertical. In such cases, the center of gravity coincides with the center of mass. It is important to keep in mind, however, the conceptual difference between the center of gravity and the center of mass because one must not confuse the two when a distinction must be made, such as in more advanced problems dealing with the longtime motion of satellites and space structures. The distinction between the center of mass and the center of gravity becomes crucial when one investigates the rotational motion of orbiting satellites for example. However, the determination of the center of gravity is neither necessary nor relevant in these or other problems! As shown in Chapters 5 and 6, what is important in rotational dynamics is the moment, about the center of mass of the body, of the forces that are applied to the body, and the *angular momentum* of the system (defined in Chapter 4).

EXAMPLE 1.10.1

The body shown in Fig. 1.9 is a dumbbell that consists of two small spheres, each of mass $m/2$, connected at their centers A and B by a rigid massless bar of length L. Point C is the center of mass of the dumbbell, and the orthogonal $\hat{x}$ and $\hat{y}$ unit vectors, with $\hat{x}$ parallel to line OC, are included in the figure for use in later calculations. The dumbbell is able to rotate in the plane of the figure, and its orientation relative to line OC is tracked by the angle $\theta(t)$, where t is time.

A gravitational *center of attraction*, represented by the sphere centered at point O, exerts the forces $\vec{F}_A$ and $\vec{F}_B$ on the spheres at A and B, respectively (see Fig. 1.9), with magnitudes $F_A = (\mu m/2)/|\overrightarrow{OA}|^2$ and $F_B = (\mu m/2)/|\overrightarrow{OB}|^2$, respectively, where μ is a constant. For $R = 4$ m, $L = 2$ m, $m = 1$ kg, and $\mu = 1.8$ m³/s², determine

a. The resultant gravitational force $\vec{F}_g = \vec{F}_A + \vec{F}_B$ acting on the dumbbell, expressed in terms of the angle θ and the unit vectors $\hat{x}$ and $\hat{y}$
b. The resultant moment, about C, due to the forces $\vec{F}_A$ and $\vec{F}_B$ acting on the dumbbell
c. The distance from B to the point of intersection of the resultant $\vec{F}_g$ with line AB, and the distance from O to that same point
d. The distance from O to the center of gravity C_g of the dumbbell

Figure 1.9 Figure for Example 1.10.1.

■ **Solution**

a. The resultant force $\vec{F}_g$ is simply

$$\vec{F}_g = \vec{F}_A + \vec{F}_B = -\frac{\mu m/2}{|\overrightarrow{OA}|^2}\underbrace{\frac{\overrightarrow{OA}}{|\overrightarrow{OA}|}}_{\substack{\text{Unit vector}\\ \text{from } O \text{ to } A}} - \frac{\mu m/2}{|\overrightarrow{OB}|^2}\underbrace{\frac{\overrightarrow{OB}}{|\overrightarrow{OB}|}}_{\substack{\text{Unit vector}\\ \text{from } O \text{ to } B}}$$

$$= -\frac{\mu m}{2|\overrightarrow{OA}|^3}\overrightarrow{OA} - \frac{\mu m}{2|\overrightarrow{OB}|^3}\overrightarrow{OB}$$

This expression for $\vec{F}_g$ shows that we now need expressions for the vectors $\overrightarrow{OA}$ and $\overrightarrow{OB}$ and for their magnitudes. This is done by examining Fig. 1.9, which leads us to write the following equations:

$$\overrightarrow{OA} = \overrightarrow{OC} + \overrightarrow{CA} = \left(R + \frac{L}{2}\cos\theta\right)\hat{x} + \frac{L}{2}(\sin\theta)\hat{y}$$

$$\overrightarrow{OB} = \overrightarrow{OC} + \overrightarrow{CB} = \left(R - \frac{L}{2}\cos\theta\right)\hat{x} - \frac{L}{2}(\sin\theta)\hat{y}$$

and from these, in turn, we can determine the magnitudes

$$|\overrightarrow{OA}| = \left[\left(R + \frac{L}{2}\cos\theta\right)^2 + \left(\frac{L}{2}\sin\theta\right)^2\right]^{1/2} = \left(R^2 + \frac{L^2}{4} + RL\cos\theta\right)^{1/2}$$

$$|\overrightarrow{OB}| = \left[\left(R - \frac{L}{2}\cos\theta\right)^2 + \left(\frac{L}{2}\sin\theta\right)^2\right]^{1/2} = \left(R^2 + \frac{L^2}{4} - RL\cos\theta\right)^{1/2}$$

Finally, this then yields the following expression for the resultant force $\vec{F}_g$:

$$\vec{F}_g = -\mu \frac{m}{2} \left[\frac{[R+(L/2)\cos\theta]\hat{x} + [(L/2)\sin\theta]\hat{y}}{(R^2 + L^2/4 + RL\cos\theta)^{3/2}} \right.$$

$$\left. + \frac{[R-(L/2)\cos\theta]\hat{x} - [(L/2)\sin\theta]\hat{y}}{(R^2 + L^2/4 - RL\cos\theta)^{3/2}} \right] \quad (1.10.1a)$$

For $R = 4$ m, $L = 2$ m, $m = 1$ kg, $\theta = 30°$, and $\mu = 1.8$ m³/s², this yields

$$\vec{F}_g \approx -0.126\hat{x} + 0.01\hat{y} \quad \text{N}$$

b. The resultant moment, about the center of mass C, due to the forces $\vec{F}_A$ and $\vec{F}_B$ acting on the dumbbell is (see Fig. 1.9)

$$\vec{M}_C = \overrightarrow{CA} \times \vec{F}_A + \overrightarrow{CB} \times \vec{F}_B = \frac{L}{2}(\hat{x}\cos\theta + \hat{y}\sin\theta) \times \vec{F}_A$$

$$- \frac{L}{2}(\hat{x}\cos\theta + \hat{y}\sin\theta) \times \vec{F}_B$$

and this gives, with $\hat{z} = \hat{x} \times \hat{y}$ as defined in Fig. 1.9,

$$\vec{M}_C = -\frac{\mu m L R}{4} \left[\frac{1}{(R^2 + L^2/4 - RL\cos\theta)^{3/2}} \right.$$

$$\left. - \frac{1}{(R^2 + L^2/4 + RL\cos\theta)^{3/2}} \right] (\sin\theta)\hat{z} \quad (1.10.1b)$$

For $R = 4$ m, $L = 2$ m, $m = 1$ kg, $\theta = 30°$, and $\mu = 1.8$ m³/s², this yields

$$\vec{M}_C \approx -0.041\hat{z} \quad \text{N} \cdot \text{m}$$

which, for $\theta = 30°$, is a clockwise moment (see the unit vectors $\hat{x}$ and $\hat{y}$ in Fig. 1.9, and then observe the direction of $\hat{z} = \hat{x} \times \hat{y}$ using the right-hand rule). Notice that such a moment causes the angle θ in Fig. 1.9 to decrease. This behavior is expected since, as seen from Fig. 1.9, the force acting at B is larger than that acting at A, and this creates the clockwise moment $\vec{M}_C$. The moment $\vec{M}_C$ will cause AB to rotate.

c. The two-force system $\vec{F}_A$ and $\vec{F}_B$ acting on the dumbbell is equivalent to the force $\vec{F}_g$ applied at the center of mass C of the dumbbell and the moment $\vec{M}_C$. That system is also equivalent to a single force $\vec{F}_g$ in some direction in space that passes through O and intersects line AB at some point C_1 for which the moment $\vec{M}_{C_1}$, about such a point C_1, is equal to zero. Such a situation is shown in Fig. 1.10, where C_1 is the point of intersection of the resultant $\vec{F}_g$ with line AB. Another point C_g on the *line of action* of such a resultant force is also shown in that figure and is used for answering the question in part d. The moment, about any point on that line of action of $\vec{F}_g$, is equal to zero.

Since point C_1 is on the line of action of $\vec{F}_g$, we then have

$$\vec{M}_{C_1} = \overrightarrow{C_1 A} \times \vec{F}_A + \overrightarrow{C_1 B} \times \vec{F}_B$$

$$= (L-D)(\hat{x}\cos\theta + \hat{y}\sin\theta) \times \vec{F}_A - D(\hat{x}\cos\theta + \hat{y}\sin\theta) \times \vec{F}_B = 0$$

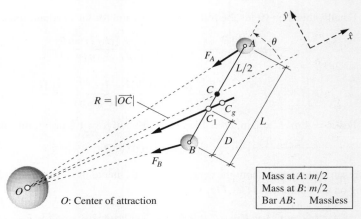

Figure 1.10 The single-force equivalent to the two forces $\vec{F}_A$ and $\vec{F}_B$ in Fig. 1.9.

and this yields

$$\vec{M}_{C_1} = -\frac{\mu m R}{2}\left[\frac{D}{(R^2 + L^2/4 - RL\cos\theta)^{3/2}} - \frac{L - D}{(R^2 + L^2/4 + RL\cos\theta)^{3/2}}\right]$$
$$\times (\sin\theta)\hat{z} = 0 \qquad (1.10.1c)$$

By solving for the distance D, we then obtain

$$D = \frac{L}{1 + \left(\dfrac{R^2 + L^2/4 + RL\cos\theta}{R^2 + L^2/4 - RL\cos\theta}\right)^{3/2}}$$

The location of point C_1 along line AB defines the orientation of the resultant force $\vec{F}_g$ shown in Fig. 1.10. With this information, any other desired point on that line can now be located.

For $R = 4$ m, $L = 2$ m, and $\theta = 30°$, we obtain

$$D \approx 0.43 \text{ m}$$

i.e., for $\theta = 30°$, C_1 is a point on line AB, between B and C (see Fig. 1.10), located 0.43 m from B and 0.57 m from the center of mass C. For the given data, its distance $|\overrightarrow{OC_1}|$ from point O, obtained by geometry from Fig. 1.10, is

$$|\overrightarrow{OC_1}| = \sqrt{\left[R - \left(\frac{L}{2} - D\right)\cos\theta\right]^2 + \left[\left(\frac{L}{2} - D\right)\sin\theta\right]^2} \approx 3.52 \text{ m}$$

Thus the answers for this part are the distances D and $|\overrightarrow{OC_1}|$.

d. Having determined the location of point C_1, one can now locate any other point on the line of action of the resultant gravitational force $\vec{F}_g$. For point C_g (see Fig. 1.10), which is defined to be the center of gravity of the dumbbell as

$$|\vec{F}_g| \triangleq \frac{\mu m}{|\overrightarrow{OC_g}|^2} \qquad (1.10.1d)$$

we obtain for the distance $|\overrightarrow{OC_g}|$, with the magnitude of $\vec{F}_g$ obtained from Eq. (1.10.1a),

$$|\overrightarrow{OC_g}| = \sqrt{\frac{\mu m}{|\vec{F}_g|}} \approx 3.78 \text{ m} \qquad (1.10.1e)$$

1.11 Free-Body Diagrams and Forces

When formulating a dynamics problem using the Newtonian approach (be it for a single mass particle, an element of a machine, or an airplane in flight), it is invaluable to draw a *free-body diagram* for the object under consideration. The free-body diagram for an object P is obtained by isolating it from its surroundings and representing, by vectors, all forces acting on P due to its interaction with the surrounding medium. The forces that are encountered in actual situations can be classified into two basic types: *field forces,* such as the force acting on an object due to the gravitational attraction of the earth or other massive objects, and *contact forces,* such as the force due to a spring connected to the object or the force due to friction. The free-body diagram for a body with finite dimensions should include these two types of forces and all the moments (or *torques*)[7] that might be applied to the body.

The moment, about a point P that coincides with a particle, of any force applied to the particle is zero. Therefore, since particles have no dimension, they can only be subjected to forces, not moments. The moment, about other points in space, of a force applied to a particle may, of course, be different from zero.

Let us look at a couple of examples to illustrate how to draw a proper free-body diagram.

EXAMPLE 1.11.1

Consider the pendulum shown in Figure 1.11. The pendulum consists of a bar AB of constant length L and mass m_1, supported by a pin at point A, and of a particle of mass m_2 located at point B. A moment M_{bar} is applied to the bar at A. That moment is due either to a motor (which is not shown in the figure) or to friction at the pin (if there is friction), or both. Moments are represented by a curved arrow (which represents the curved fingers of a hand, as in the right-hand rule) or by a straight arrow (since moments are vectors). However, since forces are always represented by straight arrows in free-body diagrams, it is wise to avoid using the same type of straight arrows to represent moments (so that they are not misunderstood as a force). The angle θ measured from the vertical to the

---

[7] The word *torque* is generally used synonymously with *moment*. It is also used by some authors for the moment of a pair of forces that are equal in magnitude and opposite in direction. Such a pair of forces is known as a *couple*.

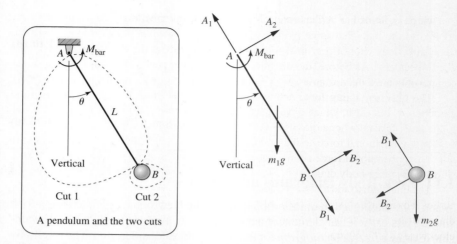

Figure 1.11 A pendulum and the free-body diagrams for each of its bodies.

bar shown in Fig. 1.11 is used for tracking the orientation of the bar. The system is close enough to the surface of the earth (i.e., L is much smaller than the radius of the earth) so that every point of the pendulum is subjected to the same constant gravitational force (which is a very reasonable approximation). It is assumed that there are no other forces applied to the system. The free-body diagrams for the bar AB and for the particle at B are obtained by isolating the bar, and the particle at B, with the cuts indicated by the dashed lines shown in Fig. 1.11.

To draw the free-body diagram for bar AB, start by drawing the bar. A vertical line and the angle θ that was chosen to track the direction of the bar at all times are also included in the diagram shown in Fig. 1.11. Their inclusion in the diagram is optional, but many authors include such information for completeness. Such a practice will prove to be convenient later, when formulating the differential equation that governs the motion of the system. The *differential equation of motion* (or equations, for the more general case when there is more than one) is the equation that allows the engineer to investigate and predict the motion of the system.

Next, put the forces in the free-body diagram of the bar. At point A on the bar there is a reaction force of unknown magnitude and direction. Forces in free-body diagrams are generally split into components along orthogonal directions. In Fig. 1.11 those directions were chosen to be parallel and perpendicular to the bar. The orthogonal arrows labeled A_1 and A_2 represent such components for the reaction force at A. The directions of the orthogonal arrows representing these components are totally arbitrary. At this stage of the formulation of the solution to a problem, it does not matter if the vector $\vec{A}_1$ is drawn pointing away from the bar (as shown in the middle part of Fig. 1.11) or toward the bar. When values for the scalar components A_1 and A_2 of $\vec{A}_1$ are eventually determined, they will be either positive or negative. A positive value for A_1, for example, simply means that the direction of the arrow labeled A_1 in Fig. 1.11 is indeed in the direction drawn in the figure, while a negative value means it is in the opposite direction.

At point B of bar AB there is also a force of unknown magnitude and direction applied to the bar. That force is also split into two orthogonal components (along the same directions chosen previously) that are represented by the arrows labeled B_1 and B_2, respectively. As before, the directions for the orthogonal arrows representing these components are totally arbitrary.

The only remaining force acting on bar AB is the gravitational force due to the attraction of the earth, which always acts along the *local vertical*. Here, and elsewhere in the book (except when noted), the gravitational field will be approximated as constant so that a gravitational force acting on a body of mass m is simply a vector of magnitude mg. The gravitational force acting on the bar[8] is represented by the arrow labeled $m_1 g$ shown in the free-body diagram in Fig. 1.11.

The final item to add to the free-body diagram for the bar is the moment M_{bar} discussed earlier. That moment is represented by the curved arrow labeled M_{bar} that is added to the diagram (the counterclockwise sense was chosen arbitrarily; it could also have been drawn clockwise). If the quantity M_{bar} is positive, the moment is counterclockwise, as drawn. If M_{bar} is negative, the negative value of M_{bar} will automatically reverse the physical interpretation of the counterclockwise arrow, making it a clockwise moment. Thus, in either case, representing the moment by that counterclockwise arrow labeled M_{bar} will account for the actual physical situation (and we do not have to draw another free-body diagram with that arrow reversed). The free-body diagram for the bar is now complete.

The free-body diagram for the point mass at B is also shown in Fig. 1.11. The forces applied to the point mass are the gravitational force due to the attraction of the earth (which is the vertical force with magnitude equal to $m_2 g$) and the reaction force applied by bar AB. The latter is represented by the negative [according to Newton's third law (Section 1.5)] of the forces B_1 and B_2 that were drawn in the free-body diagram for the bar. Therefore, the corresponding arrows for these forces should be drawn in the free-body diagram for the point mass in opposite directions to those that appear in the free-body diagram for the bar. The labels for these forces must be the same ones that were used in the free-body diagram for the bar.[9] This completes the free-body diagram for the point mass.

Since the point mass is just that, a point (i.e., it has no dimensions), no moment can ever be applied to it. That is why there is no reaction moment at point B, only forces. If the point mass were replaced by a finite body, the reactions at the connection between that

---

[8] The gravitational force acting on the bar is, of course, distributed along the bar. Each infinitesimal bar element of mass dm_1 is subjected to the gravitational force whose magnitude is $g\, dm_1$. These forces are equivalent to (i.e., they can be replaced by) a resultant force, and the moment of the distributed forces, applied at any point P chosen on the bar. The resultant force is equal to the integral, over the length of the bar, of $g\, dm_1$, or simply $m_1 g$ when g is approximated as a constant. The moment of the distributed forces depends on the choice of point P, and such a moment is zero if P is chosen to be the center of gravity of the bar. Under the assumption of a constant gravitational field, the center of gravity of the bar coincides with its center of mass. For a homogeneous bar, for example, the center of mass is at the middle of the bar.

[9] You could also keep the same directions for the arrows and attach a minus sign to the corresponding labels. Either method accomplishes the same purpose, but reversing the direction of the arrows is preferable since it makes the diagram easier to interpret.

body and bar AB would consist of the forces labeled B_1 and B_2 and a moment (which we might label M_B) if the connection were a weld or a pin with friction.

The two free-body diagrams for the pendulum shown in Fig. 1.11 involve five unknowns (assuming that M_{bar} is either a known function of time or a function of these five unknowns), namely, the angle θ and the four reaction forces A_1, A_2, B_1, and B_2. To determine them, we are going to need to generate five equations involving these five unknowns. The generation of such equations is one of the objectives of *dynamics,* the other being the analysis of such equations in order to determine the desired unknowns. This is done in examples starting in Chapter 2. The equations involving such unknowns are generated by using Newton's laws of motion and their extension to finite bodies (which was developed by Euler).

It should be noted that for electrically charged particles moving in an electromagnetic field with flux density $\vec{B}$, an electromagnetic force is also induced on the particle. It is known that the electromagnetic force is in the direction perpendicular to the relative velocity $\vec{v}$ between the two charged particles and the magnetic flux density $\vec{B}$. If q is the charge of each particle, the electromagnetic force is determined as $q\vec{v} \times \vec{B}$. Such forces are not considered in this book.

EXAMPLE 1.11.2

Consider a bar of length L, of constant mass density m/L (where m is the mass of the bar), in contact with a vertical wall and a horizontal floor as shown in Fig. 1.12a. The bar is moving [i.e., the angle $\theta = \theta(t)$ is changing with time t] on the vertical plane, and there is friction at the contacts of the bar with the wall and the floor. The initial conditions for the motion are $\theta(0) = \theta_0$ and $(d\theta/dt)_{t=0} = \omega_0$, where ω_0 may be either positive or negative, so that point B of the bar may be moving either to the left or to the right. When θ increases, point B moves to the left with velocity v_B (relative to the fixed reference represented by the floor and the wall), and point A moves up the wall with velocity v_A as indicated in the figure. The free-body diagram for the bar is shown in Fig. 1.12b.

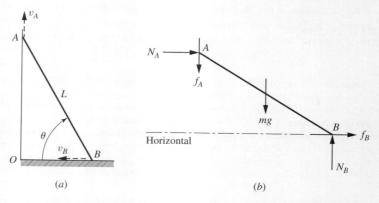

Figure 1.12 (a) A bar in contact with a wall and the floor, and (b) the free-body diagram for the bar.

The forces that appear in the free-body diagram are

- The gravitational force mg, which passes through the center of mass of the bar (located at the geometric center of the uniform bar).
- The normal reaction forces N_A (applied by the vertical wall to the bar) and N_B (applied by the floor to the bar); these forces are perpendicular to the surfaces the bar is in contact with.
- The friction forces f_A and f_B, acting at the contact points A and B, respectively.

The friction force that acts on a body that translates along a surface is *always* directed opposite to the velocity of the point of contact of the body relative to the contact surface.[10] Therefore, the friction forces f_A and f_B shown in Fig. 1.12 are directed opposite to the velocities of point A relative to the wall and of point B relative to the floor, respectively. By inspecting the figure, it is seen that the directions of f_A and f_B depend on whether θ increases or decreases. The directions indicated in Fig. 1.12b are for the case when the angle θ, measured from the floor to bar AB, is increasing, i.e., when $d\theta/dt > 0$. The directions of f_A and f_B shown in Fig. 1.12 are reversed when $d\theta/dt < 0$.

There are two types of friction, *dry friction* and *viscous friction,* and several mathematical models for them. The most common model for dry friction is *Coulomb friction,* which is named for Charles Augustin de Coulomb (1736–1806), who was a French engineer who became well known for his experimental work with friction.

For Coulomb friction between two surfaces in contact with each other, and moving relative to each other, the magnitude of the friction force is proportional to the magnitude of the normal force of contact between the surfaces. The proportionality coefficient is called the *Coulomb friction coefficient* (or the *coefficient of dry friction*). This is a reasonably good model for the friction force between two dry surfaces in contact.

Viscous friction is provided by a *viscous medium,* such as oil, when a body moves in that medium. The magnitude of the viscous friction force acting on the body is a function of the velocity of the body relative to that medium. If the dependency on the velocity is linear, the friction is called *linear viscous friction.*

The following are the expressions for the friction forces f_A and f_B that appear in the free-body diagram for the bar considered in this example.

For Coulomb friction:

$$f_A = \mu_A |N_A| \operatorname{sgn} \frac{d\theta}{dt} \qquad (1.11.1)$$

$$f_B = \mu_B |N_B| \operatorname{sgn} \frac{d\theta}{dt} \qquad (1.11.2)$$

where $\mu_A > 0$ and $\mu_B > 0$ are the kinetic coefficients of Coulomb friction at the contact points A and B, respectively, in Fig. 1.12. Notice that the use of the *sign function* (denoted either as *sign* or *sgn* in mathematical expressions) in Eqs. (1.11.1) and (1.11.2) automatically handles the change in the direction of the motion, thus making those expressions

---

[10] Rolling of a rigid body (with or without slipping) is a special case that is considered in Section 5.4.

valid for all times. Also notice that the expressions depend on $d\theta/dt$ in a nonlinear manner since the sign function does not vary linearly with its argument. The sign function is, by definition, equal to $+1$ when its argument is positive, equal to -1 when its argument is negative, and undefined when its argument is zero. One may arbitrarily define sgn(0) to be equal to zero, and this is commonly done. The sign function, with sgn(0) = 0, is one of the blocks available in Simulink.

For linear viscous friction:

$$f_A = c_A \frac{d}{dt}(L \sin \theta) = c_A L \frac{d\theta}{dt} \cos \theta \qquad (1.11.3)$$

$$f_B = c_B \frac{d}{dt}(-L \cos \theta) = c_B L \frac{d\theta}{dt} \sin \theta \qquad (1.11.4)$$

where $c_A > 0$ and $c_B > 0$ are the viscous friction coefficients at A and B, respectively.

For *linear viscous friction* acting on a point P of a body that is translating in a viscous medium, the friction force is linearly proportional to the velocity of P relative to the medium (all the necessary details about the proper formulation of velocity are presented in Chapter 2), and the direction of the friction force is opposite to that velocity. For point A in Fig. 1.12, for example, the velocity of A is the time derivative of the vector from O to A (see Example 2.5.3, p. 82), and this gives the expression for f_A given by Eq. (1.11.3) when friction is of the viscous type. The viscous friction force f_A would be provided by a *viscous damper* device (see the note on p. 41) that would be connected to points A and O.

Notice that, according to Eq. (1.11.3), f_A is positive (i.e., the viscous friction force at A points downward, as shown in Fig. 1.12) when $d\theta/dt > 0$ (because the term $c_A L \cos \theta$ in that expression is not negative for $0 \leq \theta \leq \pi/2$) and is negative when $d\theta/dt < 0$. Also, according to the expression for f_B given by Eq. (1.11.4), the viscous friction force at B is positive (i.e., that friction force points to the right, as shown in Fig. 1.12) when $d\theta/dt > 0$ and negative when $d\theta/dt < 0$. Therefore, the expressions for the viscous friction forces given by Eqs. (1.11.3) and (1.11.4) are valid for all times. The use of the sign function is not needed for the case of viscous friction.

1.12 Chain Rule of Differentiation and Exact Differentials

> The material presented in this section is indispensable when energy methods are used. Energy methods are extremely useful and versatile. Potential energy is determined using the mathematics presented in this section. You may skip this section for now, but you will need to study it later in order to fully understand how to determine potential energy (from the simplest to the most complicated cases) and the basics that constitutes the foundations of energy methods.

The chain rule of differentiation is frequently used for manipulating a differential equation of motion (basic material on the latter is presented in Section 1.13). To review the chain rule, let $x_1, x_2, \ldots, x_n$ be variables that depend on time t in an unknown manner (i.e., an expression for $x_1, x_2, \ldots$, as a function of time, is not known). Also, let $f(x_1, x_2, \ldots, x_n)$ be some known function of these variables.

For example, we might have $f = x_1^2 \sin x_2$, or $f = x_1^2 + K_1 x_2^2 + K_2 x_1 \sin x_2$, or any other function. In dynamics, $x_1, x_2, \ldots$ might stand for variables that were chosen to represent the position and velocity of a particle or the angular orientation and angular velocity of a bar, while f might stand for the sum of the kinetic and potential energies of the motion. Suppose we now want to obtain the total time derivative (which we will refer to from now on simply as the time derivative) df/dt, of the function f. According to the chain rule of differentiation, we then have

$$\boxed{\frac{d}{dt} f = \frac{\partial f}{\partial x_1} \frac{dx_1}{dt} + \frac{\partial f}{\partial x_2} \frac{dx_2}{dt} + \cdots + \frac{\partial f}{\partial x_n} \frac{dx_n}{dt}} \tag{1.12.1}$$

The coefficients of $dx_1/dt, dx_2/dt, \ldots$, are called the *partial derivatives* of f with respect to $x_1, x_2, \ldots$. As suggested by Eq. (1.12.1), the coefficient of dx_1/dt is obtained by holding $x_2, \ldots, x_n$ as constants (so that $dx_2/dt = \cdots = dx_n/dt = 0$) and then taking the derivative of f with respect to x_1. A similar procedure is followed for obtaining the coefficients of $dx_2/dt, \ldots$. For example, for $f = x_1^2 \sin x_2 + K x_2^2$, we have

$$\frac{\partial f}{\partial x_1} = 2 x_1 \sin x_2$$

$$\frac{\partial f}{\partial x_2} = x_1^2 \cos x_2 + 2 K x_2$$

which gives

$$\frac{df}{dt} = (2 x_1 \sin x_2) \frac{dx_1}{dt} + (x_1^2 \cos x_2 + 2 K x_2) \frac{dx_2}{dt}$$

Another operation that will appear in dynamics is the *differential* of a function f. That is simply df and, with $f = f(x_1, x_2, \ldots, x_n)$, it is simply obtained as

$$\boxed{df = \frac{\partial f}{\partial x_1} dx_1 + \frac{\partial f}{\partial x_2} dx_2 + \cdots + \frac{\partial f}{\partial x_n} dx_n} \tag{1.12.2}$$

The quantity df/dt is a *finite quantity*, while df is an *infinitesimal quantity*. Since df, as given by Eq. (1.12.2), was obtained by starting with a function f, it is clearly the "d of a function f," and, for this reason, it is said to be an *exact differential*. When necessary, an exact differential is written as $d(f)$, which should be read as "d of f." The integral of an exact differential $d(f)$ is, except for a constant of integration, the function f.

Suppose, now, that we have the opposite problem (which is a problem that will appear in dynamics, as you will see later): Given an expression for some infinitesimal quantity dg, as

$$dg = a_1\, dx_1 + a_2\, dx_2 + \cdots + a_n\, dx_n \qquad (1.12.3)$$

where $a_1, a_2, \ldots, a_n$ are functions of $x_1, x_2, \ldots, x_n$, we need to determine if there is a function g such that "dg is the d of a g." The answer is yes only if dg can be integrated, i.e., if dg is an exact differential. As indicated by Eq. (1.12.2), if dg is an exact differential, the coefficients of $dx_1, dx_2, \ldots$ in the expression for dg have to be of the following form:

$$a_1 = \frac{\partial g}{\partial x_1}$$

$$a_2 = \frac{\partial g}{\partial x_2}$$

$$\vdots$$

$$a_n = \frac{\partial g}{\partial x_n} \qquad (1.12.4)$$

Since $\partial^2 g/\partial x_i\, \partial x_j = \partial^2 g/\partial x_j\, \partial x_i$ for continuous functions (and we will only deal with continuous functions), the test to determine if dg is an exact differential consists of checking to see if the relationship $\partial a_i/\partial x_j = \partial a_j/\partial x_i$ is satisfied for all the coefficients that appear in the expression for the given differential dg.

As an example, consider the differential $dg = 2x_1 \sin x_2\, dx_1 + (x_1^2 \cos x_2 + 2Kx_2)\, dx_2$, where K is a constant. We already know that $dg = d(x_1^2 \sin x_2 + Kx_2^2)$ and, thus, that $g = x_1^2 \sin x_2 + Kx_2^2$ plus a constant, since this $g(x_1, x_2)$ was one of the functions used in a previous example. But if we did not have that information, the test to see whether the given dg can be integrated to yield a function $g(x_1, x_2)$ consists of checking to see if $\partial(2x_1 \sin x_2)/\partial x_2$ is equal to $\partial(x_1^2 \cos x_2 + 2Kx_2)/\partial x_1$. For this example we have

$$\frac{\partial}{\partial x_2}(2x_1 \sin x_2) = 2x_1 \cos x_2$$

$$\frac{\partial}{\partial x_1}(x_1^2 \cos x_2 + 2Kx_2) = 2x_1 \cos x_2$$

Since the given expression for dg passes the test, a function g exists. We then must have for this example,

$$dg = 2x_1 \sin x_2\, dx_1 + (x_1^2 \cos x_2 + 2Kx_2)\, dx_2 = \frac{\partial g}{\partial x_1} dx_1 + \frac{\partial g}{\partial x_2} dx_2$$

which implies that

$$\frac{\partial g}{\partial x_1} = 2x_1 \sin x_2$$

$$\frac{\partial g}{\partial x_2} = x_1^2 \cos x_2 + 2Kx_2$$

The function $g(x_1, x_2)$ is determined by integrating these two equations. By integrating the first equation with respect to x_1 (you could also start by integrating the second equation with respect to x_2; no particular order is relevant) we obtain $g = x_1^2 \sin x_2 + h(x_2)$, where $h(x_2)$ is a function of x_2 only. This then gives

$$\frac{\partial g}{\partial x_2} = x_1^2 \cos x_2 + 2Kx_2 = x_1^2 \cos x_2 + \frac{dh}{dx_2}$$

from which we obtain,

$$\frac{dh}{dx_2} = 2Kx_2$$

and, therefore,

$$h(x_2) = \int 2Kx_2 \, dx_2 = Kx_2^2 + \text{an arbitrary constant of integration}$$

1.13 Some Basics on Differential Equations

> The material and techniques presented in this section are indispensable for analyzing the motion of dynamical systems, as done in many of the examples presented in this book. This section also may be skipped for now, but you may need to study it later in order to be able to analyze the motion governed by the differential equations that you will obtain when you investigate a dynamical system.

As mentioned earlier, application of Newton's second law to a moving body yields several scalar equations that involve unknown reaction forces and the variables that were chosen to investigate the motion of the body. By appropriately combining such equations to eliminate the unknown reaction forces, one then obtains a set of scalar equations that only involve the unknown variables that describe the motion. The variables and their time derivatives (certainly the second time derivative, and maybe the first time derivative also) will appear in such equations. If only one variable is needed to describe the motion, only one of such equations is obtained; two equations are obtained if there are two variables, and so on. Such equations are *differential equations,* and the task we now have is to integrate such equations to obtain information about the motion so that we can predict how the body or bodies are going to move in space.

In general, it is impossible to integrate the differential equations of motion analytically because they are nonlinear (i.e., they may involve products of the variables and of their time derivatives, trigonometric functions of the variables, and other nonlinear relations). Finding exact analytical solutions to nonlinear equations, in general, has so far been beyond the capacity of the human brain. However, there are several cases when it is possible to extract, analytically, important information about the motion from certain types of nonlinear differential equations. There are also several methods for finding approximate solutions to such equations. An elaborate study of the types of nonlinear differential equations that appear in dynamics, and even of the approximate methods to deal with them, is beyond the level of an intermediate course in dynamics for which this book is intended. If you want to deepen your knowledge of the subject matter, look for courses and/or books that deal with more advanced matters such as nonlinear oscillations and applications such as dynamics of gyroscopic instruments, vehicle (space and otherwise) dynamics, and dynamics of flexible bodies (such as structural dynamics).

Simple analytical methods that allow one to extract important information from several types of differential equations of motion that appear in dynamics (and seen in a number of examples presented throughout this book) are presented in this section. More advanced (and, thus, more powerful) methods are covered in Chapter 7, which is included in this book for those readers who want to start to go beyond the intermediate-level material that is in the rest of the book.

> A single dot over a symbol is used throughout the book to denote the first derivative with respect to time of the variable represented by that symbol. The second derivative with respect to time is represented by two dots over the symbol, and so on. This is a very common notation that is used in books and technical journals for the derivative of a function with respect to time.

CASE 1

For some problems the acceleration of a body may be a known function of time $f(t)$, and the differential equation of motion may be of the following form:

$$\ddot{x} + f(t) = 0 \qquad (1.13.1)$$

Examples of such simple cases are a body translating along a nonrotating straight line and subjected to a known force that is proportional to $f(t)$, or a bar that is rotating on a plane and is subjected to a known moment that is also proportional to a function $f(t)$. Of course, we would not change the independent variable from t to x in Eq. (1.13.1) by using the chain rule in the $\ddot{x}$ term, followed by an integration attempt with respect to x as done in Case 4, because the resulting term $\int f(t)\,dx$ cannot be integrated [except for the simple case when $f(t)$ is a constant]. In this case, we proceed by integrating Eq. (1.13.1) with respect to time to obtain

$$\int \ddot{x}\,dt + \int f(t)\,dt = \dot{x} + \int f(t)\,dt = \text{constant} \triangleq C_0$$

$$(1.13.2)$$

Assuming that the integration that appears in this equation can be done analytically, another integration then yields the following solution for $x(t)$ (with C_1 denoting another constant of integration).

$$\int \dot{x}\,dt + \int \left[\int f(t)\,dt\right]dt = x + \int \left[\int f(t)\,dt\right]dt$$
$$= C_0 t + C_1 \qquad (1.13.3)$$

The case when $f(t)$ is a constant leads to the well-known expressions for velocity and the position of a particle moving with constant acceleration along a straight line, which are presented in introductory physics courses.

CASE 2

Another classical and relatively simple case that is sometimes encountered in dynamics involves the type of system exemplified in Case 1 but for which the force applied to the body is a function of velocity only. In such a case, the differential equation of motion is of the following form:

$$\ddot{x} + f(\dot{x}) = 0 \quad (1.13.4)$$

This equation may be rearranged in two ways:

1. Rearranging it as $d\dot{x}/f(\dot{x}) + dt = 0$ to yield (with C_0 being a constant of integration)

$$t = C_0 - \int \frac{d\dot{x}}{f(\dot{x})} \quad (1.13.5)$$

2. Making use of the chain rule (with $\ddot{x} = \dot{x}\, d\dot{x}/dx$) to rearrange it as $(\dot{x}\, d\dot{x})/f(\dot{x}) = -dx$, and then integrating the result to obtain (with C_1 being a constant of integration):

$$\int \frac{\dot{x}\, d\dot{x}}{f(\dot{x})} + x = C_1 \quad (1.13.6)$$

CASE 3 LINEAR DIFFERENTIAL EQUATIONS WITH CONSTANT COEFFICIENTS (AN INTRODUCTION TO THE LINEAR VIBRATION PROBLEM)

Now, let us review how to find the solution to linear differential equations with constant coefficients. Such equations always appear in engineering practice in the investigation of small motions of dynamical systems about an equilibrium. A linear second-order differential equation with constant coefficients is of the form

$$m\ddot{x} + c\dot{x} + kx = f(t) \quad (1.13.7)$$

where m, c, k are constants and $f(t)$ is a known function of time. An example of a system whose motion is governed by such an equation is a mass-spring-damper system suspended vertically and subjected to the force of gravity and to a known external force $F(t)$ applied to the mass. For a mass-spring-damper system, $m, c,$ and k are, respectively, the mass, the viscous damping coefficient, and the spring stiffness, and $f(t) = mg + F(t)$, with g being the acceleration of gravity.

Both sides of Eq. (1.13.7) can, of course, be divided by m to make the coefficient of $\ddot{x}$ equal to one, as in Eqs. (1.13.14) and (1.13.17) that appear later. A far better procedure is to divide Eq. (1.13.7) by k and, for $k/m > 0$, define a nondimensional time τ as $kt^2/m = \tau^2$ (thus, $\tau = t\sqrt{k/m}$) in order to rewrite the differential equation as $d^2x/d\tau^2 + c^*dx/d\tau + x = f(\tau\sqrt{m/k})$, where $c^* = (c/k)\sqrt{k/m}$. The advantage of nondimensionalization lies in the fact that the three coefficients $m, k,$ and c appear lumped together in a single nondimensional coefficient $c^* = c/\sqrt{mk}$ (thus eliminating the need for specifying individual values for $m, c,$ and k to analyze the motion). For $c/m > 0$, the quantity $c^* = (c/k)\sqrt{k/m}$ is called the *nondimensional damping coefficient* of the system.

The solution to Eq. (1.13.7) can be decomposed into two parts as $x(t) = x_p(t) + x_h(t)$, where $x_p(t)$ is called the *particular solution* and $x_h(t)$ is called the *homogeneous solution*.[11] The homogeneous solution is, by definition, the solution for $f(t) = 0$ in Eq. (1.13.7).

In this brief review, only the case for which $f(t) = \text{constant} \triangleq A$ is considered. For $f(t) = \text{constant} \triangleq A$, the particular solution is a constant.

(continued on next page)

---

[11] This decomposition, which is only valid for linear differential equations, is called the *superposition principle*.

(continued from page 37)

With $x = \text{constant} \triangleq x_e$, Eq. (1.13.7) yields $x_e = A/k$. As mentioned earlier, a particular solution that is constant is called the *equilibrium solution* (for the sake of clarity, the subscript e is used for indicating an equilibrium solution). Notice that in the case when there is no spring, i.e., when $k = 0$, the particular solution is (with $c \neq 0$) $\dot{x}_p = \text{constant} = A/c$, which gives, $x_p = x_p(t) = At/c + x_p(0)$.

With $k \neq 0$, and $f(t) = \text{constant} \triangleq A$, substitution of $x(t) = x_e + x_h(t) = A/k + x_h(t)$ in Eq. (1.13.7) yields the following linear *homogeneous* differential equation for the homogeneous solution $x_h(t)$ [which is the deviation of $x(t)$ from the equilibrium solution $x_e = A/k$].

$$m\ddot{x}_h + c\dot{x}_h + kx_h = 0 \quad (1.13.8)$$

The solution to a homogeneous linear differential equation with constant coefficients, i.e., Eq. (1.13.8), is an exponential of the form $x_h = Be^{st}$. This is so because the exponential function is the only function for which a sum involving its derivatives is proportional to the function itself. By substituting such a solution into Eq. (1.13.8), the following algebraic equation (which is called the *characteristic equation*) is obtained for the exponent s.

$$ms^2 + cs + k = 0 \quad (1.13.9)$$

The solutions to the characteristic equation are

$$s = -\frac{c}{2m} + \sqrt{\left(\frac{c}{2m}\right)^2 - \frac{k}{m}} \triangleq s_1$$

$$s = -\frac{c}{2m} - \sqrt{\left(\frac{c}{2m}\right)^2 - \frac{k}{m}} \triangleq s_2 \quad (1.13.10)$$

and this gives, when $s_1 \neq s_2$ [i.e., when $(c/2m)^2 \neq k/m$][12]

$$x_h(t) = A_1 e^{s_1 t} + A_2 e^{s_2 t} \quad (1.13.11)$$

where A_1 and A_2 are constants of integration that are determined by making use of the initial conditions $x(0)$ and $\dot{x}(0)$.

---

[12] For the special case when $(c/2m)^2$ is exactly equal to k/m, which may rarely occur, the homogeneous solution is of the form $x_h = (A_1 + A_2 t)e^{-ct/(2m)}$.

The solutions $s = s_1$ and $s = s_2$ for the exponents s are real when $4mk < c^2$ and complex conjugate numbers when $4mk > c^2$. When either the real part of a complex conjugate solution for s_1 and s_2, or one of the real solutions, is positive, the solution given by Eq. (1.13.11) grows (because of the positive exponent) exponentially as time increases. Obviously, such a motion cannot be kept in a small neighborhood of the equilibrium solution $x_e = A/k$ [i.e., the deviation $x_h(t)$ from the equilibrium $x_e = A/k$ cannot be kept arbitrarily small] by simply choosing arbitrarily small nonzero initial conditions $x_h(0)$ and $\dot{x}_h(0)$. The equilibrium $x_e = A/k$ is said to be *unstable* in these cases.

> In a linear system (i.e., a system whose motion is governed by linear differential equations) instability implies that the motion goes to infinity as time increases. However, this is not necessarily the case for nonlinear systems [see Example 1.13.3 and several of the problems (e.g., Problems 1.29 and 1.30) in Section 1.17]. Other examples also appear elsewhere in the book. That is why the definition of instability, stated earlier, says that an equilibrium is unstable if the motion $x(t)$ that is started in a small vicinity of that equilibrium cannot be kept arbitrarily close to that equilibrium, as time passes, simply by changing the initial conditions of the motion. This definition covers linear and nonlinear systems.

When $k/m > (c/2m)^2$, the exponents s_1 and s_2 given by Eqs. (1.13.10), and the constants A_1 and A_2 in Eq. (1.13.11), are complex conjugate numbers. In such a case, the solution written in the form of Eq. (1.13.11) is a complex function. When this happens, it is more convenient to write the exponents s_1 and s_2 as $s_1 = \sigma + i\omega$ and $s_2 = \sigma - i\omega$, where $i = \sqrt{-1}$, $\sigma = -c/2m$, and $\omega = \sqrt{k/m - (c/2m)^2}$. The quantity ω is real if $c^2 < 4mk$ and imaginary if $c^2 > 4mk$. Since, as discovered by Euler,

$$\begin{aligned} e^{i\omega t} &= \cos(\omega t) + i\sin(\omega t) \\ e^{-i\omega t} &= \cos(\omega t) - i\sin(\omega t) \end{aligned} \quad (1.13.12)$$

the homogeneous solution takes the form shown in Eq. (1.13.13), where $B_1 = A_1 + A_2$ and $B_2 = i(A_1 - A_2)$ are now real quantities.

$$x_h = e^{\sigma t}[B_1 \cos(\omega t) + B_2 \sin(\omega t)] \quad (1.13.13)$$

This solution shows that, when $k/m > (c/2m)^2$ (i.e., when the exponents s_1 and s_2 are complex conjugates), the homogeneous solution is oscillatory, with frequency equal to ω, and with an "amplitude" that decreases with time when $c/m > 0$ (which, is the case of an actual damper). The equilibrium $x_e = A/k$ is said to be *asymptotically stable* in this case, and such motion decays toward the equilibrium with the passage of time. Asymptotic stability is a desirable feature in practice.

For the case when $c = 0$ (i.e., no damping), the solution $x(t) = x_e + x_h(t) = x_e + B_1 \cos(\omega t) + B_2 \sin(\omega t)$ is periodic [i.e., it repeats itself every $t = 2\pi/\omega$ s, when ω is expressed in radians per second (rad/s)], with frequency equal to $\omega = \sqrt{k/m}$. The equilibrium solution $x_e = k/A$ is said to be *stable* in this case, and $x(t)$ simply consists of a perennial oscillation about the equilibrium $x = x_e$. For a stable motion, the amplitude of the oscillations $x_h(t) = x(t) - x_e$ can be made arbitrarily small by simply changing the initial conditions of the motion. The frequency $\omega = \sqrt{k/m}$ is called the *undamped natural frequency* of the oscillation about that equilibrium.

Oscillations that are governed by the linear differential equation (1.13.7) belong to a class of problems that is called *linear vibrations*. The field of vibrations is classified into two categories: *linear vibrations* and *nonlinear vibrations* (or *nonlinear oscillations*).

Although the more advanced subject of nonlinear oscillations is beyond the scope of this book, every engineer should be aware that there are circumstances for which nonlinearities in the differential equations of motion play a dominant role in the motion. Nonlinear terms in the differential equation (in general, equations) of motion are actually present in every system (even for a simple system like a pendulum). Except for rare simple cases, linear differential equations start to appear only after nonlinear differential equations are *linearized* for small motions about a particular solution (such as an equilibrium solution) exhibited by the system. The linearization process is presented in Section 1.14, and you will encounter it in several problems solved in this book.

In several examples in this book, Simulink is used for obtaining the numerical solution to differential equations of motion (which are generally nonlinear), instead of writing computer code to do so. A tutorial on the use of Simulink is given in Appendix A. Several tutorials are also available on the World Wide Web (search for them under "simulink tutorial"; include the quotation marks to restrict the search).

Summary of Main Ideas in Case 3

1. The solution of a linear differential equation with constant coefficients, $m\ddot{x} + c\dot{x} + kx = A$, where A is a constant and $k \neq 0$, is of the form $x(t) = x_e + Be^{st}$, where $x_e = \text{constant} = A/k$ is the equilibrium solution.

2. The exponents s are the roots $s = s_1$ and $s = s_2$ of the equation $ms^2 + cs + k = 0$, which is called the *characteristic equation*.

3. The equilibrium $x = x_e$ is
   - *Unstable* if the real part of *any* of the two exponents s_1 or s_2 is positive. This happens if either c/m or k/m is negative. It also happens when $c = 0$ and k/m is negative.
   - *Asymptotically stable* if the real part of the exponents s_1 and s_2 is negative. For this to happen, c/m and k/m have to be positive. The motion is a damped oscillation in this case.
   - *Stable* if $c = 0$ and k/m is positive. In this case the exponents s_1 and s_2 are purely imaginary. The motion is an undamped oscillation with frequency $\omega = \sqrt{k/m}$.

Therefore, the check for stability or instability simply consists in looking at the signs of the coefficients k/m and c/m.

EXAMPLE 1.13.1

The motion of a dynamical system is governed by the following differential equation:

$$\ddot{x} + (c-4)\dot{x} + Kx = 16$$

a. Determine the value of K if the system exhibits the equilibrium solution $x_e = 8$. For what range of values of c is the equilibrium stable, and for what range is it unstable? For what range of c is the equilibrium asymptotically stable?
b. Repeat part a for $x_e = -8$.
c. Choose your own value of c for the equilibrium to be asymptotically stable, and then determine the value of K for the system to oscillate with a frequency equal to $\omega = 1.5$. What is the value of the equilibrium x_e in this case?

■ Solution

a. Since $x_e = 16/K = 8$, it follows that $K = 2$. The equilibrium is asymptotically stable [i.e., the motion $x(t)$ approaches the equilibrium $x = \text{constant} = x_e$ as time goes to infinity] if the coefficients of $\ddot{x}, \dot{x}$, and x of the given linear differential equation have the same sign, and unstable otherwise. Therefore, since $K = 2$ is positive, the equilibrium $x_e = 8$ is asymptotically stable if $c > 4$, and unstable if $c < 4$. The case $c = 4$ is a special case. In this case, the equilibrium is classified as *stable* (instead of asymptotically stable) because, with $K > 0$, the motion is an undamped oscillation about the equilibrium.
b. For $x_e = -8 = 16/K$ we obtain $K = -2$. This does not change the characteristic equation for the given differential equation, which still is $s^2 + (c-4)s + K = 0$. However, since K is negative, the equilibrium $x_e = 16/K = -8$ is unstable. The roots of the characteristic equation are $2 - c/2 \pm \sqrt{(2-c/2)^2 + 2}$.
c. Since we want the equilibrium to be asymptotically stable, we have to have $c > 4$ and $K > 0$. Let $c = 5$, for example. The differential equation is then $\ddot{x} + \dot{x} + Kx = 16$. Since the solution to the differential equation is of the form $x = 16/K + Be^{st}$, the characteristic equation is $s^2 + s + K = 0$, and its roots are $s = -0.5 \pm \sqrt{0.25 - K}$. In order to have $s = -0.5 \pm i\omega$, with $\omega = 1.5$, we then need to have $0.25 - K = -1.5^2$, which gives $K = 2.5$. The equilibrium solution in this case is $x_e = 16/K = 6.4$. The motion in this case tends to $x = x_e = 6.4$ as time goes to infinity. For a plot of such a motion, see the example in Appendix A. More details on the linear vibration problem are presented in Chapter 8.

CASE 4

For a number of problems in dynamics (such as problems involving a frictionless pendulum and a frictionless spring-mass system) the differential equation that governs the motion of a body is of the following form:

$$\ddot{x} + f(x) = 0 \qquad (1.13.14)$$

Equation (1.13.14) can be integrated analytically if the function $f(x)$ is of the form $f(x) = Ax + B$, where A and B are constants. An example of a mechanical system whose motion is governed by such an equation with $f(x) = Ax + B$ is a mass moving along a straight line and connected to a massless linear spring, with the other end of the spring connected to a fixed point. The same equation, with $A = 0$, also appears in the investigation of the short-range motion of a projectile subjected only to the gravitational force of the earth. For the case when $A = 0$, the solution for $x(t)$ is immediately obtained as $x(t) = -Bt^2/2 + C_1 t + C_2$, where C_1 and C_2 are constants of integration that are determined by the initial conditions $x(0)$ and $\dot{x}(0)$.

For any function $f(x)$ (except for a constant), it is useless to try to integrate Eq. (1.13.14) with respect to time by first multiplying both sides of the equation by dt to obtain

$$\int \ddot{x}\, dt + \int f(x)\, dt = \text{constant}$$

(What not to do unless $f(x)$ is constant.)

The first term in this equation integrates to $\dot{x}$, but the second term cannot be integrated (unless f is a constant). The operation that leads to this equation (i.e., first, a multiplication by dt, and then the attempt to integrate) simply put us "against a corner," so to speak, from which there is no way out.

In an attempt to integrate Eq. (1.13.14) once, what if we try the chain rule to change the independent variable from t to x (which is the argument of the function f)? By recognizing that $\ddot{x}$ is simply a notation for $d(dx/dt)/dt$, we can use the chain rule to write it as

$$\frac{d}{dt}\left(\frac{dx}{dt}\right) = \left[\frac{d}{dx}\left(\frac{dx}{dt}\right)\right]\frac{dx}{dt}$$

By proceeding in this manner, Eq. (1.13.14) is then rewritten as

$$\dot{x}\frac{d\dot{x}}{dx} + f(x) = 0 \qquad (1.13.15)$$

which now yields

$$\int \dot{x}\frac{d\dot{x}}{dx}\, dx + \int f(x)\, dx = \text{constant}$$

or

$$\frac{\dot{x}^2}{2} + \int f(x)\, dx = \text{constant} \triangleq C_0 \qquad (1.13.16)$$

The integration attempt to obtain $\dot{x}$ as a function of x was now successful (at least in principle, because the expression for $f(x)$ may be so complicated that no known analytical form for the integral exists). For many dynamics problems the form of $f(x)$ is such that it can indeed be integrated analytically. In such cases, Eq. (1.13.16) is very useful since, without even knowing the solution for $x(t)$, it allows us to find the candidates x_m for the maximum and minimum values of $x(t)$. Such values are simply obtained by setting $\dot{x} = 0$ in Eq. (1.13.16) and then solving the resulting algebraic equation for x_m.

Notice that Eq. (1.13.16), which relates $\dot{x}$ to x, is a first-order differential equation for $x(t)$. The constant of integration C_0 is determined, by using the initial conditions $x(0) = x_0$ and $\dot{x}(0) = \dot{x}_0$, as $C_0 = \frac{1}{2}\dot{x}_0^2 + [\int f(x)\, dx]_{x=x_0}$.

Note: When there is *viscous friction* (or *viscous damping*) in the system (and this is very common in practice), Eq. (1.13.14) becomes (when the damping is linear)

$$\ddot{x} + c\dot{x} + f(x) = 0 \qquad (1.13.17)$$

where c is a constant. A *linear viscous damper* is a device that *reacts* to motion by applying, to the body it is connected to, a force proportional to velocity $\dot{x}$ (or a moment that is proportional to angular velocity, for the case of a torsional viscous damper). An example of such a device is the shock absorber that is present in every car (and mounted in parallel with the big springs

(continued on next page)

(continued from page 41)

that are located underneath the car). The case when $f(x)$ is of the form $Ax + B$, which is a simpler but important case that appears in practice, was considered separately. It is important to notice that the analytical technique based on the chain rule, which was presented in this section, will not lead us anywhere in the attempt to integrate Eq. (1.13.17) when $c \neq 0$. The reason for this is that application of the chain rule now yields, after the integration attempt,

$$\frac{\dot{x}^2}{2} + c\int \dot{x}\, dx + \int f(x)\, dx = \text{constant}$$

The problem with this integration attempt is that the second term in this equation cannot be integrated because it is not known how the integrand $\dot{x}$ depends on the unknown function $x(t)$. In such cases, when $f(x)$ is a nonlinear function of x, one would rely mostly on numerical integration of Eq. (1.13.17).

Consider now Examples 1.13.2 to 1.13.4. For each example, k, L, m, and g are given positive constants.

EXAMPLE 1.13.2

The motion of a vertical pendulum of length L m (such as that shown in Fig. 1.11, on p. 28, with $M_{\text{bar}} = 0$) is governed by the following differential equation:

$$\ddot{\theta} + k\sin\theta = 0$$

where $k = g/L$ and $g = 9.81$ m/s^2 is the acceleration of gravity. The variable $\theta(t)$ is the angle the pendulum makes with the vertical.

a. Determine all the equilibrium values of $\theta(t)$.
b. Determine the maximum and minimum values of $\theta(t)$ if the motion of the pendulum is started with the initial conditions $\theta(0) = 56.6°$ and $\dot{\theta}(0) = 1.53\sqrt{k}$ rad/s.

■ **Solution**

a. At an equilibrium we must have $\theta = \text{constant} \triangleq \theta_e$, with θ_e being the solutions to the following algebraic equation that is obtained from the given differential equation of motion.

$$\sin\theta_e = 0$$

The solutions for θ_e are then $\theta_e = 0, \pm\pi, 2\pi, \ldots$. (Of all these equilibrium solutions, only two are physically distinct from each other, i.e., $\theta_e = 0$ and $\theta_e = \pi$.)

b. By applying the chain rule procedure presented in this section to the given differential equation, we obtain the following result:

$$\frac{\dot{\theta}^2}{2} + k\int \sin\theta\, d\theta = \frac{\dot{\theta}^2}{2} - k\cos\theta = \text{constant} \triangleq C_1 = \frac{\dot{\theta}^2(0)}{2} - k\cos\theta(0) \approx 0.62k$$

The maximum and minimum values of θ, which we will denote as θ_m, occur when $\dot{\theta} = 0$. With $\dot{\theta} = 0$ we then obtain

$$\cos\theta_m = -0.62$$

and this yields $\theta_m \approx 128.3°$ and $\theta_m \approx -128.3°$ (or $\theta_m = 360 - 128.3 = 231.7°$). The values of $\ddot{\theta} = -k\sin\theta$ at these values of θ are approximately

- $-0.78k$ (which is negative because k is positive) when $\theta = 128.3°$. This indicates that $\theta = 128.3°$ is a maximum value.
- $0.78k$ (which is positive) when $\theta = -128.3°$ (or $231.7°$). This points to a minimum value of θ.

Therefore, we conclude the following:

$$\theta_{\max} = 128.3° \qquad \theta_{\min} = -128.3°$$

Thus, the pendulum oscillates between these two values of the angle θ. Next, Example 1.13.3 sheds more light on the motion of the pendulum (especially after observing Fig. 1.13).

EXAMPLE 1.13.3

What are the maximum and minimum values of $\theta(t)$ for Example 1.13.2 if the motion of the pendulum is started with the initial conditions $\theta(0) = 0$ and $\dot{\theta}(0) = 2.1\sqrt{k}$ rad/s? Plot $\dot{\theta}$ versus θ, and use the plot to explain the resulting motion of the pendulum.

■ **Solution**

For these new initial conditions we obtain

$$\frac{\dot{\theta}^2}{2} + k\int \sin\theta \, d\theta = \frac{\dot{\theta}^2}{2} - k\cos\theta = \text{constant} \stackrel{\Delta}{=} C_2 = \frac{\dot{\theta}^2(0)}{2} - k\cos\theta(0) = 1.205k$$

From this equation, the value of $\theta = \theta_m$ for which $\dot{\theta} = 0$ is obtained as $\cos\theta_m = -1.205$. Since the cosine of an angle is bounded between -1 and 1, we then conclude that there are no minimum and maximum values of θ for the given initial conditions.

If the angle θ does not exhibit maximum and minimum values, what is the pendulum doing then? The answer to such a question may be obtained by plotting $\dot{\theta}$ versus θ (such a plot is known in dynamics and vibrations as a *phase plane plot*), with $\dot{\theta}$ given by the following relationship.

$$\frac{\dot{\theta}^2}{2k} - \cos\theta = \text{constant} \stackrel{\Delta}{=} C$$

Such a plot, when drawn for several values of the constant C, actually provides an understanding of the types of motion the system may exhibit.

Let us plot the nondimensional angular velocity $\dot{\theta}/\sqrt{k}$ (or, simply, $d\theta/d\tau$, where $\tau = t\sqrt{k}$ is nondimensional time) versus θ so that we do not have to concern ourselves with a specific value for k. The phase plane plot for the pendulum, for θ in the range $-2\pi \leq \theta \leq 2\pi$, is shown in Fig. 1.13 (the plot repeats itself outside of that range). The three points marked with a "•" correspond to the equilibrium values $\theta_e = -2\pi$, 0, and 2π (which were determined in Example 1.13.2 and are physically the same as $\theta_e = 0$).

As indicated in Fig. 1.13, the curves $C = $ constant were generated for the following values of C: 0.62 (obtained in Example 1.13.2), 1.205 (obtained in this example), and also 1 and -0.8. Notice that, as indicated by the equation $\dot{\theta}^2/(2k) - \cos\theta = \text{constant} \stackrel{\Delta}{=} C$, the curves that exhibit maximum and minimum values of θ (i.e., the *closed curves* that exhibit a value of $\dot{\theta} = 0$) are those for which C is between -1 and 1, which is the range of $\cos\theta$ when θ is real. It is also worth observing that since a function $\theta(t)$ increases when $\dot{\theta}(t)$ is positive, and decreases when $\dot{\theta}(t)$ is negative, any curve $\dot{\theta}$ versus θ in a phase plane plot is

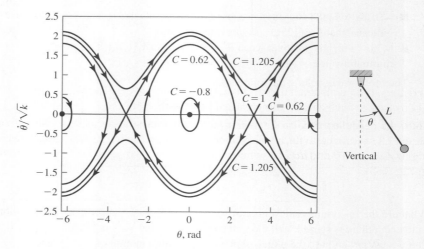

Figure 1.13 Plot of $\dot{\theta}(t)/\sqrt{k}$ versus θ for a pendulum.

always traced (with the passage of time) in the direction of the arrows shown in Fig. 1.13 (this observation is true for any other dynamical system). This can be easily observed if the differential equation of motion for the system is integrated numerically, and the result displayed as the integration proceeds with time.

Inspection of Fig. 1.13 shows that the motion $\theta(t)$ is oscillatory (thus it exhibits a maximum and a minimum value) only if the initial conditions are such that $C < 1$. This happens when the initial condition pair of values $(\dot{\theta}(0)/\sqrt{k}, \theta(0))$ fall inside the region bounded by the curves that pass through the points $(\dot{\theta}(0)/\sqrt{k} = 0, \theta(0) = -\pi)$ and $(\dot{\theta}(0)/\sqrt{k} = 0, \theta(0) = \pi)$. As determined in Example 1.13.2, $\theta = \pm\pi$ are also equilibrium points for the pendulum. Such motions are oscillations about the equilibrium $\theta_e = 0$ which is, thus, a stable equilibrium.

If the initial conditions are such that $C > 1$ [which happens when the initial condition pair of values $(\dot{\theta}(0)/\sqrt{k}, \theta(0))$ fall outside the region just mentioned], the angular velocity $\dot{\theta}$ of the pendulum never reaches zero and, as a consequence, the angle θ keeps either increasing or decreasing (i.e., the pendulum shown in the rightmost part of Fig. 1.13 keeps rotating either counterclockwise or clockwise). A motion started with initial conditions that fall on the curve $C = 1$ ends up at the equilibrium point $\theta = \pi$, but it takes an infinite time for it to reach that equilibrium. As disclosed by Fig. 1.13, the equilibrium $\theta_e = \pi$ is unstable (because the motion cannot be kept in a small neighborhood of that equilibrium when that equilibrium is perturbed). The determination of the stability of small motions about these equilibrium points for this system, using linearization, is presented in Example 1.15.3.

These conclusions about the motion can be verified by an actual experiment with a pendulum.

A more detailed and enlightening visualization concerning the motion of a system governed by the general differential equation $\ddot{x} + c\dot{x} + f(x) = 0$, where $f(x)$ is any function (linear or nonlinear) of the motion variable x, is obtained if we inspect the three-dimensional plot involving x, $\dot{x}$, and the function $C(x, \dot{x}) \triangleq \dot{x}^2/2 + \int f(x)\, dx$. Notice that,

$$\frac{d}{dt} C(x, \dot{x}) = \dot{x}\ddot{x} + \dot{x} f(x) = -c\dot{x}^2$$

and, thus, C = constant if there is no damping in the system (in which case, $c = 0$). Let us look at such a three-dimensional plot for the pendulum of Examples 1.13.2 and 1.13.3, for which $C = C(\theta, \dot{\theta}) = \dot{\theta}^2/(2k) - \cos\theta$, where $k = g/L$ (g being the acceleration of gravity, and L the length of the pendulum). A three-dimensional plot of C versus θ and $d\theta/d\tau$, where $\tau = t\sqrt{k}$ is a nondimensional time, is shown in Fig. 1.14.[13]

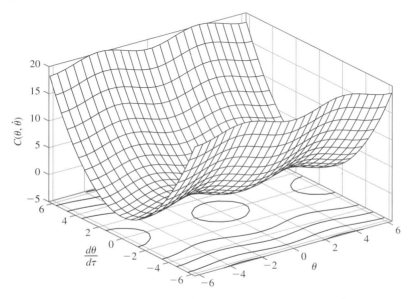

Figure 1.14 The surface $C(\theta, \dot{\theta}) = \dot{\theta}^2/2k - \cos\theta$. (Notice that $\tau = t\sqrt{k}$, and thus, $d\theta/d\tau = (1/\sqrt{k})\,\dot{\theta}$.)

The surface seen in Fig. 1.14 is like a mountain terrain that exhibits peaks and valleys, except that the plot in that figure is not a physical plot like that at all

---

[13] Figure 1.14 was generated in MATLAB using the following commands typed in the MATLAB command window (semicolons that appear in any line separate different commands in that line). Labels were inserted using the editing capabilities of the figure window.

```
X=-2*pi:0.5:2*pi; Xdot=-2*pi:0.5:2*pi; [x, xdot]=meshgrid(X, Xdot);
C=xdot.^2/2-cos(x); len=size(x); meshc(x, xdot, C, 5*ones(len));
axis([-2*pi,   2*pi,   -2*pi,   2*pi,   -5,    20]);
% Note: The .^ in the expression for C is the MATLAB exponentiation
% operator for arrays.
% A brief tutorial of MATLAB is given at the end of Appendix A.
```

because its θ and $\dot{\theta}$ axes are angular orientation and angular velocity, respectively, instead of Cartesian coordinates that specify position in space. The peaks of that surface (which appear at $\theta_e = \pm\pi, \pm 3\pi, \ldots$) correspond to unstable equilibrium points of the pendulum, while the lowest point on the valleys (which appear at $\theta_e = 0, \pm 2\pi, \ldots$) correspond to a stable equilibrium point.

The curves shown in Fig. 1.13 are cuts on the three-dimensional surface of Fig. 1.14 for different values of C. They are the level curves $C = $ constant for that surface; some of these curves are also shown in Fig. 1.14 on the $(d\theta/d\tau)$-θ plane.

If the plot shown in Fig. 1.14 were an actual physical plot of a mountain terrain, a small ball placed either at the peaks or at the bottom of a valley would stay there forever. If the ball were displaced from a peak, it would move away and it would never be able to stay close to that peak for all times. Such peaks are the physical analogy to the unstable equilibrium points in Fig. 1.13, while the bottom of the valleys of the physical terrain are the physical analogy to the stable equilibrium points in that figure; the ball in this analogy would oscillate back and forth near the bottom of a curved valley of the physical terrain. If there were friction in the physical terrain (which is equivalent to having friction in the dynamical system under investigation), the ball would roll along the physical surface on the terrain and eventually end up at the bottom of one of the curved valleys. The stable equilibrium points would then be *asymptotically stable* (instead of just stable) in such a case. Dynamicists love to come up with such analogies. They actually help to interpret difficult motions and to explain the concept of stability.

Returning to Fig. 1.14, where $C(\theta, \dot{\theta}) = \dot{\theta}^2/(2k) - \cos\theta$ is plotted versus θ and $\dot{\theta}$, keep in mind that an undamped motion takes place along a level curve $C = $ constant in that plot. If there were viscous damping in the pendulum, so that the damping moment is $-c\dot{\theta}$ where c [in N · m/(rad/s)] is the damping coefficient, we would have $d(C(\theta, \dot{\theta}))/dt = -c\dot{\theta}^2$. In such a case, the phase plane plot $\dot{\theta}$ versus θ would start at a given initial point $\dot{\theta}(0)$-$\theta(0)$ and end up at the stable equilibrium point in a spiraling manner. Use Simulink to see this, and play with different values of c to see what happens. A single trajectory can be plotted using the `XY Graph scope` described in Appendix A. Include the `animate_nbars` block in your model if you want to see the animation of the pendulum while the differential equation of motion $\ddot{\theta} + c\dot{\theta} + k\sin\theta = 0$ is being integrated.

Figure 1.15[14] shows the shape of the three-dimensional surface $C(\theta, \dot{\theta}) = \dot{\theta}^2/(2k) - \cos\theta$ near a stable equilibrium point (Fig. 1.15a) and near an unstable

---

[14] Figure 1.15 was generated in MATLAB using the following commands typed in the MATLAB command window. Labels were inserted using the editing capabilities of the figure window.

```
X=-2:0.2:2; Xdot=-2:0.2:2; [x, xdot]=meshgrid(X, Xdot);
C=xdot.^2/2-cos(x); len=size(x); subplot(1,2,1);
meshc(x, xdot, C, 5*ones(len)); axis([-2, 2, -2, 2, -1, 2.5]);
X=[-5:0.1:-1]; Xdot=[-1:0.1:1]; [x, xdot]=meshgrid(X, Xdot);
C=xdot.^2/2-cos(x); len=size(x); subplot(1,2,2);
meshc(x, xdot, C, 5*ones(len)); axis([-5, -1, -1, 1, -1, 1.5]);
```

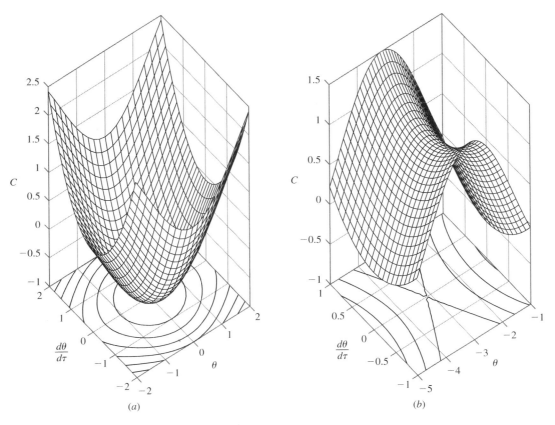

Figure 1.15 Closeup view of the surface $C(\theta, \dot{\theta})$ (a) near a center and (b) near a saddle point.

equilibrium point of the pendulum (Fig. 1.15b); some level curves on the plane $(d\theta/d\tau)$-θ, which correspond to $C =$ constant, are also shown in the figure.

As seen in Fig. 1.15, the three-dimensional surface $C(\theta, \dot{\theta}) = \dot{\theta}^2/(2k) - \cos\theta$ resembles a cup in the neighborhood of a stable equilibrium, and the cuts $C =$ constant (i.e., the level curves) are closed curves in such a case. For this reason, a stable equilibrium of an undamped system is called a *center*. It is called a stable *focus* if it is an asymptotically stable equilibrium; if c were negative, it would be called an unstable focus. The situation is quite different in the neighborhood of the unstable equilibrium point $\theta_e = \pi$ for the pendulum. The three-dimensional surface $C(\theta, \dot{\theta}) = \dot{\theta}^2/(2k) - \cos\theta$ now resembles a saddle in the vicinity of that unstable point for the pendulum. Because of such resemblance, the unstable equilibrium in this example is called a *saddle point*.

EXAMPLE 1.13.4

The motion of a mechanical system is governed by the following differential equation

$$m\ddot{x} + k(x - L) = mg$$

The system consists of a block of mass m kg connected to a spring with an unstretched length L m and a stiffness (or *spring constant*) k N/m, and is constrained to move along the vertical. The other end of the spring is connected to a fixed point. The system is shown in Fig. 1.16.

The variable $x(t)$ in Fig. 1.16 is the actual length of the stretched spring. For $mg/(kL) = 3$, determine the maximum and the minimum values of $x(t)$ when the motion of the system is started with the initial conditions $x(0) = L$ m and $\dot{x}(0) = 2L\sqrt{k/m}$ m/s.

Figure 1.16
A spring-mass system constrained to move along the vertical.

■ Solution
By making use of the chain rule to write $\ddot{x}$ as $\ddot{x} = (d\dot{x}/dx)\dot{x}$, the given differential equation can then be integrated once, as shown earlier in this section, to yield

$$\frac{m}{2k}\dot{x}^2 + \int \left(x - L - \frac{mg}{k}\right) dx = \frac{m}{2k}\dot{x}^2 + \frac{(x-L)^2}{2} - \frac{mg}{k}x = \text{constant} \triangleq C_3$$

$$= \frac{m}{2k}\dot{x}^2(0) + \frac{1}{2}[x(0) - L]^2 - \frac{mg}{k}x(0) = 2L^2 - \frac{mgL}{k}$$

The maximum and the minimum values of x occur when $\dot{x} = 0$. By letting x_m denote such values, this equation then yields, with $\dot{x}_m = 0$,

$$\frac{1}{2}(x_m - L)^2 - \frac{mg}{k}(x_m - L) = 2L^2$$

This quadratic equation can be readily solved for $x_m - L$ (and, thus, for x_m). The solutions are

$$x_m = L + \frac{mg}{k} \pm L\sqrt{\left(\frac{mg}{kL}\right)^2 + 4}$$

For $mg/(kL) = 3$, we then obtain $x_m \approx 7.61L$ and $0.39L$. These are, respectively, the maximum and minimum values of $x(t)$ (the corresponding values of $\ddot{x}$ are $\ddot{x} = -3.61kL/m$ when $x = 7.61L$, and $\ddot{x} = 3.61kL/m$ when $x = 0.39L$).

Since $x(t)$ has maximum and minimum values, the motion consists of an oscillation between $x_{\min} \approx 0.39L$ and $x_{\max} \approx 7.61L$. Since the differential equation $m\ddot{x} + k(x-L) = mg$ is linear, the frequency of the oscillation (which is an undamped oscillation) is equal to $\sqrt{k/m}$ rad/s (see Case 3, p. 37).

1.14 Taylor Series Expansion and Linearization

Consider a given continuous function $f(x)$ of a variable x. The idea of a Taylor series expansion [named after the English mathematician Brook Taylor (1685–1731)] of the function $f(x)$ is to approximate the function in the vicinity of a desired value of x, say $x = x^*$, by a polynomial series in $x_s \stackrel{\Delta}{=} x - x^*$ as

$$f(x) = f(x^* + x_s) \approx A_0 + A_1 x_s + A_2 x_s^2 + A_3 x_s^3 + \cdots \quad (1.14.1)$$

and then truncate the series at some specified degree of x_s. The new variable $x_s = x - x^*$, which is the x-deviation from the value $x = x^*$, is called the *perturbation* variable. The Taylor series approximation of the function $f(x)$ provides a reasonable approximation for the function only in a small vicinity of $x_s = 0$.

The value of the coefficient A_0 in Eq. (1.14.1) is obtained by simply setting $x_s = 0$ (i.e., $x = x^*$) in that equation, and this yields [which requires that the given function $f(x)$ be continuous at $x = x^*$]

$$A_0 = [f(x^* + x_s)]_{x_s=0} = f(x^*)$$

The value of the coefficient A_1 is determined by simply taking the derivative of both sides of Eq. (1.14.1) with respect to x_s and then evaluating the result at $x_s = 0$ to obtain [which requires that the first derivative of the given function $f(x)$ also be continuous at $x = x^*$]

$$A_1 = \left[\frac{df(x^* + x_s)}{dx_s}\right]_{x_s=0} = \left[\frac{df(x)}{dx}\right]_{x=x^*}$$

By continuing to take higher derivatives of Eq. (1.14.1), and evaluating them at $x_s = 0$, the values of the remaining coefficients $A_2, A_3, \ldots$ are then obtained as

$$A_k = \frac{1}{k!}\left[\frac{d^k f(x^* + x_s)}{dx_s^k}\right]_{x_s=0} = \frac{1}{k!}\left[\frac{d^k f(x^*)}{dx^k}\right]_{x=x^*} \quad k = 1, 2, \ldots \quad (1.14.2)$$

Thus, the approximation of $f(x)$ by an nth-degree polynomial requires that the function $f(x = x^* + x_s)$ and its first n derivatives be continuous.

Following the same procedure, the Taylor series expansion of a function of several variables may be obtained. Consider, for example, a continuous function $f(x_1, x_2)$ of two variables x_1 and x_2, with continuous partial derivatives with respect to each variable. Its expansion about a pair of values $x_1 = x_1^*$ and $x_2 = x_2^*$ is written as

$$f(x_1, x_2) = f(x_1^*, x_2^*) + A_1(x_1 - x_1^*) + A_2(x_2 - x_2^*) + B_{11}(x_1 - x_1^*)^2$$
$$+ B_{12}(x_1 - x_1^*)(x_2 - x_2^*) + B_{22}(x_2 - x_2^*)^2 + \cdots \quad (1.14.3)$$

where the "$\cdots$" indicates the cubic and higher-order terms $(x_1 - x_1^*)^3$, $(x_1 - x_1^*)^2(x_2 - x_2^*)$, $(x_1 - x_1^*)^4$, etc.

The coefficients $A_1, A_2, \ldots$, that appear in Eq. (1.14.3) are determined as

$$A_1 = \left[\frac{\partial f}{\partial x_1}\right]_{x_1=x_1^*, x_2=x_2^*} \qquad A_2 = \left[\frac{\partial f}{\partial x_2}\right]_{x_1=x_1^*, x_2=x_2^*}$$

$$B_{11} = \frac{1}{2}\left[\frac{\partial^2 f}{\partial x_1^2}\right]_{x_1=x_1^*, x_2=x_2^*} \qquad B_{12} = \left[\frac{\partial^2 f}{\partial x_1 \partial x_2}\right]_{x_1=x_1^*, x_2=x_2^*} \qquad (1.14.4)$$

$$B_{22} = \frac{1}{2}\left[\frac{\partial^2 f}{\partial x_2^2}\right]_{x_1=x_1^*, x_2=x_2^*}$$

A function $f(x)$ is said to have been linearized about $x = x^*$, when only the constant term $A_0 = f(x^*)$ and the linear term $A_1 x_s$, where $x_s = x - x^*$ is small, are kept in the Taylor series approximation given by Eq. (1.14.1). Linearization is not valid if $A_1 = 0$; the behavior of $f(x)$ in such a case is nonlinear even in a small neighborhood of $x = x^*$ (see Problem 1.31).

Similarly, linearization (about $x_1 = x_1^*, x_2 = x_2^*$, etc.) of a function $f(x_1, x_2, \ldots)$ of several variables $x_1, x_2, \ldots$ implies keeping only the constant term $f(x_1^*, x_2^*, \ldots)$ and the linear terms in $x_{s_1} \triangleq x_1 - x_1^*, x_{s_2} \triangleq x_2 - x_2^*, \ldots, x_{s_n} \triangleq x_n - x_n^*$ in its Taylor series expansion about the point under consideration.

EXAMPLE 1.14.1

Linearize $f_1 = \sin\theta$ and $f_2 = \cos\theta$ about $\theta = \theta^*$. In particular, what are the linearized approximations for $\sin\theta$ and $\cos\theta$ about $\theta = 0$ and $60°$?

■ Solution

With $\theta \triangleq \theta^* + \theta_s$, we write

$$\sin\theta = \sin(\theta^* + \theta_s) = A_0 + A_1\theta_s + \cdots$$

$$A_0 = \sin\theta^* \qquad A_1 = \left[\frac{d}{d\theta_s}\sin(\theta^* + \theta_s)\right]_{\theta_s=0} = [\cos(\theta^* + \theta_s)]_{\theta_s=0} = \cos\theta^*$$

Thus, the linearized approximation for $\sin\theta$ about any arbitrary value $\theta = \theta^*$ is $\sin\theta = \sin(\theta^* + \theta_s) = \sin\theta^* + \theta_s \cos\theta^*$.

Similarly, for $\cos\theta$ we have

$$\cos\theta = \cos(\theta^* + \theta_s) = B_0 + B_1\theta_s + \cdots$$

$$B_0 = \cos\theta^* \qquad B_1 = \left[\frac{d}{d\theta_s}\cos(\theta^* + \theta_s)\right]_{\theta_s=0} = -[\sin(\theta^* + \theta_s)]_{\theta_s=0} = -\sin\theta^*$$

For $\theta^* = 0$ and $60°$ ($\pi/3$ rad), the linearized approximations for $\sin\theta$ and $\cos\theta$ are

For $\theta^* = 0$:

$$\sin\theta = \theta + \cdots \qquad \cos\theta = 1 + \cdots$$

SECTION 1.14 Taylor Series Expansion and Linearization

For $\theta^* = \pi/3$ rad:

$$\sin\theta = \sin\frac{\pi}{3} + \left(\cos\frac{\pi}{3}\right)\left(\theta - \frac{\pi}{3}\right) = \frac{\sqrt{3}}{2} + \frac{1}{2}\left(\theta - \frac{\pi}{3}\right) + \cdots$$

$$\cos\theta = \cos\frac{\pi}{3} - \left(\sin\frac{\pi}{3}\right)\left(\theta - \frac{\pi}{3}\right) = \frac{1}{2} - \frac{\sqrt{3}}{2}\left(\theta - \frac{\pi}{3}\right) + \cdots$$

where "$\cdots$" denotes the nonlinear terms in the perturbation $\theta - \theta^*$ [i.e., the terms involving $(\theta - \theta^*)^2$, $(\theta - \theta^*)^3$, etc.] that are disregarded.

The function $\sin\theta$, and its approximations $\sin\theta^* + (\cos\theta^*)(\theta - \theta^*)$ and $\sin\theta^* + (\cos\theta^*)(\theta - \theta^*) - (\sin\theta^*)(\theta - \theta^*)^2/2$, for $\theta^* = \pi/3$, are shown in Fig. 1.17. Since $-1 \leq \sin\theta \leq 1$, the Taylor series approximations are only valid in the region where their values are between -1 and 1. It can be seen in Fig. 1.17 that the linear approximation is valid only in a very small neighborhood of $\theta^* = \pi/3$, while the quadratic approximation is valid in a little larger neighborhood of $\theta^* = \pi/3$.

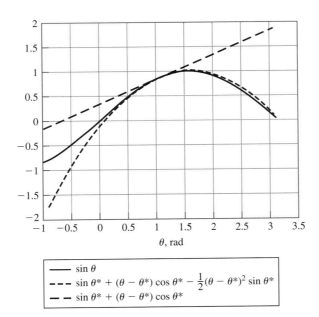

Figure 1.17 The function $\sin\theta$ and two of its Taylor series approximations about $\theta^* = \pi/3$ rad.

EXAMPLE 1.14.2

Expand $f(x) = \sqrt{1-x}$ in Taylor series about $x = 0$, and keep terms up to the quadratic order in the expansion.

■ **Solution**
The expansion of $f(x) = \sqrt{1-x} = (1-x)^{1/2}$ about $x = 0$ is

$$\sqrt{1-x} = 1 + A_1 x + A_2 x^2 + \cdots$$

where

$$A_1 = \left[\frac{df}{dx}\right]_{x=0} = \left[\frac{1}{2}(1-x)^{1/2-1}\frac{d(1-x)}{dx}\right]_{x=0} = -\left[\frac{1}{2}(1-x)^{-1/2}\right]_{x=0} = -\frac{1}{2}$$

$$A_2 = \frac{1}{2}\left[\frac{d^2 f}{dx^2}\right]_{x=0} = \frac{1}{2}\left\{\frac{d}{dx}\left[\left(-\frac{1}{2}(1-x)^{-1/2}\right)\right]\right\}_{x=0}$$

$$= \frac{1}{2}\left[-\frac{1}{4}(1-x)^{-3/2}\right]_{x=0} = -\frac{1}{8}$$

Therefore, the Taylor series of $\sqrt{1-x}$ about $x = 0$ is

$$\sqrt{1-x} = 1 - \frac{x}{2} - \frac{x^2}{8} + \cdots$$

EXAMPLE 1.14.3

Show that the Taylor series expansion of a product $F(x) = f_1(x) f_2(x)$ of two functions, about $x = x^*$, is equal to the product of the Taylor series expansions of $f_1(x)$ and $f_2(x)$ [the functions $f_1(x)$ and $f_2(x)$ are assumed to be continuous and have continuous derivatives at $x = x^*$ so that their Taylor series exist]. Then use this result to obtain the Taylor series expansion of $F(x) = (\cos x)/(1 + x)$, about $x = 0$, keeping up to quadratic terms in the final expansion.

■ **Solution**
According to Eqs. (1.14.1) and (1.14.2), the series expansion of $f(x)$ about $x = x^*$ is, with primes denoting differentiation with respect to x, and $x_s \triangleq x - x^*$,

$$f(x) = f(x^*) + [f'(x^*)]x_s + \frac{1}{2}[f''(x^*)]x_s^2 + \cdots$$

Or, with $f(x)$ being the product of two functions, i.e., $f(x) = f_1(x) f_2(x)$,

$$f(x) = f_1(x^*) f_2(x^*) + [f_1'(x^*) f_2(x^*) + f_1(x^*) f_2'(x^*)]x_s$$

$$+ \frac{1}{2}[f_1''(x^*) f_2(x^*) + 2f_1'(x^*) f_2'(x^*) + f_1(x^*) f_2''(x^*)]x_s^2 + \cdots \quad (1.14.5)$$

On the other hand, the product of the expansions of $f_1(x)$ and $f_2(x)$ yields the following result:

$$f_1(x) f_2(x) = \left\{f_1(x^*) + [f_1'(x^*)]x_s + \frac{1}{2}[f_1''(x^*)]x_s^2 + \cdots\right\}$$

$$\times \left\{f_2(x^*) + [f_2'(x^*)]x_s + \frac{1}{2}[f_2''(x^*)]x_s^2 + \cdots\right\}$$

Since each of the expansions of $f_1(x)$ and $f_2(x)$ were truncated to a certain order (i.e., the quadratic terms), we must not keep terms higher than that order when multiplying the two expansions that appear in the product. Thus, if only the quadratic terms are kept in the final result, we then obtain

$$f_1(x)f_2(x) = f_1(x^*)f_2(x^*) + [f_1'(x^*)f_2(x^*) + f_1(x^*)f_2'(x^*)]x_s$$
$$+ \frac{1}{2}[f_1''(x^*)f_2(x^*) + 2f_1'(x^*)f_2'(x^*) + f_1(x^*)f_2''(x^*)]x_s^2 + \cdots$$
(1.14.6)

By comparing Eqs. (1.14.6) and (1.14.5) it is seen that they are the same. If one desires to obtain the expansion of $f_1(x)f_2(x)$ with terms up to x_s^3, for example, then both functions $f_1(x)$ and $f_2(x)$ must be expanded to that order (i.e, to contain terms up to x_s^3) and the product truncated so that terms containing powers higher than x_s^3 are disregarded. This assures that the expansions are mathematically consistent.

The main objective of this example was to present the following result:

> The Taylor series expansion of a product $F(x) = f_1(x)f_2(x)$ of two functions, about $x = x^*$, is equal to the product of the Taylor series expansions of $f_1(x)$ and $f_2(x)$.

This result is very useful when determining the Taylor series expansion of a product of several functions. That is done simply by multiplying the expansions of the individual functions *and appropriately truncating the resulting expression to the order that is consistent with the individual expansions.*

To expand the function $f(x) = (\cos x)/(1+x)$ about $x = x^*$ we then use the result just presented. For $x^* = 0$, the expansions for $\cos x$ and for $(1+x)^{-1}$ are

$$\cos x = 1 - \frac{x^2}{2} + \cdots$$

$$(1+x)^{-1} = 1 - x + x^2 + \cdots$$

and this yields the following expansion for $(\cos x)/(1+x)$:

$$\frac{\cos x}{1+x} = \left[1 - \frac{x^2}{2} + \cdots\right][1 - x + x^2 + \cdots] = 1 - x + x^2 - \frac{x^2}{2} + \cdots = 1 - x + \frac{x^2}{2} + \cdots$$

1.15 Stability Analysis and Linearization

The differential equations of motion of dynamical systems are, in general, nonlinear. In contrast with linear differential equations, nonlinear differential equations may exhibit several equilibrium solutions, and the stability of each equilibrium needs to be addressed.

An equilibrium is considered to be *stable* if it is possible to choose pairs of initial conditions $x_s(0)$ and $\dot{x}_s(0)$, so that the perturbed motion $x_s(t) \triangleq x(t) - x_e$ can be kept in an arbitrarily small neighborhood of that equilibrium. Otherwise, the equilibrium is *unstable*. An equilibrium is *asymptotically stable* if, in addition to being stable, the perturbation $x_s(t)$ goes to zero as time goes to infinity.

The solution to nonlinear differential equations of motion cannot, in general, be determined analytically. That makes it difficult (if not impossible) to obtain general answers that can provide a physical understanding about the behavior of the system so that appropriate changes could be made to the system in order to modify that behavior to conform to specifications. Much (but not all, as alluded to in Section 1.16) can be learned about the behavior of a dynamical system from linearized equations about an equilibrium. The knowledge so acquired is always used in engineering design situations. Such situations include the design of control systems to affect the behavior of a given physical system.

In contrast, the theory of linear differential equations is well developed, especially linear differential equations with constant coefficients, whose exact solutions are easily obtained in terms of known functions. For this reason, the linearization process (which involves truncation of Taylor series up to the linear terms only) is widely used in practice for the investigation of small oscillations of dynamical systems about a particular solution of interest that is exhibited by the system.

The linearization of a differential equation $\ddot{x} = f(x, \dot{x})$ about an equilibrium solution $x = \text{constant} \stackrel{\Delta}{=} x_e$ involves two steps.

Step 1. The determination of the differential equation's constant solutions $x = x_e$. This involves solving the algebraic nonlinear equation $f(x_e, \dot{x}_e = 0)$ for x_e (which may have to be done numerically). Notice that since the resulting algebraic equation $f(x_e, \dot{x}_e = 0)$ is nonlinear, the system may exhibit several equilibrium solutions $x_{e_1}, x_{e_2}, \ldots$ (in contrast with a linear equation, which exhibits only one solution).

Step 2. The expansion of every nonlinear term in the differential equation in Taylor series about each constant solution determined in step 1.

When a nonlinear differential equation $\ddot{x} = f(x, \dot{x})$ is linearized about an equilibrium solution $x \stackrel{\Delta}{=} x_e = \text{constant}$, the linearized differential equation that is obtained is always of the form given by Eq. (1.13.7), i.e., $m\ddot{x}_s + c\dot{x}_s + kx_s = f(t)$, where $x_s(t) = x(t) - x_e$ is called the *perturbation* about the equilibrium x_e. The coefficients of the linearized differential equation depend on the value of x_e.

The information obtained from the linearized differential equation is used in the design of dynamical systems (such as the suspension system for a car or an airplane landing gear). The analysis presented in Case 3 of Section 1.13 can be applied to the linearized differential equation in order to determine the stability of each equilibrium. However, one should be aware that nonlinearities in the original differential equations may give rise to phenomena that are not predicted by the linearized differential equations (see Section 1.16 and Example 2.7.8, p. 129).

It should also be noted that stability analysis based only on the linearized differential equations may not be conclusive when any of the roots of the characteristic equation are purely imaginary. Fortunately, these cases are rare in actual practice, but if such a case does arise, one should either do a few numerical tests that involve integrating the original differential equations with different initial conditions to observe the response or resort to a method that accounts for nonlinearities such as *Lyapunov's method*. This method was devised by the Russian mathematician Aleksandr Mikhailovich Lyapunov (1857–1918) and is presented in more advanced books (see, for example, References 9, 11, and 12 of Appendix G).[15] If no further tests are done to verify the conclusions about stability in an inconclusive situation obtained from linearization, the equilibrium should be labeled *infinitesimally stable,* instead of stable. This serves as a warning that stability determination was done from a linearized analysis and no further stability test was performed with the original nonlinear differential equation(s) of motion. *The nonlinearities do not change the conclusion on stability based on linearized differential equations when the equilibrium is either asymptotically stable or unstable.* In this book, numerical integration of nonlinear differential equations with initial conditions near the equilibrium point is done in order to verify stability when linearization is inconclusive.

EXAMPLE 1.15.1

The motion of a mechanical system is governed by the differential equation

$$\ddot{x} + c\dot{x} + x + 2x^2 - 5x^3 + Kx\dot{x} = 0.2 \qquad (1.15.1a)$$

where c and K are constants, with $c \geq 0$.

a. Determine all the equilibrium values of $x(t)$, and linearize the differential equation about each equilibrium.
b. Determine which equilibrium solutions are stable, and which ones are unstable. What are the conditions, if any, to be imposed on c and K for the system equilibrium solution (or solutions, as the case may be) to be asymptotically stable?
c. What is the frequency of the small oscillations about each stable equilibrium exhibited by the system? Does the value of K affect the damping in the system?

■ Solution

a. At equilibrium, $x = \text{constant} \triangleq x_e$. The values of x_e are obtained by solving the following algebraic equation, which is obtained from the given differential equation with all the time derivative terms equated to zero.

$$x_e + 2x_e^2 - 5x_e^3 = 0.2 \qquad (1.15.1b)$$

The roots of this cubic equation may be found by trial and error with a handheld calculator. They may also be found by plotting the function $x_e + 2x_e^2 - 5x_e^3 - 0.2$ versus x_e, and the roots read from the plot. One may also find one of the roots by the iterative trial-and-error process, and then factoring out the polynomial to reduce the

---

[15] This method is based on examining the behavior of a function of x and $\dot{x}$ in the neighborhood of an equilibrium, as illustrated by the discussion involving Fig. 1.14 (p. 45). The difficulty is finding an acceptable testing function for some problems.

rest of the calculations to finding, analytically, the roots of the resulting quadratic equation. One of the roots is found to be $x_e \approx 0.167$. By dividing the polynomial $5x_e^3 - 2x_e^2 - x_e + 0.2$ by $x_e - 0.167$, we find that $5x_e^3 - 2x_e^2 - x_e + 0.2 \approx (x_e - 0.167)(5x_e^2 - 1.165x_e - 1.195)$, and the roots of the resulting quadratic polynomial are $x_e \approx -0.386$ and 0.619. The equilibrium solutions are then

$$x_e \approx 0.167, -0.386, \text{ and } 0.619$$

b. To determine the stability of each equilibrium point exhibited by the system, start by linearizing the given differential equation about each of the equilibrium solutions. For this, let

$$x(t) \triangleq x_e + x_s(t) \qquad (1.15.1c)$$

In terms of the perturbation variable $x_s(t)$, the differential equation becomes

$$\ddot{x}_s + c\dot{x}_s + x_e + x_s + 2(x_e + x_s)^2 - 5(x_e + x_s)^3 + K\dot{x}_s(x_e + x_s) - 0.2 = 0$$

$$(1.15.1d)$$

Linearization now involves disregarding the nonlinear term $\dot{x}_s x_s$ and the x_s^2 and x_s^3 terms that are obtained after the quadratic and cubic powers are expanded (which, for a polynomial is the same as obtaining the Taylor series expansion of those terms). Noticing that the constant term $x_e + 2x_e^2 - 5x_e^3 - 0.2$ is zero because that is how x_e is determined from the equilibrium equation, the following linearized equation is then obtained[16]

$$\ddot{x}_s + (c + Kx_e)\dot{x}_s + \left(1 + 4x_e - 15x_e^2\right)x_s = 0 \qquad (1.15.1e)$$

Specifically, we then have for each equilibrium:

For $x_e = 0.167$: $\qquad \ddot{x}_s + (c + 0.167K)\dot{x}_s + 1.25x_s = 0 \qquad (1.15.1f)$

For $x_e = -0.386$: $\qquad \ddot{x}_s + (c - 0.386K)\dot{x}_s - 2.779x_s = 0 \qquad (1.15.1g)$

For $x_e = 0.619$: $\qquad \ddot{x}_s + (c + 0.619K)\dot{x}_s - 2.271x_s = 0 \qquad (1.15.1h)$

The equilibrium solutions $x_e = -0.386$ and 0.619 are unstable because the coefficients of $\ddot{x}_s$ and x_s in the linearized differential equations for motion about them, namely, Eqs. (1.15.1g) and (1.15.1h), have opposite signs (see the Summary of Main Ideas in Case 3, Section 1.13, p. 39). Such unstable motions start growing exponentially with time. Although the exponential approximation is not valid for large times because the original differential equation is nonlinear (and the exponential solution is only an approximation that was obtained by linearization), such motions cannot be kept in a small neighborhood of the equilibrium solutions $x_e = -0.386$ and 0.619.

---

[16] Notice that, when more than one equilibrium solution is found, it is wise to obtain the linearized differential equation without using any specific value for x_e, and then use specific values of x_e afterward. Otherwise one would end up repeating all the same mathematical steps for each equilibrium solution.

Equation (1.15.1f) is the differential equation that governs the linearized motion about the equilibrium $x_e = 0.167$. Therefore, that equilibrium is infinitesimally stable if $c + 0.167K = 0$ (the equilibrium point $x_e = 0.167$ is a *center*, as discussed in Example 1.13.3), asymptotically stable if $c + 0.167K > 0$ (the equilibrium point that was a center is called a *stable focus* in such a case), and unstable if $c + 0.167K < 0$ (the equilibrium point $x_e = 0.167$ is called an *unstable focus* in this case). In the latter case (i.e., if $c + 0.167K < 0$), the system does not exhibit any stable equilibrium solution (and this feature is not a desirable situation in practice). It can be easily verified that the equilibrium $x_e = 0.167$ is stable by numerically integrating Eq. (1.15.1a) with initial conditions near that equilibrium, such as $x(0) = 0.15$ and 0.015 and $\dot{x}(0) = 0$.

c. Small motions about the equilibrium $x_e = 0.167$ are undamped oscillations with frequency $\omega = \sqrt{1.25} \approx 1.118$ when $c + 0.167K = 0$. Such oscillations die out with the passage of time when $c + 0.167K > 0$, and the value of K contributes to the damping of such motions because that value affects the coefficient of $\dot{x}_s$ in the linearized differential equation of motion.

When $c + 0.167K > 0$, the frequency ω of the small damped oscillations about the asymptotically stable equilibrium $x_e = 0.167$ is obtained directly from the characteristic equation for Eq. (1.15.1f), which is $s^2 + (c + 0.167K)s + 1.25 = 0$. The roots of this polynomial equation are

$$s = -\frac{1}{2}(c + 0.167K) \pm \sqrt{\frac{1}{4}(c + 0.167K)^2 - 1.25}$$

The roots are complex conjugates only when $(c + 0.167K)^2/4 < 1.25$ and, in such a case, the frequency of the damped oscillations about $x_e = 0.167$ is $\omega = \sqrt{1.25 - (c + 0.167K)^2/4}$.

All these conclusions may be verified by numerical integration of the nonlinear differential equation of motion, Eq. (1.15.1a). The numerical integration can easily be done with Simulink, which does not involve writing any computer programs. Try it yourself to see the motion of the system when you use different initial conditions to start the motion. Try initial conditions near the three equilibrium points and observe the motion $x(t)$, and also see what happens as you watch the motion in the $\dot{x}$-x plane.

EXAMPLE 1.15.2

The motion of a mechanical system is governed by the differential equation

$$\ddot{x} + \dot{x} + x(2 - \cos x) = K + \sin t \qquad (1.15.2a)$$

where K is a constant. The term $K + \sin t$ is due to an external force that is applied to the system. For $K = 0$ and -8, determine all the equilibrium values of $x(t)$, linearize the differential equation about each equilibrium, and determine whether those equilibrium solutions are stable, asymptotically stable, or unstable.

■ **Solution**

At equilibrium, $x = \text{constant} \triangleq x_e$, and x_e can be constant only in the absence of the $\sin t$ term that appears in the differential equation (i.e., only when the excitation is constant). Therefore, the values of x_e are determined by solving the following algebraic equation, which is obtained from the given differential equation with all the time derivative terms equated to zero, and leaving out the $\sin t$ excitation.

$$x_e(2 - \cos x_e) = K \tag{1.15.2b}$$

The case $K = 0$

For this case we have $x_e = 0$ and $\cos x_e = 2$. Since $x(t)$ (and, therefore, x_e, of course) must be real, there is no solution to the equation $\cos x_e = 2$. Therefore, the only equilibrium solution when $K = 0$ is $x_e = 0$.

Normally we let $x(t) = x_e + x_s(t)$ to expand the differential equation of motion for small x_s. However, since $x_e = 0$, there is no need to introduce a new symbol x_s in this case. The expansion of $\cos x$ about $x = 0$ is $\cos x = 1 + \cdots$, and the linearized differential equation in this case is then

$$\ddot{x} + \dot{x} + x = \sin t$$

Since the coefficients of the $\ddot{x}$, $\dot{x}$, and x terms have the same sign, the equilibrium $x_e = 0$ is asymptotically stable.

The case $K = -8$

For this case we have to solve the equation $x_e(2 - \cos x_e) = -8$. Since an analytical solution to this nonlinear equation cannot be obtained, we must then resort to a numerical solution. This may be done by a few iterations with a handheld calculator or by plotting the function $x_e(2 - \cos x_e) + 8$ versus x_e and then extracting the values of x_e for which the function is zero. Such a plot is shown in Fig. 1.18. The following roots of the equation $x_e(2 - \cos x_e) + 8 = 0$ are obtained from that plot: $x_e \approx -2.74, -5.19,$ and -6.87.

To linearize the given differential equation about each equilibrium solution, let

$$x(t) \triangleq x_e + x_s(t) \tag{1.15.2c}$$

In terms of the perturbation variable $x_s(t)$, the given differential equation, Eq. (1.15.2a), becomes

$$\ddot{x}_s + \dot{x}_s + (x_e + x_s)[2 - \cos(x_e + x_s)] = K + \sin t \tag{1.15.2d}$$

By expanding $\cos(x_e + x_s)$ in Taylor series for small x_s as

$$\cos(x_e + x_s) = \cos x_e + x_s \left[\frac{d}{dx_s} \cos(x_e + x_s)\right]_{x_s=0} + \cdots = \cos x_e - x_s \sin x_e + \cdots$$

linearization of Eq. (1.15.2d) for small deviations x_s from the equilibrium then yields [noticing that the constant term $x_e(2 - \cos x_e) + 8$ is zero] the following differential equation:

$$\ddot{x}_s + \dot{x}_s + (2 - \cos x_e + x_e \sin x_e)x_s = 0 \tag{1.15.2e}$$

SECTION 1.15 Stability Analysis and Linearization

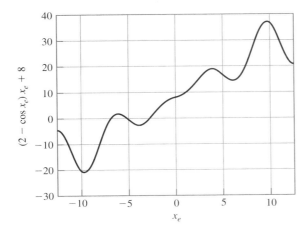

Figure 1.18 Plot for determining the roots of $(2 - \cos x_e)x_e + 8$.

We then have for each equilibrium

For $x_e = -2.74$: $\quad \ddot{x}_s + \dot{x}_s + 3.99 x_s = \sin t$ $\quad$ (1.15.2f)

For $x_e = -5.19$: $\quad \ddot{x}_s + \dot{x}_s - 3.07 x_s = \sin t$ $\quad$ (1.15.2g)

For $x_e = -6.87$: $\quad \ddot{x}_s + \dot{x}_s + 4.97 x_s = \sin t$ $\quad$ (1.15.2h)

These linearized differential equations disclose that the equilibrium solutions $x_e = -2.74$ and -6.87 are asymptotically stable. In these cases the free (i.e., without the $\sin t$ excitation) motion $x(t)$ is a decaying exponential that tends to the corresponding equilibrium $x = x_e$ when the motion is started with small values of $\dot{x}(0)$ and $x(0) - x_e$. When the $\sin t$ excitation is present, the linearized motion about each of these asymptotically stable equilibrium solutions consists of the sum of the homogeneous solution and the particular solution due to the excitation $\sin t$. The time-dependent part of the homogeneous solution dies out with the passage of time, and eventually the response $x(t)$ is the sum of the corresponding equilibrium solution and the sinusoidal function due to the $\sin t$ excitation (see Section 8.2 for a presentation on the response to a sinusoidal excitation).

The equilibrium $x_e = -5.19$ is unstable, and a motion that is started near that equilibrium does not stay near it.

EXAMPLE 1.15.3

Let us investigate further the motion of the pendulum of length L that was addressed in Examples 1.13.2 and 1.13.3. The differential equation of motion for the pendulum in terms of the angle θ measured from the vertical is repeated here, where $k = g/L$ (with $g = 9.81$ m/s² being the acceleration of gravity).

$$\ddot{\theta} + k \sin \theta = 0 \quad (1.15.3a)$$

As determined in Example 1.13.2, this system exhibits the equilibrium solutions $\theta_e = 0, \pi, 2\pi, \ldots$, of which only two are physically distinct from each other, namely, $\theta_e = 0$ and π rad. Linearize the differential equation about each of these two equilibrium solutions and determine whether they are stable, asymptotically stable, or unstable.

■ Solution

To linearize the differential equation about each equilibrium solution, let

$$\theta(t) \triangleq \theta_e + \theta_s(t) \tag{1.15.3b}$$

and expand, for small θ_s, the nonlinear term $k \sin \theta$ that appears in the differential equation. The expansion for $\sin \theta$ (see Example 1.14.1, with $\theta^* = \theta_e$ in that example) is

$$\sin \theta = \sin(\theta_e + \theta_s) = \sin \theta_e + \theta_s \cos \theta_e + \cdots$$
$$= \theta_s \cos \theta_e + \cdots \quad \text{(because } \sin \theta_e = 0\text{)} \tag{1.15.3c}$$

In terms of the perturbation variable $\theta_s(t)$, the linearized differential equation is then

$$\ddot{\theta}_s + (k \cos \theta_e)\theta_s = 0 \tag{1.15.3d}$$

Specifically, we then have for each equilibrium:

For $\theta_e = 0$: $\qquad\qquad\qquad \ddot{\theta}_s + k\theta_s = 0 \tag{1.15.3e}$

For $\theta_e = \pi$: $\qquad\qquad\qquad \ddot{\theta}_s - k\theta_s = 0 \tag{1.15.3f}$

Since $k > 0$, the equilibrium $\theta_e = 0$ is infinitesimally stable, and the equilibrium $\theta_e = \pi$ is unstable. Further investigation should be done to ascertain the stability of the equilibrium $\theta_e = 0$, with θ governed by the nonlinear equation (1.15.3a). This was actually done in Examples 1.13.2 and 1.13.3. The equilibrium $\theta_e = 0$ would be asymptotically stable if there were damping in the system.

It is worth noticing that the investigation of the same nonlinear system presented in this example, and in Examples 1.13.2 and 1.13.3, by different methods provided results that complement each other. That is what engineers do in practice when investigating the motion of dynamical systems. Generally the engineer will have to use a combination of methods (including numerical integration of the nonlinear differential equations that govern the motion) in order to obtain as much information as possible about the dynamic behavior of any particular system. Such work helps the engineer to understand the behavior of the system, to predict its motion, and to take appropriate measures to control it when it is necessary to do so.

The plot of $\dot{\theta}/\sqrt{k}$ versus θ shown in Fig. 1.13 (p. 44) for Example 1.13.3 provides a graphical visualization of the dynamic behavior of the nonlinear system investigated here. The figure shows that a motion started with initial conditions in the vicinity of the point $(\dot{\theta}(0) = 0, \theta = 0)$, which is the equilibrium point that is stable, stays in that vicinity as time increases. Such a motion is an oscillation whose amplitude can be made as small as desired. In contrast, this is not the case for a motion that is started in the vicinity of the point ($\dot{\theta}(0) = 0, \theta = \pi$), which, as determined earlier and also in this example, is the unstable equilibrium point.

Notice that the solution to the linearized differential equation about $\theta_e = \pi$ [namely, $\ddot{\theta}_s - k\theta_s = 0$, where $\theta_s = \theta - \pi$ is the deviation of $\theta(t)$ from $\theta_e = \pi$] tends to infinity as time increases. A linearized differential equation is generated under the assumption that the perturbed motion $\theta_s(t) = \theta(t) - \theta_e$ is small. Therefore, the *only* correct conclusion that the linearized differential equation $\ddot{\theta}_s - k\theta_s = 0$ provides is that the equilibrium $\theta_e = \pi$ is unstable. Important detail about the motion that is started with initial conditions near the unstable equilibrium point $\theta_e = \pi$ [i.e., the motion started with both $\dot{\theta}(0)$ and $\theta(0) - \pi$ small] is disclosed by observing Fig. 1.13. When looking at that figure, notice that no matter how small those initial conditions are in this case, $\theta(t)$ never stay in the neighborhood of the equilibrium $\theta_e = \pi$ (and that is why that equilibrium is classified as unstable).

It is important to notice that, in contrast to a linear system (i.e., a system whose motion is governed by a linear, and not just linearized, differential equation), a motion started close to an unstable equilibrium of a nonlinear system may not grow indefinitely because the nonlinearities in the system may limit the motion to finite values.

EXAMPLE 1.15.4

The motion of a mechanical system is governed by the differential equation

$$\ddot{x} + c\dot{x} + x - x^3 + Kx^2\dot{x} = 0 \tag{1.15.4a}$$

where c and $K \neq 0$ are constants, with $c \geq 0$. Determine all the equilibrium solutions $x = \text{constant} \triangleq x_e$, and specify whether they are stable, asymptotically stable, or unstable. If both c and K are zero, the quantity $H \triangleq \dot{x}^2/2 + x^2/2 - x^4/4$ is constant. Show that $\dot{H} = -(c + Kx^2)\dot{x}^2$ for all cases.

■ **Solution**

The equilibrium solutions are $x_e = 0$, $x_e = 1$, and $x_e = -1$. With $x(t) = x_e + x_s(t)$, linearization of Eq. (1.15.4a) about $x_e = 1$ and $x_e = -1$ yields $\ddot{x}_s + (c + K)\dot{x}_s - 2x_s = 0$. Therefore, since the coefficient of x_s is negative, those equilibrium solutions are unstable.

Linearization of Eq. (1.15.4a) about $x_e = 0$ yields $\ddot{x}_s + c\dot{x}_s + x_s = 0$. The equilibrium is asymptotically stable if $c > 0$. When $c = 0$, linearization indicates that the equilibrium is infinitesimally stable. However, a simple numerical integration of Eq. (1.15.4a) with a small value of $x(0)$ and with $\dot{x}(0) = 0$, say, shows that the stability of the equilibrium $x_e = 0$ depends on the sign of K when $c = 0$. In such a case, that equilibrium is actually asymptotically stable if $K > 0$, unstable if $K < 0$, and stable if $K = 0$. The explanation for this actually lies in the fact that $\dot{H} = -(c + Kx^2)\dot{x}^2$, which shows that for $K > 0$ the term Kx^2 adds damping to the system since x^2 is never negative. With $c > 0$, $c + Kx^2 \approx c$ since x^2 can be made as small as desired, in which case $\dot{H} \approx -c\dot{x}^2$, and the equilibrium is asymptotically stable. However, when $c = 0$, the term Kx^2 in the expression for $\dot{H}$ prevails. Since $\dot{H} < 0$ when a motion is started, H decreases with time. Since H is not negative when x is small (because the $x^2/2$ term is greater than the $x^4/4$ term in such a case), H must go to zero as $t \to \infty$. This then shows that the equilibrium is also asymptotically stable when $K > 0$ and $c = 0$.

1.16 An Interesting Experiment

You can build a very simple system to see the effects of nonlinearities. Just take any spring and measure its unstretched length, which we will call L_u. Now, attach a mass m at the end of the spring; keep your hands on the mass, and let the spring stretch along the vertical. Choose a proper mass until the equilibrium length of the spring (i.e., the length of the spring when there is no motion) is about $4L_u/3$. Do not worry if it is not exactly $4L_u/3$; the value $4L_u/3$ will produce the most dramatic effect in this demonstration, but any value that is close to $4L_u/3$ is good enough (how close is "close" is actually answered only by a nonlinear analysis).

To do the experiment, secure the upper end of the spring (the end not attached to the mass) to a fixed point. Now, trying to keep the spring vertical, grab the mass and stretch the spring a little bit from its equilibrium. Then release it, and watch the motion. The mass should start moving up and down, as expected. But after a little while the angle the spring makes with the vertical starts to increase (let's call it θ), while the up-and-down amplitude of the motion decreases. It decreases until the spring swings laterally only, essentially without the up-and-down motion. Then, the angle starts to decrease, and the up-and-down motion starts to appear again; and this sequence of events repeats itself.

You can repeat the experiment by changing the value of the mass from that "critical value" that you found for your spring. Change it by increasing and decreasing the value of the mass. What you will find is that there is a range of values of m for which the described phenomenon occurs. For values of mass outside that range, the motion takes place essentially along the vertical, where it was started.

This phenomenon happens to be a nonlinear phenomenon. It cannot be predicted by linearizing the differential equations of motion for small values of θ and for the up-and-down swing of the spring.

This simple experiment demonstrates a nonlinear phenomenon that can occur in many dynamical systems. Given the appropriate circumstances, it occurs with systems like satellites, ships, airplanes, buildings, and machines. In ships, for example, it can cause undesirable (and dangerous) large rolling motions due to nonlinear interactions with the pitch motion of the ship. In three-axis stabilized satellites, it can cause undesirably large roll-yaw motions also due to nonlinear interactions with pitch motions.

The phenomenon that is demonstrated with this simple spring-mass experiment is called *nonlinear resonance*. Its analysis is certainly beyond the scope of this book, but the engineer should be aware of it. You can learn more about it by taking more advanced courses in dynamics, and/or studying advanced texts (and papers published in the technical journals) that deal with nonlinear oscillations.

1.17 Problems

Unless indicated otherwise, for Problems 1.1 to 1.9, $\{\hat{x}_1, \hat{x}_2, \hat{x}_3\}$ is a unit vector triad with $\hat{x}_2 = \hat{x}_1 \times \hat{x}_3$.

1.1 It is claimed that the vectors $\vec{a} = \hat{x}_1 + 2\hat{x}_2 + 3\hat{x}_3$ and $\vec{b} = -5\hat{x}_1 + 4\hat{x}_2 - \hat{x}_3$ are perpendicular to each other. Determine whether the claim is true or false.

1.2 Determine the value of the constant K so that the vectors $\vec{a} = 2\hat{x}_1 + 5\hat{x}_2 + 3\hat{x}_3$ and $\vec{b} = 5\hat{x}_1 + K\hat{x}_2 + 3\hat{x}_3$ are perpendicular to each other.

1.3 Determine the angle between the vectors $\vec{a} = 2\hat{x}_1 + 5\hat{x}_2 + 3\hat{x}_3$ and $\vec{b} = 5\hat{x}_1 + 8\hat{x}_2 + 3\hat{x}_3$.

1.4 Determine the value of the constant K so that the angle between the vectors $\vec{a} = 4\hat{x}_1 + K\hat{x}_2 + \hat{x}_3$ and $\vec{b} = 0.5\hat{x}_1 + \hat{x}_2 + \hat{x}_3$ is $60°$.

1.5 Determine the cross product $\vec{p}$ of the vectors $\vec{a} = \hat{x}_1 + 2\hat{x}_2 + 5\hat{x}_3$ and $\vec{b} = 3\hat{x}_1 + 2.5\hat{x}_2 + 4.2\hat{x}_3$. What is the magnitude of $\vec{p}$? What would the answers be if $\hat{x}_2$ were chosen as $\hat{x}_2 = \hat{x}_3 \times \hat{x}_1$?

1.6 Determine the angle between the vectors $\vec{a}$ and $\vec{b}$ given in Problem 1.5.

1.7 Determine the vector $\vec{c}$ that is equal to the projection of the vector $\vec{a} = \hat{x}_1 + 3\hat{x}_2 + 4\hat{x}_3$ along the direction of the vector $\vec{b} = 3\hat{x}_1 + 2.5\hat{x}_2 + 4.2\hat{x}_3$.

1.8 Determine the perpendicular distance between point P whose rectangular coordinates are $x_1 = 4$, $x_2 = 3$, $x_3 = 2$ and the vector $\vec{a}$ given in Problem 1.7. The axes x_1, x_2, and x_3 are parallel to the unit vectors $\hat{x}_1$, $\hat{x}_2$, and $\hat{x}_3$, respectively, and the point $M(x_1 = 5, x_2 = 7, x_3 = 9)$ is in the line of action of $\vec{a}$.

1.9 Determine the moment of the vectors $\vec{a}$ and $\vec{b}$, given in Problem 1.7, about point P whose coordinates are given in Problem 1.8. Point $N(x_1 = 2, x_2 = 8, x_3 = 12)$ is in the line of action of $\vec{b}$.

1.10 Consider the vector $\overrightarrow{AB}$ from point $A(x = 1, y = 2)$ to point $B(x = 5, y = 0)$. Define, in a sketch, a set of convenient unit vectors that you choose to work with (clearly naming them with any letter of your choice, and using hats on the top of those letters) and then determine the moment $\vec{M}_O$ of the vector $\overrightarrow{AB}$ about the origin of the coordinate system.

1.11 Consider the vector $\overrightarrow{AB}$ from point $A(x = 1, y = 2)$ to point $B(x = 5, y = 0)$, and the vector from point $P(x = 1, y = 0)$ to point $Q(x = 5, y = 1)$. Define, in a sketch, a set of convenient unit vectors that you choose to work with (clearly naming them with any letter of your choice, and using hats on the top of those letters) and then determine the expression for the vector that is equal to the projection of $\overrightarrow{AB}$ along the unit vector from P to Q.

1.12 Consider the vector $\overrightarrow{AB}$ from point $A(x = 1, y = 2)$ to point $B(x = 5, y = 0)$. Define, in a sketch, a set of convenient unit vectors that you choose to work with (clearly naming them with any letter of your choice, and using hats on the top of those letters) and then determine the coordinates of all points Q on the y axis for which the magnitude of the moment $\vec{M}_Q$ of the vector $\overrightarrow{AB}$ about Q is equal to $|\vec{M}_Q| = 12$.

1.13 A *couple* is a pair of forces $\vec{F}$ and $-\vec{F}$ separated by a distance D. Show that the moment of a couple is independent of the point about which the moment is taken.

1.14 Determine the time derivative of the following functions $f_1 = \tan x + (1 + y^2)^{-3/2}$ and $f_2 = (\sin x)/[1 + \cos^2(xy)]$, where x and y are functions of time t.

1.15 In Problem 1.14, $x(t) = \cos(k_1 t)$ and $y(t) = k_2 t^2$, where $k_1 = 1$ rad/s and $k_2 = 1$ s^{-2}. Determine the values of $\dot{f}_1$ and $\dot{f}_2$ when $t = 12.5$ minutes.

1.16 Which of the following differentials is an exact differential? Determine the function g for those cases where dg is an exact differential.
  a. $dg = (2xy + 4)y^2\,dx + (3xy + 8)xy\,dy$
  b. $dg = x^2 \dot{x}\,dt$
  c. $dg = 6y^2\,dx + 3xy\,dy$
  d. $dg = 3y^2\,dx + 6xy\,dy$

1.17 Determine the function $P(x, \theta)$ for $dg = P\,dx - k(\sin\theta)\,d\theta$ to be an exact differential. What is the function $g(x, \theta)$ in such a case? The parameter k is a constant.

1.18 The motion $\theta(t)$ of a mechanical system is governed by the nonlinear differential equation $\ddot{\theta} - \Omega^2 (\sin\theta)\cos\theta + k\sin\theta = 0$, where $k > 0$ is a constant. An example of such a system is a pendulum that makes an angle θ with the vertical and is driven by a motor forcing it to spin at a constant angular velocity Ω about a vertical axis; for such a pendulum, $k = g/L$, where g is the acceleration of gravity and L is the length of the pendulum. For $k/\Omega^2 = 2$ and $\frac{1}{2}$, determine, analytically, the maximum and the minimum values of $\dot{\theta}(t)$ when the motion is started with the initial conditions $\theta(0) = 0.1$ rad and $\dot{\theta}(0) = 0.5\sqrt{k}$ rad/s.

1.19 Expand each function in Taylor series about $x = 0$, keeping two terms in the expansion. The parameter k is a constant.
  a. $f(x) = k\tan x$
  b. $f(x) = \dfrac{k}{\sqrt{1+x}}$
  c. $f(x) = \dfrac{k\tan x}{\sqrt{1+x}}$

1.20 Determine the equilibrium solution $x = \text{constant} \stackrel{\Delta}{=} x_e$ of each one of the given nonlinear differential equations, and then expand each equation in Taylor series about each of their equilibrium solutions by "perturbing the equilibrium" as $x(t) \stackrel{\Delta}{=} x_e + \epsilon x_s(t)$, where ϵ is just a bookkeeping constant parameter that is introduced only to keep track of orders of magnitude in the Taylor series expansion. Keep terms up to order ϵ^2 in the expansions, and then set $\epsilon = 1$ when writing your answers. The coefficient k that appears in the equations is an arbitrary constant. What are the linearized differential equations for $x_s(t)$?
  a. $\ddot{x} + \sin x + \dfrac{k\dot{x}}{1-x} = 0$
  b. $\ddot{x} + x\cos x + \dfrac{k\dot{x}}{(1-x)^2} = 0$

1.21 Determine, analytically, all the equilibrium solutions for the rotating pendulum described in Problem 1.18. How many physically different equilibrium solutions does such a pendulum exhibit when $\Omega^2 > k$ and when $\Omega^2 < k$? For each of these cases, determine the stability of each equilibrium of the system by linearizing the differential equation that governs the motion.

1.22 The motion $x(t)$ of a dynamical system is governed by the differential equation $\ddot{x} + x - x^3 = 0$, with $x(0) = 0$ and $\dot{x}(0) = 0.4$. Determine, analytically, the maximum and the minimum values of $x(t)$. Verify your answer by sketching the phase plane plot $\dot{x}$ versus x and showing the phase plane trajectory for the given initial conditions.

1.23 The motion of a dynamical system is governed by the differential equation $\ddot{x} + x + x \sin x = 0$, with $x(0) = 0$ and $\dot{x}(0) = 0.4$. Integrate the differential equation once to obtain an equation of the form $\dot{x}^2 + g(x) = \text{constant} \stackrel{\Delta}{=} C$, and then determine the maximum and minimum values of x from a plot of $g(x) - C$ versus x.

1.24 Solve, analytically, the following differential equation for $\dot{x}(t)$: $\ddot{x} + t\dot{x} = 0$. The initial conditions are $x(0) = x_0$ and $\dot{x}(0) = \dot{x}_0$.

1.25 Analyze the stability of the equilibrium (i.e., indicate whether the equilibrium is stable, asymptotically stable, or unstable) of each of the six systems whose motion is governed by the given differential equation.

  a. $\ddot{x} + 2x = 0$
  b. $\ddot{x} - 2x = 0$
  c. $\ddot{x} + 0.4\dot{x} + 2x = 0$
  d. $\ddot{x} + 0.4\dot{x} - 2x = 0$
  e. $\ddot{x} + \dot{x} + 5x = 6$
  f. $\ddot{x} - \dot{x} + 5x = 6$

1.26 Determine, analytically, the damped and the undamped frequencies of oscillation for each system in Problem 1.25. Indicate which of the systems, if any, do not oscillate.

1.27 Solve, analytically, each of the differential equations given in Problem 1.25 for $x(t)$ when the motion is started with the initial conditions $x(0) = 0.2$ and $\dot{x}(0) = 0.1$.

1.28 The motion $x(t)$ of a mechanical system is governed by the differential equation $\ddot{x} + 2\dot{x} + 17x = 0$. The motion is started with the initial conditions $x(0) = 0$ and $\dot{x}(0) = 1$. Determine, analytically, the maximum value of $x(t)$.

1.29 The motion $x(t)$ of a dynamical system is governed by the differential equation $\ddot{x} - x + x^3 = 0$.
  a. Analyze the stability of each equilibrium of the system, indicating whether the equilibrium is asymptotically stable, stable, or unstable.
  b. Determine the maximum and minimum values of $x(t)$ when the motion is started with $x(0) = 0.2$ and $\dot{x}(0) = 0$. Verify your answer by sketching the phase plane plot $\dot{x}$ versus x and showing the phase plane trajectory for the given initial conditions.

1.30 The motion $x(t)$ of a mechanical system is governed by the differential equation $\ddot{x} + x + 2x^2 - 5x^3 = 0.2$. This is the same system considered in Example 1.15.1 (p. 55), with $c = 0$ and $K = 0$ in that example, and for this case $\dot{x}^2/2 + x^2/2 + 2x^3/3 - 5x^4/4 - 0.2x = \text{constant} \stackrel{\Delta}{=} H$. Draw the phase plane plot $\dot{x}$ versus x for several values of H appropriately chosen by you, and then indicate in which region of that plot the pair of initial conditions $\dot{x}(0)$ and $x(0)$ result in an oscillatory motion about the stable equilibrium solution of the system (which is $x_e \approx 0.167$, as already determined in Example 1.15.1). For $\dot{x}(0) = 0$, for what range of $x(0)$ is the resulting motion oscillatory?

1.31 The motion $\theta(t)$ of a mechanical system is governed by the differential equation $\ddot{\theta} + \theta - \sin\theta = 0$. Show from a plot of $\theta - \sin\theta$ versus θ that $\theta = 0$ is the only equilibrium of the system. Plot $\dot{\theta}$ versus θ and show, from the plot, that the motion is oscillatory. Show that linearization of the differential equation about the equilibrium yields $\ddot{\theta} + k\theta = 0$, where the "stiffness" k is $k = 0$, and that such an equation fails, in this case, to exhibit the actual behavior disclosed by the plot of $\dot{\theta}$ versus θ for the given system.

1.32 Analyze the stability of the equilibrium of the system whose motion $x(t)$ is governed by the differential equation $\ddot{x} + x + k\dot{x}^3 = 0$. Is the stability obtained from linearization conclusive? What can you conclude by looking at the behavior of the function $H = \dot{x}^2/2 + x^2/2$ when $k \neq 0$?

The following problems involve the use of Simulink. Study the Simulink tutorial in Appendix A to prepare yourself to do them and other related problems later in the book.

1.33 Use Simulink to integrate numerically the differential equation in Problem 1.18, and then check the analytical answer you have obtained for that problem. *Hint:* Work with the nondimensional time $\tau = t\sqrt{k}$, so that the differential equation becomes $d^2\theta/d\tau^2 - (\Omega^2/k)(\sin\theta)\cos\theta + \sin\theta = 0$ (and thus, there is no need to know values for Ω and k, but only the ratio k/Ω^2).

1.34 Construct a Simulink block diagram model to integrate numerically the nonlinear differential equation for Example 1.15.1, i.e., Eq. (1.15.1a) (p. 55). Use your model to integrate the equation and to plot $\dot{x}$ versus x for $c = K = 0$, with $\dot{x}(0) = 0$ and with several values (of your choice) of the initial condition $x(0)$ near the three equilibrium points of the system (see Example 1.15.1 for the values that were determined for the equilibrium solutions $x = \text{constant} \triangleq x_e$).

1.35 Use the Simulink model you have constructed for Problem 1.34 to verify the effect the value of K has on the motion. For this, take $c = 0$ and $\dot{x}(0) = 0$ and run the simulation for two different values (of your choice) of K, one positive and the other negative, with $x(0) = 0.3$ or any other value of your choice that is near the equilibrium $x_e = 0.168$. Also, maintaining the same values of c and K, run the simulation again with a value of $x(0)$ near the other two equilibrium points (see Example 1.15.1). Do some numerical experiments of your own by running the simulation with other values of $c + 0.168K$ to verify the conclusion that the system has no stable equilibrium points when $c + 0.168K < 0$.

1.36 Check the answer to Problem 1.28 using Simulink.

1.37 The following questions should be answered without using a computer and without consulting any notes, as if you were taking a closed book exam.
  a. What does the Simulink model depicted in Fig. 1.19 do (i.e., what equation does it solve)?
  b. What is the value of the output when the simulation stops at $t = 0.2$ time units (choose your own values for any data you need).

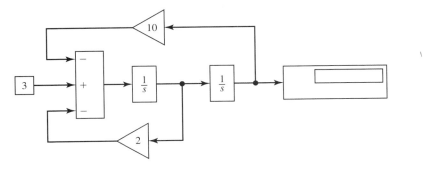

Figure 1.19 Figure for Problem 1.37.

CHAPTER 2

Kinematics and Dynamics of Point Masses

Kinematics is the study of the velocity and acceleration of particles (and of points in bodies of finite dimension) without regard to the forces that cause their motion. It is a study of the dependence of those quantities on the coordinates (and of their time derivatives) necessary to locate a point or body in space at any instant of time. In kinematics, consideration is only given to the dependence of velocity and acceleration on the variables (and their time derivatives) needed to describe the motion. In dynamics, differential equations of motion of a system must be formulated taking into account the forces acting on the system, and *integrated* so that the motion can be predicted. When the motion is known (which is the case of a "pure" kinematics problem), such as the motion of a car moving with known speed along a given track or the prescribed motion of an element of a machine, a kinematic analysis includes finding the position, the velocity, and/or the acceleration of one point relative to another.

> Implementation of numerical integration of differential equations of motion is presented in the form of *block diagrams* in this book. They are fully functional diagrams, which are Simulink *models. Their implementation and use do not involve writing any computer programs,* but only typing mathematical expressions in Simulink function blocks when they are used. Instead of using Simulink, readers and instructors using this book can, if preferred, write computer programs using any appropriate language, but that will require more effort than simply building block diagrams using Simulink. A tutorial is presented in Appendix A and, if you have not done so, it is recommended that you go over that material either now or when you encounter the use of Simulink later on in this chapter.

The Simulink models that appear throughout the book are included on the enclosed CD. All the files on the CD are listed in Appendix E. They were prepared with the latest version of MATLAB and Simulink available at the time of publication of this book but can be easily upgraded to newer versions if you wish to do so (please see the fifth paragraph in Appendix E). To use them, follow the instructions in Section A.2 in Appendix A. Assuming that you have done that, a model may be accessed by typing its name in your MATLAB command window as indicated in each example in the book. However, the easiest way is to type `simulations_menu` in the MATLAB command window and then use the left button of the computer mouse to simply choose a simulation model from a pull-down menu that appears on the computer screen.

2.1 How to Formulate a Dynamics Problem

When a particle P is free to move in three-dimensional space, three time-dependent variables are needed to locate its position at any time t. Those three variables may be, for example, the distances from a reference point O in space, taken along three orthogonal lines originating at O; or they might be the distance from O to P and two angles that specify the orientation in space of the line that passes through O and P. There are many possible choices one can make and, as far as locating the particle in space is concerned, any set of such variables is satisfactory. The number of such variables is reduced by one if the motion is constrained to a surface in space and by two if the motion is constrained to a line. For example, if a particle is constrained to move on the surface of a sphere, only two variables are needed to locate it on that surface. Possible choices for those variables are two angles.

The particle is said to have as many *degrees of freedom* as the number of time-dependent variables needed to completely specify its position in space. In dynamics, such variables are called the *motion coordinates*. Thus, a particle that is free to move in three-dimensional space, with no constraints, has three degrees of freedom.

Two particles moving independently of each other have six degrees of freedom, while the number of degrees of freedom is reduced to five if the particles are connected by a rigid bar because the six variables that are needed for two points moving independently of each other would then be related by one *equation of constraint*. The constraint equation in this case is the equation stating that the length of the bar is a constant. For a rigid bar, the five independent variables that are used for describing the motion might be three rectangular coordinates for a chosen point on the bar and two angles for specifying the orientation of the bar. If the bar were constrained to move in a plane, the motion of the bar relative to that plane would have three degrees of freedom. In such a case, the variables that one might choose to investigate the motion might be two rectangular coordinates for a point on the bar and one orientation angle.

A particle moving on a straight line has one degree of freedom. This is the simplest problem in dynamics. For two-dimensional motion the simplest problem in dynamics consists of the motion of a particle subjected to a constant force (i.e., constant in magnitude and direction). The classical projectile problem, which is widely studied in basic physics courses, is one of these problems.

> For each dynamics problem under consideration, the application of Newton's second law to a material particle of mass m, i.e.,
>
> $$\vec{F} = \frac{d}{dt}(m\vec{v}) = m\frac{d\vec{v}}{dt} \qquad (2.1.1)$$
>
> requires that
>
> 1. We isolate the particle from its surroundings and draw a free-body diagram for it, choose a set of unit vectors and variables (the *motion coordinates*) to work with, and then write an expression for the resultant $\vec{F}$ of the forces (i.e., field forces and contact forces) acting on the particle.
> 2. We develop an expression (in terms of the time-dependent variables that are chosen to describe the motion) for the *absolute velocity* $\vec{v}$ of the particle and for its *absolute acceleration* $\vec{a} = d\vec{v}/dt$.

The quantity $m\vec{v}$, with $\vec{v}$ being the absolute velocity of the particle, is called the *absolute linear momentum* of the particle.

> When use is made of Newton's second law by following the two steps just presented, a set of scalar equations involving unknown reaction forces, and the motion coordinates (and their time derivatives), is then obtained. Such scalar equations are obtained when the corresponding components along the directions of the unit vectors on both sides of the vector equation $\vec{F} = m(d\vec{v}/dt)$ are equated to each other. By a process of elimination of the unknown reaction forces in those scalar equations, a set of equations that only involve the motion coordinates and their time derivatives, and known applied forces, can be obtained. The latter equations are the *differential equations of motion* for a specific problem under consideration; they are the equations (or equation, if there is only one degree of freedom) that need to be used for investigating the motion.

The difficulty encountered in integrating the differential equations of motion for any problem being investigated will be directly related to the original choice made for the motion coordinates, i.e., for the time-dependent variables chosen to study the motion. Thus, some care should be exercised in each problem concerning the choice of such variables. For example, when analyzing the motion of a pendulum, or the motion of an orbiting satellite relative to an attracting planet, one might wish to use polar coordinates for the motion variables, while a set of rectangular coordinates might be chosen to analyze the motion of a short-range projectile. A variety of problems that illustrate the degree of difficulty inherent with different choices of motion variables is presented in Section 2.5 (see, for example, Example 2.5.10, p. 101).

Intuitively, we know that the velocity of a point P depends on one's frame of reference. Take, for example, the case of an observer who is stationary inside a moving car and looking at a point P fixed to the windshield of the car. For that observer, the velocity of that point is zero. However, for observers outside the car, the velocity of that same point is not zero.

As mentioned in Chapter 1, Newton's laws call for the existence of an inertial reference frame, and both the velocity and the acceleration used in his laws must be referred to that frame. The difficulty here is that neither an inertial point (i.e., a point with zero acceleration) nor a nonrotating line seem to exist in nature! In practice, approximations are then made. For most engineering dynamics problems the earth is treated as inertial, which means that the effect of its motion around the sun and of the spin about its own axis are neglected. Whether such approximations yield acceptable results in practice depends on how long the motion being studied is tracked. Very accurate prediction of the motion of a long-range projectile relative to the earth, for example, may require accounting for the rotational motion of the earth about its own axis when formulating the differential equations of motion of the projectile.

A classical experiment with an apparatus known as the *Foucault pendulum* is widely used for demonstrating the effect of the rotation of the earth on motion relative to the earth. The experiment was devised by Jean-Bernard-Léon Foucault (1819–1868), who was a very skilled French experimental physicist who also constructed the pendulum that now bears his name. He performed his experiment in Paris, France, in 1851, and repeated it several times after that. A Foucault pendulum can be seen in a number of science museums. Basically, it consists of a ball attached to a long, thin metal string that hangs from the ceiling of a tall room in a building. The experiment demonstrates that the oscillating pendulum rotates with respect to the earth due to the rotation of the earth about its own axis. Analysis of the motion discloses that the period of the rotation is equal to $24/|\sin \lambda|$ hours, where λ is the latitude of the pendulum's location on the earth. The rotation is clockwise in the northern hemisphere and counterclockwise in the southern hemisphere. At the poles, where $\lambda = \pm 90°$, the pendulum rotates $360/4 = 90°$ every $24/4 = 6$ hours (h). In Washington, D.C. (United States), Paris (France), and Rio de Janeiro (Brazil), a rotation of $90°$ takes approximately 9 h and 30 minutes (min), 8 h, and 15 h and 20 min, respectively. The analysis of the motion of the Foucault pendulum is presented in a separate section in Chapter 8.

A Note on Simple Cases Involving Eq. (2.1.1) If the resultant force $\vec{F}$ acting on a mass particle is equal to zero, Newton's second law yields $m\vec{v} = $ constant (or simply, $\vec{v} = $ constant), which is Newton's first law. This is also known as the *principle of conservation of linear momentum,* but this integrated form of Newton's second law is an obvious one and is of limited use when only a single particle is involved. A similar principle involving a system of particles is presented in Chapter 4.

If the resultant force $\vec{F}$ acting on a mass particle is either a constant or a known function that only involves the time t, i.e., when $\vec{F} = \vec{F}(t)$, Eq. (2.1.1) may be readily integrated to yield

$$\int_{t=t_0}^{t} \vec{F}\, dt = \int_{t=t_0}^{t} \frac{d(m\vec{v})}{dt}\, dt = [m\vec{v}]_{t=t} - [m\vec{v}]_{t=t_0} \qquad (2.1.2)$$

Equation (2.1.2) states that the change in the linear momentum $m\vec{v}$ of the particle from any time t_0 to any other time t is equal to the integral, over time, of the resultant force $\vec{F}(t)$ acting on the particle. Specific examples are found in elementary dynamics books.

An interesting integrated form of Eq. (2.1.2) is obtained when the system is subjected to a force that is applied only for a very short time Δt, short enough so that the particle essentially does not change its position during that time. Such forces are called *impulsive forces*, and the manner in which they change with time during the short time Δt is generally unknown. If the force starts to be applied at some time t_0, Eq. (2.1.2) still yields, of course,

$$\int_{t=t_0}^{t_0+\Delta t} \vec{F}\, dt = [m\vec{v}]_{t=t_0+\Delta t} - [m\vec{v}]_{t=t_0} \stackrel{\Delta}{=} m\Delta\vec{v} \qquad (2.1.3)$$

The integral on the left side of Eq. (2.1.3), in the limit as $\Delta t \to 0$, is called an *impulse*. In practice, an impulse can only be approximated, of course, by a short-duration large force. Such forces appear in impact problems that are addressed in first courses in dynamics. They are also approximations for short-duration propulsive forces that are imparted to satellites, for example, to change the velocity of their center of mass by a desired amount $\Delta\vec{v}$ in a very short time in order to correct their space trajectory or change their orbit around a celestial body.

2.2 A Note on the Reaction Force due to Springs

A spring is a particular type of flexible body and, as such, it deforms when subjected to an applied force. Many physical systems incorporate a spring attached to a mass. During the deformation process, a spring reacts to deformation with an internal force, and that force appears in the free-body diagram for the mass it is attached to. The study of deformation of flexible objects belongs to the discipline called *theory of elasticity*, which is beyond the scope of this book. Early discoveries in elasticity were made by Robert Hooke (1635–1703) who was a contemporary of Newton. In most dynamics problems, the mass of the spring is neglected, and such a practice will be followed here.

There are two types of springs: *extensional springs* (which are simply referred to as *springs*) and *torsional springs*. Both are sketched in Fig. 2.1, with a block P and a bar, respectively, connected to them. In Fig. 2.1a, L_u is the length of the spring when it is unstretched, while r is its actual length when it is stretched from that position. In Fig. 2.1b, the torsional spring is unstretched when $\theta = \theta_u$, where θ is measured from some arbitrary line taken as reference. Figure 2.1c and d shows how the spring reaction force F_s and the torsional spring reaction moment M_s appear in free-body diagrams for block P and the bar, respectively. The complete free-body diagram for a massless spring is also shown in Fig. 2.1c. For such a spring, the right-hand side of the equation $\vec{F} = d(m\vec{v})/dt$ is equal to zero when applied to the spring. That is why the free-body diagram for a massless spring simply consists of the forces $\vec{F}_s$ and $-\vec{F}_s$ acting at both ends of the spring.

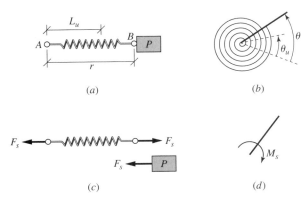

Figure 2.1 (*a*) Extensional and (*b*) torsional springs. (*c*) Free-body diagrams showing a massless spring and the spring reaction force for the extensional spring. (*d*) Free-body diagram showing the spring reaction moment for the torsional spring.

Thus, that diagram does not offer any additional useful information beyond what is already in the free-body diagram for the block that is attached to the spring.

The reaction spring force F_s is a function of the stretch $r - L_u$, and the reaction moment M_s for the torsional spring is a function of $\theta - \theta_u$. For *linear springs*, these functions are simply

For a linear extensional spring: $\quad F_s = k_1(r - L_u)$

For a linear torsional spring: $\quad M_s = M_1(\theta - \theta_u)$ (2.2.1)

where k_1 and M_1 are constants, which are determined experimentally.

The constants k_1 (in newtons per meter) and M_1 (in newton-meters per radian) are each called the *spring stiffness* (or the *spring constant*), and they are both positive quantities. Notice that the direction of the spring force F_s in Eq. (2.2.1) is automatically reversed, as expected, when the spring is compressed (i.e., when $r - L_u < 0$). The same occurs for the reaction moment due to the torsional spring.

It should be kept in mind that for real springs, the linear relationships given in Eq. (2.2.1) are only valid for some range of $|r - L_u|$ and $|\theta - \theta_u|$, respectively. For large values of $|r - L_u|$ for the extensional spring, it becomes harder to stretch (or to compress) the spring. This behavior is modeled with the better approximation $F_s = k_1(r - L_u) + k_3(r - L_u)^3$, where $k_3 > 0$. Such a spring is called a *hardening spring*.

2.3 A Moment Equation Obtained from Newton's Second Law

As indicated in Section 2.1, a number of mathematical manipulations may be required involving the scalar equations that are obtained from $\vec{F} = d(m\vec{v})/dt$ in order to eliminate unknown reaction forces to finally obtain the differential equations of motion for the particle. For some problems, such an elimination process can be shortened (and, sometimes, considerably shortened) by manipulating Newton's second law as shown in this section. For a single mass particle, the methodology presented here does not provide any new information beyond what is already obtained from $\vec{F} = d(m\vec{v})/dt$. But it may provide significant shortcuts as previously mentioned. This is illustrated in detail in a number of solved examples in the book.

If we calculate the moment $\vec{M}_O = \vec{r} \times \vec{F}$, about an inertial point O, of the resultant $\vec{F}$ of the forces acting on a particle P of mass m, and make use of Newton's second law $\vec{F} = d(m\vec{v})/dt$, where $\vec{v} = d\vec{r}/dt$, the following equality is obtained:

$$\vec{M}_O = \vec{r} \times \vec{F} = \vec{r} \times \frac{d(m\vec{v})}{dt} = \frac{d[\vec{r} \times (m\vec{v})]}{dt} \triangleq \frac{d\vec{H}_O}{dt} \qquad (2.3.1)$$

The equality $d[\vec{r} \times (m\vec{v})]/dt = \vec{r} \times [d(m\vec{v})/dt]$ is true because the extra term $(d\vec{r}/dt) \times (m\vec{v})$ in $d[\vec{r} \times (m\vec{v})]/dt = (d\vec{r}/dt) \times (m\vec{v}) + \vec{r} \times [d(m\vec{v})/dt]$ is equal to zero [since $d\vec{r}/dt = \vec{v}$, and $\vec{v} \times (m\vec{v}) = 0$].

The vector $\vec{H}_O \triangleq \vec{r} \times (m\vec{v})$, which is the moment about an inertial point O of the absolute linear momentum $m\vec{v}$ of particle P, is defined as the *absolute angular momentum* of P about the inertial point O (the prefix *absolute* refers to the fact that $\vec{v}$ is the absolute velocity of P); the quantity $m\vec{v}$ is called the *absolute linear momentum* of P.

Equation (2.3.1) shows that for those problems where the moment $\vec{M}_O$ is zero, the absolute angular momentum $\vec{H}_O$ is constant; i.e., its magnitude is constant, and its direction is fixed in inertial space (see Example 2.5.11, p. 107). In some problems, it may also happen that only one or two of the components of $\vec{M}_O$ (instead of all three), in terms of a set of inertial directions, is zero. For such cases, although the angular momentum $\vec{H}_O$ is not conserved (i.e., not constant for all times), those inertial components of $\vec{H}_O$ for which the inertial components of $\vec{M}_O$ are zero remain constant in time.

It turns out that Eq. (2.3.1), which was obtained here for a point mass by simply taking the cross product of $\vec{r}$ with both sides of the basic equation $\vec{F} = d(m\vec{v})/dt$ for that point mass, is of fundamental importance in dynamics, especially for a system of point masses, including rigid and flexible bodies. Such a mathematical operation is done again in Chapter 4 (for a system of point masses), and in Chapters 5 and 6 (for rigid bodies). The resulting equation, which involves the moment of the forces acting on the system and the absolute angular momentum of the system (defined in those chapters), is called *Euler's vector equation*.

Let us now look at the problem of how to find the expressions for the absolute velocity and acceleration of a point so that we can start analyzing specific problems in dynamics.

2.4 Absolute Position, Velocity, and Acceleration of a Point

By definition, the absolute velocity $\vec{v}$ of a point P is the time derivative, in an inertial reference frame, of its absolute position vector. As shown in Fig. 2.2, the *position vector* $\vec{r}$ of a point P, relative to a reference point O in space, is simply the vector drawn from O to P. The position vector $\vec{r}$ is said to be *absolute* when the reference point O has zero acceleration. Such a reference point is called an *inertial point,* and point O is labeled as such in the figure.

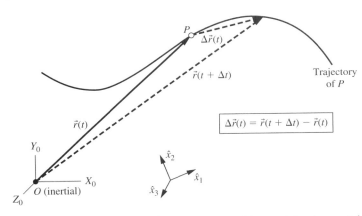

Figure 2.2 A position vector $\vec{r}$ at two instants of time and the change $\Delta\vec{r}$ of the position vector.

In order to introduce the definition of the time derivative of a vector,[1] two inertial position vectors for point P at different times are shown in Fig. 2.2, namely, $\vec{r}(t)$ (i.e., "vector r at time t"), and $\vec{r}(t + \Delta t)$ (i.e., "vector r at time t plus delta t"). They correspond to the absolute position vector of P at two instants of time t and $t + \Delta t$, after a time increment Δt. In a dynamics problem, the trajectory of P shown in Fig. 2.2 is unknown, and it is determined by integrating a set of differential equations that are obtained from Newton's second law.

An inertial reference frame is shown in Fig. 2.2 as being represented by the orthogonal lines labeled X_0, Y_0, and Z_0. A set of orthogonal unit vectors $\{\hat{x}_1, \hat{x}_2, \hat{x}_3\}$ is also shown in the figure; the need for them will start to be explained in the paragraph immediately following Eq. (2.4.2).

---

[1] We are concerned here with the time derivative of $\vec{r}(t)$ in an inertial frame because that is what is needed for correctly using Newton's second law.

The *absolute velocity* $\vec{v}(t)$ of point P at any instant of time t is, by definition, equal to the time derivative, in an inertial frame, of the absolute position vector of P. The definition of the derivative of a scalar function (which you have seen in calculus) is extended to vectors simply as

$$\boxed{\vec{v}(t) = \frac{d\vec{r}(t)}{dt}} \triangleq \lim_{\Delta t \to 0} \frac{\vec{r}(t + \Delta t) - \vec{r}(t)}{\Delta t} \qquad (2.4.1)$$

As can be seen in Fig. 2.2, the velocity $\vec{v}(t)$ of point P is always tangent to the trajectory described by the point during its motion in space. This is so because, as Δt goes to zero, the direction of the change $\Delta \vec{r} = \vec{r}(t + \Delta t) - \vec{r}(t)$ of the position vector becomes, in the limit as $\Delta t \to 0$, tangent to the trajectory at the location of P at time t.

The *absolute acceleration* $\vec{a}(t)$ of point P at any instant of time t is defined as the time derivative of the absolute velocity $\vec{v}(t)$ of P, i.e.,

$$\boxed{\vec{a}(t) = \frac{d\vec{v}(t)}{dt}} \triangleq \lim_{\Delta t \to 0} \frac{\vec{v}(t + \Delta t) - \vec{v}(t)}{\Delta t} = \frac{d^2\vec{r}(t)}{dt^2} \qquad (2.4.2)$$

Equations (2.4.1) and (2.4.2) are just the definitions of velocity $\vec{v}$ and acceleration $\vec{a}$. The more difficult task consists of carrying out the derivatives in order to obtain, for each problem, explicit expressions for the components of $\vec{v}$ and $\vec{a}$ along the three orthogonal unit vectors of a triad that is chosen to continue the work. Let us say that a triad $\{x\} = \{\hat{x}_1, \hat{x}_2, \hat{x}_3\}$, illustrated in Fig. 2.2, has been chosen. The unit vectors $\hat{x}_1, \hat{x}_2$, and $\hat{x}_3$ may be either nonrotating or drawn on a rotating body. Different choices are generally made for different problems, as will be illustrated in Section 2.5. Since Newton's second law requires using absolute velocity and absolute acceleration, a set of nonrotating axes is also shown in Fig. 2.2; the axes are labeled X_0, Y_0, and Z_0.

The absolute position vector $\vec{r}$ is of the general form

$$\vec{r} = r_{x_1}\hat{x}_1 + r_{x_2}\hat{x}_2 + r_{x_3}\hat{x}_3 \qquad (2.4.3)$$

where the scalar components r_{x_1}, r_{x_2}, and r_{x_3} are functions of the variables that are chosen to describe the motion.

The use of different coordinates to express a vector was illustrated in the review presented in Section 1.1. As indicated in that section, such variables may be rectangular, cylindrical, or spherical coordinates (or even a combination of these classical coordinates), or any other convenient set of variables. The specific choice depends on the particular problem to be analyzed.

As mentioned earlier in this section, the form of the differential equations of motion that will be obtained (and, therefore, the difficulty encountered in integrating them) will be directly related to the choice made for such variables.

The time derivative of $\vec{r}$ is simply calculated as

$$\frac{d\vec{r}}{dt} = \dot{r}_{x_1}\hat{x}_1 + \dot{r}_{x_2}\hat{x}_2 + \dot{r}_{x_3}\hat{x}_3 + r_{x_1}\frac{d\hat{x}_1}{dt} + r_{x_2}\frac{d\hat{x}_2}{dt} + r_{x_3}\frac{d\hat{x}_3}{dt}$$

$$\triangleq \left[\frac{d\vec{r}}{dt}\right]_{\{x\}} + r_{x_1}\frac{d\hat{x}_1}{dt} + r_{x_2}\frac{d\hat{x}_2}{dt} + r_{x_3}\frac{d\hat{x}_3}{dt} \qquad (2.4.4)$$

where a dot over a scalar function denotes the time derivative of the scalar function represented by that symbol. The time derivative of any unit vector (or of any vector of constant magnitude) is zero only if the direction of the unit vector is not changing with time (i.e., if the unit vector is not rotating). The translation of a unit vector in space is of no relevance at all because unit vectors are introduced only to allow us to keep track of directions.

The combination of the first three terms in the expression for $d\vec{r}/dt$ [Eq. (2.4.4)], which is renamed $[d\vec{r}/dt]_{\{x\}}$, is the time derivative of $\vec{r}$ as seen by an observer fixed at O and rotating the same way as the triad $\{x\}$. The remaining three terms in Eq. (2.4.4), i.e., the ones involving the time derivative of unit vectors, are due to the rotation of the triad $\{x\}$.

Before we make use of rotating unit vectors, let us look at some example problems that are formulated using nonrotating unit vectors. Rotating unit vectors are dealt with in Section 2.6, and a number of examples, using such unit vectors, are presented in Section 2.7.

2.5 Solved Example Problems

This section starts with the simplest type of problem in dynamics, which involves motion of a material point along a nonrotating straight line. The difficulty of the problems is then increased. As the level of difficulty is increased, the amount of mathematical manipulations involved in obtaining the differential equations that govern the motion is gradually increased. The motion is analyzed using the methods presented in Chapter 1, but often the differential equations have to be integrated numerically because no analytical solution is possible. Such situations are very typical in actual practice. Numerical integration is done in this book using Simulink so that one does not have to detract from the study of dynamics to write (and debug) computer code to solve the problems.

EXAMPLE 2.5.1

A body P of mass $m = 10$ kg is thrown from a height $h = 200$ m above the earth's surface with an initial velocity $[v]_{t=0} = v_0 = 200$ km/h directed upward. The body is modeled as a point mass and is subjected only to the gravitational force of the earth; its motion takes place along the vertical. Determine

a. The maximum height reached by the body P, measured from the ground
b. The speed (i.e., the magnitude of the velocity vector) of P when it hits the ground
c. The speed of P as a function of its height measured from the ground
d. The time it takes P to hit the ground

Figure 2.3
Free-body diagram for Example 2.5.1.

■ **Solution**

The free-body diagram for P is shown in Fig. 2.3. The position of the body is tracked by the vector from an inertial point O on the ground, using the distance $y(t)$ from O to P as the coordinate of the motion. A unit vector $\hat{r}$, for expressing the vector $\overrightarrow{OP}$, is also shown in the figure. Notice that such choices are up to the individual, and you could have made different ones. You could have chosen to work with a unit vector in the direction opposite to $\hat{r}$ and to track the body from another inertial point, such as the launching point, for example. But the choices that were made are as good as any other. Since the motion takes place along the vertical, only one unit vector is needed; it could be chosen to be pointing either upward or downward. The force $\vec{F}$ acting on P is indicated by the heavy arrowed line shown in Fig. 2.3, while the position vector $\vec{r}$ is indicated with a broken arrowed line to clearly distinguish it from a force. It is a good practice to adopt such a convention so that others do not misinterpret your free-body diagram.

In terms of the unit vector $\hat{r}$ and the motion coordinate y shown in Fig. 2.3, the expression for the absolute position vector $\vec{r} = \overrightarrow{OP}$ (with point O being inertial) is $\vec{r} = y\hat{r}$. Therefore, the absolute velocity and acceleration of P are

$$\vec{v} = \frac{d\vec{r}}{dt} = \dot{y}\hat{r} \qquad (2.5.1a)$$

$$\vec{a} = \frac{d\vec{v}}{dt} = \ddot{y}\hat{r} \qquad (2.5.1b)$$

We now need an expression for the gravitational force F acting on the particle. For this, a decision has to be made; namely, How should force F be modeled? Engineers and scientists are always facing such decisions about modeling. In this case, we have two choices. The simplest choice is to approximate F as mg, with g being the constant acceleration of gravity. Such a choice is good if the particle stays "close enough" to the earth's surface during its motion along the vertical. Given that the initial speed is only (!) 200 km/h, that is probably going to be the case (but, at this stage of the formulation, we do not know that for sure). The other choice, which is more accurate, is to use $F = GM_E m/(R_E + y)^2$, as given by Newton's law of universal gravitation, where M_E and R_E are the mass and the radius of the earth, respectively. Let's do the problem both ways, starting with the simplest choice for F as $F = mg$. When both analyses are finished, a plot is presented showing how the answers to the questions that were asked depend on how F was modeled.

With $\vec{F} = -mg\hat{r}$ (notice that the minus sign goes together with the direction of the unit vector that was chosen; if we had chosen to work with a unit vector in the opposite direction, a plus sign would show up here and a minus sign would show up in the expressions for the position vector $\vec{r}$, the velocity $\vec{v}$, and the acceleration $\vec{a}$), $\vec{F} = m\vec{a}$ then gives the following differential equation for $y(t)$:

$$\ddot{y} = -g \qquad (2.5.1c)$$

This differential equation is of the type that was reviewed in Section 1.13, and its solution is

$$y = -\frac{gt^2}{2} + t\dot{y}(0) + y(0) \qquad (2.5.1d)$$

where, for this problem, $\dot{y}(0) = v_0 = 200$ km/h and $y(0) = h = 200$ m.

Since $\ddot{y} = d\dot{y}/dt = \dot{y}(d\dot{y}/dy)$ (see the presentation in Section 1.13, about changing variables by using the chain rule), we also obtain from Eq. (2.5.1c)

$$\frac{\dot{y}^2}{2} + gy = \text{constant} = \left[\frac{\dot{y}^2}{2} + gy\right]_{t=0} = \frac{v_0^2}{2} + gh \qquad (2.5.1e)$$

All the questions that were asked are then answered from the expressions obtained for y and $\dot{y}^2$ in Eqs. (2.7.1d) and (2.7.1e).

a. We can immediately determine the maximum height P by setting $\dot{y} = 0$ in Eq. (2.5.1e) that involves $\dot{y}^2$ because, when there are no bounds imposed on the independent variable t, the maximum (or minimum) of a function $y(t)$ occurs at the point when the derivative of the function is zero. For the present problem, since the second derivative $\ddot{y} = -g$ is negative, the point where $\dot{y} = 0$ corresponds to the maximum value $y_{\max}$ of y. For the given initial conditions, Eq. (2.5.1e) yields

$$y_{\max} = h + \frac{v_0^2}{2g} \approx 357 \text{ m}$$

Since the maximum height is only 157 m above the initial point (i.e., 357 m above the earth's surface) and the radius of the earth is $R_E \approx 6378$ km, the approximation of a constant acceleration of gravity should be perfectly acceptable. But what if the answer were much higher? Would the answer be acceptable then? This question will be addressed when we use $F = GM_E m/(R_E + y)^2$.

Another way to determine the maximum height $y_{\max}$ is to use the expression $\dot{y} = -gt + \dot{y}(0)$, which is obtained by integrating $\ddot{y} = -g$ once, and set $\dot{y} = 0$ to obtain the time t when $\dot{y} = 0$, which is $t = \dot{y}(0)/g \approx 5.66$ s, and then find $y_{\max}$ from the solution $y = -gt^2/2 + t\dot{y}(0) + y(0)$ as $y_{\max} = [y]_{t=5.66} \approx 357$ m. As you see, there is more than one way to solve this problem. But no matter which way you prefer, the solution is always obtained by integrating the differential equation of motion (which is $\ddot{y} = -g$ for this particular problem).

b. We can also immediately determine the speed of P when it hits the ground by making use of the expression that involves y and $\dot{y}^2$ [Eq. (2.5.1e)]. By setting $y = 0$ in that equation we now obtain

$$[\dot{y}^2]_{\text{when } y=0} = v_0^2 + 2gh$$

or

$$[\dot{y}]_{\text{when } y=0} = \pm\sqrt{v_0^2 + 2gh} \approx \pm 83.7 \text{ m/s} \approx \pm 301 \text{ km/h}$$

Looking at the directions of the position vector $\vec{r}$ and the unit vector $\hat{r}$ in Fig. 2.3, we conclude that the velocity of P when it strikes the ground is $[\vec{v}]_{\text{when } y=0} = -301\hat{r}$ km/h (the magnitude of the velocity is 301 km/h).

c. The speed of P as a function of its height from the ground, obtained from Eq. (2.5.1e), is $v = |\pm\sqrt{v_0^2 - 2g(y-h)}|$, i.e., $v = \sqrt{v_0^2 - 2g(y-h)}$.

d. The time it takes P to hit the ground is more easily obtained when the differential equation $\ddot{y} = -g$ is integrated directly with respect to time. This was possible to do for this differential equation, and the solution is given by Eq. (2.5.1d). By setting $y = 0$ in that solution, and letting t_f denote the time when $y = 0$, we then have to solve the following quadratic equation for t_f:

$$-\frac{gt_f^2}{2} + t_f \dot{y}(0) + h = 0$$

Two solutions for the time t_f are obtained:

$$t_f = \frac{\dot{y}(0)}{g} \pm \sqrt{\left(\frac{\dot{y}(0)}{g}\right)^2 + \frac{2h}{g}}$$

Obviously, only the positive solution makes physical sense. For the given initial conditions we then obtain $t_f \approx 14.2$ s.

Notes

- If you look at the solution obtained for the maximum value of y, $y_{\max} = h + v_0^2/2g$, you see that the maximum height reached by the particle increases as we increase the initial upward speed $v_0 = [\dot{y}]_{y=h}$. This is certainly reasonable, and it is what intuition tells us. However, it is not reasonable to expect that the approximation $F = \text{constant} = mg$ remains valid as y becomes bigger and bigger.
- Another thing that the expression $y_{\max} = h + v_0^2/2g$ indicates is that for $y_{\max}$ to be infinite the initial speed $v_0 = [\dot{y}]_{t=0}$ will have to be infinite. Physically, $y_{\max} = \infty$ should be interpreted as being a very big distance, which is many times the radius of the earth. Since it certainly would take an infinite amount of time for the distance y to be infinite, that means that the particle would never return to the vicinity of the earth after it was launched. The physical interpretation of this is that the particle would have "escaped" the gravitational attraction of the earth. The minimum launch speed for this to happen is called the *escape speed*. Notice, though, that it would be physically impossible to launch a space probe to make it leave forever the neighborhood of the launching point if an infinite launching speed would be required! Satellites could never have been launched into deep space if that were the case. Let us then look at this problem again in Example 2.5.2 with the objective of determining the maximum distance $y_{\max}$ and the minimum value of the initial speed $v_0 = [\dot{y}]_{t=0}$ that is necessary to impart to the particle for it to escape the gravitational attraction of the earth.

EXAMPLE 2.5.2

Determine the maximum distance $y_{\max}$ for the problem in Example 2.5.1 when $v_0 = \dot{y}(0) = 10,000$ km/h and when $v_0 = 20,000$ km/h. What is the value of the escape speed for a body launched from the earth's surface at a height $y(0) = h = 400$ km above the earth's surface? The earth may be approximated as a homogeneous sphere with radius $R_E \approx 6378$ km.

■ Solution

With $h = 0$, the solution $y_{max} = h + v_0^2/2g$ found in Example 2.5.1 yields $y_{max} \approx 393$ km when $v_0 = 10{,}000$ km/h, and $y_{max} \approx 1573$ km when $v_0 = 20{,}000$ km/h. Let us check these answers with $F = GM_E m/(R_E + y)^2$, which is the expression that has to be used for calculating the escape speed.

With $\vec{F} = -GM_E m/(R_E + y)^2 \hat{r} = -mgR_E^2/(R_E + y)^2 \hat{r}$, $\vec{F} = m\vec{a}$ now gives the following second-order differential equation for $y(t)$:

$$\ddot{y} = -\frac{gR_E^2}{(R_E + y)^2} \tag{2.5.2a}$$

Although not impossible, it is now more difficult to integrate this differential equation with respect to time. Instead of trying to do that analytically here (you can integrate it numerically by using Simulink, for example), let us use again the method presented in Section 1.13. With $\ddot{y} = d\dot{y}/dt = \dot{y}(d\dot{y}/dy)$, we now obtain from Eq. (2.5.2a)

$$\int \dot{y} \frac{d\dot{y}}{dy} dy + \int \frac{gR_E^2\, dy}{(R_E + y)^2} = \text{constant}$$

or

$$\frac{\dot{y}^2}{2} - \frac{gR_E^2}{R_E + y} = \text{constant} = \frac{\dot{y}^2(0)}{2} - \frac{gR_E^2}{R_E + y(0)} \tag{2.5.2b}$$

Notice that this equation discloses that to have $\dot{y} = 0$ when y is infinite (i.e., to escape the gravitational field of the earth by launching the particle with the smallest required initial speed), we must impart to the body, at the launching height $y = h$, an initial speed equal to $\dot{y} = \sqrt{2gR_E/(1 + h/R_E)}$ since the constant of integration in this situation is equal to zero. Notice that the positive solution for $\dot{y}$ was used since we want the body to move away from the earth, as indicated by the convention for the directions shown in Fig. 2.3. This critical speed, which is called the *escape speed* (and is denoted as v_{esc}) at the launching height h, is the minimum initial speed that should be imparted to the body for it to escape the gravitational field of the attracting planet it is launched from. Any initial value of $\dot{y}$ that is bigger than v_{esc} will cause the body to escape the gravitational field of the attracting planet and have a nonzero "speed at infinity" (which is certainly what is desired in practice). For a body launched from the surface of the earth (i.e., $h = 0$), the escape speed is $v_{esc} \approx 40{,}270$ km/h $\approx 25{,}000$ miles per hour (mi/h). For $h = 400$ km, the escape speed is $v_{esc} \approx 39{,}000$ km/h $\approx 24{,}280$ mi/h.

The maximum value of y, y_{max}, is obtained by setting $\dot{y} = 0$ in Eq. (2.5.2b). For $y(0) = 0$, Eq. (2.5.2b) yields $y_{max} \approx 419$ km when $\dot{y}(0) = v_0 = 10{,}000$ km/h, and $y_{max} \approx 2090$ km when $\dot{y}(0) = v_0 = 20{,}000$ km/h.

Figure 2.4 shows a plot of y_{max}/R_E versus the initial speed $\dot{y}(0) = v_0$ obtained with $F = mgR_E^2/(R_E + y)^2$ and with the approximation $F = mg$, for $y(0) = h = 0$. Observe how the approximation $F = mg$ becomes worse and worse as v_0 is increased.

Figure 2.4 Maximum height y_{max} reached by a particle thrown from the earth's surface for Example 2.5.2.

EXAMPLE 2.5.3

As shown in Fig. 2.5a, a bar AB of length $L = 1.5$ m is forced to slide against a fixed (nonrotating) vertical wall so that the angle θ it makes with the horizontal changes with time as $\theta = \pi/2 - k_1 t \sin(k_2 t)$ where $k_1 = 1$ rad/s and $k_2 = 0.08$ rad/s. Point O in Fig. 2.5 is inertial; i.e., its absolute acceleration is equal to zero. Determine the absolute velocity and acceleration of points A and B when the bar strikes the ground (i.e., when $\theta = 0$), and the velocity of point B relative to point A at that instant.

■ Solution

Since the angle θ is known for all times, there are no differential equations of motion to be formulated and, therefore, this type of problem is a kinematics problem. The absolute

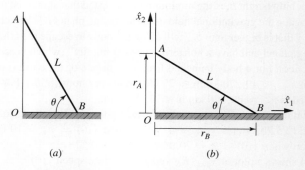

Figure 2.5 The system and the absolute position vectors of A and B for Example 2.5.3.

velocities of points A and B are equal to the time derivatives of the absolute position vectors from the inertial point O to points A and B, respectively. Also, the absolute accelerations of those points are equal to the time derivatives of their respective absolute velocities.

Since points A and B move along straight lines, the natural way to determine their absolute velocity and acceleration is to express their absolute position vectors $\vec{r}_A = \overrightarrow{OA}$ and $\vec{r}_B = \overrightarrow{OB}$ in terms of the known angle θ, and of the unit vectors $\hat{x}_1$ and $\hat{x}_2$ shown in Fig. 2.5b, as given in Eq. (2.5.3a). These expressions are simply geometrical relationships that are obtained by inspection of Fig. 2.5.

$$\vec{r}_A = (L \sin\theta)\hat{x}_2 \qquad \vec{r}_B = (L \cos\theta)\hat{x}_1 \qquad (2.5.3a)$$

Once the expressions for the absolute position vectors $\vec{r}_A$ and $\vec{r}_B$ are written down, the calculation of the absolute velocities and accelerations of points A and B simply involve taking the first and second time derivatives of $\vec{r}_A$ and $\vec{r}_B$. By proceeding in this manner, we then obtain

$$\vec{v}_A = \frac{d\vec{r}_A}{dt} = L\dot\theta(\cos\theta)\hat{x}_2 \qquad \vec{a}_A = \frac{d\vec{v}_A}{dt} = L(\ddot\theta\cos\theta - \dot\theta^2\sin\theta)\hat{x}_2 \qquad (2.5.3b)$$

$$\vec{v}_B = \frac{d\vec{r}_B}{dt} = -L\dot\theta(\sin\theta)\hat{x}_1 \qquad \vec{a}_B = \frac{d\vec{v}_B}{dt} = -L(\ddot\theta\sin\theta + \dot\theta^2\cos\theta)\hat{x}_1 \qquad (2.5.3c)$$

For $\theta = \pi/2 - t\sin(0.08t)$, the first zero-crossing, $\theta = 0$, is found to occur when $t \approx 4.48$ s. This may be obtained by a few iterations with a calculator, or from the plot of $\theta(t)$ versus t shown in Fig. 2.6. Therefore, the bar hits the floor when $t \approx 4.48$ s.

With $\theta = \pi/2 - t\sin(0.08t)$, we obtain $\dot\theta = -\sin(0.08t) - 0.08t\cos(0.08t)$. Therefore, the value of $\dot\theta(t)$ when $\theta = 0$ is

$$[\dot\theta]_{t=4.48\,\text{s}} = -\sin(0.08 \times 4.48) - 0.08 \times 4.48 \cos(0.08 \times 4.48) \approx -0.686 \text{ rad/s}$$

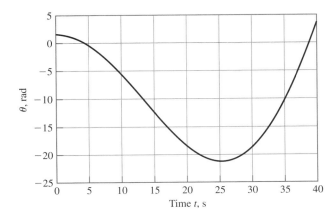

Figure 2.6 Plot of $\theta(t) = \pi/2 - t\sin(0.08t)$ for Example 2.5.3.

With $L = 1.5$ m, this gives the following velocities at the instant when $\theta = 0$:

$$[\vec{v}_A]_{\theta=0} = -[1.5 \times 0.686 \times \cos(0)]\hat{x}_2 \approx -1.03\hat{x}_2 \text{ m/s} \qquad [\vec{v}_B]_{\theta=0} = 0$$

The minus sign in the answer for $\vec{v}_A$ means that the velocity of A at that instant is opposite to the $\hat{x}_2$ unit vector; i.e., point A is moving down.

The velocity of B relative to A, denoted as $\vec{v}_{B/A}$, is, by definition, equal to $\vec{v}_B - \vec{v}_A$. Therefore,

$$[\vec{v}_{B/A}]_{\theta=0} = 1.03\hat{x}_2 \text{ m/s}$$

The respective accelerations when $\theta = 0$ are

$$[\vec{a}_A]_{\theta=0} \approx -0.21\hat{x}_2 \text{ m/s}^2 \qquad [\vec{a}_B]_{\theta=0} = -0.707\hat{x}_1 \text{ m/s}^2$$

EXAMPLE 2.5.4

A rectangular block is forced by a bar AB to move along a fixed straight groove that is inclined at a constant angle $\alpha = 30°$ with the horizontal as shown in Fig. 2.7a. Point P is the point of contact of bar AB with the block, and the distance h shown in the figure is the constant perpendicular distance from the fixed point A to the groove. The bar is driven by a motor so that the angle θ shown in the figure is changing with time t as $\theta = 2\pi/3 - k_1 \cos(k_2 t)$ rad, where $k_1 = 5\pi/12$ rad and $k_2 = \pi/2$ rad/s. Determine the absolute velocity and the absolute acceleration (magnitude and direction) of the block when $t = 1$ s, 2 s, and 4 s.

■ Solution

Since the block is only translating, each point of the body has the same velocity. The same holds for the acceleration of each of its points. We then choose to work with point P to determine the velocity and the acceleration of the block. Notice that this problem is also a kinematics problem, as the one in Example 2.5.3, since the motion of the block is already known. Also, we are not concerned here with determining the forces that are necessary to impart that specified motion to the block.

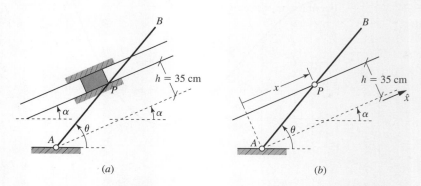

Figure 2.7 The system and the absolute position vector of P for Example 2.5.4.

Since point P moves along a straight line, this problem is also readily solved by introducing the variable x and the unit vector $\hat{x}$ shown in Fig. 2.7b, to express the absolute position vector $\vec{r}$ of point P in terms of the known angle $\theta(t)$ and the known distance $h = 35$ cm as

$$\vec{r} = x\hat{x} = \frac{h}{\tan(\theta - \alpha)}\hat{x} \quad (2.5.4a)$$

This expression is simply a geometrical relationship that is obtained by inspection of Fig. 2.7. Once the expression for the absolute position vector $\vec{r}$ is written down, the problem is readily solved simply by taking the first and second time derivatives of that expression. For this problem, we then obtain

$$\vec{v} = \frac{d\vec{r}}{dt} = \dot{x}\hat{x} = -\frac{h}{\tan^2(\theta - \alpha)}\frac{d}{dt}[\tan(\theta - \alpha)]\hat{x} = -\frac{h\dot{\theta}}{\sin^2(\theta - \alpha)}\hat{x} \quad (2.5.4b)$$

$$\vec{a} = \frac{d\vec{v}}{dt} = \ddot{x}\hat{x} = \left\{-\frac{h\ddot{\theta}}{\sin^2(\theta - \alpha)} + \frac{2h\dot{\theta}}{\sin^3(\theta - \alpha)}\frac{d}{dt}[\sin(\theta - \alpha)]\right\}\hat{x}$$

$$= \left\{-\frac{h\ddot{\theta}}{\sin^2(\theta - \alpha)} + \frac{2h\dot{\theta}^2\cos(\theta - \alpha)}{\sin^3(\theta - \alpha)}\right\}\hat{x} \quad (2.5.4c)$$

The answers to the questions that were asked for the various values of t are then calculated using these expressions. They are given in the following table, where $\dot{\theta} = [5\pi^2 \sin(\pi t/2)]/24$ and $\ddot{\theta} = [5\pi^3 \cos(\pi t/2)]/48$. The absolute velocity and the absolute acceleration of P are $\dot{x}\hat{x}$ and $\ddot{x}\hat{x}$, respectively.

t, s	θ, degrees	$\dot{\theta}$, rad/s	$\ddot{\theta}$, rad/s²	$\dot{x}$, m/s	$\ddot{x}$, m/s²
1	120	2.06	0	−0.72	0
2	195	0	−3.23	0	16.88
4	45	0	3.23	0	−16.88

EXAMPLE 2.5.5

A block of mass m kg slides on the face of a plane that is inclined at a constant angle $\alpha = 30°$ with the horizontal as shown in Fig. 2.8. The motion of the block takes place on a vertical plane, and it consists of translation only.[2] The block is initially at rest at a distance of 4 m from the bottom of the plane. Obtain the differential equation of motion for the block, and determine the time it takes the block to arrive at the bottom of the inclined plane, for the following two cases:

a. Friction in the system is viscous, with friction coefficient c N/(m/s) and $c/m = 0.8$ s^{-1}.
b. Friction in the system is of the Coulomb type; the static and kinetic friction coefficients are $\mu_s = 0.3$ and $\mu_k = 0.25$, respectively.

Figure 2.8 The system for Example 2.5.5.

---

[2] The motion of a body is only a translation if lines that are "painted" on the body do not change orientation during the motion.

Figure 2.9 Free-body diagram and the position vector $\vec{r}$ for Example 2.5.5.

■ Solution

Since there is no rotation involved, the block moves as if it were a point mass; i.e., all points of the body have the same velocity and the same acceleration. The free-body diagram and the absolute position vector $\vec{r} = \overrightarrow{OP}$ for a point P on the block are shown in Fig. 2.9. The position vector $\vec{r} = \overrightarrow{OP}$ is shown as a dashed arrowed line to distinguish it from forces, which are shown as solid arrowed lines. The inertial reference point O was arbitrarily chosen at the bottom of the inclined plane; it could have been chosen at any other fixed location on the inclined plane or elsewhere. The two orthogonal unit vectors $\hat{x}_1$ and $\hat{x}_2$ shown in that figure were also chosen for expressing all the vectors that appear in the formulation of the equations for this problem. Any other set of orthogonal unit vectors could have been chosen. The rationale behind the particular choice of $\hat{x}_1$ shown in the figure is that the expression for the absolute acceleration of P will only have a component along $\hat{x}_1$ because the motion of P takes place along a straight line on the inclined plane. The direction for $\hat{x}_1$ could have been chosen to be opposite to the one shown in the figure, and the expression obtained for the acceleration of P would still be a simple one.

In terms of $\hat{x}_1$, the absolute position vector $\vec{r}$, absolute velocity $\vec{v}$, and absolute acceleration $\vec{a}$ of P are then

$$\vec{r} = r\hat{x}_1 \qquad \vec{v} = \frac{d\vec{r}}{dt} = \dot{r}\hat{x}_1 \qquad \vec{a} = \frac{d\vec{v}}{dt} = \ddot{r}\hat{x}_1 \qquad (2.5.5a)$$

The resultant force $\vec{F}$ acting on the block is the sum of the gravitational force, the normal force $N\hat{x}_2$ applied by the plane to the block, and the friction force $\vec{f}$, which was arbitrarily drawn with an arrow pointing to the left in Fig. 2.9. Therefore, the expression for $\vec{F}$ is

$$\vec{F} = mg(-\hat{x}_1 \sin\alpha - \hat{x}_2 \cos\alpha) + N\hat{x}_2 + f\hat{x}_1 \qquad (2.5.5b)$$

From Newton's second law, $\vec{F} = m\vec{a}$, the following two equations are then obtained

$$N - mg\cos\alpha = 0 \qquad (2.5.5c)$$

$$m\ddot{r} = f - mg\sin\alpha \qquad (2.5.5d)$$

These two equations involve three unknowns, namely, $r(t)$, N, and the friction force f. A third equation is obviously necessary for completing the formulation of the problem. The missing equation involves the friction force $f\hat{x}_1$, and to generate such an equation it is necessary to know which type of friction force is involved. Let us consider both viscous and dry (Coulomb) friction. The friction force for both cases is directed opposite to the velocity $\vec{v}$ when the block translates relative to the inclined plane.

Viscous friction. $\vec{f} = -c\vec{v} = -c\dot{r}\hat{x}_1$, which gives $f = -c\dot{r}$.

Coulomb friction. Either $|f| \leq \mu_s|N|$ if there is no motion, or $f = -\mu_k|N|\operatorname{sgn}\dot{r}$ if the block is moving.

Case a: Viscous Friction

For this case the differential equation of motion, obtained from Eq. (2.5.5d), is

$$\ddot{r} + \frac{c}{m}\dot{r} = -g\sin\alpha \tag{2.5.5e}$$

This is a linear nonhomogeneous differential equation with constant coefficients, and for $c \neq 0$ its solution is (see Case 3 in Section 1.13, p. 37):

$$r = r(0) - \frac{mg\sin\alpha}{c}t + \frac{m}{c}\left[\dot{r}(0) + \frac{mg\sin\alpha}{c}\right](1 - e^{-ct/m}) \tag{2.5.5f}$$

For $r(0) = 4$ m, $\dot{r}(0) = 0$, $\alpha = 30°$, and $c/m = 0.8$ s^{-1} this solution yields, after a few iterations with a calculator with $r = 0$, $t \approx 1.54$ s.

Case b: Coulomb Friction

For this case, we first need to determine if the block will move. This is done by calculating the friction force under the assumption that the block does not move (i.e., that it is in static equilibrium) and then checking whether the assumption is true or false by comparing the magnitude of the friction force with the magnitude $\mu_s|N|$ of the maximum possible friction force. If it is found that $|f|/|N| > \mu_s$, the block is actually moving.

If the block is in static equilibrium (i.e., if $r = $ constant), Eq. (2.5.5d) yields

$$f - mg\sin\alpha = 0 \quad \text{or} \quad f = mg\sin\alpha \tag{2.5.5g}$$

which gives

$$\frac{|f|}{|N|} = \frac{mg\sin\alpha}{mg\cos\alpha} = \tan\alpha \approx 0.577$$

Since the above ratio is greater than μ_s, the block is moving. In addition, since the value found for f is positive, the block starts to move down the inclined plane (because the friction force was drawn upward in the free-body diagram shown in Fig. 2.8, and an upward friction force implies $f > 0$). With the direction convention used in Fig. 2.8, the friction force is then

$$f = -\mu_k|N|\operatorname{sgn}\dot{r} = -\mu_k mg(\cos\alpha)\operatorname{sgn}\dot{r} \tag{2.5.5h}$$

and the differential equation of motion for this case, obtained from Eq. (2.5.5d), is

$$\ddot{r} + \mu_k g(\cos\alpha)\operatorname{sgn}\dot{r} = -g\sin\alpha \tag{2.5.5i}$$

This differential equation is nonlinear because of the $\operatorname{sgn}\dot{r}$ term, which is not a linear function of its argument $\dot{r}$. However, the initial conditions of the motion are such that the block starts moving down the plane. Since the direction of the motion does not change, sgn $\dot{r} < 0$ while the block keeps moving down. Therefore, the motion of the body is governed, at all times, by the following equation, which is a linear differential equation.

$$\ddot{r} = g(\mu_k \cos \alpha - \sin \alpha)$$

The solution to this linear nonhomogeneous differential equation is simply

$$r = r(0) + t\dot{r}(0) + \frac{g(\mu_k \cos \alpha - \sin \alpha)t^2}{2}$$

For $r(0) = 4$ m, $\dot{r}(0) = 0$, $\alpha = 30°$, and $\mu_k = 0.25$ this solution yields, with $r = 0$, $t \approx 1.7$ s.

EXAMPLE 2.5.6

Solve Example 2.5.5 when the initial conditions are $r(0) = 4$ m and $\dot{r}(0) = 1.5$ m/s.

■ Solution

For the case of viscous friction, the solution (which is valid for all times) was determined in Example 2.5.5 as

$$r = r(0) - \frac{mg \sin \alpha}{c} t + \frac{m}{c} \left[\dot{r}(0) + \frac{mg \sin \alpha}{c} \right] (1 - e^{-ct/m}) \qquad (2.5.6a)$$

For $r(0) = 4$ m, $\dot{r}(0) = 1.5$ m/s, $\alpha = 30°$, and $c/m = 0.8$ s^{-1}, the solution yields (after a few iterations with a calculator with $r = 0$) $t \approx 1.85$ s.

For the case of Coulomb friction, the differential equation of motion was found to be

$$\ddot{r} + \mu_k g (\cos \alpha) \operatorname{sgn} \dot{r} = -g \sin \alpha \qquad (2.5.6b)$$

Since the function sgn $\dot{r}$ depends on $\dot{r}$ in a nonlinear manner, this differential equation is nonlinear. But, since sgn $\dot{r}$ = constant = 1 when $\dot{r} > 0$ and sgn $\dot{r}$ = constant = -1 when $\dot{r} < 0$, this differential equation can be solved analytically in a "piecewise manner," i.e., by solving it separately when $\dot{r} > 0$ and when $\dot{r} < 0$. For these two possibilities, the solutions are then

for $\dot{r} > 0$: $\qquad r = r(0) + t\dot{r}(0) - \dfrac{g(\sin \alpha + \mu_k \cos \alpha)t^2}{2} \qquad$ (2.5.6c)

for $\dot{r} < 0$: $\qquad r = r(0) + t\dot{r}(0) - \dfrac{g(\sin \alpha - \mu_k \cos \alpha)t^2}{2} \qquad$ (2.5.6d)

Since the block is forced to start moving up the plane, with $\dot{r}(0) = 1.5$ m/s, the solution for $\dot{r} > 0$ discloses that, for $r(0) = 4$ m, $\alpha = 30°$, and $\mu_k = 0.25$, the block keeps moving up the plane until $\dot{r}$ becomes zero. That happens when

$$t = \frac{\dot{r}(0)}{g(\sin \alpha + \mu_k \cos \alpha)} \approx 0.21 \text{ s}$$

The corresponding value of r is found to be

$$r(0.21) = r(0) + \frac{\dot{r}^2(0)}{2g(\sin \alpha + \mu_k \cos \alpha)} \approx 4.16 \text{ m}$$

The question to be answered now is whether the dry friction coefficient is large enough to prevent the block from moving down the plane after $t = 0.21$ s. If the block permanently stopped at that location, the friction force f shown in Fig. 2.9 would be $f = mg \sin \alpha$, as determined in Example 2.5.5. Since the normal force is $N = mg \cos \alpha$, the ratio $|f|/|N| = \tan \alpha \approx 0.577$ would then be greater than the value of the static coefficient of friction μ_s, which is physically impossible. Therefore, the block only stops momentarily (i.e., when $\dot{r}$ becomes zero at $t = 0.21$ s as previously determined), and then proceeds to travel down the plane. Notice that the block would permanently stop at the top of its trajectory up the plane if $\alpha \leq \arctan \mu_s$ (which, for $\mu_s = 0.3$, is $\alpha \leq 16.7°$).

We now have to determine how long it takes the block to travel 4.16 m down the inclined plane, starting from rest. By using the new initial conditions $r(0) = 4.16$ m and $\dot{r}(0) = 0$ in the solution previously obtained when $\dot{r}(t) < 0$, namely,

$$r = r(0) + t\dot{r}(0) - \frac{g(\sin \alpha - \mu_k \cos \alpha)t^2}{2} \tag{2.5.6e}$$

we then find that $r = 0$ occurs when

$$t = \sqrt{\frac{2r(0)}{g(\sin \alpha - \mu_k \cos \alpha)}} \approx 1.73 \text{ s}$$

The total time for the block to travel from $r(0) = 4$ m to $r = 4.16$ m up the plane and then pass the position when $r = 0$ again is therefore $t_{\text{total}} \approx 0.21 + 1.73 = 1.94$ s.

EXAMPLE 2.5.7

The same block in Example 2.5.6 is now connected to the bottom of the inclined plane by a linear spring of unstressed length $L_u = 1$ m and stiffness k N/m, where $k/m = 10$ s^{-2}. The system and its free-body diagram are shown in Fig. 2.10; the reaction forces f and f_s are due to friction and to the spring, respectively.[3]

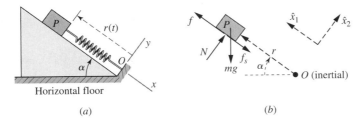

Figure 2.10 The system and its free-body diagram for Example 2.5.7.

The block is initially at rest (i.e., in static equilibrium) at a distance of $r = r_e$ m from the bottom of the plane when it is given a displacement $r(0)$. Consider, separately, both cases of viscous friction and Coulomb friction, with the friction coefficients given in

---

[3] The x-y coordinate system (centered at point O) shown in Fig. 2.10a is only used, later on, in connection with a Simulink diagram that was prepared for this system.

parts a and b of Example 2.5.5. That is, when friction is viscous, the friction coefficient is c N/(m/s), with $c/m = 0.8$ s^{-1}, and when friction is of the Coulomb type, the static and kinetic coefficients of friction are $\mu_s = 0.3$ and $\mu_k = 0.25$, respectively.

a. What is the differential equation that governs the motion of the block for each case?
b. Determine the value of the equilibrium distance r_e. Notice that for the case of viscous friction there is a single value of r_e, while for the case of Coulomb friction there is a range of values of r_e when the block does not move.
c. Use Simulink to numerically integrate the differential equation of motion for each case and to plot $r(t)$ when $r(0) = 0.86$ m.

■ Solution

a. Referring to Fig. 2.10b, there is an additional force $-k(r - L_u)\hat{x}_1$ acting on the block, which is the reaction force applied to the block by the spring. Therefore, the resultant force acting on the block is now

$$\vec{F} = mg(-\hat{x}_1 \sin\alpha - \hat{x}_2 \cos\alpha) + N\hat{x}_2 + f\hat{x}_1 - k(r - L_u)\hat{x}_1 \quad (2.5.7a)$$

and Newton's second law yields the following two equations.

$$N - mg\cos\alpha = 0 \quad (2.5.7b)$$

$$m\ddot{r} = f - k(r - L_u) - mg\sin\alpha \quad (2.5.7c)$$

Viscous friction. For this case, the expression for the friction force f shown in Fig. 2.10 is $f = -c\dot{r}$, and the differential equation of motion is then

$$\ddot{r} + \frac{c}{m}\dot{r} + \frac{k}{m}(r - L_u) + g\sin\alpha = 0 \quad (2.5.7d)$$

Equation (2.5.7d) is a linear nonhomogeneous differential equation with constant coefficients. Its solution was first presented in Section 1.13 (see Case 3).

Coulomb friction. When the block moves, the expression for the friction force f shown in Fig. 2.10 is, for this case, $f = -\mu_k |N| \text{sgn}\,\dot{r} = -(\mu_k mg \cos\alpha)\,\text{sgn}\,\dot{r}$. The differential equation of motion for this case is then

$$\ddot{r} + \frac{k}{m}(r - L_u) + g\sin\alpha + \mu_k g(\cos\alpha)\,\text{sgn}\,\dot{r} = 0 \quad (2.5.7e)$$

Equation (2.5.7e) is nonlinear because $\text{sgn}\,\dot{r}$ is a nonlinear function of $\dot{r}$.

b. *Viscous friction.* With $r = \text{constant} = r_e$, Eq. (2.5.7d) yields for this case

$$\frac{k}{m}(r_e - L_u) + g\sin\alpha = 0 \quad (2.5.7f)$$

With $L_u = 1$ m, $k/m = 10$ s^{-2}, and $\alpha = 30°$, we obtain

$$r_e = L_u - \frac{mg}{k}\sin\alpha \approx 0.51 \text{ m}$$

Coulomb friction. With $r = \text{constant} = r_e$, Eq. (2.5.7c) yields for this case

$$r_e = L_u - \frac{mg}{k}\sin\alpha + \frac{f}{k} \quad (2.5.7g)$$

For the block to remain in equilibrium the magnitude of the friction force f in Eq. (2.5.7g) cannot be greater than its maximum possible value, which is $|f_{\max}| = \mu_s |N| = \mu_s m g \cos\alpha$. Therefore, the block does not move when

$$-\mu_s m g \cos\alpha \leq f \leq \mu_s m g \cos\alpha$$

With f determined from Eq. (2.5.7g), this inequality yields

$$-\mu_s \frac{m}{k} g \cos\alpha \leq r_e - L_u + \frac{m}{k} g \sin\alpha \leq \mu_s \frac{m}{k} g \cos\alpha$$

or

$$L_u - \frac{m}{k} g \sin\alpha - \mu_s \frac{m}{k} g \cos\alpha \leq r_e \leq L_u - \frac{m}{k} g \sin\alpha + \mu_s \frac{m}{k} g \cos\alpha \quad (2.5.7h)$$

With $L_u = 1$ m, $k/m = 10$ s^{-2}, $g = 9.81$ m/s^2, $\mu_s = 0.3$, and $\alpha = 30°$, this yields

$$0.255 \text{ m} \leq r_e \leq 0.764 \text{ m}$$

c. *Viscous friction.* Figure 2.11 shows the plots of $r_s(t) \stackrel{\Delta}{=} r(t) - (r_e)_{\mu_k=0}$ versus t, where $(r_e)_{\mu_k=0} = L_u - \frac{m}{k} g \sin\alpha \approx 0.51$ m, obtained from the numerical integration of Eqs. (2.5.7d) and (2.5.7e). The integration was done using Simulink, with the initial conditions $r_s(0) = 0.35$ [which corresponds to $r(0) = (r_e)_{\mu_k=0} + r_s(0) = 0.86$, as specified in the problem statement] and $\dot{r}_s(0) = \dot{r}(0) = 0$. The plot of $\dot{r}_s(t) = \dot{r}(t)$ versus t is also shown for the case of Coulomb friction. Detailed explanations of the motion for both cases are presented.

Equation (2.5.7d), for the case of viscous friction, is a linear differential equation with constant coefficients and it can be solved analytically. As shown in Section 1.13 (see Case 3, p. 37), its solution is $r(t) = r_e + e^{\sigma t}[B_1 \cos(\omega t) + B_2 \sin(\omega t)]$, where $r_e = L_u - mg(\sin\alpha)/k$, $\sigma = -c/(2m)$, and $\omega = \sqrt{k/m - [c/(2m)]^2}$. For $\alpha = 30°$, $L_u = 1$ m, $k/m = 10$ s^{-2}, and $c/m = 0.8$ s^{-1}, we obtain $\sigma = -0.4$ s^{-1} and $\omega \approx 3.14$ rad/s. Such a motion is a damped oscillation with frequency $\omega \approx 3.14$ rad/s. For the initial conditions $r(0) = 0.86$ m and

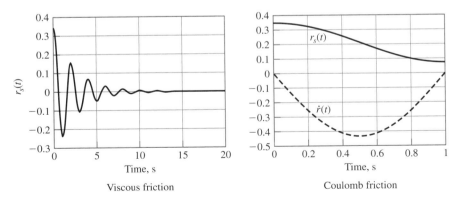

Figure 2.11 Result of the numerical integration of $\ddot{r} + (c/m)\dot{r} + (k/m)(r - L_u) + g \sin\alpha = 0$ and of $\ddot{r} + (k/m)(r - L_u) + g \sin\alpha + \mu_k g(\cos\alpha) \operatorname{sgn}\dot{r} = 0$.

$\dot{r}(0) = 0$, as specified in the problem statement, the constants B_1 and B_2 are $B_1 = r(0) - r_e \approx 0.35$ m, and $B_2 = [\dot{r}(0) - \sigma B_1]/\omega \approx 0.045$ m.

⟹ Figure 2.12 shows a Simulink block diagram model for integrating Eq. (2.5.7d), when rewritten in terms of $r_s(t) \triangleq r(t) - r_e = r(t) - [L_u - mg(\sin\alpha)/k]$. The differential equation for r_s is $\ddot{r}_s + (c/m)\dot{r}_s + (k/m)r_s = 0$. The model is the file inc_plane_viscous.mdl included on the CD that accompanies the book.

> If you have not done so yet, study the material in Appendix A to learn how to use Simulink for Dynamics. The alternative is to write computer code for solving equations, including integrating differential equations, using any language of your choice. Section A.5 provides the details on how to construct and run Simulink block diagram models.

The optional animate_nbars block is included in the inc_plane_viscous Simulink model to generate a visualization of the system in motion. The advantage of such a visualization capability is that one can immediately see how a model that is being analyzed behaves when its motion is governed by the differential equations that were formulated for it. Such a visual capability provides an additional tool to the engineer for judging how changes in initial conditions and system parameters (and even errors in the equations) will affect the model response.

For simplicity, the figure drawn by the inc_plane_viscous model is of a bar OP imitating the spring-mass system. The motion is as seen by an observer looking along the inclined plane and, for this, the x-y coordinate system shown in Fig. 2.10a, centered at point O, is used. Thus, the four inputs to the *Mux* that precedes the animate_nbars block are the $x = 0$ and $y = 0$ coordinates of point O in Fig. 2.10, followed by the x and y coordinates of point P, which are $x_P = -r = -(r_e + r_s)$ and $y_P = 0$.

As indicated at the beginning of this chapter, and in Appendix A, to use the model, either type inc_plane_viscous in the MATLAB command window, or type simulations_menu and choose the inc_plane_viscous

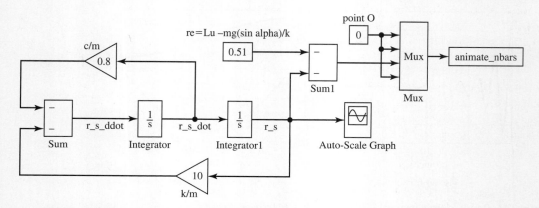

Figure 2.12 Simulink model to integrate the equation $\ddot{r}_s + (c/m)\dot{r}_s + (k/m)r_s = 0$, and the Simulation → Parameters menu showing the choices made for the simulation parameters.

model from a pull-down menu that appears on the computer screen. Click on Start in the Simulation menu of the model to start the simulation. It is assumed that you have followed the steps indicated in Section A.2 of Appendix A.

Look at the Solver part of the Simulation → Parameters menu for the model in your computer to see its default settings for the model. You may want to experiment using any different value for the maximum step size (including the string auto) of the integration scheme if your simulation runs either too slow or too fast in your computer.

Coulomb friction. As mentioned earlier, Eq. (2.5.7e), for the case of Coulomb friction, is nonlinear because of the sgn $\dot{r}$ term. Notice, however, that for each interval of the motion for which $\dot{r}$ does not change sign (i.e., sgn $\dot{r}$ is either $+1$ or -1) that differential equation is linear. For this reason, Eq. (2.5.7e) is called a *piecewise linear* differential equation, and it also could be solved analytically for the different intervals of the motion where $\dot{r} > 0$ and $\dot{r} < 0$. It is not very convenient to obtain such an analytical piecewise solution, and only the numerical solution is presented here.

Notice that for the case of Coulomb friction, the determination of $r(t)$ for all times requires more thought than the case of viscous friction. It was seen in the solution presented for part b that there is a range of values of $r(0)$ when the block does not move. For the values of the parameters given in this problem, that range is $0.255 \text{ m} \le r(0) \le 0.764 \text{ m}$.

For the given initial conditions, i.e., $\dot{r}(0) = 0$ and $r(0) = r_e + 0.35 = 0.86$ m, the particle starts to move down the inclined plane and, thus, $r(t)$ starts to decrease. It continues to decrease until it reaches a certain value when $\dot{r}(t) = 0$. If that value of $r(t)$ falls inside the range previously indicated, the body will not be able to move any longer because the friction force acting on it will be equal to $\mu_s N = \mu_s mg \cos\alpha$, which will be balanced by the spring force and by the component, along the inclined plane, of the gravitational force. The body would reverse the direction of its motion and would continue to move along the plane if the value of $r(t)$ when $\dot{r}(t) = 0$ is outside the range already indicated. Such a motion would repeat itself until the body stops when $0.255 \text{ m} \le r(0) \le 0.764 \text{ m}$. This situation must be taken into account when integrating the differential equation that governs the motion when Coulomb friction is present.

The Simulink model that was used for integrating Eq. (2.5.7e), when rewritten in terms of $r_s(t) \triangleq r(t) - r_e$, is shown in Fig. 2.13. Notice how the model is more complicated than that for viscous friction because of the logic that has to be included in order to decide whether the simulation continues or stops. The differential equation for r_s in this case is

$$\ddot{r}_s + \frac{k}{m} r_s + \mu_k g (\cos\alpha) \operatorname{sgn} \dot{r}_s = 0$$

⟹ The Simulink model of Fig. 2.13 is the file `inc_plane_coulomb.mdl` included on the CD in the back of the book. It is suggested that you run it with different values of the damping coefficient and different initial conditions in order to gain further insight about the motion of this system.

Figure 2.13 Simulink model to integrate the equation $\ddot{r}_s + (k/m)r_s + \mu_k g(\cos\alpha)\,\text{sgn}\,\dot{r}_s = 0$.

The five blocks in the lower part of the Simulink model shown in Fig. 2.13 (the ones that include the `Dead Zone` and the `Stop Simulation` blocks) were included in the model to handle the logic for deciding whether the integration should continue or stop. We want the integration to stop when $r(t)$ is neither less than 0.255 nor greater than 0.764. This is accomplished with the `Dead Zone` and the `Logical Operator` NOT blocks. These and other blocks are described in Appendix A. The output of the block labeled `Dead Zone` is equal to zero if its input $r(t)$ is in the range $0.255 \le r \le 0.764$, and equal to one otherwise (such limits are set by double clicking the block with the computer mouse, and writing their values in the window that appears on the screen). Therefore, the output of the NOT block is equal to one if $0.255 \le r \le 0.764$.

The output of the block labeled `Hit Crossing` is equal to one if its input $\dot{r}(t)$ is equal to zero (see Fig. 2.13); its output is equal to zero for any other input values. Therefore, the output of the logical operator block marked AND is equal to one if $0.255 \le r \le 0.764$ and $\dot{r}(t) = 0$. That output is connected to the input to the block labeled `Stop Simulation` since that block causes the simulation to stop when its input becomes equal to one.

Example 2.5.8 involves a simple two-degree-of-freedom system (i.e., a system for which two variables that are independent of each other are needed to describe its motion) for which Newton's second law immediately yields the two differential equations of motion for the system.

SECTION 2.5 Solved Example Problems

EXAMPLE 2.5.8

A small ball P, modeled as a mass particle, is thrown from a point A with an initial velocity as shown in Fig. 2.14. Neglecting air resistance and assuming that the only force acting on the ball is the gravitational force of attraction of the earth, determine the value of v_0 for the ball to hit a target at point B shown in the figure. What is the maximum height reached by the ball?

■ **Solution**
Because of the small distances involved, the gravitational force acting on particle P is approximated as a constant (i.e., constant direction and constant magnitude). The free-body diagram for P is shown in Fig. 2.15. Two variables are needed to describe the motion, and rectangular coordinates for point P are the most convenient ones for this problem. By arbitrarily choosing point A to be the inertial reference point, those variables are the distances $x(t)$ and $y(t)$ shown in Fig. 2.15.

In terms of the unit vectors $\hat{x}$ and $\hat{y}$ shown in the Fig. 2.15, we then have

Resultant force acting on P: $\qquad \vec{F} = -mg\hat{y}$

Absolute position vector for the mass particle P: $\qquad \vec{r} = x\hat{x} + y\hat{y}$

Absolute velocity of P: $\qquad \vec{v} = \frac{d\vec{r}}{dt} = \dot{x}\hat{x} + \dot{y}\hat{y}$

Absolute acceleration of P: $\qquad \vec{a} = \frac{d\vec{v}}{dt} = \ddot{x}\hat{x} + \ddot{y}\hat{y}$

Newton's second law $\vec{F} = m\vec{a}$ then yields the following differential equations that govern the motion of the ball:

$$\ddot{x} = 0 \quad \text{and} \quad \ddot{y} = -g$$

Figure 2.14 Figure for Example 2.5.8.

Figure 2.15 Free-body diagram for Example 2.5.8.

The solutions to these simple differential equations are

$$\dot{x} = \text{constant} = \dot{x}(0) = v_0 \cos\alpha \qquad x = (v_0 \cos\alpha)t + x(0)$$

$$\dot{y} = \dot{y}(0) - gt = v_0 \sin\alpha - gt \qquad y = (v_0 \sin\alpha)t - \frac{gt^2}{2} + y(0)$$

With $x(0) = y(0) = 0$, these solutions yield, with $x = 60$ m, $y = 2.6 - 0.6 = 2$ m, and $\alpha = 30°$,

$$v_0 \approx 26.9 \text{ m/s}$$

The maximum value of y (i.e., the maximum height) occurs when $\dot{y} = 0$; the corresponding value of $y(t)$ is maximum because $\ddot{y} = -g$ is negative. Since $\dot{y} = v_0 \sin\alpha - gt$, the maximum height is reached when $t = (v_0 \sin\alpha)/g \approx 1.37$ s, and this gives

$$y_{\max} \approx 9.2 \text{ m}$$

⇒   The enclosed CD has a Simulink block diagram model named `ballistics.mdl` that integrates the differential equations of motion. You can change that model to include the effect of aerodynamic drag on the body (see Problem 2.19 in Section 2.9). The animation block `animate_nbars` is also included in the model so that one can observe the particle during its actual motion. To use the model either type `ballistics` in the MATLAB command window or `simulations_menu` and choose the `ballistics` model as indicated in Appendix A, and click on `Start` in the `Simulation` menu, as indicated at the beginning of this chapter and in Appendix A.

Example 2.5.9 involves a more elaborate two-degree-of-freedom system that consists of a block sliding on a sliding block. For this system, unknown reaction forces have to be eliminated from the equations that are obtained from Newton's second law in order to obtain the two differential equations of motion for the system.

EXAMPLE 2.5.9

A block P_2 of mass m_2 slides on a face of a block P_1, which is an inclined plane ABC, of mass m_1. Block P_1 is able to slide on a horizontal surface as shown in Fig. 2.16

Figure 2.16 A block sliding on a sliding block for Example 2.5.9.

(the x-y coordinate system shown in the figure is used in connection with the Simulink model presented at the end of this example). Both blocks are constrained to move on the same vertical plane, which is the plane of the paper in the figure. Friction in the system is viscous, with friction coefficients c_2 between P_2 and P_1, and c_1 between P_1 and the horizontal surface. The spring DE connecting P_2 to P_1 is linear, with unstressed length L_u m and stiffness k N/m. Obtain the differential equations of motion for the system in terms of the variables $x(t)$ and $r(t)$ shown in the figure, and then construct a Simulink model that shows the system undergoing motion governed by such equations. Notice that the blocks translate only, as if they were point masses.

■ **Solution**

The free-body diagrams for P_1 and P_2 are shown in Fig. 2.17. The position vectors chosen to track the bodies, and two sets of orthogonal unit vectors $\{\hat{x}_1, \hat{x}_2\}$ and $\{\hat{y}_1, \hat{y}_2\}$, are also shown in that figure. The analysis begins by writing down expressions for the absolute position, the absolute acceleration, and the resultant force acting on each block. By inspecting Fig. 2.17, the following expressions are written.

For block P_1:

$$\vec{r}_1 = x\hat{x}_1 \qquad \vec{a}_1 = \frac{d^2\vec{r}_1}{dt^2} = \ddot{x}\hat{x}_1 \qquad (2.5.9a)$$

$$\vec{F}_1 = (N_1 - m_1 g)\hat{x}_2 - c_1\dot{x}\hat{x}_1 + [c_2\dot{r} + k(r - L_u)]\hat{y}_1 - N_2\hat{y}_2 \qquad (2.5.9b)$$

For block P_2:

$$\vec{r}_2 = x\hat{x}_1 + r\hat{y}_1 \qquad \vec{a}_2 = \frac{d^2\vec{r}_2}{dt^2} = \ddot{x}\hat{x}_1 + \ddot{r}\hat{y}_1 \qquad (2.5.9c)$$

$$\vec{F}_2 = N_2\hat{y}_2 - m_2 g\hat{x}_2 - [c_2\dot{r} + k(r - L_u)]\hat{y}_1 \qquad (2.5.9d)$$

Notice that no matter which set of unit vectors is used in this problem, i.e., either $\{\hat{x}_1, \hat{x}_2\}$ or $\{\hat{y}_1, \hat{y}_2\}$, a transformation between these two sets will always be required in order to extract the differential equations of motion from Newton's second law for each block.

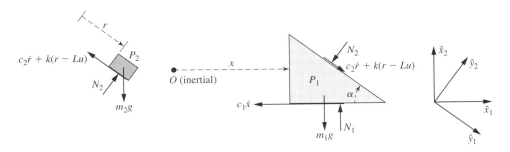

Figure 2.17 Free-body diagrams and unit vectors for Example 2.5.9.

From $\vec{F}_1 = m_1\vec{a}_1$ and $\vec{F}_2 = m_2\vec{a}_2$, the following scalar equations are obtained if we choose to work with the unit vector set $\{\hat{x}_1, \hat{x}_2\}$ for block P_1 and $\{\hat{y}_1, \hat{y}_2\}$ for block P_2.

For block P_1:

$$m_1\ddot{x} = -c_1\dot{x} + [c_2\dot{r} + k(r - L_u)]\cos\alpha - N_2\sin\alpha \qquad \text{(2.5.9e)}$$

$$N_1 - m_1 g - [c_2\dot{r} + k(r - L_u)]\sin\alpha - N_2\cos\alpha = 0 \qquad \text{(2.5.9f)}$$

For block P_2:

$$m_2(\ddot{r} + \ddot{x}\cos\alpha) = m_2 g \sin\alpha - c_2\dot{r} - k(r - L_u) \qquad \text{(2.5.9g)}$$

$$m_2\ddot{x}\sin\alpha = N_2 - m_2 g \cos\alpha \qquad \text{(2.5.9h)}$$

Notice that $x(t)$ does not appear by itself in Eqs. (2.5.9e) to (2.5.9h) and, thus, it is irrelevant where the inclined plane ABC is located when one starts the motion (as expected, by inspection of Fig. 2.16).

Equation (2.5.9g) does not involve any of the unknown reactions N_1 or N_2, and, therefore, it is one of the differential equations of motion for the system. The second differential equation that governs the motion of the system is obtained by combining Eqs. (2.5.9e) and (2.5.9h) to eliminate the unknown reaction N_2. This yields the following differential equation:

$$(m_1 + m_2\sin^2\alpha)\ddot{x} + c_1\dot{x} - [c_2\dot{r} + k(r - L_u)]\cos\alpha + m_2 g (\sin\alpha)\cos\alpha = 0 \qquad \text{(2.5.9i)}$$

Equations (2.5.9g) and (2.5.9i) are two coupled linear differential equations with constant coefficients, and they actually can be solved analytically. The solution to a set of coupled linear differential equations is shown in Chapter 8. Such material is covered in vibrations books and is included in Chapter 8 for completeness and because it is a natural extension of the material that was presented under Case 3 in Section 1.13.

Equations (2.5.9g) and (2.5.9i) are solved numerically here. For this to be done, numerical integration routines require that the expressions for $\ddot{x}$ and $\ddot{r}$ first be written in terms of x, r, $\dot{x}$, and $\dot{r}$. Inspection of Eq. (2.5.9i) discloses that $\ddot{x}$ is already in the required form. The required form for $\ddot{r}$ is then obtained by simply substituting into Eq. (2.5.9g) the expression for $\ddot{x}$ given by Eq. (2.5.9i). Take the time to verify that the resulting expression that is obtained for $\ddot{r}$ is

$$(m_1 + m_2\sin^2\alpha)\ddot{r} = c_1\dot{x}\cos\alpha + (m_1 + m_2)\left\{\left[\frac{c_2\dot{r} + k(r - L_u)}{m_2}\right] + g\sin\alpha\right\} \qquad \text{(2.5.9j)}$$

$\Rightarrow$ Figure 2.18 shows a Simulink block diagram model for numerically integrating simultaneously Eqs. (2.5.9i) and (2.5.9j). The model is the file `inclined_plane.mdl` included on the enclosed CD. It includes the optional `animate_nbars` block to animate the motion with a figure consisting of five bars: three to draw the inclined plane ABC, one to draw bar DE representing the spring, and one to draw a body at the end of the spring to represent body P_2. The figure is drawn using the x-y coordinate system shown in Fig. 2.16. Details of the required inputs to the `animate_nbars` animation block are in the boxed

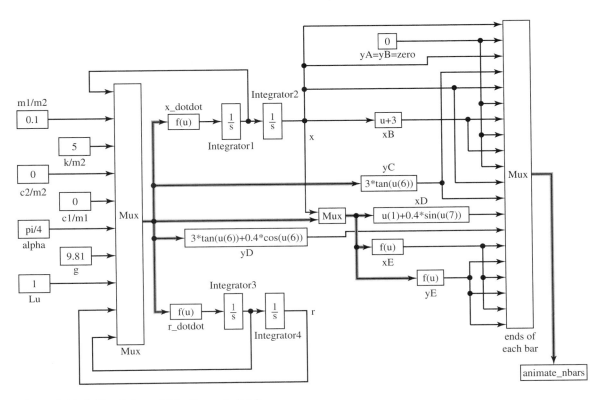

Figure 2.18 A Simulink model for Example 2.5.9.

text on p. 520 of Appendix A. The details of the 20 inputs to the `animate_nbars` block to do this in the `inclined_plane` model are shown in the sequel.

The decision on what details to include in a figure is, of course, entirely individual. The figure generated by the `animate_nbars` block in the `inclined_plane` model can be simplified, of course, and that would lead to a more manageable smaller number of inputs to that block.

The inputs to the `animate_nbars` block may also be generated with a MATLAB Function block. That would actually be preferable since the block diagram would not be so crowded with so many inputs to the animation block. Figure 2.19 shows such a diagram. The new model is the file `inclined_plane2.mdl` included on the enclosed CD. To use the models, type either `inclined_plane` or `inclined_plane2` in the MATLAB command window, or choose them from a pull-down menu that can be selected after typing `simulations_menu` as indicated earlier. Change the maximum step size of the integration scheme in the Simulation→Parameters menu if the simulation runs too fast or too slow in your computer with either model.

In the animated figure generated by the `animate_nbars` block in the `inclined_plane` and `inclined_plane2` models, the length of bar AB is set at 3 *length units* (3 m when g is input in m/s^2 as in Fig. 2.18). The generated figure (see Figs. 2.16 and 2.18) is drawn in the following sequence: bars AC, AB, BC, DE, and then a "big point" (!)

Figure 2.19 A more compact Simulink model for Example 2.5.9, producing the same animation as the model shown in Fig. 2.18.

at E (double click the `animate_nbars` block to see the default "width" of each bar). This is explained here in more detail.

Bar AC. Accounts for the first four inputs to the `MUX` that precedes the `animate_nbars` block in Fig. 2.18, i.e., the x and y coordinates of point A, which are $x_A = x$ and $y_A = 0$, followed by the x and y coordinates of point C, which are $x_C = x$ and $y_C = 3 \tan \alpha$.

Bar AB. Accounts for the next four inputs to the `MUX` labeled `ends of each bar`, i.e., the x and y coordinates of point A, followed by the x and y coordinates of point B, which are $x_B = x + 3$ and $y_B = 0$.

Bar BC. Accounts for the four inputs after the ones for bar AB, which are x_B, y_B, x_C, and y_C.

Bar DE. Accounts for the next four inputs, and then a "bar" at E with its two end points coinciding with each other. The four inputs for drawing bar DE are x_D and y_D, followed by the x and y coordinates of point E. The distance CD in Fig. 2.16 is arbitrarily set at 0.4 length units and, therefore, the x and y coordinates of point D are $x_D = x + 0.4 \sin \alpha$ and $y_D = 3 \tan \alpha + 0.4 \cos \alpha$. Similarly, the x and y coordinates of point E are $x_E = x + 0.4 \sin \alpha + r \cos \alpha$ and $y_E = 3 \tan \alpha + 0.4 \cos \alpha - r \sin \alpha$. They are labeled xE and yE in Fig. 2.18 (double click those function blocks to see these expressions). Other coordinates are labeled in that figure as xB, yC, etc., for relatively easy identification.

In the model `inclined_plane2` shown in Fig. 2.19, the 20-input *Mux* in Fig. 2.18 is replaced by a `MATLAB Function` block to make the diagram more manageable. That block provides the following set of four inputs for each bar drawn by the `animate_nbars` block (double click the `MATLAB Function` block in the model to see its syntax, which is typed in a single long line, as indicated in Appendix A):

[x,0,x,3tanα,   x,0,x+3,0,   x+3,0,x,3tanα,   x+0.4sin(α),
3tanα+0.4cosα,x+0.4sinα+rcosα,3tanα+0.4cosα-rsinα,
            x+0.4sinα+rcosα,3tanα+0.4cosα-rsinα,
x+0.4sinα+rcosα,3tanα+0.4cosα-rsinα].

> Use either model `inclined_plane` or `inclined_plane2` to investigate the effect of changes in the input parameters m_1/m_2, k/m_2, etc., on the motion of the system. Choose your own values for those parameters and initial conditions for the motion. For your chosen values, what is the equilibrium displacement for $r(t)$ and the frequencies of the oscillations? Obtain them from plots of $r(t)$ and $x(t)$.

Suggested Problem
For this example, verify that the quantity

$$E \triangleq (m_1 + m_2)\frac{\dot{x}^2}{2} + m_2\left(\frac{\dot{r}^2}{2} + \dot{x}\dot{r}\cos\alpha\right) - m_2 g r \sin\alpha + \frac{k}{2}(r - L_u)^2$$

remains constant during the motion if $c_1 = c_2 = 0$. *Hint:* Take the time derivative of E and use the expressions for $\ddot{x}$ and $\ddot{r}$ from the differential equations of motion obtained in this example to show that $\dot{E} = 0$. The quantity E happens to be equal to the sum of the *kinetic energy* and of the *potential energy* of the motion. These quantities are introduced in Section 2.8.

Example 2.5.10 illustrates the degree of difficulty associated with different choices of variables for analyzing a dynamics problem. It also shows that the number of mathematical manipulations in solving a problem depends on the unit vector triad one chooses to work with.

EXAMPLE 2.5.10

Consider the pendulum discussed in Section 1.11 and shown in Fig. 1.11. For convenience, that figure (with $m_1 = 0$) is repeated here as Fig. 2.20. Free-body diagrams for different parts of the system are shown in separate parts of the figure (explanations about the diagrams are given in the solution).

The pendulum consists of a small body (modeled as a point mass) of mass $m = 200$ g attached to one end of a bar AB of length $L = 60$ cm, which is approximated as being massless.[4] The other end of the bar is attached to a fixed frictionless pin. The motion of the system takes place in a vertical plane, and the only forces acting on the point mass are due to the gravitational attraction of the earth and the force due to contact with the

---

[4] Dynamics of rigid bodies, taking their distributed mass into account, is treated in Chapters 5 and 6.

Figure 2.20 Free-body diagrams for a pendulum with a massless bar for Example 2.5.10.

bar. The bar is subjected to a known moment M_{bar} which is applied by a motor that is not shown in the figure.

a. Obtain the differential equation of motion for the pendulum in terms of the angle θ. What is the advantage of working with the angle θ, instead of either x or y shown in Fig. 2.20?

b. Obtain the equations for calculating the reaction forces A_1, A_2, B_1, and B_2 shown in the free-body diagrams, and explain how to calculate them using such equations.

c. For $M_{\text{bar}} = 0$, determine the maximum value $\theta_{\max}$ of the angle θ if the motion is started with the initial conditions $\theta(0) = 0$ and $\dot{\theta}(0) = 2$ rad/s.

d. A sensor in the system measures the angle $\theta(t)$, and the signal that is generated is input to the motor at A so that $M_{\text{bar}} = -K\theta(t)$, where $K = 4$ N·m/rad. For the same initial conditions $\theta(0) = 0$ and $\dot{\theta}(0) = 2$ rad/s that were used in part c, determine the new value of $\theta_{\max}$.

e. With $M_{\text{bar}} = 0$ and $\theta(0) = 0$, plot the nondimensional angular velocity $\dot{\theta}(t)\sqrt{L/g}$ versus $\theta(t)$ (using $-360 \leq \theta \leq 360°$) for $\dot{\theta}(0)\sqrt{L/g} = 1.5$, $\dot{\theta}(0)\sqrt{L/g} = 2$, and $\dot{\theta}(0)\sqrt{L/g} = 2.5$. Superimpose all three plots in the same graph, and describe the motion in a few words based on the observation of the plots.

■ **Solution**

a. For this problem, point B moves along a circle centered at the inertial point A (see Fig. 2.20) and, as you have seen in basic physics courses, the absolute velocity of point B is simply $L\dot{\theta}$ directed along the perpendicular to bar AB. If, instead of working with the angle θ, we work with the variable x and also with the unit vectors

$\hat{x}_1$ and $\hat{x}_2$ shown in Fig. 2.20, inspection of that figure discloses that the expression for the absolute position vector of B is $\vec{r} = \overrightarrow{AB} = x\hat{x}_1 - \sqrt{L^2 - x^2}\,\hat{x}_2$. With such choices, the expression for the absolute velocity of B becomes $\vec{v} = d\vec{r}/dt = \dot{x}\hat{x}_1 + (x\dot{x}/\sqrt{L^2 - x^2})\hat{x}_2$. Such an expression is much more complicated than the one written in terms of the angle θ. This will yield an even more complicated expression for the acceleration $\vec{a} = d\vec{v}/dt$ if we work with the variable x, and, in such a case, the resulting differential equation of motion, obtained from $\vec{F} = m\vec{a}$, will be much more difficult to deal with than the differential equation written in terms of the angle θ. Also notice that the expression for the velocity in terms of the variable x will have a division-by-zero problem if x approaches L (i.e., if bar AB approaches the horizontal). The first problem alone, namely, the additional complexity of the differential equation, gives us enough justification to work with the variable θ. The same difficulties previously mentioned will occur if we choose to work with the variable y.

Three cuts are shown in Fig. 2.20a. Cut 1 allows us to see the internal reaction forces acting on the bar, so that we can answer the questions asked in part b about those forces. Cut 2 allows us to draw the free-body diagram for the point mass located at the end B of the bar. The free-body diagrams for the two bodies that compose the pendulum are shown in Fig. 2.20b. The two diagrams involve five unknowns: θ, A_1, A_2, B_1, and B_2. To determine all these unknowns, five equations are then needed. For a two-dimensional problem, $\vec{F} = d(m\vec{v})/dt$ for a point mass yields two scalar equations. The other three equations for a massless body are obtained as indicated in the following box (the dynamics of translating and rotating rigid bodies, taking into account their masses, is dealt with in Chapters 5 and 6).

For a *massless body* subjected to an arbitrary number n of forces $\vec{F}_1, \vec{F}_2, \ldots, \vec{F}_n$, the same equations of statics apply (even though the acceleration of points of the body is not zero), i.e.,

Summation of forces $= 0$ (in two dimensions this yields two equations)

and

Summation of moments (about any point) $= 0$ (in two dimensions this yields one equation)

Now, why is the free-body diagram for cut 3 shown in Fig. 2.20c? Strictly speaking, neither that cut nor the free-body diagram for that cut are actually needed because all the unknowns for this problem already appear in the free-body diagrams for cuts 1 and 2. However, cut 3 happens to be the most convenient cut for obtaining the differential equation of motion of the pendulum. The reason for this is that the moment about point A of all the unknown reaction forces that appear in the free-body diagram for that cut is zero. Since point A is inertial, we can apply the moment equation, Eq. (2.3.1), to the free-body diagram for cut 3 to obtain the differential equation of motion for the pendulum after performing only a line or so of mathematical manipulations. For this problem, Eq. (2.3.1) becomes

$\vec{M}_A = m\,d(\overrightarrow{AB} \times \vec{v})/dt$. We could work only with the free-body diagram for cut 3 if the reaction forces A_1, A_2, B_1, and B_2 were not asked for.

Before we proceed to obtain the equations that will allow us to calculate the four reaction forces A_1, A_2, B_1, and B_2, let us see how the free-body diagram for cut 3 yields the differential equation of motion with just a few mathematical manipulations.

The moment $\vec{M}_A$ is simply (see Fig. 2.20, and use the right-hand rule)

$$\vec{M}_A = M_{\text{bar}}\hat{x}_3 - (mgL\sin\theta)\hat{x}_3$$

Since (also, see Fig. 2.20, and use the right-hand rule)

$$\overrightarrow{AB} \times \vec{v} = L(L\dot{\theta})\hat{x}_3 = L^2\dot{\theta}\hat{x}_3$$

the equation $\vec{M}_A = m\,d(\overrightarrow{AB} \times \vec{v})/dt$ yields the following differential equation of motion:

$$\ddot{\theta} + \frac{g}{L}\sin\theta = \frac{M_{\text{bar}}}{mL^2} \qquad (2.5.10\text{a})$$

This is the equation that was asked for.

b. Let us now generate the equations for calculating the reaction forces A_1, A_2, B_1, and B_2. Referring to Fig. 2.20 we have, by working with the nonrotating unit vectors $\hat{x}_1$ and $\hat{x}_2$:

- Resultant force acting on the mass particle at B (see the free-body diagram for cut 2):

$$\vec{F} = (B_1\cos\theta - B_2\sin\theta - mg)\hat{x}_2 - (B_1\sin\theta + B_2\cos\theta)\hat{x}_1$$

- Absolute position vector for point B:

$$\vec{r} = \overrightarrow{AB} = L(\hat{x}_1\sin\theta - \hat{x}_2\cos\theta)$$

- Absolute velocity of B:

$$\vec{v} = \frac{d\vec{r}}{dt} = L\dot{\theta}(\hat{x}_1\cos\theta + \hat{x}_2\sin\theta)$$

- Absolute acceleration of B:

$$\vec{a} = \frac{d\vec{v}}{dt} = L(\ddot{\theta}\cos\theta - \dot{\theta}^2\sin\theta)\hat{x}_1 + L(\ddot{\theta}\sin\theta + \dot{\theta}^2\cos\theta)\hat{x}_2$$

Newton's second law $\vec{F} = m\vec{a}$ then yields the following scalar equations for the point mass at B:

$$mL(\ddot{\theta}\cos\theta - \dot{\theta}^2\sin\theta) = -B_1\sin\theta - B_2\cos\theta \qquad (2.5.10\text{b})$$

$$mL(\ddot{\theta}\sin\theta + \dot{\theta}^2\cos\theta) = B_1\cos\theta - B_2\sin\theta - mg \qquad (2.5.10\text{c})$$

For the massless bar AB, let us take moments about point A since this results in one simple equation that involves only the unknown force B_2. By equating to zero the summation of the forces acting on bar AB, and the summation of the moments about point A, the following equations are obtained for the free-body diagram for cut 1.

$$A_1 - B_1 = 0 \qquad (2.5.10\text{d})$$

$$A_2 + B_2 = 0 \qquad (2.5.10\text{e})$$

$$M_{\text{bar}} + B_2 L = 0 \qquad (2.5.10\text{f})$$

Notice that the moment equation, Eq. (2.5.10f), only involves one unknown because we chose to work with force components in the directions parallel and perpendicular to bar AB and took moments about point A. Because of that choice, three of the unknown reaction forces pass through point A and, thus, their moment about A is equal to zero. We could also have taken moments about point B of the massless bar to obtain a simple equation that would only involve the unknown A_2, which is $A_2 = M_{\text{bar}}/L$.

The five equations that will answer all the questions that were asked for in parts b to d are Eqs. (2.5.10b) through (2.5.10f), or, better still, Eqs. (2.5.10a), (2.5.10b) [or (2.5.10c)], (2.5.10d), (2.5.10e), (2.5.10f).

Having the differential equation of motion, we can calculate the reaction forces and also answer the questions in parts b to d. The reaction forces B_2 and A_2 are simply determined from Eqs. (2.5.6e) and (2.5.6f) as $B_2 = -A_2 = -M_{\text{bar}}/L$. To calculate the reaction forces A_1 and B_1, the nonlinear differential equation of motion, Eq. (2.5.10a), has to be integrated first in order to obtain $\theta(t)$. This can be done numerically. Once this is done (Simulink, for example, can be used for the numerical integration), the reaction force B_1 can be calculated by using either Eq. (2.5.10b) or (2.5.10c). The reaction force A_1 is then determined from Eq. (2.5.10d) as $A_1 = B_1$. The integration of the differential equation and the calculation of the reaction forces is a relatively simple process, in actual engineering practice, when it is done with software that does not involve the writing of computer programs.

c and d. Let us now determine the maximum value of the angle $\theta(t)$, as asked for in parts c and d. Since part c is the same as part d with $K = 0$, let us find the value for part d first.

With $M_{\text{bar}} = -K\theta$, integration of Eq. (2.5.10a) once with respect to θ yields, according to the analytical steps presented in Section 1.13 (see Case 4, p. 41):

$$\frac{\dot{\theta}^2}{2} - \frac{g}{L}\cos\theta + \frac{K\theta^2}{2mL^2} = \text{constant} = \frac{\dot{\theta}^2(0)}{2} - \frac{g}{L}\cos\theta(0) + \frac{K\theta^2(0)}{2mL^2} \qquad (2.5.10\text{g})$$

The maximum and minimum values of θ are obtained by setting $\dot{\theta} = 0$ in Eq. (2.5.10g). With $\theta(0) = 0$, $\dot{\theta}(0) = 2$ rad/s, $m = 0.2$ kg, $L = 0.6$ m, and $g = 9.81$ m/s^2, we obtain

$$\frac{\dot{\theta}^2}{2} - \frac{9.81}{0.6}\cos\theta + \frac{K\theta^2}{2 \times 0.2 \times 0.6^2} = \frac{2^2}{2} - \frac{9.81}{0.6}\cos(0) = -14.35 \qquad (2.5.10\text{h})$$

■ $K = 0$. For this case, Eq. (2.5.10h) yields $\theta \approx \pm 0.5$ rad $\approx \pm 28.6°$ when $\dot{\theta} = 0$. From Eq. (2.5.10a), it follows that $\ddot{\theta} = -g(\sin\theta)/L - K\theta/(mL^2)$ is negative when $\theta = 28.6°$ (thus, a maximum value for θ) and positive when $\theta = -28.6°$ (thus, a minimum value for θ). Therefore, $\theta_{max} = 28.6°$, $\theta_{min} = -\theta_{max} = -28.6°$, and θ oscillates, in this case, between θ_{min} and θ_{max}.

- $K = 4$ N·m/rad. For this case, the values of θ for which $\dot\theta = 0$ are obtained from Eq. (2.5.10a), after a few iterations with a calculator, as $\theta \approx \pm 0.236$ rad $\approx \pm 13.5°$. Again, from $\ddot\theta = -g(\sin\theta)/L - K\theta/(mL^2)$, it follows that $\ddot\theta$ is negative when $\theta = 13.5°$, and positive when $\theta = -13.5°$. Therefore, $\theta_{max} = 13.5°$ and $\theta_{min} = -\theta_{max} = -13.5°$. In this case, θ oscillates between these two values.

e. The plot of the nondimensional angular velocity $d\theta/d\tau = \dot\theta(t)\sqrt{L/g}$, where $\tau = t\sqrt{g/L}$ is nondimensional time, versus $\theta(\tau)$ for any value of the nondimensional quantity $K/(mgL)$, is generated by using Eq. (2.5.10g). For this, that equation is first rearranged as

$$\frac{L}{g}\frac{\dot\theta^2}{2} - \cos\theta + \frac{K\theta^2}{2mgL} = \text{constant} \triangleq E = \frac{L}{g}\frac{\dot\theta^2(0)}{2} - \cos\theta(0) + \frac{K\theta^2(0)}{2mgL}$$

The plots for $K = 0$ (i.e., $M_{bar} = 0$), for three values of E, namely, $E = 1.5^2/2 - \cos(0) = 0.125$, $E = 2^2/2 - \cos(0) = 1$, and $E = 2.5^2/2 - \cos(0) = 2.125$, are shown in Fig. 2.21. Although the plot can easily be done by hand, it was generated with MATLAB.[5]

Inspection of Fig. 2.21, discloses that the motion is not oscillatory when $E > 1$, with the value of the constant E being determined by the initial conditions $\theta(0)$ and $\sqrt{L/g}\dot\theta(0)$. For $K = 0$, the pendulum continuously rotates around its pivot point A (see Fig. 2.20, p. 102) when $E > 1$.

Note

Look at Eqs. (2.5.10b) and (2.5.10c) for the point mass at B. Both of those equations involve $\ddot\theta$, the reaction $B_2 = -M_{bar}/L$, and the unknown reaction B_1. More mathematical manipulations are required to eliminate B_1 to obtain the differential equation of motion for the system. This can be done by multiplying the first equation by $\cos\theta$, the second by $\sin\theta$, and then adding the results to finally obtain the differential equation of motion for this system, which is Eq. (2.5.10a) obtained earlier by a simpler procedure.

---

[5] The following MATLAB commands were used (the first command creates an array of values assigned to the variable `th`). For information about any command, such as the `plot`, `grid`, `axis`, and `xlabel` commands, type `help command_name` in the MATLAB command window, where `command_name` is the name of the desired command. The label of the vertical axis in Fig. 2.21 was not generated by MATLAB.

```
th=-360:1:360; theta_0=0;
th_prime1=sqrt(2*(cos(pi*th/180) - cos(theta_0) )+1.5^2);
th_prime2=sqrt(2*(cos(pi*th/180) - cos(theta_0) )+2.0^2);
th_prime3=sqrt(2*(cos(pi*th/180) - cos(theta_0) )+2.5^2);
plot(th, th_prime1, th, th_prime2, th, th_prime3, th, -th_prime1, th, -th_prime2, th, -th_prime3);
grid; axis([-360 360 -2.5 2.5]); xlabel('\theta, degrees');
```

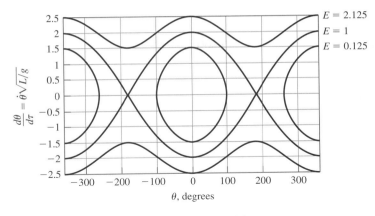

Figure 2.21 $d\theta/d\tau$ versus θ for a simple pendulum $[d^2\theta/d\tau^2 + \sin\theta + (K/mL^2)\theta = 0$, with $K = 0]$.

These extra manipulations were necessary in order to "extract" the differential equation of motion from Eqs. (2.5.10b) and (2.5.10c) because $\ddot{\theta}$ appeared in both components of the expression that was obtained for the acceleration $\vec{a}$. There would be no need for those extra steps if we could choose to work with two orthogonal unit vectors where $\ddot{\theta}$ appears only in one of the components of $\vec{a}$. Such unit vectors are actually the ones parallel and perpendicular to bar AB for this problem.[6] Since such unit vectors are rotating, it is then necessary to learn how to obtain the time derivative of rotating unit vectors when one decides to work with them. For a variety of problems, working with rotating unit vectors is advantageous because the expressions for velocity and acceleration turn out to be simpler and the number of manipulations needed to obtain the differential equations of motion is reduced. The problem of determining the time derivative of rotating unit vectors is addressed in Section 2.6.

EXAMPLE 2.5.11

To investigate the translational motion of a satellite subjected to the gravitational attraction of a planet, the satellite is modeled as a particle P of mass m, while the planet, whose mass M is much bigger than that of the satellite, is treated as a spherical inertial body centered at a point O and with radius R. If the only force acting on this two-body system is due to their mutual attraction, show that the motion of P in the three-dimensional space

---

[6] This should be expected because, as mentioned earlier, point B in Fig. 2.20 simply moves along a circle centered at point A. This can also be seen by looking at the expression $\vec{v} = L\dot{\theta}(\hat{x}_1 \cos\theta + \hat{x}_2 \sin\theta)$ that was obtained for the absolute velocity of point B and noticing that $\hat{x}_1 \cos\theta + \hat{x}_2 \sin\theta$ is the unit vector normal to line AB (and pointing in the same direction as the vector A_2 shown in Fig. 2.20). If we denote such a unit vector by $\hat{n}$, say, the absolute velocity of point B is simply $\vec{v} = L\dot{\theta}\hat{n}$, which gives $\vec{a} = (d\vec{v}/dt) = L\ddot{\theta}\hat{n} + L\dot{\theta}(d\hat{n}/dt)$. But, now, one needs to know how to determine the time derivative of a rotating unit vector.

Figure 2.22 Free-body diagram for a body P subjected to a central gravitational force for Example 2.5.11.

x-y-z shown in Fig. 2.22 takes place in a single plane. The motion of P is started with the initial conditions $\vec{r}(0) = \vec{r}_0$ and $\vec{v}(0) = \vec{v}_0$, where $\vec{r}(t) = \overrightarrow{OP}$ and $\vec{v} = d\vec{r}/dt$ (see Fig. 2.22).

■ **Solution**

The solution to this problem is strikingly simple if you notice that the moment about the inertial point O of the force $\vec{F}$ acting on P is equal to zero, i.e., $\vec{M}_O = \vec{r} \times \vec{F} = 0$. Therefore, as shown in Section 2.3,

$$\vec{H}_O = m\vec{r} \times \vec{v} = \text{constant} = m\vec{r}(0) \times \vec{v}(0) = m\vec{r}_0 \times \vec{v}_0$$

Since the cross product $\vec{r} \times \vec{v}$ is a vector that is normal to the plane defined by $\vec{r}$ and $\vec{v}$, this equation discloses that such a plane is inertial (since the normal to the plane does not change direction). Therefore, this proves that the motion of P takes place in a single plane in space.

This conclusion is a classical result in dynamics, and the system considered here is a model for the motion of any planet in our solar system around the sun (which is a good model since the mass of the sun is more than 700 times greater than the mass of all the planets together) or for the motion of an artificial satellite around a spherical celestial body. A detailed study of the two-body system is presented in Chapter 4.

2.6 Velocity and Acceleration Again and Angular Velocity

Let us now look at the problem of obtaining expressions for velocity and acceleration using rotating unit vectors. As indicated by Eq. (2.4.4) in Section 2.4, which is repeated here as Eq. (2.6.1), this requires knowing how to calculate the time derivative of a unit vector.

$$\frac{d\vec{r}}{dt} = \dot{r}_{x_1}\hat{x}_1 + \dot{r}_{x_2}\hat{x}_2 + \dot{r}_{x_3}\hat{x}_3 + r_{x_1}\frac{d\hat{x}_1}{dt} + r_{x_2}\frac{d\hat{x}_2}{dt} + r_{x_3}\frac{d\hat{x}_3}{dt}$$

$$\triangleq \left[\frac{d\vec{r}}{dt}\right]_{\{x\}} + r_{x_1}\frac{d\hat{x}_1}{dt} + r_{x_2}\frac{d\hat{x}_2}{dt} + r_{x_3}\frac{d\hat{x}_3}{dt} \quad (2.6.1)$$

SECTION 2.6 Velocity and Acceleration Again and Angular Velocity

The derivative of a rotating unit vector $\hat{x}$ (or of any vector whose magnitude is constant) is a vector that is always perpendicular to the unit vector $\hat{x}$. This can be shown by making use of the fact that the dot product $\vec{x} \bullet \vec{x}$ is constant; specifically,

$$\hat{x} \bullet \hat{x} = 1$$

By taking the time derivative of this equation and by noticing that $(d\hat{x}/dt) \bullet \hat{x} = \hat{x} \bullet (d\hat{x}/dt)$, we obtain

$$\frac{d}{dt}(\hat{x} \bullet \hat{x}) = 2\frac{d\hat{x}}{dt} \bullet \hat{x} = 0$$

which proves that $d\hat{x}/dt$ is normal to $\hat{x}$, since two vectors are normal to each other when their dot product is equal to zero.

To develop the general expression for the time derivative of any unit vector, consider a triad $\{\hat{x}, \hat{y}, \hat{z} \triangleq \hat{x} \times \hat{y}\}$ rotating about a line AA as shown in Fig. 2.23. In order to clearly specify directions, we need to bring a reference frame into the picture. For this, let the orientation of line AA be fixed in some reference frame, which, for now, is arbitrarily chosen. Such a reference frame is represented by the orthogonal lines labeled X_0, Y_0, and Z_0 in the figure. Figure 2.23a illustrates the two-dimensional case, where the axis about which the triad is rotating is perpendicular to the direction of one of the unit vectors ($\hat{z}$ in this illustration). Figure 2.23b illustrates the general case of three-dimensional motion.

To determine the time derivative of any of the unit vectors of the triad $\{\hat{x}, \hat{y}, \hat{z} \triangleq \hat{x} \times \hat{y}\}$, in the chosen reference frame, let us arbitrarily consider one

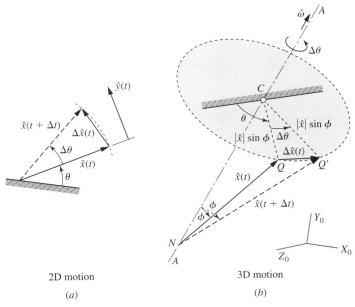

Figure 2.23 Time derivative of a unit vector and the angular velocity vector $\vec{\omega} = \dot{\theta}\hat{\omega}$.

of the unit vectors of the triad, say $\hat{x}$. Let ϕ be the angle between the unit vector $\hat{\omega}$ (which is, by definition, a unit vector parallel to the axis of rotation AA) and the unit vector $\hat{x}$. The two-dimensional case shown in Fig. 2.23a corresponds to $\phi = \pi/2$.

As shown in Fig. 2.23, the tip of the unit vector $\hat{x}$ sweeps out an arc of a circle of radius equal to $|\hat{x}| \sin \phi$, centered at a point C on the axis AA, while the unit vector $\hat{x} = \overrightarrow{NQ}$ sweeps out a portion of a right cone in space. At time t, the orientation of $\hat{x}$ is $\hat{x}(t)$ as shown in Fig. 2.23. After an infinitesimal instant of time Δt has elapsed, the radius CQ rotates by an angle $\Delta\theta$ to CQ', and the new orientation of $\hat{x}$ is along the vector $\hat{x}(t + \Delta t)$, which is also shown in Fig. 2.23. The vector $\Delta \hat{x}(t) = \hat{x}(t + \Delta t) - \hat{x}(t)$ is perpendicular to the unit vector $\hat{\omega}$ along line AA and is nearly perpendicular to $\hat{x}$ (it is perpendicular to $\hat{x}$ in the limit as $\Delta t \to 0$). As seen from Fig. 2.23, its magnitude $|\Delta \hat{x}|$ is

$$|\Delta \hat{x}| = |\hat{x}(t + \Delta t) - \hat{x}(t)|$$

In the limit as $\Delta t \to 0$, the angle of rotation $\Delta\theta$ also approaches zero, and $\Delta \hat{x}(t)$ becomes tangent to the arc swept by the tip of $\hat{x}(t)$; the length of that arc is equal to $(|\hat{x}| \sin \phi) \Delta\theta$, which is equal to the magnitude of the cross product $(\Delta\theta) \hat{\omega} \times \hat{x}$ [or the magnitude of the cross product $(\Delta\theta) \hat{x} \times \hat{\omega} = -(\Delta\theta) \hat{\omega} \times \hat{x}$]. Thus, in the limit as $\Delta t \to 0$, the magnitude of $d\hat{x}$ is

$$|d\hat{x}(t)| = \lim_{\Delta t \to 0} |\Delta \hat{x}(t)| = |(d\theta) \hat{\omega} \times \hat{x}| = |(d\theta) \hat{x} \times \hat{\omega}| \qquad (2.6.2)$$

Since, as disclosed by inspection of Fig. 2.23, the vector $d\hat{x}(t)$ is in the direction of the cross product $\hat{\omega} \times \hat{x}$ (rather than the cross product $\hat{x} \times \hat{\omega}$), we can then write from Eq. (2.6.2):

$$d\hat{x}(t) = (d\theta) \hat{\omega} \times \hat{x} \qquad (2.6.3)$$

or

$$\frac{d\hat{x}(t)}{dt} = \dot{\theta} \hat{\omega} \times \hat{x} \overset{\Delta}{=} \vec{\omega} \times \hat{x} \qquad (2.6.4)$$

where $\dot{\theta}$ denotes $d\theta/dt$. The vector $\vec{\omega}$, defined as

$$\vec{\omega} = \lim_{\Delta t \to 0} \frac{\Delta\theta}{\Delta t} \hat{\omega} = \dot{\theta} \hat{\omega} \qquad (2.6.5)$$

is called the *angular velocity* (in the chosen reference frame, as illustrated in Fig. 2.23) of the triad $\{\hat{x}, \hat{y}, \hat{z}\}$, associated with the rotation of that triad about axis AA. The direction of the vector $\vec{\omega}$ is, by definition, the direction of the infinitesimal rotation vector $(d\theta) \hat{\omega}$, given by the right-hand rule,[7] which was presented in Chapter 1.

---

[7] In dynamics, one should always use a curved line *with only one arrow* to indicate an angle. Using two arrows to indicate an angle is only fine either for angles that are constants or for statics work, where variables do not change with time. But in dynamics lines can rotate and, therefore, the rotation sense (i.e., clockwise or counterclockwise) needs to be taken into account. Two arrows would certainly make it impossible to apply the right-hand rule to determine the direction of angular velocity. *Just as one would not use two arrows for drawing a vector, two arrows should not be used for indicating an angle that can change with time.*

The magnitude $|\vec{\omega}|$ of the angular velocity $\vec{\omega}$ is called *angular speed*. The scalar component $\dot{\theta}$, in $\vec{\omega} = \dot{\theta}\hat{\omega}$, which may be positive or negative, is often also called angular velocity. It is also referred to as such throughout the book and, when it is known to be a positive quantity, it is referred to as angular speed.

Notice that for the case of planar motion, which is illustrated in Fig. 2.23a with a rotation about a fixed axis parallel to $\hat{z} \triangleq \hat{x} \times \hat{y}$ (with $\hat{y}$ being perpendicular to $\hat{x}$), one does not even have to introduce the angular velocity vector $\vec{\omega} = \dot{\theta}\hat{z}$ to obtain the result $d\hat{x}/dt = \dot{\theta}\hat{y}$. For such cases, this result can be inferred simply by inspection of a simple figure similar to Fig. 2.23.[8] Try to keep such a figure in mind, since by doing so you should always be able to write down, for planar problems, the expression for the derivative of a unit vector simply by inspection of a similar figure.

Since the unit vector $\hat{x}$ in Fig. 2.23 is only one of the three unit vectors of the $\{\hat{x}, \hat{y}, \hat{z}\}$ triad that is rotating in space with angular velocity $\vec{\omega}$ in the reference frame under consideration, we can then write the following expressions for the time derivative, in that reference frame, of the remaining unit vectors $\hat{y}$ and $\hat{z}$.

$$\frac{d\hat{y}}{dt} = \vec{\omega} \times \hat{y} \qquad \frac{d\hat{z}}{dt} = \vec{\omega} \times \hat{z} \qquad (2.6.6)$$

Therefore, Eq. (2.6.1), in which $\vec{r} = r_{x_1}\hat{x}_1 + r_{x_2}\hat{x}_2 + r_{x_3}\hat{x}_3$, can be written as follows:

$$\frac{d\vec{r}}{dt} = \left[\frac{d\vec{r}}{dt}\right]_{\{x\}} + \vec{\omega} \times (r_{x_1}\hat{x}_1 + r_{x_2}\hat{x}_2 + r_{x_3}\hat{x}_3)$$

or, simply,

$$\boxed{\frac{d\vec{r}}{dt} = \left[\frac{d\vec{r}}{dt}\right]_{\{x\}} + \vec{\omega} \times \vec{r}} \qquad (2.6.7)$$

> It is important to notice that $\vec{\omega}$ in Eq. (2.6.7) is the angular velocity, relative to the particular reference frame one is using, of the $\{x\}$ triad that was chosen to express the vector $\vec{r}$, and not the angular velocity of anything else. If the reference frame being used is inertial, the time derivative $d\vec{r}/dt$ given by Eq. (2.6.7) is called the *absolute time derivative* of the vector $\vec{r}$.

The relationship given by Eq. (2.6.7) is of fundamental importance in dynamics for vectors that are expressed in terms of rotating triads. That relationship will be used throughout the book, as it is used by all dynamicists.

---

[8] To see this, rotate the unit vector $\hat{x}$ by an infinitesimal angle $d\theta$. As you do so, the direction of the vector $d\hat{x}$ is the direction followed by the "tip of the unit vector" $\hat{x}$. Give that direction a name; the name used in Fig. 2.23 is $\hat{y}$. The length of the arc that is swept out is $(d\theta)|\hat{x}| = d\theta$. This, then, is how one can immediately conclude that $d\hat{x}/dt = (d\theta/dt)\hat{y} = \dot{\theta}\hat{y}$.

The first term in the right-hand side of Eq. (2.6.7), $[d\vec{r}/dt]_{\{x\}}$, is the time derivative of the vector $\vec{r}$ as seen by an observer fixed to the rotating triad $\{x\} = \{\hat{x}_1, \hat{x}_2, \hat{x}_3\}$. As indicated by Eq. (2.6.1), that term only involves the time derivative of the scalar components of $\vec{r}$. The second term, i.e., the one involving the omega cross operation $\vec{\omega} \times \vec{r}$ that is present because of the use of a rotating unit vector triad for expressing the vector $\vec{r}$, is the extra contribution to $d\vec{r}/dt$ seen by the observer fixed to the particular reference frame under consideration. As mentioned in Chapter 1, the velocity and acceleration that appear in Newton's second law has to be referred to an inertial reference frame.

> When does one choose to use rotating triads, and when does one choose nonrotating triads? The choice is only a matter of convenience for each particular problem. The resulting expressions for the position vector, the velocity, and the acceleration will have different forms for different triads. Some forms for such quantities will have less and simpler terms than others. More often than not, simpler forms for such expressions are obtained when rotating triads are used for expressing them. Also, some forms will require additional mathematical manipulations in the scalar equations obtained from $\vec{F} = m\vec{a}$ to obtain the differential equation (or equations, when more than one motion coordinate is involved) of motion for the system. This was illustrated in Example 2.5.10. Such additional manipulations that are required near the end of a formulation can be avoided by a judicious choice of triads made right at the beginning of the formulation.

Several examples that are formulated using rotating triads are presented in Section 2.7 and throughout the book. The determination of the angular velocity vector in the general case that involves several angles that are changing with time, even in the case of three-dimensional motion, is actually not a difficult matter (see the last sentence in the boxed text in Section B.1). Section B.2 presents the general "formula" for the acceleration of a point, which contains five different types of terms. As indicated in Appendix B, the use of that formula for solving problems is actually unnecessary and, therefore, is avoided in this book. Instead, a repetitive but basic approach is used throughout, where absolute velocity and acceleration are formulated from first principles for each problem.

2.7 More Solved Example Problems

EXAMPLE 2.7.1

The magnitude of the vector $\vec{r} = \overrightarrow{OP}$ shown in Fig. 2.24 is constant and equal to $r = 2$ m. The vector rotates in a plane, and its direction relative to a nonrotating line on that plane is tracked by the angle θ.

a. Using the nonrotating unit vectors $\hat{x}_1$ and $\hat{x}_2$, determine the time derivative of $\vec{r}$ in the reference frame defined by the plane of the paper.
b. Convert the answer obtained in part a to the unit vectors $\hat{r}$ and $\hat{n}$, where $\hat{r}$ is parallel to $\vec{r}$ and $\hat{n}$ is perpendicular to $\hat{r}$ (see Fig. 2.24) to show that the expression for $d\vec{r}/dt$

SECTION 2.7 More Solved Example Problems

Figure 2.24 Figure for Example 2.7.1.

is now a very simple one and equal to $d\vec{r}/dt = r\dot{\theta}\hat{n} = 2\dot{\theta}\hat{n}$ (see Section 1.2 if you need to review how to convert from one set of unit vectors to another).

c. Obtain the result $d\vec{r}/dt = r\dot{\theta}\hat{n} = 2\dot{\theta}\hat{n}$ by completely avoiding the nonrotating unit vectors $\hat{x}_1$ and $\hat{x}_2$ and working directly with the rotating unit vectors $\hat{r}$ and $\hat{n}$.

■ Solution

a. By inspecting Fig. 2.24, we can write for the vector $\vec{r}$,

$$\vec{r} = 2(\hat{x}_1 \cos\theta + \hat{x}_2 \sin\theta)$$

Therefore, the time derivative of $\vec{r}$ in the reference frame defined by the plane of the paper is then

$$\vec{v} = \frac{d\vec{r}}{dt} = 2\dot{\theta}(\hat{x}_2 \cos\theta - \hat{x}_1 \sin\theta) \text{ m/s}$$

b. To convert the expression found in part a to the unit vectors $\hat{r}$ and $\hat{n}$, inspect Fig. 2.24 again. As shown in Section 1.2, the relations between the unit vectors $\{\hat{x}_1, \hat{x}_2\}$ and $\{\hat{r}, \hat{n}\}$ are simply obtained by projecting the unit vectors of one set along the directions of the unit vectors of the other set. As disclosed by Fig. 2.24,

$$\hat{x}_1 = \hat{r} \cos\theta - \hat{n} \sin\theta$$

$$\hat{x}_2 = \hat{n} \cos\theta + \hat{r} \sin\theta$$

Now, substitute these relations into the expression that was obtained for $d\vec{r}/dt$. You will find

$$\vec{v} = \frac{d\vec{r}}{dt} = 2\dot{\theta}(\cos^2\theta + \sin^2\theta)\hat{n} = 2\dot{\theta}\hat{n}$$

This expression is much simpler than the one that was developed in terms of the nonrotating unit vectors $\{\hat{x}_1, \hat{x}_2\}$ in part a. It is also an expression that you could, from now on, write simply by inspecting a figure similar to Fig. 2.24. The reasons for this are

1. The time derivative $d\vec{r}/dt$ of the constant magnitude vector $\vec{r}$ must be perpendicular to $\vec{r}$.
2. $d\vec{r}$ is perpendicular to $\vec{r}$ and points in the direction followed by the tip of $\vec{r}$ as $\vec{r}$ is turned slightly in the sense indicated by the curved arrow used for the angle θ.

114   CHAPTER 2   Kinematics and Dynamics of Point Masses

3. For a vector $\vec{r}$ of constant magnitude, the magnitude of $d\vec{r}$ is equal to $|\vec{r}||d\theta|$, which is the length of an arc of a circle of radius $|\vec{r}|$.

By using the nonrotating unit vectors $\hat{x}_1$ and $\hat{x}_2$ we did not even have to mention and directly deal with angular velocity. But look at the amount of manipulations that had to be made to get the simple one-component expression for $d\vec{r}/dt$.

c. We will now see a more convenient way of obtaining the simpler (and easy to visualize) expression for $d\vec{r}/dt$. In terms of the rotating unit vectors shown in Fig. 2.24, the vector $\vec{r}$ takes the following form, which is the simplest form of a vector because it only involves one component,

$$\vec{r} = 2\hat{r}$$

and the calculation of its time derivative in the reference frame defined by the plane of the paper now involves the angular velocity vector $\vec{\omega} = \dot{\theta}\hat{z}$, where $\hat{z} = \hat{r} \times \hat{n}$. We then have

$$\vec{v} = \frac{d\vec{r}}{dt} = (\dot{\theta}\hat{z}) \times (2\hat{r}) = 2\dot{\theta}\hat{n}$$

Of course there was a price, so to speak, to be paid for the simplicity inherent in this expression. The price is that one has to know how to obtain the expression for the absolute angular velocity $\vec{\omega}$ of a rotating line and also has to know about the omega cross operation that gives the time derivative of a unit vector. For planar motion, the time derivative of a unit vector may even be written by inspection of a figure, as already pointed out. You should accustom yourself with these operations because they lead to less mathematical manipulations and are extremely convenient in dynamics.

EXAMPLE 2.7.2

Obtain the second time derivative, in the reference frame defined by the plane of the paper, of the vector $\vec{r}$ of Example 2.7.1 using the unit vectors $\{\hat{r}, \hat{n}\}$ and the unit vectors $\{\hat{x}_1, \hat{x}_2\}$ shown in Fig. 2.24.

■ Solution

In terms of the triad $\{\hat{r}, \hat{n}, \hat{z} \stackrel{\Delta}{=} \hat{r} \times \hat{n}\}$, whose angular velocity is $\vec{\omega} = \dot{\theta}\hat{z}$, we then obtain (see Fig. 2.24, and use the right-hand rule to determine the direction of the cross product $\vec{\omega} \times \hat{n}$):

$$\vec{a} = \frac{d\vec{v}}{dt} = \frac{d}{dt}(2\dot{\theta}\hat{n}) = 2\ddot{\theta}\hat{n} + 2\dot{\theta}\frac{d\hat{n}}{dt} = 2\ddot{\theta}\hat{n} + 2\dot{\theta}(\dot{\theta}\hat{z} \times \hat{n}) = 2\ddot{\theta}\hat{n} - 2\dot{\theta}^2\hat{r} \text{ m/s}^2$$

Notice that the time derivative of the unit vector $\hat{n}$, which is equal to $-\dot{\theta}\hat{r}$, might also have been obtained by direct inspection of Fig. 2.24.

Again, this expression has two components, and each one of them has a reasonably simple form. The expression for $\vec{a}$ is more complicated and, thus, less convenient, if the triad $\{\hat{x}_1, \hat{x}_2, \hat{z} = \hat{x}_1 \times \hat{x}_2\}$ is used. The angular velocity of that triad, relative to the plane of the paper, is zero, and the following expression is obtained for the vector $\vec{a} = d\vec{v}/dt$:

$$\vec{a} = \frac{d}{dt}[2\dot{\theta}(\hat{x}_2 \cos\theta - \hat{x}_1 \sin\theta)]$$
$$= 2(\ddot{\theta} \cos\theta - \dot{\theta}^2 \sin\theta)\hat{x}_2 - 2(\ddot{\theta} \sin\theta + \dot{\theta}^2 \cos\theta)\hat{x}_1$$

Usage of Polar Coordinates and of a Convenient Rotating Triad

EXAMPLE 2.7.3

For several problems in dynamics, it is convenient to use polar coordinates to locate a point in a plane. Using the polar coordinates r and θ and the unit vectors $\hat{r}$ and $\hat{n}$ shown in Fig. 2.24, obtain the general expressions for the absolute velocity and for the absolute acceleration of point P. Point O shown in the figure is an inertial point on the plane of the paper, and that plane is also inertial; i.e., it also does not rotate.

■ Solution

Since the absolute position vector $\vec{r}$ for point P is simply (see Fig. 2.24)

$$\vec{r} = r\hat{r} \qquad (2.7.3a)$$

and the absolute angular velocity $\vec{\omega}$ of the triad $\{\hat{r}, \hat{n}, \hat{z} \stackrel{\Delta}{=} \hat{r} \times \hat{n}\}$ is

$$\vec{\omega} = \dot{\theta}\hat{z} \qquad (2.7.3b)$$

the absolute velocity of point P is then

$$\vec{v} = \frac{d\vec{r}}{dt} = \dot{r}\hat{r} + r\frac{d\hat{r}}{dt} = \dot{r}\hat{r} + \vec{\omega} \times (r\hat{r}) = \dot{r}\hat{r} + r\dot{\theta}\hat{n} \qquad (2.7.3c)$$

Figure 2.25 is a graphical visualization of the time derivative $d\vec{r}/dt$ of a rotating vector $\vec{r}$. Its purpose is to help you be able to write down the expression for the velocity $\vec{v} = d\vec{r}/dt$ without having to derive it again or to consult such a derivation.

The absolute acceleration of point P is determined by taking the time derivative of Eq. (2.7.3c) as shown here.

$$\vec{a} = \frac{d\vec{v}}{dt} = \ddot{r}\hat{r} + \dot{r}\frac{d\hat{r}}{dt} + (r\ddot{\theta} + \dot{r}\dot{\theta})\hat{n} + r\dot{\theta}\frac{d\hat{n}}{dt}$$

Since $d\hat{r}/dt = \vec{\omega} \times \hat{r} = \dot{\theta}\hat{n}$ and $d\hat{n}/dt = \vec{\omega} \times \hat{n} = -\dot{\theta}\hat{r}$ (see Fig. 2.24, and use the right-hand rule), this expression for the acceleration then becomes

$$\vec{a} = \frac{d\vec{v}}{dt} = (\ddot{r} - r\dot{\theta}^2)\hat{r} + (r\ddot{\theta} + 2\dot{r}\dot{\theta})\hat{n} \qquad (2.7.3d)$$

Figure 2.25 Graphical visualization of the time derivative $(d/dt)\vec{r}$ of a rotating vector $\vec{r}$.

116 CHAPTER 2 Kinematics and Dynamics of Point Masses

As you can see, these operations are really not difficult to perform when you use polar coordinates and the unit vectors chosen in this example. The advantage of using such rotating unit vectors lies in the simplicity of the expression that is obtained for the velocity, and the relatively compact expression that is obtained for the acceleration.

Notice that the expression for the velocity given by Eq. (2.7.3c) is the sum of the following two velocities.

1. The velocity $\dot{r}\hat{r}$ that would be obtained if θ were held constant (i.e., if P were traveling along the radial line)
2. The velocity $r\dot{\theta}\hat{n}$ that would be obtained if r were held constant (i.e., if P were traveling along a circle of radius r)

The objective of these observations is to help you be able to write down Eq. (2.7.3c) for the velocity $\vec{v} = d\vec{r}/dt$ simply by inspecting a figure such as Fig. 2.24 or Fig. 2.25.

As for the expression for the acceleration, the only terms that appear in Eq. (2.7.3d) that might be viewed as obvious are the terms $\ddot{r}\hat{r}$ (which would be the acceleration if P were traveling along the radial line, with θ held constant), and $r\ddot{\theta}\hat{n}$ (which would be the acceleration if r were held constant so that P would be traveling along a circle of radius r). The other two terms $-r\dot{\theta}^2\hat{r}$ and $2\dot{r}\dot{\theta}\hat{n}$ are called the *centripetal acceleration* and the *Coriolis acceleration*, respectively. The negative of the centripetal acceleration is called centrifugal acceleration. Figure 2.26 is a graphical visualization that summarizes the results obtained here.

Look at the larger box that shows the time derivative of the radial $\hat{r}$ and of the transverse $\hat{\theta}$ components of the second time derivative of a vector $\vec{r}$. Each component of the vector $\vec{v} = d\vec{r}/dt$ yields two parts for each one of the components of the final result for $\vec{a} = d\vec{v}/dt$. The purpose of such a figure is to help you be able to write down the expression for the velocity $\vec{v} = d\vec{r}/dt$ and the acceleration $\vec{a} = d\vec{v}/dt$ from these first principles.

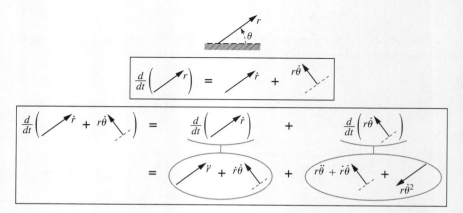

Figure 2.26 Graphical visualization of the first and second time derivatives of a rotating vector $\vec{r}$.

EXAMPLE 2.7.4

A turntable rotates about a fixed axis that passes through its center C as shown in Fig. 2.27. The angular velocity of the turntable, relative to the fixed plane of the paper, is $\Omega(t)$. Determine the absolute velocity $\vec{v}$ and the absolute acceleration $\vec{a}$ of a particle P that is able to move on the turntable. Express the answers in terms of the polar coordinates $r(t)$ and $\theta(t)$ that locate the particle relative to the turntable at any time t.

■ Solution

The absolute velocity of P is the velocity of P relative to C (since point C is inertial) taking into account the rotation of line CP relative to a nonrotating reference line (the "fixed reference line" shown in Fig. 2.27). It is the time derivative, in the inertial plane of the paper, of the vector $\vec{r} \triangleq \overrightarrow{CP}$ drawn from C to P. We now need to choose a set of unit vectors in order to be able to express the vector $\vec{r}$ and carry out the time derivative operation. Two possible choices are shown in Fig. 2.27. Suppose we decide to work with the unit vector triad $(\hat{r}, \hat{\theta}, \hat{z} \triangleq \hat{r} \times \hat{\theta})$ shown in Fig. 2.27, with $\hat{r}$ always pointing in the direction from C to P. In this case, the expression for the vector $\vec{r}$ and the absolute angular velocity of the chosen triad are, respectively,

$$\vec{r} = r\hat{r} \tag{2.7.4a}$$

$$\vec{\omega} = (\dot{\phi} + \dot{\theta})\hat{z} = (\Omega + \dot{\theta})\hat{z} \tag{2.7.4b}$$

Notice that $\dot{\phi} = \Omega$, since the angle ϕ tracks the orientation of any line fixed to the rotating table relative to the nonrotating reference line shown in Fig. 2.27. Also, notice that the term $\dot{\theta}\hat{z}$ is the angular velocity of line CP relative to the rotating table (or, in other words, the angular velocity of line CP in the reference frame fixed to the rotating table).

With $\vec{r}$ and $\vec{\omega}$ given by Eqs. (2.7.4a) and (2.7.4b), respectively, the absolute velocity and the absolute acceleration of P are then determined as follows.

$$\vec{v} = \frac{d\vec{r}}{dt} = \dot{r}\hat{r} + \vec{\omega} \times \vec{r} = \dot{r}\hat{r} + r(\Omega + \dot{\theta})\hat{\theta} \triangleq v_r\hat{r} + v_\theta\hat{\theta} \tag{2.7.4c}$$

$$\vec{a} = \frac{d\vec{v}}{dt} = \ddot{r}\hat{r} + [\dot{r}(\Omega + \dot{\theta}) + r(\dot{\Omega} + \ddot{\theta})]\hat{\theta} + \vec{\omega} \times \vec{v}$$

$$= [\ddot{r} - r(\Omega + \dot{\theta})^2]\hat{r} + [(\dot{\Omega} + \ddot{\theta})r + 2\dot{r}(\Omega + \dot{\theta})]\hat{\theta} \triangleq a_r\hat{r} + a_\theta\hat{\theta} \tag{2.7.4d}$$

These expressions for $\vec{v}$ and $\vec{a}$ are valid for any kind of motion of the particle P on the rotating table. If the motion of P were to be constrained to a radial groove fixed to

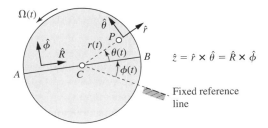

Figure 2.27 Figure for Example 2.7.4.

the rotating table, for example, you would simply have to set θ equal to a constant in Eqs. (2.7.4c) and (2.7.4d). If the motion of P were to be constrained to a circular groove of radius R and centered at C, you would, instead, set $r = \text{constant} = R$.

In formulating the expressions for the velocity and for the acceleration you could, of course, choose other unit vector triads to work with. If the triad $(\hat{R}, \hat{\phi}, \hat{z} = \hat{R} \times \hat{\phi})$ shown in Fig. 2.27 is chosen, the expression for the vector $\vec{r}$ becomes (see Fig. 2.27)

$$\vec{r} = r(\hat{R}\cos\theta + \hat{\phi}\sin\theta) \tag{2.7.4e}$$

while the absolute angular velocity of the new triad $\{\hat{R}, \hat{\phi}, \hat{z}\}$, which is denoted below as $\vec{\omega}_2$ so that it is not confused with the vector $\vec{\omega}$ used earlier, is now

$$\vec{\omega}_2 = \Omega\hat{z} \tag{2.7.4f}$$

In this case, the following expression for $\vec{v}$ is obtained:

$$\vec{v} = \frac{d\vec{r}}{dt} = \left[\frac{d\vec{r}}{dt}\right]_{\{\hat{R},\hat{\phi},\hat{z}\}} + \vec{\omega}_2 \times (\hat{R}r\cos\theta + \hat{\phi}r\sin\theta)$$

$$= (\dot{r}\cos\theta - r\dot{\theta}\sin\theta)\hat{R} + (\dot{r}\sin\theta + r\dot{\theta}\cos\theta)\hat{\phi} + \vec{\omega}_2 \times (\hat{R}r\cos\theta + \hat{\phi}r\sin\theta)$$

$$= [\dot{r}\cos\theta - r(\Omega + \dot{\theta})\sin\theta]\hat{R} + [\dot{r}\sin\theta + r(\Omega + \dot{\theta})\cos\theta]\hat{\phi}$$

$$\stackrel{\triangle}{=} v_R\hat{R} + v_\phi\hat{\phi} = (v_r\cos\theta - v_\theta\sin\theta)\hat{R} + (v_r\sin\theta + v_\theta\cos\theta)\hat{\phi} \tag{2.7.4g}$$

Notice that, as indicated in Eq. (2.7.4g) and verified by inspection of Fig. 2.27, the expression $\vec{v} = v_R\hat{R} + v_\phi\hat{\phi}$ is the same as the projection of $\vec{v} = v_r\hat{r} + v_\theta\hat{\theta}$ along the orthogonal unit vector system $\{\hat{R}, \hat{\phi}\}$ shown in that figure.

The following expression is obtained for the acceleration $\vec{a}$:

$$\vec{a} = \frac{d}{dt}\vec{v} = \left[\frac{d}{dt}\vec{v}\right]_{\{\hat{R},\hat{\phi},\hat{z}\}} + \vec{\omega}_2 \times \vec{v}$$

$$= \{[\ddot{r} - r\dot{\theta}(\Omega + \dot{\theta})]\cos\theta - [\dot{r}\dot{\theta} + \dot{r}(\Omega + \dot{\theta}) + r(\dot{\Omega} + \ddot{\theta})]\sin\theta\}\hat{R}$$

$$+ \{[\ddot{r} - r\dot{\theta}(\Omega + \dot{\theta})]\sin\theta + [\dot{r}\dot{\theta} + \dot{r}(\Omega + \dot{\theta}) + r(\dot{\Omega} + \ddot{\theta})]\cos\theta\}\hat{\phi} + \vec{\omega}_2 \times \vec{v}$$

If you perform the cross product that appears in this expression for $\vec{a}$, collect items in $\cos\theta$ and $\sin\theta$, and group the terms $r\Omega^2 + r\dot{\theta}^2 + 2r\Omega\dot{\theta}$ that appear in the resulting expression for $\vec{a}$ as $r(\Omega + \dot{\theta})^2$, you should obtain the following expression for the acceleration $\vec{a}$:

$$\vec{a} = \{[\ddot{r} - r\dot{\theta}(\Omega + \dot{\theta})^2]\cos\theta - [r(\dot{\Omega} + \ddot{\theta}) + 2\dot{r}(\Omega + \dot{\theta})]\sin\theta\}\hat{R}$$

$$+ \{[\ddot{r} - r\dot{\theta}(\Omega + \dot{\theta})^2]\sin\theta + [r(\dot{\Omega} + \ddot{\theta}) + 2\dot{r}(\Omega + \dot{\theta})]\cos\theta\}\hat{\phi}$$

$$\stackrel{\triangle}{=} a_R\hat{R} + a_\phi\hat{\phi} = (a_r\cos\theta - a_\theta\sin\theta)\hat{R} + (a_r\sin\theta + a_\theta\cos\theta)\hat{\phi} \tag{2.7.4h}$$

Also notice that, as indicated in Eq. (2.7.4h) and verified by inspection of Fig. 2.27, the expression $\vec{a} = a_R\hat{R} + a_\phi\hat{\phi}$ is the same as the projection of $\vec{a} = a_r\hat{r} + a_\theta\hat{\theta}$ along the orthogonal unit vector system $\{\hat{R}, \hat{\phi}\}$ shown in that figure.

This example showed in detail the symbolic manipulations involved in formulating expressions for velocity and for acceleration that are needed when using Newton's second law. Before we leave this example, it is worthwhile to look at the results for the velocity and for the acceleration that were obtained and reflect on them because many of the steps done here appear in a large variety of situations in practice. Let us start by making some observations about the steps that led us to the two different forms for the velocity $\vec{v}$.

> *Observation 1.* The simplest of the expressions for $\vec{v}$ and $\vec{a}$ were obtained by using the triad for which one of the unit vectors, namely, $\hat{r}$, was along the position vector $\vec{r}$, i.e., the triad $\{\hat{r}, \hat{\theta}, \hat{z}\}$ shown in Fig. 2.27. In that triad, the two scalar components of $\vec{v}$, i.e., $v_r = \dot{r}$ and $v_\theta = r(\Omega + \dot{\theta})$, have the same simple physical interpretation mentioned in Example 2.7.3 (p. 115). Those observations are repeated here.
>
> *Observation 2.* The radial $\hat{r}$-component of the velocity, $v_r \hat{r} = \dot{r}\hat{r}$, would be the velocity $\vec{v}$ if the angles θ and ϕ shown in Fig. 2.27 were held constant; in such a case, particle P would simply travel along the straight line CP.
>
> *Observation 3.* The transverse $\hat{\theta}$-component of the velocity, $v_\theta \hat{\theta} = r(\Omega + \dot{\theta})\hat{\theta}$, would be the velocity $\vec{v}$ if the polar coordinate r shown in Fig. 2.27 (which, again, is the distance measured from C to P) were kept constant. In such a case, particle P would simply travel along a circle of radius r centered at point C. The velocity $\vec{v}$ is always the vector sum of the radial $\hat{r}$-component and the transverse $\hat{\theta}$-component.
>
> *Observation 4.* As indicated in Eq. (2.7.4g), the components $v_R \hat{R}$ and $v_\phi \hat{\phi}$ of the velocity $\vec{v}$ are simply composed, as expected, of projections of the simpler components v_r and v_θ along the directions $\hat{R}$ and $\hat{\phi}$ as $v_R = v_r \cos\theta - v_\theta \sin\theta$ and $v_\phi = v_r \sin\theta + v_\theta \cos\theta$. Thus, a "better way" (meaning, one that involves less mathematical manipulations and, therefore, less calculations) to obtain the expression given by Eq. (2.7.4g) would be to find the quantities v_r and v_θ first, and then convert the result into v_R and v_ϕ by simple projections. The same observation holds for the acceleration $\vec{a}$.

EXAMPLE 2.7.5

A particle P is constrained to move on the surface of a nonrotating cylinder of radius R m as shown in Fig. 2.28. The x, y, and z axes shown in that figure are fixed to the cylinder. Point C of the cylinder has a constant acceleration equal to $a_C \hat{y}$ relative to inertial space.

a. Obtain the expression for the absolute acceleration of P by making use of the cylindrical coordinates $\phi(t)$ and $z(t)$ and the conveniently chosen triad $\{\hat{R}, \hat{\phi}, \hat{z} = \hat{R} \times \hat{\phi}\}$, all of which are shown in Fig. 2.28. The x, y, and z axes in that figure are inertial (i.e., they do not rotate relative to inertial space).

b. For $R = 1$ m and $a_C = 4$ m/s^2, plot the magnitude of the absolute acceleration $\vec{a}_P$ of point P versus the angle ϕ, if $\dot{\phi} = $ constant $= 30$ revolutions per minute (rpm) and the speed of P relative to the moving cylinder is constant and equal to 5 m/s.

■ **Solution**

a. With point O shown in Fig. 2.28 being inertial, the absolute position vector for P is

$$\overrightarrow{OP} = \overrightarrow{OC} + \overrightarrow{CP} = \overrightarrow{OC} + R\hat{R} + z\hat{z} \qquad (2.7.5a)$$

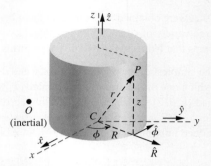

Figure 2.28 Particle constrained to move on a cylindrical surface.

Since the x axis shown in Fig. 2.28 is nonrotating, the absolute angular velocity of the triad $\{\hat{R}, \hat{\phi}, \hat{z} = \hat{R} \times \hat{\phi}\}$ is equal to[9] $\vec{\omega} = \dot{\phi}\hat{z}$. Therefore, the absolute velocity of point P is

$$\vec{v}_P = \frac{d}{dt}\overrightarrow{OP} = \underbrace{\frac{d}{dt}\overrightarrow{OC}}_{\text{call this } \vec{v}_C} + \underbrace{\dot{R}\hat{R}}_{\substack{=0 \\ \text{because } R=\text{constant}}} + \dot{z}\hat{z} + \vec{\omega} \times (R\hat{R} + z\hat{z})$$

$$= \vec{v}_C + \dot{z}\hat{z} + R\dot{\phi}\hat{\phi} \tag{2.7.5b}$$

Since we already know that $d\vec{v}_C/dt$ is equal to the given acceleration $a_C\hat{y}$, the simplest way to deal with the term $\vec{v}_C$ in Eq. (2.7.5b) is to leave it as is in that equation [i.e., it is simply not wise to express $\vec{v}_C$ in components and then take the derivative of such components, because the final expression for $d\vec{v}_C/dt$ is already known beforehand].

By differentiating $\vec{v}_P$ with respect to time, the absolute acceleration of P is then obtained as

$$\vec{a}_P = \frac{d}{dt}\vec{v}_P = \underbrace{\frac{d}{dt}\vec{v}_C}_{\text{this is } \vec{a}_C} + \ddot{z}\hat{z} + R\ddot{\phi}\hat{\phi} + \vec{\omega} \times (\dot{z}\hat{z} + R\dot{\phi}\hat{\phi})$$

$$= \vec{a}_C + \ddot{z}\hat{z} + R\ddot{\phi}\hat{\phi} - R\dot{\phi}^2\hat{R} \tag{2.7.5c}$$

Since $\vec{a}_C$ is given as $\vec{a}_C = a_C\hat{y}$, and $\hat{y} = \hat{R}\sin\phi + \hat{\phi}\cos\phi$ (see Fig. 2.28), the following result is then obtained for the acceleration $\vec{a}_P$ of point P.

$$\vec{a}_P = (a_C\sin\phi - R\dot{\phi}^2)\hat{R} + (R\ddot{\phi} + a_C\cos\phi)\hat{\phi} + \ddot{z}\hat{z} \tag{2.7.5d}$$

b. Now it is desired to plot $|\vec{a}_P|$ versus ϕ, for $R = 1$ m, $a_C = 4$ m/s^2, $\dot{\phi}$ = constant = 30 rpm = $30 \times 2\pi/60$ rad/s, and when the speed of P relative to the moving cylinder is constant and equal to 5 m/s.

---

[9] If the cylinder were rotating with angular velocity $\vec{\Omega}$ relative to inertial space, the absolute angular velocity $\vec{\omega}$ of the $\{\hat{R}, \hat{\phi}, \hat{z}\}$ triad would be equal to $\vec{\omega} = \vec{\Omega} + \dot{\phi}\hat{z}$ since the x axis, where the angle ϕ is measured from, was chosen to be fixed to the cylinder.

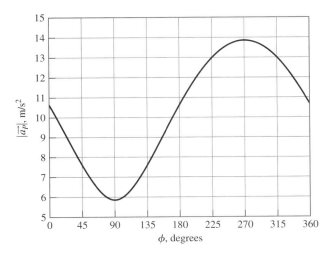

Figure 2.29 Plot of $|\vec{a}_P|$ versus ϕ for Example 2.7.5.

The magnitude of $\vec{a}_P$ is

$$|\vec{a}_P| = \sqrt{(a_C \sin\phi - R\dot\phi^2)^2 + (R\ddot\phi + a_C \cos\phi)^2 + \ddot z^2} \qquad (2.7.5e)$$

The values for $\ddot\phi$ and for $\ddot z$ are needed in this expression. Since $\dot\phi =$ constant, it follows that $\ddot\phi = 0$. To calculate $\ddot z$, use the only information that has not been used so far, i.e., that the speed of P relative to the moving cylinder is constant and equal to 5 m/s. Since the velocity of any point of the cylinder is $\vec v_C = d\overrightarrow{OC}/dt$, the speed of P relative to the moving cylinder is the magnitude of the vector $\vec v_P - \vec v_C$. By making use of Eq. (2.7.5b), $|\vec v_P - \vec v_C|$ is then determined as

$$|\vec v_P - \vec v_C| = \sqrt{\dot z^2 + (R\dot\phi)^2} = \text{constant} = 5 \text{ m/s}$$

This gives $\dot z =$ constant ≈ 3.89 m/s. Therefore, $\ddot z = 0$, and $|\vec{a}_P|$ is

$$|\vec{a}_P| = \sqrt{4^2 + \pi^4 - 2\times 4 \times \pi^2 \sin\phi} \approx \sqrt{113.4 - 79\sin\phi} \text{ m/s}^2$$

The plot of $|\vec{a}_P|$ versus ϕ is shown in Fig. 2.29. Although particle P moves with a constant speed relative to the cylinder, its acceleration is not constant (and it would not be constant even if the cylinder were not accelerating) because the velocity of P is changing direction.

A Pendulum Mounted on a Moving Base — EXAMPLE 2.7.6

The pendulum considered in Example 2.5.10 (p. 101) is connected by a bearing (represented by point A in Fig. 2.30) to a truck that is moving to the left along a straight horizontal track with a known acceleration (relative to the earth, which is approximated

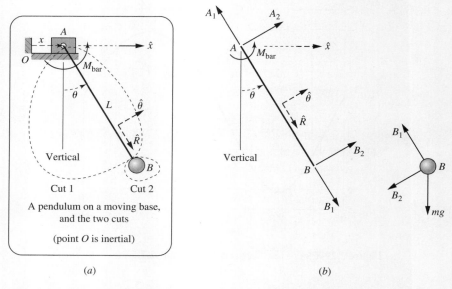

Figure 2.30 A pendulum mounted on a moving base for Example 2.7.6.

as inertial) equal to a_A m/s², also directed to the left. The motion of the pendulum is constrained to a vertical plane, and is started with the initial conditions $\theta(0) = 0$ and $\dot{\theta}(0) = 0$, and the only external force acting on the point mass at B is the gravitational force of attraction of the earth. An external moment M_{bar} is applied to the pendulum stem as shown in Fig. 2.30. The pendulum length is $L = 1$ m, and the mass of the particle at point B is $m = 1$ kg. Approximate the bar as massless.

a. Obtain the differential equation of motion for the pendulum, and then determine the maximum and minimum values of $\theta(t)$ for the case when $M_{bar} = 0$ and $a_A = $ constant $= \sqrt{3}g \approx 17$ m/s².
b. Determine the maximum and minimum values for the axial reaction force acting at end B of bar AB for the case when $M_{bar} = 0$, and $a_A = $ constant $= \sqrt{3}g \approx 17$ m/s².

  Note: The questions in parts c and d cannot be answered analytically because the differential equation of motion is nonlinear and M_{bar} is neither zero nor a function that only involves the angle θ. For both cases, $M_{bar} = -c\dot{\theta}$, where $c = 1$ N·m/(rad/s) is due to *viscous friction* in the bearing at A. To answer the questions for parts c and d, use a computer to integrate the differential equation numerically.
c. Determine the maximum value of the angle θ, and the time it occurs, for the case when $a_A = $ constant $= 2g$ m/s².
d. During the first 5 s, point A is forced to accelerate to the left (see Fig. 2.30) as $a_A(t) = kt$ where $k = 5$ m/s³. Afterward, the acceleration of A is suddenly dropped to zero, but point A is forced to continue to travel to the left at a constant velocity. Plot $\theta(t)$ versus time t for $0 \le t \le 15$ s. What is the maximum value of θ, and when does it occur?

Solution

a. The free-body diagrams for the system consisting of the massless bar and the mass particle at B are shown in Fig. 2.30b. We will use the unit vector $\hat{x}$ to express the absolute acceleration of point A, which is $\vec{a}_A = \ddot{x}\hat{x} = -a_A \hat{x}$ where $a_A = $ constant (with a_A being positive, the minus sign, which is associated with the unit vector $\hat{x}$, accounts for the fact that point A is accelerating to the left). To deal with the particle at B, we will make use of the more convenient rotating unit vectors $\hat{R}$ and $\hat{\theta}$ shown in the figure. The absolute angular velocity of the triad $\{\hat{R}, \hat{\theta}, \hat{z} \triangleq \hat{R} \times \hat{\theta}\}$ is $\vec{\omega} = \dot{\theta}\hat{z}$ (see Fig. 2.30, and use the right-hand rule to obtain $\vec{\omega}$). In terms of those unit vectors, we have for the particle at B:

- Resultant force acting on the particle at B:

$$\vec{F} = mg(\hat{R}\cos\theta - \hat{\theta}\sin\theta) - B_1 \hat{R} - B_2 \hat{\theta}$$

- Absolute position vector for the particle at B:

$$\vec{r}_B = \overrightarrow{OB} = \overrightarrow{OA} + \overrightarrow{AB} = x\hat{x} + L\hat{R}$$

- Absolute velocity of the particle at B:

$$\vec{v}_B = \frac{d\vec{r}_B}{dt} = \dot{x}\hat{x} + \frac{d}{dt}(L\hat{R}) = \dot{x}\hat{x} + \vec{\omega} \times (L\hat{R}) = \dot{x}\hat{x} + L\dot{\theta}\hat{\theta}$$

- Absolute acceleration of the particle at B:

$$\vec{a}_B = \frac{d\vec{v}_B}{dt} = \ddot{x}\hat{x} + [L\ddot{\theta}\hat{\theta} + \vec{\omega} \times (L\dot{\theta}\hat{\theta})] = \ddot{x}\hat{x} + L\ddot{\theta}\hat{\theta} - L\dot{\theta}^2 \hat{R}$$

Since $\ddot{x}\hat{x} = \ddot{x}(\hat{\theta}\cos\theta + \hat{R}\sin\theta)$, Newton's second law, $\vec{F} = m\vec{a}_B$, then yields the following scalar equations. Notice that, as indicated by Eq. (2.7.6a), only the unknown reaction force B_2 is needed for obtaining the differential equation of motion for the system.

$$m(L\ddot{\theta} + \ddot{x}\cos\theta) = -mg\sin\theta - B_2 \qquad (2.7.6a)$$

$$B_1 - mg\cos\theta = m(L\dot{\theta}^2 - \ddot{x}\sin\theta) \qquad (2.7.6b)$$

These equations involve the three unknowns θ, B_1, and B_2. Therefore, one more equation is needed. The additional equation is obtained by working with the free-body diagram for the bar. As indicated in Example 2.5.10 (p. 101), the additional equation that only involves the unknown B_2 can be readily obtained by equating to zero the summation of the moments, about any point, of the forces acting on the massless bar (the dynamics of rigid bodies, taking into account their distributed masses, is dealt with in later chapters, starting with Chapter 5). By taking moments about point A (because the unknowns A_1 and A_2 would not appear in the resulting equation), we obtain

$$B_2 = -\frac{M_{\text{bar}}}{L} \qquad (2.7.6c)$$

With B_2 as previously given, Eq. (2.7.6a) becomes

$$L\ddot{\theta} + g\sin\theta + \ddot{x}\cos\theta = \frac{M_{\text{bar}}}{mL} \qquad (2.7.6d)$$

Equation (2.7.6d) does not involve any unknown reaction forces, and, therefore, it is the differential equation of motion for the system. Once a solution for $\theta(t)$ is obtained by integrating that equation, the axial force B_1 can be calculated by making use of Eq. (2.7.6b).

As indicated in Section 1.13, application of certain techniques to a differential equation of motion may yield a variety of information about the motion of the system without having to solve the equation analytically (which is not always possible to do, since differential equations of motion are generally nonlinear). For this problem, the maximum and minimum values of θ can be determined analytically for the case when $M_{\text{bar}} = 0$ and $\vec{a}_A$ is either a constant or a function of θ only.

The maximum and minimum values for the angle θ occur when $\dot\theta = 0$. By making use of the chain rule of differentiation to express $\ddot\theta$ as

$$\ddot\theta = \frac{d\dot\theta}{dt} = \frac{d\dot\theta}{d\theta}\frac{d\theta}{dt} = \dot\theta\frac{d\dot\theta}{d\theta}$$

the differential equation of motion, Eq. (2.7.6d), may be rearranged and then integrated once, for $\ddot{x} = \text{constant} = -a_A$ and $M_{\text{bar}} = 0$.

$$L\int \dot\theta\frac{d\dot\theta}{d\theta}\, d\theta + \int (g\sin\theta - a_A\cos\theta)\, d\theta = \text{constant} \stackrel{\Delta}{=} C_1$$

or

$$\frac{L}{2}\dot\theta^2 - g\cos\theta - a_A\sin\theta = \text{constant} = C_1 \qquad (2.7.6e)$$

By making use of the given initial conditions, the constant C_1 is then $C_1 = \dot\theta^2(0)/2 - g\cos\theta(0) - a_A\sin\theta(0) = -g$, since $\dot\theta(0) = 0$ and $\theta(0) = 0$.

By setting $\dot\theta = 0$ in Eq. (2.7.6e), the maximum and minimum values of θ, $\theta \stackrel{\Delta}{=} \theta_m$, are, then, the solutions to the following algebraic equation.

$$g\cos\theta_m + a_A\sin\theta_m = g \qquad (2.7.6f)$$

Since $\sin\theta = \pm\sqrt{1-\cos^2\theta}$, this equation yields the following quadratic equation for θ_m:

$$\left(g^2 + a_A^2\right)\cos^2\theta_m - 2g^2\cos\theta_m + g^2 - a_A^2 = 0 \qquad (2.7.6g)$$

whose solutions are

$$\cos\theta_m = 1 \quad\text{and}\quad \cos\theta_m = \frac{g^2 - a_A^2}{g^2 + a_A^2}$$

Therefore, $\theta_m = 0$ and, with $a_A = \sqrt{3}g$, $\theta_m = \arccos(-0.5) = \pm 120°$. As can be verified from Eq. (2.7.6f), the positive solution $\theta_m = 120°$ is the one that corresponds to $a_A/g = \sqrt{3}$. The extra value $\theta_m = -120°$ was "created" when Eq. (2.7.6f) was squared to obtain Eq. (2.7.6g), and it does not satisfy Eq. (2.7.6f). The solution for θ_m could also have been obtained by using a calculator to do a few numerical iterations in Eq. (2.7.6f) to obtain $\theta_m = 120°$. For $a_A/g = \sqrt{3}$ we then have $\theta_{\min} = 0$ and $\theta_{\max} = 120°$.

b. For the axial force B_1, for the case when $M_{\text{bar}} = 0$ and $\ddot{x} = -a_A =$ constant $= -\sqrt{3}g$ m/s^2, the following result is obtained by combining Eqs. (2.7.6b) and (2.7.6e).

$$B_1 = 3m(g\cos\theta + a_A\sin\theta) - 2mg \qquad (2.7.6h)$$

By setting $dB_1/d\theta = 0$, with $dB_1/d\theta$ obtained from Eq. (2.7.6h), the extremum (i.e., maximum or minimum) value for B_1 is found to occur when $\tan\theta = a_A/g$. For $a_A/g = \sqrt{3}$, such a value of θ is $\theta = 60°$, which is equal to $\theta_{\max}/2$. The corresponding value for B_1 is $B_1 = 4mg \approx 39.2$ N for $m = 1$ kg. Since $(d^2B_1/d\theta^2) = -3m(g\cos\theta + a_A\sin\theta)$ is negative for $\theta = 60°$, the extremum value that was found for B_1 is a maximum. Notice that $[B_1]_{\theta_{\min}} = [B_1]_{\theta=0} = mg$. This is the minimum value of B_1.

⟹ c, d. A Simulink block diagram model for integrating the differential equation of motion [Eq. (2.7.6i) which follows], for $M_{\text{bar}} = -c\dot{\theta}$, is shown in Fig. 2.31 (see Section A.3 for a description of the Switch block and other blocks). The model is the file pend_moving_base.mdl on the enclosed CD. To use it, either type pend_moving_base in the MATLAB command window or select it by typing simulations_menu as indicated earlier. Enter either 1 or 2 in the block labeled Input selection to generate the solution to parts d and c, respectively. The animate_nbars block is optional. Look at the default settings of the Solver part of the Simulation → Parameters menu for the model in your computer, and experiment using any different value for the maximum step size, including the string auto, of the integration scheme if your simulation runs either too slow or too fast in your computer. The model may be used for answering all the questions that were asked in parts b, c, and d.

$$\ddot{\theta} + \frac{g}{L}\sin\theta = \frac{a_A}{L}\cos\theta - \frac{c}{mL^2}\dot{\theta} \qquad (2.7.6i)$$

The blocks named Ramp1 and Ramp2 in Fig. 2.31 generate a signal that varies linearly with time t as $f(t) = f(t_0) + k(t - t_0)$, with $f(t) = 0$ for $t < t_0$. The initial time t_0, the initial value $f(t_0)$, and the slope k are given as inputs to the blocks in a menu that appears when their icons are double clicked with the computer mouse. Two ramp generators and a step generator are used for producing the desired input acceleration $a_A(t)$ for part d, as shown in Fig. 2.32.

The figure generated by the animate_nbars block in the model shown in Fig. 2.31 is based on an x-y coordinate system centered at point A in Fig. 2.30, with the $+x$ axis parallel to the unit vector $\hat{x}$ in that figure and with the $+y$ axis at 90° counterclockwise from the $+x$ axis. This accounts for the four inputs to the Mux that precedes the animate_nbars block, which are

- The x and y coordinates of point A: $xA = 0$ and $yA = 0$
- The nondimensional x and y coordinates of point B: $xB/L = \sin\theta$ and $yB/L = -\cos\theta$

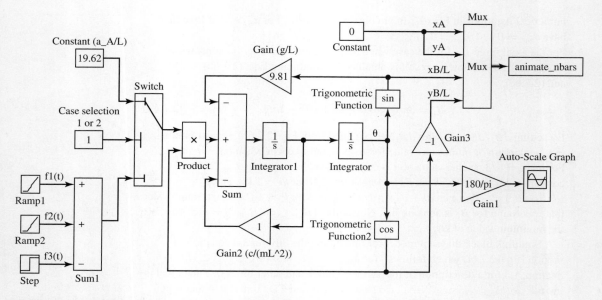

Figure 2.31 Simulation and animation of a pendulum on a moving base for Example 2.7.6.

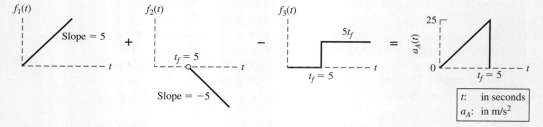

Figure 2.32 How to generate the acceleration input for part d in Example 2.7.6.

Therefore, the animation will show the pendulum as viewed by an observer located on the moving base at A. Double click the `animate_nbars` block if you want to change axes limits and/or the width of the bar in the animation.

The results of the computer simulations for $a_A = $ constant $= 2g$ m/s^2 (part c) and for the ramp input of 5 s duration and with slope $k = 5$ m/s^2, as specified in part d, (see Fig. 2.32 and the problem description), are shown in Fig. 2.33. For part c, a reading taken from Fig. 2.33a yields $\theta_{max} \approx 107°$, which occurs when $t \approx 0.7$ s. For part d, a reading from Fig. 2.33b yields $\theta_{max} \approx 68°$, which occurs when $t = 5$ s. These values can be accurately determined by zooming in near the desired location of the plot, which can be simply done by clicking the computer mouse on the zooming icon of the figure window and then selecting, also with the mouse, the desired area of the figure to zoom in. It may also be done by typing `zoom` in the MATLAB command window and then selecting the area to zoom in.

Figure 2.33 $\theta(t)$ versus t for (a) part c and (b) part d of Example 2.7.6.

EXAMPLE 2.7.7

Consider, again, the pendulum mounted on a moving base analyzed in Example 2.7.6. For $M_{bar} = 0$ and $\ddot{x} = \text{constant} = -a_A$:

a. Determine the equilibrium values for the angle θ in terms of g and a_A.
b. Linearize the differential equation of motion, Eq. (2.7.6d), about each equilibrium (in case there are more than one) determined in part a. For $a_A = 2g$, which equilibrium solutions are stable and which ones are unstable? Describe, in a few words, the motion $\theta(t)$ when it is started near the stable equilibrium.
c. Repeat part b for $a_A = 0$.

■ **Solution**

a. For convenience, the differential equation of motion for the system is repeated here.

$$L\ddot{\theta} + g\sin\theta - a_A\cos\theta = \frac{M_{bar}}{mL} = 0 \quad (2.7.7a)$$

The equilibrium values for the angle θ are, by definition, the constant values of $\theta \triangleq \theta_e$ that satisfy the differential equation of motion of the system. With $\theta = \text{constant} \triangleq \theta_e$, Eq. (2.7.7a) yields

$$g\sin\theta_e - a_A\cos\theta_e = 0 \quad (2.7.7b)$$

or, simply,

$$\theta_e = \arctan\frac{a_A}{g}$$

For a given value of a_A/g, this equation yields two physically distinct values for θ_e, with the difference between those values being equal to $180°$. Physically, if the motion were started with $\theta(0) = \theta_e$ and $\dot\theta(0) = 0$, the angle $\theta(t)$ would remain constant and equal to θ_e for $t > 0$. However, if the equilibrium were perturbed by making the initial conditions different from these values (even slightly different), a motion would result.

b. To linearize the differential equation about each equilibrium, introduce a small perturbation $\theta_s(t)$ as $\theta(t) = \theta_e + \theta_s(s)$, expand each nonlinear term in the differential equation in Taylor series in $\theta_s(t)$ (see Section 1.14), and disregard nonlinear terms in the expansion. Several examples of Taylor series expansion and linearization are given in Section 1.14, and now is the time to study that material if you have not done so yet.

The Taylor series expansions for $\sin\theta$ and $\cos\theta$ about $\theta = \theta_e$ are

$$\sin\theta = \sin\theta_e + \theta_s \cos\theta_e + \cdots \qquad (2.7.7c)$$

$$\cos\theta = \cos\theta_e - \theta_s \sin\theta_e + \cdots \qquad (2.7.7d)$$

By using these expansions, Eq. (2.7.7a) yields the following linearized differential equation for the perturbed motion $\theta_s(t)$.

$$\ddot\theta_s + \frac{g}{L}\left(\cos\theta_e + \frac{a_A}{g}\sin\theta_e\right)\theta_s = 0 \qquad (2.7.7e)$$

or, since $\tan\theta_e = a_A/g$,

$$\ddot\theta_s + \frac{g}{L\cos\theta_e}\theta_s = 0 \qquad (2.7.7f)$$

For $a_A = 2g$ we then have for each equilibrium:

for $\theta_e \approx 63.4°$:
$$\ddot\theta_s + 2.24\frac{g}{L}\theta_s = 0 \qquad (2.7.7g)$$

for $\theta_e \approx 243.4°$:
$$\ddot\theta_s - 2.24\frac{g}{L}\theta_s = 0 \qquad (2.7.7h)$$

Therefore, as discussed in Section 1.15, the equilibrium $\theta_e \approx 63.4°$ is infinitesimally stable and the equilibrium $\theta_e = 243.4°$ is unstable. It can be easily verified that the equilibrium $\theta_e \approx 63.4°$ is stable by numerically integrating Eq. (2.7.7a) with initial conditions near that equilibrium. This means that a motion $\theta(t)$ that is started near the equilibrium $\theta_e = 63.4°$ stays near that equilibrium as time passes. Such a motion consists of an undamped oscillation (as per the material that was first presented in Section 1.13) about that equilibrium, with frequency equal to $\omega = \sqrt{2.24g/L} \approx 1.49\sqrt{g/L}$ rad/s. In contrast, the angle $\theta(t)$ does not remain near the equilibrium $\theta_e = 243.4°$ even if it is started very close to that equilibrium.

Notice that the equilibrium $\theta_e = 63.4°$ becomes asymptotically stable (i.e., the motion dies out and returns to the equilibrium state as time increases) when viscous damping is added to the system. Equation (2.7.6i) (p. 125) is the differential equation of motion in such a case.

c. For $a_A = 0$, the equilibrium solutions are $\theta_e = 0$ and $\theta_e = 180°$. In this case, the linearized motion consists of small undamped oscillations only about the equilibrium $\theta_e = 0$, and the frequency of those small oscillations is $\omega = \sqrt{g/L}$ rad/s. The equilibrium $\theta_e = 180°$ is unstable.

To reinforce your learning of the material, use the Simulink model `pend_moving_base` to solve this problem numerically, and compare the results obtained from the numerical solution to the original nonlinear differential equation of motion with those obtained from the linearized equation, with and without viscous damping.

EXAMPLE 2.7.8

Analysis of the Motion of the Spring-Pendulum in Two Dimensions

The motion of the spring-pendulum shown in Fig. 2.34a takes place in a vertical plane. The x-y coordinate system centered at point O shown in Fig. 2.34 is only used, later on, in connection with a Simulink diagram that was prepared for this system.

The pendulum consists of a small body (modeled as a point mass P) of mass m attached to the end of a massless linear spring of unstressed length equal to L_u m and stiffness k N/m. The other end of the spring is connected to a fixed frictionless pin, which is an inertial point O. The only forces acting on P are the gravitational attraction of the earth and the reaction force applied by the spring. Figure 2.34b shows the free-body diagram for P.

a. Obtain the differential equations of motion for the system in terms of the polar coordinates $r = |\overrightarrow{OP}|$ and θ shown in Fig. 2.34.
b. As indicated throughout the book, a dynamical system is said to be in equilibrium when the variables that are used for describing its motion are constants. Therefore, for the case of this pendulum, equilibrium means $r = $ constant (let us call that constant r_e) and $\theta = $ constant (let us call that constant θ_e). Show that there are two physically distinct equilibrium solutions that satisfy the differential equations of motion and that they are the following two pairs of values for θ_e and r_e: $(\theta_e = 0, r_e = L_u + mg/k)$ and $(\theta_e = \pi, r_e = L_u - mg/k)$.

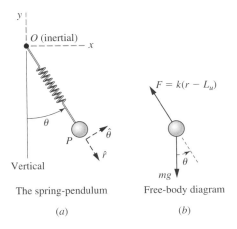

Figure 2.34 The spring-pendulum for Example 2.7.8.

c. Rewrite the differential equations obtained in part a in terms of the nondimensional variables $r_s \triangleq (r - r_e)/r_e$, which is the deviation distance from the equilibrium distance r_e (nondimensionalized by r_e) and the angle θ.
d. Linearize the differential equations of motion about the equilibrium ($\theta_e = 0$, $r_e = L_u + mg/k$), and determine the frequencies of the small oscillations of the spring-pendulum about that equilibrium (if needed, review Sections 1.14 and 1.13 for the material to do this).
e. Since the differential equations for this system are nonlinear and no exact analytical solution can be obtained for them, construct a block diagram to solve the differential equations in part c for $r_s(t)$ and $\theta(t)$. For this, use the nondimensional time τ defined as $\tau = t\sqrt{k/m}$. Integrate the differential equations with the initial conditions $r_s(0) = 0.05$, $\dot{r}_s(0) = 0$, $\theta(0) = r_s(0)/10$, $\dot{\theta}(0) = 0$, and plot $r_s(\tau)$ and $\theta(\tau)$ versus τ for $\sqrt{mg/(kr_e)} = 0.5$ and 0.3.

■ **Solution**

a. The free-body diagram for the mass particle P is shown in Fig. 2.34. For convenience, let us use the triad $\{\hat{r}, \hat{\theta}, \hat{z} \triangleq \hat{r} \times \hat{\theta}\}$, where the unit vectors $\hat{r}$ and $\hat{\theta}$ are shown in the figure. The absolute angular velocity of the triad is

$$\vec{\omega} = \dot{\theta}\hat{z} \tag{2.7.8a}$$

We then have for point P where the mass particle is located:

■ Absolute position vector of P:

$$\vec{r} = \overrightarrow{OP} = r\hat{r} \tag{2.7.8b}$$

■ Absolute velocity of P:

$$\vec{v} = \frac{d\vec{r}}{dt} = \dot{r}\hat{r} + \vec{\omega} \times \vec{r} = \dot{r}\hat{r} + r\dot{\theta}\hat{\theta} \tag{2.7.8c}$$

■ Absolute acceleration of P:

$$\vec{a} = \frac{d\vec{v}}{dt} = \ddot{r}\hat{r} + (r\ddot{\theta} + \dot{r}\dot{\theta})\hat{\theta} + \vec{\omega} \times \vec{v} = (\ddot{r} - r\dot{\theta}^2)\hat{r} + (r\ddot{\theta} + 2\dot{r}\dot{\theta})\hat{\theta} \tag{2.7.8d}$$

■ Resultant force acting on P:

$$\vec{F} = mg(\hat{r}\cos\theta - \hat{\theta}\sin\theta) - k(r - L_u)\hat{r} \tag{2.7.8e}$$

Newton's second law $\vec{F} = m\vec{a}$ then yields the following scalar equations for the spring-pendulum. These equations involve only the variables r and θ and their time derivatives. Therefore, they are the differential equations of motion for the system.

$$m(\ddot{r} - r\dot{\theta}^2) = mg\cos\theta - k(r - L_u) \tag{2.7.8f}$$

$$m(r\ddot{\theta} + 2\dot{r}\dot{\theta}) = -mg\sin\theta \tag{2.7.8g}$$

Note: Equation (2.7.8g) could also have been obtained by using the moment equation $\vec{M}_O = d\vec{H}_O/dt = d[\vec{r} \times (m\vec{v})]/dt$. This yields the following scalar equation, which is equivalent to Eq. (2.7.8g):

$$m\frac{d}{dt}(r^2\dot\theta) = -mgr\sin\theta$$

b. As indicated in the problem statement, the equilibrium solution corresponds to $r = \text{constant} \triangleq r_e$ and $\theta = \text{constant} \triangleq \theta_e$. The constant values r_e and θ_e have to satisfy the differential equations of motion. Therefore, by substituting this constant solution into Eqs. (2.7.8f) and (2.7.8g), the following equations are obtained for r_e and θ_e:

$$mg\cos\theta_e - k(r_e - L_u) = 0 \qquad (2.7.8\text{h})$$

$$mg\sin\theta_e = 0 \qquad (2.7.8\text{i})$$

The solutions to Eq. (2.7.8i) are $\theta_e = 0, \pi, 2\pi, 3\pi, \ldots$, but only two solutions $\theta_e = 0$ and $\theta_e = \pi$ are physically distinct.

For $\theta_e = 0$, the first equilibrium equation yields $r_e = L_u + mg/k$, and for $\theta_e = \pi$ we obtain $r_e = L_u - mg/k$. The second solution corresponds to an upside-down pendulum and will not be considered further. The equilibrium $\theta_e = \pi$ happens to be unstable.

c. Let us now start to investigate what the spring-pendulum does when a motion is started with some initial conditions. The main interest in practice is to find out what kind of motion results when an equilibrium of the system is perturbed. For this, it is always convenient to redefine the motion variables as

```
variable_name(t)=variable_nameₑ + new_variable_name(t)
```

where `variable_name` and `variable_nameₑ` represent, respectively, the time-dependent variables being used for describing the motion (for this problem, they are r and θ) and their constant equilibrium values. The equilibrium under consideration here is $r_e = L_u + mg/k$ and $\theta_e = 0$. The subscript e in this book stands for *equilibrium*. The time-dependent `new_variable_name` stands for any convenient symbol associated with the perturbation of the variable `variable_name`. For the problem being analyzed here we might write $r(t) = r_e + r_s(t)$, or use any other symbol for the perturbation $r(t) - r_e$ of the variable $r(t)$. Since $\theta_e = 0$, there is no need to introduce a new symbol to replace θ.

When integrating differential equations in dynamics, it is convenient to nondimensionalize all variables for the problem, including the time t (time nondimensionalization was first discussed in Section 1.13). This will actually reduce the number of parameters in a problem to a minimum. In this problem, in addition to the four initial conditions for the two second-order differential equations, Eqs. (2.7.8f) and (2.7.8g), the quantities m, k, L_u, and g appear in those equations. Nondimensionalization of r and time t can actually reduce these four quantities to only one nondimensional parameter for this problem, which, as you will see in the rest of the solution, is $kr_e/mg = 1 + kL_u/(mg)$.

Let us then proceed by redefining the variable $r(t)$ in terms of a nondimensional new variable $r_s(t)$:

$$r(t) = r_e[1 + r_s(t)] \tag{2.7.8j}$$

The physical interpretation of the variable $r_s(t)$ is simply the elongation of the spring-pendulum, measured from its equilibrium length r_e and nondimensionalized by r_e.

Substitution of Eq. (2.7.8j) into Eqs. (2.7.8f) and (2.7.8g) yields

$$\ddot{r}_s - (1 + r_s)\dot{\theta}^2 = \frac{g}{r_e}\cos\theta - \frac{k}{mr_e}(r_e + r_e r_s - L_u) = -\frac{g}{r_e}(1 - \cos\theta) - \frac{k}{m}r_s \tag{2.7.8k}$$

$$(1 + r_s)\ddot{\theta} + 2\dot{r}_s\dot{\theta} = -\frac{g}{r_e}\sin\theta \tag{2.7.8l}$$

Notice that the four quantities m, k, g, and L_u now appear only as two parameters, k/m and $g/r_e = 1/(L_u/g + m/k)$, in these differential equations. The number of parameters can be further reduced to only one by nondimensionalizing the time t. This is done in part e.

d. For the linearization process, r_s and θ are now assumed to remain small enough so that $\sin\theta \approx \theta$ and $\cos\theta \approx 1$, and products of all the higher-order small terms $r_s\ddot{\theta}$, $\dot{r}_s\dot{\theta}$, $\dot{\theta}^2$, and $r_s\dot{\theta}^2$ that appear in Eqs. (2.7.8k) and (2.7.8l) are neglected. The objective of doing this is to obtain the linearized differential equations, for which the solutions can be readily obtained (as shown in Case 3 in Section 1.13).

The differential equations that govern the linearized motions about the equilibrium that corresponds to $\theta_e = 0$ and $r_e = L_u + mg/k$ are then

$$\ddot{r}_s + \frac{k}{m}r_s = 0 \tag{2.7.8m}$$

$$\ddot{\theta} + \frac{g}{r_e}\theta = 0 \tag{2.7.8n}$$

According to the material presented in Case 3 of Section 1.13, these equations describe two independent undamped oscillations: a θ oscillation with frequency equal to $\sqrt{g/r_e}$, and an r_s oscillation with frequency equal to $\sqrt{k/m}$. In dynamics and vibrations, the frequencies of the linearized oscillations of a system are called the *natural frequencies* of oscillation. Linear oscillations are what engineers hope that systems will exhibit in practice because their character can be easily studied and predicted.

e. It is generally a good idea to nondimensionalize the time t in order to minimize the number of parameters one has to deal with when integrating the differential equations. With $\tau = t\sqrt{k/m}$, the time derivatives $\dot{r}_s$, $\ddot{r}_s$, $\dot{\theta}$, and $\ddot{\theta}$ are transformed as follows (using the chain rule)[10]:

$$\dot{r}_s = \frac{dr_s}{d\tau}\frac{d\tau}{dt} = \sqrt{\frac{k}{m}}\frac{dr_s}{d\tau} \qquad \ddot{r}_s = \frac{d\dot{r}_s}{dt} = \left[\frac{d}{d\tau}\left(\sqrt{\frac{k}{m}}\frac{dr_s}{d\tau}\right)\right]\frac{d\tau}{dt} = \frac{k}{m}\frac{d^2r_s}{d\tau^2}$$

$$\dot{\theta} = \sqrt{\frac{k}{m}}\frac{d\theta}{d\tau} \qquad \ddot{\theta} = \frac{k}{m}\frac{d^2\theta}{d\tau^2}$$

Therefore, Eqs. (2.7.8k) and (2.7.8l) become

$$\frac{d^2r_s}{d\tau^2} - (1+r_s)\left(\frac{d\theta}{d\tau}\right)^2 = -r_s - \frac{mg}{kr_e}(1-\cos\theta) \qquad (2.7.8o)$$

$$(1+r_s)\frac{d^2\theta}{d\tau^2} + 2\frac{dr_s}{d\tau}\frac{d\theta}{d\tau} = -\frac{mg}{kr_e}\sin\theta \qquad (2.7.8p)$$

Notice that now only one parameter, $mg/(kr_e)$, appears in the differential equations. This is a great advantage because the same response $r_s(\tau)$ and $\theta(\tau)$ obtained for any particular value of this parameter (which is the square of the ratio of the two natural frequencies of oscillation of the spring-pendulum) is exhibited for many spring-pendulum systems built with different springs and masses. One would have to know, though, the value of the time nondimensionalization factor $\sqrt{k/m}$ in order to relate $r_s(\tau)$ to $r_s(t)$ and $\theta(\tau)$ to $\theta(t)$ for each system.

⟹ Figure 2.35 shows a Simulink block diagram model for numerically integrating Eqs. (2.7.8o) and (2.7.8p). The model is the file spring_pendulum.mdl included on the enclosed CD. To use it, either type spring_pendulum in the MATLAB command window or select it by typing simulations_menu as indicated earlier and in Appendix A. Change the parameter $mg/(kr_e)$ by small amounts (using 0.48^2 and 0.52^2, for example) to investigate, on your own, how it affects the motion. You will see that as the value of $\sqrt{mg/(kr_e)}$ moves away from the most critical 0.5 value, the nonlinear phenomenon starts to disappear, until the motion is eventually very close to that modeled by the linearized differential equations (2.7.8m) and (2.7.8n).

The optional animate_nbars block included in the model draws a figure using the x-y coordinate system shown in Fig. 2.34 (p. 129). Its inputs, provided by a Mux block, are the x and y coordinates of points O and P, which are obtained by inspection of Fig. 2.34. The first two inputs are the $x = y = 0$ coordinates of point O, and the next two are the x and y coordinates of point P nondimensionalized by the equilibrium length r_e. By looking at Fig. 2.34, they are seen to be $x_P/r_e = (1+r_s)\sin\theta$ and $y_P/r_e = -(1+r_s)\cos\theta$. To add a "bar" with the same end points to imitate a point mass at the end of the spring, the next four inputs to that Mux block are the x and y coordinates of point P (i.e., $x = r\sin\theta$ and $y = -r\cos\theta$, as per Fig. 2.34) followed by these same quantities. Such an additional detail could, of course, be left out of the figure. Function blocks are used in the Simulink model

---

[10] The frequency of a linearized oscillation is generally a convenient time nondimensionalization factor to use. Thus, for this problem, we could also have used the nondimensional time $\tau = t\sqrt{g/r_e}$.

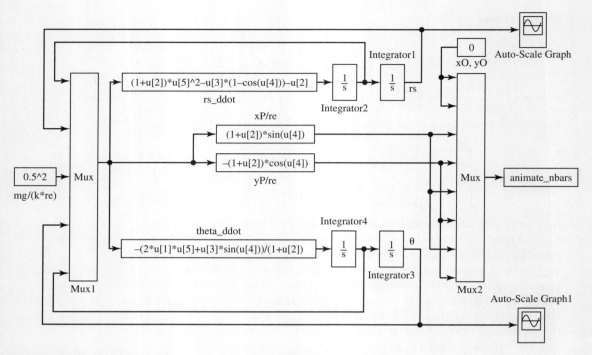

Figure 2.35 Simulation and animation of the spring-pendulum for Example 2.7.8.

shown in Fig. 2.35 in order to make the diagram more compact. This is only a matter of individual preference. The expressions for $d^2r_s/d\tau^2$ and $d^2\theta/d\tau^2$, given by Eqs. (2.7.8o) and (2.7.8p), are written in the `function blocks` labeled `rs_ddot` and `theta_ddot`, respectively, and their syntax are shown in Fig. 2.35.

Figures 2.36 and 2.37 show the results of the numerical integration of Eqs. (2.7.8o) and (2.7.8p) for $mg/(kr_e) = (0.5)^2$ and Fig. 2.38 for $mg/(kr_e) = (0.3)^2$. For Figs. 2.36 and 2.38 the integration was done with the initial conditions $r_s(0) = 0.1$, $\theta(0) = r_s(0)/10 = 0.01$, and $[dr_s/d\tau]_{\tau=0} = [d\theta/d\tau]_{\tau=0} = 0$. For Fig. 2.37 the initial conditions are $r_s(0) = 0.05$, $\theta(0) = r_s(0)/10 = 0.005$, and $[dr_s/d\tau]_{\tau=0} = [d\theta/d\tau]_{\tau=0} = 0$. Look at the `Simulation → Parameters` menu in your computer to see the settings that were used for the solver. Notice the beating phenomenon in Figs. 2.36 and 2.37, with the θ motion slowly growing to a value that is much larger than its initial value.

The beating phenomenon that is seen in Figs. 2.36 and 2.37 is a nonlinear phenomenon, and it is completely missed by the linearized differential equations $\ddot{r}_s + (k/m)r_s = 0$ and $\ddot{\theta} + (g/r_e)\theta = 0$. When such a nonlinear phenomenon occurs, the θ motion that is started with a very small value slowly grows to a much larger value that happens to be independent of the initial conditions in θ and depends only on the initial conditions in r_s. This is illustrated in Figs. 2.36 and 2.37.

SECTION 2.7 More Solved Example Problems

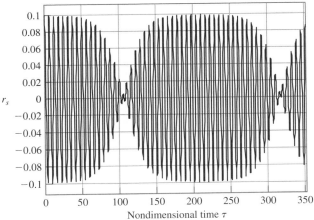

Figure 2.36 Results of the numerical integration of Eqs. (2.7.8o) and (2.7.8p) with $mg/(kr_e) = (\frac{1}{2})^2$, $r_s(0) = 0.1$, $\theta(0) = 0.01$, and $[dr_s/d\tau]_{\tau=0} = [d\theta/d\tau]_{\tau=0} = 0$ (Example 2.7.8).

The smaller the initial conditions in θ, the longer it takes for the angle θ to grow to $\theta_{\max} \approx 2r_s(0)$ when $mg/(kr_e) = (\frac{1}{2})^2$.

The nonlinear phenomenon seen in Figs. 2.36 and 2.37 is the phenomenon that can actually be observed with the experiment mentioned in Section 1.16. Such a phenomenon is known as a *nonlinear resonance phenomenon*. Since $r_e = L_u + mg/k$, the value of r_e when $mg/(kr_e) = (\frac{1}{2})^2$ is $r_e = 4L_u/3$, which is the value that was mentioned in Section 1.16.

Analytical investigations using *nonlinear perturbation techniques,* which are beyond the scope of this book, actually disclose that the nonlinear phenomenon previously mentioned occurs for a certain range of the parameter $\sqrt{mg/(kr_e)}$ (which is the ratio of the two natural frequencies of the spring-pendulum) in the

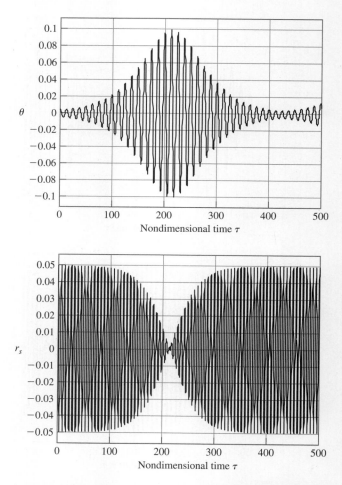

Figure 2.37 Results of the numerical integration of Eqs. (2.7.8o) and (2.7.8p) with $mg/(kr_e) = (\frac{1}{2})^2$, $r_s(0) = 0.05$, $\theta(0) = 0.005$, and $[dr_s/d\tau]_{\tau=0} = [d\theta/d\tau]_{\tau=0} = 0$ (Example 2.7.8).

neighborhood of $\sqrt{mg/(kr_e)} = \frac{1}{2}$ and disappears outside that range. The width of the nonlinear resonance range depends on the value of the initial condition $r_s(0)$. The response shown in Fig. 2.38 for $\sqrt{mg/(kr_e)} = 0.3$ does not exhibit the nonlinear resonance phenomenon that caused θ to slowly grow as seen in Figs. 2.36 and 2.37. The value of $mg/(kr_e) = 0.3^2$ happens to be outside the region for which the nonlinearities in the differential equations of motion dominate the dynamic behavior of the system. The r_s and θ motions shown in Fig. 2.38 are very close to the response predicted by the linearized differential equations. In practice, the nonlinear resonance phenomenon can be eliminated by either increasing or decreasing the mass at the end of the spring.

SECTION 2.7 More Solved Example Problems

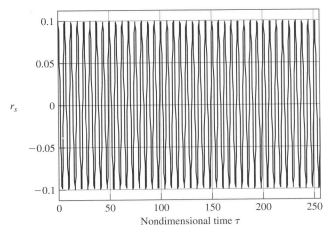

Figure 2.38 Results of the numerical integration of Eqs. (2.7.8o) and (2.7.8p) with $mg/(kr_e) = (0.3)^2$, $r_s(0) = 0.1$, $\theta(0) = 0.01$, and $[dr_s/d\tau]_{\tau=0} = [d\theta/d\tau]_{\tau=0} = 0$ (Example 2.7.8).

If you have not done so yet, try the experiment described in Section 1.16. You will probably appreciate it even more now, after having seen and studied the analysis presented in this example, and played with the computer simulation.

EXAMPLE 2.7.9

A particle P of mass $m = 0.5$ kg moves inside a 2-m horizontal thin tube filled with oil. The tube is always in contact with a fixed horizontal table and is driven by a motor with a constant clockwise angular velocity $\Omega = 5$ rad/s relative to inertial space. End O of the tube is inertial, and the particle is always in contact with a wall of the tube (see Fig. 2.39).

Because of the viscosity of the oil, P is subjected to a linear viscous friction force with friction coefficient $c = 1$ N/(m/s). If P is released from rest relative to the rotating tube (i.e., with zero velocity relative to the tube) at a distance $r_0 = 1$ m from point O, determine the time it takes for P to reach the edge of the tube at $r = 2$ m. Approximate the tube as massless.

■ Solution

By working with polar coordinates and the tube-fixed unit vectors $\hat{r}$ and $\hat{\theta}$ shown in Fig. 2.39 (whose absolute angular velocity is $\vec{\omega} = \dot{\theta}\hat{z} = \Omega\hat{z}$, where $\hat{z} = \hat{r} \times \hat{\theta}$), the absolute position vector $\vec{r}$, absolute velocity $\vec{v}$, and absolute acceleration $\vec{a}$ of the particle are (with $\Omega =$ constant, so that $\dot{\Omega} = 0$):

$$\vec{r} = r\hat{r} \tag{2.7.9a}$$

$$\vec{v} = \frac{d\vec{r}}{dt} = \dot{r}\hat{r} + \omega \times \vec{r} = \dot{r}\hat{r} + \Omega r \hat{\theta} \tag{2.7.9b}$$

$$\vec{a} = \frac{d\vec{v}}{dt} = \ddot{r}\hat{r} + \frac{d(\Omega r)}{dt}\hat{\theta} + \omega \times \vec{v} = (\ddot{r} - \Omega^2 r)\hat{r} + 2\Omega\dot{r}\hat{\theta} \tag{2.7.9c}$$

The free-body diagrams for the particle and for the tube are shown in Fig. 2.40. The forces that appear in the free-body diagram for the particle are

1. The gravitational force, equal to $-mg\hat{z}$.
2. The linear viscous friction force, which is proportional to the velocity $\dot{r}\hat{r}$ of the particle relative to the rotating tube and acts in the direction opposite to $\dot{r}\hat{r}$. It is assumed that this is the total friction force that is acting on the particle, even if the particle touches the wall of the tube. With c being the coefficient of viscous friction, that force is then $\vec{f} = -c\dot{r}\hat{r}$.
3. A force with components in the $\hat{\theta}$ and $\hat{z}$ directions due to the contact between P and the rotating tube. Those are the forces $N_\theta \hat{\theta}$ and $N_z \hat{z}$ shown in the free-body diagram.

The free-body diagram for the tube consists of the forces $-\vec{f} = c\dot{r}\hat{r}$, $-N_z\hat{z}$, $-N_\theta\hat{\theta}$, a force $N_{\text{table}}\hat{z}$ which is the resultant of the distributed force due to the contact between the tube and the table, and the support reaction forces $O_r\hat{r}$, $O_\theta\hat{\theta}$, and $O_z\hat{z}$ at point O. The moment $M\hat{z}$ applied by the motor to the tube is also shown in the diagram. It is assumed that there is no friction between the fixed table and the rotating tube (otherwise there would also be an additional force acting on the tube due to such a friction).

Figure 2.39 A constrained particle on a rotating tube for Example 2.7.9.

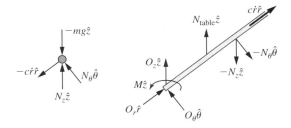

Figure 2.40 Free-body diagrams ($\hat{z} = \hat{r} \times \hat{\theta}$) for Example 2.7.9.

The vector sum of the forces acting on P is

$$\vec{F} = (N_z - mg)\hat{z} - c\dot{r}\hat{r} + N_\theta \hat{\theta} \qquad (2.7.9d)$$

By applying Newton's second law, $\vec{F} = m\vec{a}$, to the particle, the following scalar equations are then obtained.

$$\ddot{r} + \frac{c}{m}\dot{r} - \Omega^2 r = 0 \qquad (2.7.9e)$$

$$N_\theta = 2m\Omega\dot{r} \qquad (2.7.9f)$$

$$N_z = mg \qquad (2.7.9g)$$

Equation (2.7.9e) is the differential equation for the motion of P relative to the rotating tube since that equation does not involve any unknown reaction forces. That equation is a linear differential equation with constant coefficients, and its solution was first presented in Section 1.13. Equations (2.7.9f) and (2.7.9g) are simply algebraic equations that can be used for calculating, if desired, the components N_θ and N_z of the contact force between P and the tube. Notice that the calculation of N_θ requires the integration of the differential equation of motion first in order to determine $\dot{r}(t)$, which is needed in Eq. (2.7.9f) for finding that reaction force.

To determine the value of t for which $r = R = 2$ m, the differential equation of motion has to be solved for $r(t)$ first. The solution is given by either Eq. (1.13.11) or Eq. (1.13.13). With $m = 0.5$ kg, $c = 1$ N/(m/s), and $\Omega = 5$ rad/s, the two roots of the characteristic equation $s^2 + (c/m)s - \Omega^2 = 0$ are

$$s_1 = -\frac{c}{2m} + \sqrt{\Omega^2 + \left(\frac{c}{2m}\right)^2} \approx 4.1 \text{ s}^{-1}$$

$$s_2 = -\frac{c}{2m} - \sqrt{\Omega^2 + \left(\frac{c}{2m}\right)^2} \approx -6.1 \text{ s}^{-1}$$

Therefore, the general solution to Eq. (2.7.9e) is

$$r(t) = A_1 e^{s_1 t} + A_2 e^{s_2 t}$$

where A_1 and A_2 are constants determined from the initial conditions as

$$A_1 = \frac{\dot{r}_0 - s_2 r_0}{s_1 - s_2} \approx 0.6 \text{ m}$$

$$A_2 = \frac{s_1 r_0 - \dot{r}_0}{s_1 - s_2} \approx 0.4 \text{ m}$$

The value of the time t that corresponds to $r = R = 2$ m is obtained after a few numerical iterations with the earlier expression found for $r(t)$ as $t \approx 0.285$ s (≈ 0.29 s). Notice that, when $t = 0.285$ s, the tube has turned by $\Omega t = 1.425$ rad $\approx 81.6°$. It is interesting to notice that if there were no friction between P and the tube (i.e., if $c = 0$), one would obtain $r(t) = [\dot{r}_0(e^{\Omega t} - e^{-\Omega t}) + \Omega r_0(e^{\Omega t} + e^{-\Omega t})]/(2\Omega)$, and this would have yielded $\Omega t \approx 1.32$ rad $\approx 75.6°$ for $r = R = 2$ m. This shows, as one would have expected, that the friction force is responsible for keeping the particle on the tube for a longer time.

EXAMPLE 2.7.10

Two particles P_1 and P_2 of masses m_1 and m_2, respectively, are connected by a massless inextensible string of length L that passes through a small hole on a fixed frictionless horizontal table as shown in Fig. 2.41.

The x-y-z coordinate system centered at point O shown in Fig. 2.41 is only used, later on, in connection with a Simulink diagram that was prepared for this system. The motion of the system is started with the initial conditions $r(0) = r_0$, $\dot{r}(0) = \dot{r}_0$, $\theta(0)$ arbitrary, and $\dot{\theta}(0) = \dot{\theta}_0$. Particle P_2 is constrained to move along the vertical.

a. Obtain the differential equation of motion for particle P_1 in terms of the distance $r(t)$.
b. Determine the tension F_r in the string as a function of $r(t)$.
c. For $m_2 = 2m_1$, $L = 1$ m, $r(0) = 0.5$ m, $\dot{r}(0) = 1$ m/s, and $\dot{\theta}(0) = 2$ rad/s, what are the maximum and minimum values of $r(t)$ and of the tension in the string?

■ Solution

a. The free-body diagrams for both point masses are shown in Fig. 2.42. Orthogonal unit vectors $\hat{r}$, $\hat{\theta}$, and $\hat{z} = \hat{\theta} \times \hat{r}$ are also included in the figure. Line A-A, which coincides with the x axis in Fig. 2.41, is simply a nonrotating reference line that is

Figure 2.41 Dynamics of two constrained particles for Example 2.7.10.

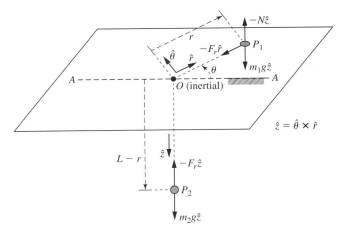

Figure 2.42 Free-body diagrams, motion variables, and unit vectors for Example 2.7.10.

used for measuring the angle $\theta(t)$ for the rotating line OP_1, while the force $-N\hat{z}$ is the normal reaction force applied by the table to particle P_1. The minus sign in $-N\hat{z}$ is associated with the unit vector $\hat{z}$. If the unit vector $\hat{z}$ were chosen to be pointing upward, instead of downward, one would write $N\hat{z}$ for the normal reaction force and $-m_1 g \hat{z}$ for the gravitational force acting on P_1. The resultant forces acting on P_1 and P_2 are then

Particle P_1: $\qquad \vec{F}_1 = -F_r \hat{r} + (m_1 g - N)\hat{z}$ (2.7.10a)

Particle P_2: $\qquad \vec{F}_2 = (m_2 g - F_r)\hat{z}$ (2.7.10b)

To obtain the expression for the absolute acceleration for each particle, start with the respective absolute position vectors which are $\vec{r}_1 = r\hat{r}$ (for particle P_1) and $\vec{r}_2 = (L - r)\hat{z}$ (for particle P_2). This gives the following expressions for the absolute velocity and for the absolute acceleration of both particles.

$$\vec{v}_1 = \frac{d\vec{r}_1}{dt} = \dot{r}\hat{r} + (-\dot{\theta}\hat{z}) \times \vec{r}_1 = \dot{r}\hat{r} + r\dot{\theta}\hat{\theta} \qquad (2.7.10c)$$

$$\vec{a}_1 = \frac{d\vec{v}_1}{dt} = \ddot{r}\hat{r} + (r\ddot{\theta} + \dot{r}\dot{\theta})\hat{\theta} + (-\dot{\theta}\hat{z}) \times \vec{v}_1$$
$$= (\ddot{r} - r\dot{\theta}^2)\hat{r} + (r\ddot{\theta} + 2\dot{r}\dot{\theta})\hat{\theta} \qquad (2.7.10d)$$

$$\vec{v}_2 = \frac{d\vec{r}_2}{dt} = -\dot{r}\hat{z} \qquad (2.7.10e)$$

$$\vec{a}_2 = \frac{d\vec{v}_2}{dt} = -\ddot{r}\hat{z} \qquad (2.7.10f)$$

By applying Newton's second law to each individual particle, the following equations are immediately obtained.

Particle P_1:

$$N = m_1 g \qquad (2.7.10\text{g})$$

$$m_1(\ddot{r} - r\dot{\theta}^2) = -F_r \qquad (2.7.10\text{h})$$

$$r\ddot{\theta} + 2\dot{r}\dot{\theta} \equiv \frac{1}{r}\frac{d}{dt}(r^2\dot{\theta}) = 0 \qquad (2.7.10\text{i})$$

Particle P_2:

$$m_2 \ddot{r} = F_r - m_2 g \qquad (2.7.10\text{j})$$

Equation (2.7.10i) is one of the two differential equations of motion for this system, and it can be readily integrated to yield

$$r^2\dot{\theta} = \text{constant} = r_0^2 \dot{\theta}_0 \qquad (2.7.10\text{k})$$

Equations (2.7.10h) and (2.7.10j) may be combined to eliminate the tension F_r to yield the second differential equation of motion for the system, which is

$$(m_1 + m_2)\ddot{r} - m_1 r \dot{\theta}^2 + m_2 g = 0 \qquad (2.7.10\text{l})$$

By making use of Eq. (2.7.10k) to eliminate $\dot{\theta}$ from this equation, we immediately obtain the differential equation that only involves the variable $r(t)$, which is

$$(m_1 + m_2)\ddot{r} - \frac{m_1 r_0^4 \dot{\theta}_0^2}{r^3} + m_2 g = 0 \qquad (2.7.10\text{m})$$

Notice that Eqs. (2.7.10k) and (2.7.10m) disclose that if the initial condition $\dot{\theta}_0$ is zero, i.e., if the initial absolute velocity of P_1 has no component along the normal to line OP_1, the angle $\theta(t)$ and the acceleration of both particles remain constant.

b. To determine the string tension F_r as a function of $r(t)$, Eqs. (2.7.10j) and (2.7.10m) are combined to obtain the following result:

$$F_r = \frac{m_1 m_2}{m_1 + m_2}\left(g + \frac{r_0^4 \dot{\theta}_0^2}{r^3}\right) \qquad (2.7.10\text{n})$$

Notice that inspection of Eq. (2.7.10n) discloses that the maximum value of F_r occurs when $r(t)$ is minimum (and vice versa).

c. Let us now determine specific values for the maximum and minimum values of $r(t)$ and the tension F_r. The extremum (i.e., maximum and minimum) values for $r(t)$ may be obtained by first making use of the chain rule of differentiation to write $\ddot{r} = (d\dot{r}/dr)\dot{r}$, and then integrating Eq. (2.7.10m) once with respect to r, according to the procedure presented in Case 4 of Section 1.13, which is used throughout the book. This yields

$$\frac{1}{2}(m_1 + m_2)\dot{r}^2 + \frac{m_1 r_0^4 \dot{\theta}_0^2}{2r^2} + m_2 g r = \text{constant} = \frac{1}{2}(m_1 + m_2)\dot{r}_0^2 + \frac{m_1 r_0^2 \dot{\theta}_0^2}{2} + m_2 g r_0$$

$$(2.7.10\text{o})$$

By letting r_m denote the extremum values of $r(t)$, the values of r_m are obtained by setting $\dot{r} = 0$ in Eq. (2.7.10o). For $m_2 = 2m_1$, $r(0) = 0.5$ m, $\dot{r}(0) = 1$ m/s, and $\dot{\theta}(0) = 2$ rad/s, this yields the following cubic equation for r_m:

$$gr_m^3 - \left(1 + \frac{g}{2}\right)r_m^2 + \frac{1}{16} = 0$$

The three roots of this cubic equation, determined with $g = 9.81$ m/s², are

$$r_m \approx 0.583 \qquad r_m \approx -0.096 \qquad r_m \approx 0.114$$

These roots may be found with a handheld calculator without too much work. Noticing that with $r_m = 0$ the left-hand side of this equation is equal to $\frac{1}{16} = 0.0625$, it is reasonable to expect one root to be close to such a value. With $g = 9.81$ m/s², a root may be found with just a few iterations as $r_m \approx 0.114$. The other two roots may be found either in a similar manner or by factoring out the root $r_m \approx 0.114$ and then solving the resulting quadratic equation to obtain them. If you wish to do so, they may also be readily determined using MATLAB, which has a `roots` command; its syntax for determining the roots of the nth-degree polynomial equation $c_n x^n + c_{n-1}x^{n-1} + \cdots + c_1 x + c_0 = 0$ is `roots([c_n, c_{n-1},..., c_1, c_0])`. For our equation, one would then use `roots([9.81,-(1+9.81/2),0,1/16])`, and the three roots would be displayed on the computer screen.

Since $r(t)$ cannot be negative, the maximum and minimum values of $r(t)$ are then

$$r_{\max} \approx 0.583 \text{ m} \qquad r_{\min} \approx 0.114 \text{ m}$$

Therefore, according to Eq. (2.7.10n), the maximum and minimum values of the tension F_r are (with m_1 in kilograms):

$$(F_r)_{\max} \approx 119 m_1 \text{ N} \qquad (F_r)_{\min} \approx 7.38 m_1 \text{ N}$$

Figure 2.43 shows a Simulink model for this system. The block `animate3d_nbars`, described in Appendix A (p. 521), generates a three-dimensional animation of the motion. For this, the string $P_1 O P_2$ in Fig. 2.41 is represented by two bars, namely, OP_1 and OP_2, and the coordinates of the two ends of each bar, using the x-y-z coordinate system shown in that figure, are then input to the animation block. The nondimensional variable r/L is used in the model. To make the diagram more compact, a `MATLAB Function` block is used for inputting the six inputs for each bar to the `animate3d_nbars` block. According to Fig. 2.41, those inputs are then

$$0, 0, 0, \frac{r}{L}\cos\theta, \frac{r}{L}\sin\theta, 0, \quad 0, 0, 0, 0, 0, -\left(1 - \frac{r}{L}\right)$$

The first three of these inputs are the x-y-z coordinates of point O, and the next three are the x-y-z coordinates of point P_1; this will draw line OP_1. This is followed by the x-y-z coordinates of point O again, and then by the x-y-z coordinates of point P_2, thus causing line OP_2 to be drawn. The syntax of this input in the MATLAB

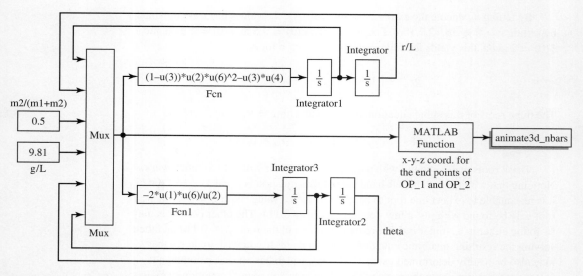

Figure 2.43 A Simulink model for the system of Example 2.7.10, with a block that shows three-dimensional animation.

⟹

Function block that precedes the animate3d_nbars is (double click the block to see it)

[0,0,0,u(2)*cos(u(5)),u(2)*sin(u(5)),0,0,0,0,0,0,u(2)-1]

The model shown in Fig. 2.43 is the file two_particles_table.mdl that is on the enclosed CD. To use it, either type two_particles_table in the MATLAB command window, or type simulations_menu and choose the two_particles_table model from a pull-down menu that appears on the computer screen. Double click the animate3d_nbars block to change the axes limits and the width of each bar, if desired. Look at the Simulation → Parameters menu on your computer to see the default settings for the simulation, and change the maximum step size of the integration scheme if the simulation runs too fast or too slow on your computer. It is suggested that you add logic blocks to the model to make the simulation stop if either P_1 or P_2 hits point O.

EXAMPLE 2.7.11 The Central Force Problem, for Modeling the Dynamics of Satellites

In Example 2.5.11 (p. 107), it was seen that the translational motion of a satellite P subjected only to the central gravitational attraction of a spherical planet takes place in a single plane. Let us now look at that problem again, also by neglecting the effect of any other force, including forces due to contact with the atmosphere. Compared to the dimensions of the earth, the satellite is modeled as a particle P of mass m and it is subjected to the central gravitational attraction of the earth, where the earth is modeled as an inertial homogeneous sphere of radius $R_E \approx 6378$ km and mass M_E. The free-body diagram for the satellite and the polar coordinates r and θ (where r is the distance from the planet's center at O to P and θ is measured from a fixed reference line in the plane of the motion) that are chosen to describe the motion are shown in Fig. 2.44. Two rotating unit vectors $\hat{r}$ and $\hat{\theta}$ are also shown in that figure. They will be used again, since it has been shown that they are very convenient for formulating the expression for the absolute acceleration of P, which is needed for obtaining the differential equations of motion for the system using Newton's second law.

a. Obtain the differential equations of motion for P in terms of the variables r and θ.
b. Determine the maximum and minimum values of the distance r when the motion of P is started with the initial conditions $r(0) = r_0 = 6578$ km, $\dot{r}(0) = 0$, and $v(0) = v_0 = 28{,}000$ km/h, where $\vec{v} = d\vec{r}/dt$ is the absolute velocity of P. Will the satellite crash on the earth's surface?

■ **Solution**

a. Since $M_E \gg m$, the effect of the mutual gravitational force of attraction $F = GM_E m/r^2$ (see Fig. 2.44) on the motion of the earth is extremely small and can certainly be neglected. If the motion of the earth in space is not considered either, point O in Fig. 2.44 is inertial, and we then have

■ Absolute position vector for particle P:
$$\vec{r} = \overrightarrow{OP} = r\hat{r}$$

■ Absolute velocity of P:
$$\vec{v} = \frac{d\vec{r}}{dt} = \dot{r}\hat{r} + r\frac{d\hat{r}}{dt} = \dot{r}\hat{r} + r\dot{\theta}\hat{\theta}$$

■ Absolute acceleration of P:
$$\vec{a} = \frac{d\vec{v}}{dt} = \ddot{r}\hat{r} + (r\ddot{\theta} + \dot{r}\dot{\theta})\hat{\theta} + \dot{r}\frac{d\hat{r}}{dt} + r\dot{\theta}\frac{d\hat{\theta}}{dt} = (\ddot{r} - r\dot{\theta}^2)\hat{r} + (r\ddot{\theta} + 2\dot{r}\dot{\theta})\hat{\theta}$$

■ Resultant force acting on P:
$$\vec{F} = -\frac{GM_E m}{r^2}\hat{r} = -\frac{mgR_E^2}{r^2}\hat{r}$$

Figure 2.44 Motion coordinates and free-body diagram for Example 2.7.11.

Newton's second law $\vec{F} = m\vec{a}$ then yields the following scalar differential equations of motion for the satellite:

$$\ddot{r} - r\dot{\theta}^2 = -\frac{gR_E^2}{r^2} \qquad (2.7.11a)$$

$$r\ddot{\theta} + 2\dot{r}\dot{\theta} = \frac{1}{r}\frac{d}{dt}(r^2\dot{\theta}) = 0 \qquad (2.7.11b)$$

b. Equation (2.7.11b) can be immediately integrated to yield

$$r^2\dot{\theta} = \text{constant} = [r^2\dot{\theta}]_{t=0} = r^2(0)\dot{\theta}(0)$$

Since $\dot{r}(0) = 0$, as stated in item (b), and $\vec{v} = \dot{r}\hat{r} + r\dot{\theta}\hat{\theta}$ for all times, it follows that $r(0)\dot{\theta}(0) = v_0$. Therefore,

$$r^2\dot{\theta} = r_0 v_0$$

This equation gives us an expression for $\dot{\theta}$ in terms of the distance r. By substituting that expression in Eq. (2.7.11a), the following differential equation that only involves the variable $r(t)$ is obtained.

$$\ddot{r} - \frac{r_0^2 v_0^2}{r^3} + \frac{gR_E^2}{r^2} = 0 \qquad (2.7.11c)$$

Since $\ddot{r} = d\dot{r}/dt = \dot{r}(d\dot{r}/dr)$, this differential equation can then be integrated once (as first shown in Case 4 in Section 1.13) as

$$\int_{\dot{r}=\dot{r}(0)}^{\dot{r}} \dot{r}\frac{d\dot{r}}{dr}\,dr - r_0^2 v_0^2 \int_{r=r_0}^{r} \frac{dr}{r^3} + gR_E^2 \int_{r=r_0}^{r} \frac{dr}{r^2} = 0$$

or

$$\frac{\dot{r}^2}{2} + \frac{r_0^2 v_0^2}{2r^2} - \frac{gR_E^2}{r} = \frac{v_0^2}{2} - \frac{gR_E^2}{r_0} \qquad (2.7.11d)$$

The maximum and the minimum values of the distance r are obtained by setting $\dot{r} = 0$ in this equation. Denoting such values by r_m, this results in a quadratic equation for r_m. Since $\dot{r}(0) = 0$, one of the solutions is $r_m = r_0 = 6578$ km. The other solution is found to be $r_m \approx 6541$ km. Therefore, $r_{\min} = 6541$ km and $r_{\max} = 6578$ km. Since $r_{\min} > R_E$, the satellite does not crash on the earth's surface.

Note: The central force problem of Example 2.7.11 is a particular case of the two-body problem of space mechanics, which is analyzed in detail in Chapter 4. A Simulink model for that problem is on the CD that accompanies the book. It is the file `two_body_simul.mdl`. However, to fully appreciate and understand the problem, it is advisable that you wait until you reach Chapter 4 and study it before you use that Simulink model.

SECTION 2.7 More Solved Example Problems

Spring-Pendulum in Three-Dimensional Space
EXAMPLE 2.7.12

The support point O of the spring-pendulum shown in Fig. 2.45 is inertial, and the orthogonal x, y, and z axes shown in that figure are fixed to the earth which, again, is taken to be inertial.

a. Obtain the expression for the absolute velocity $\vec{v}_P$ and absolute acceleration $\vec{a}_P$ of the mass particle P in terms of the spherical coordinates $r(t) = |\overrightarrow{OP}|$, $\theta(t)$, and $\phi(t)$. Express the answers in the unit vector triad $\{\hat{r}, \hat{\phi}, \hat{\theta} \triangleq \hat{r} \times \hat{\phi}\}$ shown in the figure, where $\hat{\phi}$ is perpendicular to the plane OPQ. Line OQ is the projection of OP on the x-y plane (see Fig. 2.45).
b. What are the differential equations of motion for P if the spring is linear, with unstressed length L_u and stiffness k N/m, and the only external force acting on the system is due to gravity?
c. Show that the motion of P takes place in a vertical plane if it is started with $\dot{\phi} = 0$ (i.e., if it is started with a velocity that has no component along $\hat{\phi}$).

■ **Solution**

a. From Fig. 2.45, it is seen that the absolute position vector of point P is $\vec{r}_P = \overrightarrow{OP} = r\hat{r}$ while the absolute angular velocity $\vec{\omega}$ of the $\{\hat{r}, \hat{\phi}, \hat{\theta}\}$ triad is

$$\vec{\omega} = -\dot{\theta}\hat{\phi} - \dot{\phi}\hat{z} = -\dot{\theta}\hat{\phi} - \dot{\phi}(\hat{r}\cos\theta - \hat{\theta}\sin\theta) \quad (2.7.12a)$$

The absolute velocity and acceleration of P are then obtained as

$$\vec{v}_P = \frac{d}{dt}\vec{r}_P = \dot{r}\hat{r} + \vec{\omega} \times \vec{r}_P = \dot{r}\hat{r} + r\dot{\theta}\hat{\theta} + (r\sin\theta)\dot{\phi}\,\hat{\phi}$$

$$\vec{a}_P = \frac{d}{dt}\vec{v}_P = [\ddot{r}\hat{r} + (r\ddot{\theta} + \dot{r}\dot{\theta})\hat{\theta} + (r\ddot{\phi}\sin\theta + \dot{r}\dot{\phi}\sin\theta + r\dot{\phi}\dot{\theta}\cos\theta)\hat{\phi}] + \vec{\omega} \times \vec{v}_P$$

$$= (\ddot{r} - r\dot{\theta}^2 - r\dot{\phi}^2\sin^2\theta)\hat{r} + [r\ddot{\theta} + 2\dot{r}\dot{\theta} - r\dot{\phi}^2(\sin\theta)\cos\theta]\hat{\theta}$$
$$+ [r\ddot{\phi}\sin\theta + 2\dot{r}\dot{\phi}\sin\theta + 2r\dot{\phi}\dot{\theta}\cos\theta]\hat{\phi} \quad (2.7.12b)$$

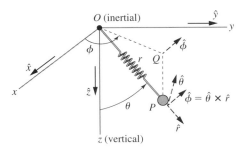

Figure 2.45 The spring-pendulum in three-dimensional space.

Notice the following properties about the expression for $\vec{v}_P$:
- The first term $\dot{r}\hat{r}$ is simply the velocity of P relative to O as if the angles θ and ϕ were constant.
- The second term $r\dot{\theta}\hat{\theta}$ is the velocity of P relative to O as if r and ϕ were constant.
- The third and last term $(r\sin\theta)\dot{\phi}\hat{\phi}$ is the velocity of P relative to O as if r and θ were held constant. If only the second term were present, with the others being zero, the motion of P relative to O would be constrained to an arc of a circle centered at O and with radius r. The third term expresses the fact that P would simply move on a circle centered on the vertical z axis and of radius $r\sin\theta$ due to the angular velocity $\dot{\phi}$ about that vertical axis.

One could now make use of the knowledge gained by these observations to write the expression for the velocity $\vec{v}_P$ by inspection from a figure for similar future situations.

It is also useful to notice that the $\hat{\phi}$ component of the acceleration $\vec{a}_P$ can be expressed in a simpler form as

$$\frac{1}{r\sin\theta}\frac{d}{dt}(r^2\dot{\phi}\sin^2\theta)$$

Such an observation is very useful as it will allow us to immediately integrate one of the differential equations of motion for this system when the moment, about O, of the resultant force acting on the particle located at P has no component in the $\hat{\phi}$ direction.

b. The free-body diagram for the mass particle at P is the same as the one shown in Fig. 2.34 (p. 129). Therefore, with the resultant force $\vec{F}$ acting on P being equal to

$$\vec{F} = mg\hat{z} - k(r - L_u)\hat{r} = mg(\hat{r}\cos\theta - \hat{\theta}\sin\theta) - k(r - L_u)\hat{r}$$

the following differential equations of motion are obtained from Newton's second law, $\vec{F} = m\vec{a}_P$.

$$m(\ddot{r} - r\dot{\theta}^2 - r\dot{\phi}^2\sin^2\theta) = mg\cos\theta - k(r - L_u) \quad \text{(2.7.12c)}$$

$$r\ddot{\theta} + 2\dot{r}\dot{\theta} - r\dot{\phi}^2(\sin\theta)\cos\theta = -g\sin\theta \quad \text{(2.7.12d)}$$

$$r\ddot{\phi}\sin\theta + 2\dot{r}\dot{\phi}\sin\theta + 2r\dot{\phi}\dot{\theta}\cos\theta = 0 \quad \text{(2.7.12e)}$$

c. To show that the motion of P takes place in a vertical plane if it is started with $\dot{\phi} = 0$, one manipulates Eq. (2.7.12e). That equation is the same as

$$\frac{1}{r\sin\theta}\frac{d}{dt}(r^2\dot{\phi}\sin^2\theta) = 0$$

or, simply,

$$r^2\dot{\phi}\sin^2\theta = \text{constant} = r^2(0)\dot{\phi}(0)\sin^2\theta(0)$$

If the motion is started with $\dot{\phi} = 0$ (i.e., with a velocity that has no component along $\hat{\phi}$), this equation yields $\phi = $ constant, since r and θ cannot be zero for all times. This means that the motion of P takes place in the same vertical plane where it was started. Such a motion was investigated in Example 2.7.8 (p. 129).

EXAMPLE 2.7.13

A point P of mass m slides on a circular frictionless groove of radius R. The groove is rotating with a known absolute angular velocity $\dot{\phi} \triangleq \Omega(t)$ about an axis AA that passes through its center C as shown in Fig. 2.46. The center C of the groove is inertial, and axis AA is vertical. The unit vector triad $\{\hat{r}, \hat{\theta}, \hat{n} \triangleq \hat{r} \times \hat{\theta}\}$ shown in the figure will be used in the solution to the problem.

a. Obtain the expressions for the absolute velocity and absolute acceleration of P by working with the variable θ, which is the angle that specifies the location of the particle on the rotating groove.
b. Obtain the expressions for the velocity and for the acceleration of P relative to the groove.
c. What is the differential equation that governs the motion of P, and what are the equations for determining the reaction forces applied on P by the groove? Explain how to determine those reactions.

■ Solution

a. In terms of the rotating triad $\{\hat{r}, \hat{\theta}, \hat{n} \triangleq \hat{r} \times \hat{\theta}\}$ shown in Fig. 2.46, the absolute position vector of P is simply $\vec{r}_P = \overrightarrow{CP} = R\hat{r}$. Since the direction of the axis AA from which the angle θ is measured is fixed in inertial space, the absolute angular velocity $\vec{\omega}$ of the $\{\hat{r}, \hat{\theta}, \hat{n}\}$ triad is equal to (see Fig. 2.46)

$$\vec{\omega} = \Omega \hat{z} + \dot{\theta} \hat{n} = \Omega(\hat{r} \cos\theta - \hat{\theta} \sin\theta) + \dot{\theta}\hat{n} \quad (2.7.13a)$$

Therefore, the absolute velocity and the absolute acceleration of P are

$$\vec{v}_P = \frac{d}{dt}\vec{r}_P = \underbrace{\dot{R}\hat{r}}_{=0} + \vec{\omega} \times \vec{r}_P = R\dot{\theta}\hat{\theta} + (\Omega R \sin\theta)\hat{n} \quad (2.7.13b)$$

$$\vec{a}_P = \frac{d}{dt}\vec{v}_P = R\ddot{\theta}\hat{\theta} + (\dot{\Omega}R\sin\theta + \Omega R\dot{\theta}\cos\theta)\hat{n} + \vec{\omega} \times \vec{v}_P$$
$$= R[\ddot{\theta} - \Omega^2(\sin\theta)\cos\theta]\hat{\theta} - R[\dot{\theta}^2 + \Omega^2 \sin^2\theta]\hat{r}$$
$$+ R[\dot{\Omega}\sin\theta + 2\Omega\dot{\theta}\cos\theta]\hat{n} \quad (2.7.13c)$$

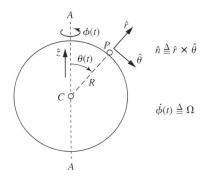

Figure 2.46 A particle moving on a rotating ring.

b. The velocity and acceleration of P relative to the rotating groove are obtained by simply freezing the rotation of the groove. Therefore, the following results are obtained directly from Eqs. (2.7.13b) and (2.7.13c).

$$\vec{v}_{P/\text{groove}} = (\vec{v}_P)_{\Omega=0} = R\dot{\theta}\hat{\theta} \tag{2.7.13d}$$

$$\vec{a}_{P/\text{groove}} = = (\vec{a}_P)_{\Omega=0} = R\ddot{\theta}\hat{\theta} - R\dot{\theta}^2\hat{r} \tag{2.7.13e}$$

Notice that the expression for the absolute velocity of P, given by Eq. (2.7.13b), is the sum of the velocity of P relative to the groove, which is the term $R\dot{\theta}\hat{\theta}$, and the absolute velocity of the point on the groove that coincides with P, which is the term $R\Omega(\sin\theta)\hat{n}$. The path of that point on the groove is simply a circle of radius $R\sin\theta$, perpendicular to axis AA.

c. The resultant $\vec{F}$ of all forces acting on particle P is the vector sum of the forces shown in Fig. 2.47. The force $\vec{f} \triangleq f\hat{n}$ is an unknown lateral reaction force that keeps P on the groove and that maintains $\dot{\phi}$ equal to the prescribed function $\Omega(t)$.

By inspecting Fig. 2.47, the following expression is obtained for $\vec{F}$.

$$\vec{F} = N\hat{r} + mg(\hat{\theta}\sin\theta - \hat{r}\cos\theta) + f\hat{n} \tag{2.7.13f}$$

Therefore, Newton's second law, $\vec{F} = m\vec{a}_P$, yields the following three equations that involve the three unknown variables θ, N, and f.

$$R[\ddot{\theta} - \Omega^2(\sin\theta)\cos\theta] = g\sin\theta \tag{2.7.13g}$$

$$N - mg\cos\theta = -mR(\dot{\theta}^2 + \Omega^2\sin^2\theta) \tag{2.7.13h}$$

$$f = mR(\dot{\Omega}\sin\theta + 2\Omega\dot{\theta}\cos\theta) \tag{2.7.13i}$$

Equation (2.7.13g), whose only unknown is $\theta(t)$, is the differential equation that governs the motion of P. It is a nonlinear differential equation because the terms $\sin\theta$ and $(\sin\theta)\cos\theta$ are nonlinear in θ.

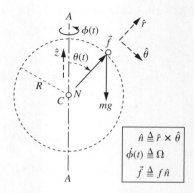

Figure 2.47 Free-body diagram for a particle moving on a rotating circular groove.

If one needs to determine the reaction forces N and f, Eqs. (2.7.13h) and (2.7.13i) are the ones that allow that to be done. To do so, one integrates Eq. (2.7.13g) numerically (because an analytical solution cannot be obtained for that nonlinear differential equation) to obtain $\theta(t)$, and then uses the result of that integration to calculate N and f from Eqs. (2.7.13h) and (2.7.13i). This can easily be done with Simulink. For more on this example, see Problems 2.33 to 2.35, and 2.41 in Section 2.9.

2.8 A Work-Energy Approach Obtained from Newton's Second Law

In some examples presented in this chapter, the fact that some of the second-order differential equations of motion could be integrated once (by making use of the chain rule of differentiation to prepare the differential equation to be integrated once) played a significant role in the analysis of the motion. As seen in Case 4 in Section 1.13, that integration was possible for differential equations of the form $\ddot{x} + f(x) = 0$, for which one obtains $\dot{x}^2/2 + \int f(x)\,dx =$ constant. The relatively trivial case for which the differential equation is of the form $\ddot{x} + f(t) = 0$ is excluded from this discussion since it can be readily solved to obtain $x(t)$. The existence of such *constants of the motion* (i.e., functions of the motion variable, and of its first time derivative, that remain constant during the motion while the variable, itself, is changing with time), or *integrals of motion* as they are more often referred to in dynamics, made it possible to reduce the problem of dealing with a second-order differential equation to one dealing with a first-order differential equation. The integrals of motion also made it possible to obtain important information about the motion of the system without even having to find a solution for the motion variable in terms of time t, which is a task that is often impossible to perform analytically.

Some examples presented in this chapter dealt with two-degree-of-freedom systems, i.e., systems that involve two independent variables that change with time in a manner that becomes known only after integrating two differential equations of motion. Under certain conditions, it is also possible to find integrals of motion for such systems and for systems involving more than two degrees of freedom. This can be done by manipulating the second-order differential equations with the chain rule, but that would, in general, involve a substantial amount of hard work.

At the end of Example 2.5.9 (p. 101) you were asked to verify that a certain function was an integral of the motion for that problem when there was no damping in the system. You might have asked yourself, How was that function found? The answer to such a question is provided in this section, in a simplistic manner suited to introductory or to many intermediate engineering dynamics courses, using the material presented in Section 1.13. It is recommended that you go back to that section and study it again any time you feel it is necessary for you to do so while studying the material that is presented in this section. A powerful and general procedure is presented in Chapter 7, which includes more advanced material.

In this section it is shown that, under certain conditions, an integrated form of the second-order differential equations of motion that results from direct application of Newton's second law, $\vec{F} = d(m\vec{v})/dt$, may readily be obtained without even having to generate those second-order differential equations. For the sake of generality, this is done here for the case of the motion of N point masses.

Consider N mass particles P_i ($i = 1, 2, \ldots, N$), each of mass m_i, subjected to several forces whose resultant is $\vec{F}_i$. Let $\vec{r}_i(t)$ be the absolute position vector for particle P_i. Now define the quantity called the "infinitesimal work done by $\vec{F}$ during an infinitesimal displacement $d\vec{r}$ of P" (or, simply, the *infinitesimal work* done by $\vec{F}$) as being the following scalar quantity dW obtained from the following dot product:

$$dW \triangleq \vec{F}_1 \bullet d\vec{r}_1 + \vec{F}_2 \bullet d\vec{r}_2 + \cdots + \vec{F}_N \bullet d\vec{r}_N \qquad (2.8.1)$$

The determination of the expression for dW as indicated here is, in general, a simple task to perform because it only involves a summation and dot products of vectors. This is illustrated in the examples at the end of this section. To obtain the expression for dW you need to write an expression for the absolute position vector $\vec{r}_i$ for each particle i (where a force is applied) in terms of a set of variables that you choose to locate each particle. For any particular problem, there will be some number n of such variables. For the motion of a single particle, for example, $n \leq 3$, with $n = 3$ being the case of unconstrained motion in three-dimensional space. By letting $q_1, q_2, \ldots, q_n$ denote those n variables, the result of the dot product operation given by Eq. (2.8.1) will always be an expression of the form shown in Eq. (2.8.2) (see the examples at the end of this section).

$$dW = \sum_{i=1}^{N} \vec{F}_i \bullet d\vec{r}_i = Q_{q_1}(q_1, q_2, \ldots, q_n, \dot{q}_1, \dot{q}_2, \ldots, \dot{q}_n; t)\, dq_1 + \cdots$$
$$+ Q_{q_n}(q_1, q_2, \ldots, q_n, \dot{q}_1, \dot{q}_2, \ldots, \dot{q}_n; t)\, dq_n$$
$$+ a_0(q_1, q_2, \ldots, q_n, \dot{q}_1, \dot{q}_2, \ldots, \dot{q}_n; t)\, dt \qquad (2.8.2)$$

The coefficients $Q_{q_1}, Q_{q_2}, \ldots, Q_{q_n}$ of each infinitesimal displacement dq_1, $dq_2, \ldots, dq_n$ are, in general, functions of the motion variables $q_1, q_2, \ldots, q_n$ and of their time derivatives $\dot{q}_1, \dot{q}_2, \ldots, \dot{q}_n$, and may be even an "explicit" function of the time t (see the following explanation). Time derivatives $\dot{q}_1, \dot{q}_2, \ldots, \dot{q}_n$ will appear in the expression obtained for dW if there are forces that are velocity dependent, such as friction forces.

The appearance of the independent variable t in a function such as $f_1(q_1, q_2; t) = q_1 q_2^2 \sin(t)$, where q_1 and q_2 are unknown functions of t, is said to be *explicit*. The unknown functions $q_1(t)$ and $q_2(t)$ in this example are said to be *implicit* functions of t. The semicolon used here in $f_1(q_1, q_2; t) = q_1 q_2^2 \sin(t)$, and in Eq. (2.8.2), is a common notation that is used for distinguishing dependent variables (such as q_1 and q_2 in this example) from the independent variable t.

SECTION 2.8 A Work-Energy Approach Obtained from Newton's Second Law

As illustrated in the examples at the end of this section, a term of the form $a_0\, dt$ may appear in the expression obtained for dW when one or more of the variables $q_1, q_2, \ldots, q_n$ is a prescribed (i.e., known) function of time.

In general, the expression that is obtained for the infinitesimal work dW *is not* the exact differential of any function W, and, thus, it should be viewed simply as an infinitesimal quantity. When necessary for the sake of clarity, an infinitesimal quantity dW that is an exact differential is written in this book as $d(W)$ and should be read as "d of W"; otherwise it is written as dW and should be read as "d W." The test to determine whether an infinitesimal quantity dW is an exact differential $d(W)$ was presented in Section 1.12. You may want to study that material now, if you feel you are not too familiar with it.

Some of the forces that are applied to any particle P_i may be perpendicular to the infinitesimal displacement $d\vec{r}_i$ for that particle. If that happens, there is no need to worry about those forces at all because the dot product of those forces with $d\vec{r}_i$ is simply equal to zero. Only the component of the forces that is parallel to $d\vec{r}_i$ contributes to the infinitesimal work dW.

In addition to the expression for dW previously obtained, an alternative expression for dW can also be obtained by making use of Newton's second law for each particle to replace the expression for each force $\vec{F}_i$ by $d(m\vec{v}_i)/dt$, where $\vec{v}_i = d\vec{r}_i/dt$. By doing so, the following expression for the infinitesimal work dW is also obtained:

$$dW \triangleq \sum_{i=1}^{N} \vec{F}_i \bullet d\vec{r}_i = \sum_{i=1}^{N} \left[\frac{d}{dt}(m_i \vec{v}_i) \right] \bullet \vec{v}_i\, dt = \sum_{i=1}^{N} [d(m_i \vec{v}_i)] \bullet \vec{v}_i$$

$$= d\left(\sum_{i=1}^{N} \frac{1}{2} m_i \vec{v}_i \bullet \vec{v}_i \right) \triangleq d(T) \tag{2.8.3}$$

In summary,

$$dW = d(T) \tag{2.8.4}$$

where the quantity

$$T \triangleq \sum_{i=1}^{N} \frac{1}{2} m_i \vec{v}_i \bullet \vec{v}_i \tag{2.8.5}$$

is called the *kinetic energy* of the motion of the entire system of point masses.

As indicated by the result expressed by Eq. (2.8.4), the infinitesimal work done by the forces that cause the displacements $d\vec{r}_i$ is equal to the change in the kinetic energy of the system during those displacements. This is known in dynamics as the *principle of work and energy*.

The right-hand side of Eq. (2.8.4) is an exact differential (i.e., it is the "d of T," as discussed in Section 1.12) and, thus, it can always be integrated to yield the function T plus a constant of integration. However, the left-hand side of that equation, with dW determined simply by performing dot product operations as indicated earlier, may not be integrated in general. It can only be integrated if the expression that is obtained for dW in that manner is an exact differential. That expression will never be an exact differential if any force $\vec{F}_i$ that contributes to dW is velocity dependent, for example. This is the case because only dq_i's, and not $d\dot{q}_i$'s, will appear in the expression that is obtained for dW as $dW = \vec{F}_1 \bullet d\vec{r}_1 + \vec{F}_2 \bullet d\vec{r}_2 + \cdots + \vec{F}_N \bullet d\vec{r}_N$ for any problem.

> In general, dW is not an exact differential $d(W)$ of any function W. For the cases when dW is an exact differential, i.e., when dW is of the form $d(W)$, Eq. (2.8.4) can be readily integrated to yield $T - W =$ constant. When dW is not an exact differential, a function W cannot be found from dW.

When dW is an exact differential, the equation $T - W =$ constant does not involve any time derivative higher than the first derivative (i.e., velocities, but not accelerations) of any variable. By making use of $dW = d(T)$, when dW can be integrated, one is then able to obtain, in a relatively simple manner, an *integral of the motion* for the system, namely, $T - W =$ constant. It should be noted that the equation $T - W =$ constant is a first-order differential equation, and important information about the motion of the system may be readily obtained from it, even without having to integrate it again in case it is possible to do so analytically.

For several problems presented in this chapter, an integral of the motion was obtained by manipulating the second-order differential equation of motion using the chain rule of differentiation to prepare that differential equation to be integrated once. For some of those problems dW is an exact differential and, thus, the integrated form of $dW = d(T)$ may be used for obtaining the same integrated equation, thus reducing the amount of mathematical manipulations involved in the process.

> **A Note on the Most General Expression for *dW***
>
> The most general expression for dW, which is of the form given by Eq. (2.8.2), may always be split into the three parts given in the following equation, with one of the parts being the term $a_0 dt$ that may or may not appear in dW.
>
> $$dW = \underbrace{- \sum_{i=1}^{n} \frac{\partial U(q_1, q_2, \ldots, q_n; t)}{\partial q_i} dq_i + \text{whatever is left out of the expression of dW}}_{\text{This is generated from the terms } Q_{q_1} dq_1 + Q_{q_2} dq_2 + \cdots + Q_{q_n} dq_n \text{ in Eq. (2.8.2)}}$$
> $$+ a_0 \, dt \qquad (2.8.6)$$
>
> The minus sign in front of the term $\frac{\partial U}{\partial q_1} dq_1 + \frac{\partial U}{\partial q_2} dq_2 + \cdots + \frac{\partial U}{\partial q_n} dq_n$ in Eq. (2.8.6) is a universal convention, and with such a convention, the function $U(q_1, q_2, \ldots, q_n; t)$ is called the *potential energy* of the motion.
>
> The development of expressions for potential energy is included in the examples at the end of this section. In dynamics, the quantity $T + U$ is called the *mechanical energy* of the motion.

SECTION 2.8 A Work-Energy Approach Obtained from Newton's Second Law

In this section, the work-energy approach was presented in an introductory to intermediate level. The work-energy approach is actually the basis for a new branch of dynamics that is called *analytical dynamics,* which is mostly based on work developed by Lagrange. In such an approach there is no need to formulate expressions for acceleration in order to obtain the differential equations of motion of dynamical systems. A number of other great masters, such as William Rowan Hamilton (1805–1865) and Karl Gustav Jacobi (1804–1851), devised very elegant and powerful theories that are part of analytical dynamics. *Hamilton's principle,* for example, is widely used by research engineers in investigations of the motion of dynamical systems, including flexible structures.

Chapter 7 is a more advanced chapter that covers the basics of *analytical dynamics* and is included in this book for those readers who want to extend their capabilities in dynamics and for instructors who introduce at least some of that material in their intermediate dynamics courses. If you study that material, you will see that using *Lagrange's equation,* and searching for *classical integrals of the motion,* as presented in that chapter, is actually simpler than working directly with $\vec{F} = m\vec{a}$. You will have a lot to gain if you do so!

Let us now look at several examples of application of the material presented in this section. Examples 2.8.1 and 2.8.2 involve a mass particle subjected to a constant gravitational force and to an inverse-square gravitational force, respectively. The differential equations of motion for both cases were formulated in Examples 2.5.8 (p. 95) and 2.7.11 (p. 144). For convenience, the two figures for those examples are repeated as Fig. 2.48.

Infinitesimal Work and Potential Energy for a Constant Gravitational Force | EXAMPLE 2.8.1

For this example refer to Fig. 2.48a (which is the same as Fig. 2.15 of Example 2.5.8). With $\vec{F} = -mg\hat{y}$ and the absolute position vector for the particle equal to $\vec{r} = x\hat{x} + y\hat{y}$ [thus, $d\vec{r} = (dx)\hat{x} + (dy)\hat{y} + x\,d\hat{x} + y\,d\hat{y} = (dx)\hat{x} + (dy)\hat{y}$ because, for the chosen unit vectors, $d\hat{x} = d\hat{y} = 0$], the following expression is obtained for the infinitesimal work dW.

$$dW = -(mg\hat{y}) \bullet d\vec{r} = -mg\,dy$$

The same expression for dW is obtained for the simpler case of motion along a vertical straight line considered in Example 2.5.1 (p. 77), or for motion along any other path.

This expression for dW is clearly an exact differential $d(W)$ and, therefore, can be integrated to yield $W = -mgy + C$, where C is an arbitrary constant of integration (which can always be set to zero without any loss of generality). The quantity $U = mgy$, obtained from $dW = -d(U)$ in this case where dW is an exact differential that involves only one variable, is called the *potential energy* associated with the constant gravitational force $-mg\hat{y}$. Thus, the potential energy associated with a constant gravitational force whose magnitude is mg is simply equal to (except for an arbitrary constant of integration) the product of mg with the particle's upward vertical distance y from a fixed reference point.

Figure 2.48 Figure for Examples 2.8.1 and 2.8.2.

For the motion of a projectile subjected only to the gravitational force, as considered in Example 2.5.8, we then obtain

$$T + U = \frac{m}{2}(\dot{x}^2 + \dot{y}^2) + mgy = \frac{1}{2}mv^2 + mgy = \text{constant} \triangleq E$$

where $v = |\vec{v}| = |\dot{x}\hat{x} + \dot{y}\hat{y}| = \sqrt{\dot{x}^2 + \dot{y}^2}$ is the magnitude of the absolute velocity of the particle and $T = mv^2/2$ is the kinetic energy of the motion. The value of the constant E is determined by the initial conditions of the motion, $\dot{x}(0)$, $\dot{y}(0)$, and $y(0)$.

EXAMPLE 2.8.2 Potential Energy for an Inverse-Square Gravitational Force

For an inverse-square gravitational force [see Fig. 2.48b, which is the same as Fig. 2.44 for Example 2.7.11 (p. 145)] we have

$$\vec{F} = -\frac{GM_E m}{r^2}\hat{r}$$

With the absolute position vector for the mass particle written as $\vec{r} = r\hat{r}$ (see Fig. 2.48b), the infinitesimal displacement vector $d\vec{r}$ is obtained as

$$d\vec{r} = (dr)\hat{r} + r\,d\hat{r} = (dr)\hat{r} + (r\,d\theta)\hat{\theta}$$

Therefore, the following expression is obtained for the infinitesimal work dW done by the force $\vec{F}$.

$$dW = \vec{F} \bullet d\vec{r} = -\frac{GM_E m}{r^2}\,dr$$

The expression for dW for this case is also an exact differential and, thus, it can be integrated to yield (with C being an arbitrary constant of integration which can be set to zero)

$$W = \frac{GM_E m}{r} + C$$

The quantity

$$U = -\frac{GM_E m}{r}$$

SECTION 2.8 A Work-Energy Approach Obtained from Newton's Second Law

which is obtained from $dW = -d(U)$ in this case, is the *potential energy* associated with the inverse-square gravitational force.

Notice that for Example 2.7.11 (p. 144), the work-energy method presented here immediately yields, without using Eqs. (2.7.11a) and (2.7.11b) obtained from $\vec{F} = m\vec{a}$ in that example,

$$T + U = \frac{1}{2}mv^2 - \frac{GM_Em}{r} = \frac{1}{2}m(\dot{r}^2 + r^2\dot{\theta}^2) - \frac{GM_Em}{r} = \text{constant} \stackrel{\triangle}{=} E$$

where v is the speed of the particle. This equation is the same as Eq. (2.7.11d). The value of the constant E is determined by the initial conditions of the motion, $r(0)$, $\dot{r}(0)$, and $\dot{\theta}(0)$.

Infinitesimal Work and Potential Energy Associated with a Spring Force   EXAMPLE 2.8.3

Figure 2.49 shows a spring connecting two bodies and the force the spring exerts on those bodies when the attachment points on the spring are at a distance $r = |\vec{r}|$ apart, where $\vec{r}$ is the vector from one connection point to the other. The spring is approximated as massless.

The spring force acting on one of the bodies is the vector $\vec{F}_s$, and, therefore, the spring force acting on the other body is $-\vec{F}_s$ since the spring is massless. The vectors $\vec{r}_A = \overrightarrow{OA}$ and $\vec{r}_B = \overrightarrow{OB}$ are the absolute position vectors for the attachment points A and B, respectively. A unit vector $\hat{u}_1$, conveniently chosen to be parallel to line AB, is also shown in the figure. The direction of line AB may be changing with time as the bodies move in space.

Letting L_u be the unstretched length of the spring, the spring force $\vec{F}_s$ is then

$$\vec{F}_s = -F_s \hat{u}_1$$

where $F_s = k(r - L_u)$ if the spring is linear. If the spring is not linear, F_s is a nonlinear function of the spring deformation $r - L_u$.

Since the spring applies the forces $\vec{F}_s$ and $-\vec{F}_s$ to both bodies, the infinitesimal work due to the spring force is then, according to the definition of infinitesimal work given by

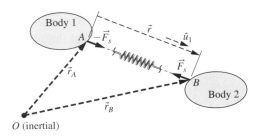

Figure 2.49 Reaction force of a spring.

Eq. (2.8.1),

$$dW = \vec{F}_s \bullet d\vec{r}_B + (-\vec{F}_s) \bullet d\vec{r}_A = \vec{F}_s \bullet d(\vec{r}_B - \vec{r}_A) = \vec{F}_s \bullet d(r\hat{u}_1)$$
$$= (-F_s\hat{u}_1) \bullet [(dr)\hat{u}_1 + r\,d\hat{u}_1]$$
$$= -F_s\,dr \qquad [\text{since } d\hat{u}_1 \text{ is perpendicular to } \hat{u}_1, \hat{u}_1 \bullet (d\hat{u}_1) = 0]$$

This expression for dW is an exact differential because the spring force depends only on r. By writing $dW = -d(U)$ where, as indicated earlier, the negative sign is always used in order to abide by international convention, the function $U = U(r) = \int F_s(r)\,dr$ is the potential energy associated with the spring force.

For a linear spring with stiffness k N/m, the potential energy is obtained as

$$U = \int k(r - L_u)\,dr = \frac{k}{2}(r - L_u)^2 + C$$

Since motion involving $r(t)$ always occurs with changes in $r(t)$ [and, thus, with changes $d(U)$ in the potential energy U of the spring], the constant of integration in the expression for U is always irrelevant. Therefore, such a constant may always be set to zero without any loss of generality.

In this example, it has been shown, in detail, *how* the expressions for dW and the potential energy of a spring were obtained. By observing the final result $dW = -F_s\,dr$ for the infinitesimal work, which led to the expression for the potential energy U after an integration, it is seen that dW only depends on the quantity $r - L_u$, which is how much the spring has been either elongated or compressed. This observation makes it possible for one to write down the expression for dW due to a linear spring in one of two simple ways:

- By simply multiplying the spring force $k(r - L_u)$ by the infinitesimal displacement dr of the spring, with a minus sign in front of the result because the spring force is always opposed to the direction the spring deforms
- By starting with the potential energy $U = \frac{1}{2}k(r - L_u)^2$, and then obtaining dW as $dW = -d(U) = -k(r - L_u)\,dr$

For a torsional spring shown in Fig. 2.1 (p. 73), the infinitesimal work done by the reaction moment M_s of the spring is $dW = -M_s\,d\theta$, which is also an exact differential because M_s only depends on the angle θ.

For a linear torsional spring with stiffness k N·m/rad, the infinitesimal work dW done by the spring moment and the potential energy U are then obtained as shown in the following equations, where θ_u is the value of θ when the torsional spring is unstretched.

$$dW = -k(\theta - \theta_u)\,d\theta = -d(U)$$
$$U = \int k(\theta - \theta_u)\,d\theta = \frac{k}{2}(\theta - \theta_u)^2 + C$$

As mentioned earlier, the quantity C in this expression is a constant of integration that is irrelevant, and may be set to zero with no loss of generality.

Infinitesimal Work and Potential Energy for the Spring-Pendulum

EXAMPLE 2.8.4

An undamped spring-pendulum is shown in Fig. 2.50. It is the same as the one that was analyzed in Example 2.7.8 using Fig. 2.34 (p. 129). By using the results obtained in Examples 2.8.1 (p. 155) and 2.8.3 (p. 157), one may write the following expression for the potential energy for the motion of this system simply by inspecting Fig. 2.50.

$$U = -mgr\cos\theta + \frac{k}{2}(r - L_u)^2$$

Since, in this example, the only forces that do work are the spring force and the gravitational force (i.e., the reaction forces at the fixed support do not do work), the expression for the infinitesimal work dW should then be

$$dW = -d(U) = [mg\cos\theta - k(r - L_u)]\,dr - (mgr\sin\theta)\,d\theta$$

If you wish to verify this result, you should obtain the expression of dW by starting with its definition, i.e. (see Fig. 2.50),

$$dW = \vec{F} \bullet d\vec{r}_P = [mg(\hat{r}\cos\theta - \hat{\theta}\sin\theta) - k(r - L_u)\hat{r}] \bullet [(dr)\hat{r} + (rd\theta)\hat{\theta}]$$

Figure 2.50 Figure for Example 2.8.4 (the unstressed length of the spring is L_u).

This gives $dW = [mg\cos\theta - k(r - L_u)]\,dr - (mgr\sin\theta)\,d\theta$, which is the same as the expression that was previously written by inspecting Fig. 2.50.

Let us now verify whether the expression for dW is an exact differential. The test for this was presented in Section 1.12.

Since $\partial[mg\cos\theta - k(r - L_u)]/\partial\theta = \partial(-mgr\sin\theta)/\partial r = -mg\sin\theta$, dW is an exact differential. By denoting it as $-d(U)$, we then have

$$\frac{\partial U}{\partial r} = k(r - L_u) - mg\cos\theta$$

$$\frac{\partial U}{\partial \theta} = mgr\sin\theta$$

The first of these partial derivatives may be integrated with respect to r to yield $U = k(r - L_u)^2/2 - mgr\cos\theta + f_1(\theta)$, where $f_1(\theta)$ is a function of θ only. The second of these partial derivatives may be integrated with respect to θ to yield $U = -mgr\cos\theta + f_2(r)$, where $f_2(r)$ is a function of r only. By comparing the two expressions obtained for U, the potential energy U is then, except for a constant of integration which, again, is arbitrarily set to zero, $U = -mgr\cos\theta + k(r - L_u)^2/2$. This is the same expression as the one that was written by inspecting Fig. 2.50 and using the knowledge gained from Examples 2.8.1 and 2.8.3.

Notice that the kinetic energy for this system is $T = m\vec{v}_P \bullet \vec{v}_P/2 = m[\dot{r}^2 + (r\dot{\theta})^2]/2$. Since dW is an exact differential $-d(U)$ for this example, it follows that the total mechanical energy $T + U$ for the undamped spring-pendulum remains constant for all times, i.e., $\frac{1}{2}mv_P^2 + k(r - L_u)^2/2 - mgr\cos\theta = \text{constant} \overset{\Delta}{=} E$, where $v_P = |\vec{v}_P| = \sqrt{\dot{r}^2 + r^2\dot{\theta}^2}$ and E is determined by the initial conditions of the motion. Notice, however, that the equation $T + U = \text{constant}$ for this system gives $\dot{r}^2 + r^2\dot{\theta}^2$ in terms of the two unknown variables r and θ and, therefore, neither the maximum nor minimum values of either r or θ can be determined from that equation alone because it is one equation in two unknowns.

To answer specific questions about r and θ, it is necessary to work with the two nonlinear differential equations of motion, Eqs. (2.7.8f) and (2.7.8g) (p. 130). Those nonlinear differential equations have to be integrated numerically so that the extremum values of r and θ can be determined,[11] and that was done in Example 2.7.8.

EXAMPLE 2.8.5 Infinitesimal Work and Potential Energy for Example 2.5.9

Referring to Fig. 2.17 of Example 2.5.9 (p. 97), the only forces that do work on that system are the gravitational force $-m_2 g \hat{x}_2$, the viscous friction force $\vec{f}_1 = -c_1 \dot{x} \hat{x}_1$ acting on block P_1, as well as the spring force and the viscous friction force acting between blocks P_1 and P_2 when the two blocks move relative to each other.

The gravitational force $-m_1 g \hat{x}_2$ and the normal forces $\vec{N}_1$ and $\vec{N}_2$ shown in Fig. 2.17 do not do work because they are always perpendicular to the displacement of the points where they are applied. The dot product of those forces with the associated displacements are equal to zero.

The infinitesimal work done by the gravitational force $-m_2 g \hat{x}_2$ and by the friction force acting on block P_1 are obtained as (see Fig. 2.17, p. 97)

$$[dW]_{\text{due to } -m_2 g \hat{x}_2} = (-m_2 g \hat{x}_2) \bullet d[x \hat{x}_1 + r \hat{y}_1] = (-m_2 g \hat{x}_2) \bullet [(dx)\hat{x}_1 + (dr)\hat{y}_1]$$
$$= (m_2 g \sin \alpha)\, dr$$

$$[dW]_{\text{due to } -c_1 \dot{x} \hat{x}_1} = -c_1 \dot{x} \hat{x}_1 \bullet (dx)\hat{x}_1 = -c_1 \dot{x}\, dx$$

Note: This expression for the infinitesimal work done by the gravitational force $-m_2 g \hat{x}_2$ could also have been obtained by making use of the knowledge acquired in Example 2.8.1 (p. 155) about the potential energy associated with a constant gravitational force. Since the vertical distance traveled by block P_2 is $r \sin \alpha$ (see Fig. 2.16, p. 96), and block P_2 is below the reference level that distance is measured from, which is a level that is independent of the motion coordinates, the potential energy associated with that force is $U = -m_2 g r \sin \alpha$. As seen in Example 2.8.1 (p. 155), the reason for the minus sign is that P_2 is below the reference level that is used. Thus, this gives the infinitesimal work as $dW = -d(U) = m_2 g (\sin \alpha)\, dr$.

The infinitesimal work done by the viscous friction force and by the spring force acting between blocks P_1 and P_2 (see Fig. 2.17, p. 97) is different from zero only when the two blocks move relative to each other. If we imagine ourselves sitting on block P_1, for example, and then look at block P_2, we then see that the infinitesimal displacement of P_2 relative to block P_1 is $(dr)\hat{y}_1$. The expression for the infinitesimal work done by

---

[11] There are advanced analytical methods, called *perturbation methods,* that go beyond linearization of the differential equations, for obtaining approximate solutions to some types of nonlinear differential equations (see, for example, Ref. 11 listed in Appendix G). Those methods are beyond the scope of this book.

SECTION 2.8 A Work-Energy Approach Obtained from Newton's Second Law

the viscous friction force and by the spring force acting between those two blocks must then be

$$[dW]_{\text{due to viscous friction between } P_1 \text{ and } P_2} = -c_2 \dot{r}\, dr$$

$$[dW]_{\text{due to spring connecting } P_1 \text{ and } P_2} = -k(r - L_u)\, dr$$

The infinitesimal work done by all forces acting on the system is equal to the sum of each dW obtained, i.e.,

$$dW = [m_2 g \sin\alpha - k(r - L_u) - c_2 \dot{r}]\, dr - c_1 \dot{x}\, dx$$

This expression for dW is of the general form given in Eq. (2.8.2) (p. 152), except that it does not contain a term in dt because neither $x(t)$ nor $r(t)$ are specified functions of time. Only the part $[m_2 g \sin\alpha - k(r - L_u)]\, dr$ in the expression for dW can be obtained from a potential energy function. The potential energy is then

$$U = \frac{1}{2} k(r - L_u)^2 - m_2 g r \sin\alpha$$

We still have, of course, $dW = d(T)$, where T is the kinetic energy of the system, whose expression follows, but dW cannot be integrated if $c_1 \neq 0$ and $c_2 \neq 0$ because neither $r(t)$ nor $x(t)$ are known functions of time.

$$T = \frac{m_1}{2}(\dot{x}\hat{x}_1) \bullet (\dot{x}\hat{x}_1) + \frac{m_2}{2}(\dot{x}\hat{x}_1 + \dot{r}\hat{y}_1) \bullet (\dot{x}\hat{x}_1 + \dot{r}\hat{y}_1)$$

$$= \frac{1}{2} m_1 \dot{x}^2 + \frac{1}{2} m_2 (\dot{x}^2 + \dot{r}^2 + 2\dot{x}\dot{r}\cos\alpha)$$

The total mechanical energy $T + U$ would be a constant during the motion only if there were no friction in the system, i.e., if $c_1 = c_2 = 0$.

The quantity $E \triangleq T + U$ in this example is the one you were asked to verify, in the suggested problem at the end of Example 2.5.9 (see p. 101), that remains constant when $c_1 = c_2 = 0$. In this example, it was shown how that quantity was obtained and why it remains constant during the motion of that system.

EXAMPLE 2.8.6 The Pendulum on a Fixed Base of Example 2.5.10

Referring to Fig. 2.20 of Example 2.5.10 (p. 102), we have for the infinitesimal work dW and for the kinetic energy T:

$$dW = M_{\text{bar}}\, d\theta - (mg \sin\theta) L\, d\theta$$

$$T = \frac{1}{2} m(L\dot{\theta})^2$$

In general, this expression for the infinitesimal work dW is not an exact differential because of the term $M_{\text{bar}}\, d\theta$. It would be an exact differential if the moment M_{bar} were either a constant or a known function that only involves the variable θ.

Note: For the case of Example 2.5.10 (p. 101), M_{bar} was given as a function of θ and, therefore, dW is an exact differential for that example. For that example, we then have

$$T - W = \frac{1}{2}m(L\dot\theta)^2 - mgL\cos\theta - \int M_{bar}(\theta)\,d\theta = \text{constant}$$

The quantity $-mgL\cos\theta - \int M_{bar}(\theta)\,d\theta$ is the potential energy of the motion for Example 2.5.10 (p. 101). For $M_{bar} = -K\theta$, as specified in Example 2.5.10, the expression $T - W = \text{constant}$ yields Eq. (2.5.10g) (p. 105). That equation was obtained in Example 2.5.10 by using the methodology based on the chain rule presented as Case 4 in Section 1.13.

The preceding integral of the motion for Example 2.5.10 allowed us to analytically determine the maximum and minimum values of the angle θ without having to obtain an explicit solution for $\theta(t)$. Two methods were presented to obtain such an integral of the motion, namely, using the methodology presented in Section 1.13 and using the work-energy methodology presented in this section. The use of either methodology is entirely a matter of personal preference. Some readers may naturally favor one of the methodologies over the other, and you should simply use the one you feel most comfortable with.

EXAMPLE 2.8.7 Potential Energy for the System in Example 2.7.6

The infinitesimal work done by the moment M_{bar} and by the gravitational force acting on the pendulum considered in Example 2.7.6 (see Fig. 2.30, p. 122) is $M_{bar}\,d\theta - (mgL\sin\theta)\,d\theta$. The only other forces for which the infinitesimal work is not zero in this example are an external force $\vec{F}_{ext}\hat{x}$ that must be applied to the block at A to make it move with a desired $x(t)$ (see Fig. 2.30), and the friction force that may exist while the block at A moves along the horizontal floor. The desired motion $x(t)$ was specified in Example 2.7.6 as a known constant acceleration $\ddot{x} \stackrel{\Delta}{=} -a_A$. Letting $F_{block}\hat{x}$ denote the vector sum of the external force $\vec{F}_{ext}\hat{x}$ and the friction force that may be acting on the block at A, the infinitesimal work done by all forces acting on the system is then determined as

$$dW = (M_{bar} - mgL\sin\theta)\,d\theta + F_{block}\,dx$$
$$= (M_{bar} - mgL\sin\theta)\,d\theta + F_{block}\dot{x}\,dt$$

Since $\dot{x}(t)$ is known, dx was expressed as $dx = \dot{x}\,dt$ in the expression for dW. The resulting expression is not an exact differential because of the contributions from the unknown quantities M_{bar} and F_{block}. The only part of the expression for dW that can be obtained for sure from a potential energy function is the exact differential $-mgL(\sin\theta)\,d\theta$, and the potential energy for this system is $U = -mgL\cos\theta$. Notice that, in this example, $T + U$ does not remain constant during the motion even if there is no friction in the system and $M_{bar} = 0$. The physical explanation for this is that energy has to be input into the system by an external force $F_{ext}\hat{x}$ in order to make the block at A move in a specified manner. If $M_{bar} = 0$ and if there is no friction in the system, the infinitesimal work dW is still $dW = -(mgL\sin\theta)\,d\theta + F_{ext}\dot{x}\,dt$.

EXAMPLE 2.8.8

A small body P, modeled as a particle of mass $m = 20$ g, is constrained to slide on a nonrotating frictionless parabolic vertical track $y = ax^2$ m (see Fig. 2.51), where $a = \frac{1}{3}$ m^{-1}. As shown in the figure, the particle is also connected to a spring, and the other end of the spring is attached to an inertial point O. The spring is linear, with stiffness $k = 9$ N/m and unstressed length $L_u = h = 1.2$ m. P is released at a point A, for which $x = x_A = -1.5$ m, with an initial velocity, down the track, of magnitude $v_A = 6$ m/s.

a. Determine, analytically, the maximum (y_{max}) and the minimum (y_{min}) values of the height y reached by the particle. What are the corresponding values of the distance x?
b. Plot the reaction force exerted on P by the track versus x. What is its maximum magnitude, and where does it occur?

■ Solution

a. The analytical determination of the maximum and minimum values of y, y_{max}, and y_{min} is possible here because there is no friction in the system. Otherwise, the nonlinear differential equation of motion for the system would have to be integrated numerically to obtain $y(t)$, from which the values y_{max} and y_{min} would be determined.

To determine y_{max} and y_{min}, we need to obtain a first-order differential equation for y, and then set $\dot{y}$ to zero in such an equation. The most expeditious way to do so is to make use of the basic equation $dW = d(T)$ because the reaction force applied to particle P by the track is normal to the infinitesimal displacement of P at all times and, therefore, does not do any work. The only other forces acting on P are the spring force and the gravitational force, both of which contribute to dW in a manner that can be obtained from a potential energy function U. Therefore, we have for this problem, $T + U = \text{constant} \triangleq E$, which is the total mechanical energy of the motion. The expressions for the kinetic energy T and for the potential energy U obtained by inspecting Fig 2.51 and by using the knowledge acquired in previous examples, are

$$T = \frac{1}{2}m(\dot{x}^2 + \dot{y}^2) = \frac{1}{2}m\dot{x}^2(1 + 4a^2x^2) \qquad (2.8.8a)$$

$$U = mgy + \frac{k}{2}\left[\sqrt{x^2 + (h-y)^2} - L_u\right]^2 \qquad (2.8.8b)$$

Figure 2.51 A mass particle constrained to move on a parabolic track.

Since $T + U =$ constant, we then have with $x^2 = y/a$,

$$\frac{1}{2}m\dot{x}^2(1 + 4a^2x^2) + mgy + \frac{k}{2}\left[\sqrt{\frac{y}{a} + (h-y)^2} - L_u\right]^2 = \text{constant} \triangleq E$$

(2.8.8c)

where

$$E = \frac{1}{2}mv_A^2 + mgy_A + \frac{k}{2}\left[\sqrt{x_A^2 + (h - ax_A^2)^2} - L_u\right]^2 \approx 1.11$$

Notice that the expressions for T and U would be more complicated to deal with if the polar coordinates $|\overrightarrow{OP}|$ and ϕ shown in Fig. 2.51 were used.

The height y reaches a maximum value $y_{\max}$ when $\dot{y} = 0$. The value of $y_{\max}$ is then determined by setting $\dot{y} = 0$ in Eq. (2.8.8c), i.e., from

$$mgy_{\max} + \frac{k}{2}\left[\sqrt{\frac{1}{a}y_{\max} + (h - y_{\max})^2} - L_u\right]^2 - 1.11 = 0$$

For the given values of m, k, a, h, and L_u, the following value for $y_{\max}$ is obtained either with a few iterations with a calculator or from a plot of the left-hand side of the preceding equation versus $y_{\max}$.

$$y_{\max} \approx 0.88 \text{ m}$$

Since $y = ax^2$ and, thus, $x = \pm\sqrt{y/a}$, the maximum and minimum values of x are

$$x_{\max} = \sqrt{y_{\max}/a} \approx 1.62 \text{ m} \qquad x_{\min} = -x_{\max}$$

b. *Reaction force exerted on P by the track.* Since there is no friction, the reaction force applied by the parabolic track to the mass particle P consists only of a force N that is normal to the track at the point of contact with P. To determine N, we now need a free-body diagram for P and to use Newton's second law to obtain the expression for N. The free-body diagram for P is shown in Fig. 2.52. Two unit vectors, $\hat{x}$ and $\hat{y}$, which are parallel to the x and the y axes, respectively, are also shown in that figure. They will be used in the following formulation.

Figure 2.52 Free-body diagram for a mass particle constrained to move on a parabolic track.

SECTION 2.8 A Work-Energy Approach Obtained from Newton's Second Law

In terms of the rectangular coordinates x and y, and of the unit vectors $\hat{x}$ and $\hat{y}$, the absolute acceleration of P and the resultant force acting on P are expressed as (see Fig. 2.52)

$$\vec{a}_P = \ddot{x}\hat{x} + \ddot{y}\hat{y} = \ddot{x}\hat{x} + 2a(x\ddot{x} + \dot{x}^2)\hat{y}$$

$$\vec{F} = k\left[\sqrt{x^2 + (h-y)^2} - L_u\right](\hat{y}\cos\phi - \hat{x}\sin\phi) - mg\hat{y} + N(\hat{y}\cos\theta - \hat{x}\sin\theta)$$

where (again, see Fig. 2.52)

$$\cos\phi = \frac{h-y}{\sqrt{x^2 + (h-y)^2}}$$

$$\sin\phi = \frac{x}{\sqrt{x^2 + (h-y)^2}}$$

$$\cos\theta = \frac{dx}{\sqrt{(dx)^2 + (dy)^2}} = \frac{1}{\sqrt{1 + (dy/dx)^2}} = \frac{1}{\sqrt{1 + 4a^2x^2}}$$

$$\sin\theta = \frac{dy/dx}{\sqrt{1 + (dy/dx)^2}} = \frac{2ax}{\sqrt{1 + 4a^2x^2}}$$

From Newton's second law, $\vec{F} = m\vec{a}_P$, the following two equations are then obtained.

$$k\left[\sqrt{x^2 + (h-y)^2} - L_u\right]\cos\phi + N\cos\theta - mg = m\ddot{y} = 2ma(x\ddot{x} + \dot{x}^2) \quad (2.8.8d)$$

$$-k\left[\sqrt{x^2 + (h-y)^2} - L_u\right]\sin\phi - N\sin\theta = m\ddot{x} \quad (2.8.8e)$$

Equations (2.8.8d) and (2.8.8e) can be readily solved for the normal force N. This can be done by multiplying Eq. (2.8.8d) by $\cos\theta$, and Eq. (2.8.8e) by $\sin\theta$, and, then, subtracting the two results. The resulting expression for N [where $\dot{x}^2$ is obtained from Eq. (2.8.8c)] is

$$N = \frac{1}{\sqrt{1 + 4a^2x^2}}\left\{mg + 2ma\dot{x}^2 - k\left[1 - \frac{L_u}{\sqrt{x^2 + (h - ax^2)^2}}\right](h+y)\right\} \quad (2.8.8f)$$

The plot of N versus x using this expression for N with the given values for a, m, k, L_u, and h, and with $\dot{x}^2$ given by Eq. (2.8.8c), is shown in Fig. 2.53. The value used for g is $g = 9.81$ m/s^2. The maximum magnitude of the reaction N, obtained from Fig. 2.53, is $N \approx 1.68$ N, and it occurs when $x = 0$, i.e., at the bottom of the track.

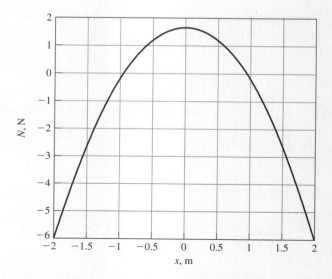

Figure 2.53 Plot of N versus x for a mass particle constrained to move on a parabolic track.

EXAMPLE 2.8.9

For the problem in Example 2.8.8, determine the maximum and the minimum values of $x(t)$ if there is friction in the system. Friction is linear and viscous, with friction coefficient $c = 0.028$ N/(m/s).

■ Solution

The linear viscous friction force acting on particle P is equal to $-c\vec{v}_P$, where $\vec{v}_P = \dot{x}\hat{x} + \dot{y}\hat{y}$ is the velocity of the particle relative to the parabolic track (which is the same as the absolute velocity of P since the track is not accelerating). Therefore, the resultant force $\vec{F}$ acting on P is now

$$\vec{F} = k\left[\sqrt{x^2 + (h-y)^2} - L_u\right](\hat{y}\cos\phi - \hat{x}\sin\phi) - mg\hat{y} + N(\hat{y}\cos\theta - \hat{x}\sin\theta) - c(\dot{x}\hat{x} + \dot{y}\hat{y})$$

From Newton's second law, $\vec{F} = m\vec{a}_P = m(\ddot{x}\hat{x} + \ddot{y}\hat{y})$, the following two equations are obtained for this case.

$$k\left[\sqrt{x^2 + (y-h)^2} - L_u\right]\cos\phi + N\cos\theta - c\dot{y} - mg = m\ddot{y} \quad \text{(2.8.9a)}$$

$$-k\left[\sqrt{x^2 + (y-h)^2} - L_u\right]\sin\phi - N\sin\theta - c\dot{x} = m\ddot{x} \quad \text{(2.8.9b)}$$

The differential equation of motion for P is obtained by eliminating the normal force N from these equations. This may be done by multiplying Eq. (2.8.9a) by $\sin\theta$ and Eq. (2.8.9b) by $\cos\theta$ and then adding the two results. With $y = ax^2$, this yields the

SECTION 2.8 A Work-Energy Approach Obtained from Newton's Second Law

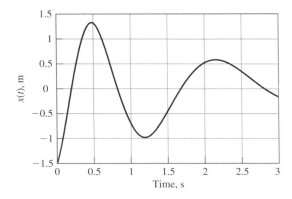

Figure 2.54 Result of the numerical integration (using Simulink) of Eq. (2.8.9c).

following nonlinear differential equation of motion when the expressions for $\cos\phi$, $\sin\phi$, $\cos\theta$, and $\sin\theta$ given in Example 2.8.8 are used.

$$(1 + 4a^2x^2)\ddot{x} + 4a^2x\dot{x}^2 + \frac{k}{m}\left[1 - \frac{L_u}{\sqrt{x^2 + (ax^2 - h)^2}}\right][2a(ax^2 - h) + 1]x$$

$$+ \frac{c}{m}\dot{x}(1 + 4a^2x^2) + 2agx = 0 \qquad (2.8.9c)$$

For $c \neq 0$, $T + U$ is not constant, and the problem cannot be solved analytically any more. Since Eq. (2.8.9c) cannot be integrated analytically to yield $\dot{x}$ as a function of x, it will have to be integrated numerically.

Figure 2.54 shows the result of the numerical integration of Eq. (2.8.9c) with the initial conditions $x(0) = -1.5$ m, $\dot{x}(0) = 3\sqrt{2}$ m/s (which corresponds to $|\vec{v}_P(0)| = 6$ m/s, as given in Example 2.8.8), and with $c = 0.028$ N/(m/s). The numerical integration was done using Simulink. The maximum and minimum values of $x(t)$ obtained from Fig. 2.54 are $x_{\max} \approx 1.24$ m and $x_{\min} = -1.5$ m, and they occur when $t \approx 0.5$ s and also $t = 0$, respectively. Notice that, because of the damping in the system, $x \to 0$ as $t \to \infty$, and that $x = 0$ is the equilibrium value for $x(t)$ for this system.

$\Longrightarrow$ The numerical integration of Eq. (2.8.9c) was done using Simulink, with the model shown in Fig. 2.55. The model is the file `parabola_track.mdl` included on the enclosed CD. To use it type, in the MATLAB command window, either `parabola_track` or `simulations_menu` to choose the model as indicated at the beginning of this chapter. The animation block `animate_nbars` is included in the model so that the motion of P can be observed as the differential equation of motion is integrated. Double click the block labeled `x_ddot` to see the syntax for the expression for $\ddot{x}$, which is a long line! The expression can be copied to an editor (with `copy` and `paste`) for better viewing.

Note: Using some creativity, the differential equation of motion for this system, and other single-degree-of-freedom systems, may be obtained more expeditiously using $dW = d(T)$ as follows. The only forces acting on P that contribute to dW are the spring force, the gravitational force, and the friction force. As mentioned in Example 2.8.8, the

Figure 2.55 Simulink block diagram for integrating Eq. (2.8.9c).

first two give a contribution to dW that can be obtained from a potential energy function U given by Eq. (2.8.8b). The contribution to dW from the friction force is

$$(dW)_{\text{friction}} = -c(\dot{x}\hat{x} + \dot{y}\hat{y}) \bullet [(dx)\hat{x} + (dy)\hat{y}] = -c\dot{x}\,dx - c\dot{y}\,dy = -c\dot{x}(1 + 4a^2x^2)\,dx$$

Therefore, $dW = d(T)$ yields, with $dW = -d(U) + (dW)_{\text{friction}}$, and T given by Eq. (2.8.8a) in Example 2.8.8:

$$-d\left(mgax^2 + \frac{k}{2}\left[\sqrt{x^2 + (h - ax^2)^2} - L_u\right]^2\right) - c\dot{x}(1 + 4a^2x^2)\,dx$$

$$= d\left[\frac{1}{2}m\dot{x}^2(1 + 4a^2x^2)\right] \tag{2.8.9d}$$

As already mentioned earlier, the expression for dW is not an exact differential because of the friction term, and, thus, it cannot be integrated. However, since the only unknown quantity that appears in Eq. (2.8.9d) is the variable $x(t)$, by manipulating it as $dW/dt = d(T)/dt$, Eq. (2.8.9c), which is the second-order differential equation that governs the motion of the system, is obtained. This is the most convenient and expeditious manner for obtaining that equation.

2.9 Problems

2.1 A point P moves along a nonrotating straight line, and the vector $\vec{r}$ from a fixed point O on that line to P is $\vec{r} = [k_1 t \sin(k_2 t)]\hat{x}$ m, where $\hat{x}$ is a unit vector pointing from O to P, $k_1 = 5$ m/s, and $k_2 = 1$ rad/s. Determine the magnitude and the direction of the absolute velocity and of the absolute acceleration of P when $t = 5$ s.

2.2 Repeat Problem 2.1 for $\vec{r} = [k_1 t^2 e^{-k_2 t} \cos(k_3 t)]\hat{x}$, where $k_1 = 1$ m/s^2, $k_2 = 1$ s^{-1}, and $k_3 = 2$ rad/s.

2.3 Consider Problem 2.1 again, but with the straight line track rotating counterclockwise on a fixed straight platform with a constant absolute angular velocity equal to 0.2 rad/s. Determine the magnitude of the following quantities when $t = 5$ s:

a. The absolute velocity and the absolute acceleration of P
b. The velocity and the acceleration of P relative to an observer fixed to the rotating track

2.4 Determine the time of the first occurrence of a maximum of the scalar component of the vector $\vec{r} = r(t)\hat{x}$ for Problems 2.1 and 2.2. You may find the answers either with a few iterations with a calculator or by generating accurate plots of $r(t)$ versus t for both cases. If you generate plots, it is suggested that you do them on your computer using either a Simulink block diagram or MATLAB commands, or any other software of your choice.

2.5 To model the effect of drag caused by air resistance, the acceleration $\ddot{r}(t)$ of a body moving along a straight line is approximated as $c_0 - c_1 \dot{r}^2$, where c_0 is a constant acceleration due to an engine on the body and c_1 is a constant. Observers at a fixed point start to measure the distance r when the velocity $\dot{r}$ is clocked at $\dot{r} = \dot{r}_0$.

a. Determine, analytically, the velocity $\dot{r}$ as a function of the distance traveled (r) for the cases when $c_1 \neq 0$ and $c_1 = 0$.
b. After its engine is turned off (i.e., $c_0 = 0$), the body travels 0.8 km when its velocity is then determined to be $\dot{r} = \dot{r}_0/2$. Determine the value of c_1; be sure to specify its units.

2.6 Consider a body P of mass m thrown upward from the earth's surface, without rotation, and with an initial velocity $\dot{y}(0) = v_0$, where y is the distance from the earth's surface to P (as in Fig. 2.3 for Example 2.5.1, p. 78). The body is also subjected to a force $\vec{f}$ that opposes its velocity $\vec{v}$ as $\vec{f} = -c\vec{v}$, where $c > 0$ is constant. Determine, analytically, $y(t)$ in terms of the quantities m, c, g, and v_0, where g is the acceleration of gravity at the surface of the earth. Also, obtain the analytical expression for the maximum height reached by the body. Approximate the gravitational force on P as a constant. Plot $y_{max}/[y_{max}]_{c=0}$ (where $[y_{max}]_{c=0} = v_0^2/(2g)$, as determined in Example 2.5.1) versus the nondimensional parameter $v_0 c/(mg)$ for $10^{-4} \leq v_0 c/(mg) \leq 1$.

2.7 Integrate numerically the differential equation of Problem 2.5 for $r(t)$, with $r(0) = 0$ and $\dot{r}(0) = 1$ m/s, and plot $r(t)$ versus t and $\dot{r}$ versus t (with $0 \leq t \leq 10$ s) for $c_1 = 0$, $c_1 = 0.01$, and $c_1 = 0.03$ m^{-1}. Use $c_0 = 5$ m/s^2 and superimpose all the $r(t)$ plots in the same graph and all the $\dot{r}(t)$ plots in another graph. *Hint:* To superimpose plots using Simulink, see the description of the `Gain` and `Auto-Scale Graph` blocks in Appendix A.

2.8 A body of mass m is launched vertically from $y = 0$ (see Fig. 2.3, p. 78) with a speed of 70 m/s. When it reaches its maximum height, a horizontal force of magnitude $f = 5mg$ is applied to it during Δt s. Some time later the body hits the horizontal plane $y = 0$ at a distance $D = 750$ m from where it was launched. Determine the time Δt, and the total time of flight counting from the instant it was launched.

2.9 Calculate the kinetic energy imparted to the body in Problem 2.8 due to the ignition of the rocket on board the body. What is the ratio between that energy and the kinetic energy that was imparted to the body when it was launched vertically? The mass of the body is $m = 40$ kg.

2.10 Determine the escape speed from the surface of the moon and from the surface of Mars. Treat both as uniform spheres, the moon with radius equal to 1740 km and Mars with radius equal to 3390 km, and disregard the gravitational attraction of any other celestial body. The acceleration of gravity at the surfaces of the moon and Mars are approximately equal to $g/6$ and $g/2.6$, where $g \approx 9.81$ m/s^2 is the acceleration of gravity at the surface of the earth.

2.11 An object of mass m kg is launched vertically from the surface of the earth with an initial velocity $v_0 = 6000$ km/h. In addition to the gravitational force of attraction of the earth, the object is also subjected to an upward constant vertical thrust T N during its entire trajectory. For $T/m = 1$ m/s^2, determine the maximum height reached by the object and the time it takes for it to reach that height. Clearly state all the approximations you decide to make when solving this problem, and justify them.

2.12 Repeat Problem 2.11 for $T/m = 4$ m/s^2 and for $T/m = 6$ m/s^2. Use the constant gravitational force approximation, and also Newton's law of universal gravitation, and comment on the results.

2.13 The object in Problem 2.11 is subjected to a variable thrust T that changes with time as $T = k_1 T_0 t e^{-k_2 t}$, where $T_0/m = 3$ m/s^2, $k_1 = 1$ s^{-1}, and $k_2 = 0.1$ s^{-1}. The initial launching speed at $y = 0$ is $v_0 = 210$ km/h. Determine the maximum value of $y(t)$ and the time it occurs. Clearly state all the approximations you make when solving this problem.

2.14 Write a brief explanation saying *how* you would determine the maximum distance $y(t)$ for Problem 2.11 (without determining it) if the thrust T were of the form $T = k_1 T_0 t y e^{-k_2 t}$, where T_0, k_1, and k_2 are constants.

2.15 A common theoretical model for the earth's atmosphere is one for which the air density changes exponentially with height y from the earth's surface as $\rho = \rho_0 e^{-ky}$, where $\rho_0 \approx 1.5$ kg/m^3 and $k \approx 0.14$ km^{-1}. Repeat Problem 2.11 if instead of the constant thrust T the object is subjected to an aerodynamic drag force $\vec{f}_{\text{drag}}$ that is always directed opposite to the velocity of the object and whose magnitude is modeled using the "exponential atmosphere" as $c\dot{y}^2 e^{-ky}$ where $k = 0.14$ km^{-1} and $c/m = 1.3$ km^{-1} for that particular object.

2.16 An object P of mass $m = 2$ kg is launched from the earth's surface with an initial velocity whose magnitude is $v_0 = 40$ m/s and whose direction makes an angle of $\alpha = 30°$ with the horizontal as shown in Fig. 2.14 (p. 95). The object is subjected to a constant thrust $T = 10$ N that makes a constant angle of $45°$ with the horizontal during the entire trajectory. Determine

a. The maximum height reached by the object
b. The time it takes to reach the maximum height

c. The distance along the horizontal plane the object strikes the ground, and the time it takes for that to happen

d. The equation $y = y(x)$ for the trajectory (see Fig. 2.15, p. 95, for the definition of the variables x and y).

Clearly state all the approximations you make when solving this problem.

2.17 A cannon A located on a platform whose height is $h = 200$ m shoots projectiles with an initial velocity whose magnitude is $v_0 = 300$ m/s. Determine the angle of firing α for the projectile to hit a stationary object that is located at a distance $R = 5000$ m from the cannon. The line of sight of the target (i.e., the line from the firing point to the target) is inclined at an angle $\beta = 20°$ with the horizontal, and the angle of firing is measured from such a line.

Suggestion: Using the constant gravitational force approximation, it is simpler to solve this problem if a rectangular x-y coordinate system is used, with x along the line of sight and y perpendicular to it. It is suggested that you also solve it with x and y along the horizontal and the vertical, respectively, so that you can see that the first choice mentioned is simpler.

2.18 Show that the relation between R, v_0, α, and β for Problem 2.17 is

$$\frac{Rg}{v_0^2} = \frac{2(\sin\alpha)\cos(\alpha + \beta)}{\cos^2\beta}$$

which is an equation that could be programmed into an onboard computer for generating a display of the required firing angle for given input values of v_0, R, and β. See the suggestion in Problem 2.17.

2.19 A projectile P of mass m is launched from the earth's surface with a velocity $\vec{v}_0$ that makes an angle α degrees with the horizontal, and whose magnitude is $v_0 = |\vec{v}_0|$ m/s. In addition to the gravitational force, the projectile is subjected to a drag force $\vec{f}_{\text{drag}}$ whose magnitude is modeled as being proportional to the square of the magnitude v of the projectile's velocity $\vec{v}$ as $|\vec{f}_{\text{drag}}| = cv^2$, with c being a constant. The direction of the drag force is opposite to $\vec{v}$.

a. Use rectangular coordinates x and y, with x along the horizontal plane and y along the vertical, and obtain the differential equations that govern the motion of the projectile using the constant acceleration of gravity model. Clearly show the free-body diagram for P, properly labeling all quantities, and show all the steps of the formulation that lead to the differential equations, as done in the book. What are the initial conditions for your differential equations?

b. Since the differential equations cannot be solved analytically when $c \neq 0$, integrate them numerically with $v_0 = 100$ m/s and $\alpha = 40°$, and determine the projectile's range (i.e., the distance, measured along the horizontal, at the instant the projectile hits the ground) for $c/m = 10^{-4}$ and $c/m = 10^{-3}$ m^{-1}.

c. For $v_0 = 100$ m/s and $\alpha = 40°$, plot the projectile's range versus c/m.

Suggestion: Modify the Simulink model `ballistics.mdl` to do this problem.

2.20 A block P of mass $m = 2$ kg is able to slide on an inclined plane that makes a constant angle α degrees with the horizontal. Friction between P and the plane is of the Coulomb type (i.e., dry friction). The static and kinetic coefficients of friction are $\mu_s = 0.25$ and $\mu_k = 0.2$, respectively. The angle α is equal to twice the value needed for P to start sliding down the plane when the block is started

from rest. The block starts its motion at a distance of $D = 10$ m from the bottom of the inclined plane and with an initial velocity of 30 m/s up the plane. Determine
a. The total distance the block travels up the plane
b. The time it takes the block to travel twice the distance determined in part a

Does the distance D affect the solution to this problem?

2.21 For the system described in Problem 2.20 determine the time it takes for P to reach the bottom of the inclined plane if its initial velocity is 30 m/s down the plane.

2.22 Determine the maximum distance the block of Problem 2.20 travels up the inclined plane if $\alpha = 15°$ and friction is viscous and linear, with friction coefficient $c = 0.2$ N/(m/s).

2.23 A block P of mass $m = 3$ kg is able to slide on an inclined plane that makes a constant angle $\alpha = 30°$ with the horizontal (see Fig. 2.56). Friction between P and the plane is of the Coulomb type (i.e., dry friction). The block is initially at rest on the inclined plane at a distance of 10 m from the bottom of the plane. The static and kinetic coefficients of friction are $\mu_s = 0.3$ and $\mu_k = 0.25$, respectively. A constant horizontal force $F_a = 10$ N is then applied to the block. Determine
a. The total distance the block travels on the plane
b. The time it takes for the block to reach the bottom of the inclined plane

2.24 Repeat Problem 2.23 for $F_a = 5$ N. Where is the block located 3 s after the application of the force F_a?

2.25 Use Simulink to simulate and animate the motion of the block in Problems 2.20 and 2.21. For the animation, use the block animate_nbars, and automatically stop the simulation if the block reaches either the top or the bottom of the inclined plane. The length of the inclined plane is 150 m.

2.26 Two blocks P_1 and P_2 of masses m_1 and m_2, respectively, are connected by an inextensible string of negligible mass, and the string passes through a small pulley at an inertial point O. Block P_1 moves on a fixed inclined vertical plane of length D, while P_2 moves along a vertical line as shown in Fig. 2.57a.

Block P_1 is connected to the inclined plane by a linear spring of stiffness k N/m and unstressed length L_u m. Friction in the system can be neglected. Obtain the differential equation of motion of the system in terms of the distance

Figure 2.56 Figure for Problem 2.23.

Figure 2.57 Figure for Problems 2.26 and 2.27.

$r(t)$, its equilibrium solution r_e, and the frequency ω of the oscillation of the blocks when they are displaced from their equilibrium. If the differential equation is nonlinear, linearize it about $r = r_e$ to find the frequency of the *linearized* oscillation, instead of the frequency of the oscillation. What are the values of r_e and ω if $\alpha = 30°$, $L_u = D$, $m_1 = 4m_2$, and $k = k_1 m_1 g$, where $k_1 = 1.25\,\text{m}^{-1}$?

2.27 In the system described in Problem 2.26 block P_2 is replaced by a constant force F N, as shown in Fig. 2.57b. Obtain the differential equation of motion of the system in terms of the distance $r(t)$, its equilibrium solution r_e, and the frequency ω of the oscillation of the block when it is displaced from its equilibrium. If the differential equation is nonlinear, linearize it about $r = r_e$ to find the frequency of the *linearized* oscillation, instead of the frequency of the oscillation. What are the values of r_e and ω if $\alpha = 30°$; $L_u = D$; $k = k_1 m_1 g$, where $k_1 = 1.25\,\text{m}^{-1}$, and $F = m_1 g/4$? Is this problem the same as Problem 2.26?

2.28 A block P of mass m kg is able to slide on a vertical inclined plane that makes a constant angle $\alpha = 30°$ with the horizontal. The block is connected by a linear spring to a fixed point A above the inclined plane as shown in Fig. 2.58. The distance h in that figure is a known constant. There is no friction in the system. The spring stiffness is K N/m, and its unstressed length is $L_u = h$ m. The block is initially motionless at point B shown in Fig. 2.58. For $Kh/(mg) = 2.5$, determine

  a. If the block loses contact with the inclined plane during its motion (*Hint:* Look at the value of the reaction force applied by the plane to the block.)

  b. The maximum value of the distance r measured from point B along the direction that points down the inclined plane

2.29 Determine the equilibrium solutions (in terms of h) for Problem 2.28, and the frequency of the small linearized oscillations about each stable equilibrium. How many equilibrium solutions does the system exhibit?

2.30 A block P of mass m kg is able to slide on a vertical inclined plane that makes a constant angle $\alpha = 30°$ with the horizontal as shown in Fig. 2.59. The plane is long enough for the block not to fall from it. Friction between the plane and the block is of the Coulomb type (i.e., dry friction), and the block is subjected to an applied force F_a that changes with time as $F_a(t) = kmt$, where $k = 0.8$ m/s^3, and is directed as shown in the figure, with $\theta = 20°$. The static and kinetic coefficients of friction are $\mu_s = 0.25$ and $\mu_k = 0.2$, respectively. The block is initially held at rest at a distance $r(0) = 40$ m from a point O on the inclined plane and then released when the force F_a is applied. Determine

  a. The time when the block starts to move, and whether it starts moving up or down the plane

Figure 2.58 Figure for Problem 2.28.

Figure 2.59 Figure for Problems 2.30 to 2.32.

b. The values of t when the block passes through point O

c. The values of $r(t)$ at the end of 20 s, and the total distance traveled by the block during that time

Describe, in detail, the motion during its first 20 s.

2.31 Use numerical integration to determine $r(t)$ and $\dot{r}(t)$ if the block in Problem 2.30 is subjected to a force $F_a = k_1 m \sin(k_2 t)$, where $k_1 = 8$ m/s^2 and $k_2 = 0.1$ rad/s, for the first 40 s of the motion. Describe, how the block moves (if it moves at all) during the first 5 s, and during the next 35 s. If it moves, how many times does it pass through point O during that time, and what are the corresponding values of t?

2.32 Solve Problem 2.30 when friction is viscous, instead of the Coulomb type. For this, consider linear viscous friction, with a friction coefficient equal to 0.2 N/(m/s). Take the mass of the block to be $m = 1$ kg. Solve this problem analytically.

2.33 A small body P of mass m slides inside an open groove on a semi-circular inertial vertical ring of radius R as shown in Fig. 2.60. Point P is initially at rest at the position when $\theta(0) = \theta_0$, where θ_0 is very small but nonzero (the particle would not move if θ_0 were equal to zero because that is a position of static equilibrium). Determine, analytically, the value of θ when the particle leaves the ring and how far away from point O it strikes the ground. The only external force acting on the particle is due to the earth's gravity. Neglect friction between the ring and the particle.

2.34 The value of θ when the particle in Problem 2.33 leaves the ring cannot be determined analytically if there is friction between the ring and the particle. Use numerical integration to determine that value of θ (in degrees) for the case when friction is of the Coulomb type with a kinetic coefficient of friction $\mu_k = 0.2$.

2.35 Repeat Problem 2.34 for the case when friction is linear and viscous (instead of the Coulomb type) and proportional to the velocity of the particle relative to the ring, with a viscous friction coefficient equal to c N/(m/s). Determine that value of θ (in degrees) when the particle leaves the ring, for $(c/m)\sqrt{R/g} = 0.2$.

Figure 2.60 Figure for Problems 2.33 to 2.35.

2.36 A block P of mass m is connected to a fixed pin O by a linear spring of stiffness k N/m and unstressed length L_u m. The block is constrained to move along a horizontal guide as shown in Fig. 2.61, and friction in the system can be neglected. The perpendicular distance from O to the guide is h m. The plane defined by the guide and the spring is vertical.

Obtain the differential equation that governs the motion of the block in terms of the variable $x(t)$ shown in the figure. For this, clearly show the free-body

Figure 2.61 Figure for Problem 2.36.

diagram for P, properly labeling all quantities, and show all the steps of the formulation that lead to the differential equation. For $L_u = h$, $x(0) = 5h$, and $\dot{x}(0) = Ah\sqrt{k/m}$, with $A = 10$, determine (analytically) the maximum and the minimum values of x/h. Also, generate a plot of $x_{\max}/h$ and $x_{\min}/h$ versus $x'(0)/h$ for $L_u = h$ and $x(0) = 5h$ for $-10 \leq A \leq 10$, where the prime denotes differentiation with respect to a nondimensional time τ defined as $\tau = t\sqrt{k/m}$.

2.37 What are the equilibrium solutions x_e for $x(t)$ for Problem 2.36 when $L_u \leq h$, and when $L_u > h$? How may equilibrium solutions does the system exhibit in each case? Is the linearized differential equation about each equilibrium you determined valid for each one of those cases? Explain in detail. If they are valid, what are such equations? Plot x_e/h versus L_u/h for $0 \leq L_u/h \leq 5$. *Hint:* see the boxed text on p. 50 in Chapter 1.

2.38 For $L_u = \sqrt{5}h$ and h, determine all the equilibrium solutions for the system described in Problem 2.36, and analyze their stability, indicating whether the equilibrium is infinitesimally stable, asymptotically stable, or unstable. If any of your linearized differential equations are not valid for describing the small motion about an equilibrium solution, plot $d(x/h)/d\tau$ versus x/h, where $\tau = t\sqrt{k/m}$, and indicate whether that particular equilibrium is stable, asymptotically stable, or unstable.

2.39 Numerically integrate the differential equation of motion for Problem 2.36, with $L_u = h$ and with linear viscous damping in the system, with damping coefficient c N/(m/s). Introduce a nondimensional time τ so that the only parameter that appears in the normalized differential equation to be integrated is $c/\sqrt{mk}$. What are the maximum and minimum values of x/h, for $c/\sqrt{mk} = 0$ and 0.2, when the motion is started with $x(0) = 5h$ and $\dot{x}(0) = 10h\sqrt{k/m}$? Compare the analytical answer obtained in Problem 2.36 (for $c = 0$) with the answer obtained from the numerical integration. Plot x/h versus normalized time and $d(x/h)/d\tau$ versus x/h for both cases. Choose an appropriate value for the final integration time that will allow you to clearly understand and describe the motion. Write a small paragraph describing the motion. *Suggestion:* Use Simulink and include the `animate_nbars` block in your model so that you can actually see how the block moves.

2.40 For the problem in Example 2.7.13 (p. 149), let $\dot{\phi} = \text{constant} \triangleq \Omega$ and the motion of P be started with the initial conditions $\theta(0) = \theta_0$ and $\dot{\theta}(0) = 0$, where θ_0 is very small (so that $\cos\theta_0 \approx 1$) but not zero. Determine, analytically, the expression for the value of θ when the particle leaves the ring, and plot that value (in degrees) versus $\Omega^2 R/g$ for $0 \leq \Omega^2 R/g \leq 1$.

2.41 Numerically integrate the differential equation of motion for Example 2.7.13, which is Eq. (2.7.13g) (p. 150), to determine the nondimensional reaction forces $N/(mg)$ and f/mg for the case when $\Omega = $ constant, where N and f are given by Eqs. (2.7.13h) and (2.7.13i), respectively. For this, work with the nondimensional time $\tau \triangleq t\sqrt{g/R}$, and rewrite those equations using τ. For the numerical integration, use $\dot\theta(0) = 0$ and a very small value for $\theta(0)$. Plot $f/(mg)$ and $N/(mg)$ versus τ for $\Omega^2 R/g = 0.5$ and 2 from $\tau = 0$ until the particle loses contact with the ring; superimpose the plots for each one of the two values of $\Omega^2 R/g$. Write a few lines about your observation on the results. *Suggestion:* Use Simulink and include the `Stop Simulation` block in your model to make the simulation stop automatically when the value of $N/(mg)$ becomes equal to zero.

2.42 Determine all the equilibrium solutions for $\theta(t)$ for the system described in Example 2.7.13 (p. 149), and analyze their stability. For each equilibrium, indicate whether it is stable, asymptotically stable, or unstable. The groove is closed so that P cannot leave it. How many equilibrium positions does the system have? Do they exist for any value of the angular velocity Ω?

Show a hand sketch of the phase plane plot $d\theta/d\tau$ (where $\tau = t\sqrt{g/R}$) versus θ for the system, for $R\Omega^2/g = 0.5$ and 2, and indicate whether the trajectories are traced clockwise or counterclockwise with the passage of time. Your sketch does not have to be an accurate plot.

2.43 The system shown in Fig. 2.62 is a double pendulum constructed of two homogeneous thin bars OA and AB of negligible mass and constant lengths L_1 and L_2, respectively, and two small bodies of masses m_1 and m_2, located at A and B as indicated in the figure. Bar OA is pivoted to an inertial point O, and the two bars are pinned at A and can rotate about that point relative to each other. The motion of the entire system takes place in the vertical plane, and friction in the system can be neglected.

a. Obtain the differential equations of motion for the system in terms of the angles $\theta_1(t)$ and $\theta_2(t)$ shown in Fig. 2.62.

b. Choose your own values for L_2/L_1 and m_2/m_1, and observe the motion by integrating the differential equations of motion numerically using Simulink and including the `animate_nbars` block in your model to animate the motion. Choose different combinations of initial conditions $\theta_1(0)$, $\theta_2(0)$, $\dot\theta_1(0)$, and $\dot\theta_2(0)$, so that you can observe different motions. Then write a few paragraphs describing your observations in detail.

Figure 2.62 Figure for Problem 2.43.

2.44 Obtain the differential equation of motion, and solve part a of Example 2.8.8 (p. 163), if the parabolic track is spun about the vertical y axis with a constant counterclockwise angular velocity $\Omega = 200$ rpm relative to inertial space. Do this without integrating the differential equation numerically. Write a paragraph describing the motion.

Suggestion: The x and y axes shown in Figs. 2.51 (p. 163) and 2.52 (p. 164) are rotating with the parabola with absolute angular velocity $\Omega \hat{y}$. Use those axes and the rotating unit vector triad $\{\hat{x}, \hat{y}, \hat{z} \stackrel{\triangle}{=} \hat{x} \times \hat{y}\}$, with $\hat{x}$ and $\hat{y}$ parallel to those axes, to formulate the differential equation of motion via Newton's second law.

Hint: Verify that the following function H is a constant of the motion (i.e., that $dH/dt = 0$), and then use this observation to do the problem analytically. The methodology to obtain such a constant of the motion is presented in Chapter 7.

$$H = \frac{1}{2}m\dot{x}^2(1 + 4a^2x^2) + mgy + \frac{k}{2}\left[\sqrt{x^2 + (h - ax^2)^2} - L_u\right]^2 - \frac{1}{2}m\Omega^2 x^2$$

2.45 Solve the problem of Example 2.8.9 (p. 166) if friction in the parabolic track is of the Coulomb type, instead of viscous. The static and kinetic coefficients of friction are $\mu_s = 0.25$ and $\mu_k = 0.2$, respectively. Determine $x_{\max}$ and $x_{\min}$ from a plot of $x(t)$ versus t. Write a paragraph describing the motion. *Suggestion:* Modify the model `parabola_track.mdl`, which is on the book CD, to integrate the differential equation of motion for this problem.

2.46 Solve the problem of Example 2.8.9 (p. 166) if the parabolic track is spun about the vertical y axis with a counterclockwise constant angular velocity $\Omega = 200$ rpm relative to inertial space and if friction between P and the rotating track is of the linear viscous type with coefficient c N/(m/s). See the suggestion for Problem 2.44. Because of the friction term, the differential equation of motion, which is nonlinear, has to be integrated numerically to determine $x_{\max}$ and $x_{\min}$, as done in Example 2.8.9. *Suggestion:* Modify the model `parabola_track.mdl`, which is on the book CD, to integrate the differential equation of motion for this problem.

2.47 Determine all the equilibrium solutions for $x(t)$ for the body on the rotating parabolic track of Problem 2.46, and analyze their stability. Choose your own value for c, and use $m = 20$ g, $a = \frac{1}{3}$ m^{-1}, $k = 9$ N/m, and $L_u = h = 1.2$ m. For each equilibrium, indicate whether it is stable, asymptotically stable, or unstable.

Show a hand sketch of the phase plane plot $\dot{x}$ versus x for the system, for $\Omega = 200$ rpm, and indicate whether the trajectories are traced clockwise or counterclockwise with the passage of time. Your sketch does not have to be an accurate plot. Show the equilibrium solutions in your sketch, indicating which ones are stable, asymptotically stable, and unstable.

2.48 Numerically integrate the differential equations of motion for the satellite problem of Example 2.7.11 (p. 144), and generate plots of $r(t)$ and $\theta(t)$ versus time t for $r(0) = 6578$ km, $\dot{r}(0) = 0$, and $r(0)\dot{\theta}(0) = v_0$. Generate the plots for $v_0 = 27{,}000$ km/h, $28{,}000$ km/h, and $30{,}000$ km/h. Determine, from the plots, the maximum and minimum values of the distance $r(t)$. Identify the cases for which the satellite crashes on the earth's surface and determine the time the crash occurs. For the cases where the satellite revolves around the earth, determine, from the plots, the time it takes the satellite to complete one revolution. The radius of the earth is $R_E \approx 6378$ km.

2.49 Consider the two-particle system described in Example 2.7.10 (p. 140) when particle P_2 shown in Fig. 2.41 is allowed to move in three-dimensional space.

a. Obtain the differential equations of motion for the system using the distance r and the angle θ shown in Fig. 2.41 and two angles defined by you to specify the position of P_2 in space.

b. Integrate numerically the equations obtained in part a and plot $r(t)$ versus t. For this, take $m_2 = 2m_1$, $L = 1$ m, $r(0) = 0.5$ m, $\dot{r}(0) = 1$ m/s, $\dot{\theta}(0) = 2$ rad/s, with OP_2 starting from rest at an angle of $20°$ with the vertical in any direction chosen by you (make sure you clearly specify that direction in your solution). What are the maximum and minimum values of r obtained from your plot? If you use Simulink, show an animation of the string P_1P_2 using the `animate3d_nbars` block described in Appendix A.

CHAPTER 3

Kinematic Analysis of Planar Mechanisms

Since a particle has no dimension, it cannot rotate in space, but only translate. Rotation is a motion that involves change of orientation of a line. The material presented in this chapter consists of the study of position, velocity, and acceleration of mechanical systems that are composed of bars or bodies interconnected to each other and moving either along a plane or parallel planes. Figures in this chapter are drawn in a single plane, which is taken to be the reference frame for velocities and accelerations that are determined here. The translational motion of some of the points of the system and/or the rotational motion of some of its bars are assumed to be known, but calculation of the forces involved will not be of concern here. Knowing the motion of some points and/or bars of the system, the objective here is to determine velocity and acceleration of other points and/or angular velocity and angular acceleration of any of the bars of the system. This material is a direct application of the kinematic formulation that was presented in Chapter 2, and several classical mechanisms are considered in separate sections of this chapter.

3.1 Instantaneous Center of Rotation

When a rigid body[1] is in motion and the directions of the velocities of any two points of the body are known at a certain instant of time, one can find the direction of the velocity of any other point of the body at that instant from a graphical construction. At that instant of time the body appears to be rotating about a certain point in space. That point is called the *instantaneous center of rotation* or the *instantaneous center of zero velocity* for the body at that instant. If, in addition, the magnitude of the velocity of one of the points of the body is known at that

---

[1] A body is said to be rigid when the distance between any two points of the body is constant.

same instant, the magnitude of the velocity of any other point of the body may also be determined for that instant by making use of the instantaneous center of rotation. In general, the instantaneous center of rotation is moving in space.

The use of the instantaneous center of rotation may provide a quick way for determining the instantaneous (i.e., at a certain instant *only*) velocity of any other point of the body. The angular velocity of the body at the instant considered is also determined from such analysis. Depending on the specific situation faced by an engineer, calculation of velocities using the instantaneous center of rotation may provide a quick and acceptable approximate answer to a problem.

The disadvantage of doing calculations using an instantaneous center of rotation is that such calculations are only valid for a specific instant of time. A specific figure that is used for calculating velocity at a certain instant using the instantaneous center of rotation at that instant is of no use for doing the same calculations at another instant. Also, the instantaneous center of rotation is not useful for determining accelerations.

The determination of the location of the instantaneous center of rotation for a body and the mathematical demonstration of its use in the calculation of velocities are now presented. Consider a rotating body for which the directions of the velocities $\vec{v}_A$ and $\vec{v}_B$ of two of its points A and B, respectively, relative to a reference point O are known as illustrated in Fig. 3.1. For the instant represented by the figure, the body is rotating clockwise and the magnitude of its angular velocity in the reference frame that is fixed to the plane of the paper is the quantity ω indicated in the figure. At that instant, the body appears to be rotating about an axis that passes through point C (thus, the name instantaneous center of rotation) shown in Fig. 3.1, which is located at the intersection of the perpendiculars to the velocity vectors $\vec{v}_A$ and $\vec{v}_B$.

To demonstrate the usefulness of point C, let $d_A = |\overrightarrow{AC}|$ and $d_B = |\overrightarrow{BC}|$ be the distances between points A and C and between points B and C, respectively.

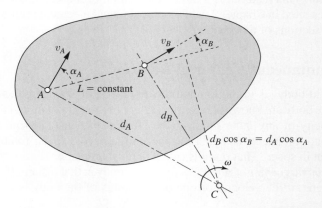

Figure 3.1 How to find the instantaneous center of rotation of a body.

Now, look at the velocity $\vec{v}_{B/A}$ of point B relative to point A, which is defined as $\vec{v}_{B/A} \triangleq \vec{v}_B - \vec{v}_A$. Since we are dealing with bodies for which the distance $L = |\overrightarrow{AB}|$ between points A and B is constant, the vector $\vec{v}_{B/A}$ is perpendicular to AB (because B has to be moving, relative to A, along an arc of a circle centered at A). Therefore, the component of $\vec{v}_{B/A}$ along line AB, which is $|\vec{v}_B|\cos\alpha_B - |\vec{v}_A|\cos\alpha_A$ has to be zero, i.e.,

$$|\vec{v}_B|\cos\alpha_B - |\vec{v}_A|\cos\alpha_A = 0 \qquad (3.1.1)$$

If we draw the perpendicular from C to line AB, we can write two expressions for the same distance, which are $d_A \cos\alpha_A$ and $d_B \cos\alpha_B$. We then have the following equation:

$$d_B \cos\alpha_B = d_A \cos\alpha_A \qquad (3.1.2)$$

Equations (3.1.1) and (3.1.2) immediately disclose that $|\vec{v}_A|/d_A = |\vec{v}_B|/d_B$ rad/s. Therefore, since points A and B are any two points on the body, we can then write

$$\frac{|\vec{v}_A|}{d_A} = \frac{|\vec{v}_B|}{d_B} = \cdots = \left(\frac{|\vec{v}_P|}{d_P}\right)_{\text{for any other point } P} = \text{constant} = \omega \qquad (3.1.3)$$

Equation (3.1.3) provides a convenient and expeditious way to determine the magnitude of the velocity of any point B on the body by the graphical construction illustrated in Fig. 3.1 when the directions of $\vec{v}_A$ and $\vec{v}_B$ and the magnitude of $\vec{v}_A$ are known. To do this, locate the instantaneous center of rotation by finding the intersection of the perpendiculars to $\vec{v}_A$ and $\vec{v}_B$ (when they are not parallel to each other), measure (or calculate by using the appropriate geometrical relations for each particular figure) the distances d_A and d_B, calculate $\omega = |\vec{v}_A|/d_A$, and then calculate $|\vec{v}_B|$ as $|\vec{v}_B| = \omega d_B$.

> It is important to remember that the instantaneous center of rotation can only be used, when convenient, for calculating angular velocity and velocity at a particular instant of time. It cannot be used for finding accelerations.

3.2 The Four-Bar Mechanism

A *four-bar mechanism* is shown in Fig. 3.2. The mechanism actually consists of three moving bars (or *links*) AB (with $|\overrightarrow{AB}| = L_2$), BC (with $|\overrightarrow{BC}| = L_3$), and CD (with $|\overrightarrow{CD}| = L_4$) joined by pins at B and C, and pinned to the ground (i.e., to the plane of the paper) at A and at D. The fourth bar is AD, with $|\overrightarrow{AD}| = L_1$, which gives rise to the name of the mechanism. The bars are approximated as rigid bodies, and their lengths are essentially approximated as L_2, L_3, and L_4, respectively. In practice, the mechanism is constructed so that the bars move along different, but parallel, planes to avoid collisions; the pins connecting the bars would be long like a nail but perpendicular to the plane of the paper. Points A

Figure 3.2 A four-bar mechanism (the angles $\theta_2, \theta_3,$ and θ_4 are changing with time).

Figure 3.3 A four-bar mechanism and the instantaneous center of rotation of bar BC.

and D are the centers of rotation of bars AB and CD, respectively. Point C_{BC} shown in Fig. 3.3 is the instantaneous center of rotation of bar BC; as shown in Section 3.1, it can be used for determining the velocity of any point on bar BC in terms of either $\dot\theta_2$ or $\dot\theta_4$. For the sake of illustration, the velocity $\vec{v}_{C/B}$ of point C relative to point B is also shown in Fig. 3.3.

Either link AB or link CD is driven by a motor (which is not shown in the figure) in order to make either θ_2 or θ_4 a desired function of time t. In such a case, the motion of the mechanism is completely determined by the knowledge of either $\theta_2(t)$ or $\theta_4(t)$. Taking the angle $\theta_2(t)$ to be the known variable, the objective of the analysis is to determine how the other angles shown in Figs. 3.2 and 3.3 and their first and second time derivatives (i.e., the angular velocity and angular acceleration of the other links) depend on $\theta_2, \dot\theta_2,$ and $\ddot\theta_2$. There are no differential equations to be integrated here since the motion is specified. This type of analysis is called *kinematic analysis,* as opposed to dynamic analysis where the objective would have been to determine the motion of the system when known forces are applied to it and/or the reaction forces that any of the bars are subjected to during the motion.

The general kinematic analysis of four-bar mechanisms, which is based on equations that are valid for all instants of time, is presented in Section 3.2.1.

3.2.1 General Analysis of the Four-Bar Mechanism

The kinematic analysis of any mechanism involves first drawing a simple sketch of the mechanism in a *general* configuration, and then writing a set of geometric relations by inspection of the sketch. Such relations consist of equations for appropriate distances along perpendicular directions. The general sketch for the four-bar mechanism is shown in Fig. 3.2. The following two equations (which are called the *loop equations* for the four-bar mechanism) can be written by inspecting either that figure or Fig. 3.3. They are simply the scalar components along line AD, and along the perpendicular to line AD, of the vector identity $\overrightarrow{AD} = \overrightarrow{AB} + \overrightarrow{BC} + \overrightarrow{CD}$.

$$L_2 \cos\theta_2 + L_3 \cos\theta_3 - L_4 \cos\theta_4 = L_1 \tag{3.2.1}$$

$$L_4 \sin\theta_4 = L_2 \sin\theta_2 + L_3 \sin\theta_3 \tag{3.2.2}$$

The left- and right-hand sides of Eq. (3.2.1) represent two expressions for the same distance $|\overrightarrow{AD}|$, and Eq. (3.2.2) represents two expressions for the perpendicular distance from point C to line AD.[2] Notice that the mechanism was drawn in a *general configuration* in Figs. 3.2 and 3.3 for generating Eqs. (3.2.1) and (3.2.2). No specific values of the angles (especially 0 and 90°, which would eliminate some terms in the equations) should be specified in those equations because, if that is done, they would be valid only for a specific time (and, thus, no time derivatives of them could be taken for continuing the analysis).

> Once Eqs. (3.2.1) and (3.2.2) are obtained, the problem of determining the angular velocities and accelerations of any of the bars of the mechanism is essentially solved. That is done simply by taking the first and the second time derivatives of those equations to obtain a set of equations that involve the angular velocities $\dot\theta_2$, $\dot\theta_3$, and $\dot\theta_4$ and the angular accelerations $\ddot\theta_2$, $\ddot\theta_3$, and $\ddot\theta_4$. This general procedure yields the equations that allow one to determine the angular velocities and the angular accelerations of all the members of a four-bar mechanism in any configuration.

The following equations that involve angular velocities are obtained by taking the first time derivative of Eqs. (3.2.1) and (3.2.2).

$$L_2 \dot\theta_2 \sin\theta_2 + L_3 \dot\theta_3 \sin\theta_3 - L_4 \dot\theta_4 \sin\theta_4 = 0 \tag{3.2.3}$$

$$L_2 \dot\theta_2 \cos\theta_2 + L_3 \dot\theta_3 \cos\theta_3 - L_4 \dot\theta_4 \cos\theta_4 = 0 \tag{3.2.4}$$

---

[2] One could also have written, instead, equations for distances along different orthogonal lines. For example, if distances along line BC and along the perpendicular to line BC are chosen, one would obtain $L_2 \cos(\theta_2 - \theta_3) + L_3 - L_4 \cos(\theta_4 - \theta_3) = L_1 \cos\theta_3$ and $L_4 \sin(\theta_4 - \theta_3) = L_2 \sin(\theta_2 - \theta_3) + L_1 \sin\theta_3$. This second choice, or any other convenient choice, yields equations that are equivalent to Eqs. (3.2.1) and (3.2.2) and accomplish the same purpose of those equations.

Note: Equations (3.2.3) and (3.2.4) could also have been obtained by making use of the following vector identity involving the velocities of points C and B.

$$\vec{v}_C = \vec{v}_B + \vec{v}_{C/B}$$

The three velocity vectors that appear in this vector relationship are shown in Fig. 3.3. As mentioned earlier, the velocities $\vec{v}_C$ and $\vec{v}_B$ are the velocities of points C and B, respectively, in the reference frame fixed to the plane of the paper. In the rest of this chapter, such a velocity of a point, say $\vec{v}_C$ (with only one subscript) of point C, will be referred to as the *absolute velocity of point C*. Since point C is constrained to move along an arc of a circle centered at D and with radius $|\overrightarrow{DC}| = L_4$, and bar CD rotates with angular velocity $\dot{\theta}_4$, the absolute velocity $\vec{v}_C$ is perpendicular to line CD as drawn in Fig. 3.3. Similarly, the absolute velocity $\vec{v}_B$ of point B is perpendicular to line AB as also shown in that figure. The velocity $\vec{v}_{C/B}$ is the velocity of point C relative to point B. Since, when viewed from point B, point C is constrained to move along an arc of a circle centered at B (because the length $|\overrightarrow{BC}| = L_3$ of bar BC is constant), and bar BC rotates in the plane of the paper with absolute angular velocity $\dot{\theta}_3$, the velocity $\vec{v}_{C/B}$ is the vector labeled $v_{C/B} = L_3 \dot{\theta}_3$ in Fig. 3.3.

The vector identity $\vec{v}_C = \vec{v}_B + \vec{v}_{C/B}$ yields two scalar equations by projecting the vectors $\vec{v}_C$, $\vec{v}_B$, and $\vec{v}_{C/B}$ along two orthogonal directions. As can be verified by inspecting Fig. 3.3, the two equations (3.2.3) and (3.2.4) simply represent the components of this vector relationship along the directions perpendicular and parallel to line AD, respectively.[3]

Equations (3.2.3) and (3.2.4) are linear in the angular velocities and, thus, can be easily solved for $\dot{\theta}_3$ and $\dot{\theta}_4$ in terms of $\dot{\theta}_2$ and the angles θ_2, θ_3, and θ_4 when their values are obtained from Eqs. (3.2.1) and (3.2.2). Having the numerical values of the angles θ_2, θ_3, and θ_4, the solution for the angular velocities $\dot{\theta}_3$ and $\dot{\theta}_4$ can be easily obtained with a simple matrix inversion and a matrix multiplication as follows. The solution can also be readily obtained by hand or with an inexpensive calculator.

$$\begin{bmatrix} L_3 \sin\theta_3 & -L_4 \sin\theta_4 \\ L_3 \cos\theta_3 & -L_4 \cos\theta_4 \end{bmatrix} \begin{bmatrix} \dot{\theta}_3 \\ \dot{\theta}_4 \end{bmatrix} = \begin{bmatrix} -L_2 \dot{\theta}_2 \sin\theta_2 \\ -L_2 \dot{\theta}_2 \cos\theta_2 \end{bmatrix}$$

or

$$\begin{bmatrix} \dot{\theta}_3 \\ \dot{\theta}_4 \end{bmatrix} = \begin{bmatrix} L_3 \sin\theta_3 & -L_4 \sin\theta_4 \\ L_3 \cos\theta_3 & -L_4 \cos\theta_4 \end{bmatrix}^{-1} \begin{bmatrix} -L_2 \dot{\theta}_2 \sin\theta_2 \\ -L_2 \dot{\theta}_2 \cos\theta_2 \end{bmatrix} \quad (3.2.5)$$

---

[3] Notice that a judicious choice of directions along which to project a vector equation may even yield an immediate solution to one of the variables in that vector equation. For example, if the vector equation $\vec{v}_C = \vec{v}_B + \vec{v}_{C/B}$ is projected along line DC (which is perpendicular to $\vec{v}_C$), the following equation is obtained (see Fig. 3.3).

$$0 = L_2 \dot{\theta}_2 \sin(\theta_4 - \theta_2) + L_3 \dot{\theta}_3 \sin(\theta_4 - \theta_3)$$

This equation immediately yields the solution for $\dot{\theta}_3$, which is the same solution that is obtained from Eqs. (3.2.3) and (3.2.4).

SECTION 3.2 The Four-Bar Mechanism

The angular accelerations $\ddot{\theta}_3$ and $\ddot{\theta}_4$ of links BC and DC, respectively, are obtained by simply taking the time derivative of Eqs. (3.2.3) and (3.2.4). This yields

$$L_2\left(\ddot{\theta}_2 \sin\theta_2 + \dot{\theta}_2^2 \cos\theta_2\right) + L_3\left(\ddot{\theta}_3 \sin\theta_3 + \dot{\theta}_3^2 \cos\theta_3\right) = L_4\left(\ddot{\theta}_4 \sin\theta_4 + \dot{\theta}_4^2 \cos\theta_4\right) \quad (3.2.6)$$

$$L_2\left(\ddot{\theta}_2 \cos\theta_2 - \dot{\theta}_2^2 \sin\theta_2\right) + L_3\left(\ddot{\theta}_3 \cos\theta_3 - \dot{\theta}_3^2 \sin\theta_3\right) = L_4\left(\ddot{\theta}_4 \cos\theta_4 - \dot{\theta}_4^2 \sin\theta_4\right) \quad (3.2.7)$$

Note: Equations (3.2.6) and (3.2.7) could also have been obtained by making use of the following vector identity involving the accelerations of points C and B.

$$\frac{d}{dt}\vec{v}_C = \frac{d}{dt}\vec{v}_B + \frac{d}{dt}\vec{v}_{C/B}$$

This vector identity yields two scalar equations when each of the three acceleration vectors $\vec{a}_C = d\vec{v}_C/dt$, $\vec{a}_B = d\vec{v}_B/dt$, and $\vec{a}_{C/B} = d\vec{v}_{C/B}/dt$ are written in terms of components along line AD and the perpendicular to line AD. Do this on your own to test yourself. If needed, look at Fig. 3.4, which is the same as the figure first presented in Example 2.7.3. The figure is a graphical visualization of the time derivative $\vec{v} = d\vec{r}/dt$ of a rotating vector $\vec{r}$ (shown in the smaller box) and also of the time derivative $\vec{a} = d\vec{v}/dt$ of $\vec{v}$ (shown in the larger box). It provides a quick reference for writing down the expressions for velocity $\vec{v} = d\vec{r}/dt$ and acceleration $\vec{a} = d\vec{v}/dt$ of a point, using rotating reference directions. For a vector $\vec{r}$ of constant magnitude, $\dot{r} = \ddot{r} = 0$ in the figure. Use it to help you, when needed, for writing expressions for $\vec{a}_C, \vec{a}_B, \vec{a}_{C/B}$, etc., in this section, in Example 3.2.2 (p. 189), and in other examples.

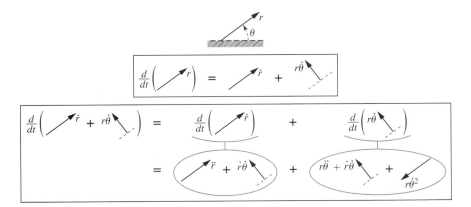

Figure 3.4 Graphical visualization of the first and second time derivatives of a rotating vector $\vec{r}$.

Equations (3.2.6) and (3.2.7) are linear in the angular accelerations and, therefore, can be easily solved for $\ddot{\theta}_3$ and $\ddot{\theta}_4$ in terms of the angular acceleration $\ddot{\theta}_2$; of the angular velocities $\dot{\theta}_2$, $\dot{\theta}_3$, and $\dot{\theta}_4$; and of the angles θ_2, θ_3, and θ_4 once their values are known. In matrix form, the solution for the angular accelerations $\ddot{\theta}_3$ and $\ddot{\theta}_4$ is obtained as follows. Again, the solution can readily be obtained by hand or with an inexpensive calculator.

$$\begin{bmatrix} \ddot{\theta}_3 \\ \ddot{\theta}_4 \end{bmatrix} = \begin{bmatrix} L_3 \sin\theta_3 & -L_4 \sin\theta_4 \\ L_3 \cos\theta_3 & -L_4 \cos\theta_4 \end{bmatrix}^{-1}$$
$$\times \begin{bmatrix} L_4 \dot{\theta}_4^2 \cos\theta_4 - L_3 \dot{\theta}_3^2 \cos\theta_3 - L_2(\ddot{\theta}_2 \sin\theta_2 + \dot{\theta}_2^2 \cos\theta_2) \\ -L_4 \dot{\theta}_4^2 \sin\theta_4 + L_3 \dot{\theta}_3^2 \sin\theta_3 - L_2(\ddot{\theta}_2 \cos\theta_2 - \dot{\theta}_2^2 \sin\theta_2) \end{bmatrix} \quad (3.2.8)$$

Of all the equations for this mechanism [Eqs. (3.2.1) and (3.2.2), (3.2.5), and (3.2.8)], Eqs. (3.2.1) and (3.2.2) are the most difficult to solve because they are nonlinear in the unknown angles θ_3 and θ_4 (with the nonlinearities being the trigonometric functions). Although, it is possible to find an analytical solution to those equations by making a number of algebraic manipulations and using trigonometric identities such as $\sin^2\alpha + \cos^2\alpha = 1$ for any angle α, a numerical solution is generally much more convenient and practical to obtain.

Numerical solutions for θ_3 and θ_4 can be obtained graphically by simply drawing the figure to scale and measuring the resulting angles. This approach is a simple one, but it is also one that may not give answers that are very accurate. Its main disadvantage, however, is that a new drawing has to be made when a new solution is desired for a different configuration of the mechanism at some other time. Another way to obtain a solution is to plot the expressions for $\sin\theta_3$ and $\cos\theta_3$ given by Eqs. (3.2.1) and (3.2.2) versus θ_4, with $0 \leq \theta_4 \leq 2\pi$, and then select, from the plot, only the values of $\sin\theta_3$ and $\cos\theta_3$ that satisfy the relationship $\sin^2\theta_3 + \cos^2\theta_3 = 1$. This strategy has the same disadvantage previously mentioned.

A more efficient and direct way to solve Eqs. (3.2.1) and (3.2.2) is given in Section 3.2.2. After the calculations presented so far are illustrated by some examples, the sequence of calculations shown in Section 3.2.2 and the calculations for angular velocities and angular accelerations are automated in a Simulink block diagram model that also animates the motion of the mechanism while the equations are being solved. The model can, of course, be used for solving any four-bar mechanism problem.

3.2.2 Solution of Eqs. (3.2.1) and (3.2.2)

The solution methodology presented here makes use of the angles α, β, and γ shown in Fig. 3.5 and of simple trigonometric relations that are obtained by inspection of that figure. First, determine the angle α by using the following expressions that are obtained by inspection of Fig. 3.5:

$$\sin\alpha = \frac{L_2 \sin\theta_2}{L} \qquad \cos\alpha = \frac{L_1 - L_2 \cos\theta_2}{L} \quad (3.2.9)$$

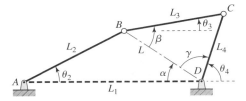

Figure 3.5 The angles α, β, and γ needed for calculating θ_3 and θ_4.

where

$$L = \sqrt{(L_2 \sin \theta_2)^2 + (L_1 - L_2 \cos \theta_2)^2} = \sqrt{L_1^2 + L_2^2 - 2L_1 L_2 \cos \theta_2} \quad (3.2.10)$$

Notice that point B in Fig. 3.5 will be below line AD, which is a physical possibility, if the value that is found for $\sin \alpha$ is negative.

> For this mechanism, the angle α can be anywhere in the range $[-\pi, \pi]$. Therefore, to calculate the value of α, both $\sin \alpha$ and $\cos \alpha$ should be used [instead of either $\sin \alpha$ or $\cos \alpha$, or $\tan \alpha = (\sin \alpha)/(\cos \alpha)$] because we need to determine which quadrant α lies on if we want to draw the actual configuration of the mechanism based on these calculations. In computer software, the function that does this is `atan2(argument_1, argument_2)`, where `argument_1` is the value for $\sin \alpha$ and `argument_2` is the value for $\cos \alpha$.

Next, determine the angles β and γ by using the following expressions that are obtained by inspection of the triangle BCD in Fig. 3.5.

$$L_3 \sin \beta = L_4 \sin \gamma \quad (3.2.11)$$

$$L_3 \cos \beta + L_4 \cos \gamma = L \quad (3.2.12)$$

To determine β, start by combining Eqs. (3.2.11) and (3.2.12) as $(L_3 \cos \beta - L)^2 + (L_3 \sin \beta)^2 = L_4^2$ to obtain

$$\cos \beta = \frac{L_3^2 + L^2 - L_4^2}{2LL_3} \quad (3.2.13)$$

Notice that a real solution for β is possible only if $|L_3^2 + L^2 - L_4^2| \leq 2LL_3$, of course. If this condition is violated for certain ranges of the angle $\theta_2(t)$, then the mechanism cannot operate in that range. The same is true when the calculation of the sine and cosine of any angle yields a value that is outside the range $[-1, 1]$.

If the value obtained for $\cos \beta$ is in the physical range $[-1, 1]$, then there are two possibilities for $\sin \beta$, each one corresponding to a possible configuration of the mechanism (i.e., with point C located either above or below line BD in

Fig. 3.5). The solution to use is determined by the initial configuration that is chosen for the mechanism to start its motion.

Solution number 1: $\qquad \sin\beta = \sqrt{1 - \cos^2\beta}$ (3.2.14)

Solution number 2: $\qquad \sin\beta = -\sqrt{1 - \cos^2\beta}$ (3.2.15)

The angle β corresponding to these two solutions is then determined by using the values that are found for $\sin\beta$ and $\cos\beta$.

Having determined α and β, the angle θ_3 is simply determined as

$$\theta_3 = \beta - \alpha \qquad (3.2.16)$$

Finally, the angle γ (which is needed for calculating θ_4) can now be determined by using Eqs. (3.2.11) and (3.2.12) to calculate $\sin\gamma = (L_3 \sin\beta)/L_4$ and $\cos\gamma = (L - L_3 \cos\beta)/L_4$ for each of the two possible solutions for β.

The angle θ_4 is then determined as

$$\theta_4 = \pi - (\alpha + \gamma) \qquad (3.2.17)$$

EXAMPLE 3.2.1

Bar AB of the mechanism shown in either Fig. 3.2 (p. 182) or Fig. 3.5 (p. 187) is driven by a motor so that $\theta_2 = \pi/6 + 1.5\pi \sin(kt)$ rad, with $k = 2$ rad/s. The lengths of the bars are $L_2 = 0.5$ m, $L_3 = 0.8$ m, and $L_4 = 0.7$ m, while $L_1 = 1$ m. Determine, for the initial instant $t = 0$, and for $t = 5$ s, the angular velocity and the angular acceleration (magnitude and direction) of bars BC and DC.

■ **Solution**

Since only θ_2 is given, we need to calculate θ_3 and θ_4 in order to be able to determine the angular velocities and angular accelerations $\dot\theta_3$, $\dot\theta_4$, $\ddot\theta_3$, and $\ddot\theta_4$ of bars BC and DC. With $\theta_2(t) = \pi/6 + 1.5\pi \sin(2t)$, we have $\theta_2(0) = \pi/6$ rad $= 30°$, and $\theta_2(5) \approx -2.04$ rad $\approx -116.9°$. The results of the sequence of calculations presented in this section for these two values of θ_2 are shown in the following table.

θ_2, degrees	L, m	$\sin\alpha$	$\cos\alpha$	$\cos\beta$	$\sin\beta$	$\sin\gamma$	$\cos\gamma$
30	0.6197	0.4034	0.9150	0.5386	±0.8426	±0.9629	0.2697
−116.9	1.305	−0.3417	0.9396	0.8875	±0.4608	±0.5270	0.85

These results yield the following solutions for θ_3 and θ_4.

Solution 1 for $\theta_2 = 30°$:

$$\theta_3 = \beta - \alpha \approx 33.6° \qquad \theta_4 = 180 - \alpha - \gamma \approx 81.8°$$

Solution 2 for $\theta_2 = 30°$:

$$\theta_3 = \beta - \alpha \approx -81.2° \qquad \theta_4 = 180 - \alpha - \gamma \approx 230.6°$$

Solution 1 for $\theta_2 = -116.9°$:

$$\theta_3 = \beta - \alpha \approx 47.4° \qquad \theta_4 = 180 - \alpha - \gamma \approx 168.2°$$

Solution 2 for $\theta_2 = -116.9°$:

$$\theta_3 = \beta - \alpha \approx -7.5° \qquad \theta_4 = 180 - \alpha - \gamma \approx 231.8°$$

The angular velocity and angular acceleration corresponding to each of the two solutions can be calculated using Eqs. (3.2.5) and (3.2.8). With $\dot{\theta}_2(t) = 3\pi \cos(2t)$ rad/s and $\ddot{\theta}_2(t) = -6\pi \sin(2t)$ rad/s^2, this yields the following results.

Solution 1 for $\theta_2(0) = 30°$ [thus, $\dot{\theta}_2(0) = 3\pi$ rad/s and $\ddot{\theta}_2(0) = 0$], $\theta_3 \approx 33.6°$, $\theta_4 \approx 81.8°$:

$$\dot{\theta}_3 \approx -6.21 \text{ rad/s} \qquad \dot{\theta}_4 \approx -0.57 \text{ rad/s}$$

$$\ddot{\theta}_3 \approx 80.0 \text{ rad/s}^2 \qquad \ddot{\theta}_4 \approx 143.7 \text{ rad/s}^2$$

Solution 2 for $\theta_2(0) = 30°$ [thus, $\dot{\theta}_2(0) = 3\pi$ rad/s and $\ddot{\theta}_2(0) = 0$], $\theta_3 \approx -81.2°$, $\theta_4 \approx 230.6°$:

$$\dot{\theta}_3 \approx -2.77 \text{ rad/s} \qquad \dot{\theta}_4 \approx -8.41 \text{ rad/s}$$

$$\ddot{\theta}_3 \approx 145.9 \text{ rad/s}^2 \qquad \ddot{\theta}_4 \approx 82.2 \text{ rad/s}^2$$

Solution 1 for $\theta_2(5) = -116.9°$ [thus, $\dot{\theta}_2(5) \approx -7.91$ rad/s and $\ddot{\theta}_2(5) = 10.25$ rad/s^2], $\theta_3 \approx 47.5°$, $\theta_4 \approx 168.2°$:

$$\dot{\theta}_3 \approx -5.55 \text{ rad/s} \qquad \dot{\theta}_4 \approx 1.77 \text{ rad/s}$$

$$\ddot{\theta}_3 \approx -2.51 \text{ rad/s}^2 \qquad \ddot{\theta}_4 \approx -9.49 \text{ rad/s}^2$$

Solution 2 for $\theta_2(5) = -116.9°$ [thus, $\dot{\theta}_2(5) \approx -7.91$ rad/s and $\ddot{\theta}_2(5) = 10.25$ rad/s^2], $\theta_3 \approx -7.5°$, $\theta_4 \approx 231.8°$:

$$\dot{\theta}_3 \approx 1.13 \text{ rad/s} \qquad \dot{\theta}_4 \approx -6.2 \text{ rad/s}$$

$$\ddot{\theta}_3 \approx -6.2 \text{ rad/s}^2 \qquad \ddot{\theta}_4 \approx 0.78 \text{ rad/s}^2$$

The angular velocities $\dot{\theta}_3$ and $\dot{\theta}_4$ may also be calculated using the graphical construction that involves the instantaneous center of rotation of bar BC, as shown in Fig. 3.3. The use of the instantaneous center of rotation to do this is illustrated in Example 3.2.2.

EXAMPLE 3.2.2

Bar AB of the mechanism shown in Fig. 3.6 is rotating clockwise with a constant angular velocity $\dot{\phi}_2 = -90$ rpm. The figure is a snapshot of the mechanism taken when $\phi_2 = 90°$. Bars AB, BC, and DC are rigid, and their lengths are $L_2 = 2$ m, $L_3 = 3$ m, and $L_4 = 2.5$ m.

190 CHAPTER 3 Kinematic Analysis of Planar Mechanisms

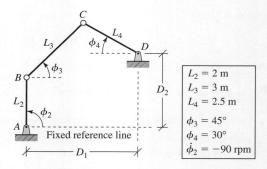

$L_2 = 2$ m
$L_3 = 3$ m
$L_4 = 2.5$ m
$\phi_3 = 45°$
$\phi_4 = 30°$
$\dot{\phi}_2 = -90$ rpm

Figure 3.6 The mechanism for Example 3.2.2.

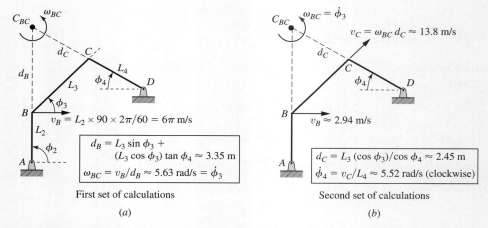

First set of calculations
(a)

Second set of calculations
(b)

Figure 3.7 Solution for the angular velocities $\omega_{BC} = \dot{\phi}_3$ and $\dot{\phi}_4$ for Example 3.2.2.

The distances D_1 and D_2 shown in the figure are also constants. Determine for the instant under consideration

a. The angular velocity and the angular acceleration of bars BC and DC. Indicate their magnitude and whether they are clockwise or counterclockwise.
b. The absolute velocity $\vec{v}_C$ and the absolute acceleration $\vec{a}_C$ of point C. Specify their magnitude and direction.

■ **Solution**

a. The sequence of calculations to obtain the angular velocity of bars BC and CD, using the instantaneous center of rotation of bar BC, is shown in Fig. 3.7. The instantaneous center of rotation of bar BC is point C_{BC}.

First, point C_{BC} is located at the intersection of the extension of bars AB and DC. The distance d_B, needed for calculating the angular velocity of bar BC, could have been measured from the figure, but it was determined analytically, using trigonometry, as shown in Fig. 3.7a. The vector $\vec{v}_B$, which is the absolute velocity of point B, is then added to the figure. It is perpendicular to bar AB and points to the right, as shown in that figure. Because $\vec{v}_B$ is pointing to the right, bar BC has to be rotating counterclockwise, which means that the angular velocity ω_{BC} of bar BC is as indicated in the figure. This, then, determines the direction of the absolute velocity $\vec{v}_C$ of point C, which is shown in Fig. 3.7b. Having the values for $v_B = |\vec{v}_B|$ and the distance d_B, the value of the angular velocity of bar BC, $\omega_{BC} = \dot{\phi}_3$, can then be determined as $\omega_{BC} = v_B/d_B \approx 5.63$ rad/s.

To determine the angular accelerations $\ddot{\phi}_3$ and $\ddot{\phi}_4$, we need to generate two equations that involve these two unknowns. The general approach is to write two equations involving the angles ϕ_2, ϕ_3, and ϕ_4 shown in Fig. 3.6 in the same manner Eqs. (3.2.1) and (3.2.2) were generated for Fig. 3.2. This could be done, for this example, simply by determining the angles θ_2, θ_3, and θ_4 between line AD and bars AB, BC, and CD, respectively, in the same manner that such angles were defined in Fig. 3.2, and then using Eqs. (3.2.6) and (3.2.7) to calculate $\ddot{\phi}_3 = \ddot{\theta}_3$ and $\ddot{\phi}_4 = -\ddot{\theta}_4$. One also could look at Fig. 3.6 and write, by inspection, the two following equations by imagining that the figure was drawn with $\phi_2 < 90°$.

$$L_2 \cos\phi_2 + L_3 \cos\phi_3 + L_4 \cos\phi_4 = \text{constant} = D_1 \quad \textbf{(3.2.2a)}$$

$$L_2 \sin\phi_2 + L_3 \sin\phi_3 - L_4 \sin\phi_4 = \text{constant} = D_2 \quad \textbf{(3.2.2b)}$$

The left- and right-hand sides of Eq. (3.2.2a) represent two expressions for the same distance D_1, and the two sides of Eq. (3.2.2b) represent two expressions for the distance D_2 in Fig. 3.6.

The following equations involve the unknown angular velocities $\dot{\phi}_3$ and $\dot{\phi}_4$. They are obtained by taking the time derivative of Eqs. (3.2.2a) and (3.2.2b).

$$L_2 \dot{\phi}_2 \sin\phi_2 + L_3 \dot{\phi}_3 \sin\phi_3 + L_4 \dot{\phi}_4 \sin\phi_4 = 0 \quad \textbf{(3.2.2c)}$$

$$L_2 \dot{\phi}_2 \cos\phi_2 + L_3 \dot{\phi}_3 \cos\phi_3 - L_4 \dot{\phi}_4 \cos\phi_4 = 0 \quad \textbf{(3.2.2d)}$$

Equations (3.2.2c) and (3.2.2d) yield, of course, the same solutions for $\dot{\phi}_3$ and $\dot{\phi}_4$ that were obtained earlier using the instantaneous center of rotation. The next set of equations involve the angular accelerations. They are obtained simply by taking the time derivative of Eqs. (3.2.2c) and (3.2.2d).

$$L_2 \left(\ddot{\phi}_2 \sin\phi_2 + \dot{\phi}_2^2 \cos\phi_2 \right) + L_3 \left(\ddot{\phi}_3 \sin\phi_3 + \dot{\phi}_3^2 \cos\phi_3 \right)$$
$$+ L_4 \left(\ddot{\phi}_4 \sin\phi_4 + \dot{\phi}_4^2 \cos\phi_4 \right) = 0 \quad \textbf{(3.2.2e)}$$

$$L_2 \left(\ddot{\phi}_2 \cos\phi_2 - \dot{\phi}_2^2 \sin\phi_2 \right) + L_3 \left(\ddot{\phi}_3 \cos\phi_3 - \dot{\phi}_3^2 \sin\phi_3 \right)$$
$$- L_4 \left(\ddot{\phi}_4 \cos\phi_4 - \dot{\phi}_4^2 \sin\phi_4 \right) = 0 \quad \textbf{(3.2.2f)}$$

For the values of L_2, L_3, and L_4 shown in Fig. 3.6, Eqs. (3.2.2e) and (3.2.2f) yield the following results, with $\dot{\phi}_2 = -90$ rpm $= -3\pi$ rad/s and $\ddot{\phi}_2 = 0$.

$$\ddot{\phi}_3 \approx -4.13 \text{ rad/s}^2 \qquad \ddot{\phi}_4 \approx -99.6 \text{ rad/s}^2$$

b. The calculation of the absolute velocity $\vec{v}_C$ of point C is shown in Fig. 3.7b. Its direction is perpendicular to bar DC. Since it was determined that bar BC is rotating counterclockwise, the direction of $\vec{v}_C$ has to be as indicated in Fig. 3.7. Its magnitude is $v_C = \omega_{BC} d_C \approx 13.8$ m/s; the calculation of the distance d_C is shown in the figure. Now that v_C is known, the angular velocity of bar DC can then be determined. Because $\vec{v}_C$ is directed as shown in Fig. 3.7b, bar DC has to be rotating clockwise. Its angular velocity is $v_C/L_4 \approx 5.52$ rad/s. Therefore, $\dot{\phi}_4 \approx 5.52$ rad/s (compare Fig. 3.7b and Fig. 3.6 to see why the sign of $\dot{\phi}_4$ is positive).

Finally, we come to the calculation of the absolute acceleration $\vec{a}_C$ of point C. The simplest way to express $\vec{a}_C$ is in terms of unit vectors $\hat{r}_4$ and $\hat{\phi}_4$ that rotate with bar DC as shown in Fig. 3.8. However, this is a matter of personal choice since any other set of orthogonal unit vectors could be used. In terms of $\hat{r}_4$ and $\hat{\phi}_4$, the absolute acceleration $\vec{a}_C$ of point C is simply (see Fig. 3.4, p. 185, if you need help with understanding how to obtain such an expression)

$$\vec{a}_C = L_4 \ddot{\phi}_4 \hat{\phi}_4 - L_4 \dot{\phi}_4^2 \hat{r}_4 \approx -249 \hat{\phi}_4 - 76 \hat{r}_4 \text{ m/s}^2$$

The magnitude of $\vec{a}_C$ is $|\vec{a}_C| = \sqrt{249^2 + 76^2} \approx 260$ m/s². The vector $\vec{a}_C$ is directed to the left of bar DC in Fig. 3.8, at an angle equal to $\pi/2 + \arctan(76/249) \approx 1.87$ rad $\approx 107°$, measured counterclockwise from line DC. As an exercise, put it in the figure to visualize it, and check all the answers obtained in this example.

One other possible way to specify the absolute acceleration $\vec{a}_C$ is to express it, for example, in terms of the unit vectors $\hat{x}$ and $\hat{y}$ that are also shown in Fig. 3.8. Again, as mentioned earlier, this is a matter of personal choice. Such an expression is given here, with the specific answer $\vec{a}_C \approx -59\hat{x} - 253\hat{y}$ being for $\phi_4 = 30°$.

$$\vec{a}_C = L_4 \left(\ddot{\phi}_4 \sin \phi_4 + \dot{\phi}_4^2 \cos \phi_4 \right) \hat{x} + L_4 \left(\ddot{\phi}_4 \cos \phi_4 - \dot{\phi}_4^2 \sin \phi_4 \right) \hat{y}$$

$$\approx -59\hat{x} - 253\hat{y} \text{ m/s}^2$$

Figure 3.8 Unit vectors chosen for expressing the acceleration of point C for Example 3.2.2.

3.2.3 Automated Solution for the Four-Bar Mechanism Using Simulink

For practical applications, the most efficient way to do all the calculations presented so far is to automate them in a computer. This can be done either by writing a program using any computer language of one's choice or by using a graphical interface software that replaces the code writing by connecting operational blocks that perform all the required calculations. The latter is what is presented in this book. The user can change parameters in the block diagram and see, almost immediately, how the mechanism would respond to such changes.

Figure 3.9 shows a Simulink block diagram model for automating the calculations of θ_3 and θ_4 presented in Section 3.2.2 and of $\dot{\theta}_3$, $\dot{\theta}_4$, $\ddot{\theta}_3$, and $\ddot{\theta}_4$ presented in Section 3.2.1 and for animating the motion. The solutions are displayed in a digital manner using digital display blocks. The model can be used for investigating and designing four-bar mechanisms. Its inputs are the lengths L_1,

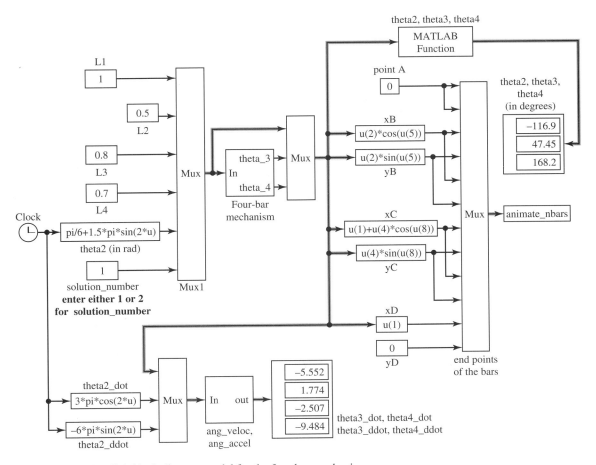

Figure 3.9 A Simulink block diagram model for the four-bar mechanism.

L_2, L_3, and L_4; the angle $\theta_2(t)$ (which may also be a constant), and the angle's first and second derivatives $\dot\theta_2$ and $\ddot\theta_2$. Details about the implementation are given next.

Notice that θ_3 and θ_4 could have been determined by using the three nondimensional quantities L_2/L_1, L_3/L_1, and L_4/L_1 instead of the four lengths L_1, L_2, L_3, and L_4. This is seen by simply dividing Eqs. (3.2.1) and (3.2.2) by L_1; such nondimensional lengths will appear. The same is true for the calculation of the angular velocities and angular accelerations.

$\Longrightarrow$ The Simulink model of Fig. 3.9 is the file `fourbar_mechanism.mdl` included on the CD. To use it, either type `fourbar_mechanism` in the MATLAB command window or type `simulations_menu` and choose the `fourbar_mechanism` model from a pull-down menu that appears on the computer screen. The default inputs of the model are the same as those for solution 1 of Example 3.2.1 (p. 188), with the default choices in the `Solver` part of the `Simulation` $\rightarrow$ `Parameters` menu shown in Fig. 3.10.

Since the model does not involve integrations, a discrete (i.e., no continuous states) solver is used. The maximum step size is set at 0.01, but you may change that and other choices in the menu. As seen in Fig. 3.9, the displayed answers are the same, except for round-off errors, as the ones obtained earlier. Run the model using a different value for the maximum step size, even using the string `auto`

Figure 3.10 The default settings of the `Solver` submenu of the `Simulation` $\rightarrow$ `Parameters` menu for the `fourbar_mechanism` model of Fig. 3.9.

(which lets Simulink automatically set its value) in case the simulation runs either too fast or too slow in your computer.

- To obtain the solution to the equations at $t = 0$ and also see the drawing of the mechanism created by the `animate_nbars` block, set the stop time and the maximum step size in the `Solver` part of the `Simulation →  Parameters` menu to a very small value, say $1.e\text{-}6$.

- To see the answers corresponding to solution number 2 of Example 3.2.1, double click the block labeled `solution_number` in the `fourbar_mechanism` model, change the default value from 1 to 2, and start the simulation again. Run the model with different values of $L_1, L_2, \ldots$ to see how your new mechanism behaves. Enter the values by first double clicking their respective blocks, which are the `Constant` blocks described in Appendix A.

    Notice that the solution number given to the `fourbar_mechanism` model must be either 1 or 2. Otherwise, the answers that are obtained will not make any sense because Eqs. (3.2.14) and (3.2.15) for $\sin\beta = \pm\sqrt{1-\cos^2\beta}$ are implemented in the model as `sinβ= (3-2*solution_number)*`$\sqrt{1\text{-cos}^2\beta}$.

- To use a new function $\theta_2(t)$, double click the `Fcn` block labeled `theta2` and then replace the default expression by the new desired one. Note that the symbol u, instead of t, must be used, as illustrated in Fig. 3.9. As indicated in Appendix A, this is an internal programming requirement in the MATLAB functions that are called by Simulink. The model will start displaying the solutions for $\theta_3, \theta_4, \dot{\theta}_3, \dot{\theta}_4, \ddot{\theta}_3$, and $\ddot{\theta}_4$ right after the simulation is started with `Simulations → Start` (i.e., by choosing `Start` in the `Simulations` menu). To plot any solution versus time t, simply add a scope to the model and connect its input to the line corresponding to the variable that is to be plotted.

- The drawing generated by the `animate_nbars` block is based on an x-y coordinate system centered at point A of Fig. 3.2, with the $+x$ axis directed along AD and the $+y$ axis at $90°$ counterclockwise from the $+x$ axis. As indicated in Appendix A and also in Chapter 2, the input to that block must be a single multisignal line. Such a line is provided at the output of either a `Mux` or a `MATLAB Function` block. The input line to the `animate_nbars` block must carry four signals for each bar to be drawn. In the `fourbar_mechanism` model shown in Fig. 3.9 only the three moving bars are drawn, which explains why the `Mux` labeled `end points of the bars` has 12 inputs.

    The first two inputs to the `animate_nbars` block in the model shown in Fig. 3.9 are the $x = 0$ and $y = 0$ coordinates of point A, and the next two are the x and y coordinates of point B, which are $x_B = L_2 \cos\theta_2$ and $y_B = L_2 \sin\theta_2$. This causes the `animate_nbars` block to draw line AB. The next four inputs are the x and y coordinates of points B and C, where $x_C = L_1 + L_4 \cos\theta_4$ and $y_C = L_4 \sin\theta_4$, and the last four are the x

Figure 3.11 The default settings of the `animate_nbars` menu in the `fourbar_mechanism` model (double click the `animate_nbars` block to get the menu).

and y coordinates of points C and D, where $x_D = L_1$ and $y_D = 0$. The coordinates $x_B, y_B, \ldots$ are calculated in the function blocks labeled `xB`, `yB`, ... in the model shown in Fig. 3.9, and the syntax for their expressions are displayed in that figure.

The default settings of the `animate_nbars` block in the `fourbar_mechanism` model are shown in Fig. 3.11. The number of bars should not be changed in this model, of course. You may want to change the axes limits, and the width of each bar that is drawn.

- All the steps presented in Section 3.2.2 for calculating θ_3 and θ_4 in the `fourbar_mechanism` model are embedded in the block labeled `Four-bar mechanism` that is seen in Fig. 3.9. That block is called a *Simulink subsystem,* and a subsystem[4] was created with the sole purpose of compacting the entire block diagram which, otherwise, might hardly fit on the computer screen. It is very simple to create such blocks, as shown in Appendix A.

The block diagram that is grouped in the subsystem labeled `Four-bar mechanism` in the `fourbar_mechanism` model can be seen by double clicking that subsystem block. It is shown in Fig. 3.12, where the syntax of the expressions in most of the function blocks in that

---

[4] Subsystems are the counterpart to *subroutines* in computer programs that most technical people are familiar with.

SECTION 3.2 The Four-Bar Mechanism

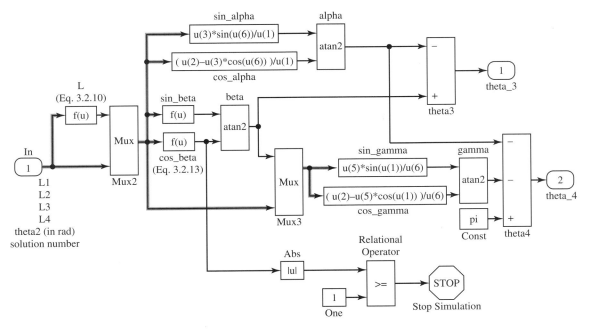

Figure 3.12 Block diagram (a subsystem) for calculating θ_3 and θ_4.

subsystem is also displayed. To save space, the syntax in other function blocks are not displayed. The $f(u)$ block labeled `L` calculates the length L using Eq. (3.2.10) (p. 187). Similarly, the $f(u)$ block labeled `cos_beta` calculates $\cos \beta$ using Eq. (3.2.13) (p. 187). Equations (3.2.14) and (3.2.15) for $\sin \beta = \pm \sqrt{1 - \cos^2 \beta}$ are implemented in the block labeled `sin_beta` as $\sin\beta = (3-2*\text{solution\_number})*\sqrt{1-\cos^2\beta}$, which must be typed in a single line in the block. Double click any function block to see its actual expression.

- Since it may happen that $|(L_3^2 + L^2 - L_4^2)/(2LL_3)| > 1$ in Eq. (3.2.13), the logic implemented by the three blocks that precede the `Stop Simulation` block shown in Fig. 3.12 is included in the `fourbar_mechanism` model to make the simulation stop if that occurs.[5] If such a condition occurs, the mechanism then "locks" itself and stops moving. Such lockings can happen in real mechanisms.

- The angular velocities and the angular accelerations are also determined in a subsystem in the `fourbar_mechanism` model. The calculations are

---

[5] If that logic is not included in the Simulink model, a warning about the square root of a negative number starts appearing in the MATLAB command window at each step of the calculation when that occurs. To prevent a warning that might also occur in the last step of the calculation, the simulation is forced to stop if $|(L_3^2 + L^2 - L_4^2)/(2LL_3)| \geq 1$.

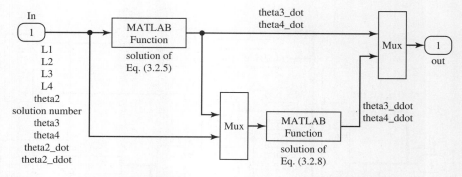

Figure 3.13 Block diagram (a subsystem) for calculating $\dot{\theta}_3$, $\dot{\theta}_4$, $\ddot{\theta}_3$, and $\ddot{\theta}_4$.

done using Eqs. (3.2.5) and (3.2.8), and that subsystem is shown in Fig. 3.13.

The syntax of Eq. (3.2.5) in the `MATLAB Function` block that solves that equation is `-u(2)*u(9)*[u(3)*sin(u(7))-u(4)* sin(u(8));u(3)*cos(u(7))-u(4)*cos(u(8))]^(-1)* [sin(u(5));cos(u(5))]` (notice that expressions have to be written in a single line in any function block). Double click the `MATLAB Function` block that solves Eq. (3.2.8) to see the syntax of the expression involving the calculation of $\ddot{\theta}_3$, and $\ddot{\theta}_4$. The expression that you will see is a very long line! You may wish to copy it to an editor for better viewing.

> The internal forces acting on each bar of a mechanism also play a role in the design of mechanisms since some operational limits for them should not be exceeded. Section 5.6 is dedicated to the calculation of such forces.

3.3 The Geneva Wheel Mechanism

This mechanism is used for creating intermittent motion. It consists of two wheels, one of which has a number of radial slots to engage a pin on the other wheel. Such a mechanism is used in mechanical clocks and watches for example. A six-slot Geneva wheel is shown in Fig. 3.14.

Wheel 1 in Fig 3.14 is the driver. It is driven by an external motor with a known angular velocity ω_1 (relative to the plane of the paper), and it has a pin B. The distance between the fixed pins A and C is such that B engages one of the slots of wheel 2 and then slides along that rotating slot without hitting the end of the slot. Thus, when B is engaged, the slotted wheel 2 is forced to rotate until B slides in and out of the slot as wheel 1 continues to turn. When B is disengaged from a slot on wheel 2, the slotted wheel will stop rotating until B engages the next slot and that wheel starts to rotate again. The slots on wheel 2 in Fig 3.14 are equally spaced, and their openings should be in position to engage pin B.

SECTION 3.3 The Geneva Wheel Mechanism

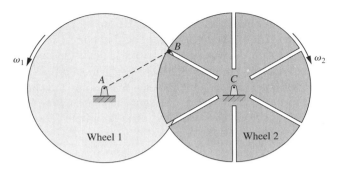

Figure 3.14 A six-slot Geneva wheel mechanism.

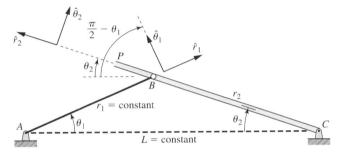

Figure 3.15 Diagram for analyzing the kinematics of a Geneva wheel mechanism.

Given the angular velocity ω_1 of wheel 1 and the position of B at any instant of time t, the kinematic analysis of the Geneva wheel consists of determining the angular velocity and the angular acceleration of the slotted wheel and the velocity and acceleration of pin B relative to the slot in which it slides. The analysis is done with the aid of Fig. 3.15.

Notice that the angular velocity ω_1 shown in Fig. 3.14, drawn with a counter-clockwise arrow, is the same as $\dot{\theta}_1$ because θ_1 in Fig. 3.15 is also drawn counter-clockwise. Similarly, the angular velocity ω_2 shown in Fig. 3.14, drawn with a clockwise arrow, is the same as $\dot{\theta}_2$ because θ_2 in Fig. 3.15 is also drawn clockwise. Bar AB is driven to make the angle $\theta_1(t)$ a known function of time.

As in the four-bar mechanism, the kinematics problem is solved after two independent equations involving the two unknowns $r_2 = |\overrightarrow{CB}|$ and θ_2 shown in Fig. 3.15 are generated by inspection of the figure. Such an inspection yields

$$r_1 \cos\theta_1 + r_2 \cos\theta_2 = \text{constant} = L \qquad (3.3.1)$$

$$r_1 \sin\theta_1 = r_2 \sin\theta_2 \qquad (3.3.2)$$

These are simply the scalar components along line AC and along the perpendicular to line AC of the vector identity $\overrightarrow{AC} = \overrightarrow{AB} + \overrightarrow{BC}$. For simplicity, intermittent motion is not considered, and it is assumed that the length $|\overrightarrow{CP}|$ of bar CP is large enough for pin B to be always engaged on that bar.

The left- and the right-hand sides of Eq. (3.3.1) represent two expressions for the same distance $|\overrightarrow{AC}|$, and the left- and right-hand sides of Eq. (3.3.2) represent two expressions for the perpendicular distance from point B to line AC.

Unlike Eqs. (3.2.1) and (3.2.2) for the four-bar mechanism, Eqs. (3.3.1) and (3.3.2) can be easily solved for r_2 and θ_2. The solution for r_2 is obtained by combining Eqs. (3.2.1) and (3.2.2) to yield Eq. (3.3.3):

$$r_2 = \sqrt{(L - r_1 \cos\theta_1)^2 + (r_1 \sin\theta_1)^2} = \sqrt{L^2 + r_1^2 - 2Lr_1 \cos\theta_1} \qquad \textbf{(3.3.3)}$$

> With r_2 given by Eq. (3.3.3), the only unknown in Eqs. (3.3.1) and (3.3.2) is now the angle θ_2 and, for this mechanism, θ_2 can take any value in any quadrant. Therefore, one should resist the temptation to use only one of those equations to determine θ_2 based only on the value of either $\sin\theta_2$ or $\cos\theta_2$ (or even on $\theta_2 = \arctan[(\sin\theta_2)/(\cos\theta_2)] = \arctan[(r_1 \sin\theta_1)/(L - r_1 \cos\theta_1)]$, for example, as obtained by combining those equations), because the answer that is obtained from such a calculation is an angle that is only in the range $[-\pi/2, \pi/2]$ (i.e., only in two quadrants, instead of all four). A proper calculation of θ_2 for the mechanism should give a result in any of the four possible quadrants so that the actual position of bar CP of the mechanism can be determined.

Generally, having determined r_2, as given by Eq. (3.3.3), one then uses *both* Eqs. (3.3.1) and (3.3.2) to calculate $\sin\theta_2$ and $\cos\theta_2$ as

$$\sin\theta_2 = \frac{r_1 \sin\theta_1}{r_2} \quad \text{and} \quad \cos\theta_2 = \frac{L - r_1 \cos\theta_1}{r_2}$$

The angle θ_2 for bar CP is then determined based on the values of both $\sin\theta_2$ and $\cos\theta_2$. This and other calculations are illustrated in Example 3.3.1.

The problem of determining $\dot{r}_2$, which is the velocity of pin B as seen by an observer rotating with bar CP, and the angular velocity[6] $\dot{\theta}_2(t)$ of bar CP is readily solved by using the following two equations that are obtained by simply taking the time derivative of Eqs. (3.3.1) and (3.3.2):

$$\dot{r}_2 \cos\theta_2 - r_2 \dot{\theta}_2 \sin\theta_2 = r_1 \dot{\theta}_1 \sin\theta_1 \qquad \textbf{(3.3.4)}$$

$$\dot{r}_2 \sin\theta_2 + r_2 \dot{\theta}_2 \cos\theta_2 = r_1 \dot{\theta}_1 \cos\theta_1 \qquad \textbf{(3.3.5)}$$

---

[6] Notice that these scalar quantities are also being referred to as *velocity*, instead of *speed*, since they may be negative. Recall that the word speed is used for the magnitude of velocity.

Equations (3.3.4) and (3.3.5) may be put in matrix form and readily solved for $\dot{r}_2$ and $\dot{\theta}_2$ by using a handheld calculator that handles matrices. They can also be solved by hand, and their step-by-step solution is as follows.[7]

$$\begin{bmatrix} \dot{r}_2 \\ \dot{\theta}_2 \end{bmatrix} = r_1 \dot{\theta}_1 \begin{bmatrix} \cos\theta_2 & -r_2 \sin\theta_2 \\ \sin\theta_2 & r_2 \cos\theta_2 \end{bmatrix}^{-1} \begin{bmatrix} \sin\theta_1 \\ \cos\theta_1 \end{bmatrix}$$

$$= \frac{r_1 \dot{\theta}_1}{r_2} \begin{bmatrix} r_2 \cos\theta_2 & r_2 \sin\theta_2 \\ -\sin\theta_2 & \cos\theta_2 \end{bmatrix} \begin{bmatrix} \sin\theta_1 \\ \cos\theta_1 \end{bmatrix}$$

$$= \begin{bmatrix} r_1 \dot{\theta}_1 (\sin\theta_1 \cos\theta_2 + \sin\theta_2 \cos\theta_1) \\ r_1 \dot{\theta}_1 (\cos\theta_1 \cos\theta_2 - \sin\theta_1 \sin\theta_2)/r_2 \end{bmatrix} \quad (3.3.6)$$

The velocity $\dot{r}_2$ may also be determined by simply taking the time derivative of Eq. (3.3.3) to obtain

$$\dot{r}_2 = \frac{L r_1 \dot{\theta}_1 \sin\theta_1}{\sqrt{L^2 + r_1^2 - 2Lr_1 \cos\theta_1}} \quad (3.3.7)$$

Finally, equations that involve the accelerations $\ddot{r}_2$ (which is the acceleration of pin B as seen by an observer rotating with bar CP) and $\ddot{\theta}_2$ are obtained by taking the time derivative of Eqs. (3.3.4) and (3.3.5). The resulting equations that follow can be simultaneously solved for $\ddot{r}_2$ and $\ddot{\theta}_2$ to obtain numerical values for those quantities once r_2, θ_2, $\dot{r}_2$, and $\dot{\theta}_2$ have been determined.

$$\ddot{r}_2 \cos\theta_2 - r_2 \ddot{\theta}_2 \sin\theta_2 = 2\dot{r}_2 \dot{\theta}_2 \sin\theta_2 + r_2 \dot{\theta}_2^2 \cos\theta_2 + r_1 \left(\ddot{\theta}_1 \sin\theta_1 + \dot{\theta}_1^2 \cos\theta_1 \right) \quad (3.3.8)$$

$$\ddot{r}_2 \sin\theta_2 + r_2 \ddot{\theta}_2 \cos\theta_2 = r_2 \dot{\theta}_2^2 \sin\theta_2 - 2\dot{r}_2 \dot{\theta}_2 \cos\theta_2 + r_1 \left(\ddot{\theta}_1 \cos\theta_1 - \dot{\theta}_1^2 \sin\theta_1 \right) \quad (3.3.9)$$

An Alternative Way to Determine $\dot{r}_2$, $\dot{\theta}_2$, $\ddot{r}_2$, and $\ddot{\theta}_2$ The solution for the velocities $\dot{r}_2$ and $\dot{\theta}_2$ may also be obtained by making use of two different expressions for the absolute velocity $\vec{v}_B$ of point B, which are obtained by inspection of Fig. 3.15 (p. 199) as

$$\vec{v}_B = \frac{d}{dt}(r_1 \hat{r}_1) = r_1 \dot{\theta}_1 \hat{\theta}_1 \quad (3.3.10)$$

$$\vec{v}_B = \frac{d}{dt}(r_2 \hat{r}_2) = \dot{r}_2 \hat{r}_2 + r_2 \dot{\theta}_2 \hat{\theta}_2 \quad (3.3.11)$$

---

[7] The analytical solution for $\dot{r}_2$ seen in Eq. (3.3.6) may also be obtained by multiplying Eq. (3.3.4) by $\cos\theta_2$, multiplying Eq. (3.3.5) by $\sin\theta_2$, adding the results, and then noticing that $\sin^2\theta_2 + \cos^2\theta_2 = 1$. Similarly, the analytical solution for $\dot{\theta}_2$ seen in Eq. (3.3.6) may also be obtained by multiplying Eq. (3.3.5) by $\cos\theta_2$, multiplying Eq. (3.3.4) by $\sin\theta_2$, and then subtracting the results.

> The solutions for $\dot{r}_2$ and $\dot{\theta}_2$ in the resulting vector equation $r_1\dot{\theta}_1\hat{\theta}_1 = \dot{r}_2\hat{r}_2 + r_2\dot{\theta}_2\hat{\theta}_2$ are simply obtained by projecting the known velocity $\vec{v}_B = r_1\dot{\theta}_1\hat{\theta}_1$ in the $\hat{r}_2$ and $\hat{\theta}_2$ directions (see such directions in Fig. 3.15, p. 199), and then equating the projections to $\dot{r}_2$ and $r_2\dot{\theta}_2$, respectively, since those projections must be the ones given by Eq. (3.3.11).

To project $\vec{v}_B = r_1\dot{\theta}_1\hat{\theta}_1$ in the $\hat{r}_2$ and $\hat{\theta}_2$ directions, we need to express $\hat{\theta}_1$ in terms of $\hat{r}_2$ and $\hat{\theta}_2$, and this is done simply by inspection of Fig. 3.15. This yields the following expression

$$\hat{\theta}_1 = \hat{r}_2 \cos\left(\frac{\pi}{2} - \theta_1 - \theta_2\right) + \hat{\theta}_2 \sin\left(\frac{\pi}{2} - \theta_1 - \theta_2\right)$$
$$= \hat{r}_2 \sin(\theta_1 + \theta_2) + \hat{\theta}_2 \cos(\theta_1 + \theta_2) \qquad (3.3.12)$$

Therefore, Eqs. (3.3.10) and (3.3.11) yield the following expressions for $\dot{r}_2$ and $\dot{\theta}_2$,

$$\dot{r}_2 = r_1\dot{\theta}_1 \sin(\theta_1 + \theta_2) \qquad (3.3.13)$$
$$r_2\dot{\theta}_2 = r_1\dot{\theta}_1 \cos(\theta_1 + \theta_2) \qquad (3.3.14)$$

which are the same as those seen in Eq. (3.3.6) if one remembers that $\sin(\theta_1 + \theta_2) = \sin\theta_1\cos\theta_2 + \sin\theta_2\cos\theta_1$ and that $\cos(\theta_1 + \theta_2) = \cos\theta_1\cos\theta_2 - \sin\theta_1\sin\theta_2$.

The solutions for the accelerations $\ddot{r}_2$ and $\ddot{\theta}_2$ may be obtained in the same manner by working with the following two expressions for the acceleration $\vec{a}_B$ (see Fig. 3.15):

$$\vec{a}_B = \frac{d}{dt}(r_1\dot{\theta}_1\hat{\theta}_1) = r_1(\ddot{\theta}_1\hat{\theta}_1 - \dot{\theta}_1^2\hat{r}_1) \qquad (3.3.15)$$
$$\vec{a}_B = (\ddot{r}_2 - r_2\dot{\theta}_2^2)\hat{r}_2 + (r_2\ddot{\theta}_2 + 2\dot{r}_2\dot{\theta}_2)\hat{\theta}_2 \qquad (3.3.16)$$

The solutions for $\ddot{r}_2$ and $\ddot{\theta}_2$ may be obtained by projecting the vector $r_1(\ddot{\theta}_1\hat{\theta}_1 - \dot{\theta}_1^2\hat{r}_1)$ along the $\hat{r}_2$ and $\hat{\theta}_2$ directions, and then equating those projections to $\ddot{r}_2 - r_2\dot{\theta}_2^2$ and $r_2\ddot{\theta}_2 + 2\dot{r}_2\dot{\theta}_2$, respectively. The results obtained are the same (as they should be!) as those obtained from Eqs. (3.3.8) and (3.3.9).

> The use of any of the methodologies presented here is entirely a matter of personal preference. Some readers may naturally favor one of the methodologies over the other, and you should simply use the one you feel most comfortable with.

EXAMPLE 3.3.1

For the mechanism shown in Fig. 3.15 (p. 199), $r_1 = 0.3$ m and $L = 1$ m. Bar AB is driven by a motor so that $\theta_1 = \pi/2 + 3\pi \sin(kt)$ rad, where $k = 1$ rad/s. For $t = 5$ s, determine the absolute velocity and acceleration of point B, the angular velocity $\dot{\theta}_2$, and angular acceleration $\ddot{\theta}_2$ of bar PC, and the acceleration of point B relative to an observer fixed to the rotating bar PC.

■ Solution
We have,

$$\theta_1(5) = \frac{\pi}{2} + 3\pi \sin(5) \approx -7.47 \text{ rad} \qquad (\approx -427.8° \implies -67.8°)$$

$$\dot{\theta}_1(t) = 3k\pi \cos(kt) \qquad \text{thus,} \qquad \dot{\theta}_1(5) \approx 2.67 \text{ rad/s}$$

$$\ddot{\theta}_1(t) = -3k^2\pi \sin(kt) \qquad \text{thus,} \qquad \ddot{\theta}_1(5) \approx 9.04 \text{ rad/s}^2$$

$$r_2(t) = \sqrt{L^2 + r_1^2 - 2Lr_1 \cos\theta_1(t)} \qquad \text{thus,} \qquad r_2(5) \approx 0.93 \text{ m}$$

$$\sin\theta_2 = \frac{r_1 \sin\theta_1}{r_2} \qquad \text{thus,} \qquad \sin\theta_2(5) \approx -0.299$$

$$\cos\theta_2 = \frac{L - r_1 \cos\theta_1}{r_2} \qquad \text{thus,} \qquad \cos\theta_2(5) \approx 0.954$$

These values for $\sin\theta_2(5)$ and for $\cos\theta_2(5)$ then yield

$$\theta_2(5) \approx -17.4°$$

The absolute velocity and acceleration of point B are simply

$$\vec{v}_B = r_1 \dot{\theta}_1 \hat{\theta}_1 \underbrace{\approx}_{\text{for } t=5} 0.8 \hat{\theta}_1 \text{ m/s}$$

$$\vec{a}_B = \frac{d}{dt}\vec{v}_B = r_1 \ddot{\theta}_1 \hat{\theta}_1 - r_1 \dot{\theta}_1^2 \hat{r}_1 \underbrace{\approx}_{\text{for } t=5} 2.71 \hat{\theta}_1 - 2.14 \hat{r}_1 \text{ m/s}^2$$

The angular velocity $\dot{\theta}_2$ of bar PC and the angular acceleration $\ddot{\theta}_2$ may be calculated using any of the methodologies presented in this section. If we work with the projection of the velocity $\vec{v}_B$ along bar PC and along the perpendicular to bar PC, we obtain

$$\dot{r}_2 = r_1 \dot{\theta}_1 \sin(\theta_1 + \theta_2) \qquad \text{which gives} \qquad \dot{r}_2 \underbrace{\approx}_{\text{for } t=5} -0.8 \text{ m/s}$$

$$r_2 \dot{\theta}_2 = r_1 \dot{\theta}_1 \cos(\theta_1 + \theta_2) \qquad \text{which gives} \qquad \dot{\theta}_2 \underbrace{\approx}_{\text{for } t=5} 0.07 \text{ rad/s}$$

Since $\dot{\theta}_2(5)$ is positive, and θ_2 is measured clockwise in Fig. 3.15, bar PC is rotating clockwise at $t = 5$ s.

The acceleration of point B as viewed by an observer fixed to the rotating bar PC is $\ddot{r}_2 \hat{r}_2$. As mentioned earlier, one way to determine $\ddot{r}_2$ and the angular acceleration $\ddot{\theta}_2$ is by taking the time derivative of the given expressions that involve $\dot{r}_2$ and $r_2 \dot{\theta}_2$. The results are

$$\ddot{r}_2 = r_1[\ddot{\theta}_1 \sin(\theta_1 + \theta_2) + \dot{\theta}_1(\dot{\theta}_1 + \dot{\theta}_2)\cos(\theta_1 + \theta_2)]$$

which gives
$$\ddot{r}_2 \underbrace{\approx}_{\text{for } t=5} -2.52 \text{ m/s}^2$$

$$r_2 \ddot{\theta}_2 + \dot{r}_2 \dot{\theta}_2 = r_1[\ddot{\theta}_1 \cos(\theta_1 + \theta_2) - \dot{\theta}_1(\dot{\theta}_1 + \dot{\theta}_2)\sin(\theta_1 + \theta_2)]$$

which gives
$$\ddot{\theta}_2 \underbrace{\approx}_{\text{for } t=5} 2.67 \text{ rad/s}^2 \qquad \text{i.e., } \dot{\theta}_2 \text{ is increasing}$$

EXAMPLE 3.3.2

Bar AB of the mechanism shown in Fig. 3.15 (p. 199) is driven with a constant angular velocity $\dot{\theta}_1 = 2$ rad/s, and its motion is started with $\theta_1 = 0$. For $r_1 = 0.3$ m and $L = 1$ m, determine the first instant of time the angular velocity $\dot{\theta}_2$ of bar PC changes direction. Also determine, for that instant of time, the velocity and acceleration of pin B as viewed by an observer fixed to bar PC and the angular acceleration $\ddot{\theta}_2$ of bar PC.

■ Solution

With $\dot{\theta}_1(t) = $ constant $= 2$ rad/s and $\theta_1(0) = 0$, we obtain $\theta_1(t) = 2t$ rad, with t in seconds. This determines $\theta_1(t)$ at all times.

At all times t we can also write the following expression either by inspecting Fig. 3.15 again or by using Eqs. (3.3.1) and (3.3.2), which were also obtained by inspection of that figure:

$$\frac{\sin\theta_2(t)}{\cos\theta_2(t)} = \frac{r_1 \sin\theta_1}{L - r_1 \cos\theta_1}$$

The rotation of bar PC starts changing direction at some time $t = t^*$ when its angular velocity $\dot{\theta}_2(t^*)$ is $\dot{\theta}_2(t^*) = 0$. With $\dot{\theta}_2(t^*) = 0$, we can also write from Eqs. (3.3.4) and (3.3.5)

$$\frac{\sin\theta_2(t^*)}{\cos\theta_2(t^*)} = \frac{\cos\theta_1(t^*)}{\sin\theta_1(t^*)}$$

Notice that this trigonometric result indicates that $\theta_1(t^*) + \theta_2(t^*) = \pm 90°$ and bars AB and PC are perpendicular to each other at $t = t^*$. This physical situation could also have been inferred by inspection of Fig. 3.15, where bar PC is long enough for it to be always in contact with bar AB. We then have

$$\frac{r_1 \sin\theta_1(t^*)}{L - r_1 \cos\theta_1(t^*)} = \frac{\cos\theta_1(t^*)}{\sin\theta_1(t^*)}$$

This equation may be rearranged to yield, after making use of the fact that $\sin^2\theta_1 + \cos^2\theta_1 = 1$,

$$\cos\theta_1(t^*) = \frac{r_1}{L}$$

A real solution for $\theta_1(t^*)$ exists only when $r_1 \leq L$. For $r_1/L = 0.3$, as specified in the problem statement, and $\theta_1 = 2t$ as already determined, the smallest value of t^* that satisfies this equation is $t^* \approx 0.633$ s. This is the time of the first change in the direction of the rotation of bar PC, and $\theta_1(t^*) = 2t^* = 1.266$ rad $\approx 72.5°$. Since it was concluded that bar PC in Fig. 3.15 is perpendicular to bar AB at $t = t^*$, and since $r_1/L < 1$ in this example, we must then have $\theta_2(t^*) = 90 - 72.5 = 17.5°$. With $r_2(t) = \sqrt{L^2 + r_1^2 - 2Lr_1 \cos\theta_1}$, we then obtain $r_2(t^*) \approx 0.954$ m.

The rest of the calculation has been presented in detail in this section, and one could follow those same steps to obtain the answers to the questions that were asked in this problem. However, for the sake of diversity, let us make use of the fact that bars AB and PC are perpendicular to each other to solve the rest of this problem.

At the instant bars AB and PC are perpendicular to each other, we must have for $\dot{r}_2(t)$ [see Fig. 3.15, which is best for the learning process, or use Eq. (3.3.13)]

$$\dot{r}_2(t) = r_1\dot{\theta}_1(t) \quad \text{which gives} \quad \dot{r}_2(t^*) = 2r_1 = 0.6 \text{ m/s}$$

The acceleration $\ddot{r}_2(t^*)$ and the angular acceleration $\ddot{\theta}_2(t^*)$ can be determined by using the following expressions for the acceleration $\vec{a}_B$ of point B in Fig. 3.15:

$$\vec{a}_B = r_1\ddot{\theta}_1\hat{\theta}_1 - r_1\dot{\theta}_1^2\hat{r}_1$$
$$\vec{a}_B = \left(\ddot{r}_2 - r_2\dot{\theta}_2^2\right)\hat{r}_2 + (r_2\ddot{\theta}_2 + 2\dot{r}_2\dot{\theta}_2)\hat{\theta}_2$$

These are Eqs. (3.3.15) and (3.3.16) whose basic derivations were first presented in Chapter 2 and are summarized in Fig. 3.4 (p. 185).

When bars AB and PC are perpendicular to each other, we must have (see Fig. 3.15) $\hat{\theta}_1 = \hat{r}_2$ and $\hat{r}_1 = \hat{\theta}_2$. Therefore, with $\dot{\theta}_2(t^*) = 0$, the two expressions for $\vec{a}_B$ yield the following results:

$$\ddot{r}_2(t^*) = r_1\ddot{\theta}_1 = 0$$
$$r_2(t^*)\ddot{\theta}_2(t^*) = -r_1\dot{\theta}_1^2 \quad \text{which gives} \quad \ddot{\theta}_2(t^*) \approx -1.26 \text{ m/s}^2$$

Notice that these two preceding expressions may be written right away by inspection of Fig. 3.15 (especially when bars AB and PC are perpendicular to each other) and with the knowledge of the results summarized in Fig. 3.4 (p. 185). To test yourself, do this on your own by looking at Fig. 3.15, as that might allow you to solve many of these type of problems in just a few lines.

3.3.1 Automated Solution for the Geneva Wheel Mechanism using Simulink

The calculations to determine r_2 and θ_2, $\dot{r}_2$ and $\dot{\theta}_2$, and $\ddot{r}_2$ and $\ddot{\theta}_2$, based on Eqs. (3.3.1) to (3.3.9) for the mechanism shown in Fig. 3.15 (p. 199), under the assumption that the two bars AB and CP are engaged at all times, are implemented in the Simulink model shown in Fig. 3.16.

⟹ The model shown in Fig. 3.16 is the file `geneva_wheel.mdl` included on the CD. To use it, either type `geneva_wheel` in the MATLAB command window or choose the model from a menu that appears when you type `simulations_menu`. The user inputs to the model are $L, r_1, \theta_1(t), \dot{\theta}_1(t)$, and $\ddot{\theta}_1(t)$.

■ The default settings for r_1 and for $\theta_1(t)$ in the `geneva_wheel` model are the same as those of Example 3.3.1 (p. 202), with the default choices in the `Solver` part of the `Simulation` → `Parameters` menu being the same as those shown in Fig. 3.10 for the four-bar mechanism. As seen in Fig. 3.16, the answers displayed by the model are the same (except for round-off errors) as the ones obtained in Example 3.3.1. Run the model using a different value for the maximum step size in case the simulation runs either too fast or too slow in your computer. Try the model for finding the solution to Example 3.3.2; you should get a very accurate answer when

206 CHAPTER 3 Kinematic Analysis of Planar Mechanisms

Figure 3.16 A Simulink block diagram model for two engaged bars of the Geneva wheel mechanism (see Fig. 3.15, p. 199).

$t = t^*$ as determined in that example, by setting the stop time in the `Solver` part of the `Simulation` → `Parameters` menu to 0.633. Alternatively, you may want to set the stop time to a larger value and include a logic in the model to make the simulation stop when $|\theta_1 + \theta_2| = \pi/2$.

- The drawing generated by the `animate_nbars` block in the `geneva_wheel` model is based on an x-y coordinate system centered at point A of Fig. 3.15, with the $+x$ axis directed along AC and the $+y$ axis at $90°$ counterclockwise from the $+x$ axis. Thus, as shown in Fig. 3.16, the coordinates of points A and C are $x_A = y_A = 0$, $x_C = L$, and $y_C = 0$. The length of bar CP is set to $L + r_1$ in the Simulink model, which makes the x and y coordinates of point P of bar CP equal to $x_P = L - (L + r_1) \cos \theta_2$ and $y_P = (L + r_1) \sin \theta_2$, respectively. The expressions for the x and y coordinates of points B and P are displayed in the blocks labeled xB, yB, xP, and yP in Fig. 3.16.

- The calculations for determining r_2 and θ_2 in the `geneva_wheel` model are done in the subsystem block labeled `Geneva-wheel`. The diagram that is grouped in that subsystem is shown in Fig. 3.17.

SECTION 3.3 The Geneva Wheel Mechanism

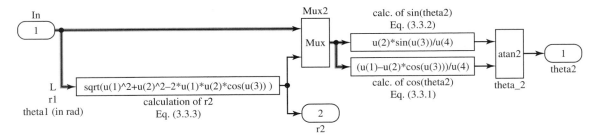

Figure 3.17 Block diagram (a subsystem) for calculating θ_2 and r_2 in the `geneva_wheel` model.

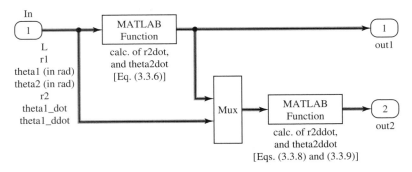

Figure 3.18 Block diagram (a subsystem) for calculating $\dot{r}_2$, $\dot{\theta}_2$, $\ddot{r}_2$, and $\ddot{\theta}_2$ in the `geneva_wheel` model.

- The calculations of $\dot{r}_2$ and $\dot{\theta}_2$, using Eq. (3.3.6), and of $\ddot{r}_2$ and $\ddot{\theta}_2$, using Eqs. (3.3.8) and (3.3.9), are done in the subsystem block labeled `velocities and accelerations` shown in Fig. 3.16. The details of that block are shown in Fig. 3.18.

The syntax of Eq. (3.3.6) in the `MATLAB Function` block that solves it is `[cos(u(4)),-u(5)*sin(u(4));sin(u(4)),u(5)*cos(u(4))]^(-1)*u(2)*u(6)*[sin(u(3));cos(u(3))]`, typed in a single line. Double click the `MATLAB Function` block that solves Eqs. (3.3.8) and (3.3.9) to see the long line of syntax of the matrix solution of those equations in that block.

EXAMPLE 3.3.3

For the mechanism of Example 3.3.2 (see Fig. 3.15, p. 199), for which $r_1 = 0.3$ m, $L = 1$ m, and $\theta_1(t) = \pi/2 + 3\pi \sin(kt)$ rad, where $k = 1$ rad/s, determine the maximum and minimum values of the angular velocity $\dot{\theta}_2(t)$ of bar CP.

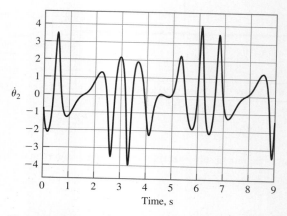

Figure 3.19 Plot of $\dot{\theta}_2(t)$ versus t for Example 3.3.3.

■ **Solution**

The maximum and minimum values of $\dot{\theta}_2(t)$ occur at the instants $t = t^*$ when $[d\dot{\theta}_2/dt]_{t=t^*} = \ddot{\theta}_2(t^*) = 0$; $\dot{\theta}_2(t^*)$ is a maximum if $[d^2\dot{\theta}_2/dt^2]_{t=t^*} < 0$ and a minimum if $[d^2\dot{\theta}_2/dt^2]_{t=t^*} > 0$. If $[d^2\dot{\theta}_2/dt^2]_{t=t^*} = 0$, the sign of the next derivative $[d^3\dot{\theta}_2/dt^3]_{t=t^*}$ has to be checked; the process continues if $[d^3\dot{\theta}_2/dt^3]_{t=t^*}$ is also zero. Since the expression that is obtained for $\dot{\theta}_2(t)$, and even for $\theta_2(t)$, is somewhat complicated, a numerical solution for the maximum and minimum values of $\dot{\theta}_2$ is determined from a plot of $\dot{\theta}_2$ versus t. The plot is shown in Fig. 3.19, and it was obtained by using the Simulink model `geneva_wheel` and adding the `auto_scale_scope` block (described in Appendix A) to it.

The plot discloses that the function $\dot{\theta}_2(t)$ exhibits several maximum and minimum values. The *absolute maximum* (i.e., the largest of the maximum values) and *absolute minimum* (i.e., the smallest of the minimum values) values for $\dot{\theta}_2(t)$ are approximately 4 rad/s, and −4 rad/s, respectively.

3.4 The Slider Crank Mechanism

The slider crank mechanism is shown in Fig. 3.20. The small insert in the figure shows the mechanism in its more common configuration, for $D = 0$ in the larger drawing. The constant lengths L_1 and L_2 of links OA and AB and the design parameter D are known, and the "slider" at B is constrained to move along a straight line. In terms of the unit vectors $\hat{x}$ and $\hat{y}$ shown in the figure, the position vector of point B relative to the fixed point O is $\overrightarrow{OB} = x_B\hat{x} + D\hat{y}$; the constant D is positive for the configuration illustrated in the figure, but negative values may also be chosen.

In practice, the mechanism is constructed so that bars OA and AB and the slider at B move along different, but parallel, planes to avoid collisions. The pin at A, for example, would be long like a nail, but still perpendicular to the plane of

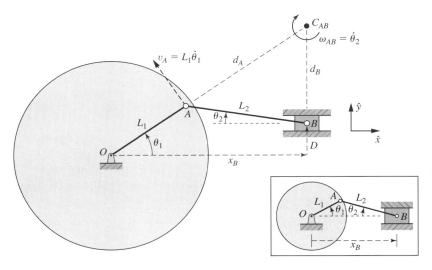

Figure 3.20 The slider crank mechanism.

the paper. If link OA is driven with a known angular velocity $\dot{\theta}_1(t)$, the objective of the kinematic analysis is to determine the angular velocity $\dot{\theta}_2$ and the angular acceleration $\ddot{\theta}_2$ of link AB and the velocity $\dot{x}_B \hat{x}$ and acceleration $\ddot{x}_B \hat{x}$ of the slider at B (see Fig. 3.20). Since the length L_2 of link AB is constant, the angular velocity $\dot{\theta}_2$ of that link may be determined by using its instantaneous center of rotation (which is point C_{AB} in Fig. 3.20), if one wishes to do so. This is illustrated in Fig. 3.20 and will now be discussed.

Analysis 1: Using the Instantaneous Center of Rotation C_{AB} of Link AB

The location of the instantaneous center of rotation C_{AB} for link AB is determined as shown in Fig. 3.20. Since $v_A = L_1 \dot{\theta}_1$ is known, and since it is also equal to $\omega_{AB} d_A$, where d_A is determined from the figure (either analytically, as presented here, or by drawing the figure to scale and measuring it), ω_{AB} is then determined as $\omega_{AB} = L_1 \dot{\theta}_1 / d_A$. If ω_{AB} is positive, as illustrated in Fig. 3.20, the velocity $\dot{x}_B \hat{x}$ of point B will be directed along $-\hat{x}$ since $\dot{x}_B$ will be negative; its magnitude is found as $|\dot{x}_B| = \omega_{AB} d_B$, where d_B is also determined from the figure. The sequence of calculations is then (see Fig. 3.20)

$$L_1 \sin\theta_1 = D + L_2 \sin\theta_2 \quad \text{(this determines } \theta_2 \text{ in the figure)} \quad (3.4.1)$$

$$d_A \cos\theta_1 = L_2 \cos\theta_2 \quad \text{(this determines } d_A) \quad (3.4.2)$$

$$\omega_{AB} d_A = L_1 \dot{\theta}_1 \quad \text{(this determines } \omega_{AB} = \dot{\theta}_2) \quad (3.4.3)$$

$$d_B + D = (L_1 + d_A) \sin\theta_1 \quad \text{(this determines } d_B) \quad (3.4.4)$$

$$\dot{x}_B = -\omega_{AB} d_B \quad \text{(this determines } \vec{v}_B = \dot{x}_B \hat{x}) \quad (3.4.5)$$

Analysis 2: Starting with Two Basic Geometric Relations from Fig. 3.20 For this type of solution, start with the following relations, which are written simply by inspecting Fig. 3.20:

$$L_1 \sin\theta_1 = D + L_2 \sin\theta_2 \tag{3.4.6}$$

$$x_B = L_1 \cos\theta_1 + L_2 \cos\theta_2 \tag{3.4.7}$$

A solution for θ_2 exists only if $|L_1 \sin\theta_1 - D| \leq L_2$, since one must have $|\sin\theta_2| \leq 1$. If the value obtained for $\cos\theta_2$ is in the physical range $[-1, 1]$, we then have two possibilities for $\cos\theta_2$, each one corresponding to a possible configuration of the mechanism. The solution to use is determined by the initial configuration that is chosen for the mechanism to start its motion.

Solution number 1: $\qquad \cos\theta_2 = \sqrt{1 - \sin^2\theta_2} \tag{3.4.8}$

Solution number 2: $\qquad \cos\theta_2 = -\sqrt{1 - \sin^2\theta_2} \tag{3.4.9}$

The angle θ_2 corresponding to the two solutions is then determined by using the values found for $\sin\theta_2$ and for $\cos\theta_2$.

The angular velocity $\dot\theta_2$ of bar AB and the velocity $\vec{v}_B = \dot{x}_B \hat{x}$ of point B are readily calculated from

$$L_1 \dot\theta_1 \cos\theta_1 = L_2 \dot\theta_2 \cos\theta_2 \tag{3.4.10}$$

$$\dot{x}_B = -L_1 \dot\theta_1 \sin\theta_1 - L_2 \dot\theta_2 \sin\theta_2 \tag{3.4.11}$$

These equations are obtained by taking the time derivative of Eqs. (3.4.6) and (3.4.7). It is a useful exercise for you to also try to obtain such equations by inspecting Fig. 3.20 and using the vector identity $\vec{v}_B = \vec{v}_A + \vec{v}_{B/A}$, as explained at the beginning of this chapter.

Observe Eq. (3.4.10) and notice that $\dot\theta_2 = 0$ when $\theta_1 = \pm 90°$. Now look at Fig. 3.20 and notice that, when this happens, the instantaneous center of rotation of bar AB is at infinity. Also, notice, again by observing Eq. (3.4.10), that $\dot\theta_2 \to \infty$ as $\cos\theta_2 \to 0$, i.e., as $\theta_2 \to \pm 90°$. As indicated by Eq. (3.4.6), this situation can be avoided by having $|L_1 \sin\theta_1 - D| < L_2$. For $D = 0$, for example, a mechanism constructed with $L_1 < L_2$ will not encounter such a situation.

With the general approach used in this method of analysis, one can also immediately determine the angular acceleration $\ddot\theta_2$ of link AB, and the acceleration $\vec{a}_B = \ddot{x}_B \hat{x}$ of point B, from the two equations that are obtained by taking the time derivative of Eqs. (3.4.10) and (3.4.11). The resulting equations are

$$L_1(\ddot\theta_1 \cos\theta_1 - \dot\theta_1^2 \sin\theta_1) = L_2(\ddot\theta_2 \cos\theta_2 - \dot\theta_2^2 \sin\theta_2) \tag{3.4.12}$$

$$\ddot{x}_B = -L_1(\ddot\theta_1 \sin\theta_1 + \dot\theta_1^2 \cos\theta_1) - L_2(\ddot\theta_2 \sin\theta_2 + \dot\theta_2^2 \cos\theta_2) \tag{3.4.13}$$

By using Eqs. (3.4.12) and (3.4.13), the numerical values for the accelerations $\ddot{\theta}_2$ and $\ddot{x}_B$ can be readily obtained.[8]

EXAMPLE 3.4.1

For the slider crank mechanism shown in Fig. 3.20 (p. 209), $L_1 = 0.5$ m, $L_2 = 1$ m, and $D = 0.25$ m (i.e., since D is positive, the slider moves in a slot *above* point O as shown in the figure). At a certain instant of time, $\theta_2 = 0$, $\dot{\theta}_1 = 60$ rpm, and $\ddot{\theta}_1 = 0$ (see Fig. 3.20). Determine the angular velocity $\dot{\theta}_2$ of bar AB and the velocity $\dot{x}_B$ of the slider at that instant. Also, determine the distance point B is from point O at that instant.

■ Solution

The steps leading to all the equations that are necessary for determining the angular velocity $\dot{\theta}_2$ and angular acceleration $\ddot{\theta}_2$ of bar AB, as well as the velocity $\dot{x}_B$ and acceleration $\ddot{x}_B$ of the slider for a general slider crank mechanism were presented in detail in this section.

At the instant under consideration, for which $\theta_2 = 0$, we have (see Fig. 3.20 and/or refer to the equations already formulated)

$$\sin\theta_1 = \frac{D}{L_1} = 0.5 \qquad \text{therefore,} \qquad \theta_1 = 30°$$

This yields, $x_B = L_1 \cos\theta_1 + L_2 \approx 1.43$ m.

The distance from point O to point B at the instant under consideration is then

$$\sqrt{x_B^2 + D^2} \approx 1.45 \text{ m}$$

The angular velocity $\dot{\theta}_2$ of bar AB when $\theta_2 = 0$ is determined as

$$L_1\dot{\theta}_1 \cos\theta_1 = L_2\dot{\theta}_2 \qquad \text{which yields} \qquad \dot{\theta}_2 = 15\sqrt{3} \text{ rpm} \approx 26 \text{ rpm}$$

Notice that there is no need to convert revolutions per minute to radians per second for this calculation, and an answer in revolutions per minute is perfectly acceptable. With $\theta_2 = 0$, the preceding equation can be quickly written by inspecting Fig. 3.20 (p. 209) and using the vector identity $\vec{v}_B = \vec{v}_A + \vec{v}_{B/A}$, as suggested earlier. It is also obtained by using Eq. (3.4.10). Notice that the preceding equation is valid only when $\theta_2 = 0$, and, thus, one should never take its derivative to try to determine $\ddot{\theta}_2$, for example.

When $\theta_2 = 0$, the expression for the velocity $\dot{x}_B$ of point B given by Eq. (3.4.11) becomes

$$\dot{x}_B = -L_1\dot{\theta}_1 \sin\theta_1$$

With $\dot{\theta}_1 = 60$ rpm $= 60 \times 2\pi/60$ rad/s (so that a proper answer in meters per second is obtained for $\dot{x}_B$), the preceding result yields $\dot{x}_B = -\pi/2 \approx -1.57$ m/s.

---

[8] As a matter of minor importance, notice that one might prefer to express $\dot{\theta}_2$ as $\dot{\theta}_2 = \pm L_1\dot{\theta}_1(\cos\theta_1)/\sqrt{L_2^2 - (L_1\sin\theta_1 - D)^2}$, which is Eq. (3.4.10) with $\cos\theta_2$ obtained from Eq. (3.4.6) as $\cos\theta_2 = \pm\sqrt{1 - [(L_1\sin\theta_1 - D)/L_2]^2}$. The $+$ sign should be used for solution 1, and the $-$ sign for solution 2. This is a matter of personal choice. With such a choice, one might also prefer to determine $\ddot{\theta}_2$ using the expression that is obtained by taking the time derivative of this equation.

The angular velocity $\dot\theta_2$ and the velocity $\dot x_B$ may also be determined, if one prefers to do so, using the instantaneous center of rotation of bar AB. This is actually the quickest way to determine these quantities for this problem. The sequence of calculations in such a solution is shown here; it is based on the drawing shown in Fig. 3.20 (p. 209), with $\theta_2 = 0$.

Angle θ_1: $\qquad\qquad \theta_1 = \arctan \dfrac{D}{L_1} = 30°$

Distance d_B: $\qquad\quad d_B = L_2 \tan\theta_1 = \dfrac{1}{\sqrt{3}}$ m

Distance d_A: $\qquad\quad d_A = \sqrt{L_2^2 + d_B^2} = \dfrac{2}{\sqrt{3}}$ m

$$d_A \dot\theta_2 = L_1 \dot\theta_1 \quad \text{which yields} \quad \dot\theta_2 = 15\sqrt{3} \text{ rpm} \approx 26 \text{ rpm}$$

$$\dot x_B = -d_B \omega_{AB} = -d_B \dot\theta_2 = -15 \times \dfrac{2\pi}{60} \approx -1.57 \text{ m/s}$$

EXAMPLE 3.4.2

Point B of the mechanism shown in Fig. 3.21 is constrained to move along an inclined slot. The location of the slot is specified by the distance $L = 1$ m from point O and by the angle $\alpha = \text{constant} = 30°$. For this mechanism, $L_1 = 1$ m, $L_2 = 2.5$ m, and bar OA is driven by a motor so that $\phi_1 = kt$ rad, where $k = 2\pi$ rad/s. Determine, for the instants of time when bar OA is parallel to the slot where B moves

a. The angular velocity $\dot\phi_2$ of bar AB and the velocity $\dot r_B$ of point B
b. The angular acceleration $\ddot\phi_2$ of bar AB and the acceleration $\ddot r_B$ of point B

■ **Solution**

To solve this problem, it is better to look at this mechanism from the point of view shown in Fig. 3.22, which makes it clear that this is the same as the general slider crank mechanism of Fig. 3.20 (p. 209), that was already analyzed in this section. Thus, instead of working with the angles ϕ_1 and ϕ_2 and the variable r_B shown in Fig. 3.21, let us work with the angles θ_1 and θ_2 and the variable x_B shown in Fig. 3.22. Since $\phi_1 = \theta_1 + \alpha$, and $\phi_2 = \theta_2 - \alpha$, and α is constant, we then have $\dot\phi_1 = \dot\theta_1$ and $\dot\phi_2 = \dot\theta_2$. Also, since x_B and r_B only differ by a constant, we have $\dot r_B = \dot x_B$.

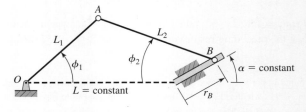

Figure 3.21 The slider crank mechanism for Example 3.4.2.

SECTION 3.4 The Slider Crank Mechanism

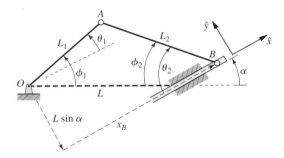

Figure 3.22 The slider crank mechanism for Example 3.4.2.

Comparing Figs. 3.22 and 3.20, we have

$$\overrightarrow{OB} = -(L \sin \alpha)\hat{y} + x_B \hat{x} = x_B \hat{x} + D\hat{y}$$

and, therefore, the variable D in the generic figure 3.20 is, for this problem,

$$D = -L \sin \alpha = -\sin \frac{\pi}{6} = -0.5 \text{ m}$$

The basic equations for the mechanism of Fig. 3.22 are

$$L_1 \sin \theta_1 = L_2 \sin \theta_2 - L \sin \alpha$$

$$x_B = L_1 \cos \theta_1 + L_2 \cos \theta_2$$

and these are, of course, the same as Eqs. (3.4.6) and (3.4.7). Therefore, there is no need to repeat the rest of the analysis here.

As seen in Fig. 3.21, bar OA is parallel to the slot where B moves when $\theta_1 = 0$ and $\theta_1 = 180°$. The results of the calculations of $\theta_2, x_B, \dot{\theta}_2, \dot{x}_B, \ddot{\theta}_2,$ and $\ddot{x}_B$, using the procedure that is presented starting on p. 210 are shown in the following table.

t, s	Solution no.	θ_1, degree	θ_2, degree	x_B, m	$\dot{\theta}_2$, rad/s	$\dot{x}_B$, m/s	$\ddot{\theta}_2$, rad/s²	$\ddot{x}_B$, m/s²
$\frac{1}{12}$	1	0	11.5	3.45	2.56	−1.28	1.34	−56.3
	2	0	168.5	−1.45	−2.56	1.28	−1.34	−22.7
$\frac{7}{12}$	1	180	11.5	1.45	−2.56	1.28	1.34	22.7
	2	180	168.5	−3.45	2.56	−1.28	−1.34	56.3

The entire analysis of the general slider crank mechanism shown in Fig. 3.20 (p. 209), based on the equations developed in this section, is automated in Section 3.4.1. This provides a convenient tool for analyzing such mechanisms.

Figure 3.23 A Simulink block diagram model for the slider crank mechanism.

3.4.1 Automated Solution for the Slider Crank Mechanism Using Simulink

Figure 3.23 shows a Simulink block diagram model for automating the calculations presented in this section for the slider crank mechanism shown in Fig. 3.20 (p. 209). The user inputs to the model are the quantity D (which can be positive, negative, or zero) for the vector $\bar{D}$ shown in Fig. 3.20, the lengths L_1 and L_2, the function $\theta_1(t)$ (which may also be just a constant, if desired) as well as $\dot{\theta}_1$ and $\ddot{\theta}_1$, and the *solution number* introduced in Eqs. (3.4.8) and (3.4.9) (p. 210).

⟹ The Simulink model of Fig. 3.23 is the file `slider_crank.mdl` included on the CD. To use it, either type `slider_crank` in the MATLAB command window or choose the model from a menu that appears when you type `simulations_menu`. The default inputs of the model are the same as those for solution 1 of Example 3.4.2. The default settings of the `Solver` submenu of the `Simulation → Parameters` menu are the same as those shown in Fig. 3.10 (p. 194), except for the final time, which is set to $\frac{1}{12}$.

As seen in Fig. 3.23, the answers displayed by the model are the same (except for round-off errors) as the ones obtained in Example 3.4.2. Try running the model using a different value for the maximum step size (even using the string `auto`, which lets Simulink automatically set its value) in case the simulation runs either too fast or too slow in your computer.

- The drawing generated by the `animate_nbars` block in the `slider_crank` model is based on an x-y coordinate system centered at point O of Fig. 3.20 (p. 209), with the $+x$ and $+y$ axes pointing, respectively, in the same direction as the $\hat{x}$ and $\hat{y}$ unit vectors in that figure. The figure consists of bars OA and AB only. Thus, for drawing bar OA, the first four inputs to the `Mux` labeled `end points of the bars` in Fig. 3.23 are the x and y coordinates of the end points of bar OA, which are $x_O = 0$, $y_O = 0$, $x_A = L_2 \cos \theta_2$, and $y_A = L_2 \sin \theta_2$, respectively. To draw bar AB, the next four inputs to that same `Mux` are the x and y coordinates of the end points of bar AB, which are x_A and y_A again, and x_B and $y_B = D$.

- The calculations for determining θ_2 and x_B, as shown in this section, are performed in the subsystem block labeled `slider-crank mechanism` in the `slider_crank` model. The diagram that is seen when that block is double clicked is shown in Fig. 3.24. Notice the inclusion of the logic for stopping the simulation based on the value obtained for $(L_1 \sin \theta_1 - D)/L_2$. As per Eq. (3.4.6), such a quantity is equal to $\sin \theta_2$, but only if it is in the range $[-1, 1]$. Since $|(L_1 \sin \theta_1 - D)/L_2| = 1$ causes a division by zero in Eq. (3.4.10), which is used for calculating $\dot{\theta}_2$ (see Fig. 3.25), it was decided to make the logic stop the simulation if $|(L_1 \sin \theta_1 - D)/L_2| \geq 1$, instead of $|(L_1 \sin \theta_1 - D)/L_2| > 1$.

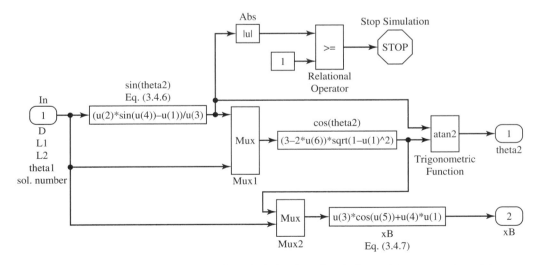

Figure 3.24 Block diagram (a subsystem) for calculating θ_2 and x_B in the `slider_crank` model.

Figure 3.25 Block diagram (a subsystem) for calculating $\dot{\theta}_2$, $\dot{x}_B$, $\ddot{\theta}_2$, and $\ddot{x}_B$ in the `slider_crank` model.

- The calculations of the velocities $\dot{\theta}_2$ and $\dot{x}_B$ and accelerations $\ddot{\theta}_2$ and $\ddot{x}_B$ in the `slider_crank` model are done in the subsystem block labeled `velocities and accelerations` in that model. The details of that subsystem are shown in Fig. 3.25.

EXAMPLE 3.4.3

Bar OA of the mechanism shown in Fig. 3.20 on (p. 209) is driven by a motor so that $\theta_1 = 2\sin(kt)$ rad, where $k = 1$ rad/s. For this mechanism, $D = 0$ and the lengths of the bars are $L_2 = 1$ m and $L_1 = 0.6$ m. Determine

a. The maximum and minimum values of the angular velocity $\dot{\theta}_2$ of bar AB and the maximum and minimum values of the velocity $\dot{x}_B$ of point B
b. The maximum and minimum values of the angular acceleration $\ddot{\theta}_2(t)$ of bar AB

■ **Solution**

a. Since analytical solutions for $\dot{\theta}_2(t)$, $\dot{x}_B(t)$, and $\ddot{\theta}_2(t)$ are somewhat complicated, numerical solutions are sought here. Plots are then generated from the numerical solutions, and the maximum and minimum values of those variables are obtained from such plots.

A plot of $\dot{\theta}_2(t)$ and $\dot{x}_B$ versus time t, obtained simply by adding the `auto_scale_scope` block (described in Appendix A) to the `slider_crank` Simulink model, is shown in Fig. 3.26. The plot discloses that these variables exhibit several maximum and minimum values. Their absolute maximum (i.e., the largest of the maximum values) and the absolute minimum (i.e., the smallest of the

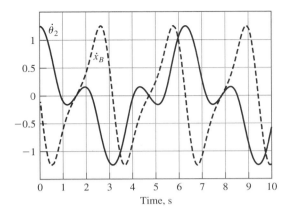

Figure 3.26 Plots of $\dot{\theta}_2(t)$ in radians per second versus t, and $\dot{x}_B(t)$ in meters per second for Example 3.4.3.

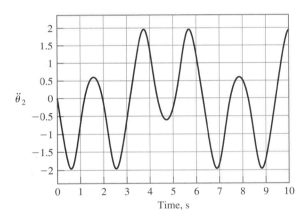

Figure 3.27 Plot of $\ddot{\theta}_2(t)$ in radians per second squared versus t for Example 3.4.3.

minimum values), obtained from Fig. 3.26, are

$$(\dot{\theta}_2)_{min} \approx -1.2 \text{ rad/s} \qquad (\dot{\theta}_2)_{max} \approx 1.2 \text{ rad/s}$$
$$(\dot{x}_B)_{min} \approx -1.2 \text{ m/s} \qquad (\dot{x}_B)_{max} \approx 1.2 \text{ m/s}$$

b. Figure 3.27 shows a plot of $\ddot{\theta}_2(t)$ versus t, also obtained from an `auto_scale_scope` block added to the `slider_crank` Simulink model. The absolute maximum and minimum values of the angular acceleration $\ddot{\theta}_2$ of bar AB, obtained from Fig. 3.27, are

$$(\ddot{\theta}_2)_{min} \approx -1.9 \text{ rad/s}^2 \qquad (\ddot{\theta}_2)_{max} \approx 1.9 \text{ rad/s}^2$$

3.5 The Scotch Yoke Mechanism

A scotch yoke mechanism is shown in Fig. 3.28. It is a very simple mechanism to analyze, and it is used for converting rotational motion into translational motion, or vice versa.

The mechanism consists of a slotted body, which is constrained to move along a straight guide perpendicular to the slot, and of a rotating arm OP of length R, at the end of which there is a pin P that is constrained to move along the slot in the slotted body. By imparting a rotation to arm OP, the slotted body is forced to move along its guide, and the variable x_A that locates point A on the body relative to the fixed point O in Fig. 3.28 is simply $x_A = R\cos\theta$. If arm OP is driven by a motor with a constant angular velocity $\dot\theta = \omega$ [thus, $\theta = \omega t + \theta(0)$], this mechanism can be used, for example, to impart a simple harmonic motion to the slotted body, which could be a table shaker that can be used for vibration tests in a laboratory or in field experiments.

The objective of the kinematic analysis of this mechanism is to determine the velocity $\vec{v}_A = \dot{x}_A \hat{x}_1$ (see Fig. 3.28) and the acceleration $\vec{a}_A = \ddot{x}_A \hat{x}_1$ of the slotted body for any given angular motion $\theta(t)$ of arm OP. The only unknown in this kinematics problem is the variable x_A that tracks the motion of the slotted body, and the basic relationship for this mechanism is simply

$$x_A = R\cos\theta \tag{3.5.1}$$

Therefore, the velocity $\dot{x}_A$ and the acceleration $\ddot{x}_A$ are simply obtained as

$$\dot{x}_A = -R\dot\theta \sin\theta \tag{3.5.2}$$

$$\ddot{x}_A = -R(\ddot\theta \sin\theta + \dot\theta^2 \cos\theta) \tag{3.5.3}$$

Note: The absolute velocity $\vec{v}_P$ and absolute acceleration $\vec{a}_P$ of pin P of bar OP, expressed in terms of the unit vectors $\hat{u}_1$ and $\hat{u}_2$ that rotate with bar OP (see Fig. 3.28), are

$$\vec{v}_P = R\dot\theta \hat{u}_2 \tag{3.5.4}$$

$$\vec{a}_P = \frac{d}{dt}\vec{v}_P = R(\ddot\theta \hat{u}_2 - \dot\theta^2 \hat{u}_1) \tag{3.5.5}$$

Figure 3.28 The scotch yoke mechanism.

SECTION 3.5 The Scotch Yoke Mechanism

Notice that the velocity $\vec{v}_A = \dot{x}_A \hat{x}_1$ and the acceleration $\vec{a}_A = \ddot{x}_A \hat{x}_1$ of point A of the slotted body are the same as the projection of the vectors $\vec{v}_P$ and $\vec{a}_P$, respectively, along the fixed direction $\hat{x}_1$. They are, respectively, the absolute velocity and absolute acceleration of point A.

$\Longrightarrow$ A Simulink model that performs the calculations presented here and animates the scotch yoke mechanism is shown in Fig. 3.29. The model is the file scotch_yoke.mdl on the CD. To use it, either type scotch_yoke in the MATLAB command window or choose the model from a menu that appears when you type simulations_menu. The user inputs to the model are R, $\theta(t)$, $\dot{\theta}(t)$, and $\ddot{\theta}(t)$, and the default settings for R and θ are $R = 1$ and $\theta = \pi/2 + kt$, with $k = 2$ rad/s, as shown in Fig. 3.29.

The default settings of the Solver part of the Simulation → Parameters menu of the scotch_yoke model are the same as those for the other models in this chapter, except for the final time, which is set to 10. Run the model using a different value for the maximum step size in case the simulation runs either too fast or too slow in your computer.

The drawing generated by the animate_nbars block in the scotch_yoke model is based on an x-y coordinate system centered at point O of Fig. 3.28, with the $+x$ and $+y$ axes pointing, respectively, in the same direction

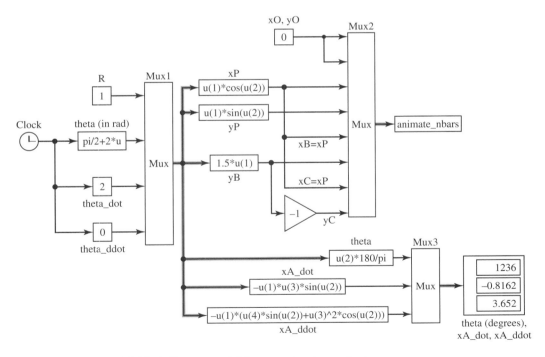

Figure 3.29 A Simulink block diagram model for a scotch yoke mechanism.

as the $\hat{x}_1$ and $\hat{x}_2$ unit vectors shown in that figure. Two bars are drawn, namely, OP and a bar BC, to represent the slot in the mechanism. As displayed in Fig. 3.29, the half-length AB of the slot BC is set to $1.5R$ in the Simulink model, which makes the y coordinates of points B and C equal to $y_B = 1.5R$ and $y_C = -y_B$, respectively.

3.6 Problems

3.1 At a certain instant of time, bar AB of a four-bar mechanism is perpendicular to line AD and bar CD is along AD as shown in Fig. 3.30. The absolute angular velocity of bar AB at that time is 10 rad/s counterclockwise. Determine the absolute velocity of point C (magnitude and direction) and the absolute angular velocity (magnitude and direction) of bars BC and CD at that instant. Do not use a computer to solve this problem.

Figure 3.30 Figure for Problem 3.1.

3.2 At a certain instant of time, the bars of a mechanism are aligned along a straight line as shown in Fig. 3.31. At that instant, the absolute angular velocity and angular acceleration of bar AB are $\omega_{AB} = 15$ rad/s and $\dot{\omega}_{AB} = 24$ rad/s^2, respectively, both counterclockwise. Determine, for that instant, the velocity and acceleration of point B (magnitude and direction) relative to the slot it slides in and the absolute angular velocity and angular acceleration of that slot (magnitude and direction). Do not use a computer to solve this problem.

Figure 3.31 Figure for Problem 3.2.

3.3 At a certain instant of time, bar AB of a slider crank mechanism is perpendicular to the direction of the motion of the slider at C as shown in Fig. 3.32. That bar is rotating counterclockwise with constant absolute angular velocity $\omega_{AB} = 20$ rad/s. Determine, for that instant (without using a computer)

a. The absolute velocity of point C (magnitude and direction) and the absolute angular velocity (magnitude and direction) of bar BC

b. The absolute acceleration of point C (magnitude and direction) and the absolute angular acceleration (magnitude and direction) of bar BC

Figure 3.32 Figure for Problem 3.3.

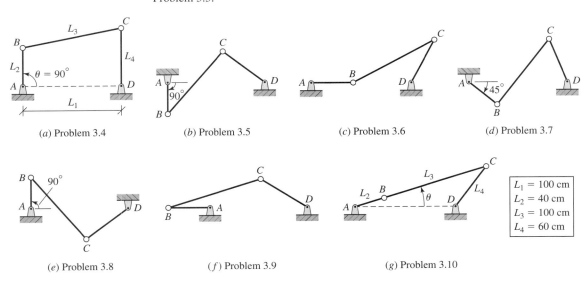

Figure 3.33 Figures for Problems 3.4 to 3.10.

Figure 3.33 shows snapshots of a four-bar mechanism at different times. For all the snapshots, the distance L_1 between the fixed supports at A and D and the lengths of the rotating bars are indicated in the figure. At the instant represented by each snapshot, the absolute angular velocity and angular acceleration of bar AB are $\omega_{AB} = 100$ rpm and $\dot{\omega}_{AB} = 0.3$ rpm/s, respectively, both counterclockwise.

For each problem 3.4 to 3.10, determine, without using the instantaneous center of rotation method, the absolute angular velocity of bars BC and CD and the absolute velocity of point C (specifying its magnitude and direction), at the instant represented by each snapshot in Fig. 3.33. Specify the magnitude of those angular velocities and indicate whether they are clockwise or counterclockwise. Do not use Simulink.

3.4 See Fig. 3.33a.
3.5 See Fig. 3.33b.
3.6 See Fig. 3.33c. Bar AB is along line AD at that instant.
3.7 See Fig. 3.33d.
3.8 See Fig. 3.33e.
3.9 See Fig. 3.33f. Bar AB is along line AD at that instant.
3.10 See Fig. 3.33g. Bars AB and BC are collinear at that instant.

For Problems 3.11 to 3.17, solve Problems 3.4 to 3.10 using instantaneous centers of rotation. *Suggestions:* determine the angles between bars AD and DC and bars AD and BC using the loop equations for those mechanisms; eliminate one of the unknown angles by combining those equations to obtain an equation that involves only one of the unknowns. When necessary, use a calculator to do a few numerical iterations to determine the unknown angle in the resulting equations, as you might do in a closed-book exam. *Hint:* $\sin^2 \alpha + \cos^2 \alpha = 1$ for any α.

3.11 See Problem 3.4 and Fig. 3.33a.
3.12 See Problem 3.5 and Fig. 3.33b.
3.13 See Problem 3.6 and Fig. 3.33c.
3.14 See Problem 3.7 and Fig. 3.33d.
3.15 See Problem 3.8 and Fig. 3.33e.
3.16 See Problem 3.9 and Fig. 3.33f.
3.17 See Problem 3.10 and Fig. 3.33g.

For Problems 3.18 to 3.24, determine the absolute angular acceleration of bars BC and CD and the absolute acceleration of point C (specifying its magnitude and direction), at the instant represented by each snapshot in Fig. 3.33. Specify the magnitude of those angular accelerations, and indicate whether they are clockwise or counterclockwise. Do not use Simulink.

3.18 See Fig. 3.33a.
3.19 See Fig. 3.33b.
3.20 See Fig. 3.33c. Bar AB is along line AD at that instant.
3.21 See Fig. 3.33d.
3.22 See Fig. 3.33e.
3.23 See Fig. 3.33f. Bar AB is along line AD at that instant.
3.24 See Fig. 3.33g. Bars AB and BC are collinear at that instant.

3.25 Determine the angular velocity $\dot{\theta}_2$ and the angular acceleration $\ddot{\theta}_2$ of bar CP for the mechanism shown in Fig. 3.15 (p. 199), at the instants when $\theta_1 = 45°$ and $\theta_1 = 90°$. Also, determine the velocity $\dot{r}_2$ and acceleration $\ddot{r}_2$ (i.e., the velocity and acceleration of point B in a reference frame fixed to the rotating bar CP) at those same instants. When $\theta_1 = 45°$ (see Fig. 3.15), $\dot{\theta}_1 = -0.74$ rad/s and $\ddot{\theta}_1 = -0.55$ rad/s^2. When $\theta_1 = 90°$, $\dot{\theta}_1 = -1.05$ rad/s and $\ddot{\theta}_1 = 0$. For that mechanism, $|\overrightarrow{AB}| = 1.2$ m and $|\overrightarrow{AC}| = 1.5$ m. Do not use a computer to solve this problem.

3.26 Determine the angular velocity $\dot{\theta}_2$ and the angular acceleration $\ddot{\theta}_2$ of bar AB and the absolute velocity and acceleration of point B for the slider crank mechanism shown in Fig. 3.20 (p. 209), at the instants when $\theta_1 = 45°$ and $\theta_1 = 90°$. When $\theta_1 = 45°$, $\dot{\theta}_1 = 0.74$ rad/s and $\ddot{\theta}_1 = 0.55$ rad/s^2. When $\theta_1 = 90°$, $\dot{\theta}_1 = 1.05$ rad/s and $\ddot{\theta}_1 = 0$. For that mechanism, $L_1 = 0.3$ m, $L_2 = 1.5$ m, and $D = 0$. Do not use a computer to solve this problem.

3.27 The mechanism shown in Fig. 3.34 consists of a bar AB, with a groove that is welded at its end B making a constant angle α with AB; a bar PC that is forced to pass through the groove that is welded to bar AB; and a bar CD. Points A and D are pinned to the ground, and bar AB is driven by a motor so that the angle

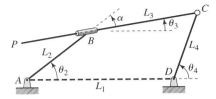

Figure 3.34 Figure for Problem 3.27.

$\theta_2(t)$ is a known function of time t. A pin at C allows PC to rotate relative to CD. For $L_1 = 1$ m, $\alpha = 30°$, $\theta_2(t) = \pi/2 + 3\pi \sin(kt)$ rad, where $k = 1$ rad/s, $L_2 = 0.4$ m, and $L_4 = 1.5$ m, determine the velocity $\dot{L}_3$ and the angular velocities $\dot{\theta}_3$ and $\dot{\theta}_4$, the acceleration $\ddot{L}_3$ (where $L_3 = |\overrightarrow{CB}|$), and the angular accelerations $\ddot{\theta}_3$ and $\ddot{\theta}_4$, when $t = 5$ and 10 s.

Hint: Start, as done in Section 3.2.1, by writing the loop equations for this mechanism and the constraint equation between the appropriate angles labeled in Fig. 3.34 that guarantees that the angle α remains constant.

3.28 At a certain instant $t = t^*$, $\theta_2(t^*) = -45°$, $\dot{\theta}_2(t^*) = 100$ rpm, and $\ddot{\theta}_2(t^*) = 0.3$ rpm/s for a mechanism constructed in an identical manner to the one shown in Fig. 3.34 but driven with a different motor. Determine the angular velocity of bars CP and CD for that same instant.

3.29 What is the angular acceleration of bars CP and CD, at $t = t^*$, for the mechanism of Problem 3.28?

3.30 The distance x_A for the mechanism shown in Fig. 3.35 is changing with time as $x_A = 40 + 15 \sin[\omega(t - 5)]$ cm, where $\omega = 2$ rad/s. For that mechanism, $L_{BE} = 30$ cm, $L_{BC} = L_{ED} = 100$ cm, $L_{CD} = 60$ cm, and $L_{AB} = 50$ cm. Determine, for $t = 5$ s, the absolute angular velocity (magnitude and direction) of all the bars of the mechanism and the magnitude of the absolute velocity of points C and B.

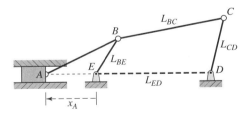

Figure 3.35 Figure for Problem 3.30.

3.31 For the mechanism in Problem 3.30, determine, for $t = 5$ s, the absolute angular acceleration (magnitude and direction) of all the bars of the mechanism and the magnitude of the absolute acceleration of points C and B.

3.32 If you have not done so yet, use the Simulink model `fourbar_mechanism` on the CD to solve Problems 3.4 to 3.10 and 3.18 to 3.24. Show hand calculation of the angle θ for the configuration in Fig. 3.33g.

3.33 Use the Simulink model `fourbar_mechanism` on the CD to solve Example 3.2.2 (p. 189). The mechanism is shown in Fig. 3.6 (p. 190).

3.34 It turns out that the mechanism of Example 3.2.2 (p. 189) is only able to operate without locking when ϕ_2 is in a certain range $\phi_{2\min} \leq \phi_2 \leq \phi_{2\max}$. Determine that range. Then, redesign the mechanism by changing the length of either AB, BC, or CD (without changing the location of the supports at A and D) to prevent locking, and then answer the questions in parts a and b of that example for your redesigned mechanism when $\phi_2 = $ constant $= -90$ rpm. The Simulink model `fourbar_mechanism` is a convenient engineering tool to use to solve this problem.

3.35 Solve Problem 3.25 using the Simulink model `geneva_wheel` that is on the CD, and check the answers you obtained in your solution to that problem.

3.36 Solve Problem 3.26 using the Simulink model `slider_crank` that is on the CD, and check the answers you obtained in your solution to that problem.

3.37 Construct a Simulink model for solving the equations you formulated for the mechanism in Problems 3.27, 3.28, and 3.29, and for animating the motion of that mechanism. Use your model to check the answers you obtained in your solution to those problems.

3.38 Construct a Simulink model for solving the equations you formulated for the mechanism in Problems 3.30 and 3.31 and for animating the motion of that mechanism. Use your model to check the answers you obtained in your solution to those problems.

CHAPTER 4

System of Point Masses: Special Dynamical Properties

To analyze the motion of each particle in a system of N point masses one starts with a free-body diagram for the particles, which is done by isolating them from their surroundings. From then on, the problem is no different than that presented in Chapter 2 for a single particle, where one of the forces acting on the particle is the gravitational force of attraction of the planet the particles are subjected to (say, the earth!).[1] For many applications, the mutual force of attraction between the particles, given by Newton's law of universal gravitation, may be neglected when compared to other forces acting on them (as in Example 2.7.10, p. 140).

The motion of a system of particles exhibits some special properties when the system is viewed as a whole. Such properties are of fundamental importance to the understanding of the motion and are essential to the formulation of the differential equations of motion of the system. They are essential to the analysis of the motion of continuum systems such as rigid bodies. Specifically, these properties deal with the *motion of the center of mass* of the system under the action of the external forces acting on the entire system and with the relationship between the moment of the external forces acting on the system and the time rate of change of the absolute *angular momentum* associated with the motion of the system. The latter relationship provides the means for analyzing the rotational motion of the system, which is the motion associated with the change in direction of a line that joins any two particles of the system. Such properties of the motion of a system of particles are addressed and studied in detail in this chapter.

---

[1] Actually, such a single-particle problem really involves two particles, where the second particle represents the faraway center of mass C of the earth. If the arc length traveled by the particle is small compared to the radius of the earth, the direction of the gravitational force of attraction due to the earth remains essentially constant, and the earth acts as if it were flat. Examples of this were given in Section 2.5.

Two classical problems dealing with the motion of unconstrained particles are analyzed in Sections 4.4 and 4.6. These are the two-body problem of space mechanics and the restricted three-body problem of space mechanics.

4.1 Motion of the Center of Mass of a System of Point Masses

Consider a system of N point masses P_i, each of mass m_i ($i = 1, 2, \ldots, N$), as shown in Fig. 4.1. The position vector of an arbitrary particle P_i, relative to a reference point O, is the vector $\vec{r}_i = \overrightarrow{OP_i}$ shown in the figure.

For convenience, the forces acting on each particle are separated into two types, referred to as *internal* and *external* forces. The former are due to the mutual attraction between any two particles, and the latter constitute all the other forces acting on the particle (such as gravitational forces, friction, and externally applied forces). Each particle P_i in Fig. 4.1 is, thus, subjected to $N - 1$ internal forces $\vec{f}_{ij}$ (which stands for the force of attraction on particle P_i due to particle P_j) and to the resultant $\vec{F}_i$ of the external forces acting on that particle.

It will prove convenient in the development that follows, to define a point with position vector $\vec{r}_C$ from O (point C in Fig. 4.1), determined as

$$\boxed{\vec{r}_C \triangleq \frac{1}{m} \sum_{i=1}^{N} m_i \vec{r}_i} \qquad (4.1.1)$$

where $m = \sum_{i=1}^{N} m_i$ is the total mass of the system. Point C is called the *center of mass* of the system. Notice that if the position vector $\vec{r}_i$ of an arbitrary particle P_i is decomposed as $\vec{r}_i = \vec{r}_C + \vec{\rho}_i$, where $\vec{\rho}_i = \overrightarrow{CP_i}$ is the position vector of P_i relative to C (see Fig. 4.1), Eq. (4.1.1) yields $\sum_{i=1}^{N} m_i \vec{\rho}_i = 0$. This is an alternative equation, which is equivalent to Eq. (4.1.1), for locating the center of mass C.

The center of mass plays a significant role in the dynamics of a system of point masses, especially on the dynamics of bodies with finite dimensions (which is the case for all bodies in nature!). As already indicated in Section 1.10, it is important not to confuse the concepts of *center of mass* and *center of gravity*. It is recommended that you read that section again at this time.

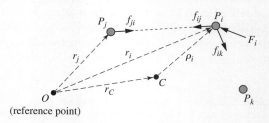

Figure 4.1 A system of N point masses (C is the *center of mass* of the system).

By letting the reference point O in Fig. 4.1 be inertial (i.e., with no acceleration), application of Newton's second law to each particle P_i yields

$$\vec{F}_i + \sum_{\substack{j=1 \\ j \neq i}}^{N} \vec{f}_{ij} = m_i \frac{d}{dt} \vec{v}_i = m_i \frac{d^2}{dt^2} \vec{r}_i \qquad i = 1, 2, \ldots, N \qquad (4.1.2)$$

Equation (4.1.2) represents N vector equations (or $3N$ scalar equations if each particle is allowed to move in three-dimensional space). An important result is obtained by adding all the N vector equations obtained from Eq. (4.1.2). By making use of Eq. (4.1.1), which defines the position vector $\vec{r}_C$ of the center of mass C, Eq. (4.1.2) yields

$$\sum_{i=1}^{N} \vec{F}_i + \sum_{i=1}^{N} \sum_{\substack{j=1 \\ j \neq i}}^{N} \vec{f}_{ij} = \frac{d^2}{dt^2} \sum_{i=1}^{N} (m_i \vec{r}_i) = m \frac{d^2}{dt^2} \vec{r}_C \qquad (4.1.3)$$

By noticing that $\sum_{i=1}^{N} \vec{F}_i$ is the resultant $\vec{F}$ of all the external forces acting on the system of particles, and that the double summation in Eq. (4.1.3) vanishes since $\vec{f}_{ij} = -\vec{f}_{ji}$ (as a consequence of Newton's third law), Eq. (4.1.3) is reduced to

$$\boxed{\vec{F} \triangleq \sum_{i=1}^{N} \vec{F}_i = m \frac{d^2}{dt^2} \vec{r}_C = m \frac{d}{dt} \vec{v}_C} \qquad (4.1.4)$$

The equation $\sum_{i=1}^{N}[\vec{F}_i - m_i(d^2\vec{r}_i/dt^2)] = 0$ is called *d'Alembert's principle,* and it is due to Jean Le Rond d'Alembert (1717–1783).

Equation (4.1.4) discloses that the motion of the center of mass of a system of N particles is governed by the differential equation of motion of a point C of mass equal to the total mass of the system and subjected to the resultant force $\vec{F}$ of all the external forces acting on the system. This is true whether or not the resultant $\vec{F}$ passes through the center of mass C. This is referred to as the *principle of the motion of the center of mass,* and it is one of the properties that were mentioned in the introduction to this chapter. The total absolute linear momentum of the system of particles, defined as $\vec{p} = \sum_{i=1}^{N} m_i \vec{v}_i$, is equal to the absolute linear momentum $m \vec{v}_C$ of the center of mass C. This conclusion is obtained from the time derivative of Eq. (4.1.1) that defines the position vector for C.

4.2 Time Rate of Change of Angular Momentum and Moment

An important relation that is essential for analyzing the motion of a body, either rigid or flexible, is obtained by first defining a vector $\vec{H}_A$ for the system of

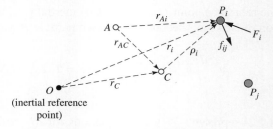

Figure 4.2 A system of N point masses again.

particles as

$$\vec{H}_A \triangleq \sum_{i=1}^{N} \vec{r}_{Ai} \times \frac{d}{dt}(m_i \vec{r}_i) \qquad (4.2.1)$$

where $\vec{r}_{Ai}$ is the vector from an arbitrary point A to particle P_i as shown in Fig. 4.2. The vector $\vec{H}_A$ is known as the *absolute angular momentum, about point A, of the system of particles*. It is the sum of the moment, about an arbitrary point A, of the absolute linear momentum for each particle of the system. The angular momentum $\vec{H}_A$, referred to an arbitrary point A, is a generalization, for a system of particles, of the quantity $\vec{H}_O$ that was introduced in Section 2.3 for a single point mass. The usefulness of the quantity $\vec{H}_A$, and the reason for using an arbitrary point A (instead of simply an inertial point O as done in Section 2.3 for a single point mass), are disclosed when we take its time derivative. By making use of Eq. (4.1.2) we obtain

$$\frac{d}{dt}\vec{H}_A = \sum_{i=1}^{N} \left(\frac{d}{dt}\vec{r}_{Ai}\right) \times \frac{d}{dt}(m_i \vec{r}_i) + \sum_{i=1}^{N} \vec{r}_{Ai} \times \left[\vec{F}_i + \sum_{\substack{j=1 \\ j \neq i}}^{N} \vec{f}_{ij}\right] \qquad (4.2.2)$$

With $\vec{f}_{ij} = -\vec{f}_{ji}$, the term involving the double summation in Eq. (4.2.2) vanishes. By expressing the position vectors $\vec{r}_{Ai}$ and $\vec{r}_i$ as (see Fig. 4.2)

$$\vec{r}_{Ai} = \vec{r}_{AC} + \vec{\rho}_i \qquad (4.2.3)$$

$$\vec{r}_i = \vec{r}_C + \vec{\rho}_i \qquad (4.2.4)$$

and by noticing that the second summation term in Eq. (4.2.2) is equal to the total moment $\vec{M}_A$ of the external forces $\vec{F}_i$, about point A, i.e., $\sum_{i=1}^{N} \vec{r}_{Ai} \times \vec{F}_i = \vec{M}_A$, Eq. (4.2.2) yields

$$\frac{d}{dt}\vec{H}_A = \sum_{i=1}^{N} \frac{d\vec{r}_{AC}}{dt} \times \frac{d(m_i \vec{r}_C)}{dt} + \vec{M}_A + \sum_{i=1}^{N} \frac{d\vec{r}_{AC}}{dt} \times \frac{d(m_i \vec{\rho}_i)}{dt}$$

$$+ \sum_{i=1}^{N} \frac{m_i \, d\vec{\rho}_i}{dt} \times \frac{d(\vec{r}_C)}{dt} \qquad (4.2.5)$$

SECTION 4.3 An Integrated Form of Newton's Second Law and of the Moment Equation

The last two summation terms in Eq. (4.2.5) vanish because $\sum_{i=1}^{N} m_i \vec{\rho}_i = 0$. Therefore, Eq. (4.2.5) simply reduces to

$$\frac{d}{dt}\vec{H}_A = \vec{M}_A + m \frac{d\vec{r}_{AC}}{dt} \times \frac{d\vec{r}_C}{dt} \qquad (4.2.6)$$

In obtaining Eq. (4.2.6), no restrictions were imposed on the choice of point A about which the moment of the external forces and the absolute angular momentum of the system of particles were taken. Thus, that equation is completely general. The choice of working with an arbitrary point A now puts us in a privileged situation of having an equation that discloses that a proper choice of A can eliminate the cross product term in that equation, thus yielding a simpler relationship between the moment $\vec{M}_A$ and the time derivative of the absolute angular momentum $\vec{H}_A$. This can be accomplished by either choosing point A to be inertial, in which case $\vec{r}_{AC} = \vec{r}_C$, or to be the center of mass C, in which case $\vec{r}_{AC} = 0$. The following simpler relations, which still can be applied to any problem, are then obtained:

$$\boxed{\begin{aligned} \frac{d}{dt}\vec{H}_O &= \vec{M}_O \qquad \text{when } O \text{ is an inertial point} \\ \text{and} & \\ \frac{d}{dt}\vec{H}_C &= \vec{M}_C \qquad \text{when } C \text{ is the center of mass of the system} \end{aligned}} \qquad (4.2.7)$$

The second equation of (4.2.7) is particularly significant since it is valid even when the center of mass C of the system of particles is undergoing any arbitrary motion. When analyzing the motion of a particular system, it is more convenient, in general, to work with the moment/angular-momentum relationship relative to a point that is moving with the system rather than relative to an inertial point. In such cases, the simpler form of that relationship, given by the second equation of (4.2.7), evidences the advantage of working with the center of mass.

The equations of (4.2.7) constitute what was referred to as the second property of the motion of a system of point masses mentioned in the introduction to this chapter.

> The material in Sections 4.1 and 4.2 leads, in a natural way, to the study of rigid body dynamics, which is started in Chapter 5.

4.3 An Integrated Form of Newton's Second Law and of the Moment Equation

When the resultant force $\vec{F}$ acting on a system of mass particles is equal to zero, Eq. (4.1.4) discloses that the absolute velocity $\vec{v}_C$ of the center of mass of the system is constant. This means that point C moves along a straight line in space, with constant speed along that line. For this to happen, the system would have to be immune from the attracting force of any celestial body in the universe! What is more realistic, though, is that a component, along an inertial direction, of the resultant force $\vec{F}$ might be equal to zero. In such a case, the component, along that direction, of the absolute velocity $\vec{v}_C$ of the center of mass of the system of particles remains constant.

Since the position vector $\vec{r}_C$ of the center of mass C (see Fig. 4.2) is defined by Eq. (4.1.1), that equation yields $m\vec{v}_C = (m_1 + m_2 + \cdots + m_N)\vec{v}_C = m_1\vec{v}_1 + m_2\vec{v}_2 + \cdots + m_N\vec{v}_N$, where $\vec{v}_1, \vec{v}_2, \ldots$ are the absolute velocity of particle 1, particle 2, etc. The result mentioned in the previous paragraph, i.e., $\vec{v}_C =$ constant when $\vec{F} = 0$, is known in dynamics as the *principle of conservation of linear momentum* for the system of N point masses.

Another conservation principle in dynamics is disclosed by either equation of (4.2.7), which relates the absolute angular momentum of the system of particles to the moment of the forces applied to the system, i.e., $d\vec{H}_O/dt = \vec{M}_O$ or $d\vec{H}_C/dt = \vec{M}_C$ where O is any inertial point. Either the absolute angular momentum $\vec{H}_O$ is constant if $\vec{M}_O = 0$ or the absolute angular momentum $\vec{H}_C$ is constant if $\vec{M}_C = 0$. This is known as the *principle of conservation of angular momentum*. The latter result (i.e., $\vec{H}_C =$ constant if $\vec{M}_C = 0$) is particularly useful because the center of mass of the system may be moving in any manner in space. For many problems, a component, along an inertial direction, of the moment $\vec{M}_C$ (or $\vec{M}_O$) might be equal to zero, instead of the moment itself being zero. In such a case, the component, along that direction, of the absolute angular momentum $\vec{H}_C$ (or $\vec{H}_O$) remains constant.

EXAMPLE 4.3.1

A person of mass $m_1 = 70$ kg is standing on a stationary horizontal platform of mass $m_2 = 400$ kg that is on the smooth surface of a frozen lake, as illustrated in Fig. 4.3. Point C is the center of mass of that person. Neglecting friction between the platform and the frozen surface, determine the absolute velocity of the platform when the person starts to walk along the platform with a constant velocity $\dot{x}_1 = 4$ km/h relative to the platform.

■ Solution

When walking, one holds a foot on the surface, lifts the other foot off the surface, and moves it forward. The same procedure is then repeated with the other foot. If there is no sliding, the friction force acting on the foot that is in contact with the surface is static.

Figure 4.3 Motion on a platform that is free to slide (Example 4.3.1).

SECTION 4.3 An Integrated Form of Newton's Second Law and of the Moment Equation

Figure 4.4 Free-body diagrams for Example 4.3.1.

The free-body diagrams for the person and platform are shown in Fig. 4.4; $\vec{f}_1$ and $\vec{f}_2$ are friction forces, with $\vec{f}_2 = 0$ since there is no friction between the platform and the stationary ground (the earth).[2] The absolute position vectors $x_2 \hat{x}$ and $(x_1 + x_2)\hat{x}$ for the platform and for the center of mass C of the person are also shown in the figure; the unit vectors $\hat{x}$ and $\hat{y}$ are inertial (i.e., they maintain the same directions relative to inertial space).

Application of Newton's second law to the center of mass of the person and to the platform yield the following equations.

For the person:
$$N_1 = m_1 g \tag{4.3.1a}$$
$$f_1 = m_1(\ddot{x}_1 + \ddot{x}_2) \tag{4.3.1b}$$

For the platform:
$$N_2 = N_1 + m_2 g \tag{4.3.1c}$$
$$f_1 = -m_2 \ddot{x}_2 \tag{4.3.1d}$$

The following equation is obtained by combining Eqs. (4.3.1b) and (4.3.1d):
$$m_2 \ddot{x}_2 + m_1(\ddot{x}_1 + \ddot{x}_2) = 0$$

This equation can easily be integrated with respect to time. Since values for $\dot{x}_1$ and $\dot{x}_2$ are known just before the person starts to walk on the platform (and not just after), let us integrate that equation from $t = 0^-$ (where $t = 0^-$ denotes the instant just before the person started walking) to an arbitrary value of t. This then yields

$$\int_{t=0^-}^{t} [m_2 \ddot{x}_2 + m_1(\ddot{x}_1 + \ddot{x}_2)]\, dt = m_2 \dot{x}_2 + m_1(\dot{x}_1 + \dot{x}_2) - [m_2 \dot{x}_2 + m_1(\dot{x}_1 + \dot{x}_2)]_{t=0^-} = 0 \tag{4.3.1e}$$

---

[2] The earth would be a third body in this problem (and other similar ones). However, because the mass of the earth is very large (essentially infinite) compared to the other masses involved and because of the small amount of time involved in such a motion, the earth behaves as an inertial body when its long time motion around the sun is not considered.

The result expressed by Eq. (4.3.1e) discloses that the linear momentum of the entire system, which is the quantity $[m_2\dot{x}_2 + m_1(\dot{x}_1 + \dot{x}_2)]\hat{x}$, is conserved (i.e., remains constant for all times). Such results, which are generally referred to as the principle of conservation of linear momentum, are obtained in a natural manner simply by application of Newton's second law to the appropriate parts of the system and by working with the resulting equations, as illustrated in this example.

Equation (4.3.1e) is easily solved for the desired unknown $\dot{x}_2$. With the given initial conditions, i.e., $\dot{x}_1(0^-) = \dot{x}_2(0^-) = 0$, we then obtain

$$\dot{x}_2 = -\frac{m_1}{m_1 + m_2}\dot{x}_1$$

Right after the person starts to walk along the platform (i.e., at $t = 0^+$) with $\dot{x}_1 =$ constant $= 4$ km/h, the platform starts moving in the opposite direction and $\dot{x}_2 =$ constant ≈ -0.596 km/h ≈ -9.93 m/min.

EXAMPLE 4.3.2 The *Two-Body* Problem of Space Mechanics (or of Spaceflight Dynamics)

As indicated in footnote 3, the system under consideration consists of two spherical bodies with masses m_1 and m_2, with geometric centers at P_1 and P_2, respectively, subjected only to their mutual forces of attraction. The free-body diagrams for the two bodies are shown in Fig. 4.5b, where $\vec{f}_{12}$ is the force on P_1 due to P_2 and $\vec{f}_{21} = -\vec{f}_{12}$ is the force on P_2 due to P_1. The vectors $\vec{r}_1$ and $\vec{r}_2$ shown in Fig. 4.5a are the position vectors of P_1 and P_2 relative to the center of mass C of the system while the vector $\vec{r} = \vec{r}_2 - \vec{r}_1$ is the position vector of P_2 relative to P_1. The x-y coordinate system shown in the figure is inertial (i.e., the orientation of the x and y axes does not change), and the angle θ is chosen for tracking the orientation of line $P_1 P_2$. By working with the polar coordinates $r \stackrel{\Delta}{=} |\overrightarrow{P_1 P_2}|$ (which is the distance from P_1 to P_2) and θ (see Fig. 4.5),

a. Obtain the differential equations that govern the motion of the system.
b. Verify that the differential equations of motion obtained in part a are of the same form as those for the central force problem analyzed in Example 2.7.11 (p. 144) and, thus, that all the conclusions and results obtained for that example are valid for the two-body problem under consideration here.

■ **Solution**

a. Two important conclusions about the motion can be obtained right away.
   ■ Since the resultant of the forces acting on the system $\vec{F} = \vec{f}_{12} + \vec{f}_{21}$ is equal to zero (see Fig. 4.5), Eq. (4.1.4) discloses that the center of mass C of the system is

---

[3] This is one of the classical problems in dynamics. It consists of determining the motion of two bodies subjected only to their mutual force of attraction. The solution to such a problem provides a first approximation for the motion of the center of mass of a satellite around a planet or the motion of any planet around the sun. In this classical problem, both bodies are treated as homogeneous spherical bodies. As shown in Section 1.7, the gravitational force of attraction due to a homogeneous sphere is the same as that of a particle with mass equal to the mass of the sphere, located at its center.

SECTION 4.3 An Integrated Form of Newton's Second Law and of the Moment Equation

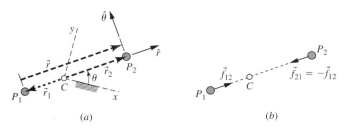

Figure 4.5 The two-body system and the free-body diagrams for Example 4.3.2.

inertial (i.e., its absolute acceleration is zero). Therefore, the position vectors $\vec{r}_1$ and $\vec{r}_2$ shown in Fig. 4.5 are, respectively, the absolute position vectors for P_1 and P_2.

- Since the moment about C of the forces $\vec{f}_{12}$ and $\vec{f}_{21}$ is equal to zero, the second equation of (4.2.7) discloses that the angular momentum $\vec{H}_C$ is a constant (because $d\vec{H}_C/dt = \vec{M}_C = 0$). Thus, as concluded in Example 2.5.11, the motion takes place in a single plane in space. Therefore, only two variables are needed to describe the motion. With $\vec{r} \stackrel{\Delta}{=} r\hat{r}$ (see Fig. 4.5), the variables chosen here are the relative distance r between P_1 and P_2, and the angle θ, which is measured from a nonrotating reference line in the plane of the motion and tracks the direction of line $P_1 P_2$ in that plane.[4]

Since Newton's second law requires the use of the absolute position vectors $\vec{r}_1$ for P_1, and $\vec{r}_2$ for P_2, and since we want to use the relative position vector $\vec{r} = r\hat{r}$, it is necessary to determine the expressions that relate $\vec{r}_1$ and $\vec{r}_2$ to $\vec{r}$. The first relationship, Eq. (4.3.2a), is simply obtained from the definition of the center of mass C of the system, and the second one, Eq. (4.3.2b), is obtained by inspecting Fig. 4.5.

$$m_1 \vec{r}_1 + m_2 \vec{r}_2 = 0 \qquad (4.3.2a)$$

$$\vec{r} = \vec{r}_2 - \vec{r}_1 \qquad (4.3.2b)$$

From Eqs. (4.3.2a) and (4.3.2b), the following expressions for the absolute position vectors $\vec{r}_1$ and $\vec{r}_2$ are then obtained in terms of the relative position vector $\vec{r}$.

$$\vec{r}_1 = -\frac{m_2}{m_1 + m_2}\vec{r} \qquad \vec{r}_2 = \frac{m_1}{m_1 + m_2}\vec{r} \qquad (4.3.2c)$$

Notice that when $m_1 \gg m_2$, one obtains $\vec{r}_2 \approx \vec{r}$ and $\vec{r}_1 \approx 0$, which means that P_1 is essentially inertial. This was the central force problem considered in Examples 2.5.11 and 2.7.11.

---

[4] In practical terms, it is more convenient to work with the relative distance r instead of either distance $r_1 = |\overrightarrow{CP_1}|$ or $r_2 = |\overrightarrow{CP_2}|$. The desirability of doing so is seen if we think of one body as being the earth, with a person on the earth observing the motion of the other body relative to the earth.

Since we are dealing with relative motion, the same differential equations involving r and θ will be obtained if we consider either P_1 or P_2 for formulating those equations. By arbitrarily using P_2 we then have

- Absolute position vector for P_2:

$$\vec{r}_2 = \frac{m_1}{m_1 + m_2}\vec{r} = \frac{m_1}{m_1 + m_2}r\hat{r} \qquad (4.3.2d)$$

- Absolute velocity of P_2:

$$\vec{v}_2 = \frac{d\vec{r}_2}{dt} = \frac{m_1}{m_1 + m_2}\left[\dot{r}\hat{r} + r\frac{d\hat{r}}{dt}\right] = \frac{m_1}{m_1 + m_2}[\dot{r}\hat{r} + r\dot{\theta}\hat{\theta}] \qquad (4.3.2e)$$

- Absolute acceleration of P_2:

$$\vec{a}_2 = \frac{d\vec{v}_2}{dt} = \frac{m_1}{m_1 + m_2}\left[\ddot{r}\hat{r} + (r\ddot{\theta} + \dot{r}\dot{\theta})\hat{\theta} + \dot{r}\frac{d\hat{r}}{dt} + r\dot{\theta}\frac{d\hat{\theta}}{dt}\right]$$

$$= \frac{m_1}{m_1 + m_2}[(\ddot{r} - r\dot{\theta}^2)\hat{r} + (r\ddot{\theta} + 2\dot{r}\dot{\theta})\hat{\theta}] \qquad (4.3.2f)$$

- Resultant force acting on P_2 due to the attraction of P_1:

$$\vec{f}_{21} = -\frac{Gm_1 m_2}{r^2}\hat{r} \qquad (4.3.2g)$$

Newton's second law, $\vec{f}_{21} = m_2 \vec{a}_2$, then yields the following scalar differential equations of motion:

$$\ddot{r} - r\dot{\theta}^2 = -\frac{G(m_1 + m_2)}{r^2} \qquad (4.3.2h)$$

$$r\ddot{\theta} + 2\dot{r}\dot{\theta} = \frac{1}{r}\frac{d}{dt}(r^2\dot{\theta}) = 0 \quad \text{or} \quad r^2\dot{\theta} = \text{constant} \stackrel{\Delta}{=} h \qquad (4.3.2i)$$

b. Equations (4.3.2h) and (4.3.2i) are the differential equations that govern the relative motion between the two bodies (the trajectory of one body relative to the other is called an *orbit*). These differential equations are exactly of the same form as Eqs. (2.7.11a) and (2.7.11b) obtained for the central force problem analyzed in Example 2.7.11, with $GM_E = gR_E^2$ in those equations replaced by $G(m_1 + m_2)$. Therefore, all the conclusions and results obtained for the central force problem are valid for the two-body problem, for which both bodies are moving.

It is worth noticing that Eq. (4.3.2i) also follows directly from the fact that the angular momentum $\vec{H}_C$ is constant. The angular momentum $\vec{H}_C$ for this system is [see Fig. 4.5, and use the expressions for $\vec{r}_1$ and $\vec{r}_2$ given by Eqs. (4.3.2c)]:

$$\vec{H}_C = m_1 \vec{r}_1 \times \frac{d}{dt}\vec{r}_1 + m_2 \vec{r}_2 \times \frac{d}{dt}\vec{r}_2 = \frac{m_1 m_2}{m_1 + m_2}\vec{r} \times \frac{d}{dt}\vec{r}$$

$$= \frac{m_1 m_2}{m_1 + m_2}r\hat{r} \times (\dot{r}\hat{r} + r\dot{\theta}\hat{\theta}) = \frac{m_1 m_2}{m_1 + m_2}r^2\dot{\theta}\hat{r} \times \hat{\theta} = \text{constant} \qquad (4.3.2j)$$

The mass $m \stackrel{\Delta}{=} m_1 m_2/(m_1 + m_2)$ is called the *reduced mass* of the system. The quantity $h = r^2\dot{\theta} = [r^2\dot{\theta}]_{t=0}$ is called the *angular momentum per unit reduced mass* of the system.

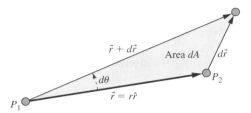

Figure 4.6 Infinitesimal area element swept by the position vector $\vec{r}$ for Example 4.3.2.

Such a quantity has a nice physical interpretation, which is arrived at with the aid of Fig. 4.6. The figure shows an infinitesimal portion of the trajectory of P_2 relative to P_1. The infinitesimal area dA swept by the position vector $\vec{r}$ during an infinitesimally small elapsed time dt is equal to half the area of a parallelogram of sides $r = |\vec{r}|$ and $r\,d\theta$. The area dA may be calculated as (refer to Fig. 4.5 when needed)

$$dA = \frac{1}{2}|\vec{r} \times d\vec{r}| = \frac{1}{2}|r\hat{r} \times [(dr)\hat{r} + r(d\theta)\hat{\theta}]| = \frac{1}{2}|r^2\,d\theta| \qquad (4.3.2\text{k})$$

Equation (4.3.2k) discloses that $dA/dt = \frac{1}{2}r^2|\dot{\theta}|$ remains constant during the motion since $r^2\dot{\theta}$ is constant. This is the same conclusion obtained by Kepler after many years of exhaustive analysis of the data gathered by Tycho Brahe concerning the motion of the planets. Kepler's second law states that *the motion of each planet is such that the radius vector from the sun to the planet sweeps equal areas in equal times*.

4.4 Detailed Analysis of the Two-Body Problem

Because of the importance of the two-body problem, a detailed investigation of the motion is presented in this section. As mentioned earlier, the solution to such a problem provides a first approximation for the motion of a satellite around a planet or for the motion of any planet around the sun. A more detailed model for the earth-satellite problem, for example, involves taking into account the oblateness of the earth. Such an effect, and other effects, are treated in advanced books on *celestial mechanics*.

When the expression for $\dot{\theta}$ obtained from Eq. (4.3.2i) is substituted into Eq. (4.3.2h), the resulting equation is of the form $\ddot{r} = f(r)$ and, thus, it can be integrated once with respect to r to obtain the following result. Such a result is the same as that given by Eq. (2.7.11d) of Example 2.7.11.

$$\frac{1}{2}\underbrace{\left(\dot{r}^2 + \frac{h^2}{r^2}\right)}_{=v^2} - \frac{\mu}{r} = \text{constant} \stackrel{\Delta}{=} E \qquad \text{where } \mu = G(m_1 + m_2) \qquad (4.4.1)$$

As indicated in Eq. (4.4.1), the quantity $\dot{r}^2 + h^2/r^2 = \dot{r}^2 + r^2\dot{\theta}^2$ is the square of the speed of P_2 relative to P_1. As indicated throughout the book, an expression of the form of Eq. (4.4.1), which only involves $\dot{r}$ and r, yields important information about the motion even without determining how $r(t)$ changes with time. Such an investigation is presented in Section 4.4.2. An analytical integration of Eq. (4.4.1) is presented in Section 4.4.3 with the objective of generating additional important information about the motion of P_2 relative to P_1. The knowledge gained from such analyses is of crucial importance in the planning of spacecraft missions. Before this is done, the physical interpretation of the terms that appear in Eq. (4.4.1) is presented in Section 4.4.1.

4.4.1 The Meaning of the Quantity E: Kinetic and Potential Energies of the Motion

The quantity E in Eq. (4.4.1) has a physical interpretation, which is arrived at if we apply the work-energy relationship $dW = d(T)$ to this problem. Such an approach is presented in Section 2.8. By referring to Fig. 4.5, and by using the expressions for $\vec{r}_1$ and $\vec{r}_2$ given by Eqs. (4.3.2c), the expressions for the infinitesimal work dW done by the forces acting on the system, and for the kinetic energy T of the motion, are developed as follows:

Infinitesimal work:

$$dW = \vec{f}_{21} \bullet d\vec{r}_2 + \vec{f}_{12} \bullet d\vec{r}_1 = \vec{f}_{21} \bullet d\vec{r}_2 - \vec{f}_{21} \bullet d\vec{r}_1 = \vec{f}_{21} \bullet d(\vec{r}_2 - \vec{r}_1)$$

$$= \vec{f}_{21} \bullet d\vec{r} = -\frac{Gm_1 m_2}{r^2} dr \qquad (4.4.2)$$

Kinetic energy:

$$T = \frac{1}{2}m_1 \frac{d\vec{r}_1}{dt} \bullet \frac{d\vec{r}_1}{dt} + \frac{1}{2}m_2 \frac{d\vec{r}_2}{dt} \bullet \frac{d\vec{r}_2}{dt} = \frac{1}{2}\frac{m_1 m_2}{m_1 + m_2} \frac{d\vec{r}}{dt} \bullet \frac{d\vec{r}}{dt}$$

$$= \frac{1}{2}\frac{m_1 m_2}{m_1 + m_2}(\dot{r}^2 + r^2\dot{\theta}^2) = \frac{1}{2}\frac{m_1 m_2}{m_1 + m_2}\left(\dot{r}^2 + \frac{h^2}{r^2}\right) \qquad (4.4.3)$$

Equation (4.4.3) discloses that the quantity $\frac{1}{2}(\dot{r}^2 + h^2/r^2)$ is the *kinetic energy per unit reduced mass* of the system.

Since the expression for dW is an exact differential, integration of $dW = d(T)$ yields

$$T - W = \frac{m_1 m_2}{m_1 + m_2}\left[\frac{1}{2}\left(\dot{r}^2 + \frac{h^2}{r^2}\right) - \frac{\mu}{r}\right] = \text{constant} \triangleq \frac{m_1 m_2}{m_1 + m_2} E \qquad (4.4.4)$$

which is the total energy of the motion. The quantity $-\mu/r$ is called the *potential energy per unit reduced mass* of the system, and the quantity E is the *total energy per unit reduced mass* of the system. The potential energy of the system is the quantity U obtained by integrating $d(U) \triangleq -dW = (Gm_1m_2/r^2)\,dr$, where the minus sign is for conforming to universal convention. Thus, except for an arbitrary constant of integration that can be set to zero without any loss of generality, as discussed in Section 2.8, we then obtain $U = -Gm_1m_2/r$.

Notice that, since $T + U = [m_1m_2/(m_1+m_2)](v^2/2 - \mu/r)$ always remains constant (i.e., a decrease in the kinetic energy is compensated by an increase in potential energy, and vice versa), the speed of P_2 relative to P_1 is maximum when P_2 is closest to P_1 and minimum when P_2 is farthest.

Note: The use of the work-energy equation $dW = d(T)$, together with the moment and angular momentum relationship $d\bar{H}_C/dt = \bar{M}_C$, yielded the integral of the motion given by Eq. (4.4.4) without generating Eq. (4.3.2h) (which required the use of acceleration to do so). This observation is made here to reinforce the notion that one actually has a choice on how to approach the formulation of a dynamics problem, and that different choices lead to different amounts of effort to arrive at the same place.

4.4.2 Determination of the Maximum and Minimum Values of $r(t)$

The maximum and minimum values of $r(t)$ are obtained by setting $\dot{r} = 0$ in Eq. (4.4.1). This yields the following equation, where r_m is being used for denoting such values.

$$\frac{h^2}{2r_m^2} - \frac{\mu}{r_m} - E = 0 \tag{4.4.5}$$

Equation (4.4.5) is a quadratic equation in r_m. When $E \neq 0$, its roots are

$$r_m = -\frac{\mu}{2E}(1 \pm e) \tag{4.4.6}$$

where

$$e = \sqrt{1 + \frac{2Eh^2}{\mu^2}} \tag{4.4.7}$$

The trajectory r versus θ of P_2 relative to P_1 is called an *orbit*, and the quantity e defined by Eq. (4.4.7) is called the *orbit eccentricity*. All the different cases involving the possible values of e are presented.

CASE 1 $0 \leq e < 1$

E Is Negative

In this case both values given by Eq. (4.4.6) are positive and, thus, they are the maximum and the minimum values of $r(t)$:

$$r_{\max} = -\frac{\mu}{2E}(1+e) \qquad r_{\min} = -\frac{\mu}{2E}(1-e) \qquad (4.4.8)$$

The trajectory of P_2 relative to P_1 is then a closed curve in space. The points where the minimum and the maximum values of $r(t)$ occur are called, respectively, the *pericenter* and the *apocenter* of the orbit.[5]

Notice that the analysis so far still did not reveal what kind of closed trajectory P_2 is executing around P_1. To determine that, it is necessary to integrate Eq. (4.4.1) analytically. This is done in Section 4.4.3. When $r_{\min} = r_{\max} \triangleq r_{\text{circ}}$, which corresponds to $E = -\mu^2/(2h^2)$, the closed trajectory is a circle. The speed v_{circ} of a body P_2 in circular orbit of a radius r_{circ} around P_1 is obtained from Eq. (4.3.2h) and is given below. Notice that the constant value $r = r_{\text{circ}}$ is the same as the equilibrium solution to Eq. (4.3.2h). Equilibrium solutions were first presented in Chapter 1.

$$v_{\text{circ}} = \sqrt{r_{\text{circ}}^2 \dot{\theta}_{\text{circ}}^2} = \sqrt{\frac{\mu}{r_{\text{circ}}}} = \sqrt{\frac{G(m_1+m_2)}{r_{\text{circ}}}}$$

The constant orbital angular velocity for a circular orbit is $\dot{\theta}_{\text{circ}} = v_{\text{circ}}/r_{\text{circ}} = \sqrt{\mu/r_{\text{circ}}^3}$, and the orbital period, i.e., the time it takes for P_2 to complete one revolution around P_1, is equal to $2\pi/\dot{\theta}_{\text{circ}}$.

Notice that to place a satellite in a circular orbit of radius r_{circ} around a planet, the satellite has to be "injected into orbit" so that its initial velocity relative to the attracting planet has no radial component, and its magnitude must be equal to v_{circ}.

---

[5] For an orbit around the earth, such points are called the *perigee* and *apogee*, respectively. For an orbit around the sun, they are called the *perihelion* and *apohelion*.

CASE 2 $e = 1$

$E = 0$

If the initial conditions are such that $E = 0$, the solutions to Eq. (4.4.5) are $r_m = h^2/(2\mu)$ and $r_m = \infty$. Since $h^2/(2\mu)$ is a positive quantity, we then have for this case

$$r_{\min} = \frac{h^2}{2\mu} \qquad r_{\max} = \infty \qquad (4.4.9)$$

This case represents a very interesting situation. If point P_1 represents the center of the earth for example, and P_2 represents the center of mass of a spacecraft launched from the vicinity of the earth at a distance $r = r_0$ from P_1 and with a speed v_0 relative to the earth, the spacecraft would leave the vicinity of the earth and never return, unless it is equipped with a power system to do so. The required initial speed v_0 is the minimum launching speed for that to happen and, as indicated by Eq. (4.4.1), as r tends to infinity, the speed v of P_2 relative to P_1 tends to zero. The minimum launching speed $v_0 \triangleq v_{\text{esc}}$ for P_2 to escape the gravitational field of P_1 is known either as the *escape speed* or *escape velocity*. For a launching at $r = r_0$, it is determined as

$$v_{\text{esc}} = \sqrt{\frac{2\mu}{r_0}} = \sqrt{\frac{2G(m_1+m_2)}{r_0}} \qquad (4.4.10)$$

Notice that the escape speed at $r = r_0$ is equal to $\sqrt{2}$ times the speed on a circular orbit of radius r_0.

The escape speed from the earth's surface for a spacecraft with mass $m_2 \ll m_1$ is about 40,270 km/h ($\approx$ 25,000 mi/h). The escape speed from the surface of the moon for the same spacecraft is about 8550 km/h ($\approx$ 5300 mi/h).

CASE 3 $e > 1$

E Is Positive

In this case only one of the solutions given by Eq. (4.4.5) is positive, and that solution is $r_m = \mu(e-1)/(2E) = (h^2/\mu)/(1+e)$. As in the case for $E < 0$, this solution corresponds to the minimum[6] value of $r(t)$. When $E > 0$, there is no other real and positive value of r for which $\dot{r} = 0$, which is also confirmed by observation of Eq. (4.4.1). Therefore, as in the case $E = 0$, the trajectory is also open, and we then have

$$r_{\min} = \frac{h^2/\mu}{1+e} \qquad r_{\max} = \infty \qquad (4.4.11)$$

According to Eq. (4.4.1), one obtains for the speed v, in the limit as r tends to infinity,

$$\lim_{r \to \infty} v = \sqrt{2E}$$

In this case, P_2 escapes the gravitational field of P_1 and, provided it does not come "close" to any other body in space, it continues to travel in space with a constant velocity (i.e., along a straight line and with constant speed).

---

[6] This may also be verified by making use of Eq. (4.3.2h) to obtain, for $r = (h^2/\mu)/(1+e)$, $\ddot{r} = \mu^3 e(1+e)^3/h^4$. Since this value of $\ddot{r}$ is positive, the solution in question is a minimum.

Here is an important feature of the analysis presented in this section:

> Given the initial conditions of the motion of P_2 relative to P_1, one can immediately determine whether the trajectory is closed or open simply by verifying whether the specific energy $E = \frac{1}{2}v^2 - \mu/r$ is negative or not.

4.4.3 Solution of the Orbital Differential Equation of Motion for r in Terms of θ

Equation (4.4.1) may be rearranged and then integrated analytically to determine the time t as a function of distance traveled along an orbit to yield the following,

$$\int_{t=0}^{t} dt = t = \left| \int_{r(0)}^{r} \frac{r\, dr}{\sqrt{2Er^2 + 2\mu r - h^2}} \right|$$

Although the integral that appears in this equation can be found in a table of integrals, the determination of $r(t)$ as a function of time is done more conveniently by numerical integration of Eq. (4.4.1). The absolute value in the preceding equation is introduced to ensure that the value found for t is positive.

It turns out that a variety of important information about the motion is obtained if Eq. (4.4.1) is integrated with respect to θ, which happens to be possible to do. This is done by using the chain rule of differentiation, i.e., $d(\ )/dt = [d(\ )/d\theta]\, d\theta/dt$ which yields the following expression for $\dot{r}$:

$$\dot{r} = \frac{dr}{d\theta} \dot{\theta} = \frac{h}{r^2} \frac{dr}{d\theta} = -h \frac{d}{d\theta}\left(\frac{1}{r}\right) \qquad (4.4.12)$$

By making use of this expression for $\dot{r}$, Eq. (4.4.1) can then be rearranged and integrated as

$$\int \frac{h\, d(1/r)}{\sqrt{2E + 2\mu/r - h^2/r^2}} = \pm \int d\theta$$

The integral that appears in this equation can be found in a table of integrals, and the integration yields the result shown in Eq. (4.4.13), where e is the same nondimensional parameter defined in Eq. (4.4.7).

$$\arcsin \frac{1/r - \mu/h^2}{\sqrt{(\mu/h^2)^2 + 2E/h^2}} = \pm\theta - \theta_0$$

or

$$\frac{1}{r} = \frac{\mu}{h^2}[1 + e\sin(\pm\theta - \theta_0)] \qquad (4.4.13)$$

The quantity θ_0 in the solution for $1/r$ given by Eq. (4.4.13) is an arbitrary constant of integration. It may be set to any desired value without any loss of generality. The choice $\theta_0 = -\pi/2$ is a convenient one since it transforms the term $\sin(\pm\theta - \theta_0)$ into the more convenient form $\cos(\pm\theta) = \cos\theta$, thus eliminating the appearing of the $\pm$ signs. In practical terms, the choice $\theta_0 = -\pi/2$ implies that the tracking angle $\theta(t)$ (which is called the *true anomaly* of the orbit) shown in Fig. 4.5 is measured from a reference line that passes through the attracting body at P_1 and the point that corresponds to the minimum distance between P_1 and P_2. The resulting expression for $r(\theta)$, which is a form that is adopted universally, is

$$r(\theta) = \frac{h^2/\mu}{1 + e\cos\theta} \qquad (4.4.14)$$

The solution given by Eq. (4.4.14) shows that for a closed trajectory (i.e., one for which $e < 1$), the maximum value of r, $r_{max} = (h^2/\mu)/(1 - e)$, occurs when $\theta = \pi$, i.e., at a point on the line joining P_1 and P_2 that is diametrically opposed to the point where $r_{min} = (h^2/\mu)/(1 + e)$ occurs.

Notice that if the values of r_{max} and r_{min} for a closed orbit are measured in practice by *distance measuring equipment* such as radar, the expressions for r_{max} and r_{min} disclose that the orbit eccentricity can be readily determined as

$$e = \frac{r_{max} - r_{min}}{r_{max} + r_{min}} \qquad (4.4.15)$$

In addition, when the value of r is measured when $\theta = 90°$, the orbit is completely determined since that value is equal to h^2/μ, as disclosed by Eq. (4.4.14).

It is known from calculus that Eq. (4.4.14) is the polar equation of a conic section (i.e., a circle, ellipse, parabola, or hyperbola) with the origin located at a focus of that conic. The value of the parameter e, which only depends on the initial conditions $r(0)$, $\dot{r}(0)$, and $\dot{\theta}(0)$, determines the type of conic. For $e < 1$, the trajectory is a closed curve, as it has been previously concluded even without having to find the solution for $r(\theta)$. As determined in this section, the closed curve is an ellipse (which becomes a circle when $e = 0$). The trajectory is open for $e \geq 1$, as also concluded earlier. It is a parabola if $e = 1$, and a hyperbola if $e > 1$. This is summarized in the following table.

Type of orbit	Total specific energy E	Eccentricity e
Circular	$E = -\dfrac{\mu^2}{2h^2}$	$e = 0$
Elliptical	$-\dfrac{\mu^2}{2h^2} < E < 0$	$0 < e < 1$
Parabolic	$E = 0$	$e = 1$
Hyperbolic	$E > 0$	$e > 1$

As previously concluded, if an artificial satellite P_2, of mass m_2, is to be launched so that it escapes the gravitational field of the attracting body P_1, of mass m_1, the initial conditions of the motion of P_2 relative to P_1 have to be such that the eccentricity e is no less than unity, with $e = 1$ accomplishing the desired task.

> These results show that the open orbit that requires the smallest energy for P_2 to escape the gravitational attraction of P_1 is a parabola.

Figure 4.7 illustrates the types of conic sections that are obtained when a right cone is cut by a plane, thus justifying their names. Figure 4.8 shows the types of orbits and some of their geometric properties. In each case, the point mass P_1 is located at the focus F_1 of any of the conics, as Kepler predicted in his second law. The focus F_2 of the conic section is known as the *vacant focus*. For an ellipse, the *semi-major axis*, of length a, and the *semi-minor axis*, of length b, shown in Fig. 4.8, are related to the orbital parameters h^2/μ and e as

$$a = \frac{1}{2}(r_{\min} + r_{\max}) = \frac{h^2/\mu}{1 - e^2} = \frac{r_{\min}}{1 - e} = \frac{r_{\max}}{1 + e} \quad (4.4.16)$$

$$b = \sqrt{a^2 - (a - r_{\min})^2} = \sqrt{r_{\max} r_{\min}} = a\sqrt{1 - e^2} = r_{\min}\sqrt{\frac{1 + e}{1 - e}} \quad (4.4.17)$$

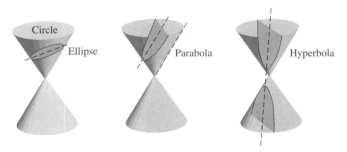

Figure 4.7 The three types of conic sections and how to obtain them.

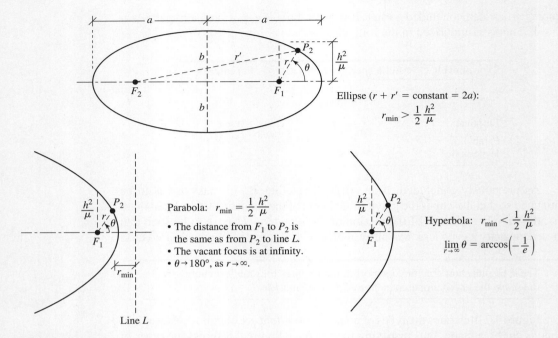

Figure 4.8 The three types of trajectories and some of their geometric properties.

Also, as disclosed by Eq. (4.4.14), the quantity h^2/μ is equal to the distance r when $\theta = 90°$. This provides a practical way for determining such a quantity from actual observation data involving a satellite.

The orbital period $t_{2\pi}$ for an elliptical orbit (i.e., the time it takes P_2 to complete one revolution around P_1) may be calculated by knowing, from calculus, that the area A of the ellipse is equal to $A = \pi a b$ (see Fig. 4.8), and by making use of Eq. (4.3.2k), which yields

$$\frac{dA}{dt} = \frac{1}{2}|r^2\dot\theta| = \text{constant} = \frac{\text{area of ellipse}}{t_{2\pi}} = \frac{\pi a b}{t_{2\pi}}$$

With $b = a\sqrt{1-e^2}$, as per Eq. (4.4.17), and $|r^2\dot\theta| = |h| = \sqrt{\mu a(1-e^2)}$, as per Eq. (4.4.16), the following result is obtained for the period $t_{2\pi}$:

$$t_{2\pi} = 2\pi\sqrt{\frac{a^3}{\mu}}$$

or

$$t_{2\pi} = 2\pi\sqrt{\frac{[(r_{\max}+r_{\min})/2]^3}{\mu}} \tag{4.4.18}$$

The result given by Eq. (4.4.18) agrees with Kepler's prediction stated in his third law, i.e., the square of the period of a planet in its motion around the sun is proportional to the cube of the semi-major axis of its elliptical orbit around the sun.

4.4.4 Solved Example Problems

EXAMPLE 4.4.1

Detection equipment provided the following data at a time $t = t_0$ for the position and velocity, relative to the earth, of a small object traveling near the earth (the unit vectors $\hat{x}$, $\hat{y}$, and $\hat{z}$, with $\hat{z} = \hat{x} \times \hat{y}$, are inertial):

$$\vec{r}_0 = 5000(\hat{x} + \hat{y} + \hat{z}) \text{ km} \qquad \vec{v}_0 = 16{,}000(\hat{x} - \hat{y} + \hat{z}) \text{ km/h}$$

By approximating the earth as a homogeneous sphere of radius $R_E \approx 6378$ km, assuming that the object is unpowered, and disregarding the effect of the earth's atmosphere during the entire trajectory of the body:

a. Determine the specific angular momentum of the body.
b. Determine whether the body's trajectory is open or closed. What are the maximum and minimum values of the distance from the earth's center to the body?
c. Determine whether the body hits or misses the earth.

■ **Solution**

a. The specific angular momentum is simply

$$\vec{h} = \vec{r} \times \vec{v} = \text{constant} = \vec{r}_0 \times \vec{v}_0 = 16 \times 10^7(\hat{x} - \hat{z}) \text{ km}^2/\text{h}$$

b. The specific energy $E = v^2/2 - \mu/r$ of a two-body system determines whether the trajectory is open or closed. The trajectory is open if $E \geq 0$ and closed otherwise. With $\mu = G(m_{\text{earth}} + m_{\text{object}}) \approx Gm_{\text{earth}} = gR_E^2$, we then obtain with $g \approx 9.81$ m/s^2 = $9.81 \times (3600^2/1000)$ km/h^2 and $R_E \approx 6378$ km:

$$E = \frac{v^2}{2} - \frac{\mu}{r} = \frac{\vec{v} \bullet \vec{v}}{2} - \frac{\mu}{|\vec{r}|} = \frac{\vec{v}_0 \bullet \vec{v}_0}{2} - \frac{\mu}{|\vec{r}_0|} \approx -2.132 \times 10^8 \text{ km}^2/\text{h}^2$$

Since $E < 0$, the trajectory is closed.

c. Since the energy during the motion is constant, and since $|\vec{h}| = r^2|\dot{\theta}|$ is also constant, we have

$$\frac{v^2}{2} - \frac{\mu}{r} = \frac{1}{2}(\dot{r}^2 + r^2\dot{\theta}^2) - \frac{\mu}{r} = \frac{1}{2}\left(\dot{r}^2 + \frac{h^2}{r^2}\right) - \frac{\mu}{r} = E$$

With $h^2 = \vec{h} \bullet \vec{h} = (16\sqrt{2} \times 10^7)^2$ km^4/h^2, we then obtain for the values of r when $\dot{r} = 0$:

$$r_{\max} \approx 17{,}330 \text{ km} \qquad r_{\min} \approx 6929 \text{ km}$$

Since $r_{\min}$ is greater than the radius of the earth, the body will not hit the earth. The minimum height from the earth's surface will be $6929 - 6378 = 551$ km. The body that was observed traveling near the earth is then in orbit around the earth. We have already seen in this section that the orbit is an ellipse; its period is

$$t_{2\pi} = 2\pi \sqrt{\frac{[(r_{\max} + r_{\min})/2]^3}{\mu}} \approx 221 \text{ min}$$

EXAMPLE 4.4.2

For the satellite considered in Example 4.4.1, determine the angle θ_p between the line of perigee (i.e., the line joining the center of the earth and the perigee) and the initial position vector $\vec{r}_0$. Also, determine the speed of the satellite at perigee and at apogee.

■ **Solution**

To determine θ_p one needs to use the orbital equation

$$r = \frac{h^2/\mu}{1 + e \cos \theta}$$

With $e = (r_{\max} - r_{\min})/(r_{\max} + r_{\min}) \approx 0.429$, we then obtain, with $r = |\vec{r}_0| = 5000\sqrt{3}$ km,

$$\cos \theta \approx 0.334$$

Thus, θ is either 70.5 or $-70.5°$.

To determine the speed of the satellite at perigee and at apogee, notice that at both of those points the position vector $\vec{r}$ is normal to the velocity $\vec{v}$. Therefore, the speeds v_p (at perigee) and v_a (at apogee) may be obtained as

$$|\vec{h}| = v_p r_{\min} = v_a r_{\max} = 16\sqrt{2} \times 10^7 \text{ km}^2/\text{h}$$

This yields

$$v_p \approx 32{,}656 \text{ km/h} \qquad v_a \approx 13{,}057 \text{ km/h}$$

EXAMPLE 4.4.3

The *flight path angle* of a flying object is defined as the angle the velocity of the object with respect to the attracting planet makes with the local horizontal. Determine the flight path angle of the orbiting body in Example 4.4.1 at the instant it was sighted. Also determine, for that instant, the values of $\dot{r}$ and the absolute angular velocity of the line $P_1 P_2$.

■ **Solution**

To solve problems of this type start by drawing a figure such as Fig. 4.9. The *local horizontal* is, by definition, the line perpendicular to line $P_1 P_2$, and the flight path angle is the angle γ shown in the figure. To obtain the solution, observe Fig. 4.9 and notice the following items.

- To "see" the flight path angle γ, the plane of the orbit needs to be determined. This is done by determining the direction of the unit vector $\hat{n}$ that is normal to that plane. Since $\vec{r}$ and $\vec{v}$ are known at a given instant, we then have $\hat{n} = (\vec{r} \times \vec{v})/|\vec{r} \times \vec{v}|$.

SECTION 4.4 Detailed Analysis of the Two-Body Problem

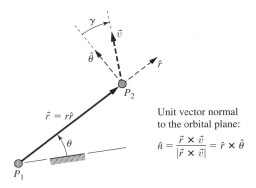

Figure 4.9 Flight path angle γ for Example 4.4.3.

■ Having determined $\hat{n}$, which is the same as $\hat{r} \times \hat{\theta}$, the unit vector $\hat{\theta}$, parallel to the orbital plane and normal to $\hat{r}$, is then $\hat{\theta} = \hat{n} \times \hat{r}$. The flight path angle γ is then found from $\cos \gamma = \hat{\theta} \bullet (\vec{v}/|\vec{v}|)$

At the instant the satellite was sighted, its position vector $\vec{r}_0$ and velocity $\vec{v}_0$ relative to the earth were given as $\vec{r} = 5000(\hat{x} + \hat{y} + \hat{z})$ km, and $\vec{v} = 16{,}000(\hat{x} - \hat{y} + \hat{z})$ km/h. This yields, at that instant,

$$\hat{r} = \frac{\vec{r}}{|\vec{r}|} = \frac{1}{\sqrt{3}}(\hat{x} + \hat{y} + \hat{z})$$

$$\hat{n} = \frac{\vec{r} \times \vec{v}}{|\vec{r} \times \vec{v}|} = \frac{(\hat{x} + \hat{y} + \hat{z}) \times (\hat{x} - \hat{y} + \hat{z})}{|(\hat{x} + \hat{y} + \hat{z}) \times (\hat{x} - \hat{y} + \hat{z})|} = \frac{\hat{x} - \hat{z}}{\sqrt{2}}$$

$$\hat{\theta} = \hat{n} \times \hat{r} = \frac{\hat{x} - 2\hat{y} + \hat{z}}{\sqrt{6}}$$

$$\cos \gamma = \frac{\vec{v}}{|\vec{v}|} \bullet \hat{\theta} = \frac{4}{\sqrt{18}} \qquad \text{which yields } \gamma \approx \pm 19.47°$$

$$\dot{r} = \vec{v} \bullet \hat{r} \approx \frac{16{,}000}{\sqrt{3}} \approx 9237.6 \text{ km/h}$$

$$|\vec{r}|\dot{\theta} = \vec{v} \bullet \hat{\theta} \approx 26{,}128 \text{ km/h} \qquad \text{which yields } \dot{\theta} \approx 3.017 \text{ rad/s}$$

Since the value of $\dot{r}$ is positive, inspection of Fig. 4.9 discloses that γ has to be in the range $0 < \gamma < \pi$. Therefore, the flight path angle (which, based on the value of $\cos \gamma$ was found to be $\pm 19.47°$) is $\gamma \approx 19.47°$.

4.5 Automated Solution for the Two-Body Problem

As it turned out, the nonlinear differential equations of motion for the two-body problem were in a form that made analytical investigations possible. Those investigations revealed a substantial amount of extremely useful information about the motion. Specifically, it allowed us to

- Determine whether the trajectory is open or closed simply by looking at the sign of the total energy E of the motion
- Determine the maximum and minimum values of $r(t)$ without having to solve the differential equation of motion for $r(t)$
- Obtain the expression for r as a function of the angle θ, which allowed for the recognition of the different types of orbits
- Obtain the period of a closed orbit

This section presents the numerical determination of the solution of the differential equations of motion for the two-body system. This is done with the Simulink block diagram model shown in Fig. 4.10. Such a model is a convenient

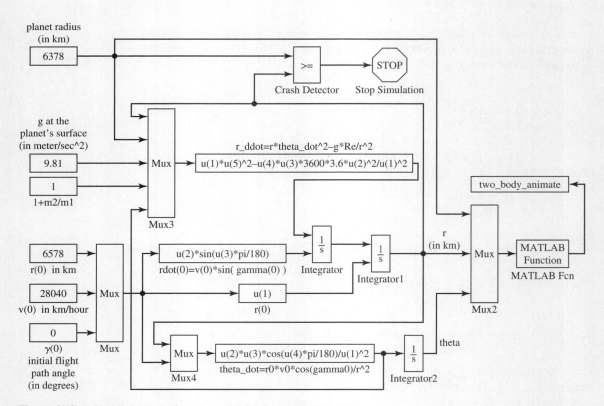

Figure 4.10 A Simulink block diagram model for the two-body problem.

SECTION 4.5 Automated Solution for the Two-Body Problem

tool for solving related problems and for investigating the motion of the system. The model integrates Eqs. (4.3.2h) and (4.3.2i) to obtain $r(t)$ and $\theta(t)$. Details about the implementation are given in the rest of this section.

⟹ The model shown in Fig. 4.10 is the file `two_body_simul.mdl` included on the CD. To use it, either type `two_body_simul` in the MATLAB command window or type `simulations_menu` and choose the `two_body_simul` model from a pull-down menu that appears on the computer screen. Change the maximum step size of the `Solver` in the `Simulation → Parameters` menu if the simulation runs either too fast or too slow in your computer. (See the caution note in Problem 4.11, p. 261.)

There are six user-defined inputs to the model (inputs 4 to 6 are the initial conditions for the motion).

1. The planet (i.e., body P_1 with mass m_1) radius R_E (in kilometers)
2. The acceleration of gravity g at the planet's surface
3. The quantity $1 + m_2/m_1$ (where m_2 is the mass of body P_2), which is essentially equal to 1 if body P_2 is a spacecraft
4. The initial value $r(0)$ for $r(t)$, measured from the center of the attracting planet, as in the analysis
5. The initial speed $v(0)$ of body P_2 relative to body P_1
6. The initial value $\gamma(0)$ of the flight path angle $\gamma(t)$, which is the angle shown in Fig. 4.9 (p. 245).

The initial velocity $\vec{v}(0) = [\dot{r}\hat{r} + r\dot{\theta}\hat{\theta}]_{t=0}$ is specified in the `two_body_simul` model in terms of its magnitude $v(0) = |\vec{v}(0)|$ and of the initial flight path angle $\gamma(0)$ because these are the quantities that are preferred in practice, instead of the scalar components $\dot{r}(0)$ and $r(0)\dot{\theta}(0)$. Notice that inspection of Fig 4.9 discloses that $\dot{r}$ and $r\dot{\theta}$ are related to v and γ as $\dot{r} = v \sin \gamma$ and $r\dot{\theta} = v \cos \gamma$.

As you examine Fig. 4.10, notice that the expressions $\ddot{r} = r\dot{\theta}^2 - gR_E^2(1 + m_2/m_1)/r^2$ and $\dot{\theta} = h/r^2$, from Eqs. (4.3.2h) and (4.3.2i), are shown in the model. They are in the function blocks labeled `r_ddot` and `theta_dot`, respectively, for easy identification. For the user's convenience, the two integrators for Eq. (4.3.2h) are configured so that the initial conditions $\dot{r}(0) = v_0 \sin \gamma(0)$ and $r(0)$ are automatically input externally to them (i.e., one does not double click those integrators to enter initial conditions) using the user-defined inputs to the model. See Section A.3 for a description of such a feature of the integrator.

The quantity $\ddot{r}$ is integrated with the initial condition $\dot{r}(0) = v_0 \sin \gamma(0)$ to yield $\dot{r}$ which, in turn, is integrated by the integrator labeled `Integrator1` to yield $r(t)$. Notice that advantage was taken of the fact that $r^2\dot{\theta} = \text{constant} \stackrel{\Delta}{=} h = r(0)v(0)\cos\gamma(0)$, to obtain θ by integrating the first-order nonlinear differential equation $\dot{\theta} = h/r^2$. The integration of $\dot{\theta}$, with the default initial condition

$\theta(0) = 0$, is done by the integrator labeled `Integrator2`. Double click that integrator to see the default initial condition.

Logic is included in the model to force the simulation to stop if $r(t) \leq R_E$. The default input settings for the motion puts the center of mass of a satellite in a circular orbit 200 km above the earth's surface.

The block named `two_body_animate`, which was prepared by the author, is optional in the model. It is a nice and useful feature to the model, though. Its inclusion produces an animation of the motion using the solutions $r(t)$ and $\theta(t)$ and the radius R_E of the main attracting planet (i.e., body P_1). A snapshot of the default animation figure that appears on the computer screen is shown in Fig. 4.11. To change the minimum and maximum settings of the x and y axes of the figure ($x_{\min}$, $x_{\max}$, $y_{\min}$, and $y_{\max}$), double click the `two_body_animate` block and enter your desired values.

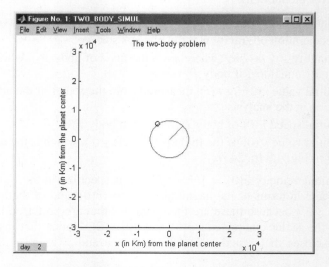

Figure 4.11 The animation figure created by the `two_body_animate` block in the Simulink `two_body_simul` model.

For the default input values in the model, the big circle seen in Fig. 4.11 represents the earth (of radius $R_E = 6378$ km), and the small circle represents the satellite in a circular orbit 200 km above the earth's surface. Change the initial flight path angle to 2°, for example, and you will see the satellite crash on the earth. Keep $\gamma(0) = 0$, and change the initial speed to 38,040 km/h, for example, and you will see the satellite go into a long elliptical orbit, and appear to go off the figure window; stop the simulation, change $x_{\min}$ to $-100{,}000$ km, say, and you will see the satellite reappear in the figure window. Restart the simulation, and you will see the satellite travel in its entire orbit, slowing down as it goes away from the earth and speeding up when it approaches it on its way back. The earth's rotation in the simulation is imitated by a radial line that turns with an angular velocity of 2π rad/day.[7] The day number, counting from the start of the simulation, is automatically updated during the simulation every time that line rotates $360°$.

Plots of r versus θ and of r versus t, for example, can be obtained simply by adding scopes to the model, as shown in other parts of the book. Experiment using the model to obtain solutions to the examples presented in this chapter and to other problems related to the two-body problem. If you modify the differential equations of motion to include aerodynamic drag, for example, and include the modifications in the Simulink model, you can then deal with either reasonably realistic short-range or long-range ballistic problems.

The problem that is analyzed in Section 4.6 is another important classical problem in dynamics, and it has attracted the attention of engineers and scientists for a long time. Its analysis is the most involved one up to this point in the book, and the type of work involved is typical of what is encountered in professional practice. The entire analysis involves the formulation of the differential equations of motion, determination of the equilibrium solutions, linearization, and analysis of the stability of each equilibrium.

4.6 The Restricted Three-Body Problem of Space Mechanics

A spacecraft is subjected to the gravitational attraction of several celestial bodies, and relatively little can be done analytically for the general case involving any number N of bodies attracting each other. Numerical integration of the equations is the main recourse in such a case. The problem that is analyzed in this section involves two main attracting bodies and a third body whose mass is much smaller than either one of the other two so that its presence essentially does not affect their motions. The problem consists of investigating the motion of that smaller body when the only forces acting on it are the gravitational forces of attraction of the other two massive bodies. This problem is known as the *restricted three-body problem* of celestial mechanics (or of spacecraft dynamics).

This problem represents a first approximation to the motion of a spacecraft where the main external forces acting on it are the gravitational attraction of two celestial bodies (such as the earth and the moon, for example). As shown in Section 4.6.2, such a system exhibits several equilibrium points where the smaller body does not move relative to the other two bodies. Such equilibrium points have attracted the attention of the scientific community and might be used in the future for placing artificial satellites in their neighborhood for communication purposes, for example. The stability of those equilibrium points is also investigated in this section.

---

[7] This is done in a MATLAB file called by the Simulink block `two_body_animate` as the simulation progresses, both of which were prepared by the author. The user does not need to deal with that file at all, just like any Simulink user does not need to deal with a number of MATLAB files that are automatically called by Simulink blocks in a simulation.

4.6.1 Differential Equations of Motion

The system under consideration consists of two massive spherical bodies in orbit around each other, both attracting a third body that is much smaller than the other two. Such a system is shown in Fig. 4.12a and b, where the two massive spherical bodies of masses m_1 and m_2 are represented by their geometric centers P_1 and P_2, respectively, and the smaller body, with center of mass at P_3 and with mass m_3, is subjected to the forces $\vec{F}_{31}$ and $\vec{F}_{32}$ due to the gravitational attraction of P_1 and P_2, respectively. The mass m_3 of P_3 is assumed to be very small compared to m_1 and m_2 so that the forces acting on P_1 and P_2 are essentially due to their mutual attraction. Thus, their motion is not affected by the presence of the smaller body. Therefore, P_1 and P_2 constitute a two-body problem and, as such, they move in a plane, which is the orbital plane shown in Fig. 4.12.

Different coordinate systems, such as spherical, cylindrical, or rectangular, may be chosen to analyze the motion of P_3. Cylindrical coordinates r-θ-z, with r

Figure 4.12 The three-body system and the free-body diagrams. (a) Using cylindrical (r, θ, z) coordinates to specify the position of P_3. (b) Using rectangular (x, y, z) coordinates to specify the position of P_3.

measured from P_1, are shown in Fig. 4.12a, while rectangular coordinates x-y-z, measured from the center of mass C of the system, are shown in Fig. 4.12b. Both of these coordinates have been used in the technical literature to investigate this system. Spherical coordinates would involve a distance (say, from either P_1 or P_2 to P_3) and two angles. No matter what choice is made, the differential equations of motion for the system will certainly not be very simple. They rarely are, either for the system under consideration or for other systems!

In Fig. 4.12a, $r \triangleq |\overrightarrow{P_1 Q}|$, where $\overrightarrow{P_1 Q}$ is the projection of the vector $\overrightarrow{P_1 P_3}$ on the orbital plane of $P_1 P_2$ and z is the *elevation* of P_3 relative to that orbital plane. The mutually orthogonal x, y, and z axes shown in Fig. 4.12a and b with x and y in the orbital plane and rotating in space, are centered at the center of mass C of the P_1-P_2 system. From the analysis of the two-body problem, it is already known that point C is inertial, i.e., that its absolute acceleration is equal to zero.

Several unit vectors, which will be used later, are also shown in Fig. 4.12a and b. The unit vector $\hat{z}$ is perpendicular to the orbital plane of $P_1 P_2$. At any time t the distance L between P_1 and P_2 is determined from the solution of the two-body problem. Line $P_1 P_2$ rotates in space with absolute angular velocity $\vec{\Omega} = \dot{\phi}\hat{z} \triangleq \Omega\hat{z}$ where, as shown in Fig. 4.12, ϕ is the angle measured from a nonrotating reference line in the orbital plane to line $P_1 P_2$.

The analysis that follows is restricted to the case where P_1 and P_2 are in a circular orbit around C, which is known as the *restricted circular three-body problem*. In such a case, the distance $L = |\overrightarrow{P_1 P_2}|$ is constant. As determined from the analysis of the two-body problem, the angular velocity $\dot{\phi} = \Omega$ is also constant, and it is determined as

$$\Omega = \text{constant} = \sqrt{\frac{G(m_1 + m_2)}{L^3}} \qquad (4.6.1)$$

where G, as before, is the universal gravitational constant.

The motion is analyzed below using the cylindrical coordinates r, θ, and z shown in Fig. 4.12a. The formulation of the differential equations involving the rectangular coordinates x, y, and z shown in Fig. 4.12b is left as an exercise (see Problem 4.13).

Differential Equations Using the Cylindrical Coordinates $r = |\overrightarrow{P_1 Q}|$, θ, and z
Referring to Fig. 4.12 and to Eqs. (4.3.2c) (p. 233) for the two-body problem, the absolute position vectors of P_1 and P_2 are

$$\vec{r}_1 = \overrightarrow{CP_1} = -\frac{m_2 L}{m_1 + m_2}\hat{x} \qquad (4.6.2)$$

$$\vec{r}_2 = \overrightarrow{CP_2} = \frac{m_1 L}{m_1 + m_2}\hat{x} \qquad (4.6.3)$$

The following can also be written after inspecting Fig. 4.12.

- Absolute position vector of P_3:

$$\vec{r}_3 = \overrightarrow{CP_3} = \overrightarrow{CP_1} + \overrightarrow{P_1Q} + \overrightarrow{QP_3} = -\frac{m_2 L}{m_1 + m_2}\hat{x} + r\hat{r} + z\hat{z} \quad (4.6.4)$$

- Absolute velocity of P_3:

$$\vec{v}_3 = \frac{d\vec{r}_3}{dt} = \underbrace{\dot{r}\hat{r} + [(\dot{\phi}+\dot{\theta})\hat{z}] \times (r\hat{r})}_{\text{This is } \frac{d}{dt}(r\hat{r})} + \dot{z}\hat{z} - \frac{m_2 L}{m_1 + m_2}\underbrace{(\dot{\phi}\hat{z}) \times \hat{x}}_{\text{This is } \frac{d}{dt}\hat{x}}$$

$$= \dot{r}\hat{r} + r(\dot{\phi}+\dot{\theta})\hat{\theta} + \dot{z}\hat{z} - \frac{m_2 L}{m_1 + m_2}\dot{\phi}\hat{y} \quad (4.6.5)$$

since $\hat{z} \times \hat{x} = \hat{y}$ and $\hat{z} \times \hat{r} = \hat{\theta}$ (see Fig. 4.12, and use the right-hand rule to do these cross products).

- Absolute acceleration of P_3:

$$\vec{a}_3 = \frac{d\vec{v}_3}{dt} = \underbrace{\ddot{r}\hat{r} + [(\dot{\phi}+\dot{\theta})\hat{z}] \times (\dot{r}\hat{r})}_{\text{This is } \frac{d}{dt}(\dot{r}\hat{r})}$$

$$+ \underbrace{[r(\ddot{\phi}+\ddot{\theta}) + \dot{r}(\dot{\phi}+\dot{\theta})]\hat{\theta} + (\dot{\phi}+\dot{\theta})\hat{z} \times [r(\dot{\phi}+\dot{\theta})\hat{\theta}]}_{\text{This is } \frac{d}{dt}[r(\dot{\phi}+\dot{\theta})\hat{\theta}]}$$

$$+ \ddot{z}\hat{z} - \frac{m_2 L}{m_1 + m_2}\underbrace{[\ddot{\phi}\hat{y} + (\dot{\phi}\hat{z}) \times (\dot{\phi}\hat{y})]}_{\text{This is } \frac{d}{dt}(\dot{\phi}\hat{y})}$$

$$= [\ddot{r} - r(\dot{\phi}+\dot{\theta})^2]\hat{r} + [r(\ddot{\phi}+\ddot{\theta}) + 2\dot{r}(\dot{\phi}+\dot{\theta})]\hat{\theta}$$

$$+ \ddot{z}\hat{z} - \frac{m_2 L}{m_1 + m_2}[\ddot{\phi}\hat{y} - \dot{\phi}^2\hat{x}] \quad (4.6.6)$$

Since $\dot{\phi} = \text{constant} \triangleq \Omega$, $\ddot{\phi} = 0$ in this expression for $\vec{a}_3$.

- Resultant force $\vec{F}$ acting on P_3 due to the gravitational attractions of P_1 and P_2, as per Newton's law of universal gravitation:

$$\vec{F} = \vec{F}_{31} + \vec{F}_{32} = \frac{Gm_1 m_3}{r_{31}^2}\frac{\overrightarrow{P_3 P_1}}{|\overrightarrow{P_3 P_1}|} + \frac{Gm_2 m_3}{r_{32}^2}\frac{\overrightarrow{P_3 P_2}}{|\overrightarrow{P_3 P_2}|} \quad (4.6.7)$$

where $r_{31} = |\overrightarrow{P_3 P_1}|$ is the distance between P_3 and P_1, and $r_{32} = |\overrightarrow{P_3 P_2}|$ is the distance between P_3 and P_2. The vectors $\overrightarrow{P_3 P_1}/|\overrightarrow{P_3 P_1}|$ and $\overrightarrow{P_3 P_2}/|\overrightarrow{P_3 P_2}|$ are the unit vectors from P_3 to P_1 and from P_3 to P_2, respectively.

By inspecting Fig. 4.12, the following expressions can be written for $\overrightarrow{P_3 P_1}$ and $\overrightarrow{P_3 P_2}$ in terms of the quantities r, θ, z, and L:

$$\overrightarrow{P_3 P_1} = -z\hat{z} - r\hat{r} \quad (4.6.8)$$

$$\overrightarrow{P_3 P_2} = -z\hat{z} - r\hat{r} + L\hat{x} = -z\hat{z} - r\hat{r} + L(\hat{r}\cos\theta - \hat{\theta}\sin\theta) \quad (4.6.9)$$

SECTION 4.6 The Restricted Three-Body Problem of Space Mechanics

and this yields

$$r_{31} = |\overrightarrow{P_3 P_1}| = \sqrt{r^2 + z^2} \tag{4.6.10}$$

$$r_{32} = |\overrightarrow{P_3 P_2}| = \sqrt{L^2 + r^2 + z^2 - 2Lr\cos\theta} \tag{4.6.11}$$

We now have all the ingredients to obtain the differential equations of motion for the system based on Newton's second law $\vec{F} = m_3 \vec{a}_3$. With $\vec{a}_3$ given by Eq. (4.6.6) and $\vec{F}$ by Eq. (4.6.7), $\overrightarrow{P_3 P_1}$ and $\overrightarrow{P_3 P_2}$ given by Eqs. (4.6.8) and (4.6.9), and $\hat{x}$ and $\hat{y}$ that appear in Eq. (4.6.6) obtained by inspection of Fig. 4.12 as

$$\hat{x} = \hat{r}\cos\theta - \hat{\theta}\sin\theta \qquad \hat{y} = \hat{r}\sin\theta + \hat{\theta}\cos\theta \tag{4.6.12}$$

you can verify that $\vec{F} = m_3 \vec{a}_3$ yields the following scalar differential equations of motion:

$$\ddot{r} - r(\dot{\theta} + \Omega)^2 + \Omega^2 \frac{m_2 L}{m_1 + m_2} \cos\theta = \frac{Gm_2}{r_{32}^3}(L\cos\theta - r) - \frac{Gm_1}{r_{31}^3} r$$

$$= \frac{\Omega^2 L^3}{m_1 + m_2} \left[\frac{m_2(L\cos\theta - r)}{r_{32}^3} - \frac{m_1 r}{r_{31}^3} \right] \tag{4.6.13}$$

$$r\ddot{\theta} + 2\dot{r}(\dot{\theta} + \Omega) - \Omega^2 \frac{m_2 L}{m_1 + m_2} \sin\theta = -\frac{Gm_2}{r_{32}^3} L\sin\theta = -\frac{\Omega^2 L^3}{m_1 + m_2} \frac{m_2 L \sin\theta}{r_{32}^3} \tag{4.6.14}$$

$$\ddot{z} = -\left[\frac{Gm_1}{r_{31}^3} + \frac{Gm_2}{r_{32}^3} \right] z \tag{4.6.15}$$

4.6.2 Equilibrium Solutions

The equilibrium solutions to Eqs. (4.6.13), (4.6.14), and (4.6.15) correspond to $r = \text{constant} \triangleq r_e$, $\theta = \text{constant} \triangleq \theta_e$, and $z = \text{constant} \triangleq z_e$, respectively. Equation (4.6.15) immediately discloses that

$$z_e = 0 \tag{4.6.16}$$

Since $z_e = 0$, any equilibrium point lies in the orbital plane of $P_1 P_2$. Thus, at equilibrium, point Q in Fig. 4.12 is the same as P_3, which means that the distance $r_{31} = \overrightarrow{P_3 P_1}$ is $r_{31} = r_e$ (which still needs to be determined). By inspecting Eq. (4.6.14), it follows that either

$$r_{32} = \text{constant} \triangleq r_{32e} = L \qquad \text{or} \qquad \sin\theta_e = 0 \tag{4.6.17}$$

Both cases are considered. The notations r_{31e} and r_{32e} denote the equilibrium values of the distances r_{31} and r_{32}, respectively.

CASE 1 $r_{32e} = L$

With $r_{31e} = r_e$, Eq. (4.6.13) yields

$$r_e = L \qquad (4.6.18)$$

Since $r_{32e} = L$ and also $r_{31e} = r_e = L$, points P_1, P_2, and P_3 form an equilateral triangle for this solution.

There are two such configurations, with $\theta_e = 60°$ and $-60°$, respectively. These two solutions are known as the L_4 and L_5 equilibrium points, respectively, and they were first found by Lagrange.

CASE 2 $\sin\theta_e = 0$

There are two possibilities for this case, namely, $\theta_e = 0$ and $180°$. For both cases, Eq. (4.6.13) yields

$$-(m_1 + m_2)r_e + m_2 L \cos\theta_e$$
$$= L^3 \left[\frac{m_2(L\cos\theta_e - r_e)}{r_{32e}^3} - \frac{m_1}{r_e^2} \right] \qquad (4.6.19)$$

where

$$r_{32e} = \sqrt{L^2 + r_e^2 - 2r_e L \cos\theta_e}$$
$$= \begin{cases} r_e + L & \text{when } \theta_e = 180° \\ |r_e - L| & \text{when } \theta_e = 0° \end{cases} \qquad (4.6.20)$$

Equation (4.6.19) is nonlinear in r_e. It was found many years ago that, by separately considering the two cases $\theta_e = 0$ and $180°$, it yields a fifth-degree polynomial equation in r_e/L when it is reduced to a single denominator for each one of these two cases. Numerical solutions for r_e/L can then be obtained for each value of either m_2/m_1 or m_1/m_2, with only the real and positive solutions being the acceptable ones. An alternative and much easier solution procedure is shown next.

Inspection of Eq. (4.6.19) discloses that it is certainly nonlinear in r_e, but it is linear in either m_2 or m_1. Therefore, why not solve it either for m_2/m_1 or m_1/m_2 in terms of r_e/L? A plot involving these variables can then be generated so that values of r_e/L versus either m_2/m_1 or m_1/m_2 are then taken from the plot. The solution for m_2/m_1 is

$$\frac{m_2}{m_1} = \frac{r_e/L - (L/r_e)^2}{(\cos\theta_e - r_e/L)\{1 - 1/[1 + (r_e/L)^2 - 2(r_e/L)\cos\theta_e]^{3/2}\}} \qquad (4.6.21)$$

Without any loss of generality, one can choose $m_2/m_1 \leq 1$. A plot of m_2/m_1 versus r_e/L (or vice versa, if preferred) can then be generated for any range of values of r_e/L using Eq. (4.6.21), and then limiting the range of the plot to the one that yields $m_2/m_1 \leq 1$. Such a plot can be generated with Simulink without having to resort to knowing language-dependent plotting commands. A block diagram for doing this is shown in Fig. 4.13, and the plot it generates is shown in

Figure 4.13 A Simulink block diagram for plotting the solutions to Eq. (4.6.21).

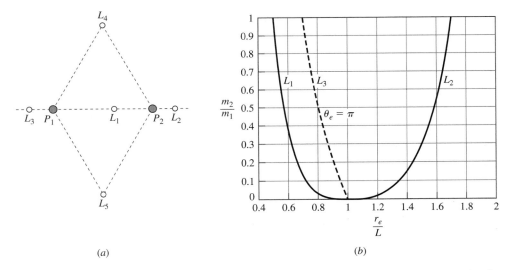

Figure 4.14 The equilibrium points L_1 to L_5 for the restricted circular three-body problem, and a plot of the nondimensional distances r_e/L for L_1, L_2, and L_3 (r_e is measured from P_1).

Fig. 4.14b. If one prefers, the plot shown in Fig. 4.14b could also be generated by hand, with MATLAB commands, or with any other appropriate software.[8]

The three equilibrium points obtained from Eq. (4.6.21), which are along a straight line, are known as the L_1, L_2, and L_3 points. They were first found by

---

[8] In MATLAB, the following commands could be used to plot m_2/m_1 (represented by the variable m) versus r_2/L (represented by the variable r).

```
fig=figure; hold on; grid; r=0.4:0.01:2; cos_theta=1;
m=(r-1./r.^2)./((cos_theta-r).*(1-1./(1+r.^2-2*cos_theta*r).^(3/2)));
plot(r,m); cos_theta=-1;
m=(r-1./r.^2)./((cos_theta-r).*(1-1./(1+r.^2-2*cos_theta*r).^(3/2)));
plot(r,m, '- -'); axis([0.4, 2, 0, 1]);
```

Euler. As an example, if P_1 represents the center of the earth and P_2 represents the center of the moon, the L_3 and L_2 equilibrium points are always "behind" the earth and the moon, respectively, and L_1 is between the earth and the moon. The five points $L_1, \ldots, L_5$ are known as the *libration points*, and that is what the L in their notation stands for. They are shown in Fig. 4.14a.

The question that needs to be answered now is whether a body P_3, with mass m_3 that is much smaller than m_1 and m_2, placed in the neighborhood of any of the equilibrium points $L_1, \ldots, L_5$ will remain in that neighborhood with the passage of time. The answer to such a question, which is important for space applications, involves the analysis of the stability of each equilibrium. Such an analysis is presented in Section 4.6.3.

4.6.3 Stability of the Equilibrium Solutions

If a spacecraft is placed in the vicinity of any of the equilibrium points $L_1, \ldots, L_5$, it is important to know if it will stay near that equilibrium point for all times. If that happens, the equilibrium is said to be stable. Otherwise, it is unstable. As done in Section 1.15 and in other parts of the book (see, for example, the spring-pendulum analysis on p. 129 in Chapter 2), stability of each equilibrium is analyzed here by linearizing the differential equations of motion about each equilibrium. To do this, the equilibrium solutions are "perturbed" by introducing small perturbations $r_s(t)$, $\theta_s(t)$, and $z_s(t)$, respectively, as

$$r(t) \triangleq r_e + r_s(t) \qquad \theta(t) \triangleq \theta_e + \theta_s(t) \qquad z(t) \triangleq \underbrace{z_e}_{=0} + z_s(t) = z_s(t)$$
(4.6.22)

These expressions are substituted into the differential equations of motion—Eqs. (4.6.13), (4.6.14), and (4.6.15)—and all terms are expanded in Taylor series in the small perturbations. Linearization of the equations is done by neglecting any product of small quantities in the Taylor series. The series expansion of the terms $r(\dot{\theta} + \Omega)^2$, $\sin\theta$, $\cos\theta$, r_{31}^{-3}, and r_{32}^{-3} that appear in those equations yields

$$r(\dot{\theta} + \Omega)^2 = (r_e + r_s)(\dot{\theta}_s + \Omega)^2 = (r_e + r_s)(\Omega^2 + 2\Omega\dot{\theta}_s + \dot{\theta}_s^2)$$
$$= \Omega^2 r_e + \Omega^2 r_s + 2\Omega r_e \dot{\theta}_s + \cdots \qquad (4.6.23)$$

$$\sin\theta = \sin(\theta_e + \theta_s) = \sin\theta_e + \theta_s \cos\theta_e + \cdots \qquad (4.6.24)$$

$$\cos\theta = \cos(\theta_e + \theta_s) = \cos\theta_e - \theta_s \sin\theta_e + \cdots \qquad (4.6.25)$$

$$r_{31}^{-3} = \left[(r_e + r_s)^2 + z_s^2\right]^{-3/2} = r_e^{-3}\left(1 - 3\frac{r_s}{r_e}\right) + \cdots \qquad (4.6.26)$$

$$r_{32}^{-3} = \left[L^2 + (r_e + r_s)^2 + z_s^2 - 2L(r_e + r_s)\cos(\theta_e + \theta_s)\right]^{-3/2}$$
$$= r_{32e}^{-3}\left\{1 - \frac{3}{r_{32e}^2}[Lr_e(\sin\theta_e)\theta_s + (r_e - L\cos\theta_e)r_s]\right\} + \cdots$$
(4.6.27)

where "$\ldots$" denote all neglected products of small quantities.

SECTION 4.6 The Restricted Three-Body Problem of Space Mechanics

By proceeding in this manner with the rest of the terms in the differential equations, and by noticing that the nonlinear term $\dot{r}\dot{\theta} = \dot{r}_s \dot{\theta}_s$ that appears in Eq. (4.6.14) is neglected in the linearization, you should obtain the following linearized differential equations of motion.

$$\ddot{r}_s - 2\Omega r_e \dot{\theta}_s - \Omega^2 r_s - \frac{\Omega^2 m_2 L \sin\theta_e}{m_1 + m_2}\theta_s$$

$$= -\frac{\Omega^2 L^3 m_2}{(m_1 + m_2)r_{32e}^3}\left\{r_s + L(\sin\theta_e)\theta_s \right.$$

$$\left. + \frac{3(L\cos\theta_e - r_e)[Lr_e(\sin\theta_e)\theta_s + (r_e - L\cos\theta_e)r_s]}{r_{32e}^2}\right\} + \frac{2\Omega^2 L^3 m_1 r_s}{(m_1 + m_2)r_{32e}^3}$$

$$\tag{4.6.28}$$

$$r_e \ddot{\theta}_s + 2\Omega \dot{r}_s - \frac{\Omega^2 m_2 L \cos\theta_e}{m_1 + m_2}\theta_s$$

$$= -\frac{\Omega^2 L^4 m_2}{(m_1 + m_2)r_{32e}^3}\left\{\theta_s \cos\theta_e - 3\frac{Lr_e(\sin\theta_e)\theta_s + (r_e - L\cos\theta_e)r_s}{r_{32e}^2}\sin\theta_e\right\}$$

$$\tag{4.6.29}$$

$$\ddot{z}_s = -\left[\frac{Gm_1}{r_e^3} + \frac{Gm_2}{r_{32e}^3}\right]z_s \tag{4.6.30}$$

These linearized differential equations are valid for small perturbed motions about all five equilibrium solutions. Equation (4.6.30) immediately discloses that the out-of-plane z-motion is oscillatory because the quantity $Gm_1/r_e^3 + Gm_2/r_{32e}^3$ is always positive. Therefore, the "linearized $z_s(t)$ motion" simply consists of an oscillation of frequency equal to $\sqrt{Gm_1/r_e^3 + Gm_2/r_{32e}^3}$. We now have to deal only with Eqs. (4.6.28) and (4.6.29), which are independent of $z_s(t)$. For convenience, the analysis is separated into two parts: the equilateral triangle and the straight-line configurations.

The Equilateral Triangle Configuration (Points L_4 and L_5 in Fig. 4.14) For these equilibrium points, $r_e = r_{32e} = L$ and $\theta = \pm 60°$. In this case, $\cos\theta_e = \frac{1}{2}$, $\sin\theta_e = \pm\frac{\sqrt{3}}{2}$ for the L_4 and L_5 points, respectively, and you can verify that Eqs. (4.6.28) and (4.6.29) simplify to the following, where $k \triangleq m_2/m_1$

$$\ddot{r}_s - 2\Omega L \dot{\theta}_s = \frac{3(1+k/4)r_s \pm \frac{3\sqrt{3}}{4}kL\theta_s}{1+k}\Omega^2 \tag{4.6.31}$$

$$L\ddot{\theta}_s + 2\Omega \dot{r}_s = \frac{\Omega^2 k(9L\theta_s \pm 3\sqrt{3}r_s)}{4(1+k)} \tag{4.6.32}$$

Similarly to the material that was presented in Chapter 1, the general solutions to the linear differential equations (4.6.31) and (4.6.32) are also exponentials of the form

$$r_s(t) = Ae^{st} \qquad \theta_s(t) = Be^{st}$$

where A, B, and s are constants. By substituting this general solution into Eqs. (4.6.31) and (4.6.32), the exponential function e^{st} will be common to all terms. This is the reason why the exponential function is the general solution to a set of "homogeneous" linear differential equations with constant coefficients. The following algebraic equations involving A, B, and s are then obtained:

$$\left[s^2 - \frac{3\Omega^2(1+k/4)}{1+k}\right]A - \left[2s\Omega L \pm \frac{3\sqrt{3}L\Omega^2 k}{4(1+k)}\right]B = 0 \quad (4.6.33)$$

$$\left[2s\Omega \mp \frac{3\sqrt{3}\Omega^2 k}{4(1+k)}\right]A + \left[Ls^2 - \frac{9\Omega^2 Lk}{4(1+k)}\right]B = 0 \quad (4.6.34)$$

By solving Eq. (4.6.33), say, for either A or B and by substituting the result into Eq. (4.6.34), the following equation is obtained:

$$\left[s^4 + \Omega^2 s^2 + \frac{27k\Omega^4}{4(1+k)^2}\right]B = 0 \quad (4.6.35)$$

By imposing the requirement that $B \neq 0$ (otherwise one would just have the obvious solution $r_s = \theta_s = 0$, i.e., the equilibrium is not perturbed!), the coefficient of B in the equation has to be equal to zero. As in Chapter 1, the resulting equation is also called the *characteristic equation* associated with this equilibrium. The solution to the characteristic equation for s^2 is

$$s^2 = -\frac{\Omega^2}{2}\left[1 \pm \sqrt{1 - \frac{27k}{(1+k)^2}}\right] \quad (4.6.36)$$

If the roots of Eq. (4.6.36) are complex, the square root of such a root will also be complex, but with a positive real part.[9] The equilibrium is unstable in such a case. Therefore, based on the linearized differential equation, the equilibrium is stable, only if Eq. (4.6.36) yields two negative real roots, and this happens only if $(1+k)^2 - 27k > 0$. The roots of this quadratic equation are $k = m_2/m_1 = (25 \pm 3\sqrt{69})/2 \approx 24.96$ and also $k = 1/24.96$. Taking $m_2 \leq m_1$, without any loss of generality, the L_4 and L_5 equilibrium points are unstable if $m_2/m_1 > 1/24.96$, and infinitesimally stable otherwise. It can be easily verified that such an equilibrium is stable when $m_2/m_1 \leq 1/24.96$ by numerically integrating Eqs. (4.6.13), (4.6.14), and (4.6.15) with initial conditions near the equilibrium. When this is done with $m_1 = 80m_2$, $\dot{r}(0) = \dot{\theta}(0) = \dot{z}(0) = 0$, $\theta(0) = 59°$,

---

[9] This is so because any complex number is of the form $C(\cos\alpha \pm i\sin\alpha) = Ce^{\pm i\alpha}$, where C and α are real numbers, with $C > 0$ and $0 \leq \alpha \leq \pi$, and $i = \sqrt{-1}$. The square root of this is then $\sqrt{C}e^{\pm i\alpha/2} = \sqrt{C}[\cos(\alpha/2) \pm i\sin(\alpha/2)]$, whose real part $\sqrt{C}\cos(\alpha/2)$ is never negative for any α.

and $z(0)/L = 0.01$, for example, the maximum values obtained for r/L are approximately 1.01 and 1.006 when $r(0)/L = 1.001$ and $r(0)/L = 1.0005$, respectively.

$\implies$ The Simulink model named `restricted_3body.mdl` for integrating Eqs. (4.6.13), (4.6.14), and (4.6.15) is included on the CD.

For the earth-moon system, $m_2/m_1 \approx 1/81.3$, and for the sun-earth system, $m_2/m_1 \approx 3 \times 10^{-6}$. The L_4 and L_5 points for both of these systems are stable.

The Straight-Line Configuration (Points L_1, L_2, and L_3 in Fig. 4.14) For this configuration $\theta_e = 0$ for the equilibrium points L_1 and L_2, and $\theta_e = 180°$ for the L_3 point (see Fig. 4.14, p. 255). For these cases, $r_{32e} = |r_e \pm L|$, as per Eq. (4.6.20) (p. 254). The term $r_e - L\cos\theta_e$ that appears in Eqs. (4.6.28) and (4.6.29) becomes $|r_e \pm L|$, and those linearized differential equations reduce to

$$\ddot{r}_s - 2\Omega r_e \dot{\theta}_s = \Omega^2 \underbrace{\left[1 + \frac{2m_1 L^3}{(m_1 + m_2)r_e^3} + \frac{2m_2 L^3}{(m_1 + m_2)r_{32e}^3}\right]}_{\text{Call this } f_1} r_s \triangleq \Omega^2 f_1 r_s$$

(4.6.37)

$$r_e \ddot{\theta}_s + 2\Omega \dot{r}_s = \Omega^2 \underbrace{\frac{m_2 L \cos\theta_e}{m_1 + m_2} \left[1 - \left(\frac{L}{r_{32e}}\right)^3\right]}_{\text{Call this } f_2} \theta_s \triangleq \Omega^2 f_2 \theta_s \quad (4.6.38)$$

The characteristic equation in this case is readily found to be

$$r_e s^4 + \Omega^2 (4r_e - f_2 - r_e f_1) s^2 + \Omega^4 f_1 f_2 = 0 \quad (4.6.39)$$

This equation is of the form $s^4 + a_2 s^2 + a_0 = 0$, which yields the roots

$$s^2 = \frac{-a_2 \pm \sqrt{a_2^2 - 4a_0}}{2} \quad (4.6.40)$$

Two different roots for s^2 that are real and negative are obtained only when $a_2 > 0$, $a_0 > 0$, and $a_2^2 - 4a_0 > 0$. Thus, all these three conditions must be satisfied for the equilibrium to be infinitesimally stable. The equilibrium is unstable if $a_2 < 0$ or $a_0 < 0$ or $a_2^2 - 4a_0 < 0$.

Looking at Eq. (4.6.39), the quickest way to perform the stability test for this case is to start examining the term $f_1 f_2$. Since $f_1 > 0$, as seen from its definition in Eq. (4.6.37), one of the conditions for an infinitesimally stable equilibrium is for f_2 to be positive. From the definition of f_2 in Eq. (4.6.38), this requires $[1 - (L/r_{32e})^3]\cos\theta_e$ to be positive, where, as per Eq. (4.6.20) (p. 254), $r_{32e} = r_e + L$ when $\theta_e = 180°$, and $r_{32e} = |r_e - L|$ when $\theta_e = 0$. However, for both cases, inspection of Fig. 4.14 (p. 255) discloses that $[1 - (L/r_{32e})^3]\cos\theta_e$ is negative. Therefore all three equilibrium points L_1, L_2, and L_3 are unstable. In practice, a propulsion device activated by a control system is required if a spacecraft is to be maintained near any of these three equilibrium points.

4.7 Problems

Some of the problems have to be solved numerically since the equations for them cannot be solved analytically.

4.1 A geosynchronous satellite is one whose orbital period is equal to the period of the rotation of the earth about its own axis and whose orbit is on the equatorial plane. The satellite is geostationary (i.e., appears to be at the same point in the sky when it is viewed from the earth) if the orbit is a perfect circle. Small errors in orbit injection will cause small drifts in the satellite position relative to the earth. Determine the height above the surface of the earth a satellite should be placed for it to be geostationary.

4.2 An astronomical observation revealed that a small moon of a distant planet is in a nearly circular orbit about that planet. The diameter of the orbit and the orbital period (measured in earth time) are 80,000 km and 10 h and 35 min, respectively. Calculate the mass of the planet, in terms of the mass of the earth, by treating the planet and its moon as a two-body system.

4.3 If the maximum and minimum distances between an inertial center of attraction O and the center of mass C of a spacecraft in an elliptical orbit around O are related as $r_{max}/r_{min} = k$, show that the maximum and minimum magnitudes of the absolute velocity of C are $v_{max} = \sqrt{2(\mu/r_{min})k/(k+1)}$ and $v_{min} = v_{max}/k$.

4.4 The center of mass C of a spacecraft is in a circular orbit 300 km above the earth's surface, which is well above the earth's atmosphere. The mass of the spacecraft is $m = 2500$ kg. The spacecraft is equipped with an engine that, when ignited, generates a constant thrust $\vec{F}$ of magnitude $F = 2000$ N. To change the orbit to an elliptical orbit, the spacecraft is oriented with small thrusters so that $\vec{F}$ is directed along the local horizontal. The small thrusters are then turned off, and the engine is fired for a short time Δt s. Estimate the value of Δt so that the minimum and maximum heights of the new orbit are 300 and 450 km, respectively. Should the initial speed of C have been increased or decreased by the engine firing to achieve the new orbit? Approximate the earth as a homogeneous sphere of radius $R_E = 6378$ km. *Hint:* See the note on p. 71 in Chapter 2; use the relationship given in Problem 4.3.

4.5 The spacecraft in Problem 4.4 is in an elliptical orbit of minimum and maximum heights of 300 and 450 km, respectively, above the earth's surface after the reorientation and firing maneuver described in that problem. It is now desired to fire the engines a second time, for a short time Δt_2 s, when the spacecraft reaches its apogee, in order to transfer the spacecraft to a circular orbit 450 km above the surface of the earth. Estimate the value of Δt_2 to accomplish the task. Should the speed of C at apogee be increased or decreased to achieve the circular orbit? Approximate the earth as a homogeneous sphere of radius $R_E = 6378$ km. See the hint in Problem 4.4.

4.6 The space shuttle is in a circular orbit at a height of 300 km above the earth's surface when it releases a satellite with the following initial conditions: velocity relative to the Shuttle is 20 km/h and the flight path angle is $10°$. Determine the maximum and minimum values of $r(t)$ for the satellite and the eccentricity of its resulting orbit.

4.7 A small space probe approaches a distant planet P_1 as shown in Fig. 4.15. The distant planet is Uranus, and its mass is 14.536 times the mass of the earth, and its

Figure 4.15 Figure for Problem 4.7.

radius is $R \approx 4.007 R_E$, where $R_E \approx 6378$ km is the radius of the earth. When the probe is far away from P_1, it is traveling essentially along a straight line with velocity $v_0 = 60{,}000$ km/h relative to P_1, and the perpendicular distance from the center of P_1 to the asymptote of the flight path is $D = 1.8R$. By treating the probe and Uranus as a two-body system, determine the minimum value $r_{\min}$ of the distance $r(t)$ from the center of the planet to the probe. Does the probe crash on P_1?

4.8 For Problem 4.7, plot $r_{\min}$ versus the distance D. What is the minimum distance D needed for the probe not to crash on the planet?

4.9 A small body P of mass m is launched, along the equator, from a height of 1 km above the earth's surface (treating the earth as done in this chapter) with a speed of $v_0 = 20{,}000$ km/h and a flight path angle of $\gamma_0 = 30°$. Determine the range R of its trajectory, i.e., the distance, measured along the earth's surface, from the launching point to the point it hits the earth, and the total time of flight. Consider only the mutual force of attraction between the earth and P. Estimate the error in the answers when the earth's rotation about its own axis is taken into account. For simplicity, an equatorial trajectory is assumed, so that the plane of the trajectory is perpendicular to the axis of rotation of the earth. Use $R_E = 6378$ km for the radius of the earth. What would the range be if the constant gravitational force approximation were used? *Suggestion:* Use the Simulink model `two_body_simul` to do part of this problem.

4.10 For Problem 4.9, plot the range R versus γ_0 [with $20 \leq \gamma_0 \leq 60°$] for $v_0 = 18{,}000$ km/h and $v_0 = 20{,}000$ km/h, preferably superimposing the plots on the same graph (such types of plots are prepared, for a number of other problems, for engineering manuals). For $v_0 = 20{,}000$ km/h, what value of γ_0 maximizes the range R, and what should be the values of γ_0 so that $R = 3500$ km?

4.11 In addition to the gravitational force of attraction due to the earth, a satellite of mass m is subjected to a force $\vec{f}_1 = -c_1 \vec{v}$, and to a drag force modeled as $\vec{f}_2 = -c_2 v^2 (\vec{v}/|\vec{v}|)$, where $\vec{v} = \dot{r}\hat{r} + r\dot{\theta}\hat{\theta}$ is the velocity of the center of mass C of the satellite relative to the earth, and c_1 and c_2 are constants.

  a. What are the differential equations that govern the motion of the center of mass of the satellite? Approximate the earth as an inertial homogeneous sphere of radius $R_E \approx 6378$ km.

  b. Use Simulink to integrate the differential equations of motion, to animate the motion, and to record the total flight time. Automatically stop the simulation if $r \leq R_E$. The satellite was put in space at an altitude of 102 km above the surface of the earth, with a velocity of 29,500 km/h directed along the local horizontal. For $c_1 = 0$, $c_1/m = 0.0001$ s^{-1}, and $c_1/m = 0.001$ s^{-1}, record the flight path and observe what happens during the first 72 h of flight if $c_2/m = 5 \times 10^{-9}$ m^{-1}, and write a few paragraphs about your observations. Also, indicate what happens when $c_1 = c_2 = 0$. *Suggestion:* Modify the Simulink model `two_body_simul` to do this problem. *Caution:* The flight times are long and can be sensitive to the maximum step size $(\Delta t)_{\max}$ of the integration and also to the integration method. Therefore, one should experiment with smaller values of $(\Delta t)_{\max}$ (in the `Simulation` → `Parameters` menu of the model window) until the answers become independent of $(\Delta t)_{\max}$.

4.12 A common theoretical model for the earth's atmosphere is one for which the air density changes exponentially with height H from the earth's surface as $\rho = \rho_0 e^{-kH}$. Such a model is an approximation for the *standard atmosphere,* for which you can find details on the World Wide Web from the sites www.digitaldutch.com/atmoscalc/ and www.lerc.nasa.gov/Other_Groups/K-12/airplane/atmosmet.html, for example. For altitudes up to about 90 km or so, a fair approximation for the air density from the standard atmosphere model is obtained with $\rho_0 = 1.5$ kg/m^3 and $k = 0.14$ km^{-1}.

It is extremely difficult to accurately account for the effect of air drag on satellites when they enter the atmosphere since the drag depends on the shape of the vehicle and on its constantly changing orientation in space, and also on changes of atmospheric conditions. The purpose of this problem is to estimate the effect of air drag on the motion of the center of mass of a spacecraft using the exponential model.

a. Obtain the differential equations that govern the motion of the center of mass C of a satellite of mass m taking into account the drag force modeled as $\vec{F}_{\mathrm{drag}} = -\frac{1}{2} C_D \rho A v^2 \hat{v}$, where $\vec{v} = \dot{r}\hat{r} + r\dot{\theta}\hat{\theta}$ is the velocity of C relative to the earth, and $\hat{v} = \vec{v}/|\vec{v}|$. The nondimensional quantity C_D is called the *drag coefficient,* and A is the area of the satellite that is exposed to the airflow. Approximate the earth as an inertial homogeneous sphere of radius $R_E \approx 6378$ km, treat C_D and A as constants, and assume $\vec{F}_{\mathrm{drag}}$ acts on the satellite all the time.

b. At a certain time t_0 the speed of a satellite circling the earth is $v_0 = 29{,}500$ km/h, its height from the earth's surface is 102 km, and its flight path angle is $\gamma = 0$. With $C_D A/m = 5 \times 10^{-3}$ m^2/kg, numerically integrate the differential equations of motion and then determine how much time it takes for the satellite to crash on the earth's surface (assuming it survives the heat generated due to travel in the atmosphere). What are the values of the crash times when $C_D A/m$ is changed by ± 10 percent? Write a short paragraph commenting on your observations of the motion.

Suggestion: Modify the Simulink model `two_body_simul` to do this problem. Also, see the caution note in Problem 4.11.

4.13 Show that, in terms of the x, y, and z rectangular coordinates shown in Fig. 4.12 (p. 250), the differential equations of motion for the three-body problem investigated in Section 4.6 are

$$\ddot{x} - 2\Omega\dot{y} - \Omega^2 x = -\left(\frac{Gm_1}{r_{31}^3} + \frac{Gm_2}{r_{32}^3}\right) x \qquad (4.7.1)$$

$$\ddot{y} + 2\Omega\dot{y} - \Omega^2 y = -\left(\frac{Gm_1}{r_{31}^3} + \frac{Gm_2}{r_{32}^3}\right) y - \frac{Gm_1 m_2 L}{m_1 + m_2}\left(\frac{1}{r_{31}^3} - \frac{1}{r_{32}^3}\right) \qquad (4.7.2)$$

$$\ddot{z} = -\left(\frac{Gm_1}{r_{31}^3} + \frac{Gm_2}{r_{32}^3}\right) z \qquad (4.7.3)$$

Figure 4.16 Figure for Problem 4.14.

4.14 Body AB shown in Fig. 4.16 consists of two mass particles, each of mass $m/2$, connected by a rigid bar of negligible mass and length L. Body AB is able to rotate freely, in the plane of the figure, about its center of mass C, which is connected to a fixed pin. The system is attracted by an inertial center of attraction at O that exerts a force whose magnitude is equal to $\mu M/|\overrightarrow{OP}|^2$, where μ is a constant, at any point P of mass M. Show that the system exhibits the equilibrium orientations $\theta = \text{constant} \triangleq \theta_e$, where $\theta_e = 0$ and $90°$. Which one of these is stable? For $R = 3$ m, $L = 2$ m, and $\mu = 1.8$ m³/s², determine the frequency of the small oscillations about the stable equilibrium orientation of the system. *Hint:* The moment $\vec{M}_C$ was formulated in Example 1.10.1 (p. 23).

4.15 For the system shown in Fig. 4.16, point O represents the center of the earth, and the earth is approximated as a homogeneous sphere of radius $R_E \approx 6378$ km. Instead of being stationary as in Problem 4.14, point C is in a circular orbit of radius R about O, above the earth's atmosphere. For $L \ll R_E$, show, either analytically or numerically, that the frequency ω of the small oscillations of the system about the equilibrium $\theta = \theta_e = 0$ is $\omega = \sqrt{3}\Omega$, where $\Omega = \sqrt{gR_E^2/R^3}$ (with g being the acceleration of gravity at the surface of the earth) is the absolute orbital angular velocity of the circular orbit. *Hint:* The unit vectors $\hat{x}$ and $\hat{y}$ shown in Fig. 1.9 (p. 24) are no longer inertial. Introduce an angle ϕ, where $\dot{\phi} = \Omega$, to keep track of their rotation.

4.16 For the system in Problem 4.15, show that the linearized differential equation of motion about the equilibrium $\theta_e = 0$ is $\ddot{\theta} + 3\Omega^2\theta = 0$, where Ω is the orbital angular velocity of the circular orbit.

CHAPTER 5

Dynamics of Rigid Bodies in "Simpler" Planar Motion

The motion of each particle of a system of N point masses is investigated simply by applying Newton's second law to each particle. As shown in Chapter 4, the resulting vector equations can also be manipulated to yield a vector equation that governs the translational motion of the center of mass of the system, and also a vector equation that involves the time rate of change of the *absolute angular momentum* of the system. These vector equations, which yield six scalar equations for a general motion in three-dimensional space, provide the transition from a system composed of discrete mass particles to a continuum of matter (which is achieved in the limit as $N \to \infty$), which includes either rigid or flexible bodies.

A rigid body is a continuum of matter for which the distance between any two of its points is constant. A rigid body may be viewed as a very large, in fact infinite, number of particles satisfying such a distance constraint. The fundamental vector equations for analyzing the motion of a rigid body consist of the two equations previously mentioned, namely, the equation that governs the translational motion of the center of mass of the system and a vector equation that involves the time rate of change of the absolute angular momentum of the system. As shown in this chapter, the latter equation yields a set of scalar equations that govern the *rotational motion* (i.e., motion that involves the change of orientation of a line) of the rigid body. Internal *action* and *reaction forces* between individual particles of the body do not appear in the differential equations of motion for the system. Those sets of equations are the fundamental equations that govern the most general motion of a rigid body.

This chapter starts with a section showing that a rigid body that is able to move in three-dimensional space has six degrees of freedom; i.e., six independent coordinates are needed to specify the position and orientation of the body in three-dimensional space. The rest of the chapter is restricted to *"simpler" planar motion* (defined at the end of Section 5.1) of rigid bodies. Section 5.6 is dedicated to the calculation of the reaction forces acting on the members of mechanisms. A Simulink model named `fourbarforces.mdl` that performs such calculations based on the equations developed in that section is included on the CD that accompanies the book.

5.1 Degrees of Freedom of a Rigid Body

A particle P_1 that is able to move in three-dimensional space has three degrees of freedom since three time-dependent variables, or motion *coordinates,* are necessary and sufficient to locate the particle in that space and, thus, to describe its motion. Similarly, two unconstrained particles P_1 and P_2 have six degrees of freedom, while such a number is reduced to five if the distance between the particles is constrained to be constant. This is so because if six variables were used, say six rectangular coordinates $x_1, y_1, z_1, x_2, y_2, z_2$, those variables would be related by one equation of constraint, namely, $(x_2 - x_1)^2 + (y_2 - y_1)^2 + (z_2 - z_1)^2 = $ constant $= L^2$, where L is the constant distance between the two particles. Thus, only five of those six variables would be independent of each other. This is illustrated in Fig. 5.1. The addition of one more particle P_3 to the constrained P_1-P_2 system increases the number of degrees of freedom of the entire system from five to eight. However, if the distances between P_3 and the other two particles is constant, so that the particles form a rigid body, the number of degrees of freedom of such a system is reduced from eight to six. This happens because each one of the two additional rigid connections that are needed to transform the system into a rigid body introduces one constraint between the three rectangular coordinates of each particle.

As illustrated in Fig. 5.1, further addition of any number of particles to a rigid three-particle system, in such a way that the resulting N-particle system is rigid, will not change the number of degrees of freedom of the resulting system. In the limit, as $N \to \infty$, the system becomes a continuum body.

As the number of particles is increased, three new variables are added for each particle that is added to the system. However, three new constraint equations are also introduced when the distance constraints are imposed, and the number of degrees of freedom of the system remains equal to six. Thus, although a rigid body has an infinite number of particles, its number of degrees of freedom is still $n = 6$ for motion in three-dimensional space. This is an important result, and it implies that six scalar differential equations are necessary

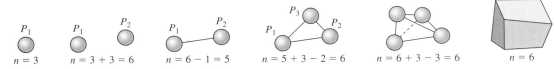

Figure 5.1 From a single particle to a rigid body in three-dimensional motion (n is the number of degrees of freedom).

and sufficient to analyze the most general motion of a rigid body in three-dimensional space. If the motion of the body is constrained to a single plane in space, the number of degrees of freedom for such motion is equal to $n = 3$. In such cases, the motion of the body is governed by only three scalar differential equations.

In this chapter, only "simpler" planar motion of rigid bodies is considered. A *simpler* (but not necessarily "simple" by any means!) planar motion is defined in this book as follows: For thin bodies that may be approximated as being planar, the motion is restricted to a plane that is parallel to the plane of the body; for general three-dimensional bodies, only bodies with an axis of symmetry and that are rotating either about that axis or about an axis that is parallel to it are considered. Such cases still encompass a large class of problems that are encountered in engineering practice. They do not involve *products of inertia,* which are quantities that appear in Chapter 6 in connection with the general motion of rigid bodies. In that chapter you will be introduced to the basic material that is necessary for dealing with the general case of three-dimensional motion for more advanced studies involving such motions (such as studies involving satellite dynamics, aircraft and ship dynamics, and gyroscopic instruments). The material presented in this chapter provides a transition to the more general material presented in Chapter 6.

5.2 Types of Rigid Body Displacements

A rigid body can undergo two types of basic motion: *translation* and *rotation*. In translation, the orientation of any line fixed to the body does not change as the body moves. This is illustrated in Fig. 5.2a where a top is simply moved from one location to another without spinning it or changing its orientation. In *rotation,* there is at least one line painted on the body that changes orientation as the body moves. The rotation of a line is illustrated in Fig. 5.2b. If the top in Fig. 5.2a were rotated, the line AB would change direction during its motion.

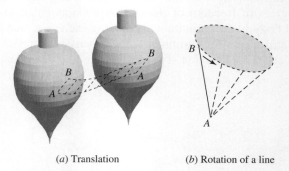

(*a*) Translation (*b*) Rotation of a line

Figure 5.2 The two types of motion of a rigid body (translation and rotation).

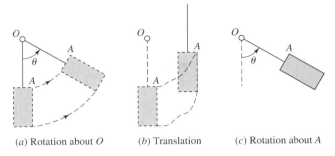

(a) Rotation about O (b) Translation (c) Rotation about A

Figure 5.3 Decomposition of a motion (a) into a translation of a point (b), and a rotation (c).

The most general motion of a rigid body can be "decomposed" into a translation of an arbitrarily chosen point A on the body and a rotation about an axis that passes through A. Each one of these two basic motions accounts for three of the six degrees of freedom of the motion of a rigid body. The choice of a point A is arbitrary. Such a decomposition is illustrated by the planar motion of the pendulum shown in Fig. 5.3. The motion may be thought of as either one of the following:

- A rotation θ about an axis that passes through O and is perpendicular to the plane of the paper, as shown in Fig. 5.3a.
- A translation of an arbitrary point A on the body (i.e., no rotation is involved in this step, as shown in Fig. 5.3b), followed by a rotation θ about an axis that passes through A and is also perpendicular to the plane of the paper (as shown in Fig. 5.3c). Such a rotation brings point O to its actual position at all times in space, without changing the position of point A (which is already in the right place after the translation shown in Fig. 5.3b occurs).

The rotation shown in Fig. 5.3c brings the body shown in Fig. 5.3b to its correct orientation in space as shown in Fig. 5.3a. Notice that the angle the body rotates (namely, the angle θ) is independent of the choice of point A. Thus, the rotation θ is the same for any line drawn on the body shown in Fig. 5.3.

The motion of a rigid body that is only translating in space is similar to that of a point mass since each point of the body is moving in parallel paths. Thus, for such cases, the motion of any point on the body describes the motion of the entire body.

> The motion of a rigid body will be purely translational only if the forces acting on the body are equivalent to a single resultant force $\vec{F}$ that passes through its center of mass C (i.e., the moment $\vec{M}_C$ about C of all forces acting on the body is equal to zero) *and* if the motion is started with zero angular velocity.

If the moment $\vec{M}_C$, about the center of mass C, of the forces acting on a body is not zero, such a moment will cause the body to rotate.

The formulation of the equations that are needed for analyzing the motion of a rigid body in simpler planar motion is presented in Section 5.3. This is then followed by Section 5.4 where several solved examples are presented.

5.3 Dynamics of Simpler Planar Motion of a Rigid Body

If a rigid body is viewed as a system of a large number of point masses connected by massless bars of constant lengths, it is seen that the results obtained in Sections 4.1 and 4.2 are applicable to such bodies. For a continuum body, the summations in the equations in that chapter are replaced by an integration that spans the volume of the entire body.

Figure 5.4 shows a planar rigid body subjected to a set of forces whose resultant is $\vec{F}$. Some of the applied forces may appear as a couple (i.e., two forces that are equal in magnitude and opposite in direction), and their effect is a moment applied to the body (such as the moment M_{bar} in Example 2.7.6, p. 121). In general, the body will be translating and rotating. Specific details about Fig. 5.4 are listed here.

■ Point P (whose absolute position vector is $\vec{r} = \overrightarrow{OP}$, where O is an inertial point) represents an infinitesimal mass element of the body, with mass dm. The total mass of the body is $m = \int_{\text{body}} dm$, where the integration spans the entire body.

■ Point C is the center of mass of the body, and its absolute position vector is $\vec{r}_C = \overrightarrow{OC}$. The absolute position vector $\vec{r}_C$ was defined in Chapter 4 for a system of point masses. For continuum bodies, either rigid or flexible, it is

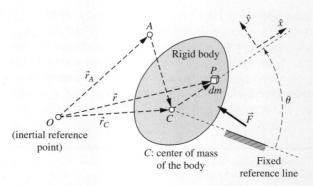

Figure 5.4 A rigid body subjected to a resultant force $\vec{F}$.

SECTION 5.3 Dynamics of Simpler Planar Motion of a Rigid Body

determined as

$$\vec{r}_C = \frac{1}{m} \int_{\text{body}} \vec{r} \, dm \qquad (5.3.1)$$

Notice that the center of mass of the body may not coincide with any point on the body itself, as in the case of a ring, for example. In such cases the center of mass may be simply viewed as belonging to an extension of the body.

- The angle θ is the variable that is used for keeping track of the rotational motion of the body in inertial space. The angle measured from a line that is fixed in space to *any* line that is fixed to the body will accomplish such a purpose. Since θ is measured from a nonrotating reference line, the absolute angular velocity of the body is, according to the right-hand rule, $\vec{\omega} = \dot{\theta}\hat{z}$, where $\hat{z} = \hat{x} \times \hat{y}$ is a unit vector perpendicular to the plane of Fig. 5.4, and pointing toward you, the reader.

- Point A is an arbitrary point in space. Its absolute position vector is $\vec{r}_A = \overrightarrow{OA}$.

It was shown in Chapter 4 that the motion of the center of mass C of a system of particles is governed by Newton's second law as if the entire mass of the system were concentrated at C. The same result is applicable, of course, to a rigid body. If the body is subjected to several forces (say, N forces $\vec{F}_1, \vec{F}_2, \ldots, \vec{F}_N$), the motion of its center of mass is then governed by Newton's second law as [notice that m is assumed to be constant so that $d(m\vec{v}_C)/dt = m\,d\vec{v}_C/dt$]:

$$\boxed{\vec{F} \triangleq \sum_{i=1}^{N} \vec{F}_i = m \frac{d\vec{v}_C}{dt} = m \frac{d^2 \vec{r}_C}{dt^2}} \qquad (5.3.2)$$

The motion governed by this equation was considered in detail in Chapter 2.

We now need to analyze the rotational motion of the rigid body and, for this, it is necessary to look at the *absolute angular momentum* of the body. Returning to Fig. 5.4, let us look at the absolute angular momentum referred to an arbitrary point A, which is denoted by $\vec{H}_A$. For a continuum body, $\vec{H}_A$ is still defined by Eq. (4.2.1), with the summation replaced by an integral that should span every infinitesimal mass element of the body. Thus, $\vec{H}_A$, which involves the moment about A of the absolute linear momentum $(d\vec{r}/dt)\,dm$, is defined as being the following quantity:

$$\vec{H}_A \triangleq \int_{\text{body}} \left[\overrightarrow{AP} \times \frac{d\vec{r}}{dt}\right] dm \qquad (5.3.3)$$

The rotational motion of the body is governed by Eq. (4.2.6). That equation is repeated here.

$$\frac{d}{dt}\vec{H}_A = \vec{M}_A + m\left(\frac{d}{dt}\overrightarrow{AC}\right) \times \frac{d\vec{r}_C}{dt} \qquad (5.3.4)$$

The most convenient way to "read" Eq. (5.3.4) is, "The moment $\vec{M}_A$ about a point A of the resultant force $\vec{F}$ acting on a body is equal to the time rate of change of the absolute angular momentum $\vec{H}_A$, plus a correction." The "correction" is zero for two particularly convenient choices of point A, as indicated in the next paragraph (which is essentially repeated from Chapter 4). When read as indicated earlier, Eq. (5.3.4) is the "counterpart" to the fundamental equation $\vec{F} = m\, d\vec{v}_C/dt$. It is also a fundamental equation in dynamics, an equation whose form is simple and not difficult to remember.[1] This simpler form of Eq. (5.3.4) is the one that is used in the solution to examples in this book.

Equation (5.3.4) is completely general since no restrictions were imposed on the choice of point A about which the moment of the external forces and the absolute angular momentum of the system of particles were taken. However, by proper choice of a point A (as discussed in Section 4.2), the cross product term in Eq. (5.3.4) can be eliminated, thus yielding a simpler relationship (that still can be applied to any problem) between the moment $\vec{M}_A$ and the time derivative of the angular momentum $\vec{H}_A$. This can be accomplished by either choosing point A to be inertial, in which case $\overrightarrow{AC} = \vec{r}_C$, or to be the center of mass C of the body, in which case $\overrightarrow{AC} = 0$. The following simpler relations are then obtained. These equations were originally developed by Euler, and they are appropriately called *Euler's vector equations,* or simply Euler's equations.

$$\boxed{\begin{aligned} \frac{d}{dt}\vec{H}_O &= \vec{M}_O \qquad \text{(when } O \text{ is inertial)} \\ \frac{d}{dt}\vec{H}_C &= \vec{M}_C \qquad \text{(when } C \text{ is the center of mass)} \end{aligned}} \qquad (5.3.5)$$

The second equation of (5.3.5) is particularly significant since it is valid even when the center of mass C is not inertial and, thus, may be moving in an arbitrary manner in space. The first equation of (5.3.5) is useful in problems when a body is rotating about a fixed point O that is either a point on the body or on a "rigid extension" of the body.

To apply the second equation of (5.3.5) to specific problems, an expression for the angular momentum $\vec{H}_C$ is needed. With $\vec{r} = \vec{r}_C + \overrightarrow{CP}$ (see Fig. 5.4), such

---

[1] The similarity between the two equations is seen by replacing the force $\vec{F}$ in $\vec{F} = m\, d\vec{v}_C/dt$ by the moment $\vec{M}_C$, and the linear momentum $m\vec{v}_C$ by the angular momentum $\vec{H}_C$.

an expression is developed as follows:

$$\vec{H}_C = \int_{body} \left[\overrightarrow{CP} \times \frac{d}{dt}\vec{r}\right] dm = \int_{body} \left[\overrightarrow{CP} \times \frac{d}{dt}\vec{r}_C\right] dm + \int_{body} \left[\overrightarrow{CP} \times \frac{d}{dt}\overrightarrow{CP}\right] dm$$

$$= \underbrace{\left[\int_{body} \overrightarrow{CP}\, dm\right]}_{=0} \times \frac{d}{dt}\vec{r}_C + \int_{body} \left[\overrightarrow{CP} \times \frac{d}{dt}\overrightarrow{CP}\right] dm$$

The first integral in the right-hand side of this expression is zero because that is how the location of the center of mass C is defined. For the second integral, it is convenient to express the vector $\overrightarrow{CP}$ in terms of the unit vector $\hat{x}$ fixed to the rotating rigid body as $\overrightarrow{CP} = |\overrightarrow{CP}|\hat{x}$ (see Fig. 5.4), where $|\overrightarrow{CP}|=$ constant because the body is rigid. Therefore, since $d\hat{x}/dt = \dot{\theta}\hat{y}$ (again, see Fig. 5.4), it follows that

$$\frac{d}{dt}\overrightarrow{CP} = |\overrightarrow{CP}|\dot{\theta}\hat{y}$$

This immediately yields

$$\boxed{\vec{H}_C = I_C \dot{\theta} \hat{z}} \tag{5.3.6}$$

where

$$\boxed{I_C \triangleq \int_{body} |\overrightarrow{CP}|^2\, dm} \tag{5.3.7}$$

is called the *mass moment of inertia* (or simply, the *moment of inertia*) of the rigid body about the line that passes through the center of mass C and is perpendicular to the plane of Fig. 5.4. Notice that the integrand in the expression for I_C is simply the square of the distance from point C to a mass element dm of the body and, thus, I_C is never a negative quantity. The unit vector $\hat{z} = \hat{x} \times \hat{y}$ is perpendicular to the plane of the motion in Fig. 5.4, and the quantity $\vec{\omega} \triangleq \dot{\theta}\hat{z}$ that appears in Eq. (5.3.6) is the absolute angular velocity of the body in planar motion.

> The definition of the moment of inertia about a line that passes through the center of mass, given by Eq. (5.3.7), is extended for any other line in space, say a line AA, as $I_A = \int_{body} |\overrightarrow{AP}|^2\, dm$, where $|\overrightarrow{AP}|$ is the perpendicular distance from the mass element dm of the body to line AA. This general definition appears in the analysis of three-dimensional motion of rigid bodies.

The moments of inertia of a body about two parallel lines, one of which passes through the center of mass of the body, are related to each other in a relatively simple manner. Such a relation, which is presented next, is known as the *parallel axis theorem*. When the moment of inertia I_C about an axis that passes through C is known, use of the parallel axis theorem provides a simple means for calculating the moment of inertia I_A about an axis that passes thorough A and is parallel to the one that passes through C.

> **The Parallel Axis Theorem**
> By making use of Fig. 5.4, and of the definition of the moment of inertia, it can be shown that
>
> $$\boxed{I_A = I_C + m|\overrightarrow{AC}|^2} \tag{5.3.8}$$
>
> This relationship between the moments of inertia about two parallel lines, one of which passes through the center of mass of the body, is known as the *parallel axis theorem*.
>
> The relationship between I_A and I_C given by Eq. (5.3.8) is readily developed by referring to Fig. 5.4. From the definition of the moment of inertia, we have for I_A and for I_C:
>
> $$I_A = \int_{\text{body}} |\overrightarrow{AP}|^2 \, dm \qquad I_C = \int_{\text{body}} |\overrightarrow{CP}|^2 \, dm$$
>
> Since $\overrightarrow{AP} = \overrightarrow{AC} + \overrightarrow{CP}$ (see Fig. 5.4), we can then write
>
> $$|\overrightarrow{AP}|^2 = \overrightarrow{AP} \cdot \overrightarrow{AP} = \overrightarrow{AC} \cdot \overrightarrow{AC} + 2\overrightarrow{AC} \cdot \overrightarrow{CP} + \overrightarrow{CP} \cdot \overrightarrow{CP}$$
> $$= |\overrightarrow{AC}|^2 + 2\overrightarrow{AC} \cdot \overrightarrow{CP} + |\overrightarrow{CP}|^2$$
>
> and this gives
>
> $$I_A = \int_{\text{body}} |\overrightarrow{AC}|^2 \, dm + 2 \underbrace{\int_{\text{body}} \overrightarrow{AC} \cdot \overrightarrow{CP} \, dm}_{= \, \overrightarrow{AC} \, \cdot \int_{\text{body}} \overrightarrow{CP} \, dm \, = \, 0} + \int_{\text{body}} |\overrightarrow{CP}|^2 \, dm = I_C + m|\overrightarrow{AC}|^2$$
>
> The second integral in this expression is zero because of the definition of the center of mass C.

Notes

1. The absolute angular momentum $\vec{H}_A$ about an arbitrary point A is related to the absolute angular momentum $\vec{H}_C$ about the center of mass C as (see Problem 5.4)

$$\vec{H}_A = \vec{H}_C + m\overrightarrow{AC} \times \vec{v}_C$$

2. In general, the relationship between the absolute angular momentum $\vec{H}_A$ about an arbitrary point A, and the body's absolute angular velocity $\vec{\omega} = \dot{\theta}\hat{z}$, is not as simple as Eq. (5.3.6) (with I_C replaced by I_A). One would have $\vec{H}_A = I_A \dot{\theta} \hat{z}$ only when point A is inertial and the distance between A and the center of mass C of the body is constant (see Problem 5.5).

Several examples of the application of the material presented so far are given next. Examples 5.3.1, 5.3.2, and 5.3.3 illustrate the calculations involved in finding the location of the center of mass of a body, and Examples 5.3.4 through 5.3.8 illustrate the calculation of mass moments of inertia. Examples 5.3.9 through 5.3.13 involve the dynamics of rigid bodies undergoing the plane motions addressed in this chapter.

EXAMPLE 5.3.1

Three wires AB, BD, and AD, of lengths $L_1 = 3$ m, $L_2 = 4$ m, and $L_3 = \sqrt{3^2 + 4^2} = 5$ m, respectively, are welded together to form a rectangular triangle as shown in Fig. 5.5a. The wires are made of the same material (i.e., they have the same material density ρ kg/m). Determine the location of the center of mass of the wire triangle.

■ **Solution**

The additional details needed for the calculations are shown in Fig. 5.5b. They are

- A rectangular coordinate system (x, y) that is arbitrarily chosen to be centered at A.
- Two orthogonal unit vectors $\hat{x}$ and $\hat{y}$ that will allow us to keep track of directions.
- Three infinitesimal line segments along the wires AD, AB, and BD. The lengths of the first two are dx and dy, respectively. Since the equation for the line that passes through points B and D is $y = L_1 - (L_1/L_2)x$, the length of the infinitesimal line segment along BD is $ds = \sqrt{(dx)^2 + [(L_1/L_2)\,dx]^2} = (L_3/L_2)\,dx$.

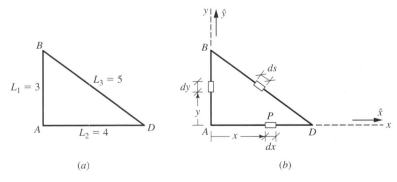

(a)   (b)

Figure 5.5 The wire triangle and its infinitesimal mass elements for Example 5.3.1.

Solution 1: Direct Integration

Since the mass of the wires AD, AB, and BD are $\rho\,dx$, $\rho\,dy$, and $\rho\,ds$, respectively, the position vector $\overrightarrow{AC}$ for the center of mass of the bent wire is then

$$\underbrace{\rho(L_1+L_2+L_3)}_{\text{total mass}}\overrightarrow{AC}$$

$$= \underbrace{\int_{x=0}^{L_2}(x\hat{x})\rho\,dx}_{\text{for wire }AD} + \underbrace{\int_{y=0}^{L_1}(y\hat{y})\rho\,dy}_{\text{for wire }AB} + \underbrace{\int_{x=0}^{L_2}\left[x\hat{x}+\left(L_1-\frac{L_1}{L_2}x\right)\hat{y}\right]\rho\frac{L_3}{L_2}\,dx}_{\text{for wire }BD}$$

$$= \frac{\rho L_2^2}{2}\hat{x} + \frac{\rho L_1^2}{2}\hat{y} + \frac{\rho L_3}{L_2}\left[\frac{L_2^2}{2}\hat{x}+\left(L_1 L_2 - \frac{L_1}{L_2}\frac{L_2^2}{2}\right)\hat{y}\right] \quad (5.3.1a)$$

From this equation, we then obtain

$$\overrightarrow{AC} = 1.5\hat{x} + \hat{y}\text{ m}$$

Solution 2: Reducing the Problem to Three Point Masses (Indirect Integration)

Since each wire has a constant mass density, the center of mass of each wire is located at its geometric center (as also obtained by direct integration). With this observation, the problem is then reduced to finding the center of mass of a system of three mass particles: a particle of mass ρL_2 located in the middle of wire AD, a particle of mass ρL_1 located in the middle of wire AB, and a particle of mass ρL_3 located in the middle of wire BD. The position vector $\overrightarrow{AC}$ of the center of mass of this system is then given by

$$(\rho L_1 + \rho L_2 + \rho L_3)\overrightarrow{AC} = \rho L_1\frac{L_1}{2}\hat{y} + \rho L_2\frac{L_2}{2}\hat{x} + \rho L_3\left(\frac{L_2}{2}\hat{x}+\frac{L_1}{2}\hat{y}\right) \quad (5.3.1b)$$

which is the same as the expression obtained in solution 1.

EXAMPLE 5.3.2

Determine the location of the center of mass C of the homogeneous triangular plate ABD shown in Fig. 5.6a. The mass of the plate is m, and its thickness is small enough (when compared to either L_1 or L_2) to be neglected in the calculations.

■ **Solution**

As shown in Fig. 5.6b, let $\overrightarrow{AP} = x\hat{x}+y\hat{y}$ be the position vector from the arbitrarily chosen reference point A to an infinitesimal mass element whose mass is equal to $dm = \rho\,dx\,dy$, where $\rho = 2m/(ab)$ is the mass density of the plate. By noticing that the boundary BD of the plate is the straight line $y = L_1 - (L_1/L_2)x$, we then obtain for the position vector $\overrightarrow{AC}$ of the center of mass of the plate,

$$m\overrightarrow{AC} = \frac{2m}{L_1 L_2}\int_{x=0}^{L_2}\int_{y=0}^{L_1-(L_1/L_2)x}(x\hat{x}+y\hat{y})\,dy\,dx \quad (5.3.2a)$$

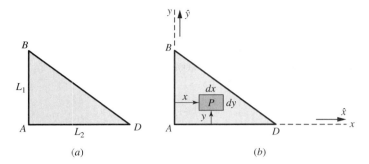

Figure 5.6 The triangular plate and the infinitesimal mass element for Example 5.3.2.

This gives

$$\overrightarrow{AC} = \frac{L_2}{3}\hat{x} + \frac{L_1}{3}\hat{y}$$

A Composite Body — EXAMPLE 5.3.3

Determine the location of the center of mass C of the *composite* body (i.e., a relatively complex body that consists of several simple bodies joined to each other) shown in Fig. 5.7, based on the x-y rectangular coordinate system (centered at point A) that is also shown in the figure. If needed, use the $\hat{x}$ and $\hat{y}$ unit vectors shown in the figure. The body consists of a homogeneous triangular plate ABD welded to a homogeneous rectangular plate $DEFG$. The mass of the entire body is m kg, and its thickness is small enough to be neglected in the calculations. All dimensions shown in Fig. 5.7 are in meters.

■ **Solution**

The easiest way to calculate the coordinates of the center of mass for a composite body is to reduce the problem to the calculation of the center of mass of a system of point masses (as done in solution 2 of Example 5.3.1). For this, one needs to determine the location of

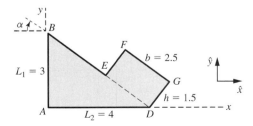

Figure 5.7 The composite body for Example 5.3.3.

the center of mass of each individual body that makes up the composite body. Here, the composite body can be divided into a homogeneous triangular plate and a homogeneous rectangular plate.

The x and y coordinates of the center of mass C_1 of the homogeneous triangular plate ABD are (see Example 5.3.2)

$$x_{C_1} = \frac{L_2}{3} = \frac{4}{3} \text{ m}$$

$$y_{C_1} = \frac{L_1}{3} = 1 \text{ m}$$

The center of mass C_2 of the homogeneous rectangular plate $DEFG$ is located at the middle of the plate. Since the angle α shown in Fig. 5.7 is $\alpha = \arctan \frac{3}{4}$, the x and y coordinates of C_2 are then (see Fig. 5.7 to understand the calculations)

$$x_{C_2} = L_2 - \frac{b}{2} \cos \alpha + \frac{h}{2} \sin \alpha = 4 - \frac{2.5}{2} \times \frac{4}{\sqrt{3^2 + 4^2}} + \frac{1.5}{2} \times \frac{3}{\sqrt{3^2 + 4^2}} = 3.45 \text{ m}$$

$$y_{C_2} = \frac{h}{2} \cos \alpha + \frac{b}{2} \sin \alpha = \frac{1.5}{2} \times \frac{4}{\sqrt{3^2 + 4^2}} + \frac{2.5}{2} \times \frac{3}{\sqrt{3^2 + 4^2}} = 1.35 \text{ m}$$

The area of the triangular plate is $A_1 = L_1 L_2/2 = 6 \text{ m}^2$, and the area of the rectangular plate is $A_2 = bh = 3.75 \text{ m}^2$.

The problem is now reduced to finding the center of mass of two mass particles: a particle of mass $m_1 = mA_1/(A_1 + A_2)$, located at C_1, and a particle of mass $m_2 = mA_2/(A_1 + A_2)$, located at C_2. The position vector $\overrightarrow{AC}$ of the center of mass of the composite plate is then given by

$$m\overrightarrow{AC} = m_1 \overrightarrow{AC_1} + m_2 \overrightarrow{AC_2} \tag{5.3.3a}$$

This gives,

$$m\overrightarrow{AC} = \frac{6}{9.75} m \left(\frac{4}{3} \hat{x} + \hat{y} \right) + \frac{3.75}{9.75} m (3.45 \hat{x} + 1.35 \hat{y})$$

or

$$\overrightarrow{AC} \approx 2.15 \hat{x} + 1.13 \hat{y} \text{ m}$$

The x and y coordinates of the center of mass of the composite plate are then $x_C \approx 2.15$ m and $y_C \approx 1.13$ m.

EXAMPLE 5.3.4

For the three-wire triangle of Example 5.3.1, determine the moments of inertia I_A and I_C about the axes that are perpendicular to the plane of Fig. 5.5 and that pass through point A and the center of mass C of the system, respectively.

SECTION 5.3 Dynamics of Simpler Planar Motion of a Rigid Body

■ **Solution**

Solution 1

By definition, the moment of inertia I_A is given by the following expression (see the boxed text on p. 271):

$$I_A = \int_{\text{body}} (x^2 + y^2)\, dm$$

$$= \int_{\text{body } AB} (x^2 + y^2)\, dm + \int_{\text{body } AD} (x^2 + y^2)\, dm + \int_{\text{body } BD} (x^2 + y^2)\, dm \quad \text{(5.3.4a)}$$

Therefore, referring to Fig. 5.5, we obtain for I_A [where $ds = (L_3/L_2)\, dx$],

$$I_A = \int_{y=0}^{L_1} y^2 \rho\, dy + \int_{x=0}^{L_2} x^2 \rho\, dx + \int_{x=0}^{L_2} \left[x^2 + \left(L_1 - \frac{L_1}{L_2}x\right)^2 \right] \rho\, ds$$

$$= \frac{\rho}{3}\left[L_1^3 + L_2^3 + L_3(L_1^2 + L_2^2)\right] = 72\rho \text{ kg} \cdot \text{m}^2$$

The moment of inertia I_C may be readily calculated by making use of the parallel axis theorem as

$$I_C = I_A - m|\overrightarrow{AC}|^2 \quad \text{(5.3.4b)}$$

We then obtain

$$I_C = 72\rho - (3 + 4 + 5)\rho(1.5^2 + 1) = 33\rho \text{ kg} \cdot \text{m}^2$$

Solution 2

The moment of inertia $(I_C)_{\text{bar}}$ of a homogeneous bar of mass density ρ kg/m and length L is $(I_C)_{\text{bar}} = \int_{-L/2}^{L/2} \rho x^2\, dx = \rho L^3/12 = mL^2/12$ (where $m = \rho L$ is the mass of the bar). Therefore, the moment of inertia I_A for the three-wire triangle may also be calculated as

$$I_A = \underbrace{\frac{\rho L_1^3}{12} + (\rho L_1)\left(\frac{L_1}{2}\right)^2}_{\text{for bar } AB} + \underbrace{\frac{\rho L_2^3}{12} + (\rho L_2)\left(\frac{L_2}{2}\right)^2}_{\text{for bar } AD}$$

$$+ \underbrace{\frac{\rho L_3^3}{12} + (\rho L_3)\left[\left(\frac{L_1}{2}\right)^2 + \left(\frac{L_2}{2}\right)^2\right]}_{\text{for bar } BD}$$

$$= 72\rho \text{ kg} \cdot \text{m}^2$$

Notice that the first two underbraced terms in this result, which reduce to $\rho L_1^3/3$ and $\rho L_2^3/3$, are the moments of inertia about A of bars AB and AD, respectively. The third underbraced term is the sum of the moment of inertia $\rho L_3^3/12$ of bar BD about its center of mass, with a term that is equal to the product of the mass ρL_3 of bar BD and the square of the distance from A to the center of mass of BD.

EXAMPLE 5.3.5

For the triangular plate of Example 5.3.2, determine the moments of inertia I_A and I_C about the axes that are perpendicular to the plane of the figure and that pass through point A and the center of mass of the plate, respectively.

■ **Solution**

Referring to Fig. 5.6 and the definition of the moment of inertia, we obtain for I_A,

$$I_A = \int_{body} (x^2 + y^2)\, dm$$

$$= \frac{m}{L_1 L_2/2} \left(\int_{x=0}^{L_2} \int_{y=0}^{L_1-(L_1/L_2)x} x^2\, dy\, dx + \int_{x=0}^{L_2} \int_{y=0}^{L_1-(L_1/L_2)x} y^2\, dy\, dx \right) = \frac{m}{6}\left(L_1^2 + L_2^2\right)$$

The moment of inertia I_C may be calculated by using the parallel axis theorem as

$$I_C = I_A - m|\overrightarrow{AC}|^2 = I_A - m\left[\left(\frac{L_1}{3}\right)^2 + \left(\frac{L_2}{3}\right)^2\right] = \frac{m}{18}\left(L_1^2 + L_2^2\right)$$

EXAMPLE 5.3.6

A thin rectangular plate with sides $a = 4$ m and $b = 3$ m is welded to a thin triangular plate with sides $b = 3$ m and $L = 2$ m as shown in Fig. 5.8. Both plates are made of steel and have the same thickness. The total mass of the system is $m = 115.5$ kg. Points C_1 and C_2 shown in the figure are the centers of mass of the two plates. Determine the following quantities for the entire plate.

a. The x and y coordinates of its center of mass using the rectangular coordinate system shown in Fig. 5.8, which is centered at point A. If needed, use the two unit vectors $\hat{x}$ and $\hat{y}$ that are shown in the figure.
b. The moment of inertia I_A, about point A (more precisely, about the line that is perpendicular to the plane of the figure and that passes through point A). The moment of inertia of a rectangular plate of mass m_1 and sides a and b, about its center of mass, is $m_1(a^2 + b^2)/12$.

Figure 5.8 The composite plate for Example 5.3.6.

■ Solution

a. The location of the center of mass of a homogeneous triangular plate was determined in Example 5.3.2), and the center of mass of a homogeneous rectangular plate is at the middle of the plate. By choosing point A as the reference point (notice that such a choice is arbitrary), and also choosing to work with the unit vectors $\hat{x}$ and $\hat{y}$ shown in Fig. 5.8, the position vectors for C_1 and C_2 relative to point A are then

$$\vec{r}_{C_1} = \overrightarrow{AC_1} = \frac{a}{2}\hat{x} + \frac{b}{2}\hat{y} = 2\hat{x} + 1.5\hat{y} \text{ m}$$

$$\vec{r}_{C_2} = \overrightarrow{AC_2} = \left(a + \frac{L}{3}\right)\hat{x} + \frac{b}{3}\hat{y} \approx 4.67\hat{x} + \hat{y} \text{ m}$$

The position vector $\vec{r}_C$ of the center of mass C of the entire system, relative to point A, is determined by using the equation that defines a center of mass, i.e.,

$$(m_1 + m_2)\vec{r}_C = m_1\vec{r}_{C_1} + m_2\vec{r}_{C_2} \tag{5.3.6a}$$

where m_1 and m_2 are the masses of the rectangular and triangular plates, respectively.

Since both plates have the same thickness and the total mass of the system is $m = m_1 + m_2 = 115.5$ kg, the mass of each plate is then proportional to the plate's area:

$$m_1 = \frac{ab}{ab + bL/2}m = 92.4 \text{ kg} \qquad m_2 = \frac{bL/2}{ab + bL/2}m = 23.1 \text{ kg}$$

This then gives for $\vec{r}_C$,

$$\vec{r}_C \approx 2.53\hat{x} + 1.4\hat{y} \text{ m}$$

b. As determined in Example 5.3.5, the moment of inertia of the triangular plate about its center of mass is $m_2(b^2 + L^2)/18$. Since the moment of inertia of the plate about its center of mass is already given as $m_1(a^2 + b^2)/12$ (as an exercise for yourself, do the appropriate integrations to obtain this result), we can use the parallel axis theorem to determine the moment of inertia I_A. We then have

$$I_A = [I_A]_{\text{rectangle}} + [I_A]_{\text{triangle}} \tag{5.3.6b}$$

where (see Fig. 5.8)

$$[I_A]_{\text{rectangle}} = [I_{C_1}]_{\text{rectangle}} + m_1|\overrightarrow{AC_1}|^2 = [I_{C_1}]_{\text{rectangle}} + m_1\left[\left(\frac{a}{2}\right)^2 + \left(\frac{b}{2}\right)^2\right]$$

$$[I_A]_{\text{triangle}} = [I_{C_2}]_{\text{triangle}} + m_2|\overrightarrow{AC_2}|^2 = [I_{C_2}]_{\text{triangle}} + m_2\left[\left(a + \frac{L}{3}\right)^2 + \left(\frac{b}{3}\right)^2\right]$$

With $[I_{C_2}]_{\text{triangle}} = m_2(b^2 + L^2)/18$, as determined in Example 5.3.5, and $[I_{C_1}]_{\text{rectangle}} = m_1(a^2 + b^2)/12$, we obtain

$$I_A \approx 1313 \text{ kg} \cdot \text{m}^2$$

EXAMPLE 5.3.7

A homogeneous bar of mass m is bent as shown in Fig. 5.9 (the unit vectors $\hat{x}$ and $\hat{y}$ shown in the figure are used in the calculation of the location of the center of mass C). The thickness of the cross section of the bar is very small and may be neglected.

a. Verify that the center of mass of the bar is at its geometric center C shown in the figure.
b. Determine the moments of inertia I_B and I_C about the axes that are perpendicular to the plane of the figure and that pass through points B and C, respectively.

Figure 5.9 The bent bar for Example 5.3.7.

■ Solution

a. Since the bar is homogeneous, its density (mass per unit length) is equal to $\rho = m/(L + 2\pi R)$. Therefore, by dividing the bar into the curved segments AB and DE, and the straight segment BD (see Fig. 5.9), the masses of the infinitesimal mass elements for each segment are equal to $dm_1 = \rho R \, d\theta_1$, $dm_2 = \rho R \, d\theta_2$, and $dm_3 = \rho \, dx$, respectively.

By arbitrarily taking point B as the reference point, we obtain from the definition of the center of mass C,

$$m\overrightarrow{BC} = \underbrace{\int_0^\pi [(-R\sin\theta_1)\hat{x} - (R + R\cos\theta_1)\hat{y}]\rho R \, d\theta_1}_{\text{for wire } AB}$$

$$+ \underbrace{\int_0^L (x\hat{x})\rho \, dx}_{\text{for wire } BD} + \underbrace{\int_0^\pi [(L + R\sin\theta_2)\hat{x} + (R + R\cos\theta_2)\hat{y}]\rho R \, d\theta_2}_{\text{for wire } DE}$$

$$= \rho(L + 2\pi R)\frac{L}{2}\hat{x} = m\frac{L}{2}\hat{x}$$

This result confirms that $\vec{BC} = (L/2)\hat{x}$ (i.e., that the center of mass of the homogeneous symmetric body shown in Fig. 5.9 is located at its geometric center).

b. For the moment of inertia I_B we obtain

$$I_B = \underbrace{\int_{\theta_1=0}^{\pi} [(R + R\cos\theta_1)^2 + (R\sin\theta_1)^2]\rho R\, d\theta_1}_{\text{for } AB}$$

$$+ \underbrace{\int_{\theta_2=0}^{\pi} [(L + R\sin\theta_2)^2 + (R + R\cos\theta_2)^2]\rho R\, d\theta_2}_{\text{for } DE} + \underbrace{\int_{x=0}^{L} x^2 \rho\, dx}_{\text{for } BD}$$

$$= \left[\pi R(L^2 + 4R^2) + 4LR^2 + \frac{L^3}{3}\right]\rho$$

$$= \frac{m}{2\pi R + L}\left[\pi R(L^2 + 4R^2) + 4LR^2 + \frac{L^3}{3}\right]$$

For the moment of inertia I_C, since $I_B = I_C + m(L/2)^2$ (as per the parallel axis theorem), we then have

$$I_C = I_B - m\left(\frac{L}{2}\right)^2$$

EXAMPLE 5.3.8

Figure 5.10a shows a hollow cylinder of mass m, inner radius R_1 and outer radius R_2, and length L. The cylinder is homogeneous, and its mass is m kg. Determine its mass moment of inertia about the axial axis that passes through the center C of the cylinder cross section, as shown in the figure.

■ **Solution**

A cross section of the cylinder, perpendicular to the axial axis, is shown in Fig. 5.10. By definition, the moment of inertia I_C is equal to the integral, over the entire volume of the cylinder (i.e., a triple integral), of the product of the infinitesimal mass dm and the square

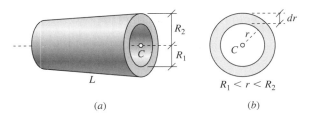

(a) (b)

Figure 5.10 The hollow cylinder for Example 5.3.8.

of the perpendicular distance from the axial axis to dm, i.e.,

$$I_C = \int_{\text{cylinder}} r^2 \, dm \tag{5.3.8a}$$

Since the cylinder is homogeneous, the triple integral that spans the entire cylinder can be replaced by a single integral if we work with a larger mass element of radius r, width dr, and length L. The side view of this mass element is shown in Fig. 5.10b. The volume dV of such a larger element is $dV = (2\pi r \, dr) L$, and its mass density (i.e., its mass per unit volume) is $\rho = m/(\pi R_2^2 L - \pi R_1^2 L)$. Therefore, since $dm = \rho \, dV$, the moment of inertia I_C can be obtained as

$$I_C = \frac{m}{\left(\pi R_2^2 L - \pi R_1^2 L\right)} \int_{r=R_1}^{R_2} r^2 \times (2\pi r L \, dr) = \frac{m}{2\left(R_2^2 - R_1^2\right)} \left(R_2^4 - R_1^4\right) = \frac{m}{2}\left(R_2^2 + R_1^2\right)$$

Notice that two particularly interesting cases can be extracted from this result, namely, a ring and a massive cylinder.

A ring or a cylindrical shell. For this case $R_1 = R_2$, but the mass of the body is still m (i.e., the mass density is the mass per unit length, which is m/L). Thus, $I_C = m R_2^2$. This result should be expected since the perpendicular distance between any mass element of the ring and the axis through C is equal to R_2.

A massive cylinder. For this case, $R_1 = 0$ and, thus, $I_C = m R_2^2/2$ (which happens to be one-half of the moment of inertia of a ring with the same mass).

The following examples deal with dynamics of simpler planar motion of rigid bodies. The basic equations that need to be used were developed in Section 5.3. They are Eq. (5.3.2) (p. 269) and either equation in Eq. (5.3.5) (p. 270) (whichever is simpler to apply to any particular problem under consideration). Equation (5.3.2) governs the motion of the center of mass of the body and either equation in Eq. (5.3.5) governs the rotational motion of the body. Together, those equations yield the differential equations of motion of the body.

EXAMPLE 5.3.9

The system shown in Fig. 5.11 consists of a rigid body of mass m kg, pinned to a fixed point O. Motion takes place in the vertical plane, and the body is free to rotate about O, with no friction. The distance between O and the center of mass C of the body is r_C m. The body is subjected to an external force $P(t)$ that is always perpendicular to line OC, applied at a distance D from O. The motion is started with given initial conditions $\theta(0) = \theta_0$ and $\dot{\theta}(0) = \dot{\theta}_0$.

a. Obtain the equations for determining the support reaction forces at O, along OC and perpendicular to OC, and the differential equation that governs the resulting motion of the system. Describe, in a few words, the necessary steps for calculating the support reactions.

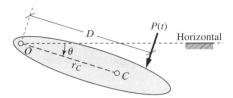

Figure 5.11 A pinned rigid body subjected to a force $P(t)$ for Example 5.3.9.

b. Is it possible to predict how the component of the support reaction perpendicular to OC changes with the angle θ? If so, determine how.

■ Solution

a. The free-body diagram for the system, using reaction force components along line OC and along the perpendicular to OC, is shown in Fig. 5.12. Two rotating unit vectors, $\hat{r}$ and $\hat{\theta}$, are also shown in the figure. In terms of these unit vectors, the expression for the absolute acceleration of the center of mass C is $\vec{a}_C = r_C(\ddot{\theta}\hat{\theta} - \dot{\theta}^2\hat{r})$.

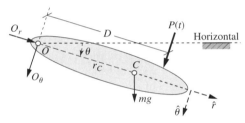

Figure 5.12 Free-body diagram for Example 5.3.9.

From Newton's second law $\vec{F} = m\vec{a}_C$, the following equations are obtained (see Fig. 5.12):

$$O_r + mg\sin\theta = -mr_C\dot{\theta}^2 \quad (5.3.9a)$$

$$O_\theta + mg\cos\theta + P = mr_C\ddot{\theta} \quad (5.3.9b)$$

These equations involve three unknowns, namely, $\theta(t)$, $O_r(t)$, and $O_\theta(t)$. Therefore, one more equation is needed. The third equation is the moment equation. By taking moments about the inertial point O, the following equation that does not involve the support reactions is obtained (see Fig. 5.12, and use the right-hand rule).

$$PD + mgr_C\cos\theta = \frac{d}{dt}(I_O\dot{\theta}) = I_O\ddot{\theta} \quad (5.3.9c)$$

This is the differential equation that governs the motion of the system. The quantity I_O that appears in this equation is the moment of inertia of the body about an axis that passes through point O and is perpendicular to the plane of Fig. 5.12. For a homogeneous thin bar of length L, for example, $I_O = mL^2/3$.

To calculate the support reactions, Eq. (5.3.9c) needs to be integrated with the known initial conditions to determine $\theta(t)$. This is done numerically since that equation is nonlinear in $\theta(t)$. Equations (5.3.9a) and (5.3.9b) are then used, simultaneously with the numerical integration, for calculating O_r and O_θ.

b. By combining Eqs. (5.3.9b) and (5.3.9c), the following expression is obtained for the component O_θ of the support reaction.

$$O_\theta = \left(\frac{mr_C^2}{I_O} - 1\right) mg \cos\theta + \left(\frac{mDr_C}{I_O} - 1\right) P \qquad (5.3.9d)$$

Equation (5.3.9d) discloses that O_θ changes with θ as $\cos\theta$.

EXAMPLE 5.3.10

The system considered in Example 2.7.6 (p. 121) consisted of a particle of mass m attached to the end B of a massless bar AB. What is the differential equation of motion of the system if the massless bar and the point mass are replaced by a thin homogeneous bent bar ABD, of mass m and total length $L = L_1 + L_2$, as shown in Fig. 5.13? The support point A is given a known displacement $x(t)$ m as indicated in the figure.

■ **Solution**

Two free-body diagrams for the bent bar are shown in Fig. 5.14. They are equivalent to each other, and one can choose to work with either one of them. Two rotating unit vectors $\hat{r}$ and $\hat{\theta}$ and a nonrotating unit vector $\hat{x}$ are also shown in the figure.

Point C in Fig. 5.14 is the center of mass of the bent bar. With

$$m_1 = \frac{mL_1}{L_1 + L_2} \qquad m_2 = \frac{mL_2}{L_1 + L_2} \qquad (5.3.10a)$$

the location of point C is determined as follows:

$$m\overrightarrow{AC} = m_1 \frac{L_1}{2}\hat{r} + m_2\left(L_1 \hat{r} + \frac{L_2}{2}\hat{\theta}\right) \triangleq m(x_C \hat{r} + y_C \hat{\theta})$$

Figure 5.13 A pendulum on a moving support for Example 5.3.10.

SECTION 5.3 Dynamics of Simpler Planar Motion of a Rigid Body

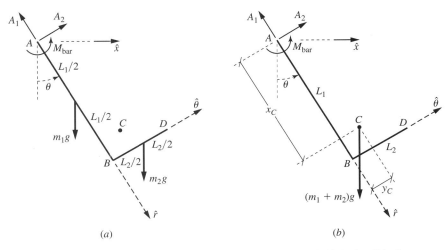

Figure 5.14 Two equivalent free-body diagrams for Example 5.3.10 (point C is the center of mass of the system).

This gives for the distances x_C and y_C shown in Fig. 5.14,

$$x_C = \frac{L_1}{L_1 + L_2}\left(L_2 + \frac{L_1}{2}\right) \tag{5.3.10b}$$

$$y_C = \frac{L_2^2}{2(L_1 + L_2)} \tag{5.3.10c}$$

The support point A shown in Fig. 5.14 is not inertial and, thus, $\vec{M}_A \neq d\vec{H}_A/dt$. If we take moments about the center of mass C, and use the basic equation $\vec{M}_C = d\vec{H}_C/dt$, where $\vec{H}_C = I_C \dot{\theta}\, \hat{r} \times \hat{\theta}$, we obtain the following equation (see Fig. 5.14, and use the right-hand rule to obtain the correct signs as they appear in the equation)

$$I_C \ddot{\theta} = M_{\text{bar}} - A_1 y_C - A_2 x_C \tag{5.3.10d}$$

The expression for the moment of inertia I_C is obtained as follows:

$$I_C = \underbrace{m_2 \frac{L_2^2}{12}}_{\substack{\text{moment of inertia of bar } BD \\ \text{about its center of mass}}} + \underbrace{m_2 \left[(L_1 - x_C)^2 + \left(\frac{L_2}{2} - y_C\right)^2\right]}_{\substack{m_2 \text{ times the square of the distance from } C \\ \text{to the center of mass of line } BD}}$$

$$\underbrace{\phantom{m_2 \frac{L_2^2}{12} + m_2 \left[(L_1 - x_C)^2 + \left(\frac{L_2}{2} - y_C\right)^2\right]}}_{\text{The sum of these two terms is } I_C \text{ for bar } BD}$$

$$+ \underbrace{m_1 \frac{L_1^2}{12}}_{\substack{\text{moment of inertia of bar } AB \\ \text{about its center of mass}}} + \underbrace{m_1 \left[\left(x_C - \frac{L_1}{2}\right)^2 + y_C^2\right]}_{\substack{m_1 \text{ times the square of the distance from } C \\ \text{to the center of mass of line } AB}}$$

$$\underbrace{\phantom{m_1 \frac{L_1^2}{12} + m_1 \left[\left(x_C - \frac{L_1}{2}\right)^2 + y_C^2\right]}}_{\text{The sum of these two terms is } I_C \text{ for bar } AB}$$

where use is made of the parallel axis theorem for bars BD and AB in each step of the calculation. The expressions $m_1 L_1^2/12$ and $m_2 L_2^2/12$ were first obtained in solution 2 of Example 5.3.4 (p. 277).

This expression is good enough for generating a numerical value for I_C, but a much simpler expression for I_C is obtained if the expressions for m_1 (which is the mass of bar AB) and m_2 (which is the mass of bar BD) given by the two equations in (5.3.10a) are substituted in this equation. The following simpler expression for I_C is obtained after some algebraic manipulations:

$$I_C = \frac{m}{12}(L_1 + L_2)^2 - \frac{m}{2}\left(\frac{L_1 L_2}{L_1 + L_2}\right)^2 \qquad (5.3.10e)$$

Now, use Newton's second law for the motion of the center of mass, $\vec{F} = m\vec{a}_C = m\,d\vec{v}_C/dt$, where $\vec{v}_C = \dot{x}\hat{x} + d(x_C \hat{r} + y_C \hat{\theta})/dt$ as seen from Fig. 5.14. With $\hat{z} \triangleq \hat{r} \times \hat{\theta}$ and noticing that $\hat{z} \times \hat{r} = \hat{\theta}$ and $\hat{z} \times \hat{\theta} = -\hat{r}$, the following expressions are then obtained for $\vec{v}_C$ and $\vec{a}_C$:

$$\vec{v}_C = \dot{x}\hat{x} + \frac{d}{dt}(x_C \hat{r} + y_C \hat{\theta}) = \dot{x}\hat{x} + \dot{\theta}\hat{z} \times (x_C \hat{r} + y_C \hat{\theta})$$
$$= \dot{x}\hat{x} + \dot{\theta}(x_C \hat{\theta} - y_C \hat{r}) \qquad (5.3.10f)$$
$$\vec{a}_C = \frac{d}{dt}\vec{v}_C = \ddot{x}\hat{x} + \ddot{\theta}(x_C \hat{\theta} - y_C \hat{r}) + \dot{\theta}\hat{z} \times (x_C \dot{\theta}\hat{\theta} - y_C \dot{\theta}\hat{r})$$
$$= \ddot{x}\hat{x} + (x_C \ddot{\theta} - y_C \dot{\theta}^2)\hat{\theta} - (y_C \ddot{\theta} + x_C \dot{\theta}^2)\hat{r} \qquad (5.3.10g)$$

Since $\hat{x} = \hat{r}\sin\theta + \hat{\theta}\cos\theta$ (as disclosed by Fig. 5.14), $\vec{F} = m\vec{a}_C$ yields the following additional equations.

$$m(\ddot{x}\sin\theta - y_C \ddot{\theta} - x_C \dot{\theta}^2) = mg\cos\theta - A_1 \qquad (5.3.10h)$$
$$m(\ddot{x}\cos\theta + x_C \ddot{\theta} - y_C \dot{\theta}^2) = A_2 - mg\sin\theta \qquad (5.3.10i)$$

Finally, by combining Eqs. (5.3.10d), (5.3.10h), and (5.3.10i), the following differential equation that governs the motion of the bent bar is obtained (where $\ddot{x}$ is a prescribed function, such as $\ddot{x} = \text{constant} \triangleq -a_A$ or any known function of time t).

$$\left[I_C + m\left(x_C^2 + y_C^2\right)\right]\ddot{\theta} = M_{\text{bar}} + m(y_C \sin\theta - x_C \cos\theta)\ddot{x} - mg(x_C \sin\theta + y_C \cos\theta) \qquad (5.3.10j)$$

Equation (5.3.10j), which only involves θ and the applied moment M_{bar}, is the differential equation of motion of the system. The analysis of the motion is done in Example 5.3.11.

EXAMPLE 5.3.11

The support point A of the system considered in Example 5.3.10 is given an acceleration $\ddot{x} = \text{constant} = -2g$ m/s². For $M_{\text{bar}} = 0$ and $L_2 = 2L_1$,

a. Plot the angular velocity $\dot{\theta}$ versus θ and describe, by observing the plot, the types of motion the system may exhibit.

b. Determine the maximum and the minimum values of the angle $\theta(t)$ if the motion is started with the initial conditions $\theta(0) = 90°$ and $\dot\theta(0) = 0$.
c. Determine the equilibrium value (or values if there is more than one equilibrium solution) for the angle θ.
d. Determine the frequency of the small oscillations of the system about the stable equilibrium values of the angle θ.

■ **Solution**

a. From Eqs. (5.3.10b), (5.3.10c), and (5.3.10e) developed in Example 5.3.10 we obtain

$$x_C = \frac{5}{6}L_1 \qquad y_C = \frac{2}{3}L_1 \qquad I_C = \frac{19}{36}mL_1^2$$

The differential equation that governs the motion of the system, Eq. (5.3.10j), then becomes

$$\frac{5}{3}\ddot\theta + \frac{g}{L_1}\left(\frac{13}{6}\sin\theta - \cos\theta\right) = M_{\text{bar}} = 0 \qquad (5.3.11a)$$

Since Eq. (5.3.11a) is of the form $\ddot\theta + f(\theta) = 0$, it can be integrated analytically once (as shown in Chapter 1 and in several examples in Chapter 2) to yield

$$\dot\theta^2 - \frac{6g}{5L_1}\left(\frac{13}{6}\cos\theta + \sin\theta\right) = \text{constant} \triangleq \frac{g}{L_1}H$$

$$= \left[\dot\theta^2 - \frac{6g}{5L_1}\left(\frac{13}{6}\cos\theta + \sin\theta\right)\right]_{t=0} \qquad (5.3.11b)$$

For the initial conditions $\dot\theta = 0$ and $\theta(0) = 90°$, the constant H in Eq. (5.3.11b) is equal to $-\frac{6}{5}$. The plot of the nondimensional angular velocity $d\theta/d\tau$ (where $\tau = t\sqrt{g/L_1}$ is a nondimensional time) versus the angle θ is shown in Fig. 5.15.

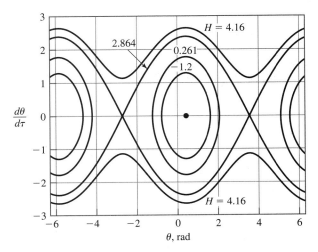

Figure 5.15 Plot of $d\theta/d\tau$ versus θ, where $\tau = t\sqrt{g/L_1}$ is a nondimensional time for Example 5.3.11.

Nondimensional time is used in order to generate a plot that is valid for any value of L_1. Otherwise, plots would have to be generated for each desired value of L_1. The corresponding values for the constant H associated with each one of the four curves shown in the plot are indicated in the figure.

b. Observation of Fig. 5.15 discloses that the system may exhibit two types of motion depending on the initial conditions of the motion:

1. One type of motion, which corresponds to the closed curves in Fig. 5.15, is an oscillation about the point on the θ axis marked with a •. That point, which corresponds to $\theta = \text{constant} \triangleq \theta_e \approx 24.77°$, is an equilibrium solution, and that equilibrium is, therefore, stable (because a motion that starts in its neighborhood stays in that neighborhood). Other stable equilibrium points are seen to be $24.77 \pm 360n$ degrees, where $n = 1, 2, \ldots$, but these new points correspond to the same physical orientation of line AC in Fig. 5.14.

    The maximum and minimum values of the angle θ for the motion just described correspond to $\dot\theta = 0$ in Eq. (5.3.11b). For the initial conditions given in the problem statement, which correspond to $H = -\frac{6}{5}$ as obtained earlier, that equation yields $\theta_{\max} = \pi/2\text{ rad} = 90°$ and $\theta_{\min} \approx -0.71\text{ rad} \approx -40.7°$. Reasonably accurate values for $\theta_{\max}$ and $\theta_{\min}$ could, of course, be extracted from the plot in Fig. 5.15. That figure also discloses that $\theta \approx 24.77 + 180 = 204.77°$ is an equilibrium point, and that equilibrium is unstable.

2. The other type of motion the system can exhibit corresponds to open curves in Fig. 5.15. Two such curves (for which $H = 4.16$) are shown in the figure. These motions occur when the system is given a big enough initial angular velocity $\dot\theta$, either clockwise or counterclockwise, causing the system to rotate about the support point A shown in Figs. 5.13 and 5.14.

c. The equilibrium solutions for the angle θ are obtained by setting $\theta = \text{constant} \triangleq \theta_e$ in the differential equation of motion, Eq. (5.3.11a). They correspond to

$$\frac{13}{6}\sin\theta_e - \cos\theta_e = 0 \quad\text{or}\quad \tan\theta_e = \frac{6}{13}$$

This gives two physically distinct solutions $\theta_e \approx 24.77$ and $204.77°$ which are the same values (as they should be) that were extracted from Fig. 5.15.

d. To determine the frequency of small oscillations about the stable equilibrium, let

$$\theta(t) \triangleq \theta_e + \theta_s(t)$$

and expand in Taylor series, for small θ_s, the nonlinear terms $\sin\theta = \sin(\theta_e + \theta_s)$ and $\cos\theta = \cos(\theta_e + \theta_s)$ that appear in the differential equation of motion, Eq. (5.3.11a), and disregard powers in the small quantities that appear in the expansion. The resulting approximate differential equation is the linearized differential equation of motion:

$$\frac{5}{3}\ddot\theta_s + \frac{g}{L_1}\left(\frac{13}{6}\cos\theta_e + \sin\theta_e\right)\theta_s = 0$$

SECTION 5.3 Dynamics of Simpler Planar Motion of a Rigid Body

Specifically, we then have for each equilibrium:

for $\theta_e = 24.77°$:
$$\frac{5}{3}\ddot{\theta}_s + 2.39\frac{g}{L_1}\theta_s = 0$$

for $\theta_e = 204.77°$:
$$\frac{5}{3}\ddot{\theta}_s - 2.39\frac{g}{L_1}\theta_s = 0$$

These linearized differential equations disclose that the equilibrium $\theta_e = 204.77°$ is unstable (as already concluded from Fig. 5.15) and the equilibrium $\theta_e = 24.77°$ is infinitesimally stable (it would be asymptotically stable if there were damping in the system). Actually, inspection of the $\dot{\theta}$ versus θ plot shown in Fig. 5.15 already disclosed that the equilibrium $\theta_e = 24.77°$ is stable.

As disclosed by the linearized differential equation $\frac{5}{3}\ddot{\theta}_s + 2.39(g/L_1)\theta_s = 0$, the frequency of the small oscillations about the stable equilibrium $\theta_e = 24.77°$ is equal to

$$\omega = \sqrt{\frac{3 \times 2.39}{5}\frac{g}{L_1}} \approx 1.2\sqrt{\frac{g}{L_1}} \text{ rad/s}$$

For $L_1 = 1$ m, for example, the frequency is $\omega \approx 3.75$ rad/s.

The motion of the bent bar, which is governed by Eq. (5.3.10j) (p. 286), can be easily simulated, and animated, using Simulink. A block diagram model to do this is shown in Fig. 5.16. A linear viscous damping moment in the support at point A, equal to $-c\dot{\theta}\hat{z}$ (where $c > 0$ is the viscous damping coefficient), which opposes the angular velocity $\dot{\theta}\hat{z}$ (see Fig. 5.14) is included in the model. The inclusion of friction, as well as the applied moment M_{bar}, can easily be handled by the numerical

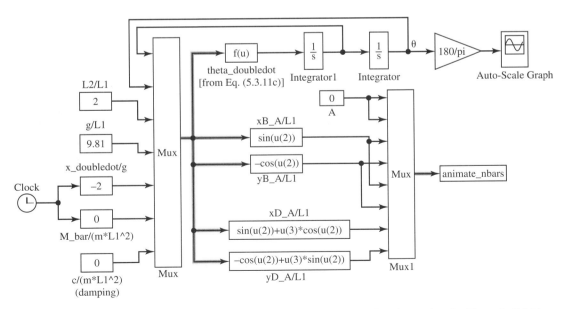

Figure 5.16 A Simulink model for integrating Eq. (5.3.11c) and for animating the motion, for Example 5.3.11.

integration that is implemented in the model. The animation of the motion, as viewed by an observer fixed to the moving support at A (which is accomplished by simply setting the rectangular coordinates of A to zero), is included in the model shown in Fig. 5.16. The inputs to the `animate_nbars` animation block are the rectangular x-y coordinates (nondimensionalized by L_1) of point A, and those of points B and D relative to point A, with the $+x$ axis pointing along the $\hat{x}$ direction shown in Fig. 5.13 and the $+y$ axis at 90° counterclockwise from the x axis. The nondimensional coordinates of points B and D in Fig. 5.13 are in the function blocks labeled `xB_A/L1, yB_A/L1, xD_A/L1,` and `yD_/L1` in the model shown in Fig. 5.16.

$\Longrightarrow$ The Simulink model of Fig. 5.16 is the file `sim_model_ex5_3_11.mdl` included on the CD that accompanies the book. To use it, either type `sim_model_ex5_3_11` in the MATLAB command window or choose the model from a menu that appears when you type `simulations_menu`. Look at the `Solver` part of the `Simulation` $\rightarrow$ `Parameters` menu for the model to see its default settings. You may want to experiment using a different value for the maximum step size (including the string `auto`) of the integration scheme if your simulation runs either too slow or too fast in your computer.

The differential equation of motion that is integrated by the model shown in Fig. 5.16 is

$$\left[1 + 3\frac{L_2}{L_1} + \left(\frac{L_2}{L_1}\right)^3\right]\ddot{\theta} + \frac{3g}{2L_1}\left\{\left[1 + 2\frac{L_2}{L_1} - \frac{\ddot{x}}{g}\left(\frac{L_2}{L_1}\right)^2\right]\sin\theta\right.$$
$$\left. + \left[\left(1 + 2\frac{L_2}{L_1}\right)\frac{\ddot{x}}{g} + \left(\frac{L_2}{L_1}\right)^2\right]\cos\theta\right\}$$
$$= 3\left(1 + \frac{L_2}{L_1}\right)\left(\frac{M_{\text{bar}}}{mL_1^2} - \frac{c}{mL_1^2}\dot{\theta}\right) \qquad (5.3.11c)$$

Equation (5.3.11c) is the same, of course, as Eq. (5.3.10j) (p. 286). Equations (5.3.10j) and (5.3.11c) only appear to be different because the author chose to simplify the resulting equation by making direct use of the expressions for x_C, y_C, and I_C given, respectively, by Eqs. (5.3.10b), (5.3.10c), and (5.3.10e), before putting the resulting expression for $\ddot{\theta}$ in the Simulink model. Also, the author chose to rearrange the final equation[2] into the form shown in Eq. (5.3.11c) so that only five external inputs L_2/L_1, g/L_1, $c/(m_1L_1^2)$, $\ddot{x}(t)/g$, and $M_{\text{bar}}(t)/(mL_1^2)$ [instead of the six external inputs L_1, L_2, c/m, g, $\ddot{x}(t)$, and $M_{\text{bar}}(t)/m$ that would be required otherwise] are used in the model shown in Fig. 5.16 (this is, of course, an optional "construction" style that varies from one individual to another).

---

[2] Before rearranging, the resulting differential equation is

$$\frac{m}{3}\left(L_1^3 + 3L_1^2L_2 + L_2^3\right)\ddot{\theta} + c_1\dot{\theta}$$
$$+ \frac{1}{2}m\left\{\left[gL_1(L_1 + 2L_2) - \ddot{x}L_2^2\right]\sin\theta + \left[\ddot{x}L_1(L_1 + 2L_2) + gL_2^2\right]\cos\theta\right\} = M_{\text{bar}}$$

EXAMPLE 5.3.12

A thin homogeneous rectangular plate of mass m_2 kg, with dimensions b m and h m, and a thin homogeneous bar of mass m_1 kg and length L m, are welded at the center of mass C_2 of the plate. The other side of the bar is connected to a fixed bearing at point O as shown in Fig. 5.17. Friction in the bearing is of the linear viscous type with friction coefficient c_1 N·m/(rad/s). The mass moments of inertia for the plate and for the bar are given in the figure. The motion of the system takes place in a vertical plane. The angle ϕ, measured from a fixed vertical line, is used for describing the orientation of the system. The angle θ is constant since the plate and the bar are welded to each other.

a. Obtain the angular momentum of the entire system about C_2 and about O.
b. What is the differential equation of motion for the system?

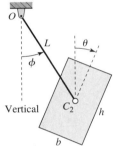

	I_{C_2}	I_O
Plate	$\dfrac{m_2}{12}(b^2+h^2)$	$I_{C_2} + m_2 L^2$
Bar	$\dfrac{m_1 L^2}{3}$	$\dfrac{m_1 L^2}{3}$

Figure 5.17 Figure for Example 5.3.12, and mass moments of inertia.

■ Solution

a. Let $\hat{z}$ be a unit vector perpendicular to the plane of Fig. 5.17 and coming out of the paper toward you, the reader. The absolute angular velocity of the rigid system is then

$$\vec{\omega} = \dot{\phi}\hat{z} \qquad (5.3.12a)$$

Therefore, by noticing that $(I_O)_{\text{bar}} = (I_{C_2})_{\text{bar}}$ (because both C_2 and O are end points of the bar), the angular momentum vectors $\vec{H}_{C_2}$ and $\vec{H}_O$ for the entire rigid system is calculated as follows:

$$\vec{H}_{C_2} = (I_{C_2})_{\text{plate}}\,\vec{\omega} + (I_{C_2})_{\text{bar}}\,\vec{\omega} = \left[\frac{m_2(b^2+h^2)}{12} + \frac{m_1 L^2}{3}\right]\dot{\phi}\hat{z} \qquad (5.3.12b)$$

$$\vec{H}_O = (I_O)_{\text{plate}}\,\vec{\omega} + (I_O)_{\text{bar}}\,\vec{\omega} = \left[\underbrace{\frac{m_2(b^2+h^2)}{12} + m_2 L^2}_{\text{as per the parallel axis theorem}} + \frac{m_1 L^2}{3}\right]\dot{\phi}\hat{z}$$

$$(5.3.12c)$$

b. The differential equation of motion for the system may be obtained either from $\vec{M}_O = d\vec{H}_O/dt$ or from $\vec{M}_{C_2} = d\vec{H}_{C_2}/dt$. The support reactions at O consist of a force $\vec{F}_O$ and of the moment $-c_1\dot\phi\hat{z}$ due to viscous friction in the bearing at O. Since the moment about C_2 of force $\vec{F}_O$ is not zero, the equation $\vec{M}_{C_2} = d\vec{H}_{C_2}/dt$ will involve three unknowns, namely, the angle ϕ and two components of the reaction force $\vec{F}_O$. Thus, using the equation $\vec{M}_{C_2} = d\vec{H}_{C_2}/dt$ will also require the use of Newton's second law for the center of mass of the system in order to obtain two additional equations involving those same unknowns. Therefore, to obtain the differential equation of motion for the system, it is much simpler to use, instead, the equation $\vec{M}_O = d\vec{H}_O/dt$ since this will only involve the desired unknown $\phi(t)$.

The moment $\vec{M}_O$ (which consists of the reaction moment at the bearings, and of the moment due to the gravitational forces) and the differential equation of motion that is immediately obtained from $\vec{M}_O = d\vec{H}_O/dt$ are

$$\vec{M}_O = -c_1\dot\phi\hat{z} - m_1 g \frac{L}{2}(\sin\phi)\hat{z} - m_2 g L(\sin\phi)\hat{z} \tag{5.3.12d}$$

$$\left[\frac{m_2(b^2 + h^2)}{12} + m_2 L^2 + \frac{m_1 L^2}{3}\right]\ddot\phi = -c_1\dot\phi - \left(m_1 g \frac{L}{2} + m_2 g L\right)\sin\phi \tag{5.3.12e}$$

If the system were a simple pendulum consisting of a particle of mass m_2 connected at the end C_2 of a massless bar of length L, the differential equation of motion would simply be

$$m_2 L^2 \ddot\phi + c_1\dot\phi + m_2 g L \sin\phi = 0 \tag{5.3.12f}$$

The effect of having a finite body of mass m_2 connected at C_2 and the bar OC_2 having mass was simply to change the coefficients of $\ddot\phi$ and $\sin\phi$ in Eq. (5.3.12f). The coefficient of $\dot\phi$ was not changed because the term $c_1\dot\phi$ is dependent only on the viscous damper, and not on the physical characteristics of any of the bodies.

Suggested problem: Choose your own values for the system parameters (plate dimensions, masses, etc.) and determine the frequencies of the damped and undamped oscillations about the stable equilibrium $\phi = 0$ (you already know it is stable because of similar problems in Chapter 2). Compare the answers with those you would have if the plate were to be replaced by a point mass at C_2.

EXAMPLE 5.3.13

Instead of being welded to bar OC_2, the plate of Problem 5.3.12 is now connected at C_2 by a bearing, and friction in that bearing is also of the linear viscous type, with friction coefficient c_2 N · m/(rad/s). The motion of the system is started with the initial conditions

$\phi(0) = 20°$, $\dot\phi(0) = \dot\phi_0$, $\theta(0) = 32°$, and $\dot\theta(0) = \dot\theta_0$. Obtain the differential equations of motion for the system, and

a. Determine the angle $\theta(t)$ for the case when $c_2 = 0$ and $\dot\theta_0 = 0$.
b. Verify that $\phi = 0$ is a stable equilibrium.
c. Determine the frequency Ω of the undamped linearized oscillations of the system about the equilibrium $\phi = 0$, in terms of m_1, m_2, L, b, h.
d. Rewrite the differential equations of motion in terms of a nondimensional time $\tau \triangleq \Omega t$, with Ω being the frequency obtained in part c. The damping coefficients c_1 and c_2 should appear in the resulting equations as the nondimensional quantities $c_1^* = \Omega c_1/(m_1 g L/2 + m_2 g L)$ and $c_2^* = \Omega c_2/(m_1 g L/2 + m_2 g L)$, respectively. Integrate the resulting equations numerically, and plot the angles $\phi(\tau)$ and $\theta(\tau)$ versus τ, for $0 \leq \tau \leq 50$. To do this, use $(b^2 + h^2)/L^2 = \frac{7}{6}$, $m_1/m_2 = \frac{1}{2}$, $c_1^* = 0.1$, $c_2^* = 0.1$, and the initial conditions $\phi(0) = 20°$, $\theta(0) = 32°$, $[d\phi/d\tau]_{\tau=0} = 0$, and $[d\theta/d\tau]_{\tau=0} = 50$. Do the integration using a Simulink block diagram model, observe what the system does by including an animation of the motion in the model, and comment on the resulting motion.

■ **Solution**

a. Unlike Example 5.3.12, the system is no longer a rigid body. The free-body diagram for each rigid body component of the system is shown in Fig. 5.18. The figure also shows a convenient unit vector triad for expressing the absolute velocity and acceleration of point C_2 and the absolute angular velocity of each body. The moments M_1 and M_2 shown in the free-body diagrams are the reaction moments due to the friction in the system. The friction moment $\vec{M}_2$ that is acting on the plate is proportional to the angular velocity of the plate relative to the bar. Thus, $\vec{M}_2 = -c_2[\dot\theta \hat z - (-\dot\phi \hat z)] = -c_2(\dot\theta + \dot\phi)\hat z$, which accounts for the way it is shown in Fig. 5.18a.

Notice that the two free-body diagrams involve six unknowns, namely, $\phi, \theta, F_1, F_2, O_1$, and O_2, and six scalar equations can be generated by using both

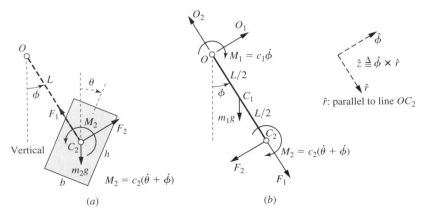

Figure 5.18 Free-body diagrams for the system of Example 5.3.13.

free-body diagrams (i.e., each diagram yields one moment equation and two equations obtained from $\vec{F} = m\vec{a}$ for the center of mass of the body).

The following equations are obtained for the free-body diagram of the plate, shown in Fig. 5.18a.

$$\frac{m_2}{12}(b^2 + h^2)\ddot{\theta} = -M_2 = -c_2(\dot{\theta} + \dot{\phi}) \tag{5.3.13a}$$

$$F_1 - m_2 g \cos\phi = m_2 L \dot{\phi}^2 \tag{5.3.13b}$$

$$F_2 - m_2 g \sin\phi = m_2 L \ddot{\phi} \tag{5.3.13c}$$

Since the absolute angular velocity of the plate is $\dot{\theta}\hat{z}$ (see Fig. 5.18), the moment equation for the plate is $\vec{M}_{C_2} = d[(I_{C_2})_{\text{plate}}\dot{\theta}\hat{z}]/dt$, where $\vec{M}_{C_2} = -c_2(\dot{\theta} + \dot{\phi})\hat{z}$ and $(I_{C_2})_{\text{plate}} = m_2(b^2 + h^2)/12$, as given in Example 5.3.12.

Equation (5.3.13a) involves only the unknown coordinates θ and ϕ and, thus, is one of the two differential equations of motion for the system. It can be analytically integrated once to yield

$$\frac{m_2}{12}(b^2 + h^2)\dot{\theta} + c_2(\theta + \phi) = \text{constant} = \left[\frac{m_2}{12}(b^2 + h^2)\dot{\theta} + c_2(\theta + \phi)\right]_{t=0} \tag{5.3.13d}$$

Equation (5.3.13d) immediately discloses that the absolute angular momentum of the plate, $(I_{C_2})_{\text{plate}}\dot{\theta}$, is conserved [i.e., $(I_{C_2})_{\text{plate}}\dot{\theta}(t) = \text{constant} = (I_{C_2})_{\text{plate}}\dot{\theta}(0)$] if there is no damping in the pin at point C_2 (i.e., if $c_2 = 0$). Thus, the absolute angular velocity of the plate is constant in such a case and equal to its initial absolute angular velocity.

For $\dot{\theta} = 0$ (i.e., if the initial absolute angular velocity of the plate is zero), as specified in part a, the plate maintains a fixed direction in space for all times if $c_2 = 0$ [i.e., the angle θ shown in Fig. 5.18 remains constant and equal to $\theta(0) = 32°$ during the entire motion of the system]. Notice that the solution to part a, with $\dot{\theta}(0) = 0$, only involved Eq. (5.3.13a).

b. Another differential equation is needed to solve this part since the problem involves two unknown coordinates, i.e., θ and ϕ. The easiest and quickest way to obtain the second differential equation of motion is to use the free-body diagram for the bar shown in Fig. 5.18b, take moments about point O to obtain an expression for F_2, and then use the resulting expression in Eq. (5.3.13c).

The moment equation for the bar is $\vec{M}_O = d[-(I_O)_{\text{bar}}\dot{\phi}\hat{z}]/dt$, where $(I_O)_{\text{bar}} = m_1 L^2/3$ as determined in Example 5.3.4 and also given in Example 5.3.12, and $\vec{M}_O$ is given as follows:

$$\vec{M}_O = \left[c_1\dot{\phi} + c_2(\dot{\theta} + \dot{\phi}) + m_1 g \frac{L}{2}\sin\phi + LF_2\right]\hat{z} \tag{5.3.13e}$$

Such an expression is obtained by inspecting Fig. 5.18b. Therefore, the approach just described yields

$$\frac{m_1 L^2}{3}\ddot{\phi} = -c_1\dot{\phi} - c_2(\dot{\theta} + \dot{\phi}) - m_1 g \frac{L}{2}\sin\phi - LF_2 \tag{5.3.13f}$$

By solving Eq. (5.3.13f) for F_2 and substituting the result into Eq. (5.3.13c), the following differential equation is obtained.

$$\left(\frac{m_1 L^2}{3} + m_2 L^2\right)\ddot{\phi} + \left(m_1 g \frac{L}{2} + m_2 g L\right)\sin\phi + c_1\dot{\phi} + c_2(\dot{\theta} + \dot{\phi}) = 0 \quad (5.3.13g)$$

Equations (5.3.13a) and (5.3.13g) [or (5.3.13d) and (5.3.13g)] are the differential equations that govern the motion of the system, in terms of the two unknowns $\phi(t)$ and $\theta(t)$. Notice that there was no need to apply Newton's second law to the free-body diagram for the bar OC_1 in order to answer the questions that were asked in this problem. However, we would need to do so if we wanted to find the support reactions O_1 and O_2 at point O. The determination of those reactions is left as an exercise (see Problem 5.16).

c. The equilibrium solution to Eq. (5.3.13g) is $\theta = \text{constant} \triangleq \theta_e$ and $\phi = \text{constant} \triangleq \phi_e$, with $\phi_e = 0$ and $\phi_e = \pi$. Linearization of Eq. (5.3.13g) about $\phi_e = 0$ yields

$$\left(\frac{m_1 L^2}{3} + m_2 L^2\right)\ddot{\phi} + \left(m_1 g \frac{L}{2} + m_2 g L\right)\phi + c_1\dot{\phi} + c_2(\dot{\theta} + \dot{\phi}) = 0 \quad (5.3.13h)$$

which discloses that the equilibrium is asymptotically stable if c_1 and c_2 are positive and infinitesimally stable if $c_1 = c_2 = 0$. From other similar problems that have already been presented, it is concluded that the equilibrium is stable, instead of just infinitesimally stable. The frequency Ω of the undamped linearized oscillations is

$$\Omega = \sqrt{\frac{m_2 + m_1/2}{m_2 + m_1/3} \frac{g}{L}} \quad (5.3.13i)$$

Notice that as $m_1/m_2 \to \infty$ the above expression yields $\Omega \to \sqrt{3g/(2L)}$, which agrees with the frequency of a pendulum that consists of a homogeneous bar length L. On the other hand, as $m_2/m_1 \to \infty$, $\Omega \to \sqrt{g/L}$, which is the frequency of a pendulum that consists of a massless bar of length L with a mass particle at the end of the bar.

d. Equations (5.3.13a) and (5.3.13g) [or (5.3.13d) and (5.3.13g)] for the motion coordinates $\phi(t)$ and $\theta(t)$ involve five dimensional coefficients $m_1 L^2/3$, $(m_2 + m_1/2)gL$, c_1, c_2, and $m_2(b^2 + h^2)$, and the time t. As indicated in Chapter 2, the number of coefficients in the equations can be reduced if nondimensional variables are used (notice, however, that this is an optional step in any analysis). Since the variables ϕ and θ are already nondimensional, time is the only variable to be nondimensionalized. The frequency Ω of the undamped oscillation is a convenient nondimensionalization factor to use.

With $\tau \triangleq \Omega t$, the time derivatives $\dot{\phi}, \ddot{\phi}, \dot{\theta}$, and $\ddot{\theta}$ are transformed as follows (using the chain rule):

$$\dot{\phi} = \frac{d\phi}{dt} = \frac{d\phi}{d\tau}\frac{d\tau}{dt} = \Omega\frac{d\phi}{d\tau} \qquad \ddot{\phi} = \frac{d\dot{\phi}}{dt} = \left[\frac{d}{d\tau}\left(\Omega\frac{d\phi}{d\tau}\right)\right]\frac{d\tau}{dt} = \Omega^2\frac{d^2\phi}{d\tau^2}$$

$$\dot{\theta} = \Omega\frac{d\theta}{d\tau} \qquad\qquad\qquad \ddot{\theta} = \Omega^2\frac{d^2\theta}{d\tau^2}$$

By arbitrarily choosing to work with Eq. (5.3.13a), instead of Eq. (5.3.13d), Eqs. (5.3.13h) and (5.3.13a) become, after minor manipulations that can be readily checked,

$$\frac{d^2\phi}{d\tau^2} + \sin\phi + c_1^* \frac{d\phi}{d\tau} + c_2^* \left(\frac{d\theta}{d\tau} + \frac{d\phi}{d\tau}\right) = 0 \quad \text{(5.3.13j)}$$

$$\frac{d^2\theta}{d\tau^2} + \frac{12[1 + m_1/(3m_2)]}{(b/L)^2 + (h/L)^2} c_2^* \left(\frac{d\theta}{d\tau} + \frac{d\phi}{d\tau}\right) = 0 \quad \text{(5.3.13k)}$$

In these equations c_1^* and c_2^* are the nondimensional damping coefficients as follows:

$$c_1^* = \frac{\Omega c_1}{(m_2 + m_1/2)gL} = \frac{c_1}{L\sqrt{gL(m_2 + m_1/2)(m_2 + m_1/3)}} \quad \text{(5.3.13l)}$$

$$c_2^* = \frac{\Omega c_2}{(m_2 + m_1/2)gL} = \frac{c_2}{L\sqrt{gL(m_2 + m_1/2)(m_2 + m_1/3)}} \quad \text{(5.3.13m)}$$

Notice that the new equations involve only three different nondimensional coefficients instead of the five dimensional ones that appeared before.

$\implies$ Figure 5.19 shows a Simulink model for integrating Eqs. (5.3.13j) and (5.3.13k). The model is the file `sim_model_ex5_3_13.mdl` included on the CD. To use it, either type `sim_model_ex5_3_13` in the MATLAB command window or choose the model from a menu that appears when you type `simulations_menu`. Look at the `Solver` part of the `Simulation` $\rightarrow$ `Parameters` menu for

Figure 5.19 The Simulink model `sim_model_ex5_3_13` for Example 5.3.13.

the model to see its default settings, and change the maximum step size of the integration scheme if your simulation runs either too slow or too fast in your computer. The outputs seen in Fig. 5.19 are essentially the equilibrium solutions for θ and ϕ when $\theta(0) = 32°$.

Figure 5.20 shows the details that were used for creating the inputs to the `animate_nbars` block for generating only a simple figure to imitate the system considered in this example. Instead of a plate, only a bar MN of length $h = 0.25L$ is drawn. The x-y coordinate system that is used and the coordinates of points C_2, M, and N are shown in Fig. 5.20. Those expressions, nondimensionalized by L, are the ones that appear in the blocks labeled `x_C2/L`, `y_C2/L`, `x_N/L`, etc., in Fig. 5.19.

Figure 5.21 shows plots of $\theta(t)$ and $\phi(t)$ versus τ, obtained by simply putting two `auto_scale_scope` blocks (described in Appendix A) in the `sim_model_ex5_3_13` model. As seen in the plots, $\theta(t)$ practically reaches an equilibrium when $\tau \approx 50$ while it takes a little longer for $\phi(t)$ to reach an equilibrium.

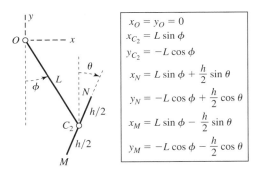

Figure 5.20 Details for the creation of the figure generated by the `animate_nbars` block in the simulation model `sim_model_ex5_3_13`.

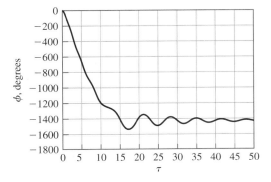

Figure 5.21 Plot of $\theta(\tau)$ versus $\tau = t\sqrt{g/L}$, and $\phi(\tau)$ versus τ, in degrees, for Example 5.3.13, with $\phi(0) = 20°$, $[d\phi/d\tau]_{\tau=0} = 0$, $\theta(0) = 32°$, and $[d\theta/d\tau]_{\tau=0} = 50$.

Comments on the Observed Motion

When the simulation is started with the `sim_model_ex5_3_13` model shown in Fig. 5.19, with the system parameters shown in that figure, and with the given initial conditions $\phi(0) = 20°$, $[d\phi/d\tau]_{\tau=0} = 0$, $\theta(0) = 32°$, and $[d\theta/d\tau]_{\tau=0} = 50$, the system responds as described in the following. Run the `sim_model_ex5_3_13` model on your computer to see the response.

The plate starts to spin, and bar OC_2 starts to revolve around point O. This happens because both were given a relatively high angular velocity. As time progresses, the damping in both pins (at O and at C_2) starts to slow down the motion until bar OC_2 and the plate stop spinning and then oscillate. The oscillations are damped and eventually become unnoticeable; i.e., the system essentially reached an equilibrium state $\phi = \text{constant} \stackrel{\Delta}{=} \phi_e$ and $\theta = \text{constant} \stackrel{\Delta}{=} \theta_e$, where $\phi_e = 0$ (which is the expected stable equilibrium; i.e., the pendulum orients itself upside down). Bar MN that imitates the plate in the simulation ends up inclined with the horizontal. Here is a curious observation then. Why does that happen? To investigate this further, change the initial conditions of the motion and see what happens. If the motion is started with $\theta(0) = 0$ and the other initial conditions are kept the same, bar MN ends up in a different orientation. The results of these two observations are shown in Fig. 5.22, which is two snapshots of the animation generated by the `animate_nbars` block in the Simulink model.

The snapshot in Fig. 5.22a is for $\theta(0) = 32°$, and that in Fig. 5.22b is for $\theta(0) = 0$. The values of the angle θ_e for both cases are listed in the following table.

$\theta(0)$	θ_e
32°	3879°, which is equivalent to $3879 - 10 \times 360 = 279°$
0	3847°, which is equivalent to $3847 - 10 \times 360 = 247°$

To understand what is happening, look at the differential equations of motion and determine their equilibrium solutions. With $\theta = \text{constant} \stackrel{\Delta}{=} \theta_e$ and $\phi = \text{constant} \stackrel{\Delta}{=} \phi_e$,

(a)

(b)

Figure 5.22 The equilibrium orientation of bars OC_2 and MN: (a) with $\theta(0) = 32°$ and (b) with $\theta(0) = 0$.

Eq. (5.3.13a) is automatically satisfied, and Eq. (5.3.13g) yields $\sin\phi_e = 0$. Thus, $\phi_e = 0$ and $\phi_e = 180°$ are the two physically different solutions, with the solution $\phi_e = 0$ being the stable equilibrium as you have seen in Chapters 1 and 2. This still does not explain why there should be an equilibrium value for θ, as observed in the simulation. The explanation lies in the fact that Eq. (5.3.13a) can be integrated once to yield Eq. (5.3.13d), and that equation explains why the angle θ tends to an equilibrium value that depends on the initial condition θ_0. Using Eq. (5.3.13k) with $\phi_e = 0$, which is the nondimensional form of Eq. (5.3.13d), the equilibrium angle θ_e is obtained as follows:

$$\theta_e = \theta(0) + \phi(0) + \frac{1}{12c_2^*[1 + m_1/(3m_2)]}\left[\left(\frac{b}{L}\right)^2 + \left(\frac{h}{L}\right)^2\right]\left[\frac{d\theta}{d\tau}\right]_{\tau=0} \quad (5.3.13n)$$

Equation (5.3.13n) yields values of θ_e that are equivalent to the ones shown in the table presented earlier. For example, for $\phi(0) = 20°$, $\theta(0) = 32°$, $\phi_e = 0$, $(b^2 + h^2)/L^2 = \frac{7}{6}$, $m_1/m_2 = \frac{1}{2}$, $c_2^* = 0.1$, and $[d\theta/d\tau]_{\tau=0} = 50$, Eq. (5.3.13n) yields $\theta_e \approx 2439°$. Such a value is equivalent to $\theta_e = 2439 - 6 \times 360 = 279°$, which is shown in the table on p. 298.

5.4 Rolling of a Rigid Body

A special case of rigid body motion consists of a cylinder rolling on a surface, with friction between the cylinder and the surface modeled as Coulomb friction. Two typical cases are shown in Fig. 5.23, namely, rolling on a flat surface and rolling on the inside of another cylinder.

The analysis of the motion of the cylinder in either case shown in Fig. 5.23 involves four unknowns, namely, two motion coordinates and two unknown reactions. The unknown reactions are a normal force and the friction force of contact between the surfaces. The motion coordinates are x and θ for Fig. 5.23a (i.e., for the cylinder rolling on a flat surface), and θ and ϕ for Fig. 5.23b. However, the equations of dynamics [i.e., $\vec{F} = m\vec{a}$ for the motion of the center of mass C, and the moment equation $\vec{M}_C = d\vec{H}_C/dt$] yield only three scalar equations. Thus, one more equation is needed since four unknowns are involved. The fourth equation in this type of problem has to do with whether the cylinder rolls with or without slipping, and it is obtained as described in the following.

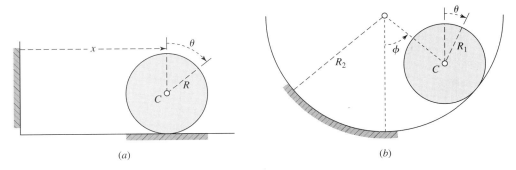

Figure 5.23 A cylinder rolling on a flat surface, and on a cylinder of radius R_2.

Rolling is said to be without slipping when the length of the infinitesimal path traveled by the center of the rolling cylinder during an infinitesimal time dt is equal to the length of the infinitesimal path traveled along the surface of contact during the same time. Thus, for each one of the two cases shown in Fig. 5.23, rolling occurs without slipping only when the condition indicated for each case is satisfied. The variables x and θ shown in Fig. 5.23a and ϕ and θ shown in Fig. 5.23b are related to each other in these cases.

Case 1. For a cylinder of radius R, rolling without slipping on a flat surface: $dx = R\,d\theta$. This implies

$$\dot{x} = R\dot{\theta} \tag{5.4.1}$$

Case 2. For a cylinder of radius R_1, rolling without slipping, on a cylinder of radius $R_2 > R_1$: $(R_2 - R_1)\,d\phi = R_1\,d\theta$. This implies

$$(R_2 - R_1)\dot{\phi} = R_1\dot{\theta} \tag{5.4.2}$$

When the cylinder rolls without slipping, the fourth equation needed for determining all the unknowns is either Eq. (5.4.1) for rolling on a flat surface or Eq. (5.4.2) for rolling on another cylinder. In such a case, the friction force is determined from one of the equations obtained from $\vec{F} = m\vec{a}$. Its magnitude is smaller than the maximum possible value $\mu_s|N|$, where μ_s is the static coefficient of friction between the surfaces in contact and $|N|$ is the magnitude of the normal force of contact. If the magnitude of the friction force happens to be equal to $\mu_s|N|$, then slipping is impending. The magnitude of the friction force becomes smaller than $\mu_s|N|$ when slipping occurs, and such a motion is treated separately.

Rolling occurs with slipping if the no-slipping condition given by either Eq. (5.4.1) or Eq. (5.4.2) is violated. In such a case, the magnitude of the Coulomb friction force acting on the body is equal to $\mu_k|N|$, where μ_k is the kinetic coefficient of friction between the surfaces in contact. In this case, this is the fourth equation that is needed for solving the problem. The variables x and θ shown in Fig. 5.23a, and ϕ and θ shown in Fig. 5.23b are independent of each other in this case.

Problems involving dry friction are solved by guessing whether the motion occurs with or without slipping and then checking whether the guess is true or false at the end of the calculations. Examples 5.4.1 through 5.4.3 illustrate the calculations that are typical of such problems. For the sake of completeness, details of the calculations are repeated in the solution to each example.

EXAMPLE 5.4.1

A homogeneous cylinder of mass m and radius R rolls on a fixed inclined plane that makes an angle α with the horizontal. The static and kinetic coefficients of Coulomb friction between the cylinder and the plane are μ_s and μ_k, respectively.

a. What are the differential equations that govern the motion of the cylinder? For this, consider the case when the cylinder rolls without slipping and the case when it slips.

b. Determine, in terms of μ_s, the maximum value of the angle α, $\alpha \triangleq \alpha_{\text{slip}}$, for the cylinder to roll on the plane without slipping. What is the direction of the friction force acting on the cylinder? What is its magnitude? Sketch the magnitude of the friction force versus α (this is a good way to understand whether slipping occurs or not).

c. What are the values of α_{slip} (in terms of μ_s) in the case where the cylinder is a ring and the case where it is a solid homogeneous body?

■ **Solution**

a. Figure 5.24 shows the cylinder on the inclined plane and the free-body diagram for the cylinder. Since the cylinder is homogeneous, its center of mass coincides with its geometric center C. The absolute position vector $\overrightarrow{OC}$ of the center of mass C, and the angle θ that was chosen to track the rotation of the cylinder, are also shown in the figure.

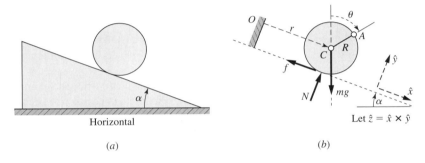

(a) (b)

Figure 5.24 Planar rolling motion of a rigid body for Example 5.4.1.

The forces acting on the cylinder are the gravitational force, the normal force (which is perpendicular to the inclined plane), and the friction force (which is tangent to the inclined plane). The direction of these forces, indicated by the arrows, were arbitrarily drawn as shown in Fig. 5.24. If those directions are correct, the calculations will yield positive values for N and f. Otherwise, they will yield a negative value for either f or N or both. Two mutually orthogonal unit vectors that were conveniently chosen for expressing the vector quantities that will appear in the formulation presented in the following are also shown in the figure.

In terms of the unit vector triad $\{\hat{x}, \hat{y}, \hat{z}\}$ shown in Fig. 5.24, we then have

Resultant force acting on the cylinder: $\quad \vec{F} = (mg \sin\alpha - f)\hat{x} + (N - mg\cos\alpha)\hat{y}$

Absolute position vector for the center of mass C of the cylinder: $\quad \vec{r}_C = \overrightarrow{OC} = r\hat{x}$

Absolute velocity of C: $\quad \vec{v}_C = \dfrac{d}{dt}\vec{r}_C = \dot{r}\hat{x}$

Absolute acceleration of C: $\quad \vec{a}_C = \dfrac{d}{dt}\vec{v}_C = \ddot{r}\hat{x}$

$\vec{F} = m\vec{a}_C$ then yields the following scalar equations

$$m\ddot{r} = mg\sin\alpha - f \tag{5.4.1a}$$

$$N = mg\cos\alpha \tag{5.4.1b}$$

Next, take moments about the center of mass C of the cylinder to obtain the third equation that governs the motion. With $\vec{M}_C = d\vec{H}_C/dt$, where $\vec{M}_C = -fR\hat{z}$, and $\vec{H}_C = -I_C\dot{\theta}\hat{z}$, with I_C being the mass moment of inertia of the cylinder about the axis that passes through C and is perpendicular to the plane of the figure, we then obtain

$$I_c\ddot{\theta} = fR \tag{5.4.1c}$$

The three equations (5.4.1a), (5.4.1b), and (5.4.1c) involve four unknowns, namely, r, θ, f, and N, which, according to Eq. (5.4.1b), is simply $N = mg\cos\alpha$. Therefore, one more equation is needed in order to determine the remaining unknowns. The difficulty we are now facing is that all the equations from dynamics, namely $\vec{F} = m\vec{a}_C$ and $\vec{M}_C = d\vec{H}_C/dt$, have already been used in the formulation, and those vector equations already yielded all the scalar equations they can yield. A fourth equation must then be generated by some other means. Such a difficulty is typical of problems that involve dry friction.

To generate a fourth equation for such a problem with dry friction we must make a guess about the motion, and then check whether the guess is true or false at the end of the calculations. The guess to be made is that the body rolls either with or without slipping. It is simpler to assume that the cylinder rolls without slipping, and then check whether such an assumption is valid or not based on the magnitude of the friction force.

If the body rolls without slipping, the displacement dr of the center C of the cylinder must be equal to the displacement $R\,d\theta$ of the point of contact of the cylinder with the inclined plane. The no-slipping condition then relates the variables $r(t)$ and $\theta(t)$ as

$$dr = R\,d\theta$$

which yields

$$\dot{r} = R\dot{\theta} \tag{5.4.1d}$$

By combining Eqs. (5.4.1a), (5.4.1c), and (5.4.1d), one obtains for the angular acceleration $\ddot{\theta}$ and for the friction force f,

$$(I_C + mR^2)\ddot{\theta} = mgR\sin\alpha \tag{5.4.1e}$$

$$f = \frac{mg\sin\alpha}{1 + mR^2/I_c} \tag{5.4.1f}$$

If the cylinder is not slipping, its motion is governed only by one of the following two variables: $\theta(t)$ or $r(t) = R\theta(t)$. In such a case, Eq. (5.4.1e) is the differential equation of motion in terms of $\theta(t)$. That equation, which discloses that $\ddot{\theta}$ is constant, can be easily integrated. For the differential equation of motion when there is slipping, see the note at the end of this problem.

Since the scalar quantity f given by Eq. (5.4.1f) is a positive quantity, it can now be said with certainty that the direction of the friction force is indeed as shown in Fig. 5.24.

b. It is now necessary to check whether these solutions are valid or not, which then ascertains whether the cylinder is slipping or not. The way to do this is to determine whether the solution for the friction force f given by Eq. (5.4.1f) is valid or not. This is done by comparing the magnitude of f with the maximum possible magnitude of the friction force. If μ_s is the static coefficient of friction between the cylinder and the plane, the maximum magnitude of the friction force, $|f_{\max}|$, must be equal to

$$|f_{\max}| = \mu_s |N|$$

Or, since $N = mg \cos \alpha$ is a positive quantity, and f given by Eq. (5.4.1f) is also positive, we can then write

$$f_{\max} = \mu_s mg \cos \alpha \qquad (5.4.1g)$$

The body rolls without slipping if the friction force calculated by using Eq. (5.4.1f) is at most equal to $f_{\max}$ as given by Eq. (5.4.1g). Figure 5.25 shows a sketch of $f/(mg)$ versus the angle α. Such a sketch is a good way to understand whether slipping occurs or not.

As α is increased from $\alpha = 0$, the friction force follows the sinusoidal path shown in Fig. 5.25, as given by Eq. (5.4.1f), while the maximum value of f follows a cosine path that starts at $f = mg$ when $\alpha = 0$. When α reaches a value $\alpha = \alpha_{\text{slip}}$, the two paths intersect each other and the friction force becomes equal to $f_{\max} = \mu_s |N| = \mu_s mg \cos \alpha_{\text{slip}}$ when $\alpha = \alpha_{\text{slip}}$. If α is increased further, slipping occurs and the friction force becomes $f = \mu_k |N| = \mu_k mg \cos \alpha$. The value of α_{slip} above which slipping occurs is determined as follows:

$$\frac{\sin \alpha_{\text{slip}}}{1 + mR^2/I_C} = \mu_s \cos \alpha_{\text{slip}} \quad \text{or} \quad \tan \alpha_{\text{slip}} = \mu_s \left(1 + \frac{mR^2}{I_C}\right) \qquad (5.4.1h)$$

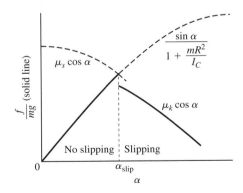

Figure 5.25 Nondimensional friction force $f/(mg)$ versus the angle α for a rolling rigid body.

Table 5.1 Values of $\tan\alpha_{\text{slip}}$ for two rolling bodies

Type of body	I_C	$\tan\alpha_{\text{slip}}$
Ring or cylindrical shell of radius R	mR^2	$2\mu_s$
Solid homogeneous cylinder of radius R	$mR^2/2$	$3\mu_s$

c. Table 5.1 shows the relationship between $\tan\alpha_{\text{slip}}$ and the static coefficient of friction for a ring or a cylindrical shell and for a solid homogeneous cylinder. For a hollow cylinder of mass m, inner radius R_1, and outer radius R_2, the moment of inertia I_C is $I_C = m(R_2^2 + R_1^2)/2$ (see Example 5.3.8, p. 281), and the value of the angle of slip is between $\arctan(2\mu_s)$ and $\arctan(3\mu_s)$.

Note: For $\alpha > \alpha_{\text{slip}}$, the equations that govern the motion are Eqs. (5.4.1a) and (5.4.1c), with the friction force f given by Eq. (5.4.1g), i.e., $f = \mu_k N = \mu_k mg\cos\alpha$. For such a case, those equations then yield the following differential equations for the independent variables $r(t)$ and $\theta(t)$:

$$\ddot{r} = g(\sin\alpha - \mu_k\cos\alpha) \tag{5.4.1i}$$

$$I_C\ddot{\theta} = \mu_k mgR\cos\alpha \tag{5.4.1j}$$

The right-hand sides of Eqs. (5.4.1i) and (5.4.1j) are constants, and thus, they can be easily integrated to obtain $r(t)$ and $\theta(t)$.

EXAMPLE 5.4.2

A wheel, modeled as a solid homogeneous cylinder of mass m and radius R, is able to roll on a horizontal surface. The wheel is subjected to an applied clockwise moment $M(t)$ that is equal to $M(t) = kmgRt$ N·m (where $k = 0.1$ s^{-1}) for $0 \leq t \leq 6$ s. The power source that produces the moment is disengaged from the wheel right after 6 s so that $M = 0$ for $t > 6$ s. Friction between the wheel and the horizontal surface is of the Coulomb type; the static and kinetic coefficients of friction are, respectively, $\mu_s = 0.3$ and $\mu_k = 0.25$. The wheel is initially at rest on the horizontal surface.

a. Obtain the differential equations of motion for the wheel for each phase of the motion, namely, when the wheel does not slip and when it does slip.
b. Integrate, analytically, the differential equations of motion for each phase of the motion, and determine the times when the wheel starts and stops to slip. How far does the center of mass of the wheel travel after 10 s has elapsed since the application of the moment $M(t)$?
c. Plot the velocity of the center of mass of the wheel versus time t for $0 \leq t \leq 10$ s, and indicate, in the plot, the time intervals when the wheel slips and when it does not.

■ Solution

a. The free-body diagram for the wheel is shown in Fig. 5.26. The figure also shows the absolute position vector $\overrightarrow{OC}$ for the center of mass C of the wheel, and an angle θ that is introduced for tracking the rotation of the wheel.

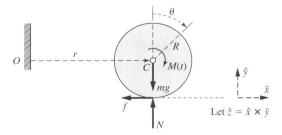

Figure 5.26 A driven wheel rolling on a horizontal surface for Example 5.4.2.

The steps that should be followed to formulate the equations that govern the motion of the wheel are the same as those presented in Example 5.4.1. With the absolute acceleration of point C written as $\vec{a}_C = \ddot{r}\hat{x}$, and the resultant force acting on the wheel being equal to $\vec{F} = -f\hat{x} + (N - mg)\hat{y}$, Newton's second law $\vec{F} = m\vec{a}_C$ then yields

$$m\ddot{r} = -f \tag{5.4.2a}$$

$$N = mg \tag{5.4.2b}$$

Also, by taking moments about the center of mass C, the equation $\vec{M}_C = d\vec{H}_C/dt$ yields, with $\vec{M}_C = -(M + fR)\hat{z}$ (see Fig. 5.26) and $\vec{H}_C = -I_C\dot{\theta}\hat{z}$, where $I_C = mR^2/2$:

$$I_C\ddot{\theta} = M(t) + fR \tag{5.4.2c}$$

There are two possibilities to investigate: rolling without slipping and with slipping.

1. *Rolling with slipping.* For rolling with slipping, r and θ are independent of each other. Thus, the motion is governed by two differential equations, and they are Eqs. (5.4.2a) and (5.4.2c). The magnitude of the friction force for this type of motion is $|f| = \mu_k|N| = \mu_k mg$.
2. *Rolling without slipping.* For rolling without slipping, we have $\dot{r} = R\dot{\theta}$. In this case, Eqs. (5.4.2a) and (5.4.2c) can be combined to yield, with $M(t) = 0.1mgRt$ and $I_C = mR^2/2$,

$$\ddot{\theta} = \frac{\ddot{r}}{R} = \frac{M(t)}{I_C + mR^2} = \frac{2M(t)}{3mR^2} = \frac{0.2gt}{3R} \tag{5.4.2d}$$

$$f = -\frac{mRM(t)}{I_C + mR^2} = -\frac{2M(t)}{3R} = -\frac{0.2mgt}{3} \tag{5.4.2e}$$

Notice that, since $f = -0.2mgt/3$ is a negative quantity, the direction of the friction force is opposite to the one shown in Fig. 5.26.

b. The wheel will roll without slipping as long as $|f| < \mu_s|N|$, and slipping will be impending when $|f| = \mu_s|N|$. Since $N = mg$, this then occurs when

$0.2mgt/3 = \mu_s mg$. With $\mu_s = 0.3$, this gives $t = 4.5$ s. Thus, the wheel starts its motion by rolling without slipping.

The analysis of the motion now has to be separated into several parts:

$0 \leq t \leq 4.5$ s: $M(t) = 0.1mgRt$ N, *and the wheel rolls without slipping.* The differential equation of motion is then Eq. (5.4.2d), and the friction force is given by Eq. (5.4.2e). Since $\ddot{r} = R\ddot{\theta}$ during this time interval, successive integrations yield

$$\int_{\dot{r}=0}^{\dot{r}} \ddot{r}\, dt = \int_{t=0}^{t} \frac{0.2gt}{3}\, dt \quad \text{i.e.,} \quad \dot{r}(t) = \frac{0.1gt^2}{3}$$

$$\int_{r=r(0)}^{r} \dot{r}\, dt = \int_{t=0}^{t} \frac{0.1gt^2}{3}\, dt \quad \text{i.e.,} \quad r(t) = r(0) + \frac{0.1gt^3}{9}$$

At the end of 4.5 s, the velocity $\dot{r}$ of the center of mass C of the wheel, and the distance $r(4.5) - r(0)$ traveled by C, are then

$$\dot{r}(4.5) = \frac{0.1g \times 4.5^2}{3} \approx 6.6 \text{ m/s}$$

$$r(4.5) - r(0) = \frac{0.1g \times 4.5^3}{9} \approx 9.9 \text{ m}$$

$4.5 \leq t \leq 6$ s: $M(t) = 0.1mgRt$ N, *and slipping occurs.* The friction force f was negative for $0 \leq t \leq 4.5$ s, and continues to be negative after that. However, due to slipping, it now becomes $f = -\mu_k|N| = -\mu_k mg = -0.25mg$. The variables $r(t)$ and $\theta(t)$ are now independent of each other, and the differential equations that govern the motion are now Eq. (5.4.2a) and (5.4.2c). With $I_C = mR^2/2$, and $M = 0.1mgRt$, successive integrations now yield

$$\int_{\dot{r}=\dot{r}(4.5)}^{\dot{r}} \ddot{r}\, dt = 0.25g \int_{t=4.5}^{t} dt \quad \text{i.e.,} \quad \dot{r}(t) = \dot{r}(4.5) + 0.25g(t - 4.5)$$

$$\int_{r=r(4.5)}^{r} \dot{r}\, dt = \int_{t=4.5}^{t} [\dot{r}(4.5) + 0.25g(t - 4.5)]\, dt$$

$$\text{i.e.,} \quad r(t) = r(4.5) + [\dot{r}(4.5)](t - 4.5) + 0.25g \frac{(t - 4.5)^2}{2}$$

$$\int_{\dot{\theta}=\dot{\theta}(4.5)}^{\dot{\theta}} \ddot{\theta}\, dt = \int_{t=4.5}^{t} \left(\frac{0.1mgRt - 0.25mgR}{mR^2/2} \right) dt$$

$$\text{i.e.,} \quad \dot{\theta}(t) = \frac{\dot{r}(4.5)}{R} + \frac{0.1g}{R}(t^2 - 4.5^2) - \frac{0.5g}{R}(t - 4.5)$$

At the end of 6 s, these solutions give for the velocity $\dot{r}$ of the center of mass C of the wheel, and for the distance $r(6) - r(0)$ traveled by C,

$$\dot{r}(6) \approx 10.3 \text{ m/s}$$

$$r(6) - r(0) \approx 22.6 \text{ m}$$

$t > 6$ s until slipping stops: $M(t) = 0$. Since neither the angular velocity of the wheel nor the velocity of its center of mass can change instantaneously,[3] the wheel continues to roll and slip after $t = 6$ s. Since slipping continues, the subsequent motion is still governed by Eqs. (5.4.2a) and (5.4.2c), but with $M(t) = 0$. Integration of the differential equations of motion now yields

$$\int_{\dot{r}=\dot{r}(6)}^{\dot{r}} \ddot{r}\, dt = 0.25g \int_{t=6}^{t} dt \qquad \text{i.e.,} \qquad \dot{r}(t) = \dot{r}(6) + 0.25g(t-6)$$

$$\int_{r=r(6)}^{r} \dot{r}\, dt = \int_{r=r(6)}^{r} [\dot{r}(6) + 0.25g(t-6)]\, dt$$

$$\text{i.e.,} \qquad r(t) = r(6) + [\dot{r}(6)](t-6) + 0.25g \frac{(t-6)^2}{2}$$

$$\int_{\dot{\theta}=\dot{\theta}(6)}^{\dot{\theta}} \ddot{\theta}\, dt = \int_{t=6}^{t} \left(\frac{-0.25mgR}{mR^2/2} \right) dt \qquad \text{i.e.,} \qquad \dot{\theta}(t) = \frac{\dot{r}(6)}{R} - \frac{0.5g}{R}(t-6)$$

After $t = 6$ s, slipping stops when $\dot{r} = R\dot{\theta}$, and this occurs when

$$\dot{r}(6) + 0.25g(t-6) = R\dot{\theta}(6) - 0.5g(t-6) \qquad \text{which gives} \qquad t = 6.6 \text{ s}$$

As obtained from these equations, the velocity of the center of mass C of the wheel and the distance traveled by C at the end of 6.6 s are

$$\dot{r}(6.6) \approx 11.8 \text{ m/s}$$

$$r(6.6) - r(0) \approx 29.2 \text{ m}$$

$t > 6.6$ s: $M(t) = 0$, and there is no slipping. During this last phase of the motion, $\dot{r} = R\dot{\theta}$. The motion is governed by Eq. (5.4.2d) again, and the friction force is given by Eq. (5.4.2e), with $M(t) = 0$. In this case, both the angular acceleration $\ddot{\theta}$

---

[3] This can be shown as follows. Since $\vec{F} = m\, d\vec{v}_C/dt$, an instantaneous change of $\vec{v}_C$ implies that $d\vec{v}_C/dt$ is infinite, which requires an abrupt, infinite change, in $\vec{F}$ for that to happen. By the same argument, since $\vec{M}_C = I_C\, d\vec{\omega}/dt$, an instantaneous change of the angular velocity $\vec{\omega}$ of the wheel requires an abrupt, infinite change in the moment $\vec{M}_C$.

and the friction force are zero. Since $\dot{r} = R\dot{\theta}$, and $\dot{\theta} = $ constant (because $\ddot{\theta} = 0$), we then have

$$\dot{r} = \text{constant} = \dot{r}(6.6) = 11.8 \text{ m/s}$$

$$\int_{r=r(6.6)}^{r} \dot{r} \, dt = r(t) - r(6.6) = \int_{t=6.6}^{t} 11.8 \, dt = 11.8(t - 6.6)$$

Since $r(6.6) = r(0) + 29.2$ m, the total distance traveled by the center of mass C after 10 s is then

$$r_{\text{total}} = r(10) - r(0) = 69.3 \text{ m}$$

c. Figure 5.27 shows the plot of $\dot{r}(t)$ versus time t for $0 \leq t \leq 10$ s, obtained by piecing together the solutions that were obtained for $\dot{r}$ at the various time intervals the problem had to be divided into. The time interval where the wheel slips, and those where it does not, are indicated in the plot.

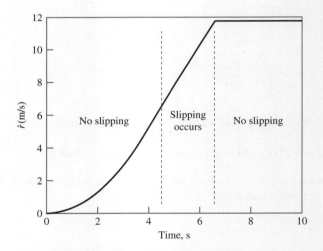

Figure 5.27 Velocity of point C (in meters per second) in Fig. 5.26 versus time t for Example 5.4.2.

EXAMPLE 5.4.3

A solid cylinder of mass m and radius $R_1 = 8$ cm rolls on a vertical semicircle of radius $R_2 = 24$ cm as shown in the free-body diagram in Fig. 5.28 (the unit vectors $\hat{r}$, $\hat{\phi}$, and $\hat{z} \triangleq \hat{\phi} \times \hat{r}$ included in the figure are used in the analysis of the problem). The angles ϕ and θ are measured from a fixed vertical line. Friction between the surfaces that are in contact is of the Coulomb type, and the static and kinetic coefficients of friction are, respectively,

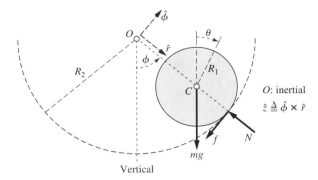

Figure 5.28 Free-body diagram for Example 5.4.3 (point O is inertial).

$\mu_s = 0.3$ and $\mu_k = 0.25$. The cylinder is released from rest at an angle $\phi = \phi_0$ so that it rolls without slipping.

a. Obtain the differential equation of motion for the cylinder in terms of the angle ϕ. How does the magnitude and the direction of the friction force change with ϕ?
b. Show that the differential equation of motion predicts that $\phi(t)$ is an oscillatory motion about the equilibrium $\phi = 0$. What is the frequency of the small (linearized) oscillation about that equilibrium?
c. Determine the magnitude of the angular velocity of the cylinder in terms of ϕ_0, at the time when $\phi = 0$.
d. Determine the maximum value ϕ_0 of the angle ϕ for the cylinder to roll without slipping.

■ **Solution**

a. In terms of the unit vectors $\hat{r}$ and $\hat{\phi}$ shown in Fig. 5.28, the absolute position vector for the center of mass C of the cylinder is

$$\vec{r}_C = \overrightarrow{OC} = (R_2 - R_1)\hat{r}$$

Therefore, the following expressions are obtained for the absolute velocity $\vec{v}_C$ and for the absolute acceleration $\vec{a}_C$ of point C.

$$\vec{v}_C = \frac{d}{dt}\vec{r}_C = (R_2 - R_1)\frac{d}{dt}\hat{r} = (R_2 - R_1)\dot{\phi}\hat{\phi}$$

$$\vec{a}_C = \frac{d}{dt}\vec{v}_C = (R_2 - R_1)\ddot{\phi}\hat{\phi} + (R_2 - R_1)\dot{\phi}\frac{d}{dt}\hat{\phi} = (R_2 - R_1)\ddot{\phi}\hat{\phi} - (R_2 - R_1)\dot{\phi}^2\hat{r}$$

The resultant force acting on the cylinder is

$$\vec{F} = (mg\cos\phi - N)\hat{r} - (f + mg\sin\phi)\hat{\phi}$$

Therefore, $\vec{F} = m\vec{a}_C$ yields the following scalar equations:

$$m(R_2 - R_1)\ddot{\phi} = -mg\sin\phi - f \quad \text{(5.4.3a)}$$

$$N - mg\cos\phi = m(R_2 - R_1)\dot{\phi}^2 \quad \text{(5.4.3b)}$$

As indicated by Eqs. (5.3.5) (p. 270), the third equation that governs the motion is obtained either from $\vec{M}_C = d\vec{H}_C/dt$ or, since point O is inertial, from $\vec{M}_O = d\vec{H}_O/dt$, where $\vec{M}_C$ and $\vec{M}_O$ are the moment of the resultant $\vec{F}$ about the center of mass C and the inertial point O, respectively. Since the absolute angular velocity of the cylinder is $\vec{\omega} = \dot{\theta}\hat{z}$ (see Fig. 5.28 and use the right-hand rule), the absolute angular momentum $\vec{H}_C$ is $\vec{H}_C = I_C\dot{\theta}\hat{z} = (mR_1^2/2)\dot{\theta}\hat{z}$.

By making use of the equation $\vec{M}_C = d\vec{H}_C/dt$, the following scalar equation is obtained:

$$\frac{mR_1^2}{2}\ddot{\theta} = fR_1 \tag{5.4.3c}$$

We now have three equations, namely, Eqs. (5.4.3a), (5.4.3b), and (5.4.3c), but they involve the four unknowns θ, ϕ, N, and f. Another equation is needed, and it is the constraint relation between $\dot{\theta}$ and $\dot{\phi}$ which, as shown earlier in this section, is

$$(R_2 - R_1)\dot{\phi} = R_1\dot{\theta} \tag{5.4.3d}$$

By combining Eqs. (5.4.3a), (5.4.3c), and (5.4.3d), the following differential equation for the angle ϕ is obtained:

$$\ddot{\phi} + \frac{2g}{3(R_2 - R_1)}\sin\phi = 0 \tag{5.4.3e}$$

Equation (5.4.3e) is the differential equation that governs the motion of the cylinder. The friction force f can now be obtained by combining Eqs. (5.4.3a) and (5.4.3e) to yield

$$f = -\frac{mg}{3}\sin\phi \tag{5.4.3f}$$

This solution for f shows that the direction of the friction force is opposite to that shown in Fig. 5.28 when $\phi > 0$. The friction force is equal to zero when $\phi = 0$ (i.e., when C crosses the vertical line), and it will point to the left in Fig. 5.28 when $\phi < 0$.

b. Equation (5.4.3e) is a nonlinear differential equation that is of the same form as the differential equation for a pendulum, as considered in Example 1.13.2 (p. 42). Its equilibrium solutions $\phi = \text{constant} \triangleq \phi_e$ satisfy the equation $\sin\phi_e = 0$, and the only solution of interest here is $\phi_e = 0$. As shown in Example 1.13.2, the equilibrium $\phi_e = 0$ is stable. Here, the motion of the cylinder is started with $\dot{\phi} = 0$, and the subsequent motion is an oscillation about the equilibrium $\phi = 0$ (see Fig. 1.13, p. 44).

To determine the frequency of the small oscillations of the system, linearize Eq. (5.4.3e) about the stable equilibrium $\phi = 0$. Linearization was first presented in Section 1.14, and used in a number of examples in subsequent chapters. For small values of ϕ, $\sin\phi \approx \phi$ and linearization of Eq. (5.4.3e) yields

$$\ddot{\phi} + \frac{2g}{3(R_2 - R_1)}\phi = 0$$

Thus, the frequency of the linearized oscillations is $\sqrt{2g/[3(R_2 - R_1)]}$, which, for $R_1 = 8$ cm and $R_2 = 24$ cm, is approximately 6.4 rad/s.

c. The nonlinear differential equation (5.4.3e) is also of the same form as Eq. (1.13.14) (p. 41), and it can be integrated once with respect to ϕ by using the identity $\ddot{\phi} = [d\dot{\phi}/d\phi]\dot{\phi}$, and the given initial conditions, as

$$\int_{\dot{\phi}=0}^{\dot{\phi}} \frac{d\dot{\phi}}{d\phi} \dot{\phi}\, d\phi + \int_{\phi=\phi_0}^{\phi} \frac{2g\sin\phi}{3(R_2 - R_1)}\, d\phi = 0$$

This gives

$$\frac{1}{2}\dot{\phi}^2 - \frac{2g\cos\phi}{3(R_2 - R_1)} = -\frac{2g\cos\phi_0}{3(R_2 - R_1)} \qquad (5.4.3g)$$

or

$$\dot{\phi} = \pm\sqrt{\frac{4g(\cos\phi - \cos\phi_0)}{3(R_2 - R_1)}}$$

The angular velocity of the cylinder is $\vec{\omega} = \dot{\theta}\hat{z}$, and its magnitude $|\dot{\theta}| = (R_2 - R_1)|\dot{\phi}|/R_1$ can now be readily obtained by using the solution obtained for $\dot{\phi}$. With $R_1 = 8$ cm and $R_2 = 24$ cm, we obtain, when $\phi = 0$,

$$|\dot{\theta}|_{\phi=0} \approx 18.1\sqrt{1 - \cos\phi_0}\ \text{rad/s}$$

d. The last question that was asked was to determine the maximum value of ϕ_0 for the cylinder to roll without slipping. This is done by setting the magnitude of the friction force, $|f|$, to its maximum possible value $\mu_s|N|$.

By combining Eqs. (5.4.3b) and (5.4.3g), the normal force N is obtained as

$$N = \frac{mg}{3}(7\cos\phi - 4\cos\phi_0) \qquad (5.4.3h)$$

With f given by Eq. (5.4.3f), the condition $|f| = \mu_s|N|$ yields

$$|\sin\phi| = \mu_s|7\cos\phi - 4\cos\phi_0|$$

By setting $\phi = \phi_0$ in this equation, it follows that $|\tan\phi_0| = 3\mu_s$. For $\mu_s = 0.3$, this gives $|\phi_0| \approx 42°$. For the cylinder to be able to roll without slipping when the motion is started from rest, we need to have $-42° \leq \phi_0 \leq 42°$.

5.5 A Work-Energy Approach for Rigid Body Dynamics

A work-energy approach involving the relationship $dW = d(T)$ for a system of point masses was presented in Section 2.8. The same relationship and all the results presented in that section are also valid for a rigid body. The infinitesimal work dW done by a set of N forces $\vec{F}_1, \vec{F}_2, \ldots, \vec{F}_N$ acting on the rigid body is still given as

$$dW \triangleq \sum_{i=1}^{N} \vec{F}_i \bullet d\vec{r}_i \qquad (5.5.1)$$

where $\vec{r}_i$ is the absolute position vector of the point on the body where the force $\vec{F}_i$ is applied. The dot product $\vec{F}_i \bullet d\vec{r}_i$ is a simple operation to perform. It is even simpler when it involves a pair of forces that are equal in magnitude and opposite in direction, which is called a *couple*. The infinitesimal work done by a couple is equal to the dot product of the moment of the couple and the infinitesimal rotation $\vec{\omega}\,dt$ of the body, where $\vec{\omega}$ is the absolute angular velocity of the body.

For the kinetic energy associated with the motion of the rigid body, one simply needs to replace the summation in Eq. 2.8.5 (p. 153) in Chapter 2 by an integral that spans the entire body. Therefore, the kinetic energy of the motion of a rigid body is defined as

$$T \triangleq \frac{1}{2} \int_{\text{body}} \vec{v}_P \bullet \vec{v}_P \, dm \tag{5.5.2}$$

where $\vec{v}_P$ is the absolute velocity of an infinitesimal mass element of the body, represented by point P in Fig. 5.4 (p. 268).

Let us now take a closer look at the expression for the kinetic energy T, and manipulate it to obtain a more useful expression for calculating T. Referring to Fig. 5.4, by writing the absolute position vector of P as $\vec{r} = \vec{r}_C + \overrightarrow{CP}$, where $\vec{r}_C$ is the absolute position vector for the body's center of mass, the expression for the kinetic energy T becomes

$$T = \frac{1}{2} \int_{\text{body}} \left(\vec{v}_C + \frac{d}{dt}\overrightarrow{CP} \right) \bullet \left(\vec{v}_C + \frac{d}{dt}\overrightarrow{CP} \right) dm$$

$$= \frac{1}{2} \int_{\text{body}} \vec{v}_C \bullet \vec{v}_C \, dm + \int_{\text{body}} \vec{v}_C \bullet \left(\frac{d}{dt}\overrightarrow{CP} \right) dm + \frac{1}{2} \int_{\text{body}} \left(\frac{d}{dt}\overrightarrow{CP} \right) \bullet \left(\frac{d}{dt}\overrightarrow{CP} \right) dm$$

$$= \frac{1}{2} m \vec{v}_C \bullet \vec{v}_C + \vec{v}_C \bullet \frac{d}{dt} \underbrace{\left(\int_{\text{body}} \overrightarrow{CP} \, dm \right)}_{=0} + \frac{1}{2} \int_{\text{body}} \left(\frac{d}{dt}\overrightarrow{CP} \right) \bullet \left(\frac{d}{dt}\overrightarrow{CP} \right) dm$$

The second integral in this expression is zero because of the definition of the center of mass C. Since the magnitude of $\overrightarrow{CP}$ is constant because the body is rigid, and thus, $|d\overrightarrow{CP}/dt| = |\dot{\theta}\,\overrightarrow{CP}|$ (see Fig. 5.4), the expression for T then becomes

$$T = \frac{1}{2} m \vec{v}_C \bullet \vec{v}_C + \frac{1}{2} I_C \dot{\theta}^2 \tag{5.5.3}$$

where

$$I_C = \int_{\text{body}} |\overrightarrow{CP}|^2 \, dm$$

is recognized as the body's moment of inertia about the axis that passes through the center of mass C and is perpendicular to the plane of Fig. 5.4.

> Notice that, as indicated by Eq. (5.5.3), the kinetic energy of the motion of a rigid body is simply the sum of two simple terms that actually have a physical interpretation. The first term in that equation is the kinetic energy associated only with the translational motion of the center of mass of the body. The second term is the kinetic energy due only to rotation of the body about an axis that passes through the body's center of mass.

As indicated in Chapter 2, when dW is an exact differential, the equation $T - W = $ constant does not involve any time derivative higher than the first derivative (i.e., velocities, but not accelerations) of any variable. In such cases, $dW = d(T)$ yields $T - W = $ constant, which is a first-order differential equation. This is the differential equation of motion if the motion is governed by only one variable (i.e., a one-degree-of-freedom system). Its time derivative, $d(T - W)/dt = 0$, gives the same second-order differential equation that one would obtain by direct use of a combination of Newton's second law $\vec{F} = m\vec{a}_C$ and the moment equation $\vec{M}_C = d\vec{H}_C/dt$ (or $\vec{M}_O = d\vec{H}_O/dt$). The advantage of the work-energy approach, as described earlier, is that the differential equation of motion is obtained with much less effort. When dW is not an exact differential, one could still obtain a second-order differential equation via $dW/dt = d(T)/dt$.

In this section, the work-energy approach was presented in an introductory/intermediate level. Such an approach is very powerful. It is actually the basis for a new approach to dynamics that is called *analytical dynamics* and is mostly based on work developed by Lagrange. A number of other great masters, such as Hamilton and Jacobi, were also creators of very elegant and powerful theories that are part of analytical dynamics. The reader will gain much by broadening his or her knowledge of dynamics to include analytical dynamics. An introduction to the subject matter is presented in Chapter 7, which is the most advanced chapter in the book.

5.6 Forces on Mechanisms

This section is devoted to the calculation of the reaction forces acting on each member of a mechanism. These calculations are typically very lengthy and require the use of the free-body diagram for each member of the mechanism. However, the formulation of the required equations is trivial and it only involves the use of the basic equation $\vec{F} = m\vec{a}_C$, for the motion of the center of mass C of a rigid body, and either one of the moment equations $\vec{M}_O = d\vec{H}_O/dt$ (if point O is inertial) or $\vec{M}_C = d\vec{H}_C/dt$ for the rotational motion of the body. The kinematic analysis of mechanisms was presented in detail in Chapter 3, and

Figure 5.29 A four-bar mechanism (the angles θ_2, θ_3, and θ_4 are changing with time).

Figure 5.30 Free-body diagram for each bar of the four-bar mechanism of Fig. 5.29.

the results obtained in that chapter are needed here. A four-bar mechanism is used for illustrating the calculations involved in determining the forces acting on each bar of the mechanism, but the procedure shown here is the same for other mechanisms.

A four-bar mechanism is shown in Fig. 5.29. It is assumed that bar AB is driven by an external motor so that the angle $\theta_2(t)$ is know at all times. The reaction forces acting on each bar of the mechanism will be determined in terms of the mass, length, angular position, angular velocity, and angular acceleration of each of its members.

Figure 5.30 shows free-body diagrams for each bar of the four-bar mechanism for the simpler case where the mechanism is either in a horizontal plane or the effect of the gravitational forces acting on each member is disregarded. The reaction forces in the plane perpendicular to the plane of the motion of the mechanism are not shown since they do not affect the solution to the problem. The moment M_{ext} is applied to bar AB by a motor in order to make $\theta_2(t)$ a desired function of time. Points C_2, C_3, and C_4 shown in Fig. 5.30 are the centers of mass of bars AB, BC, and CD, respectively. Three sets of orthogonal unit vectors, which will be used in the formulation of the equations for obtaining the reaction forces, are also shown in the figure.

SECTION 5.6 Forces on Mechanisms

In terms of the unit vector triad $\{\hat{r}_2, \hat{\theta}_2, \hat{z} \triangleq \hat{r}_2 \times \hat{\theta}_2\}$ that rotates with bar AB shown in Fig. 5.30, we then have

Absolute position vector for the center of mass C_2 of bar AB:

$$\vec{r}_{C_2} = \overrightarrow{AC_2} = \frac{L_2}{2}\hat{r}_2$$

Absolute velocity of C_2:

$$\vec{v}_{C_2} = \frac{d}{dt}\vec{r}_{C_2} = \dot{\theta}_2 \hat{z} \times \left(\frac{L_2}{2}\hat{r}_2\right) = \frac{L_2}{2}\dot{\theta}_2 \hat{\theta}_2$$

Absolute acceleration of C_2:

$$\vec{a}_{C_2} = \frac{d}{dt}\vec{v}_{C_2} = \frac{L_2}{2}\ddot{\theta}_2 \hat{\theta}_2 + \dot{\theta}_2 \hat{z} \times \vec{v}_{C_2} = \frac{L_2}{2}\left(\ddot{\theta}_2 \hat{\theta}_2 - \dot{\theta}_2^2 \hat{r}_2\right)$$

Resultant force acting on bar AB:

$$\vec{F}_{AB} = (A_{r_2} + B_{r_2})\hat{r}_2 + (A_{n_2} + B_{n_2})\hat{\theta}_2$$

The motion of the center of mass of the bar is governed by Newton's second law $\vec{F} = m_2 \vec{a}_{C_2}$, and this yields the following scalar equations:

$$A_{r_2} + B_{r_2} = -m_2 \frac{L_2}{2}\dot{\theta}_2^2 \qquad (5.6.1)$$

$$A_{n_2} + B_{n_2} = m_2 \frac{L_2}{2}\ddot{\theta}_2 \qquad (5.6.2)$$

Since point A is inertial, let us take moments about A and use the simple moment equation $\vec{M}_A = d\vec{H}_A/dt$, where $\vec{H}_A = I_A \dot{\theta}_2 \hat{z}$ and $I_A = m_2 L_2^2/3$. This moment equation yields the following scalar equation:

$$M_{\text{ext}} + L_2 B_{n_2} = \frac{1}{3}m_2 L_2^2 \ddot{\theta}_2 \qquad (5.6.3)$$

So far we have three equations, but they involve four unknown reaction forces and one unknown moment M_{ext}. To be able to solve the problem, more equations have to be generated until the number of equations and unknowns are the same.

Before proceeding, notice that the form of Eqs. (5.6.1) to (5.6.3) are relatively simple because force components A_{r_2}, B_{r_2}, A_{n_2}, and B_{n_2} in the directions $\hat{r}_2$ along bar AB, and $\hat{\theta}_2$ normal to AB, are being used in the analysis. Because of such a choice, the coefficients of these force components in Eqs. (5.6.1) and (5.6.2) are equal to one, and the expressions for the acceleration components of $\vec{a}_{C_2}$ are simply $(L_2/2)\ddot{\theta}_2 \hat{\theta}_2$ and $-(L_2/2)\dot{\theta}_2^2 \hat{r}_2$.

Therefore, in order to obtain more equations that are as simple as Eqs. (5.6.1) to (5.6.3), consider bar CD next, for which point D is inertial. Treat that bar the same way bar AB was treated, but using the unit vectors $\hat{r}_4$ and $\hat{\theta}_4$ that rotate with bar CD. With the absolute position vector of the center of mass C_4 of bar CD

being $\overrightarrow{DC_4} = (L_4/2)\hat{r}_4$, the following equations are then obtained for that bar:

$$C_{r_4} + D_{r_4} = -m_4 \frac{L_4}{2}\dot{\theta}_4^2 \quad (5.6.4)$$

$$C_{n_4} + D_{n_4} = m_4 \frac{L_4}{2}\ddot{\theta}_4 \quad (5.6.5)$$

$$L_4 C_{n_4} = \frac{1}{3}m_4 L_4^2 \ddot{\theta}_4 \quad (5.6.6)$$

Finally, consider the last bar, which is bar BC. The reaction forces acting on that bar are B_{r_2}, B_{n_2}, C_{r_4}, and C_{n_4}, and they appear as shown in the free-body diagram in Fig. 5.30c. Since these forces are not in the directions along the bar and perpendicular to the bar, and since neither point B nor point C is inertial, the equations in which the unknown reaction forces B_{r_2}, B_{n_2}, C_{r_4}, and C_{n_4} will appear will be a little more complicated than the equations that were obtained for the other two bars. The formulation of the equations involve obtaining expressions for the absolute acceleration of the center of mass of bar BC and the resultant force acting on the bar.

Absolute position vector for the center of mass C_3 of bar BC:

$$\vec{r}_{C_3} = \overrightarrow{AC_3} = \overrightarrow{AB} + \frac{L_3}{2}\hat{r}_3$$

Absolute velocity of C_3:

$$\vec{v}_{C_3} = \frac{d}{dt}\vec{r}_{C_3} = \underbrace{\frac{d}{dt}\overrightarrow{AB}}_{=L_2\dot{\theta}_2\hat{\theta}_2} + \dot{\theta}_3\hat{z} \times \left(\frac{L_3}{2}\hat{r}_3\right) = L_2\dot{\theta}_2\hat{\theta}_2 + \frac{L_3}{2}\dot{\theta}_3\hat{\theta}_3$$

Absolute acceleration of C_3:

$$\vec{a}_{C_3} = \frac{d}{dt}\vec{v}_{C_3} = L_2\ddot{\theta}_2\hat{\theta}_2 + \dot{\theta}_2\hat{z} \times (L_2\dot{\theta}_2\hat{\theta}_2) + \frac{L_3}{2}\ddot{\theta}_3\hat{\theta}_3 + \dot{\theta}_3\hat{z} \times \left(\frac{L_3}{2}\dot{\theta}_3\hat{\theta}_3\right)$$

$$= L_2\left(\ddot{\theta}_2\hat{\theta}_2 - \dot{\theta}_2^2\hat{r}_2\right) + \frac{L_3}{2}\left(\ddot{\theta}_3\hat{\theta}_3 - \dot{\theta}_3^2\hat{r}_3\right)$$

Resultant force acting on bar BC:

$$\vec{F}_{BC} = -B_{r_2}\hat{r}_2 - B_{n_2}\hat{\theta}_2 - C_{r_4}\hat{r}_4 - C_{n_4}\hat{\theta}_4$$

Newton's second law is now applied to bar BC, i.e., $\vec{F}_{BC} = m_3\vec{a}_{C_3}$. To do this, first express $\vec{F}_{BC}$ and $\vec{a}_{C_3}$ in the same orthogonal set of unit vectors. By arbitrarily choosing to work with vector components along $\hat{r}_2$ and $\hat{\theta}_2$, the transformations that relate $\{\hat{r}_3, \hat{\theta}_3\}$ (which appear in the expression for $\vec{a}_{C_3}$) and $\{\hat{r}_4, \hat{\theta}_4\}$ (which appear

Figure 5.31 The three unit vector systems used with Fig. 5.29.

in the expression for $\vec{F}_{BC}$) to $\{\hat{r}_3, \hat{\theta}_3\}$ are obtained by inspection of Fig. 5.31. These relations are as follows:

$$\hat{r}_3 = \hat{r}_2 \cos(\theta_2 - \theta_3) - \hat{\theta}_2 \sin(\theta_2 - \theta_3) \qquad \hat{\theta}_3 = \hat{r}_2 \sin(\theta_2 - \theta_3) + \hat{\theta}_2 \cos(\theta_2 - \theta_3)$$

$$\hat{r}_4 = \hat{r}_2 \cos(\theta_4 - \theta_2) + \hat{\theta}_2 \sin(\theta_4 - \theta_2) \qquad \hat{\theta}_4 = \hat{\theta}_2 \cos(\theta_4 - \theta_2) - \hat{r}_2 \sin(\theta_4 - \theta_2)$$

With $\hat{r}_3, \hat{\theta}_3, \hat{r}_4$, and $\hat{\theta}_4$ as expressed in these equations, the following scalar equations are obtained from the vector equation $\vec{F}_{BC} = m_3 \vec{a}_{C_3}$:

$$B_{r_2} + C_{r_4} \cos(\theta_4 - \theta_2) - C_{n_4} \sin(\theta_4 - \theta_2)$$
$$= m_3 \frac{L_3}{2} [\dot{\theta}_3^2 \cos(\theta_2 - \theta_3) - \ddot{\theta}_3 \sin(\theta_2 - \theta_3)] + m_3 L_2 \dot{\theta}_2^2 \qquad (5.6.7)$$

$$B_{n_2} + C_{r_4} \sin(\theta_4 - \theta_2) + C_{n_4} \cos(\theta_4 - \theta_2)$$
$$= -m_3 \frac{L_3}{2} [\dot{\theta}_3^2 \sin(\theta_2 - \theta_3) + \ddot{\theta}_3 \cos(\theta_2 - \theta_3)] - m_3 L_2 \ddot{\theta}_2 \qquad (5.6.8)$$

Since neither point B nor point C is inertial, let us take moments about the center of mass C_3 of bar BC and use the moment equation $\vec{M}_{C_3} = d\vec{H}_{C_3}/dt$, where $\vec{H}_{C_3} = I_{C_3} \dot{\theta}_3 \hat{z}$ and $I_{C_3} = m_3 L_3^2/12$. This moment equation yields the following scalar equation:

$$\frac{L_3}{2} [B_{r_2} \sin(\theta_2 - \theta_3) + B_{n_2} \cos(\theta_2 - \theta_3)$$
$$- C_{r_4} \sin(\theta_4 - \theta_3) - C_{n_4} \cos(\theta_4 - \theta_3)] = I_{C_3} \ddot{\theta}_3 \qquad (5.6.9)$$

where the moment $\vec{M}_{C_3}$ is obtained by inspecting Fig. 5.30c.

Equations (5.6.1) to (5.6.9) constitute a system of nine equations involving the nine unknowns $M_{\text{ext}}, A_{r_2}, A_{n_2}, B_{r_2}, B_{n_2}, C_{r_4}, C_{n_4}, D_{r_4}$, and D_{n_4}. They are linear in these variables and can be easily solved for. They could, for example, be put in matrix form and then solved by inverting a 9×9 matrix, as done in Chapter 3 for a 2×2 matrix. However, they can be solved in a simpler manner.

Equations (5.6.5) and (5.6.6) can immediately be solved for the reaction forces C_{n_4} and D_{n_4} to obtain the following results:

$$C_{n_4} = \frac{1}{3}m_4 L_4 \ddot{\theta}_4 \qquad (5.6.10)$$

$$D_{n_4} = \frac{1}{2}C_{n_4} \qquad (5.6.11)$$

with $\ddot{\theta}_4$ determined as shown in Chapter 3.

Having solved for C_{n_4} and D_{n_4}, Eqs. (5.6.7) to (5.6.9) now only involve the unknowns B_{r_2}, B_{n_2}, and C_{r_4}. They may be solved with a simple matrix inversion and a matrix multiplication as follows, and numerical values may be readily obtained either by hand or with an inexpensive calculator.

$$\begin{bmatrix} B_{r_2} \\ B_{n_2} \\ C_{r_4} \end{bmatrix} = \begin{bmatrix} 1 & 0 & \cos(\theta_4 - \theta_2) \\ 0 & 1 & \sin(\theta_4 - \theta_2) \\ \sin(\theta_2 - \theta_3) & \cos(\theta_2 - \theta_3) & -\sin(\theta_4 - \theta_3) \end{bmatrix}^{-1}$$

$$\times \begin{bmatrix} \frac{1}{2}m_3 L_3 \left(\dot{\theta}_3^2 \cos(\theta_2 - \theta_3) - \ddot{\theta}_3 \sin(\theta_2 - \theta_3) \right) + m_3 L_2 \dot{\theta}_2^2 + C_{n_4} \sin(\theta_4 - \theta_2) \\ -\frac{1}{2}m_3 L_3 \left(\dot{\theta}_3^2 \sin(\theta_2 - \theta_3) + \ddot{\theta}_3 \cos(\theta_2 - \theta_3) \right) - m_3 L_2 \ddot{\theta}_2 - C_{n_4} \cos(\theta_4 - \theta_2) \\ \frac{1}{6}m_3 L_3 \ddot{\theta}_3 + C_{n_4} \cos(\theta_4 - \theta_3) \end{bmatrix}$$

(5.6.12)

The remaining unknowns A_{r_2}, A_{n_2}, M_{ext}, and D_{r_4} are readily obtained from Eqs. (5.6.1) to (5.6.4).

⟹ Figure 5.32 shows a Simulink block diagram model for automating the calculations just presented and for animating the motion of the mechanism. The model is the file `fourbarforces.mdl` included on the CD. To use it, either type `fourbarforces` in the MATLAB command window or type `simulations_menu` and choose the `fourbarforces` model from a pull-down menu that appears on the computer screen.

■ The model shown in Fig. 5.32 is based on the `fourbar_mechanism` model shown in Fig. 3.9 (p. 193). It contains everything that is in that model. The only difference between the two models is the addition of the blocks seen to the right of the block that displays the angular velocities $\dot{\theta}_3$ and $\dot{\theta}_4$ and angular accelerations $\ddot{\theta}_3$ and $\ddot{\theta}_4$.

■ The default inputs of the `fourbarforces` model are the same as those for Example 3.2.2 (p. 189) (see Fig. 3.6, p. 190), for which the material density of each bar is set to $\rho = 0.075$ kg/m, and $L_1 = \sqrt{D_1^2 + D_2^2}$ (where D_1 and D_2 are identified in Fig. 3.6). The default stop time in the `Solver` part of the `Simulation → Parameters` menu of the model is $1.e - 6$, and the other default choices in that menu are the same as those shown in Fig. 3.10 (p. 194). The input variable `solution_number` is described in Section 3.2.3.

SECTION 5.6 Forces on Mechanisms

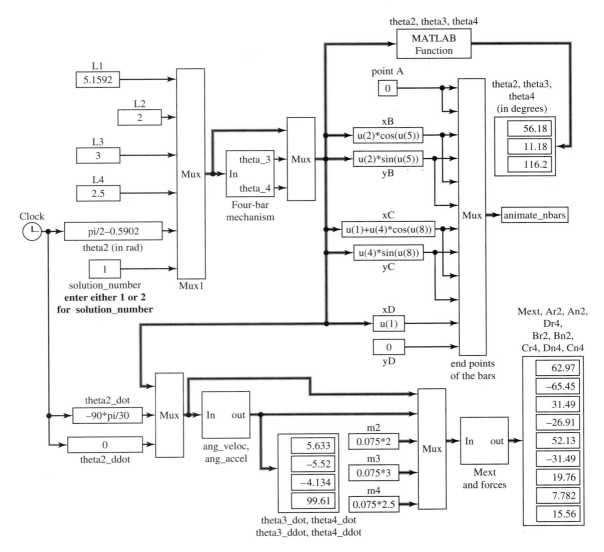

Figure 5.32 Simulink model `fourbarforces.mdl` for a four-bar mechanism (input data for Example 3.2.2 is used).

- All the steps presented for calculating the applied moment M_{ext} and the reaction forces on each bar of a four-bar mechanism are embedded in the subsystem block labeled `Mext and forces` that is seen in Fig. 5.32. A subsystem was created with the sole purpose of compacting the entire block diagram which, otherwise, would hardly fit on the computer screen. The block diagram that is grouped in that subsystem is shown in Fig. 5.33.

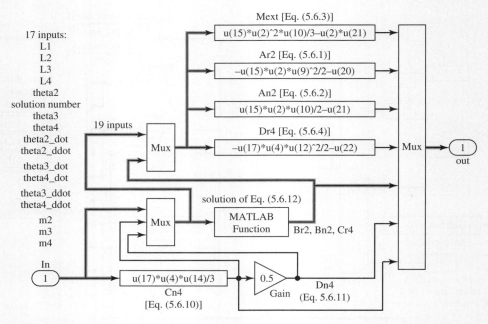

Figure 5.33 Block diagram (a subsystem) for calculating the reaction forces and the moment M_{ext}. The 17 inputs to the subsystem are listed on the left.

Double click that subsystem block to see those same details in your computer. The calculation of the reactions B_{r_2}, B_{n_2}, and C_{r_4} using Eq. (5.6.12) is done in the `MATLAB Function` block that is seen in Fig. 5.33. The syntax for that matrix operation is a very long line! It can be seen by double clicking that block. You may wish to copy that long expression to an editor for better viewing.

■ The solutions displayed in Fig. 5.32 are for a single time instant, which is essentially $t = 0$. To run the model with a different function $\theta_2(t)$, double click the blocks labeled `theta2`, `theta2_dot`, and `theta2_ddot` in the model and enter the desired functions in the windows that appear on your computer screen (using the letter u for the time t in the expressions you enter). You may want to set the final time in the `Simulation → Parameters` menu to a large value (say, 9999) so that the simulation is stopped by you. Try different values for the maximum step size if the animation runs either too fast or too slow in your computer. Change the length of some of the bars if the mechanism locks (a message saying there is a square root of a negative number may appear on the screen in such cases).

5.7 Problems

5.1 The body in Fig. 5.34 is a thin homogeneous plate with dimensions $b = 100$ cm, $h = 50$ cm, and $L = 20$ cm. The plate would be rectangular if it were not for the cutout. Its mass density is $\rho = 15.4$ kg/m^2, which is for a steel sheet of approximately 2 mm of thickness. Determine the location of its center of mass C relative to the corner at point O shown in the figure. Also, determine the moments of inertia I_O and I_C about the axis that are perpendicular to the plate and pass through O and the center of mass C of the plate, respectively. Disregard the thickness in the calculations. *Suggestion:* Solve the problem for a plate without the cutout, and then subtract the cutout (which is the same as treating the cutout as if its mass were negative).

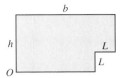

Figure 5.34 Figure for Problem 5.1.

5.2 The body shown in Fig. 5.35 consists of a homogeneous solid ring of inner and outer radius $R_1 = 90$ cm and $R_2 = 100$ cm, respectively. To reinforce the ring, three crossbars with rectangular cross section are symmetrically welded to its inside. The ring and the crossbars are made of steel, and their thickness is 2 mm, which is very small. The density of steel is 7700 kg/m^3. Determine the body's moment of inertia I_C about the axis that is perpendicular to the plane of the ring and that passes through its center of mass C. What percentage do the crossbars contribute to the answer? Disregard the thickness in the calculations.

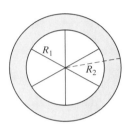

Figure 5.35 Figure for Problem 5.2.

5.3 The body in Fig. 5.36 is a thin homogeneous triangular plate with a square cutout and $L = 100$ cm. Its mass density is $\rho = 15.4$ kg/m^2, which is for a steel sheet of approximately 2 mm of thickness. Determine the location of its center of mass C relative to point O shown in the figure, which is in the middle of the bottom part of the plate. Also, determine the moments of inertia I_O and I_C about the axis that is perpendicular to the plate and passes through O and through its center of mass C, respectively. Disregard the thickness in the calculations.

5.4 Show that the absolute angular momentum $\vec{H}_A$ of a body, about an arbitrary point A, is related to the absolute angular momentum $\vec{H}_C$ about the center of mass C of the body as $\vec{H}_A = \vec{H}_C + m\overrightarrow{AC} \times \vec{v}_C$, where $\vec{v}_C$ is the absolute velocity of C.

Figure 5.36 Figure for Problem 5.3.

5.5 A rigid body of mass m, such as the one in Fig. 5.4 (p. 268), is rotating, in a plane, with absolute angular velocity $\dot{\theta}\hat{z}$, where $\hat{z}$ is a unit vector normal to the plane of rotation. The rotation is about an axis that passes through the inertial point O shown in that figure, and line OC is a massless bar rigidly attached to the body. Show that the absolute angular momentum $\vec{H}_O$ is equal to $\vec{H}_O = I_O \dot{\theta}\hat{z}$, where $I_O = I_C + m|\overrightarrow{OC}|^2$. *Suggestion:* Start by using the result of Problem 5.4, i.e., $\vec{H}_O = \vec{H}_C + m\overrightarrow{OC} \times \vec{v}_C$.

5.6 An engineering company needs to determine the location of the center of mass C of the thin plate shown in Fig. 5.37. The moment of inertia I_C of the plate, about an axis perpendicular to the plate and passing through C, also needs to be determined. The problem is that the plate is inhomogeneous, in addition to its shape being somewhat complicated. Experiments were then devised to determine those quantities. First, two points A and B were marked on the plate, with the distance between them equal to $D = 1.1$ m, as shown in Fig. 5.37. Thin pins were attached at those points so that the plate could be suspended from them in such a

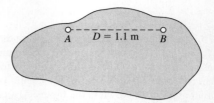

Figure 5.37 Figure for Problem 5.6.

way that it remained vertical. The plate was also weighted, and its mass was found to be equal to 30.5 kg.

a. The plate was hung from A, and line AB remained at $19°$ with the vertical when the plate was in static equilibrium (i.e., no motion). It was then given a small angular displacement from that equilibrium, and the period of the resulting oscillation was found to be 3.3 s. The engineers made sure that friction in the system was small enough to be disregarded in their subsequent calculations.

b. The plate was also hung from B and, in static equilibrium, line AB remained at $48°$ with the vertical.

Is the data gathered sufficient to perform the task at hand? If so, determine the location of C, and the moment of inertia I_C. What would be the period of the resulting small oscillations if the plate were suspended from B?

5.7 Figure 5.38 is a side view of a solid homogeneous cylinder of radius R and mass m, rolling on a plane that is inclined α degrees with the horizontal. The center of mass C of the cylinder is on its axial axis. That axis is connected to two identical linear springs (one on each face of the cylinder) of unstressed length L_u. The other end of each spring is connected to fixed points on the inclined plane. The total stiffness of the parallel springs is k N/m. Friction between the cylinder and the plane is of the Coulomb type, and the static and kinetic friction coefficients are μ_s and μ_k, respectively. The cylinder rolls without slipping on the inclined plane.

a. Obtain the differential equation of motion for the system in terms of the variable $x(t)$ shown in Fig. 5.38. Also, determine the equilibrium $x = \text{constant} \triangleq x_e$.

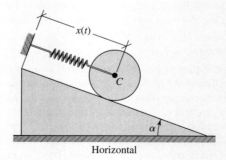

Figure 5.38 Figure for Problem 5.7.

b. Is the differential equation obtained in part a linear or nonlinear? If it is nonlinear, linearize it about $x = x_e$ and then obtain the frequency of the small oscillations of the system about each stable equilibrium. If it is linear, obtain the frequency of the oscillations of the system about its equilibrium. The system constants are $k/m = 54 \text{ s}^{-2}$, $\mu_s = 0.3$, $\mu_k = 0.25$, $L_u = 50$ cm, and $\alpha = 30°$.

5.8 The motion of the system described in Problem 5.7, with its parameter values given in part b of that problem, is started with the initial conditions $x(0) = x_0$ and $\dot{x}(0) = 0$. Physically one would expect that the cylinder will roll with slipping if the absolute value of $x(0)$ is "too big." Determine the range of x_0 for the cylinder to roll without slipping.

5.9 The system shown in Fig. 5.39 consists of a homogeneous circular plate of radius R with a rectangular cutout, and a homogeneous rectangular plate of length R and $h = (\sqrt{3}/2)R$. The two plates are glued to each other, are of small thickness, and their mass density is ρ kg/m². Point A is the center of mass of the circular plate, and line AB is parallel to one of the sides of the rectangular plate. The plate rolls without slipping on a horizontal surface, and the system is constrained to move on a vertical plane by two horizontal frictionless guides that are not shown in the figure. Friction between the plate and the horizontal surface shown is of the Coulomb type, and the static and kinetic coefficients of friction are μ_s and μ_k, respectively.

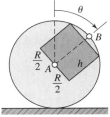

Figure 5.39 Figure for Problem 5.9.

a. Obtain the differential equation of motion for the system in terms of the angle $\theta(t)$ shown in Fig. 5.39, and all the equilibrium solutions $\theta = \text{constant} \stackrel{\Delta}{=} \theta_e$.

b. Analyze the stability of each equilibrium solution, and then determine the frequency of the small oscillations of the system about each stable equilibrium for $R = 15$ cm.

5.10 It is desired to determine if the wheel described in Problem 5.9 slips during its motion. This is to be done by numerically solving the differential equation of motion of the system and then calculating the friction and normal forces acting on the wheel. Obtain the differential equation of motion of the system and integrate it numerically to solve the problem. First, verify that no slipping occurs if the motion is started from rest and using $\theta(0) = 0.01$ rad. Then, by numerical experimentation, determine the maximum value of $|\dot{\theta}(0)/\sqrt{g/R}|$ to prevent slipping. The static and kinetic coefficients of Coulomb friction are $\mu_s = 0.3$ and $\mu_k = 0.25$, respectively (it is simpler to do all these tasks if you use Simulink since there is no computer program to write). *Hint:* Slipping occurs when $|f/N| \geq \mu_s$; use the normalized time $\tau = t\sqrt{g/R}$ and stop the simulation when slipping starts to occur.

5.11 Consider the system of Examples 5.3.10 (p. 284) and 5.3.11 (p. 286), with the value of L_2/L_1 being such that the equilibrium value of the angle θ is $\theta_e = 30°$ when $M_{\text{bar}} = -c\dot{\theta}$, and the support point A is given a constant acceleration $\ddot{x} = -7$ m/s² (see Fig. 5.13, p. 284). For $L_1 = 1$ m and $c/m = 4.5$ m²/s, determine the damped and the undamped frequencies of the small oscillations of the system about its equilibrium.

5.12 Consider the system of Examples 5.3.10 (p. 284) and 5.3.11 (p. 286), with $L_1 = 1$ m, $L_2 = 0.5$ m, $M_{\text{bar}} = 0$, and when the support point A is given a constant acceleration $\ddot{x} = -7$m/s² (see Fig. 5.13, p. 284). If the motion $\theta(t)$ is started with the initial conditions $\theta(0) = 0$ and $\dot{\theta}(0) = \dot{\theta}_0$, determine the range of values of $\dot{\theta}_0$ for $\theta(t)$ to be oscillatory.

5.13 Solve Problems 5.11 and 5.12 by integrating numerically the differential equation that governs the motion of the system. For Problem 5.11, use $L_2 = 0.42L_1$, which is the approximate answer for L_2 for that problem. *Suggestion:* The model `sim_model_ex5_3_11.mdl` may be used for this.

5.14 The system shown in Fig. 5.40a is a double pendulum constructed of two homogeneous thin bars OA and AB of constant length L_1 and L_2, respectively, and having the same mass density ρ kg/m. Bar OA is pivoted to an inertial point O, and the two bars are pinned at A and can rotate about that point relative to each other. The motion of the entire system takes place in the vertical plane, and friction in the system can be neglected.

a. Obtain the differential equations of motion for the system in terms of the angles θ_1 and θ_2 shown in Fig. 5.40a.

b. Use the energy method $dW = d(T)$ to show that $dW = -d(U)$ and that the mechanical energy $T + U$ of this system remains constant for all times.

c. Numerically integrate the differential equations of motion. For this, use the nondimensional time $\tau = t\sqrt{g/L_1}$. Do the integration with $L_2/L_1 = 2$, and with the initial conditions $\theta_1(0) = 30°$, $\theta_2(0) = 0$, $\dot{\theta}_1(0) = \dot{\theta}_2(0) = 0$, and plot $\theta_1(\tau)$ and $\theta_2(\tau)$ versus τ. Also, numerically verify that $T + U = $ constant. It is suggested that you use Simulink and include the animation block `animate_nbars` in your simulation in order to observe how the two bodies move. Do some numerical experiments by changing the initial conditions of the motion, and then write a few paragraphs describing the resulting motion.

Figure 5.40 Figure for Problems 5.14 and 5.15.

5.15 The double pendulum of Problem 5.14 is modified as shown in Fig. 5.40b. The modification consists of adding the torsional spring and the torsional viscous damper at the connection between the two bars. The spring and the damper are linear, with coefficients equal to k N·m/rad and c N·m/(rad/s), respectively. The spring is unstressed when $\theta_2 = 0$. Obtain the differential equations of motion for this system, and integrate them numerically using the nondimensional time $\tau = t\sqrt{g/L_1}$. Do the integration with $L_2/L_1 = 2$, and with the initial conditions $\theta_1(0) = 30°$, $\theta_2(0) = 0$, $\dot{\theta}_1(0) = \dot{\theta}_2(0) = 0$, and plot $\theta_1(\tau)$ and $\theta_2(\tau)$ versus τ. Choose your own nonzero values for the nondimensional spring and damping coefficients $k^* = k/(m_1 g L_1)$ and $c^* = c/(m_1 L_1 \sqrt{gL_1})$. Animate the motion in order to observe how the two bodies move. Do some numerical experiments by changing the initial conditions of the motion, and then write a few paragraphs describing the resulting motion.

5.16 Determine the maximum values of the nondimensional force reactions $F_1/(m_2 g)$ and $F_2/(m_2 g)$ acting on the plate of Example 5.3.13 (p. 292). Use the data and the initial conditions given in part d of that example, and plot those reactions versus the normalized time τ that is also defined in that example. What are the maximum values of those nondimensional reaction forces if the nondimensional damping coefficients are changed to $c_1^* = c_2^* = 2$? Write a brief comment on your observations about those reaction forces, including your conclusion about what their steady-state values should be. *Suggestion:* Modify the model `sim_model_ex5_3_13` to do the calculations.

5.17 For the four-bar mechanism shown in Fig. 3.2 (p. 182), $\dot{\theta}_2 = \text{constant} = -100$ rpm. Friction in the system can be neglected. The motion of the mechanism takes place in a horizontal plane. Each bar is homogeneous and is approximated as a slender line with mass density equal to 0.75 g/cm. For that mechanism, $L_2 = 40$ cm, $L_3 = 100$ cm, $L_4 = 60$ cm, and $L_1 = 100$ cm. Determine the magnitude of the reaction forces acting at the ends of each bar of the mechanism, and on the plane of the motion, when the mechanism is in the position labeled "Problem 3.6" in Fig. 3.33 (p. 221). Also one needs to buy the right motor to drive the mechanism and, for that, one needs to know the maximum torque to be supplied by the motor. Plot the torque M_{ext} versus time and then specify the maximum torque the motor must supply. Make sure that your answers have appropriate units, and clearly specify those units for each answer. *Suggestion:* Modify the model `fourbarforces.mdl` to solve the problem.

5.18 Repeat Problem 5.17 if motion takes place in a *vertical* plane. Use the equations you have developed for this problem to verify that the answers to Problem 5.17 may be obtained by simply setting the value of the acceleration of gravity in this problem to $g = 0$. If that does not happen, you can conclude that there is an error in your calculations.

5.19 The four-bar mechanism shown in Fig. 5.41 moves in a vertical plane, and its motion is started when $\theta_2 = 0$ and point C is above line AD. Friction in the pins can be neglected. Bar AB is driven by a motor so that $\dot{\theta}_2 = \text{constant} = 100$ rpm.

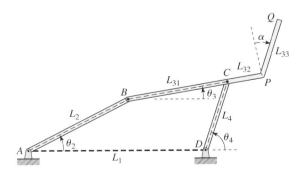

Figure 5.41 Figure for Problem 5.19.

Each bar (including the bent bar $BCPQ$) is homogeneous, with mass density of 0.75 g/cm, and may be approximated as a line. For that mechanism, $L_2 = 25$ cm, $L_{31} = 60$ cm, $L_{32} = 26$ cm, $L_{33} = 14$ cm, $L_4 = 70$ cm, $L_1 = 100$ cm, and $\alpha = \text{constant} = 30°$. Determine the reaction forces acting on each bar of the mechanism (clearly indicating them in a free-body diagram) and the external moment (magnitude and direction) applied to bar AB, when θ_2 reaches $45°$ after bar AB turns two revolutions. While performing your calculations, approximate each bar as a slender line. Make sure that your answers have appropriate units, and clearly specify those units for each answer.

5.20 If you have not done so in Problem 5.19, prepare a Simulink model that will do all the calculations for solving that problem and also display all the answers. Using the model, plot the applied moment versus time t while bar AB turns only three revolutions. What is the maximum value of the magnitude of that moment?

5.21 Bar AB of the mechanism shown in Fig. 3.15 (p. 199), is driven by a motor that applies to it a moment M_{ext} so that $\theta_1 = \pi/6 + kt$ rad, where $k = -8\pi$ rad/s. Friction in the pins and in the slotted bar can be neglected. The motion of the mechanism takes place in a horizontal plane. Each bar is homogeneous, with mass density equal to 0.75 g/cm, and may be approximated as a slender line. For that mechanism, $r_1 = 40$ cm, $L = 100$ cm, and the length of the slotted bar is equal to $L + 1.2 r_1$.

Determine the reaction forces acting on each bar of the mechanism and on the plane of the motion, the reaction moments (if any), and the external moment M_{ext} (magnitude and direction) applied to bar AB (clearly indicating them in a free-body diagram and showing the details of each step of the calculations) immediately after bar AB turns four complete revolutions.

5.22 For the slider crank mechanism shown in Fig. 3.20 (p. 209), $D = 0$, and $\theta_1 = kt - \pi/6$ rad, where $k = 2\pi$ rad/s. The motion of the mechanism takes place in a horizontal plane, and friction in the system can be neglected. Each bar of the mechanism is homogeneous, with mass density equal to 0.75 g/cm, and may be approximated as a slender line. For that mechanism, $L_1 = 40$ cm, $L_2 = 100$ cm, and the mass of the block at B is 1 kg. The frictionless pin that connects bar AB to the block at B passes through the center of mass of the block.

Determine the reaction forces acting on each bar of the mechanism and on the plane of the motion, the reaction moments (if any), and the external moment M_{ext} (magnitude and direction) applied to bar OA (clearly indicating them in a free-body diagram and showing the details of each step of the calculations) immediately after bar OA turns two complete revolutions. Treat the block as a point mass. *Note:* The problem has two solutions, one for $x_B > 0$ and another for $x_B < 0$.

5.23 Repeat Problem 5.21 for the case when the mechanism is vertical.

5.24 Repeat Problem 5.22 for the case when the mechanism is vertical. Automate the solution of your equations in a Simulink block diagram, display all your answers, and generate a plot of the external moment M_{ext} during the entire time interval. Include an animation of the mechanism in the simulation. The simulation should stop automatically at the end of the specified time. What is the maximum value of the magnitude of the moment M_{ext}? Make sure that your answers have appropriate units, and clearly specify those units for each answer.

5.25 Bar AB of the mechanism in Problem 3.27, shown in Fig. 3.34 (p. 223), is driven by a motor that applies to it a moment M_{ext} so that $\theta_2(t) = \pi/2 + k_1 \sin(k_2 t)$ rad, where $k_1 = 3\pi$ rad and $k_2 = 1$ rad/s. For that mechanism, $L_1 = 1$ m, $L_2 = 0.4$ m, $|\vec{CP}| = 3$ m, $L_4 = |\vec{CD}| = 1.5$ m, and $\alpha = 30°$. Motion takes place on a horizontal plane, and friction can be neglected. Each bar is uniform and may be approximated as a line with density $\rho = 0.75$ g/cm. Determine the reaction forces acting on each bar of the mechanism and on the plane of the motion, the reaction moments (if any), and the external moment M_{ext} when $t = 20$ s.

5.26 Repeat Problem 5.25 when the mechanism is in a vertical plane.

5.27 In the system shown in Fig. 5.42 a thin homogeneous bar of mass m_{bar} and length L is able to rotate about a pin located at a fixed point O. Two bodies (approximated as point masses), each of mass m kg, are attached to the ends of the bar, and those ends are connected by linear springs, each of stiffness k N/m and unstressed length L_u, to the fixed points P and Q. The springs are unstretched when $\theta = 0$. A motor applies a moment M_{motor} to the bar, and there is viscous friction at the pin contact at O, with friction coefficient c N · m/(rad/s). The motion of the system takes place in a horizontal plane, and the moment M_{motor} is a known function of time.

  a. Obtain the differential equation of motion for the system using the angle θ to describe the motion. Notice that the equations may be obtained either by the vectorial formulation of Newton and Euler or by the energy method. Use whatever one you choose.

  b. With $M_{\text{motor}} = 0$, determine *all* the equilibrium values θ_e of the angle θ for $h = L$.

  c. Analyze the stability of each equilibrium obtained in part b by linearizing the differential equation of motion about each equilibrium. What is the frequency of the small undamped oscillations of the system about each stable equilibrium?

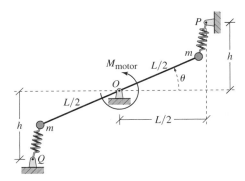

Figure 5.42 Figure for Problem 5.27.

5.28 Repeat Problem 5.27 if the plane of Fig. 5.42 is *vertical* and the motion takes place in that vertical plane. Disregard friction in the system. Does the mechanical energy $T + U$ of this system remain constant for all times? Explain. What is the expression for U? In addition, consider the case when the motor moment is a function of θ, instead of being a known function of time, i.e., $M_{\text{motor}} = Kf(\theta)$

where $K =$ constant. What is the expression for U in this case, and what can be said about the change of $T + U$ with time?

5.29 The system shown in Fig. 5.43 consists of a rigid homogeneous T-beam of dimensions L, $L/4$, and $L/4$, where $L = 70$ cm, with a pin support at point A of the T. Motion takes place on a horizontal plane. An external force is applied at A to make that point move with a constant absolute velocity of magnitude $|\vec{v}_A| = 2$ m/s, along a straight track on the horizontal plane. The torsional spring that is attached to the pin at A and to the beam is linear with stiffness $k = 3$ N·m/rad. The spring is unstressed when $\theta = 0$. Determine the frequency of the undamped oscillations of the beam about its equilibrium position by approximating its two parts as thin bars of density $\rho = 0.075$ kg/m.

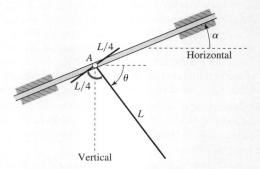

Figure 5.43 Figure for Problems 5.29 to 5.32.

5.30 Repeat Problem 5.29 if the smaller part of the T-beam is a homogeneous thin bar of constant density $\rho_1 = 0.075$ kg/m and the larger part of length L has variable density $\rho_2 = 0.05 + 0.03r/L$ kg/m, with r being distance, measured from point A, along that part of the beam.

5.31 The system shown in Fig. 5.43 consists of a rigid homogeneous T-beam of dimensions L, $L/4$, and $L/4$, where $L = 70$ cm, with a pin support at point A of the T. An external force is applied at A to make that point move with a constant absolute velocity of magnitude $|\vec{v}_A| = 2$ m/s, along a straight track that is inclined $\alpha = 30°$ with the horizontal. Motion of the beam takes place in a vertical plane. The torsional spring that is attached to the pin at A and to the beam is linear with stiffness k N·m/rad, and k is such that the beam is in equilibrium when $\theta \stackrel{\Delta}{=} \theta_e = 30°$. The spring is unstressed when $\theta \stackrel{\Delta}{=} \theta_u = 0$. Determine the frequency of the small oscillations of the beam about its equilibrium position by approximating its two parts as thin bars of density ρ kg/m. Neglect friction in the support at A.

5.32 Repeat Problem 5.31 if the smaller part of the T-beam is a homogeneous thin bar of constant density $\rho_1 = 0.075$ kg/m, and the larger part of length L has variable density $\rho_2 = 0.05 + 0.03r/L$ kg/m, with r being distance, measured from point A, along that part of the beam.

5.33 A thin homogeneous rectangular plate of dimensions $a = 30$ cm and $b = 2a$ is welded to a thin homogeneous triangular plate of dimensions a, b, and $c = \sqrt{a^2 + b^2}$. Both plates are made of the same material and have the same

Figure 5.44 Figure for Problem 5.33.

thickness. The system is pinned to a fixed point A as shown in Fig. 5.44, and motion takes place in the vertical plane, which is the plane of the figure. Neglecting friction, obtain the differential equation of motion of the system in terms of the angle θ, the equilibrium values (in degrees) for θ, and the frequency of the small oscillations about the stable equilibrium of the system. Disregard the thickness of the plates in your calculations.

5.34 A thin bar AB of length $L = 1.8$ m and mass m_1 is welded to a thin triangular plate of dimensions $a = 30$ cm, $b = 45$ cm, $c = \sqrt{a^2 + b^2}$, and mass $m_2 = 3m_1$, and the system is pinned to a fixed point A as shown in Fig. 5.45. The motion of the system takes place in the vertical plane, which is the plane of the figure. Neglecting friction, obtain the differential equation of motion of the system in terms of the angle θ, the equilibrium solutions (in degrees) for θ, and the frequency of the small oscillations about the stable equilibrium of the system.

5.35 The system shown in Fig. 5.46 consists of a homogeneous disk of mass m_1 and radius R and a homogeneous rectangular plate of dimensions a and $b > a$ that is pinned, with a frictionless pin, at the center of one of its sides to the center C of the disk. The disk rolls without slipping on a plane that is inclined at an angle α

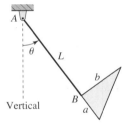

Figure 5.45 Figure for Problem 5.34.

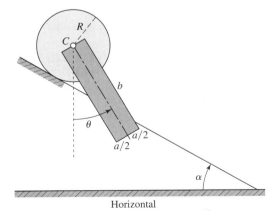

Figure 5.46 Figure for Problem 5.35.

with the horizontal, and motion of the system takes place in the vertical plane. For $m_1 = m_2$, $b = 2a$, and $\alpha = 30°$, determine

a. The equilibrium values $\theta = \text{constant} \triangleq \theta_e$, in degrees, and the corresponding acceleration of the center of mass C of the disk.

b. The frequency of the small oscillations of the plate about its stable equilibrium θ_e for $b = 35$ cm and $b = 25$ cm. Does the frequency increase or decrease with the dimension b?

5.36 The system shown in Fig. 5.47 consists of a thin block of mass m, with a pin through its center of mass C that is able to slide along a vertical circular groove of radius R. The system also has a spring that is attached to a fixed point O and to C so that the distance from O to C is equal to $2R$ when line OC is horizontal. The spring is linear with stiffness k N/m, and it is unstressed when $\phi = 0$ (i.e., when C is along the vertical). The block can rotate about C, in the vertical plane, as indicated by the angle θ shown in the figure. Neglect friction in the system.

a. Obtain the differential equations of motion of the system and, based on such equations, explain what $\theta(t)$ does if the motion is started with the initial conditions $\theta(0) = \theta_0$ (arbitrary), $\dot{\theta}(0) = 20$ rpm, and with any $\phi(0)$ and any $\dot{\phi}(0)$.

b. Verify that $\phi = 0$ and $\phi = 180°$ are two of the equilibrium solutions. There is a third equilibrium solution $\phi = \text{constant} \triangleq \phi_e$. Plot it versus $kR/(mg)$ (where $g = 9.81$ m/s^2 is the acceleration of gravity) for ϕ_e in the range $-60° \leq \phi_e \leq 60°$.

c. When $kR/(mg) = \sqrt{3} + \sqrt{2}$, $\phi_e = 30°$ is one equilibrium of the system. Determine all the other equilibrium solutions ϕ_e and analyze the stability of all of them (including the equilibrium $\phi_e = 30°$), specifying whether they are stable, asymptotically stable, or unstable.

d. For $kR/(mg) = \sqrt{3} + \sqrt{2}$, plot $d\phi/d\tau$ versus $\phi(\tau)$, where $\tau = t\sqrt{g/R}$ is nondimensional time, and identify in your plot the region of $\phi(0)$ and $[d\phi/d\tau]_{\tau=0}$ that result in an oscillatory motion $\phi(t)$ about $\phi_e = 0$.

Suggestion: It is simpler to use the approach $dW = d(T)$, instead of $\vec{F} = m\vec{a}_C$.

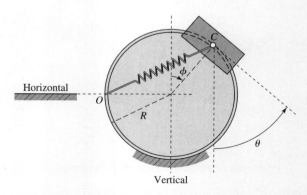

Figure 5.47 Figure for Problem 5.36.

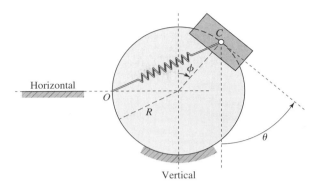

Figure 5.48 Figure for Problem 5.37.

5.37 The system shown in Fig. 5.48 consists of a thin block of mass m, with a pin through its center of mass C that is able to slide along a vertical circular plate of radius R. The block may leave the surface of the plate because no provision is made to keep it there at all times. A spring is attached from a fixed point O to C, and the distance from O to C is equal to $2R$ when line OC is horizontal. The spring is linear with stiffness k N/m, is unstressed when $\phi = 0$ (i.e., when C is along the vertical), and contacts are approximated as frictionless. The block can rotate about C, in the vertical plane, as indicated by the angle θ shown in the figure.

a. With g being the acceleration of gravity, show that the normal reaction force N applied by the plate to the block is (the direction of the force is left for you to determine)

$$\frac{N}{mg} = 3\cos\phi - 2H + \frac{kR}{mg}\left(5 + 3\sin\phi - 5\sqrt{1+\sin\phi}\right)$$

where (with $\tau = t\sqrt{g/R}$)

$$H = \frac{1}{2}\left(\frac{d\phi}{d\tau}\right)^2 + \cos\phi + \frac{kR}{mg}\left(\sqrt{1+\sin\phi} - 1\right)^2 = \text{constant} = [H]_{\tau=0}$$

b. With $kR/(mg) = 5.5$, $\phi(0) = 30°$, and $\dot\phi(0) = 0$, integrate numerically the differential equation of motion stopping the integration automatically if the block starts to leave the plate, and plotting $\phi(\tau)$ and $N/(mg)$ versus τ and $d\phi/d\tau$ versus $\phi(\tau)$. Then, repeat the integration using slightly larger values of $\phi(0)$ in the range $30° < \phi(0) \leq 40°$. Observe the results of the integration, and write a few paragraphs with your comments and conclusions. If the block ever leaves the surface of the plate, determine the value of $\phi(0)$ for which that happens.

5.38 The body shown in Fig. 5.49 consists of a homogeneous half cylinder of radius R, mass m, and negligible thickness, that can roll on a horizontal surface. Motion takes place in a vertical plane (i.e., the body neither tips over nor rotates laterally)

Figure 5.49 Figure for Problem 5.38.

and without slipping. Show that the center of mass C of the body is located at a distance equal to $4R/(3\pi)$ from the flat portion of the body and that its moment of inertia I_C about the axis that passes through C and is perpendicular to the plane of the body is equal to $I_C = [\frac{1}{2} - 16/(9\pi^2)]mR^2$. Using these results, obtain the differential equation of motion for the body in terms of the angle θ shown in Fig. 5.49. Verify that your differential equation exhibits the equilibrium $\theta = \text{constant} \triangleq \theta_e = 0$ and that the equilibrium is stable. For $R = 1$ m, what is the frequency of the small oscillations of the system about its equilibrium?

CHAPTER 6

Introduction to Dynamics of Rigid Bodies in General Motion

The general motion of a rigid body consists of the translation of a chosen reference point on the body, and a rotation of the body about an axis that passes through that reference point. In general, the direction of that axis is also changing with time, and, for that reason, such an axis is called an *instantaneous axis of rotation*. The simpler case when the motion is in two-dimensional space was dealt with in Chapter 5. Two-dimensional motion involves only three degrees of freedom, namely, two for the motion of any reference point on the body (normally taken to be the center of mass of the body) and one that describes the rotational motion (i.e., change in angular orientation) of the body. Three-dimensional motion involves six degrees of freedom; i.e., six variables are necessary to completely describe the position and the orientation of the body in space. Three such variables locate the position of a reference point of the body (again, normally taken to be the center of mass of the body), and the other three specify the angular orientation of the body in space. The basic vector equations that govern the motion are the same as Eqs. (5.3.2) and (5.3.5). Those equations govern the motion of the center of mass of the body and the rotational motion, respectively.

For rotational motion in three-dimensional space, the absolute angular momentum and absolute angular velocity of the body are not necessarily parallel to each other. For such a motion, the expression for the angular momentum involves three *mass moments of inertia* and three *mass products of inertia* for the body. These are essential quantities that appear in the formulation of the differential equations that govern the most general rotational motion of a body.

Section 6.1 introduces a mathematical operation with vectors that was not presented in Chapter 1. The operation is the *juxtaposition* of two vectors side by side to form what is called a *dyadic* (or second-order *tensor*). The matrix representation of a tensor is also presented and used in this chapter. The use of dyadics is optional (since one could work, instead, with the matrix representation of a dyadic), but they are convenient quantities that naturally appear when the explicit expression for the angular momentum of a body is formulated. The studying of Section 6.1 may be skipped for now since the use of dyadics appears for the first time in the book in Eq. (6.3.7) (p. 340). If you skip Section 6.1 for now, return to study it when you encounter that equation in Section 6.3.

Several examples involving kinematics and dynamics of rigid bodies in three dimensions are included in this chapter. Also included are several classical problems involving rotational motion in three-dimensional space.

6.1 Juxtaposition of Two Vectors: Dyadics

To see how dyadics appear naturally, consider two parallel vectors $\vec{v}$ and $\vec{c}$. Since the vectors are parallel to each other, they are related by some scalar quantity k as $\vec{v} = k\vec{c}$. Now, express k as a dot product of two other vectors, say $\vec{a}$ and $\vec{b}$, i.e., $k = \vec{a} \bullet \vec{b}$. Thus, we can always write $\vec{v} = (\vec{a} \bullet \vec{b})\vec{c}$. The right-hand side of this expression simply reads, "a scalar times $\vec{c}$," which results in a vector, of course. If the parentheses are left out, such an expression is written as $\vec{v} = \vec{a} \bullet \vec{b}\vec{c}$, and we may now stare at the quantity $\vec{b}\vec{c}$ and ask, "What is that?" The quantity $\vec{b}\vec{c}$, which is just the juxtaposition of two vectors, i.e., two vectors written side by side, is called a *dyad*. If we have, for example, the vectors $2\vec{v}_1$ and $3\vec{v}_2$, their juxtaposition is, by definition, $6\vec{v}_1\vec{v}_2$; i.e., the scalars associated with them are, by definition of the juxtaposition operation, multiplied by each other.

It is seen that the dot product of a vector with a dyad, or vice versa, yields a vector. Notice that $\vec{a} \bullet (\vec{b}\vec{c}) \neq (\vec{b}\vec{c}) \bullet \vec{a}$ in general. Also notice that $\vec{b}\vec{c} \neq \vec{c}\vec{b}$. The use of dyads is especially convenient in rigid body kinematics for transforming the inertia matrix of a body from a set of axes into another (see Example 6.3.5, p. 349).

A *dyadic* (which is also called a *second-order tensor*) is a quantity defined as a sum of dyads, such as $\vec{a}_1\vec{a}_2 + \vec{b}_1\vec{b}_2 + \cdots$. In this book dyadics are denoted by a letter with two arrows over it, such as $\vec{\vec{c}}$.

Since any vector $\vec{a}$ may be written in terms of three components in any triad $\{\hat{e}_1, \hat{e}_2, \hat{e}_3\}$ as $\vec{a} = a_{e_1}\hat{e}_1 + a_{e_2}\hat{e}_2 + a_{e_3}\hat{e}_3$, and since $a_{e_i} = \vec{a} \bullet \hat{e}_i$ for $i = 1, 2, 3$, it follows that

$$\vec{a} = (\vec{a} \bullet \hat{e}_1)\hat{e}_1 + (\vec{a} \bullet \hat{e}_2)\hat{e}_2 + (\vec{a} \bullet \hat{e}_3)\hat{e}_3$$
$$= \vec{a} \bullet (\hat{e}_1\hat{e}_1 + \hat{e}_2\hat{e}_2 + \hat{e}_3\hat{e}_3) \stackrel{\Delta}{=} \vec{a} \bullet \hat{U} \quad (6.1.1)$$

The quantity $\hat{U} \stackrel{\Delta}{=} \hat{e}_1\hat{e}_1 + \hat{e}_2\hat{e}_2 + \hat{e}_3\hat{e}_3$ is called the *unit dyadic*. Notice that $\vec{a} \bullet \hat{U} = \hat{U} \bullet \vec{a} = \vec{a}$. Also, if $\vec{b}$ is any other vector expressed in the same $\{\hat{e}_1, \hat{e}_2, \hat{e}_3\}$ triad as $\vec{a}$, i.e., $\vec{b} = b_{e_1}\hat{e}_1 + b_{e_2}\hat{e}_2 + b_{e_3}\hat{e}_3$, it is seen that any dyadic $\vec{a}\vec{b}$ has nine elements, as follows:

$$\vec{a}\vec{b} = (a_{e_1}\hat{e}_1 + a_{e_2}\hat{e}_2 + a_{e_3}\hat{e}_3)(b_{e_1}\hat{e}_1 + b_{e_2}\hat{e}_2 + b_{e_3}\hat{e}_3)$$
$$= a_{e_1}b_{e_1}\hat{e}_1\hat{e}_1 + a_{e_1}b_{e_2}\hat{e}_1\hat{e}_2 + a_{e_1}b_{e_3}\hat{e}_1\hat{e}_3$$
$$+ a_{e_2}b_{e_1}\hat{e}_2\hat{e}_1 + a_{e_2}b_{e_2}\hat{e}_2\hat{e}_2 + a_{e_2}b_{e_3}\hat{e}_2\hat{e}_3$$
$$+ a_{e_3}b_{e_1}\hat{e}_3\hat{e}_1 + a_{e_3}b_{e_2}\hat{e}_3\hat{e}_2 + a_{e_3}b_{e_3}\hat{e}_3\hat{e}_3 \quad (6.1.2)$$

The dot product of a dyadic $\vec{\vec{I}} = I_{e_1 e_1} \hat{e}_1 \hat{e}_1 + \cdots + I_{e_3 e_3} \hat{e}_3 \hat{e}_3$ with a vector $\vec{a} = a_{e_1} \hat{e}_1 + a_{e_2} \hat{e}_2 + a_{e_3} \hat{e}_3$ is another vector $\vec{b} = \vec{\vec{I}} \bullet \vec{a}$ that is obtained as follows:

$$\vec{b} = \vec{\vec{I}} \bullet \vec{a} = \left(I_{e_1 e_1} a_{e_1} + I_{e_1 e_2} a_{e_2} + I_{e_1 e_3} a_{e_3} \right) \hat{e}_1$$
$$+ \left(I_{e_2 e_1} a_{e_1} + I_{e_2 e_2} a_{e_2} + I_{e_2 e_3} a_{e_3} \right) \hat{e}_2$$
$$+ \left(I_{e_3 e_1} a_{e_1} + I_{e_3 e_2} a_{e_2} + I_{e_3 e_3} a_{e_3} \right) \hat{e}_3 \quad \quad (6.1.3)$$

As suggested by Eqs. (6.1.2) and (6.1.3), all nine elements of the dyadic $\vec{\vec{I}}$ may be represented by a matrix $[I]$ as

$$[I] = \begin{bmatrix} I_{e_1 e_1} & I_{e_1 e_2} & I_{e_1 e_3} \\ I_{e_2 e_1} & I_{e_2 e_2} & I_{e_2 e_3} \\ I_{e_3 e_1} & I_{e_3 e_2} & I_{e_3 e_3} \end{bmatrix} \quad \quad (6.1.4)$$

Thus, in matrix notation, the representation of the dot product $\vec{b} = \vec{\vec{I}} \bullet \vec{a}$ given by Eq. (6.1.3), with both $\vec{a}$ and $\vec{b}$ expressed in the same unit vector triad $\{e\} = \{\hat{e}_1, \hat{e}_2, \hat{e}_3\}$, is simply obtained by a matrix multiplication as follows:

$$\underline{b} \triangleq \begin{bmatrix} b_{e_1} \\ b_{e_2} \\ b_{e_3} \end{bmatrix} = [I] \underline{a} = \begin{bmatrix} I_{e_1 e_1} & I_{e_1 e_2} & I_{e_1 e_3} \\ I_{e_2 e_1} & I_{e_2 e_2} & I_{e_2 e_3} \\ I_{e_3 e_1} & I_{e_3 e_2} & I_{e_3 e_3} \end{bmatrix} \begin{bmatrix} a_{e_1} \\ a_{e_2} \\ a_{e_3} \end{bmatrix} \quad \quad (6.1.5)$$

Notice that the *matrix representation of a vector* is a column matrix whose elements are the scalar components of the vector. A column matrix is represented in this book with an underlined symbol, such as

$$\underline{v} = \begin{bmatrix} v_1 \\ v_2 \\ v_3 \end{bmatrix} = [v_1 \quad v_2 \quad v_3]^T$$

with the superscript T denoting the transpose operation.

A dyadic $\vec{\vec{I}}$ is said to be symmetric if its matrix representation $[I]$ is symmetric, i.e, if $I_{ij} = I_{ji}$.

6.2 The Two Basic Vector Equations of Rigid Body Dynamics

Figure 6.1 illustrates an arbitrary rigid body subjected to a set of forces $\vec{F}_1, \vec{F}_2, \ldots$. Point C is the center of mass of the body, O is an inertial point, and A is a point that may move in any manner in space.

The body may or may not be given an initial motion, and the basic dynamics problem consists of determining the subsequent motion of the body, i.e., its position and its angular orientation in space at all times. As mentioned in the introduction to this chapter, motion in three-dimensional space involves six degrees of freedom; i.e., six variables are needed to describe the motion. Three such

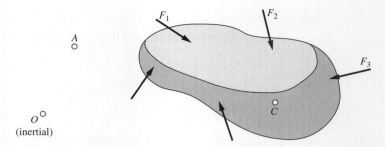

Figure 6.1 A rigid body subjected to a set of forces (C is the center of mass of the body).

variables describe the translation of a reference point that is either on the body or on a rigid extension of the body (such as the case of a doughnut, for example). The other three variables describe the rotational motion of the body, i.e., motion that is characterized by the change of orientation of a set of axes fixed to the body.

The basic vector equations that describe the general motion of a rigid body are the same as those already obtained in Chapter 4 and used in Chapter 5. They are the following.

■ For the motion of the center of mass (see Fig. 6.1):

$$\vec{F} \triangleq \sum_{i=1}^{N} \vec{F}_i = m \frac{d}{dt} \vec{v}_C = m \frac{d^2}{dt^2} \overrightarrow{OC} \tag{6.2.1}$$

■ For the rotational motion of the body:

$$\frac{d}{dt} \vec{H}_A = \vec{M}_A + m \left(\frac{d}{dt} \overrightarrow{AC} \right) \times \left(\frac{d}{dt} \overrightarrow{OC} \right) \tag{6.2.2}$$

where $\vec{M}_A$ is the moment, about point A (see Fig. 6.1) of all forces applied to the body, and $\vec{H}_A$ is the absolute angular momentum (referred to point A) of the body. The absolute angular momentum $\vec{H}_A$ was defined in Chapters 4 and 5. For a rigid body it is the integral over the body (i.e., a triple integral that spans the entire volume of the body) of the moment about point A of the absolute linear momentum of a mass element dm of the body (where dm is equal to the body mass density times a volume element of the body). The integral over the body assures that the mass element dm spans the entire body. Mathematically, $\vec{H}_A$ is defined as follows (where P is the location of an infinitesimal element of the body, with mass dm):

$$\vec{H}_A \triangleq \int_{\text{body}} \left(\overrightarrow{AP} \times \frac{d}{dt} \overrightarrow{OP} \right) dm \tag{6.2.3}$$

Equation (6.2.2) reduces to the simpler forms as follows, when point A is either an inertial point O or the center of mass C of the body.

$$\frac{d}{dt}\vec{H}_O = \vec{M}_O \qquad \text{(when } O \text{ is inertial)} \qquad (6.2.4)$$

$$\frac{d}{dt}\vec{H}_C = \vec{M}_C \qquad \text{(when } C \text{ is the center of mass)} \qquad (6.2.5)$$

These are the forms that are used in this book. When point A is an inertial point O, $\overrightarrow{AP}$ is, of course, $\overrightarrow{OP}$ in Eq. (6.2.3). Similarly, $\overrightarrow{AP} = \overrightarrow{CP}$ when point A is the center of mass C.

To apply either Eq. (6.2.5) or Eq. (6.2.4) to specific problems, an explicit expression for the angular momentum $\vec{H}_C$ (or $\vec{H}_O$) needs to be developed. This is done in Section 6.3. Several problems are then presented afterward.

6.3 Rotational Kinematics of Rigid Bodies in General Motion

As will be seen in this section, the absolute angular momentum of a rigid body depends on the angular velocity $\vec{\omega}^B$ of the body with respect to inertial space, and on a set of quantities that are called moments of inertia and products of inertia of the body. It will also be seen that the angular momentum vector and the angular velocity vector are not necessarily parallel to each other, as was the case for the simpler type of motion considered in Chapter 5. To analyze the rotational motion of the body it is necessary to determine the relationship between the absolute angular momentum and the absolute angular velocity of the body before Eq. (6.2.5) or Eq. (6.2.4) can be used for solving problems. This is done in Section 6.3.1.

6.3.1 Absolute Angular Momentum

Consider a rigid body undergoing general motion. As illustrated in Fig. 6.2, its center of mass C is translating in space, and the body is also rotating with absolute (i.e., relative to inertial space) angular velocity $\vec{\omega}^B(t)$. Both the magnitude and

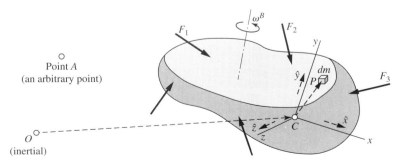

Figure 6.2 A rigid body able to move in space with six degrees of freedom.

the direction of $\vec{\omega}^B(t)$ may be changing with time. The x-y-z axes shown in that figure are centered at C, but are not necessarily fixed to the body. The unit vectors $\hat{x}$, $\hat{y}$, and $\hat{z}$ are parallel to the x, y, and z axes, respectively. If one chooses the x-y-z axes not to be fixed to the rigid body, their absolute angular velocity is different from $\vec{\omega}^B$ and the coordinates of any vector $\overrightarrow{CP}$ change with time (but the magnitude of $\overrightarrow{CP}$ still remains constant because the body is rigid).

The absolute angular momentum $\vec{H}_A$ of the body, referred to any point A, is given by Eq. (6.2.3). By letting point A in Eq. (6.2.3) be the center of mass C, that equation yields the following expression for the absolute angular momentum $\vec{H}_C$:

$$\vec{H}_C = \int_{\text{body}} \left(\overrightarrow{CP} \times \frac{d}{dt} \overrightarrow{OP} \right) dm$$

$$= \underbrace{\left(\int_{\text{body}} \overrightarrow{CP}\, dm \right)}_{=0} \times \frac{d}{dt} \overrightarrow{OC} + \int_{\text{body}} \left(\overrightarrow{CP} \times \frac{d}{dt} \overrightarrow{CP} \right) dm \quad (6.3.1)$$

The first integral in Eq. (6.3.1) is equal to zero because of the definition of the center of mass.

Note: For any other point A, the angular momentum $\vec{H}_A$ is the sum of the angular momentum $\vec{H}_C$ and the moment, about A, of the linear momentum $m\vec{v}_C$ (where $\vec{v}_C$ is the absolute velocity of C) of the body, i.e.,

$$\vec{H}_A = \vec{H}_C + \overrightarrow{AC} \times (m\vec{v}_C)$$

This relationship between $\vec{H}_A$ and $\vec{H}_C$ is obtained as follows:

$$\vec{H}_A = \int_{\text{body}} \left(\overrightarrow{AP} \times \frac{d}{dt} \overrightarrow{OP} \right) dm = \int_{\text{body}} \left(\overrightarrow{AC} \times \frac{d}{dt} \overrightarrow{OP} \right) dm$$

$$= \int_{\text{body}} \left(\overrightarrow{AC} \times \frac{d}{dt} \overrightarrow{OC} \right) dm + \int_{\text{body}} \left(\overrightarrow{AC} \times \frac{d}{dt} \overrightarrow{CP} \right) dm$$

$$+ \underbrace{\int_{\text{body}} \left(\overrightarrow{CP} \times \frac{d}{dt} \overrightarrow{OP} \right) dm}_{=\vec{H}_C}$$

$$= \overrightarrow{AC} \times \underbrace{\frac{d}{dt} \overrightarrow{OC}}_{=\vec{v}_C} \underbrace{\left(\int_{\text{body}} dm \right)}_{=m} + \overrightarrow{AC} \times \left(\frac{d}{dt} \underbrace{\int_{\text{body}} \overrightarrow{CP}\, dm}_{=0} \right) + \vec{H}_C$$

$$= \vec{H}_C + \overrightarrow{AC} \times (m\vec{v}_C)$$

SECTION 6.3 Rotational Kinematics of Rigid Bodies in General Motion

The vector identities $\overrightarrow{AP} = \overrightarrow{AC} + \overrightarrow{CP}$ and $\overrightarrow{OP} = \overrightarrow{OC} + \overrightarrow{CP}$ used in this equation are obtained by inspection of Fig. 6.2. Also, the last integral that appears in the equation is equal to zero because of the definition of the center of mass C.

To develop a more useful expression for the angular momentum $\vec{H}_C$ from Eq. (6.3.1), it is convenient to express the vector $\overrightarrow{CP}$ and the absolute angular velocity $\vec{\omega}^B$ of the body in terms of orthogonal unit vectors $\hat{x}$, $\hat{y}$, and $\hat{z}$ that are parallel to the x, y, and z axes, respectively, centered at C as shown in Fig. 6.2. At this point of this development, those axes are not necessarily fixed to the body, and they may be chosen to be rotating in any manner whatsoever. It will be seen later on, however, that it is advantageous to choose such axes to be fixed to the rotating rigid body. In general, the expressions for $\overrightarrow{CP}$ and for $\vec{\omega}^B$ are of the following forms:

$$\overrightarrow{CP} \triangleq x\hat{x} + y\hat{y} + z\hat{z} \tag{6.3.2}$$

$$\vec{\omega}^B \triangleq \omega_x \hat{x} + \omega_y \hat{y} + \omega_z \hat{z} \tag{6.3.3}$$

Since the magnitude of the vector $\overrightarrow{CP}$ from C to an element of mass dm (located at point P in Fig. 6.2) is constant, the time derivative of the vector $\overrightarrow{CP}$ that appears in Eq. (6.3.1) for the angular momentum $\vec{H}_C$ is then $d\overrightarrow{CP}/dt = \vec{\omega}^B \times \overrightarrow{CP}$ (because the angular velocity $\vec{\omega}^B$ is what causes the direction of $\overrightarrow{CP}$ to change).

The cross product $\overrightarrow{CP} \times (\vec{\omega}^B \times \overrightarrow{CP})$ is then

$$\overrightarrow{CP} \times (\vec{\omega}^B \times \overrightarrow{CP}) = \overrightarrow{CP} \times \begin{vmatrix} \hat{x} & \hat{y} & \hat{z} \\ \omega_x & \omega_y & \omega_z \\ x & y & z \end{vmatrix}$$

$$= \begin{vmatrix} \hat{x} & \hat{y} & \hat{z} \\ x & y & z \\ \omega_y z - \omega_z y & \omega_z x - \omega_x z & \omega_x y - \omega_y x \end{vmatrix}$$

$$= [(y^2 + z^2)\omega_x - xy\omega_y - xz\omega_z]\hat{x}$$
$$+ [-yx\omega_x + (x^2 + z^2)\omega_y - yz\omega_z]\hat{y}$$
$$+ [-zx\omega_x - zy\omega_y + (x^2 + y^2)\omega_z]\hat{z} \tag{6.3.4}$$

By substituting this expression for $\overrightarrow{CP} \times (\vec{\omega}^B \times \overrightarrow{CP})$ into Eq. (6.3.1), the following expression is then obtained for the angular momentum $\vec{H}_C$:

$$\vec{H}_C = (I_{xx}\omega_x + I_{xy}\omega_y + I_{xz}\omega_z)\hat{x}$$
$$+ (I_{yx}\omega_x + I_{yy}\omega_y + I_{yz}\omega_z)\hat{y}$$
$$+ (I_{zx}\omega_x + I_{zy}\omega_y + I_{zz}\omega_z)\hat{z}$$
$$\triangleq H_x\hat{x} + H_y\hat{y} + H_z\hat{z} \tag{6.3.5}$$

where

$$I_{xx} = \int_{\text{body}} (y^2 + z^2)\, dm \qquad I_{yy} = \int_{\text{body}} (x^2 + z^2)\, dm \qquad I_{zz} = \int_{\text{body}} (x^2 + y^2)\, dm$$

$$I_{xy} = I_{yx} = -\int_{\text{body}} xy\, dm \qquad I_{xz} = I_{zx} = -\int_{\text{body}} xz\, dm \qquad I_{yz} = I_{zy} = -\int_{\text{body}} yz\, dm$$

(6.3.6)

The expression for the angular momentum given in Eq. (6.3.5) may be written in the following form, which resembles the one that was obtained in Chapter 5 for the case of simpler planar motion of rigid bodies.

$$\boxed{\vec{H}_C = \vec{\bar{I}}_C \bullet \vec{\omega}^B} \qquad (6.3.7)$$

In this equation, the quantity $\vec{\bar{I}}_C$ is called the *inertia dyadic,* or *inertia tensor* of the body, and it is equal to the following nine-element dyadic:

$$\begin{aligned}\vec{\bar{I}}_C = {}& I_{xx}\hat{x}\hat{x} + I_{xy}\hat{x}\hat{y} + I_{xz}\hat{x}\hat{z} \\ & + I_{yx}\hat{y}\hat{x} + I_{yy}\hat{y}\hat{y} + I_{yz}\hat{y}\hat{z} \\ & + I_{zx}\hat{z}\hat{x} + I_{zy}\hat{z}\hat{y} + I_{zz}\hat{z}\hat{z}\end{aligned} \qquad (6.3.8)$$

The relationship between the angular momentum $\vec{H}_C$ and the absolute angular velocity $\vec{\omega}^B$ of the body may also be represented in matrix form as follows:

$$\begin{bmatrix} H_x \\ H_y \\ H_z \end{bmatrix} = [I_C] \begin{bmatrix} \omega_x \\ \omega_y \\ \omega_z \end{bmatrix}$$

or, in short notation,

$$\boxed{\underline{H}_C = [I_C]\underline{\omega}^B} \qquad (6.3.9)$$

where $\underline{H}_C \triangleq [H_x \ H_y \ H_z]^T$, $\underline{\omega}^B = [\omega_x \ \omega_y \ \omega_z]^T$, and $[I_C]$ is the following *symmetric* 3×3 matrix,

$$[I_C] = \begin{bmatrix} I_{xx} & I_{xy} & I_{xz} \\ I_{yx} & I_{yy} & I_{yz} \\ I_{zx} & I_{zy} & I_{zz} \end{bmatrix} \qquad (6.3.10)$$

which is called the *inertia matrix* of the body, referred to the center of mass C.

> The inertia matrix $[I_C]$ is the *matrix representation* of the inertia dyadic $\vec{\vec{I}}_C$. Whether one uses dyadics or matrices and whether one writes the angular momentum as $\vec{H}_C = \vec{\vec{I}}_C \bullet \vec{\omega}^B$ or $\underline{H}_C = [I_C]\underline{\omega}^B$ is entirely a matter of individual preference. But one should never equate $\vec{H}_C$ to $\underline{H}_C$ because the former is a vector $\vec{H}_C = H_x\hat{x} + H_y\hat{y} + H_z\hat{z}$ and the latter is the column matrix $\underline{H}_C = [H_x \ H_y \ H_z]^T$ (which is the *matrix representation* of the vector $\vec{H}_C$).

- The elements I_{xx}, I_{yy}, and I_{zz} that appear in the diagonal of the inertia matrix are called the *mass moments of inertia* (or, for short, the *moments of inertia*) of the body about the axes x, y, and z, respectively.
- The quantities I_{xy}, I_{xz}, and I_{yz} defined by the last three equations of (6.3.6) are called the *products of inertia* of the body with respect to the x-y-z axes.
- Notice that each one of the quantities $y^2 + z^2$, $x^2 + z^2$, and $x^2 + y^2$ that appear in the definitions of the moments of inertia are simply the squares of the perpendicular distances from the mass element dm to the axes x, y, and z, respectively. Although the value of each one of the moments of inertia of the body depends on the particular orientation of the x-y-z axes, it is of interest to notice that the sum $I_{xx} + I_{yy} + I_{zz}$ does not since, as disclosed by the equations of (6.3.6),

$$I_{xx} + I_{yy} + I_{zz} = 2\int_{\text{body}} (x^2 + y^2 + z^2)\,dm = 2\int_{\text{body}} \overrightarrow{CP} \bullet \overrightarrow{CP}\,dm$$

$$= 2\int_{\text{body}} |\overrightarrow{CP}|^2\,dm \qquad (6.3.11)$$

where $|\overrightarrow{CP}|$ is the constant distance from C to the mass element dm. This is one of the properties of any inertia matrix.

Inspection of the equations of (6.3.6), which define the moments and products of inertia, also reveals the two properties of the inertia matrix.

Properties of the Inertia Matrix

- The equations of (6.3.6) also disclose that any of the three moments of inertia of a body is never greater than the sum of the other two. This is an important property of inertia matrices. For example, $I_{zz} \leq I_{xx} + I_{yy}$ (since $I_{xx} + I_{yy} = I_{zz} + 2\int_{\text{body}} z^2\,dm$ and the integral term is never negative), with the equal sign holding for a flat body for which $z = 0$.
- Also notice that $I_{xx} \pm 2I_{yz} = \int_{\text{body}} (y^2 + z^2 \mp 2yz)\,dm = \int_{\text{body}} (y \mp z)^2\,dm$ is never negative. The other two similar combinations involve I_{yy} and I_{xz},

and also I_{zz} and I_{xy}. Therefore, the elements of any inertia matrix must also satisfy the following conditions:

$$I_{xx} \pm 2I_{yz} \geq 0 \qquad I_{yy} \pm 2I_{xz} \geq 0 \qquad I_{zz} \pm 2I_{xy} \geq 0$$

■ The other properties of the inertia matrix are given in Appendix C.

As mentioned earlier, the quantities I_{xy}, I_{xz}, and I_{yz} defined by the last three equations of (6.3.6) are called the *products of inertia* of the body with respect to the chosen x, y, and z axes. Depending on the orientation of such axes, each one of those quantities may be positive, negative, or equal to zero.

■ Notice that the negative signs that you see in Eq. (6.3.4) were just embedded in the definitions of products of inertia given by the last three equations of (6.3.6). Many authors (perhaps most) define products of inertia in this manner, and several others do not embed that negative sign in their definitions. Actually, there is no universally adopted sign convention for the definition of products of inertia, and this can cause confusion in engineering practice. If one defines I_{xy} as $\int_{\text{body}} xy\, dm$ (with similar sign definitions for I_{xz} and I_{yz}), the off-diagonal elements of the matrix $[I_C]$ in Eq. (6.3.9) for the angular momentum will be the negative of such products of inertia, instead of being the products of inertia themselves. However, regardless of the manner in which they are defined, the actual values for products of inertia may be positive, negative, or zero, while none of the moments of inertia of a body can be negative.

If the x-y-z axes are not fixed to the rigid body, the moments and products of inertia of the body referred to those axes will, in general, change with time as the body rotates, which is inconvenient. To formulate and analyze dynamics problems, it is convenient to work with axes for which the moments and products of inertia are constants.

It is shown in Section 6.3.4 that it is always possible to find an orientation for any set of orthogonal x-y-z axes, centered at any point on the body, so that the three products of inertia of the body are zero when referred to such axes. Such axes are called the *principal axes of inertia* of the body. Thus, if the x-y-z axes are principal axes of inertia, the expression for the angular momentum $\vec{H}_C$ given by Eq. (6.3.5) reduces to the simpler expression $\vec{H}_C = I_{xx}\omega_x \hat{x} + I_{yy}\omega_y \hat{y} + I_{zz}\omega_z \hat{z}$, which is another advantage of working with such axes once they are located. The moments of inertia of a body about its principal axes are called the *principal moments of inertia* (or, more precisely, the *central principal moments of inertia*).

6.3.2 The Parallel Axis Theorem

Some applications require calculation of the moments and products of inertia of a body about two sets of orthogonal axes, one of which is centered at the center of mass of the body, and with each axis of one set being parallel to a corresponding axis of the other set. Once the moments and products of inertia of the body relative

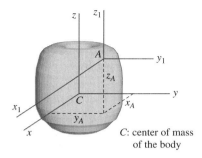

Figure 6.3 Figure for calculations involving the parallel axes theorem.

to one such set of axes are known, the moments and products of inertia relative to the other set of axes can be calculated in a simple manner. The relationships between the inertias of the body about those two sets of orthogonal axes are called the *parallel axis theorem*. This is presented in this section; for an application, see Example 6.3.2.

Figure 6.3 shows a rigid body and two sets of parallel axes; the x-y-z axes are centered at the center of mass C of the body, and the x_1-y_1-z_1 axes are centered at some other point A, whose rectangular coordinates relative to C are x_A, y_A, and z_A.

The elements of the inertia matrix $[I_C]$ about the axes centered at the center of mass C are calculated by performing the integrations shown in the equations of (6.3.6). The elements $I_{x_1 x_1}$, $I_{x_1 y_1}$, etc., of the inertia matrix $[I_A]$ about the axes centered at the origin A of the x_1-y_1-z_1 axes are calculated in the same manner, with x, y, and z replaced by x_1, y_1, and z_1, respectively. The steps involved in the calculation of $I_{x_1 x_1}$ and $I_{x_1 y_1}$ are presented in the following. The calculation of the remaining moments and products of inertia follows the same procedure, and the results are similar to the ones obtained for $I_{x_1 x_1}$ and $I_{x_1 y_1}$.

By letting $x = x_A + x_1$, $y = y_A + y_1$, and $z = z_A + z_1$ (see Fig. 6.3) we obtain:

$$I_{x_1 x_1} = \int_{body} (y_1^2 + z_1^2) \, dm = \int_{body} [(y - y_A)^2 + (z - z_A)^2] \, dm$$

$$= \underbrace{\int_{body} (y^2 + z^2) \, dm}_{=I_{xx}} + m(y_A^2 + z_A^2) - 2 y_A \underbrace{\int_{body} y \, dm}_{=0} - 2 z_A \underbrace{\int_{body} z \, dm}_{=0}$$

$$I_{x_1 y_1} = I_{y_1 x_1} = -\int_{body} x_1 y_1 \, dm = -\int_{body} (x - x_A)(y - y_A) \, dm$$

$$= \underbrace{-\int_{body} xy \, dm}_{=I_{xy}} - m x_A y_A + y_A \underbrace{\int_{body} x \, dm}_{=0} + x_A \underbrace{\int_{body} y \, dm}_{=0}$$

The last two integrals in the expressions for $I_{x_1 x_1}$ and $I_{x_1 y_1}$ are zero because of the definition of the center of mass C. They would not be zero if the x-y-z axes were not centered at the center of mass of the body.

These results can be summarized, with similar expressions involving z as follows[1]:

$$I_{x_1 x_1} = I_{xx} + m(y_A^2 + z_A^2) \qquad (6.3.12)$$

$$I_{x_1 y_1} = I_{xy} - m x_A y_A \qquad (6.3.13)$$

> The relationship given by Eq. (6.3.12) between the moment of inertia I_{xx} about an x axis centered at the center of mass, and the moment of inertia $I_{x_1 x_1}$ about an x_1 axis parallel to the x axis, is of the same form as Eq. (5.3.8) (p. 272). The quantity $m(y_A^2 + z_A^2)$ is simply the product of the mass of the body and the square of the perpendicular distance between the two parallel axes x and x_1.
>
> Equation (6.3.13) is the parallel axis theorem for products of inertia. The quantity $m x_A y_A$ that appears in the expression for $I_{x_1 y_1}$ is simply the product of the mass of the body and the x-y rectangular coordinates of the new origin A relative to the center of mass C. Thus, for $I_{y_1 z_1}$, for example, one would have $I_{y_1 z_1} = I_{yz} - m y_A z_A$.

6.3.3 Examples Involving Moments and Products of Inertia

EXAMPLE 6.3.1

A homogeneous bar of mass m is bent as shown in Fig. 6.4. The dimensions of the cross section of the bar are very small and may be neglected. For the x-y-z axes shown, which are centered at the center of mass C of the bar and with the z axis being perpendicular to the plane of the figure, determine all the elements of the inertia matrix $[I_C]$ for the bent bar.

Figure 6.4 The bar for Example 6.3.1.

---

[1] Equations (6.3.12) and (6.3.13) would also have been obtained if one used, instead, $x = x_A - x_1$, $y = y_A - y_1$, and $z = z_A - z_1$.

■ Solution

Since the bar is homogeneous, its density (i.e., its mass per unit length) is equal to $\rho = m/(L + 2\pi R)$. Therefore, by dividing the bar into the segments AB, BD, and DE shown in Fig. 6.4, the masses of the infinitesimal mass elements for each segment are equal to $dm_1 = \rho R\, d\theta_1$, $dm_2 = \rho R\, d\theta_2$ and $dm_3 = \rho\, dx$, respectively. With $z = 0$, we then obtain

$$I_{xx} = \iiint y^2\, dm = \int_{\theta_1=0}^{\pi} (R + R\cos\theta_1)^2 \rho R\, d\theta_1 + \int_{\theta_2=0}^{\pi} (R + R\cos\theta_2)^2 \rho R\, d\theta_2$$

$$= 2\int_{\theta=0}^{\pi} (R + R\cos\theta)^2 \rho R\, d\theta = 3\pi\rho R^3$$

$$I_{yy} = \iiint x^2\, dm = 2\int_{\theta=0}^{\pi} \left(\frac{L}{2} + R\sin\theta\right)^2 \rho R\, d\theta + \int_{x=-L/2}^{L/2} x^2 \rho\, dx$$

$$= \rho R\left[\left(R^2 + \frac{L^2}{2}\right)\pi + 4LR\right] + \rho\frac{L^3}{12}$$

$$I_{zz} = I_{xx} + I_{yy}$$

$$I_{xy} = -\iiint xy\, dm = -2\int_{\theta=0}^{\pi} \left(\frac{L}{2} + R\sin\theta\right)(R + R\cos\theta)\rho R\, d\theta = -\rho R^2(\pi L + 4R)$$

$$I_{xz} = I_{yz} = 0$$

EXAMPLE 6.3.2

Figure 6.5 shows a homogeneous steel triangular plate ABD of mass $m = 52$ kg and sides $L_1 = 3$ m and $L_2 = 4.5$ m. Determine the moments and products of inertia of the plate about the x-y-z axes and the x_1-y_1-z_1 axes shown in the figure. The x-y-z axes are centered at the center of mass C of the plate. The thickness of the plate, which is 2 mm, is small enough to be neglected in the calculations.

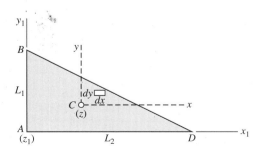

Figure 6.5 The triangular plate for Example 6.3.2.

■ **Solution**

As shown in Example 5.3.2 (p. 274), the center of mass of this triangular plate is located at $x_1 = L_2/3 = 1.5$ m and $y_1 = L_1/3 = 1$ m relative to the corner A. To calculate the inertias, let us use the infinitesimal mass element with mass $dm = [2m/(L_1 L_2)]\,dx\,dy = [2m/(L_1 L_2)]\,dx_1\,dy_1$, shown in Fig. 6.5. Since the boundary BD of the plate is the straight line $y_1 = L_1 - L_1 x_1/L_2$, and $z_1 = z = 0$, we then obtain for the inertias relative to the x_1-y_1-z_1 axes:

$$I_{x_1 x_1} = \frac{2m}{L_1 L_2} \int_{x_1=0}^{L_2} \int_{y_1=0}^{L_1 - L_1 x_1/L_2} y_1^2\, dy_1\, dx_1 = \frac{1}{6} m L_1^2 = 78 \text{ kg} \cdot \text{m}^2$$

$$I_{y_1 y_1} = \frac{2m}{L_1 L_2} \int_{x_1=0}^{L_2} \int_{y_1=0}^{L_1 - L_1 x_1/L_2} x_1^2\, dy_1\, dx_1 = \frac{1}{6} m L_2^2 = 175.5 \text{ kg} \cdot \text{m}^2$$

$$I_{z_1 z_1} = \frac{2m}{L_1 L_2} \int_{x_1=0}^{L_2} \int_{y_1=0}^{L_1 - L_1 x_1/L_2} \left(x_1^2 + y_1^2\right) dy_1\, dx_1 = I_{x_1 x_1} + I_{y_1 y_1}$$

$$I_{x_1 y_1} = -\frac{2m}{L_1 L_2} \int_{x_1=0}^{L_2} \int_{y_1=0}^{L_1 - L_1 x_1/L_2} x_1 y_1\, dy_1\, dx_1 = -\frac{1}{12} m L_1 L_2 = -58.5 \text{ kg} \cdot \text{m}^2$$

$$I_{x_1 z_1} = I_{z_1 x_1} = 0 \qquad I_{y_1 z_1} = I_{z_1 y_1} = 0 \qquad \text{(because } z_1 = 0\text{)}$$

The moments and products of inertia referred to the x-y-z axes centered at the center of mass C may be calculated in one of the following manners to obtain the given results.

■ Let $x_1 = L_2/3 + x$ and $y_1 = L_1/3 + y$ in the earlier integrals and proceed as done in Section 6.3.2.
■ Simply notice that the rectangular coordinates of A relative to C are $x_A = -L_2/3$, $y_A = -L_1/3$, $z_A = 0$, and use the simple and general results obtained in Section 6.3.2.

$$I_{x_1 x_1} = I_{xx} + m y_A^2 \quad \therefore \quad I_{xx} = \frac{1}{18} m L_1^2 = 26 \text{ kg} \cdot \text{m}^2$$

$$I_{y_1 y_1} = I_{yy} + m x_A^2 \quad \therefore \quad I_{yy} = \frac{1}{18} m L_2^2 = 58.5 \text{ kg} \cdot \text{m}^2$$

$$I_{z_1 z_1} = I_{zz} + m \left(x_A^2 + y_A^2\right) \quad \therefore \quad I_{zz} = \frac{1}{18} m \left(L_1^2 + L_2^2\right) = 84.5 \text{ kg} \cdot \text{m}^2$$

$$I_{x_1 y_1} = I_{xy} - m x_A y_A \quad \therefore \quad I_{xy} = \frac{1}{36} m L_1 L_2 = 19.5 \text{ kg} \cdot \text{m}^2$$

$$I_{x_1 z_1} = I_{z_1 x_1} = I_{xz} - m x_A z_A = 0$$

$$I_{y_1 z_1} = I_{z_1 y_1} = I_{yz} - m y_A z_A = 0$$

EXAMPLE 6.3.3

Figure 6.6 shows a homogeneous rectangular plate of mass m and dimensions a, b, and L. Determine the moments and products of inertia of the plate about the x-y-z axes that are parallel to the sides of the plate and are centered at its center of mass C.

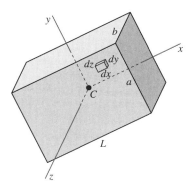

Figure 6.6 The rectangular plate for Example 6.3.3.

■ Solution

Since the plate is homogeneous, its mass density (mass per unit volume) is equal to $\rho = m/(abL)$. Therefore, the mass of the infinitesimal volume element with sides dx, dy, and dz, shown in Fig. 6.6, is $\rho\,dx\,dy\,dz$. The moment of inertia I_{xx} and the product of inertia I_{xy} are then determined as follows:

$$I_{xx} = \int_{z=-b/2}^{b/2} \int_{y=-a/2}^{a/2} \int_{x=-L/2}^{L/2} (y^2 + z^2)\,\rho\,dx\,dy\,dz = Lb\left[\frac{y^3}{3}\right]_{y=-a/2}^{a/2} + La\left[\frac{z^3}{3}\right]_{z=-b/2}^{b/2}$$

$$= \underbrace{\frac{\rho Lab}{12}}_{=m/12}(a^2 + b^2) = \frac{m}{12}(a^2 + b^2)$$

$$I_{xy} = -\int_{z=-b/2}^{b/2} \int_{y=-a/2}^{a/2} \int_{x=-L/2}^{L/2} \rho xy\,dx\,dy\,dz = 0$$

By inspecting Fig. 6.6 and these results, the following can be inferred:

$$I_{yy} = \frac{m}{12}(L^2 + b^2)$$

$$I_{zz} = \frac{m}{12}(L^2 + a^2)$$

$$I_{xz} = I_{yz} = 0$$

EXAMPLE 6.3.4

The inertia matrix $[I]_{x_1,x_2,x_3}$ of a body, referred to a set of orthogonal axes x_1, x_2, x_3, is given below. Determine the inertia matrix of the body referred to a set of orthogonal axes y_1, y_2, y_3 that are rotated θ degrees about the x_3 axis.

$$[I]_{x_1,x_2,x_3} = \begin{bmatrix} 4 & -3 & 0 \\ -3 & 5 & 0 \\ 0 & 0 & 9 \end{bmatrix} \text{ kg} \cdot \text{m}^2$$

■ **Solution**

As shown in the following, this type of calculation is readily done if dyadics and the transformations of triads presented in Section 1.2 are used. Two sets of unit vectors, one parallel to the x_1-x_2-x_3 axes and the other parallel to the y_1-y_2-y_3 axes, are shown in Fig. 6.7. Using those unit vectors, the dyadic representation $\vec{\vec{I}}$ of the given inertia matrix is expressed as follows:

Figure 6.7 The triads $\{\hat{x}_1, \hat{x}_2, \hat{x}_3\}$ and $\{\hat{y}_1, \hat{y}_2, \hat{y}_3\}$ for Example 6.3.4.

$$\vec{\vec{I}} = I_{x_1x_1}\hat{x}_1\hat{x}_1 + I_{x_1x_2}\hat{x}_1\hat{x}_2 + I_{x_1x_3}\hat{x}_1\hat{x}_3 + I_{x_2x_1}\hat{x}_2\hat{x}_1 + I_{x_2x_2}\hat{x}_2\hat{x}_2 + I_{x_2x_3}\hat{x}_2\hat{x}_3 + I_{x_3x_1}\hat{x}_3\hat{x}_1$$
$$+ I_{x_3x_2}\hat{x}_3\hat{x}_2 + I_{x_3x_3}\hat{x}_3\hat{x}_3 = 4\hat{x}_1\hat{x}_1 - 3\hat{x}_1\hat{x}_2 - 3\hat{x}_2\hat{x}_1 + 5\hat{x}_2\hat{x}_2 + 9\hat{x}_3\hat{x}_3$$

Inspection of Fig. 6.7 discloses the following relations between the unit vectors x_1, x_2, x_3 and y_1, y_2, y_3:

$$\hat{x}_1 = \hat{y}_1 \cos\theta - \hat{y}_2 \sin\theta$$
$$\hat{x}_2 = \hat{y}_1 \sin\theta + \hat{y}_2 \cos\theta$$
$$\hat{x}_3 = \hat{y}_3$$

By making use of these relations, the inertia dyadic $\vec{\vec{I}}$ becomes

$$\vec{\vec{I}} = 4(\hat{y}_1\cos\theta - \hat{y}_2\sin\theta)(\hat{y}_1\cos\theta - \hat{y}_2\sin\theta) - 3(\hat{y}_1\cos\theta - \hat{y}_2\sin\theta)(\hat{y}_1\sin\theta + \hat{y}_2\cos\theta)$$
$$- 3(\hat{y}_1\sin\theta + \hat{y}_2\cos\theta)(\hat{y}_1\cos\theta - \hat{y}_2\sin\theta)$$
$$+ 5(\hat{y}_1\sin\theta + \hat{y}_2\cos\theta)(\hat{y}_1\sin\theta + \hat{y}_2\cos\theta) + 9\hat{y}_3\hat{y}_3$$

Or, by carrying out the multiplications such as $(\cos\theta)\cos\theta$, $(\cos\theta)\sin\theta$, etc., we then obtain

$$\vec{\vec{I}} = [4\cos^2\theta - 6(\cos\theta)\sin\theta + 5\sin^2\theta]\hat{y}_1\hat{y}_1 + [3\sin^2\theta - 3\cos^2\theta + (\cos\theta)\sin\theta]\hat{y}_1\hat{y}_2$$
$$+ [3\sin^2\theta - 3\cos^2\theta + (\cos\theta)\sin\theta]\hat{y}_2\hat{y}_1$$
$$+ [4\sin^2\theta + 6(\cos\theta)\sin\theta + 5\cos^2\theta]\hat{y}_2\hat{y}_2 + 9\hat{y}_3\hat{y}_3$$

This expression discloses that the elements of the inertia matrix of the body, referred to the y_1-y_2-y_3 axes, are

$I_{y_1 y_1} = 4\cos^2\theta - 6(\cos\theta)\sin\theta + 5\sin^2\theta$	$I_{y_1 y_2} = 3\sin^2\theta - 3\cos^2\theta + (\cos\theta)\sin\theta$	$I_{y_1 y_3} = 0$
$I_{y_2 y_1} = I_{y_1 y_2}$	$I_{y_2 y_2} = 4\sin^2\theta + 6(\cos\theta)\sin\theta + 5\cos^2\theta$	$I_{y_2 y_3} = 0$
$I_{y_3 y_1} = 0$	$I_{y_3 y_2} = 0$	$I_{y_3 y_3} = 9$

EXAMPLE 6.3.5

A rigid body has the following moments and products of inertia (in kg·m²) about a set of orthogonal axes x, y, z that are centered at a point A of the body and are fixed to the body:

$$I_{xx} = 3 \qquad I_{xy} = \sqrt{2} \qquad I_{xz} = -\sqrt{1.5}$$
$$I_{yy} = 4 \qquad I_{yz} = -1 \qquad I_{zz} = 6$$

The body is rigidly mounted on a horizontal shaft as shown in Fig. 6.8a. Point A is in the middle of the shaft and is inertial, and the angle θ between x and x_s (and between y and y_s) is constant. The x_s axis shown in the figure is aligned with the shaft, while the y_s and z_s axes (with $z_s = z$) are perpendicular to the shaft.

a. Determine the inertia matrix of the body about the x_s-y_s-z_s axes for $\theta = 30°$.
b. Determine the angular momentum $\vec{H}_A$ of the body if the shaft is spun counterclockwise with an angular velocity of 120 rpm about the x_s axis. Express the answer in terms of the unit vectors $\hat{x}_s$, $\hat{y}_s$, and $\hat{z}_s$ shown in Fig. 6.8b.
c. What is the time derivative of $\vec{H}_A$ if the only moment applied to the body is an internal moment $\vec{M} = M_{xs}\hat{x}_s + M_{ys}\hat{y}_s + M_{zs}\hat{z}_s$ N·m due to the contact with the shaft?

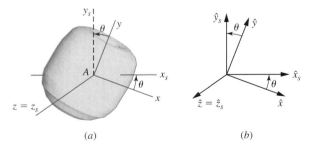

Figure 6.8 A simple transformation of triads for Example 6.3.5.

■ Solution

Figure 6.8b shows a set of unit vectors that will be used in the following calculations.

a. As shown in Example 6.3.4, this type of calculation is readily done if we use dyadics and the transformations of triads presented in Section 1.2. Start by writing the following expression for the inertia dyadic $\vec{\vec{I}}_A$ for the body, based on the given information:

$$\vec{\vec{I}}_A = I_{xx}\hat{x}\hat{x} + I_{xy}\hat{x}\hat{y} + I_{xz}\hat{x}\hat{z} + I_{yx}\hat{y}\hat{x} + I_{yy}\hat{y}\hat{y} + I_{yz}\hat{y}\hat{z} + I_{zx}\hat{z}\hat{x} + I_{zy}\hat{z}\hat{y} + I_{zz}\hat{z}\hat{z}$$
$$= 3\hat{x}\hat{x} + \sqrt{2}\hat{x}\hat{y} - \sqrt{1.5}\hat{x}\hat{z} + \sqrt{2}\hat{y}\hat{x} + 4\hat{y}\hat{y} - \hat{y}\hat{z} - \sqrt{1.5}\hat{z}\hat{x} - \hat{z}\hat{y} + 6\hat{z}\hat{z}$$

Now, express the unit vectors $\hat{x}$, $\hat{y}$, and $\hat{z}$ in terms of the unit vectors $\hat{x}_s$, $\hat{y}_s$, and $\hat{z}_s$ that are parallel to the x_s-y_s-z_s axes. As indicated in Chapter 1, the relationships between these two sets of unit vectors are obtained by inspecting Fig. 6.8b. The following relations are written by such an inspection:

$$\hat{x} = \hat{x}_s \cos\theta - \hat{y}_s \sin\theta = \frac{\sqrt{3}}{2}\hat{x}_s - \frac{1}{2}\hat{y}_s$$

$$\hat{y} = \hat{x}_s \sin\theta + \hat{y}_s \cos\theta = \frac{1}{2}\hat{x}_s + \frac{\sqrt{3}}{2}\hat{y}_s$$

$$\hat{z} = \hat{z}_s$$

When these relationships are substituted into the expression for $\vec{\vec{I}}_A$ and all the multiplications are performed [such as $3\hat{x}\hat{x} = 3(\hat{x}_s \cos\theta - \hat{y}_s \sin\theta)(\hat{x}_s \cos\theta - \hat{y}_s \sin\theta) = (3\cos^2\theta)\hat{x}_s\hat{x}_s - 3(\sin\theta)(\cos\theta)\hat{x}_s\hat{y}_s - 3(\sin\theta)(\cos\theta)\hat{y}_s\hat{x}_s + (3\sin^2\theta)\hat{y}_s\hat{y}_s$, etc.], an expression of the following type is obtained for the inertia dyadic $\vec{\vec{I}}_A$ in terms of the unit vectors $\hat{x}_s$, $\hat{y}_s$, and $\hat{z}_s$:

$$\vec{\vec{I}}_A = I_{x_s x_s}\hat{x}_s\hat{x}_s + I_{x_s y_s}\hat{x}_s\hat{y}_s + I_{x_s z_s}\hat{x}_s\hat{z}_s + I_{y_s x_s}\hat{y}_s\hat{x}_s + I_{y_s y_s}\hat{y}_s\hat{y}_s + I_{y_s z_s}\hat{y}_s\hat{z}_s$$
$$+ I_{z_s x_s}\hat{z}_s\hat{x}_s + I_{z_s y_s}\hat{z}_s\hat{y}_s + I_{z_s z_s}\hat{z}_s\hat{z}_s$$

The expressions that are obtained for all the moments and products of inertia $I_{x_s x_s}$, $I_{x_s y_s}$, etc., and their values for $\theta = 30°$, are as follows:

$$I_{x_s x_s} = I_{xx}\cos^2\theta + 2I_{xy}(\sin\theta)\cos\theta + I_{yy}\sin^2\theta \approx 4.47 \text{ kg} \cdot \text{m}^2$$

$$I_{x_s y_s} = (I_{yy} - I_{xx})(\sin\theta)\cos\theta + I_{xy}(\cos^2\theta - \sin^2\theta) \approx 1.14 \text{ kg} \cdot \text{m}^2$$

$$I_{x_s z_s} = I_{xz}\cos\theta + I_{yz}\sin\theta \approx -1.56 \text{ kg} \cdot \text{m}^2$$

$$I_{y_s x_s} = I_{x_s y_s}$$

$$I_{y_s y_s} = I_{xx}\sin^2\theta - 2I_{xy}(\sin\theta)\cos\theta + I_{yy}\cos^2\theta \approx 2.53 \text{ kg} \cdot \text{m}^2$$

$$I_{y_s z_s} = I_{yz}\cos\theta - I_{xz}\sin\theta \approx -0.25 \text{ kg} \cdot \text{m}^2$$

$$I_{z_s x_s} = I_{x_s z_s}$$

$$I_{z_s y_s} = I_{y_s z_s}$$

$$I_{z_s z_s} = I_{zz} = 6 \text{ kg} \cdot \text{m}^2$$

The algebraic steps that lead to these expressions are left for you, the reader, to do.

b. The angular velocity $\vec{\omega}^B$ of the body is the same as the angular velocity of the shaft because the two are welded to each other. Therefore, $\vec{\omega}^B = 120\hat{x}_s$ rpm $= 120 \times (2\pi/60)\hat{x}_s$ rad/s $= 4\pi\hat{x}_s$ rad/s. In terms of the unit vectors $\hat{x}_s$, $\hat{y}_s$, and $\hat{z}_s$, the angular momentum $\vec{H}_A$ is then

$$\vec{H}_A = \vec{\vec{I}}_A \bullet \vec{\omega}^B = \left(I_{x_s x_s}\hat{x}_s\hat{x}_s + I_{x_s y_s}\hat{x}_s\hat{y}_s + I_{x_s z_s}\hat{x}_s\hat{z}_s + I_{y_s x_s}\hat{y}_s\hat{x}_s + I_{y_s y_s}\hat{y}_s\hat{y}_s + I_{y_s z_s}\hat{y}_s\hat{z}_s \right.$$
$$\left. + I_{z_s x_s}\hat{z}_s\hat{x}_s + I_{z_s y_s}\hat{z}_s\hat{y}_s + I_{z_s z_s}\hat{z}_s\hat{z}_s\right) \bullet (4\pi\hat{x}_s)$$
$$= 4\pi\left(I_{x_s x_s}\hat{x}_s + I_{y_s x_s}\hat{y}_s + I_{z_s x_s}\hat{z}_s\right)$$
$$\approx 56.23\hat{x}_s + 14.33\hat{y}_s - 19.61\hat{z}_s \text{ kg} \cdot \text{m}^2/\text{s}$$

c. Since point A is inertial, it follows that (see Chapter 5 or Section 6.2) $d\vec{H}_A/dt = \vec{M}_A = \vec{M}$. Therefore, no calculations are needed to answer the question in this item!

6.3.4 Principal Axes and Principal Moments of Inertia

It was mentioned in Section 6.3.1 that it is always possible to find the orientation for any set of orthogonal x-y-z axes so that the three products of inertia of the body are zero when referred to such axes. Such axes are called the *principal axes of inertia* of the body.

The principal axes of inertia of a body are, in general, fixed to the body. As will be seen, the principal axes need not be fixed to the body when the body exhibits symmetries, as in the case of a homogeneous cylinder or a sphere for example. For the purpose of calculating the six inertias of a body, the x-y-z axes may be centered anywhere on the body. When they are centered at the center of mass of the body, the corresponding principal axes are called the *central principal axes of inertia*. It is frequent practice, however, to refer to the latter simply as the *principal axes* of a body, with the word *central* added at times for more clarity.

To locate the central principal axes of a body, use is made of Eq. (6.3.7) (p. 340), which relates the angular momentum $\vec{H}_C$ to the absolute angular velocity $\vec{\omega}^B$ of the body, i.e., $\vec{H}_C = \vec{\vec{I}}_C \bullet \vec{\omega}^B$. However, the calculations that will follow are more conveniently done if one works with the matrix form given by Eq. (6.3.9), i.e., $\underline{H}_C = [I_C]\underline{\omega}^B$, where $[I_C]$ is the symmetric 3×3 inertia matrix of the body (referred to the orthogonal x-y-z axes centered at C) as in Eq. (6.3.10). Recall that $\underline{H}_C$ is the 3×1 column matrix of the components of $\vec{H}_C$, i.e., $\underline{H}_C = [H_x \; H_y \; H_z]^T$, and that $\underline{\omega}^B$ is the 3×1 column matrix of the components of $\vec{\omega}^B$, i.e., $\underline{\omega}^B = [\omega_x \; \omega_y \; \omega_z]^T$.

In general, the angular momentum $\vec{H}_C$ and the angular velocity of the body, $\vec{\omega}^B$, are not parallel to each other since they are not necessarily proportional to each other. If the x-y-z axes were the principal axes of the body, the angular momentum $\vec{H}_C$ would be simply equal to $\vec{H}_C = I_{xx}\omega_x\hat{x} + I_{yy}\omega_y\hat{y} + I_{zz}\omega_z\hat{z}$. In such a case, if the body were given a spin about any of its principal axes, say the x axis, the resulting angular momentum would be a vector along that axis. In other words, if $\omega_y = \omega_z = 0$, the angular momentum would simply be a vector parallel to $\omega_x\hat{x}$ and equal to $\vec{H}_C = I_{xx}\omega_x\hat{x}$. This observation is the key for the mathematical determination of the direction of the principal axes of the body with respect to an arbitrarily chosen set of x-y-z axes.

To determine the direction of the central principal axes of a body, choose a set of orthogonal axes centered at the center of mass C of the body and calculate all the moments and products of inertia of the body relative to such axes. Next, consider the following question: If the body is given an angular velocity $\vec{\omega} = \omega_x \hat{x} + \omega_y \hat{y} + \omega_z \hat{z}$, what should be the direction of $\vec{\omega}$ so that the angular momentum of the body is parallel to $\vec{\omega}$? The answer to such a question gives the direction of a principal axis of the body. Since a vector parallel to $\vec{\omega}$ must be equal to $\lambda \vec{\omega}$, where λ is a scalar, the determination of the direction of any of the principal axes of a body is accomplished by finding the solution to the matrix equation $[I_C]\underline{\omega} = \lambda \underline{\omega}$, i.e.,

$$\begin{bmatrix} I_{xx} - \lambda & I_{xy} & I_{xz} \\ I_{yx} & I_{yy} - \lambda & I_{yz} \\ I_{zx} & I_{zy} & I_{zz} - \lambda \end{bmatrix} \begin{bmatrix} \omega_x \\ \omega_y \\ \omega_z \end{bmatrix} \triangleq [B] \begin{bmatrix} \omega_x \\ \omega_y \\ \omega_z \end{bmatrix} = \begin{bmatrix} 0 \\ 0 \\ 0 \end{bmatrix} \quad (6.3.14)$$

This is a classical eigenvalue-eigenvector problem of matrix algebra.

If the matrix $[B]$ has an inverse, the only solution to Eq. (6.3.14) is $\omega_x = \omega_y = \omega_z = 0$. For Eq. (6.3.14) to have a nonzero solution, the determinant of the matrix $[B]$ has to be equal to zero. This condition yields the following cubic polynomial equation for λ, which is called the *characteristic equation* for the matrix $[B]$:

$$\det \begin{bmatrix} I_{xx} - \lambda & I_{xy} & I_{xz} \\ I_{yx} & I_{yy} - \lambda & I_{yz} \\ I_{zx} & I_{zy} & I_{zz} - \lambda \end{bmatrix} \triangleq P(\lambda) = -[\lambda^3 - a_2 \lambda^2 + a_1 \lambda - a_0] = 0$$

$$(6.3.15)$$

In this equation,

$$a_2 = I_{xx} + I_{yy} + I_{zz} \quad (6.3.16)$$

$$a_1 = \begin{vmatrix} I_{xx} & I_{xy} \\ I_{yx} & I_{yy} \end{vmatrix} + \begin{vmatrix} I_{xx} & I_{xz} \\ I_{zx} & I_{zz} \end{vmatrix} + \begin{vmatrix} I_{yy} & I_{yz} \\ I_{zy} & I_{zz} \end{vmatrix} \quad (6.3.17)$$

$$a_0 = \det([I_C]) \quad (6.3.18)$$

Equation (6.3.15) is a cubic polynomial equation and, thus, admits three solutions $\lambda = \lambda_1$, $\lambda = \lambda_2$, and $\lambda = \lambda_3$. These solutions, which are called the *eigenvalues* of the inertia matrix $[I_C]$, are the central principal moments of inertia of the body.

When the solutions of the characteristic equation are found, each individual eigenvalue λ_i should be substituted into Eq. (6.3.14) in order to determine the components ω_x, ω_y, and ω_z of the vector $\vec{\omega}$ that are associated with each λ_i. The column matrix $[\omega_x \ \omega_y \ \omega_z]^T$ is an *eigenvector* of the inertia matrix $[I_C]$. However, as a consequence of Eq. (6.3.15), at most two of the three equations involving the quantities ω_x, ω_y, and ω_z obtained for each λ_i ($i = 1, 2, 3$) will be independent of each other. Thus, since we are left with more unknowns than equations to determine the components of $\vec{\omega}$, none of the eigenvectors $[\omega_x \ \omega_y \ \omega_z]^T$ are uniquely determined. In other words, what we are left with is an equation of a straight line in terms of the originally chosen x-y-z axes. But these straight lines, which are the central principal axes of the body, are precisely what we want and need. These straight lines may be specified by arbitrarily choosing a nonzero value for any of the components of $\vec{\omega}$, say ω_x, and solving for the other unknowns ω_y and ω_y in terms of ω_x. Any column matrix $[\omega_x \ \omega_y \ \omega_z]$ representing a vector $\omega_x \hat{x} + \omega_y \hat{y} + \omega_z \hat{z}$ along such a straight line is, of course, an eigenvector of the inertia matrix $[I_C]$. Although not necessary, these eigenvectors may be "normalized," if desired, so that their magnitude $\sqrt{\omega_x^2 + \omega_y^2 + \omega_z^2}$ is equal to 1. If this is done, the values obtained for ω_x, ω_y, and ω_z are the direction cosines of the principal axes of the body relative to the originally chosen x-y-z axes.

> It turns out that the eigenvectors of a symmetric matrix, corresponding to distinct eigenvalues, are orthogonal to each other. The proof of this and of other properties of the inertia matrix are presented in Appendix C. Thus, since the inertia matrix is symmetric, the principal axes of any body having three distinct principal moments of inertia are orthogonal to each other. In contrast, the angle between the principal directions corresponding to equal principal inertias is arbitrary. For example, for a homogeneous cylinder, the central principal axes are the axial axis and any transverse axis that passes through the center of mass of the cylinder. For a homogeneous sphere, any axis through the center of the sphere is a central principal axis. Also, if a homogeneous body has a plane of symmetry, any axis through its center of mass and perpendicular to that plane of symmetry is a central principal axis.

Several examples of the determination of principal axes are given in Section 6.3.5.

6.3.5 Examples of the Determination of Principal Axes and Principal Inertias

EXAMPLE 6.3.6

Determine the central principal moments of inertia and the direction of the principal axes of the triangular plate of Example 6.3.2 (p. 345).

■ **Solution**

The elements I_{xx}, I_{xy}, etc., of the inertia matrix for that plate were calculated in Example 6.3.2, and the inertia matrix $[I_C]$, relative to the center of mass C of the plate, was found to be the following:

$$[I_C] = \begin{bmatrix} I_{xx} & I_{xy} & I_{xz} \\ I_{yx} & I_{yy} & I_{yz} \\ I_{zx} & I_{zy} & I_{zz} \end{bmatrix} = \begin{bmatrix} 26 & 19.5 & 0 \\ 19.5 & 58.5 & 0 \\ 0 & 0 & 84.5 \end{bmatrix} \quad \textbf{(6.3.6a)}$$

The principal moments of inertia of the plate and the corresponding principal axes are determined by solving the eigenvector problem given by the following matrix equation:

$$\begin{bmatrix} 26 - \lambda & 19.5 & 0 \\ 19.5 & 58.5 - \lambda & 0 \\ 0 & 0 & 84.5 - \lambda \end{bmatrix} \begin{bmatrix} \omega_x \\ \omega_y \\ \omega_z \end{bmatrix} = \begin{bmatrix} 0 \\ 0 \\ 0 \end{bmatrix} \quad (6.3.6b)$$

The characteristic equation for Eq. (6.3.6b) is

$$(\lambda - 84.5)(\lambda^2 - 84.5\lambda + 1140.75) = 0$$

whose roots λ_1, λ_2, and λ_3 are

$$\lambda_1 = 84.5 \text{ kg} \cdot \text{m}^2$$

$$\lambda_2 = \frac{84.5 + \sqrt{84.5^2 - 4 \times 1140.75}}{2} \approx 67.63 \text{ kg} \cdot \text{m}^2$$

$$\lambda_2 = \frac{84.5 - \sqrt{84.5^2 - 4 \times 1140.75}}{2} \approx 16.87 \text{ kg} \cdot \text{m}^2$$

Notice that the z axis is a principal axis of the plate because $I_{xz} = I_{yz} = 0$. The two eigenvectors in the x-y plane, corresponding to the eigenvalues $\lambda = \lambda_2$ and $\lambda = \lambda_3$, are determined by using one of the two equations obtained from Eq. (6.3.6b), i.e., either from $(I_{xx} - \lambda)\omega_x + I_{xy}\omega_y = 0$ or from $(I_{yy} - \lambda)\omega_y + I_{yx}\omega_x = 0$. These two equations are actually the same since the determinant of the matrix shown in Eq. (6.3.6b) is equal to zero. For example, for $\lambda = 67.63$, they give $-41.63\omega_x + 19.5\omega_y = 0$ and $19.5\omega_x - 9.13\omega_y = 0$. Except for the round-off error in $\lambda \approx 67.63$, these equations are the same as $\omega_y - 2.135\omega_x = 0$.

By arbitrarily working with the first row of Eq. (6.3.6b), we obtain the following eigenvectors $\underline{\omega}_2$ and $\underline{\omega}_3$ corresponding to the eigenvalues λ_2 and λ_3, respectively:

For $\lambda = \lambda_2 = 67.63$: $\quad\quad\quad \underline{\omega}_2 = \omega_x \begin{bmatrix} 1 \\ 2.135 \end{bmatrix}$

For $\lambda = \lambda_3 = 16.87$: $\quad\quad\quad \underline{\omega}_3 = \omega_x \begin{bmatrix} 1 \\ -1/2.135 \end{bmatrix}$

The angle between the x axis shown in Fig. 6.5 (p. 345) and the principal axis for which the moment of inertia is equal to $\lambda_2 = 67.63$ kg$\cdot$m^2 is $\arctan(2.135) \approx 64.9°$. The angle between that same x axis and the third principal axis (for which the moment of inertia is $\lambda_3 = 16.87$ kg$\cdot$m^2) is $\arctan(-1/2.135) \approx -25.1°$.

Notice that the three principal axes of the triangular plate, which are associated with three different principal moments of inertia, are perpendicular to each other. As mentioned earlier, such a property of the inertia matrix is proved in Appendix C.

EXAMPLE 6.3.7

Determine the central principal moments of inertia of the homogeneous solid rotor of mass m, radius R, and height h shown in Fig. 6.9. Because of symmetry, the center of mass C of the rotor is located at its geometric center. The x-y-z axes shown in the figure,

SECTION 6.3 Rotational Kinematics of Rigid Bodies in General Motion

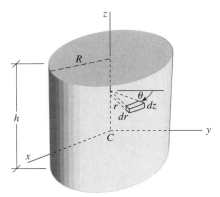

Figure 6.9 The solid rotor of Example 6.3.7.

which are centered at C, are the central principal axes of the rotor. The figure also shows a mass element of the rotor (conveniently using cylindrical coordinates) to be used in the calculations.

■ **Solution**
Because of the symmetry, the z axis shown in Fig. 6.9 is a principal axis, which means that $I_{zx} = I_{zy} = 0$. Also, any axis through the center of mass C, which is the geometric center of the rotor, and perpendicular to the z axis is a principal axis. This is confirmed by the results of the following calculations.

With the mass density of the rotor being equal to $\rho = m/(\pi R^2 h)$, the mass of the infinitesimal volume element shown in Fig. 6.9 is equal to $dm = \rho(r\, d\theta)\, dr\, dz$. Therefore, with $x = r \sin\theta$ and $y = r \cos\theta$, the principal moments of inertia of the rotor are obtained as follows:

$$I_{zz} = \frac{m}{\pi R^2 h} \int_{r=0}^{R} \int_{\theta=0}^{2\pi} \int_{z=-h/2}^{h/2} r^2 (r\, dz\, d\theta\, dr) = m\frac{R^2}{2}$$

$$I_{yy} = \frac{m}{\pi R^2 h} \int_{r=0}^{R} \int_{\theta=0}^{2\pi} \int_{z=-h/2}^{h/2} (r^2 \sin^2\theta + z^2) r\, dz\, d\theta\, dr$$

$$= \frac{m}{\pi R^2 h} \int_{r=0}^{R} r^3\, dr \int_{\theta=0}^{2\pi} \frac{h}{2}(1 - \cos 2\theta)\, d\theta + \frac{mh^2}{4} = m\left(\frac{R^2}{4} + \frac{h^2}{12}\right)$$

Because of the symmetry in the x-y plane, the x axis is really indistinguishable from the y axis and, thus, $I_{xx} = I_{yy}$.

For the product of inertia I_{xz} (which is equal to I_{yz} because the x axis is indistinguishable from the y axis) we have, with $x = r\sin\theta$,

$$I_{xz} = -\frac{m}{\pi R^2 h}\int_{r=0}^{R}\int_{z=-h/2}^{h/2}\int_{\theta=0}^{2\pi}(zr\sin\theta)r\,d\theta\,dz\,dr = 0$$

For the product of inertia I_{xy} we have, with $xy = (r\sin\theta)r\cos\theta$,

$$I_{xy} = -\frac{m}{\pi R^2 h}\left[\int_{r=0}^{R}\int_{z=-h/2}^{h/2}r^3\,dr\,dz\right]\times\underbrace{\int_{\theta=0}^{2\pi}\sin\theta\cos\theta\,d\theta}_{=\left[\frac{1}{4}(\sin^2\theta-\cos^2\theta)\right]_{\theta=0}^{2\pi}=0} = 0$$

EXAMPLE 6.3.8

A rigid body has the following moments and products of inertia (in kg · m²) about a set of orthogonal x-y-z axes centered at its center of mass:

$$I_{xx} = 3 \qquad I_{yy} = 4 \qquad I_{zz} = 6$$
$$I_{xy} = \sqrt{2} \qquad I_{xz} = -\sqrt{1.5} \qquad I_{yz} = -1$$

Determine the central principal moments of inertia of the body and the direction cosines of its central principal axes.

■ **Solution**

Let λ be the moment of inertia of the body about one of its principal axes, whose orientation is still to be determined. Imagine now that we spin the body about that axis, and let ω_x, ω_y, and ω_z denote, respectively, the components of the angular velocity $\vec{\omega}^B$ of the body along unit vectors $\hat{x}$, $\hat{y}$, and $\hat{z}$ that are parallel to those axes. If the spin axis is a principal axis, the angular momentum of the body $\vec{\vec{I}}_C \bullet \vec{\omega}^B$ (or $[I_C]\underline{\omega}^B$ in matrix form) must be parallel to its angular velocity, which means that it must be equal to $\lambda\vec{\omega}^B$ (or $\lambda\underline{\omega}^B$ in matrix notation). Therefore, to determine λ and the direction of a principal axis, we must solve the eigenvector problem given by the following matrix equation:

$$\begin{bmatrix} 3-\lambda & \sqrt{2} & -\sqrt{1.5} \\ \sqrt{2} & 4-\lambda & -1 \\ -\sqrt{1.5} & -1 & 6-\lambda \end{bmatrix}\begin{bmatrix} \omega_x \\ \omega_y \\ \omega_z \end{bmatrix} = \begin{bmatrix} 0 \\ 0 \\ 0 \end{bmatrix} \qquad (6.3.8a)$$

To have a nonzero solution for the angular velocity $\vec{\omega}^B$, the determinant of the matrix shown in Eq. (6.3.8a) must be equal to zero. This gives the following polynomial equation for λ:

$$\lambda^3 - 13\lambda^2 + 49.5\lambda - (51 + 2\sqrt{3}) = 0 \qquad (6.3.8b)$$

The following roots are obtained for Eq. (6.3.8b) using either the MATLAB command `roots[1, -13, 49.5, -(51+2*sqrt(3))]` or by a trial-and-error process (done with a pocket calculator):

$$\lambda_1 \approx 7.126 \qquad \lambda_2 \approx 3.928 \qquad \lambda_3 \approx 1.946$$

SECTION 6.3 Rotational Kinematics of Rigid Bodies in General Motion

These roots are the central principal moments of inertia of the body. Notice that $\lambda_1 + \lambda_2 + \lambda_3 = 13$, which is the *trace* of the inertia matrix (i.e., the sum of its diagonal elements). This provides a spot check on the validity of the numbers obtained for the three roots. One might also use, of course, approximations as $\lambda_1 \approx 7.13$, $\lambda_2 \approx 3.93$, and $\lambda_3 \approx 1.95$, which yields $\lambda_1 + \lambda_2 + \lambda_3 = 13.01$ instead of 13. The difference is simply due to round-off errors.

Let us denote the three central principal axes of the body by x_1 (corresponding to the moment of inertia λ_1), x_2 (corresponding to the moment of inertia λ_2), and x_3 (corresponding to the moment of inertia λ_3). To determine the direction of any of these axes, Eq. (6.3.8a) needs to be solved for ω_y and ω_z, say, in terms of ω_x for each one of the eigenvalues λ. Since the direction cosines of a line are the scalar components of a unit vector parallel to that line, it follows that the quantities ω_x, ω_y, and ω_z are the direction cosines of a principal axis if they are normalized such that $\sqrt{\omega_x^2 + \omega_y^2 + \omega_z^2} = 1$.

By arbitrarily choosing to work with the first two rows of the matrix shown in Eq. (6.3.8a), the following eigenvector $\underline{\omega}_1 = [\omega_{1x} \ \omega_{1y} \ \omega_{1z}]^T$ corresponding to the eigenvalue $\lambda_1 \approx 7.126$ is obtained:

$$\underline{\omega}_1 \approx \omega_x \begin{bmatrix} 1 \\ 1.117 \\ -2.078 \end{bmatrix}$$

Or, after normalization so that $(1 + 1.117^2 + 2.078^2)\omega_x^2 = 1$,

$$\underline{\omega}_1 \approx \begin{bmatrix} 0.390 \\ 0.436 \\ -0.811 \end{bmatrix} \triangleq \begin{bmatrix} \cos\theta_{1x} \\ \cos\theta_{1y} \\ \cos\theta_{1z} \end{bmatrix}$$

The angles this principal axis makes with the x, y, and z axes are, respectively, $\theta_{1x} \approx 67°$, $\theta_{1y} \approx 64.2°$, and $\theta_{1z} \approx 144.2°$. Notice that the components obtained for the eigenvector $\underline{\omega}_1$ do satisfy, within the accuracy of the numerical computations, the equation obtained from the third row of the matrix in Eq. (6.3.8a).

By repeating this procedure for the other two eigenvalues, the following normalized eigenvectors are obtained.

For the eigenvalue $\lambda = \lambda_2 \approx 3.928$:

$$\underline{\omega}_2 \approx \begin{bmatrix} 0.366 \\ 0.735 \\ 0.571 \end{bmatrix} \triangleq \begin{bmatrix} \cos\theta_{2x} \\ \cos\theta_{2y} \\ \cos\theta_{2z} \end{bmatrix}$$

For the eigenvalue $\lambda = \lambda_3 \approx 1.946$:

$$\underline{\omega}_2 \approx \begin{bmatrix} 0.845 \\ -0.520 \\ 0.126 \end{bmatrix} \triangleq \begin{bmatrix} \cos\theta_{3x} \\ \cos\theta_{3y} \\ \cos\theta_{3z} \end{bmatrix}$$

For the principal axis x_2 we obtain $\theta_{2x} \approx 68.5°$, $\theta_{2y} \approx 42.7°$, and $\theta_{2z} \approx 55.2°$, and for the principal axis x_3, $\theta_{3x} \approx 32.3°$, $\theta_{3y} \approx 121.3°$, and $\theta_{3z} \approx 82.8°$. Notice that the products $\underline{\omega}_1^T \underline{\omega}_2$, $\underline{\omega}_1^T \underline{\omega}_3$, and $\underline{\omega}_2^T \underline{\omega}_3$, are all zero, which indicates that the three principal axes are orthogonal to each other, as expected.

6.4 Euler's Differential Equations of Rotational Motion

It was seen in Section 6.2 that the rotational motion of a rigid body is governed by the vector equation

$$\frac{d}{dt}\vec{H}_C = \vec{M}_C \qquad (6.4.1)$$

where the expression for the angular momentum $\vec{H}_C$ is given by Eq. (6.3.5). For convenience, that expression for $\vec{H}_C$ is repeated here.

$$\begin{aligned}\vec{H}_C &= (I_{xx}\omega_x + I_{xy}\omega_y + I_{xz}\omega_z)\hat{x} \\ &+ (I_{yx}\omega_x + I_{yy}\omega_y + I_{yz}\omega_z)\hat{y} \\ &+ (I_{zx}\omega_x + I_{zy}\omega_y + I_{zz}\omega_z)\hat{z} \\ &\triangleq H_x\hat{x} + H_y\hat{y} + H_z\hat{z} \end{aligned} \qquad (6.4.2)$$

If the x-y-z axes shown in Fig. 6.2 (p. 337) are not fixed to the rigid body, the moments and products of inertia I_{xx}, I_{xy}, etc., change with time. In such a case, Eq. (6.4.1) yields the following result:

$$\frac{d}{dt}\vec{H}_C = \dot{H}_x\hat{x} + \dot{H}_y\hat{y} + \dot{H}_z\hat{z} + \vec{\omega} \times \vec{H}_C = \vec{M}_C \qquad (6.4.3)$$

where

$$\begin{aligned} \dot{H}_x &= \dot{I}_{xx}\omega_x + \dot{I}_{xy}\omega_y + \dot{I}_{xz}\omega_z + I_{xx}\dot{\omega}_x + I_{xy}\dot{\omega}_y + I_{xz}\dot{\omega}_z \\ \dot{H}_y &= \dot{I}_{yx}\omega_x + \dot{I}_{yy}\omega_y + \dot{I}_{yz}\omega_z + I_{yx}\dot{\omega}_x + I_{yy}\dot{\omega}_y + I_{yz}\dot{\omega}_z \\ \dot{H}_z &= \dot{I}_{zx}\omega_x + \dot{I}_{zy}\omega_y + \dot{I}_{zz}\omega_z + I_{zx}\dot{\omega}_x + I_{zy}\dot{\omega}_y + I_{zz}\dot{\omega}_z \end{aligned} \qquad (6.4.4)$$

It is convenient to avoid the additional complication associated with time-varying inertias by choosing to work with a set of axes that are fixed to the rigid body. In such a case, the absolute angular velocity $\vec{\omega}$ of the unit vector triad $\{\hat{x}, \hat{y}, \hat{z}\}$ is equal to the absolute angular velocity $\vec{\omega}^B \triangleq \omega_x\hat{x} + \omega_y\hat{y} + \omega_z\hat{z}$ of the rigid body. By expressing the moment $\vec{M}_C$, acting on the body, in terms of the unit vectors $\{\hat{x}, \hat{y}, \hat{z}\}$, an expression of the following form is obtained for each particular problem under consideration.

$$\vec{M}_C = M_x\hat{x} + M_y\hat{y} + M_z\hat{z} \qquad (6.4.5)$$

In such a case, Eq. (6.4.3) yields the following scalar differential equations for the body-fixed angular velocity components ω_x, ω_y, and ω_z:

$$I_{xx}\dot{\omega}_x + I_{xy}\dot{\omega}_y + I_{xz}\dot{\omega}_z + \omega_y H_z - \omega_z H_y = M_x \qquad (6.4.6)$$

$$I_{yx}\dot{\omega}_x + I_{yy}\dot{\omega}_y + I_{yz}\dot{\omega}_z + \omega_z H_x - \omega_x H_z = M_y \qquad (6.4.7)$$

$$I_{zx}\dot{\omega}_x + I_{zy}\dot{\omega}_y + I_{zz}\dot{\omega}_z + \omega_x H_y - \omega_y H_x = M_z \qquad (6.4.8)$$

where H_x, H_y, and H_z are the $\hat{x}$, $\hat{y}$, and $\hat{z}$ components, respectively, of the angular momentum $\vec{H}_C$, defined by Eq. (6.4.2). Equations (6.4.6) to (6.4.8) are called *Euler's equations for nonprincipal axes*. They are used in some applications such as aircraft and ship dynamics where the moments due to the aerodynamic forces and to the hydrodynamic forces, as the case may be, are more conveniently expressed in terms of some nonprincipal axes that are chosen as a trade-off in complexity between the left-hand sides of Euler's equations and the expressions for the moment components that appear on their right-hand sides.

In numerous applications, working with nonprincipal axes means an unnecessary complication of the differential equations that govern the rotational motion of the body. In principal axes, one simply has $H_x = I_{xx}\omega_x$, $H_y = I_{yy}\omega_y$, and $H_z = I_{zz}\omega_z$. Thus, when all the products of inertia are equal to zero, the following simpler (but not simple!) differential equations are obtained from Eqs. (6.4.6) to (6.4.8). In such cases, it is more convenient to use a single subscript for denoting the three principal moments of inertia, i.e., to use I_x instead of I_{xx}, etc. Such a convention will be adopted, from this point on, when working with principal axes.

$$I_x \dot{\omega}_x + (I_z - I_y)\omega_y\omega_z = M_x \quad (6.4.9)$$

$$I_y \dot{\omega}_y + (I_x - I_z)\omega_x\omega_z = M_y \quad (6.4.10)$$

$$I_z \dot{\omega}_z + (I_y - I_x)\omega_x\omega_y = M_z \quad (6.4.11)$$

These three equations (which are nonlinear and coupled) are called *Euler's equations for principal axes*.

Notice the symmetry that appears in Eqs. (6.4.9) to (6.4.11), which is evidenced when each equation is divided by the moment of inertia that multiplies the time derivative of the respective component of the angular velocity: the moments of inertia in the second term in each equation appear in a counterclockwise x-y-z sequence (with $\hat{z} = \hat{x} \times \hat{y}$), such as $(I_z - I_y)/I_x$ in the first equation.

When the components of the moment $\vec{M}_C$ are not functions of the angular orientation of the body, integration of Eqs. (6.4.9) to (6.4.11) yields the components ω_x, ω_y, and ω_z of the absolute angular velocity $\vec{\omega}^B$ of the body as a function of time. Otherwise, the left-hand side of those equations would also have to be expressed in terms of the variables that are chosen to describe the body's orientation in space, and this results in a set of three nonlinear second-order differential equations for those variables. As seen in this book, numerical integration of the differential equations of motion can easily be done with Simulink without having to write computer code.

Before dealing with several examples of the direct application of Euler's equations for investigating rotational motion, the work-energy approach is presented in Section 6.5. Then several important classical problems in dynamics, such as the stability of a body that is spun about one of its principal axes, dynamic balancing of a spinning rotor with rigid bodies attached to it, and the motion of a gyroscope (or a top) are presented in a separate section (Section 6.6) due to their special significance.

6.5 Kinetic Energy and the Work-Energy Approach

The determination of the kinetic energy for a rigid body undergoing a simpler plane motion as defined in Section 5.1, was presented in Section 5.5. Here, the material in that section is extended to rigid bodies that can undergo any type of three-dimensional motion. The calculations involve the same steps involved in that chapter. However, the final expression for the kinetic energy for a general motion involves other elements of the inertia matrix of the body, instead of only one for the case of the type of simpler planar motion that was dealt with in Chapter 5.

Referring to Fig. 6.2 (p. 337), recall that the kinetic energy of the motion of a rigid body is defined as

$$T \triangleq \frac{1}{2} \int_{\text{body}} \vec{v}_P \bullet \vec{v}_P \, dm \qquad (6.5.1)$$

where $\vec{v}_P = d\overrightarrow{OP}/dt$ is the absolute velocity of an infinitesimal mass element dm of the body, which is represented by point P in Fig. 6.2. It is important to remember that time derivatives such as $d\overrightarrow{OP}/dt$ have to take into account any change of orientation of $\overrightarrow{OP}$ relative to an inertial reference frame, and that point O has to be an inertial reference point. Such a fact is embedded in the phrase "where $\vec{v}_P$ is the absolute velocity ...."

With $\overrightarrow{OP} = \overrightarrow{OC} + \overrightarrow{CP}$, where C is the center of mass of the body (see Fig. 6.2), it was seen in Chapter 5 that Eq. (6.5.1) reduces to

$$T = \frac{1}{2} \int_{\text{body}} \left(\vec{v}_C + \frac{d}{dt}\overrightarrow{CP} \right) \bullet \left(\vec{v}_C + \frac{d}{dt}\overrightarrow{CP} \right) dm$$

$$= \frac{1}{2}m\vec{v}_C \bullet \vec{v}_C + \frac{1}{2} \int_{\text{body}} \left(\frac{d}{dt}\overrightarrow{CP} \right) \bullet \left(\frac{d}{dt}\overrightarrow{CP} \right) dm \qquad (6.5.2)$$

where $\vec{v}_C = d\overrightarrow{OC}/dt$ is the absolute velocity of the center of mass C of the body.

It is at this point that the development of the final expression for the kinetic energy T for a body moving in three-dimensional space starts to be a little more involved than the planar motion case considered in Chapter 5. Basically, however, the only thing that needs to be done is to take the time derivative of the rotating vector $\overrightarrow{CP}$ (as already done in Section 6.3.1, p. 337), use the result in Eq. (6.5.2), and perform the integration to come up with the final expression for the kinetic energy T. By expressing $\overrightarrow{CP}$, and also the absolute angular velocity $\vec{\omega}^B$, as given

SECTION 6.5 Kinetic Energy and the Work-Energy Approach

in Eqs. (6.3.2) and (6.3.3) (p. 339), the following result is obtained for the kinetic energy T, where $(\underline{\omega}^B)^T$ is the transpose of the 3×1 column matrix $\underline{\omega}^B$.

$$T = \frac{1}{2} m \vec{v}_C \bullet \vec{v}_C$$
$$+ \underbrace{\frac{1}{2}\left(I_{xx}\omega_x^2 + I_{yy}\omega_y^2 + I_{zz}\omega_z^2 + 2I_{xy}\omega_x\omega_y + 2I_{xz}\omega_x\omega_z + 2I_{yz}\omega_y\omega_z\right)}_{\text{same as } \frac{1}{2}\vec{\omega}^B \bullet \vec{H}_C = \frac{1}{2}\vec{\omega}^B \bullet \vec{\vec{I}}_C \bullet \vec{\omega}^B \text{ or } \frac{1}{2}(\underline{\omega}^B)^T[I_C]\underline{\omega}^B}$$

In summary, the kinetic energy is determined as

$$T = \frac{1}{2} m \vec{v}_C \bullet \vec{v}_C + \frac{1}{2} \vec{\omega}^B \bullet \vec{H}_C = \frac{1}{2} m \vec{v}_C \bullet \vec{v}_C + \frac{1}{2} \vec{\omega}^B \bullet \vec{\vec{I}}_C \bullet \vec{\omega}^B \quad (6.5.3)$$

or, in matrix notation for the last term,

$$T = \frac{1}{2} m \vec{v}_C \bullet \vec{v}_C + \frac{1}{2} (\underline{\omega}^B)^T [I_C] \underline{\omega}^B \quad (6.5.4)$$

The first term in these expressions is the kinetic energy due to translation of the center of mass C of the body, and the second term is the kinetic energy due to rotation of the body about an axis that passes through the center of mass of the body. Notice that it is not inconsistent to "mix" the vector and matrix representations in the expression for T because each one of the two terms in that expression is a scalar.

Notes

1. When principal axes of inertia are used, the kinetic energy due to rotation about the center of mass reduces to the simpler expression

$$T_{\text{rot}} = \frac{1}{2}\left(I_x \omega_x^2 + I_y \omega_y^2 + I_z \omega_z^2\right)$$

2. If the rigid body is rotating about a fixed point O so that the distance from O to any point of the body is constant, the kinetic energy given by Eq. (6.5.4) reduces to the simpler form

$$T = \frac{1}{2} \vec{\omega}^B \bullet \vec{\vec{I}}_O \bullet \vec{\omega}^B \quad \text{or} \quad T = \frac{1}{2}(\underline{\omega}^B)^T [I_O]\underline{\omega}^B$$

In such an expression, $\vec{\vec{I}}_O$ is the inertia dyadic of the body, referred to point O, and $[I_O]$ is its matrix representation, i.e., the inertia matrix of the body about orthogonal axes that pass through O.

The work-energy approach involving the relationship $dW = d(T)$ for a rigid body is the same as presented in Section 5.5. Referring again to Fig. 6.2 (p. 337), recall that the infinitesimal work dW done by a set of N forces $\vec{F}_1, \vec{F}_2, \ldots, \vec{F}_N$ acting on the rigid body is determined by a relatively simple operation involving dot products of vectors, i.e.,

$$dW \triangleq \sum_{i=1}^{N} \vec{F}_i \bullet d\vec{r}_i \qquad (6.5.5)$$

In Eq. (6.5.5), $\vec{r}_i$ is the absolute position vector of the point on the body where the force $\vec{F}_i$ is applied.

As indicated in Chapters 2 and 5, when dW is an exact differential, Eq. (6.5.5) can be readily integrated to yield $T - W =$ constant, which is a first-order differential equation. In other words, such a differential equation only involves the first derivative of any motion variable. As mentioned in Chapter 5, that is the differential equation of motion if the motion is governed by only one variable (i.e., a one-degree-of-freedom system). Generally, for motion involving more than one variable, additional work is needed to make full use of the work-energy approach. Such work has been developed by Lagrange and other great masters of dynamics and mathematics. That branch of dynamics is best known as *analytical dynamics, Lagrangian dynamics,* or *Lagrangian and Hamiltonian Dynamics*. An introduction to the subject matter is presented in Chapter 7. That material is an extremely powerful, versatile, and convenient methodology for formulating the differential equations of motion of dynamical systems and for searching for integrals of motion.

6.6 Classical Problems

The problems in this section are classical problems that have repercussions in a number of important practical applications, and they involve direct application of Euler differential equations of motion.

6.6.1 Stability of a Body Spun about One of Its Principal Axes

> This problem has repercussions in a number of important applications involving rotating bodies, such as balancing of rotors and satellite dynamics.

Problem Description A rigid body of mass m and principal moments of inertia I_x, I_y, and I_z is thrown upward with an angular velocity Ω about one of its principal axes, as shown in Fig. 6.10a. The angle α shown in that figure is supposedly equal to zero, but it is essentially impossible for that to happen in practice. Therefore, α is interpreted to be a very small "pointing error" due to the practical inability to make it exactly equal to zero. All forces acting on the body other than the force due to gravity are neglected.

- What are the differential equations of motion for the system in terms of the angular velocity components ω_x, ω_y, and ω_z, of the subsequent motion, shown in Fig. 6.10b?

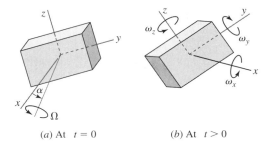

(a) At $t = 0$ (b) At $t > 0$

Figure 6.10 Motion of a body given a spin that is nominally about a principal axis.

■ The differential equations of motion admit a constant solution that corresponds to the body spinning about one of its principal axes of inertia. Linearize the differential equations of motion about that solution, and then investigate the stability of the motion using the linearized differential equations.

Solution Since the resultant of the gravitational force acting on the body passes through the center of mass C of the body, the moment $\vec{M}_C$ of that force is equal to zero. The rotational motion of the body is then governed by Eqs. (6.4.9) to (6.4.11), with $M_x = M_y = M_z = 0$.

With $M_x = M_y = M_z = 0$, Eqs. (6.4.9) to (6.4.11) admit the following three particular solutions. They also admit the solution $\omega_x = \omega_y = \omega_z = 0$, but such a solution is of no interest here because it only corresponds to translation of the body, with no rotation.

Solution 1: $\qquad \omega_x = \text{constant} \triangleq \Omega \qquad \omega_y = \omega_z = 0$

Solution 2: $\qquad \omega_y = \text{constant} \triangleq \Omega \qquad \omega_x = \omega_z = 0$

Solution 3: $\qquad \omega_z = \text{constant} \triangleq \Omega \qquad \omega_x = \omega_y = 0$

All three solutions have the same physical meaning, namely, a body spinning with a constant angular velocity about one of its three principal axes.

Thus, if the body is given a constant spin about any of its x, y, or z principal axis of inertia, it continues to spin about that axis with the same angular velocity it started with. The question that arises now is the following:

> What happens if that motion is perturbed, i.e., if the spin is about an axis that makes a slight angle α (as illustrated in Fig. 6.10a) with the principal axis we were supposed to spin the body about?

To answer this question, we can take, without any loss of generality, the particular solution

$$\omega_x = \text{constant} \triangleq \Omega \qquad \omega_y = \omega_z = 0 \qquad (6.6.1a)$$

and, then, introduce a small perturbation (as presented in Chapter 1) $\omega_{xs}(t)$ as

$$\omega_x(t) = \Omega + \omega_{xs}(t) \tag{6.6.1b}$$

In terms of the variables $\omega_{xs}(t)$, $\omega_y(t)$, and $\omega_z(t)$, the differential equations of motion become

$$I_x\dot{\omega}_{xs} + (I_z - I_y)\omega_y\omega_z = 0 \tag{6.6.1c}$$

$$I_y\dot{\omega}_y + (I_x - I_z)(\Omega + \omega_{xs})\omega_z = 0 \tag{6.6.1d}$$

$$I_z\dot{\omega}_z + (I_y - I_x)(\Omega + \omega_{xs})\omega_y = 0 \tag{6.6.1e}$$

To analyze the stability of the motion, we linearize these differential equations. This is done by simply neglecting products of small quantities to obtain the following equations:

$$I_x\dot{\omega}_{xs} \approx 0 \tag{6.6.1f}$$

$$I_y\dot{\omega}_y + (I_x - I_z)\Omega\omega_z \approx 0 \tag{6.6.1g}$$

$$I_z\dot{\omega}_z + (I_y - I_x)\Omega\omega_y \approx 0 \tag{6.6.1h}$$

The first one of these equations discloses that $\omega_{xs} = $ constant and, without loss of generality, the perturbation ω_{xs} may be equated to zero.

Notice that Eqs. (6.6.1f) to (6.6.1h) are also obtained in the situation where an external control moment $\vec{M} = M_x\hat{x}$ is applied to the body in order to maintain the x component of its angular velocity constant for all times, i.e., $\omega_x = $ constant $\triangleq \Omega$. This would be the case if the body were rigidly attached to a fixed spinning shaft, for example.

By taking the first time derivative of Eq. (6.6.1g) and combining the result with Eq. (6.6.1h), the following second-order linear differential equation with constant coefficients is obtained for ω_y (the same differential equation is also obtained for ω_z):

$$\ddot{\omega}_y + \frac{(I_x - I_z)(I_x - I_y)}{I_yI_z}\Omega^2\omega_y = 0 \tag{6.6.1i}$$

As shown in Section 1.13, the particular solution $\omega_{\hat{x}} = $ constant $\triangleq \Omega$ is unstable if the coefficient of ω_y in Eq. (6.6.1i) is negative (because, in such a case, the solution for ω_y grows with time). This occurs when $I_x > I_z$ and $I_x < I_y$ or when $I_x < I_z$ and $I_x > I_y$, i.e., if the axis about which the body is spun is the axis of intermediate inertia. This is an extremely important result in practice, and it is emphasized in the following boxed text.

> The rotational motion of a moment-free rigid body is unstable if the body is spun about its axis of intermediate inertia.

For an infinitesimally stable particular solution $\omega_x = $ constant, the coefficients of $\ddot{\omega}_y$ and ω_y must have the same sign. Therefore, for an infinitesimally stable particular solution, we must have one of the following:

- $I_x > I_z$ and $I_x > I_y$, which means that the body is spun about its axis of maximum inertia
- $I_x < I_z$ and $I_x < I_y$, which means that the body is spun about its axis of minimum inertia

It can be easily verified by numerical integration of Eqs. (6.6.1c) to (6.6.1e) that the particular solution under consideration is stable when these conditions are satisfied. In such a case, the frequency of the small oscillations governed by Eq. (6.6.1i) is equal to $\Omega\sqrt{(I_x/I_z - 1)(I_x/I_y - 1)}$.

An Interesting Experiment The conclusions about the stability of a body spun about one of its principal axes are very easy to verify in practice with a simple experiment with a solid rectangular plate (such as one made of wood) or with a book that is solidly wrapped with a few rubber bands. The principal axes of inertia of such a body are easily identified. Just toss the rectangle into space by trying to spin it about one of its principal axes, and observe the resulting motion. Do this in three separate instances, by spinning the body about a principal axis in each case. You will see that the body will always tumble when the spin is about its intermediate axis of inertia. For the other two cases (i.e., a spin about either the axis of maximum or minimum moment of inertia), the body continues to spin about that axis, with an unnoticeable deviation of the direction of that axis as the body moves in space.

6.6.2 Dynamic Balancing of a Rotating Part of a Machine

Problem Description A rigid body of mass m is welded to a horizontal symmetric shaft of mass m_{shaft} as shown in Fig. 6.11. Point O is the center of mass of the body and is on the centerline of the shaft, and line AB is inertial. The x-y-z axes shown in the figure are fixed to the body, with the y axis coinciding with the centerline of the horizontal shaft. Thus, the x and z axes are in a vertical plane. The orthogonal x-y-z axes that are fixed to the body are arbitrarily chosen and,

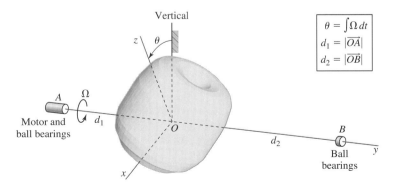

Figure 6.11 A body welded to a spinning shaft (balancing of rotors).

therefore, are not necessarily the principal axes of that body. The shaft is forced by a motor to rotate with a known angular velocity $\dot{\theta} \triangleq \Omega(t)$, where t denotes time and θ is the angle shown in Fig. 6.11. The quantities d_1 and d_2 that are also shown in the figure are the distances from O to A and O to B, respectively, i.e., $d_1 = |\overrightarrow{OA}|$ and $d_2 = |\overrightarrow{OB}|$.

The following need to be determined:

- The moment $\vec{M}_{\text{motor}}$ applied to the shaft by the motor
- The support reactions at the bearings at A and B

It is also necessary to know whether it is possible to weld the body to the shaft so that the support reactions at A and B are not affected by the angular velocity $\Omega(t)$. If that is possible to do, how should the body be mounted on the shaft for that to happen?

Solution The free-body diagram in Fig. 6.12 shows the details that are necessary for solving the problem. The diagram is for the rigid system composed of the shaft and the body that is welded to it. Such a system is rigid because of the weld connection. We could also have split the system into two parts, namely, the shaft and the body, to work with the free-body diagram for each part, thus generating a set of equations for the dynamics of both parts. However, that would introduce additional unknowns into the problem and would lead to additional unnecessary work to eliminate them by combining the resulting equations.

Note the following from Fig. 6.12:

- $\vec{F}_A$ and $\vec{F}_B$ are the bearing reactions at A and B, respectively, and m_{shaft} is the mass of the shaft. The gravitational force with magnitude $m_{\text{shaft}}g$ acts at the middle of the homogeneous shaft.
- $\vec{M}_{\text{motor}} \triangleq M_{\text{motor}} \hat{y}$ is the moment applied to the system by the motor.

The moment M_{motor} is a vector directed along the shaft, but it is indicated

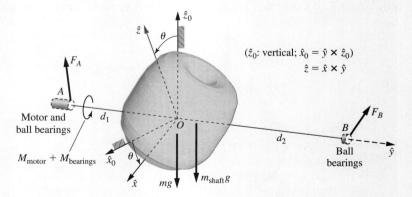

Figure 6.12 Details needed for finding the bearing reactions and the moment M_{motor} applied by the motor.

in the figure by a curved arrow to distinguish it from a force. The counterclockwise sense shown in the figure was arbitrarily chosen.

- $\vec{M}_{\text{bearings}} \triangleq M_{\text{bearings}} \hat{y}$ is a reaction moment due to friction in the two bearings. It is also a vector directed along the shaft, but it is indicated in the figure as part of the same curved arrow mentioned previously. Again, the counterclockwise sense shown in the figure was arbitrarily chosen. $\vec{M}_{\text{bearings}} = 0$ if the bearings are frictionless. If friction in the bearings is of the linear viscous type, with coefficients c_A and c_B at the bearings at A and B, respectively, then $M_{\text{bearings}} = -(c_A + c_B)\Omega$.

For convenience, five unit vectors are also included in Fig. 6.12, namely, $\hat{z}_0$; three orthogonal unit vectors $\hat{x}$, $\hat{y}$, and $\hat{z} \triangleq \hat{x} \times \hat{y}$; and $\hat{x}_0 \triangleq \hat{y} \times \hat{z}_0$. The unit vectors $\hat{z}_0$ and $\hat{x}_0$, with $\hat{z}_0$ vertical, do not change direction, while the unit vectors $\hat{x}$, $\hat{y}$, and $\hat{z}$ are parallel to the x-y-z axes that are fixed to the rotating body. The unit vector $\hat{y}$ is horizontal and, therefore, does not change direction either.

Let us now start the mathematical formulation of the problem in order to answer the questions that were asked. Three scalar equations for the system are obtained from Euler's equation for the rotational motion, and three more from Newton's second law for the motion of the center of mass C of the entire system. Point C, which is not shown in Fig. 6.12, lies on the centerline of the shaft. Since point O is inertial, and it is also fixed to the rotating system, let us use the expression $\vec{M}_O = d\vec{H}_O/dt$.

The inertia dyadic $\vec{\vec{I}}_O$ of the system is

$$\vec{\vec{I}}_O = (\vec{\vec{I}}_O)_{\text{body}} + (\vec{\vec{I}}_O)_{\text{shaft}} \qquad (6.6.2a)$$

For the body that is welded to the shaft,

$$(\vec{\vec{I}}_O)_{\text{body}} = I_{xx}\hat{x}\hat{x} + I_{xy}\hat{x}\hat{y} + I_{xz}\hat{x}\hat{z} + I_{yx}\hat{y}\hat{x} + I_{yy}\hat{y}\hat{y} + I_{yz}\hat{y}\hat{z}$$
$$+ I_{zx}\hat{z}\hat{x} + I_{zy}\hat{z}\hat{y} + I_{zz}\hat{z}\hat{z} \qquad (6.6.2b)$$

and for the symmetric shaft (for which all the products of inertia referred to the x-y-z axes are zero),

$$(\vec{\vec{I}}_O)_{\text{shaft}} = (I_{xx})_{\text{shaft}}(\hat{x}\hat{x} + \hat{z}\hat{z}) + (I_{yy})_{\text{shaft}}\hat{y}\hat{y} \qquad (6.6.2c)$$

Referring to Fig. 6.12, the absolute angular velocity of the system is simply

$$\vec{\omega} = \Omega \hat{y} \qquad (6.6.2d)$$

Since the x-y-z axes are not principal axes of the system, the scalar equations that are obtained from $d\vec{H}_O/dt$ are Eqs. (6.4.6) to (6.4.8) that were developed in Section 6.4, with $\omega_x = \omega_z = 0$ and $\omega_y = \Omega$, as determined earlier. One could then formulate the expressions for the x-y-z components of the moment $\vec{M}_O$ and directly apply those equations to the problem at hand. One could also reformulate the equations for the rotational motion starting from the basic equation $\vec{M}_O = d\vec{H}_O/dt$, without having to consult the reasonably complicated Eqs. (6.4.6) to (6.4.8), as shown in the following.

With the inertia dyadic and the absolute angular velocity of the body given as determined earlier, the absolute angular momentum $\vec{H}_O$ and its time derivative are

$$\vec{H}_O = \vec{\vec{I}}_O \bullet \vec{\omega} = I_{xy}\Omega\hat{x} + [I_{yy} + (I_{yy})_{\text{shaft}}]\Omega\hat{y} + I_{zy}\Omega\hat{z} \quad (6.6.2e)$$

$$\frac{d}{dt}\vec{H}_O = I_{xy}\dot{\Omega}\hat{x} + [I_{yy} + (I_{yy})_{\text{shaft}}]\dot{\Omega}\hat{y} + I_{zy}\dot{\Omega}\hat{z} + \vec{\omega} \times \vec{H}_O$$

$$= (I_{xy}\dot{\Omega} + \Omega^2 I_{zy})\hat{x} + [I_{yy} + (I_{yy})_{\text{shaft}}]\dot{\Omega}\hat{y} + (I_{zy}\dot{\Omega} - \Omega^2 I_{xy})\hat{z} \quad (6.6.2f)$$

Referring again to Fig. 6.12, the moment $\vec{M}_O$ that is applied to the system is calculated as follows:

$$\vec{M}_O = d_2\hat{y} \times \vec{F}_B - d_1\hat{y} \times \vec{F}_A$$
$$+ \underbrace{\left(\frac{d_1+d_2}{2} - d_1\right)\hat{y} \times (-m_{\text{shaft}}g\hat{z}_0)}_{= -\left(\frac{d_2-d_1}{2}m_{\text{shaft}}g\right)\hat{x}_0} + \vec{M}_{\text{motor}} + \vec{M}_{\text{bearings}} \quad (6.6.2g)$$

The expression for the reaction forces $\vec{F}_A$ and $\vec{F}_B$ at the bearings are of the following form:

$$\vec{F}_A = F_{A_x}\hat{x} + F_{A_y}\hat{y} + F_{A_z}\hat{z} \quad (6.6.2h)$$

$$\vec{F}_B = F_{B_x}\hat{x} + F_{B_y}\hat{y} + F_{B_z}\hat{z} \quad (6.6.2i)$$

Therefore, since $\hat{x}_0 = \hat{x}\cos\theta + \hat{z}\sin\theta$ (see Fig. 6.12), Eq. (6.6.2g) yields for the moment $\vec{M}_O$:

$$\vec{M}_O = \left(d_2 F_{B_z} - d_1 F_{A_z} - \frac{d_2-d_1}{2}m_{\text{shaft}}g\cos\theta\right)\hat{x} + (M_{\text{motor}} + M_{\text{bearings}})\hat{y}$$
$$+ \left(d_1 F_{A_x} - d_2 F_{B_x} - \frac{d_2-d_1}{2}m_{\text{shaft}}g\sin\theta\right)\hat{z} \quad (6.6.2j)$$

With $\vec{M}_O$ given by Eq. (6.6.2j), and $d\vec{H}_O/dt$ by Eq. (6.6.2f), the following scalar equations are then obtained from Euler's equation $d\vec{H}_O/dt = \vec{M}_O$:

$$d_2 F_{B_z} - d_1 F_{A_z} - \frac{d_2-d_1}{2}m_{\text{shaft}}g\cos\theta = I_{xy}\dot{\Omega} + \Omega^2 I_{zy} \quad (6.6.2k)$$

$$M_{\text{motor}} + M_{\text{bearings}} = [I_{yy} + (I_{yy})_{\text{shaft}}]\dot{\Omega} \quad (6.6.2l)$$

$$d_1 F_{A_x} - d_2 F_{B_x} - \frac{d_2-d_1}{2}m_{\text{shaft}}g\sin\theta = I_{zy}\dot{\Omega} - \Omega^2 I_{xy} \quad (6.6.2m)$$

It can be readily verified that Eqs. (6.6.2k), (6.6.2l), and (6.6.2m) are the same as Eqs. (6.4.6), (6.4.7), and (6.4.8), respectively.

> Equation (6.6.2l) gives the moment M_{motor} that the motor has to apply to the shaft in order to drive the system with a desired angular velocity $\Omega(t)$. Notice that if the angular velocity is to be maintained constant (which is achieved by a motor that is called a synchronous motor), the motor only needs to apply a moment to counteract the friction in the bearings, if any.

Equations (6.6.2k) and (6.6.2m) involve four unknowns, namely, F_{A_x}, F_{A_z}, F_{B_x}, and F_{B_z}. Therefore, two more equations involving those same unknowns are needed in order to determine them. The additional equations are obtained by applying Newton's second law to the motion of the center of mass of the system. Since the centerline of the shaft is inertial, and the center of mass C of the system is in that centerline, it follows that the absolute acceleration $\vec{a}_C$ of C is equal to zero. Newton's second law governing the motion of C then reduces to $\vec{F} = 0$, where (see the free-body diagram in Fig. 6.12)

$$\vec{F} = \vec{F}_A + \vec{F}_B - (m + m_{\text{shaft}})g\hat{z}_0$$
$$= (F_{A_x} + F_{B_x})\hat{x} + (F_{A_y} + F_{B_y})\hat{y} + (F_{A_z} + F_{B_z})\hat{z} - (m + m_{\text{shaft}})g\hat{z}_0 = 0$$
(6.6.2n)

Since (again, see Fig. 6.12)

$$\hat{z}_0 = \hat{z}\cos\theta - \hat{x}\sin\theta \qquad (6.6.2o)$$

Equation (6.6.2n) yields the following scalar equations involving the force components F_{A_x}, F_{A_y}, F_{A_z}, F_{B_x}, F_{B_y}, and F_{B_z}:

$$F_{A_x} + F_{B_x} = -(m + m_{\text{shaft}})g\sin\theta \qquad (6.6.2p)$$
$$F_{A_y} + F_{B_y} = 0 \qquad (6.6.2q)$$
$$F_{A_z} + F_{B_z} = (m + m_{\text{shaft}})g\cos\theta \qquad (6.6.2r)$$

The four equations (6.6.2k), (6.6.2m), (6.6.2p), and (6.6.2r) can now be solved for F_{A_x}, F_{A_z}, F_{B_x}, and F_{B_z}. The components F_{A_y} and F_{B_y} have opposite signs, and that is all that can be said about them. The following expressions are then obtained for the support reactions $\vec{F}_A = F_{A_x}\hat{x} + F_{A_y}\hat{y} + F_{A_z}\hat{z}$ and $\vec{F}_B = F_{B_x}\hat{x} + F_{B_y}\hat{y} + F_{B_z}\hat{z}$ at the bearings:

$$\vec{F}_A = F_{A_y}\hat{y} + \left(\frac{md_2}{d_1 + d_2} + \frac{1}{2}m_{\text{shaft}}\right)g\underbrace{(\hat{z}\cos\theta - \hat{x}\sin\theta)}_{=\hat{z}_0}$$
$$+ \frac{(\dot{\Omega}I_{zy} - \Omega^2 I_{xy})\hat{x} - (\dot{\Omega}I_{xy} + \Omega^2 I_{zy})\hat{z}}{d_1 + d_2} \qquad (6.6.2s)$$

$$\vec{F}_B = -F_{A_y}\hat{y} + \left(\frac{md_1}{d_1 + d_2} + \frac{1}{2}m_{\text{shaft}}\right)g\underbrace{(\hat{z}\cos\theta - \hat{x}\sin\theta)}_{=\hat{z}_0}$$
$$- \frac{(\dot{\Omega}I_{zy} - \Omega^2 I_{xy})\hat{x} - (\dot{\Omega}I_{xy} + \Omega^2 I_{zy})\hat{z}}{d_1 + d_2} \qquad (6.6.2t)$$

As disclosed by Eqs. (6.6.2s) and (6.6.2t), the bearing reactions $\vec{F}_A$ and $\vec{F}_B$ consist of constant vertical forces (i.e., in the $\hat{z}_0$ direction shown in Fig. 6.12) whose sum simply counteracts the gravitational force $-(m + m_{\text{shaft}})g\hat{z}_0$ acting on the system, of the forces $F_{A_y}\hat{y}$ and $F_{B_y}\hat{y} = -F_{A_y}\hat{y}$ along the shaft, and of a pair of forces $\vec{F}'_A$ and $\vec{F}'_B = -\vec{F}'_A$, acting at A and at B, respectively, where

$$\vec{F}'_A \triangleq \frac{(\dot{\Omega} I_{zy} - \Omega^2 I_{xy})\hat{x} - (\dot{\Omega} I_{xy} + \Omega^2 I_{zy})\hat{z}}{d_1 + d_2} \tag{6.6.2u}$$

The forces $\vec{F}'_A$ and $\vec{F}'_B$ rotate with the shaft with angular velocity $\vec{\omega} = \Omega \hat{y}$ (since both $\hat{x}$ and $\hat{y}$ rotate with that angular velocity). The magnitude of these rotating "imbalance forces" is

$$|\vec{F}'_A| = |\vec{F}'_B| = \frac{\sqrt{I_{xy}^2 + I_{zy}^2}}{d_1 + d_2}\sqrt{\Omega^4 + \dot{\Omega}^2} \tag{6.6.2v}$$

Notice that the magnitude of the rotating imbalance forces $\vec{F}'_A$ and $F'_B = -\vec{F}'_A$ is proportional to the quantity $\sqrt{\Omega^4 + \dot{\Omega}^2}$. Such a quantity is greater than or equal to the square of the shaft angular velocity Ω, and $|\vec{F}'_A|$ is proportional to Ω^2 if Ω is constant. These imbalance forces will cause, in practice, major damage to the bearings. Fortunately such an undesirable pair of forces may be eliminated by welding the body to the shaft so that I_{xy} and I_{zy} are equal to zero. Such a condition means that the axis about which the body is rotated (which is the y axis in Fig. 6.12) is a principal axis of the body. When this is done, the rotating system is said to be *dynamically balanced*.

> In summary, to minimize damage to the bearings due to the rotating imbalance forces, it is important that the rotor be *dynamically balanced*. This is accomplished by assuring that the axis of rotation is a principal axis of the system composed of the rotor and of the body (or bodies) that are attached to it.

In practice, one should ensure that the system is rotated about an axis that is as close as possible to one of its principal axes in order to make the imbalance rotating forces acting on the bearings as close to zero as possible for any angular velocity $\Omega(t)$.

As indicated earlier, careful alignment is required in order to keep the magnitude of the imbalance forces $\vec{F}'_A$ and $\vec{F}'_B = -\vec{F}'_A$ below acceptable levels. The manufacturer of a part that is meant to be spun (such as a vehicle tire) tries to make the body symmetric about one of its principal axes, but small deviations from symmetry are always bound to occur in practice. To counteract such small unavoidable deviations, small masses are placed at strategic locations on the body in order to achieve dynamic balancing. In the balancing process, the imbalance forces $\vec{F}'_A$ and $\vec{F}'_B$ are measured by sensors that are part of a "balancing machine."

A Note on Balancing of Rotating Parts of Machines: In addition to dynamic balancing a system consisting of a rotating shaft attached to one or more bodies, the system should also be *statically balanced*. Static balancing involves only the balancing of the gravitational forces acting on the shaft so that the shaft does not turn when it is not driven. This is achieved by connecting the bodies on the shaft so that their centers of mass coincide with the axis of rotation of the shaft. Undesirable rotating bearing reactions will also appear if the perpendicular distance e between the center of mass of any of such bodies and the centerline of the shaft in Fig. 6.11 is not equal to zero (see Problem 6.13).

6.6.3 Motion of a Symmetric Top (or Gyroscope)

Problem Description A system consisting of a homogeneous rotor of radius R, height h, and mass m is welded to an arm of negligible mass that coincides with the axial axis of the rotor. The whole assembly is connected by a ball-and-socket joint to a point O as shown in Fig. 6.13, where $L = |\overrightarrow{OC}|$ is the distance between O and the center of mass C of the rotor. The system is a model for a top and for a gyroscope (in which case, $L = 0$). The unit vectors $= \hat{z}, \hat{n}$, and $\hat{\phi} \stackrel{\Delta}{=} \hat{z} \times \hat{n}$ that are included in the figure will be used in the analysis that follows. Point O is inertial, the x_0-y_0-z_0 axes are fixed in inertial space, and friction in the system is neglected. The angles $\theta(t)$ and $\phi(t)$ locate arm OC in inertial space. The rotor is given an initial spin $[\dot{\psi}(t)]_{t=0} = \dot{\psi}(0)$ rad/s, while arm OC is held still with $[\theta(t)]_{t=0} = \theta(0)$ and $[\phi(t)]_{t=0} = \phi(0)$. The system is then released with $[\dot{\theta}(t)]_{t=0} = \dot{\theta}(0)$ rad/s and $[\dot{\phi}(t)]_{t=0} = \dot{\phi}(0)$ rad/s.

a. Obtain the differential equations that govern the rotational motion of the system in terms of the angles $\theta(t), \phi(t)$, and, if needed, an angle $\psi(t) \stackrel{\Delta}{=} \int \dot{\psi}\, dt$, which is an angle measured from a nonspinning line parallel to the plane of the flat part of the rotor.

b. Obtain all the equilibrium solutions for the system. Show that one equilibrium corresponds to $\theta = \text{constant} \stackrel{\Delta}{=} \theta_e$, with $\theta_e \neq 0$ and $\theta_e \neq \pi$, and that for such an equilibrium to exist the rotor must be spun about its axial

x_0-y_0-z_0 and O: inertial
$\hat{\phi}$: perpendicular to the OQC plane
$\hat{n}$: parallel to the OQC plane

Figure 6.13 A symmetric top (or a gyroscope) subjected to a gravitational force that creates a moment $\vec{M}_O$.

axis with an angular velocity greater than a certain minimum value. Determine the expression for that minimum. For $R = 6$ cm and $R = 10$ cm, what is that minimum angular velocity so that $\theta_e = 30°$ when $L/R = 1$ and $h/R = \frac{1}{3}$?

Note: In addition to answering the questions (which could be typical of a problem in a formal course), the solution includes a number of additional instructive comments and observations about the motion.

Solution To formulate the differential equations of motion, let us work with the rotating unit vector triad $\{\hat{z}, \hat{n}, \hat{\phi} \triangleq \hat{z} \times \hat{n}\}$ shown in Fig. 6.13, with $\hat{n}$ being normal to arm OC, and in the plane defined by OC and the vertical z_0 axis (i.e., plane OQC shown in the figure). Both $\hat{n}$ and $\hat{\phi}$ are perpendicular to the axial axis of the rotor. Although the axes that are centered at the center of mass C of the rotor and are parallel to the unit vectors $\hat{n}$ and $\hat{\phi}$ are not fixed to the spinning rotor, they are still principal axes of the system because the rotor is symmetric about its axial axis.

After the system is released, the only external moment acting on it is due to the gravitational force $-mg\hat{z}_0$ (see Fig. 6.13). The moment $\vec{M}_O$, about point O, of this force is

$$\vec{M}_O = \overrightarrow{OC} \times (-mg\hat{z}_0) = \overrightarrow{OQ} \times (-mg\hat{z}_0) + \underbrace{\overrightarrow{QC} \times (-mg\hat{z}_0)}_{=0} = mgL(\sin\theta)\hat{\phi}$$

(6.6.3a)

The differential equations that govern the motion of the system, which are needed for investigating the motions, are obtained from $\vec{M}_O = d\vec{H}_O/dt$. There is no need to consider the reaction forces at O because they do not contribute to $\vec{M}_O$. Notice that the inertial $\hat{z}_0$ component of the angular momentum $\vec{H}_O$ is constant because the component of the moment $\vec{M}_O$ in that inertial direction is equal to zero.

Before proceeding further, notice that since $\vec{M}_O$ is a vector that is always perpendicular to the plane OQC, the equation $d\vec{H}_O/dt = \vec{M}_O$ indicates that the infinitesimal change of the angular momentum of the body is in the same direction of the applied moment $\vec{M}_O$. Thus, if the motion is started with $\theta(0) \neq 0$ and $\theta(0) \neq \pi$ (i.e., the initial angular momentum is not along the vertical direction), line OC should rotate about the vertical z_0 axis (which is something anyone who played with tops has observed). But, as will be seen, there is more to the resulting motion, as the analysis of the differential equations of motion (and experiments) reveal!

The expression for the absolute angular momentum $\vec{H}_O$ is developed next. With the axial and transverse principal moments of inertia of the rotor taken from Example 6.3.7 (p. 354), the inertia dyadic of the rotor, referred to its center of mass C, is

$$\vec{\vec{I}}_C = \frac{mR^2}{2}\hat{z}\hat{z} + m\left(\frac{R^2}{4} + \frac{h^2}{12}\right)(\hat{n}\hat{n} + \hat{\phi}\hat{\phi})$$

(6.6.3b)

Using the parallel axis theorem, the inertia dyadic $\vec{\vec{I}}_O$ is then

$$\vec{\vec{I}}_O = \vec{\vec{I}}_C + mL^2(\hat{n}\hat{n} + \hat{\phi}\hat{\phi})$$
$$= m\frac{R^2}{2}\hat{z}\hat{z} + m\left(\frac{R^2}{4} + \frac{h^2}{12} + L^2\right)(\hat{n}\hat{n} + \hat{\phi}\hat{\phi}) \triangleq I_z\hat{z}\hat{z} + I_n(\hat{n}\hat{n} + \hat{\phi}\hat{\phi})$$

(6.6.3c)

This expression for $\vec{\vec{I}}_O$ is the inertia dyadic for the entire system since the bar that is welded to the rotor is being treated as massless.

Inspection of Fig. 6.13 discloses that the absolute angular velocity $\vec{\omega}_{\text{rotor}}$ of the rotor (which is the same as the absolute angular velocity of the entire system because of the rigid connection between the rotor and the bar) is

$$\vec{\omega}_{\text{rotor}} = \dot{\phi}\hat{z}_0 + \dot{\theta}\hat{\phi} + \dot{\psi}\hat{z} = \dot{\phi}(\hat{z}\cos\theta - \hat{n}\sin\theta) + \dot{\theta}\hat{\phi} + \dot{\psi}\hat{z} \quad \text{(6.6.3d)}$$

Note: Having the three components of the absolute angular velocity $\vec{\omega}_{\text{rotor}}$ and the three principal inertias I_z, I_n, and I_n again, it is natural to ask whether they could be used directly in Euler's equations [i.e., Eqs. (6.4.9) to (6.4.11), p. 359] to obtain the differential equations of motion for the system. It is important to notice that doing so would yield wrong equations. The reason is that those equations are for axes that are fixed to the rigid body, which is not the case for the axes that are being used here. If the body were not symmetric (i.e., if the three principal moments of inertia were different), then one would work with body-fixed axes and those equations could be used directly. One could also use body-fixed axes for the symmetric body, of course, and that would require introducing an angle $\psi = \int \dot{\psi}\, dt$ in Fig. 6.13. It is suggested that, in addition to using the formulation that follows, you should also derive the equations using body-fixed axes (see Problem 6.16). You will then see that the angle ψ can be eliminated by combining the resulting equations to yield those that are formulated in the following (with a little less effort!).

Using the expression for $\vec{\omega}_{\text{rotor}}$ given by Eq. (6.6.3d), the absolute angular momentum $\vec{H}_O$ of the system is then obtained as follows:

$$\vec{H}_O = \vec{\vec{I}}_O \bullet \vec{\omega}_{\text{rotor}} = I_z(\dot{\psi} + \dot{\phi}\cos\theta)\hat{z} + I_n[\dot{\theta}\hat{\phi} - (\dot{\phi}\sin\theta)\hat{n}] \triangleq H_z\hat{z} + H_\phi\hat{\phi} + H_n\hat{n}$$

(6.6.3e)

The time derivative of $\vec{H}_O$ is obtained as, where $(\ )^\bullet$ denotes $d(\ )/dt$:

$$\frac{d}{dt}\vec{H}_O = I_z(\dot{\psi} + \dot{\phi}\cos\theta)^\bullet \hat{z} + I_n[\dot{\theta}\hat{\phi} - (\dot{\phi}\sin\theta)^\bullet\hat{n}] + \vec{\omega} \times \vec{H}_O \quad \text{(6.6.3f)}$$

In this expression, $\vec{\omega}$ is the absolute angular velocity of the unit vector triad $\{\hat{n}, \hat{z}, \hat{\phi} = \hat{z} \times \hat{n}\}$ in which the absolute angular momentum $\vec{H}_O$ has been expressed. This is a fine point to pay attention to since one could inadvertently make a mistake here! Since that unit vector triad is not fixed to the spinning rotor, its absolute angular velocity $\vec{\omega}$ is not equal to $\vec{\omega}_{\text{rotor}}$, but is rather

$$\vec{\omega} = \dot{\phi}\hat{z}_0 + \dot{\theta}\hat{\phi} = \dot{\phi}(\hat{z}\cos\theta - \hat{n}\sin\theta) + \dot{\theta}\hat{\phi} \qquad (6.6.3\text{g})$$

This yields the following expression for $\vec{\omega} \times \vec{H}_O$:

$$\vec{\omega} \times \vec{H}_O = [I_z(\dot{\psi} + \dot{\phi}\cos\theta) - I_n\dot{\phi}\cos\theta][\dot{\theta}\,\hat{n} + (\dot{\phi}\sin\theta)\hat{\phi}] \qquad (6.6.3\text{h})$$

Finally, the equation $\vec{M}_O = d\vec{H}_O/dt$ yields the following differential equations involving the variables ψ, θ, and ϕ:

$$I_z(\dot{\psi} + \dot{\phi}\cos\theta)^\bullet = 0 \quad \text{or} \quad \dot{\psi} + \dot{\phi}\cos\theta = \text{constant} \stackrel{\Delta}{=} \omega_z \qquad (6.6.3\text{i})$$

$$I_n(\dot{\phi}\sin\theta)^\bullet + (I_n\dot{\phi}\cos\theta - I_z\omega_z)\dot{\theta} = 0 \qquad (6.6.3\text{j})$$

$$I_n\ddot{\theta} + (I_z\omega_z - I_n\dot{\phi}\cos\theta)\dot{\phi}\sin\theta = mgL\sin\theta \qquad (6.6.3\text{k})$$

The constant $\omega_z = \dot{\psi} + \dot{\phi}\cos\theta = [\dot{\psi} + \dot{\phi}\cos\theta]_{t=0}$ is the axial $\hat{z}$ component (see Fig. 6.13) of the absolute angular velocity $\vec{\omega}_{\text{rotor}}$ of the system.

The two equations (6.6.3j) and (6.6.3k) are the differential equations that govern the motion of the system. They involve only the two unknown angles θ and ϕ, and the constant angular velocity ω_z, which is determined by the initial conditions $\dot{\psi}(0)$, $\dot{\phi}(0)$, and $\theta(0)$ of the motion. They admit several equilibrium solutions, which are now determined.

Equilibrium Solutions Equations (6.6.3j) and (6.6.3k) are identically satisfied when $\theta = \text{constant} \stackrel{\Delta}{=} \theta_e$. One such equilibrium corresponds to $\sin\theta_e = 0$, which has two physically distinct solutions, namely, $\theta_e = 0$ and $\theta_e = \pi$. For these equilibrium solutions, the top is vertical and spinning about that axis with a constant angular velocity equal to ω_z. The top is called a *sleeping top* when it is spinning in the equilibrium $\theta_e = 0$.

Another type of equilibrium with $\theta_e = \text{constant}$ corresponds to

$$(I_z\omega_z - I_n\dot{\phi}\cos\theta_e)\dot{\phi} = mgL \qquad (6.6.3\text{l})$$

which implies that $\dot{\phi} = \text{constant} \stackrel{\Delta}{=} \dot{\phi}_e$. It also implies that $\dot{\psi} = \text{constant} \stackrel{\Delta}{=} \dot{\psi}_e$, as per Eq. (6.6.3i). In this case, Eq. (6.6.3k) yields the following relation for calculating the angle θ_e for given angular velocities ω_z and $\dot{\phi}_e$:

$$\cos\theta_e = \frac{1}{I_n\dot{\phi}_e}\left(I_z\omega_z - \frac{mgL}{\dot{\phi}_e}\right) \qquad (6.6.3\text{m})$$

In the equilibrium solution given by Eq. (6.6.3m) the top is spinning about its axial axis with a constant angular velocity ω_z. While doing so, that axis is inclined at a constant angle θ_e with the vertical and is turning about the vertical with a constant angular velocity $\dot{\phi}_e$. Such a motion is called *precession*.

Equation (6.6.3l) is also a quadratic equation in $\dot{\phi}_e$, i.e.,

$$I_n(\cos\theta_e)\dot{\phi}_e^2 - I_z\omega_z\dot{\phi}_e + mgL = 0 \qquad (6.6.3\text{n})$$

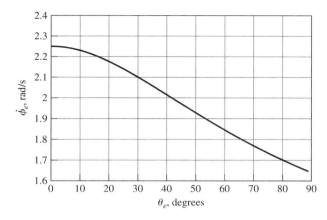

Figure 6.14 Plot of $\dot\phi_e$ versus θ_e for $L/R = 1$, $h/R = \frac{1}{3}$, with $R = 10$ cm and $\omega_z = 50$ rad/s.

This quadratic equation in $\dot\phi_e$ has two solutions when $\cos\theta_e \neq 0$ (i.e., when the spin axis of the top is not horizontal), which are

$$\dot\phi_e = \frac{I_z\omega_z \pm \sqrt{I_z^2\omega_z^2 - 4mgLI_n\cos\theta_e}}{2I_n\cos\theta_e} \quad (6.6.3\text{o})$$

Figure 6.14 shows a plot of the smallest root of Eq. (6.6.3o), versus θ_e, for $L/R = 1$, $h/R = \frac{1}{3}$, with $R = 10$ cm and $\omega_z = 50$ rad/s (see observation 2, which follows).

The top exhibits an equilibrium with $\theta = \text{constant} \overset{\Delta}{=} \theta_e$ only if the solutions to Eq. (6.6.3o) are real. This happens only if

$$|\omega_z| \geq \frac{\sqrt{4mgLI_n\cos\theta_e}}{I_z} \quad (6.6.3\text{p})$$

In other words, the top has to be given a minimum spin rate $|\omega_{z_{\min}}| = \sqrt{4mgLI_n\cos\theta_e}/I_z$ in order for it to exhibit a motion with $\theta = \text{constant} \overset{\Delta}{=} \theta_e$. For $\theta_e = 30°$, $L/R = 1$, and $h/R = \frac{1}{3}$, it follows that $|\omega_{z_{\min}}| \approx 53.4$ rad/s (≈ 510 rpm) for $R = 6$ cm and $|\omega_{z_{\min}}| \approx 41.4$ rad/s (≈ 395 rpm) for $R = 10$ cm. In practice, when friction is present, one should spin the rotor with an angular speed $|\omega_z|$ much greater than its minimum value. Friction would cause the value of $|\omega_z|$ to decrease, and eventually the top would fall down as $|\omega_z|$ becomes smaller than $\omega_{z_{\min}} = \sqrt{4mgLI_n\cos\theta_e}/I_z$.

Observation 1. When $\cos\theta_e = 0$, i.e., when the spin axis of the top is horizontal, the solution to Eq. (6.6.3n) is

$$\dot\phi_e = \frac{mgL}{I_z\omega_z}$$

which again shows that the higher the value of the spin rate $|\omega_z|$, the slower the top rotates around the vertical.

Observation 2. When $I_z|\omega_z| \gg mgL|\cos\theta_e|$, the roots of Eq. (6.6.3o) are

$$\dot{\phi}_e = \frac{I_z\omega_z}{2I_n\cos\theta_e}\left(1 \pm \sqrt{1 - \frac{4mgLI_n\cos\theta_e}{(I_z\omega_z)^2}}\right)$$

$$\approx \frac{I_z\omega_z}{2I_n\cos\theta_e}\left[1 \pm \left(1 - \frac{2mgLI_n\cos\theta_e}{(I_z\omega_z)^2}\right)\right]$$

$$= \frac{I_z\omega_z}{I_n\cos\theta_e} - \frac{mgL}{I_z\omega_z} \quad \text{and} \quad \frac{mgL}{I_z\omega_z}$$

Of the two roots, the one that matches the result for $\cos\theta_e = 0$ is the smallest root $\dot{\phi}_e = mgL/(I_z\omega_z)$. Such a root is the one that gives the actual value of the angular velocity $\dot{\phi}_e$ that is exhibited by the top. For the general case where $\dot{\phi}_e$ and θ_e are related by Eq. (6.6.3o), the actual value of $\dot{\phi}_e$ exhibited by a top is also the smallest root of that equation.

Stability Analysis by Linearization To linearize the differential Eqs. (6.6.3j) and (6.6.3k) about $\theta = \text{constant} \stackrel{\Delta}{=} \theta_e$ and $\dot{\phi} = \text{constant} \stackrel{\Delta}{=} \dot{\phi}_e$, start by introducing small perturbations $\theta_s(t)$ and $\dot{\phi}_s(t)$ and letting

$$\theta(t) = \theta_e + \theta_s(t) \qquad \dot{\phi}(t) = \dot{\phi}_e + \dot{\phi}_s(t)$$

By substituting these expressions into Eq. (6.6.3j), and expanding the resulting equation in Taylor series about $\dot{\phi}_s = 0$ and $\theta_s = 0$, the following linearized equation is obtained after products of small quantities in the expansion are disregarded:

$$I_n\ddot{\phi}_s \sin\theta_e + (2I_n\dot{\phi}_e \cos\theta_e - I_z\omega_z)\dot{\theta}_s = 0 \qquad \textbf{(6.6.3q)}$$

Following the same procedure with Eq. (6.6.3k), the following linearized equation is obtained:

$$I_n\ddot{\theta}_s + \overbrace{[I_z\omega_z\dot{\phi}_e \cos\theta_e - I_n\dot{\phi}_e^2(\cos^2\theta_e - \sin^2\theta_e) - mgL\cos\theta_e]}^{=I_n\dot{\phi}_e^2 \sin^2\theta_e \text{ since } I_z\omega_z\dot{\phi}_e - mgL - I_n\dot{\phi}_e^2\cos\theta_e = 0 \text{ as per Eq. (6.6.3n)}} \theta_s$$
$$+ [I_z\omega_z \sin\theta_e - 2I_n\dot{\phi}_e(\sin\theta_e)\cos\theta_e]\dot{\phi}_s = 0 \qquad \textbf{(6.6.3r)}$$

The characteristic equation for Eqs. (6.6.3q) and (6.6.3r) turns out to be

$$s^2\left[s^2I_n^2 + (I_z\omega_z - 2I_n\dot{\phi}_e)^2\right]\sin\theta_e = 0$$

The s^2 factor is due to the fact that Eq. (6.6.3q) can be integrated once to yield a differential equation that only involves $\dot{\phi}_s$ and θ_s. The equilibrium in question is infinitesimally stable.

The following procedure is more appropriate to use because Eq. (6.6.3j) can be integrated once to eliminate $\dot{\phi}$ from the formulation. Start by noticing that the moment $\vec{M}_O$ given by Eq. (6.6.3a) has no component along the inertial $\hat{z}_0$ direction shown in Fig. 6.13. Therefore, since $d\vec{H}_O/dt = \vec{M}_O$, the $\hat{z}_0$ component of $\vec{H}_O$ is constant, i.e., $\vec{H}_O \bullet \hat{z}_0 = \text{constant} \triangleq H_{z_0}$. By making use of the expression for $\vec{H}_O = H_z\hat{z} + H_\phi\hat{\phi} + H_n\hat{n}$ given by Eq. (6.6.3e), and by inspecting Fig. 6.13, the following is obtained[2]:

$$I_n\dot{\phi}\sin^2\theta + I_z\omega_z\cos\theta = \text{constant} \triangleq H_{z_0} \quad (6.6.3s)$$

Notice that the time derivative of this equation yields Eq. (6.6.3j).

The expression for $\dot{\phi}$ obtained from Eq. (6.6.3s) is then substituted into Eq. (6.6.3k) to obtain an equation that involves θ only. When the resulting equation is expanded about its equilibrium $\theta = \text{constant} \triangleq \theta_e$, the following linearized differential equation is obtained:

$$I_n\ddot{\theta}_s + \overbrace{\left[I_z\omega_z\dot{\phi}_e\cos\theta_e - I_n\dot{\phi}_e^2(\cos^2\theta_e - \sin^2\theta_e) - mgL\cos\theta_e\right]}^{=I_n\dot{\phi}_e^2\sin^2\theta_e \text{ since } I_z\omega_z\dot{\phi}_e - mgL - I_n\dot{\phi}_e^2\cos\theta_e = 0 \text{ as per Eq. (6.6.3n)}}\theta_s$$
$$+ \underbrace{[I_z\omega_z\sin\theta_e - 2I_n\dot{\phi}_e(\sin\theta_e)\cos\theta_e]\dot{\phi}_s}_{=[I_z\omega_z - 2I_n\dot{\phi}_e\cos\theta_e]\left\{\frac{[I_z\omega_z - 2I_n\dot{\phi}_e\cos\theta_e]\theta_s}{I_n}\right\}} = 0 \quad (6.6.3t)$$

Since the coefficient of $\ddot{\theta}_s$ in this equation is positive, the equilibrium is infinitesimally stable.

Additional Comments

- It can be easily verified by numerical integration of the nonlinear differential equations of motion that the equilibrium $\theta = \text{constant} \triangleq \theta_e$ and $\dot{\phi} = \text{constant} \triangleq \dot{\phi}_e$ is stable when the conditions obtained from the stability analysis done by linearization are satisfied.

- The precession rate of a top may be readily obtained from a very simple approximate calculation when the spin rate $|\omega_z|$ is very large, so the absolute angular momentum $\vec{H}_O = I_z\omega_z\hat{z} + I_n\dot{\theta}\hat{\phi} - I_n(\dot{\phi}\sin\theta)\hat{n}$ may be approximated as $H_O \approx I_z\omega_z\hat{z}$. In such a case, application of the equation $d\vec{H}_O/dt = \vec{M}_O = (mgL\sin\theta_e)\hat{\phi}$ is readily visualized as shown in Fig. 6.15. The figure shows the large angular momentum $\vec{H}_O \approx I_z\omega_z\hat{z}$ and its change $d\vec{H}_O$ after an infinitesimal time dt is elapsed while $\theta = \text{constant} = \theta_e$. That change is in the same direction of the moment $\vec{M}_O$. Thus, it occurs in a horizontal plane. It is equal to $(I_z\omega_z\sin\theta_e\, d\phi)\hat{\phi}$, and it is represented by the arrow labeled "$M_O\, dt = mgL\sin\theta_e\, dt$" in the figure. As shown in the insert of the figure, this yields $\dot{\phi}_e = mgL/(I_z\omega_z)$, which is the smallest root of Eq. (6.6.3o) as explained earlier.

---

[2] One could also obtain Eq. (6.6.3s) by multiplying Eq. (6.6.3j) by $\sin\theta$, which allows the terms involving $\dot{\phi}$ to be combined as $d(\dot{\phi}\sin^2\theta)/dt$, and then integrating the result with respect to time.

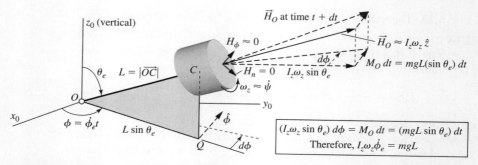

Figure 6.15 Visualization of $d\vec{H}_O/dt = \vec{M}_O$ when $\vec{H}_O = I_z\omega_z\hat{z} + I_n\dot{\theta}\hat{\phi} + I_n(\dot{\phi}\sin\theta)\hat{n} \approx I_z\omega_z\hat{z}$.

- The top spins about its axial axis OC (see Fig. 6.13) with a constant angular velocity ω_z. When in its equilibrium state $\theta = \text{constant} = \theta_e$, that axis rotates about the vertical with a constant angular velocity $\dot{\phi}_e$ causing points on the axial axis of the top to describe a vertical cone. When ω_z is counterclockwise (i.e., $\omega_z > 0$), the cone is traced in a counterclockwise manner, and clockwise otherwise. It was seen that higher values of the spin rate $|\omega_z|$ yield smaller values of the angular velocity $\dot{\phi}_e$; i.e., it takes longer for the spin axis of the top to change direction. That is the type of behavior one wants in a gyroscope for, in such a case, its spin axis would serve as a nearly fixed reference direction in space. As long as Eq. (6.6.3o) yields a real solution for θ_e, the motion with $\theta = \text{constant}$ occurs when it is started with the initial conditions $\theta(0) = \theta_e$, $\dot{\theta}(0) = 0$, and $\dot{\phi}(0) = \dot{\phi}_e$. The stability analysis of the equilibrium disclosed that when such an equilibrium exists (i.e., when $|\omega_z| \geq \sqrt{4mgLI_n\cos\theta_e}/I_z$), it is stable. For a stable equilibrium, the angle $\theta(t)$ oscillates about θ_e when the initial conditions of the motion are slightly deviated from the ones previously indicated.

- The type of motion where the spin axis of the top revolves around the vertical is called *precession*. When $\theta = \text{constant} \triangleq \theta_e$, the precession is said to be *steady*. If the angle θ_e is perturbed (say, by hitting the spin axis a little bit or by adding a little additional mass to the tip of the top), the axial axis of the top still revolves around the vertical but in a manner that neither $\theta(t)$ nor $\dot{\phi}(t)$ remain constant. The precession is unsteady in such a situation.

- The oscillation of line OC about the equilibrium $\theta = \theta_e$ is called *nutation*. Precession and nutation are two important responses exhibited by spinning bodies such as a top or a gyroscope. For a gyroscope, one wants to have $L = 0$ in order to eliminate the gravitational moment. In other words, in a gyroscope the spinning body is attached to a set of gimbals that allow for the direction of the spin axis to change, as illustrated in Fig. 6.16.

An ideal gyroscope is one that is not subjected to a moment $\vec{M}_C$, and, thus, its angular momentum would be constant. In such a case, the magnitude of $\vec{H}_C$ would be constant, and $\vec{H}_C$ would maintain the same direction in inertial space. In practice, there is always a drift of the spin axis

Figure 6.16 A typical gyroscope, showing its gimbals.

due to small disturbing moments that could be caused by small mass asymmetry for example. High-precision gyroscopes are widely used for inertial navigation, such as in commercial aircraft instrumentation, and in many other applications.

Simulink Model for Integrating the Differential Equations of Motion   The differential equations of motion can be integrated numerically with no major difficulty, especially when one makes use of Simulink since it does not involve writing any computer code to do so. The Simulink model can be simplified a little by using the integrated form of Eq. (6.6.3j), which is Eq. (6.6.3s).

⟹ Figure 6.17 shows a Simulink model for integrating Eqs. (6.6.3i), (6.6.3s), and (6.6.3k) using the five user-defined inputs shown in that figure. Explanations about the model construction are given in the following. The model is the file `top_dynamics.mdl` included on the CD. To use it, either type `top_dynamics` in the MATLAB command window or select it by typing `simulations_menu` and selecting the model as indicated throughout the book. Look at the default settings of the `Solver` part of the `Simulation → Parameters` menu for the model in your computer, and experiment using any different value for the maximum step size of the integration scheme if your simulation runs either too slow or too fast in your computer.

The default values for the $Iz/In \approx 1/2.52$ and $mgL/In \approx 196.2/2.52$ inputs correspond to $L/R = 1$ and $h/R = \frac{1}{3}$ for the rotor in Fig. 6.13, with $R = 10$ cm. The default values for the initial conditions $\dot{\phi}(0)$ and $\dot{\psi}(0)$ (which are in the blocks labeled `phi_dot(0)` and `psi_dot(0)` seen in Fig. 6.17) yield $\omega_z = 50$ rad/s for $\theta_e = 30°$, which is the default value for $\theta(0)$ in the model. The value $\dot{\phi}_e \approx 5.03$ rad/s is obtained from Eq. (6.6.3o) with $\theta_e = 30°$. Experiment running the simulation with a smaller value of either $\dot{\phi}(0)$ or $\dot{\psi}(0)$, including values that yield $\omega_z < \omega_{z_{\min}}$, and watch the animation to see what happens to the top.

Figure 6.17 A Simulink model (with animation) for the top dynamics.

The fifth user-defined input in the `top_dynamics` model is the initial value $\theta(0)$ for the integrator whose output is the angle $\theta(t)$. That initial condition is supplied "externally" to that integrator only for convenience. After the values of the constants ω_z and H_{z_0}/I_z are calculated using the equations indicated in Fig. 6.17 (double click the appropriate function blocks to see the syntax for their expressions), the expression for $\dot{\phi}$, $\dot{\psi}$, and $\ddot{\theta}$ are generated using Eqs. (6.6.3s), (6.6.3i), and (6.6.3k), respectively. Since values of $\psi(t)$ are not needed for the animation, only $\dot{\phi}$, $\ddot{\theta}$, and $\dot{\theta}$ are integrated to obtain the angles $\phi(t)$ and $\theta(t)$. They are, then, used in the `MATLAB Function` block shown in Fig. 6.17 for generating a simple animation of bar OC to illustrate the motion of the top, using nondimensional coordinates. The x_0-y_0-z_0 nondimensional coordinates of the end points of that bar are $(0, 0, 0)$ for the support at O, and $(\sin\theta \cos\phi, \sin\theta \sin\phi, \cos\theta)$ for point C (see Fig. 6.13, p. 371). Double click that `MATLAB Function` block to see the syntax for such expressions. Double click the respective integrators to see their initial conditions and to change them when desired.

6.6.4 On the Nutational Motion of a Symmetric Top

The Simulink model presented in Section 6.6.3 is very convenient for investigating and observing the motion of the top. One can also modify the model to include damping at the support and investigate the resulting motion.

Further insight into the motion can be obtained by pushing analytical techniques as far as possible, and it turns out that this can be done if we solve

Eq. (6.6.3s) for $\dot{\phi}$ in terms of θ and use the solution to eliminate $\dot{\phi}$ from Eq. (6.6.3k) to obtain a single differential equation whose only unknown is the angle $\theta(t)$. The final equation is of the general form $I_n \ddot{\theta} + f(\theta) = 0$, which is a form that can be integrated once to obtain an equation that involves $\dot{\theta}$ and θ, as shown in Chapter 1 using the chain rule. However, it turns out that such an integration is much more easily done if we make use of the basic work-energy equation $dW = d(T)$, where T is the kinetic energy of the motion and dW is the infinitesimal work done by all forces acting on the system.

The infinitesimal work done by the gravitational force $-mg\hat{z}_0$ (see Fig. 6.13) is simply

$$dW = -mg\hat{z}_0 \bullet \vec{v}_C \, dt = -mg\hat{z}_0 \bullet [(L\dot{\theta}\, dt)\hat{n} + (L\sin\theta)(\dot{\phi}\, dt)\hat{\phi}]$$
$$= -mg\hat{z}_0 \bullet [(L\, d\theta)\hat{n} + (L\sin\theta)(d\phi)\hat{\phi}] \qquad (6.6.4a)$$

and since $\hat{z}_0 = \hat{z}\cos\theta - \hat{n}\sin\theta$, we obtain:

$$dW = mgL\sin\theta \, d\theta \qquad (6.6.4b)$$

The expression $dW = mgL\sin\theta \, d\theta$ is the exact differential $d(-mgl\cos\theta)$, plus a constant, and since $dW = d(T)$, this can be immediately integrated (and this is the reason why we used the work-energy approach here) to yield

$$T + mgL\cos\theta = \text{constant} \stackrel{\Delta}{=} E \qquad (6.6.4c)$$

The quantity $mgL\cos\theta$ happens to be the potential energy of the motion, and it could have been written by inspection of Fig. 6.13, as explained in Chapter 2.

The expression for the kinetic energy T is obtained by following the steps shown in the boxed text on p. 361 in Section 6.5. Using the expressions for the absolute angular velocity of the rotor and for the inertia dyadic $\vec{\vec{I}}_O$, we obtain

$$T = \frac{1}{2}\vec{\omega}_{\text{rotor}} \bullet \vec{\vec{I}}_O \bullet \vec{\omega}_{\text{rotor}} = \frac{1}{2}I_n\dot{\theta}^2 + \frac{1}{2}I_n\dot{\phi}^2 \sin^2\theta + \frac{1}{2}I_z(\dot{\psi} + \dot{\phi}\cos\theta)^2 \qquad (6.6.4d)$$

Since $\dot{\psi} + \dot{\phi}\cos\theta$ was found to be constant, $T + mgL\cos\theta = \text{constant}$ then yields, also using the expression for $\dot{\phi}$ obtained from Eq. (6.6.3s),

$$\frac{1}{2}I_n(\dot{\theta}^2 + \dot{\phi}^2\sin^2\theta) + mgL\cos\theta = \frac{1}{2}I_n\left[\dot{\theta}^2 + \frac{(H_z - I_z\omega_z\cos\theta)^2}{I_n^2\sin^2\theta}\right] + mgL\cos\theta$$
$$= \text{constant} \stackrel{\Delta}{=} C \qquad (6.6.4e)$$

This equation looks more complicated than it actually is! By premultiplying it by $\sin^2\theta$, because the quantity $\dot{\theta}^2\sin^2\theta$ is the same as $[d(\cos\theta)/dt]^2$, we can give preference to working with $\cos\theta$ to reduce it to

$$\left(\frac{d}{dt}\cos\theta\right)^2 = \frac{2C}{I_n}(1-\cos^2\theta) - \left(\frac{H_z}{I_n} - \frac{I_z}{I_n}\omega_z\cos\theta\right)^2$$
$$- \frac{2mgL}{I_n}(\cos\theta)(1-\cos^2\theta) \qquad (6.6.4f)$$

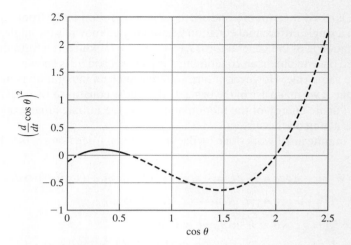

Figure 6.18 A sketch of $[d(\cos\theta)/dt]^2$ versus $\cos\theta$ for Eq. (6.6.4f).

The right-hand side of Eq. (6.6.4f) is a cubic function of $\cos\theta$ and, thus, can easily be plotted versus $\cos\theta$ for given values of the system parameters and initial conditions of the motion. A sketch of a plot of $[d(\cos\theta)/dt]^2$ versus $\cos\theta$ is shown in Fig. 6.18.

Since $[d(\cos\theta)/dt]^2$ is never negative and $|\cos\theta| \leq 1$, the plot is valid only in the region defined by the solid part of the curve. Since $[d(\cos\theta)/dt]^2 = \dot\theta^2 \sin^2\theta = 0$ when $\dot\theta = 0$ (i.e., when $\theta(t)$ reaches a maximum or minimum value), the sketch in Fig. 6.18 is disclosing that the spin axis of the top oscillates between a minimum and maximum value of $\theta(t)$. The minimum and maximum values of $\theta(t)$ can coalesce into a single value for certain initial conditions. When that happens, the equilibrium solution determined in Section 6.6.3 is then observed. The solid part of the curve in Fig. 6.18 is shifted down and just touches the $\cos\theta$ axis in such a case. A further shift down of that curve implies that the spin rate $\dot\psi_e$ of the top is too small for the spin axis to nutate between two values, and the top keeps "falling" as it precesses.

6.7 Solved Example Problems

EXAMPLE 6.7.1

A homogeneous rectangular plate of mass m and dimensions L, a, and b is hinged to a rotating horizontal circular platform as shown in Fig. 6.19. The center O of the platform is inertial. The plate is connected to the platform by a hinge and, thus, is forced to turn with the platform. The hinge allows the angle θ shown in Fig. 6.19 to change with time. The platform is driven by a motor with a known absolute angular velocity $\dot\phi$ (i.e., ϕ is measured

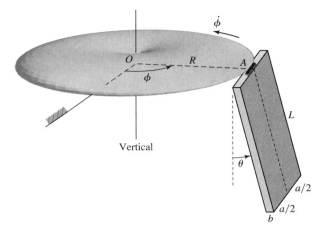

Figure 6.19 A plate on a rotating platform for Example 6.7.1.

from a line fixed in space and in the plane of rotation of line OA of the platform). Friction in the hinge may be neglected.

a. Obtain the differential equation for the plate in terms of the angle θ.
b. For $\dot\phi = \text{constant} \triangleq \Omega$, with $\Omega^2 L/g = 1.5$, $L = 3R$, $a = L/4$, and b so small that it can be neglected:
   1. Determine the equilibrium solutions $\theta = \text{constant} \triangleq \theta_e$, and analyze the stability of each equilibrium by linearizing the differential equation obtained in part a. What is the frequency of the small oscillations of the plate about its stable equilibrium?
   2. Plot the magnitude of the reaction force and the magnitude of the reaction moment at the hinge versus nondimensional time $\tau = t\sqrt{g/L}$, and then determine, from the plots, their maximum values when the motion is started with the initial conditions $\theta(0) = 30°$ and $\dot\theta(0) = 0$.

■ **Solution**

a. Figure 6.20 shows the free-body diagram for the plate in Fig. 6.19. A unit vector triad $\{\hat{x}, \hat{y}, \hat{z}\}$, parallel to the principal x-y-z axes of the plate, and a vertical unit vector $\hat{z}_0 = \hat{z}\sin\theta - \hat{x}\cos\theta$, are also superimposed on the diagram. The principal x-y-z axes of the plate are centered at its center of mass C. The vectors labeled A_1, A_2, and A_3 are the $\hat{x}$, $\hat{y}$, and $\hat{z}$ components of the reaction force applied at point A to the plate. The curved arrows labeled M_1 and M_3 represent the reaction moments $\vec{M}_1 = M_1\hat{x}$ and $\vec{M}_3 = M_3\hat{z}$ applied by the hinge to the plate. Those reaction moments are nonzero because the hinge, by its construction, prevents the plate from rotating along axes parallel to $\hat{x}$ and $\hat{z}$, respectively. The component of the reaction moment parallel to $\hat{y}$ is equal to zero because the hinge is frictionless.

Six scalar equations involving the six unknowns θ, A_1, A_2, A_3, M_1, and M_3 are obtained from the equation $\vec{F} = m\vec{a}_C$, where $\vec{a}_C$ is the absolute acceleration of the center of mass C of the plate, and from $\vec{M}_C = d\vec{H}_C/dt$, where $\vec{H}_C$ is the absolute

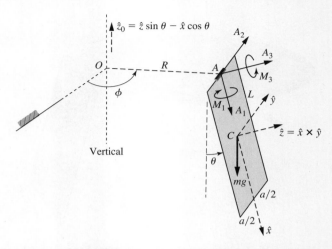

Figure 6.20 Free-body diagram (also showing a unit vector triad $\{\hat{x}, \hat{y}, \hat{z}\}$) for the plate in Fig. 6.19.

angular momentum of the plate. Of those six equations, only one of them is a differential equation of motion since θ is the only unknown motion variable.

The resultant $\vec{F}$ of the forces acting on the plate, and the moment $\vec{M}_C$, about C, of those forces are (see Fig. 6.20)

$$\vec{F} = A_1\hat{x} + A_2\hat{y} + A_3\hat{z} - mg\hat{z}_0$$
$$= A_1\hat{x} + A_2\hat{y} + A_3\hat{z} - mg(\hat{z}\sin\theta - \hat{x}\cos\theta) \qquad (6.7.1a)$$

$$\vec{M}_C = M_1\hat{x} + M_3\hat{z} - \frac{L}{2}\hat{x} \times (A_1\hat{x} + A_2\hat{y} + A_3\hat{z})$$
$$= M_1\hat{x} + \frac{L}{2}A_3\hat{y} + \left(M_3 - \frac{L}{2}A_2\right)\hat{z} \stackrel{\Delta}{=} M_x\hat{x} + M_y\hat{y} + M_z\hat{z} \qquad (6.7.1b)$$

We now need to get expressions for the absolute angular momentum $\vec{H}_C$ and for the absolute acceleration $\vec{a}_C$ to use them in the equations $\vec{F} = m\vec{a}_C$ and $\vec{M}_C = d\vec{H}_C/dt$. Let us start with the formulation of the expression for $\vec{H}_C$.

In terms of the unit vectors $\hat{x}$, $\hat{y}$, and $\hat{z}$ that are parallel to the principal axes of inertia of the plate (see Fig. 6.20), the inertia dyadic $\vec{\bar{I}}_C$ of the plate is

$$\vec{\bar{I}}_C = I_x\hat{x}\hat{x} + I_y\hat{y}\hat{y} + I_z\hat{z}\hat{z} \qquad (6.7.1c)$$

where

$$I_x = \frac{m(a^2 + b^2)}{12} \qquad I_y = \frac{m(L^2 + b^2)}{12} \qquad I_z = \frac{m(L^2 + a^2)}{12} \qquad (6.7.1d)$$

The absolute angular velocity $\vec{\omega}$ of the plate is (see Fig. 6.20)

$$\vec{\omega} = \dot{\phi}\hat{z}_0 - \dot{\theta}\hat{y} = \dot{\phi}(\hat{z}\sin\theta - \hat{x}\cos\theta) - \dot{\theta}\hat{y} \stackrel{\Delta}{=} \omega_x\hat{x} + \omega_y\hat{y} + \omega_z\hat{z} \qquad (6.7.1e)$$

Euler's equations for principal axes, which are Eqs. (6.4.9) to (6.4.11) in Section 6.4, then yield the following equations for the present problem:

$$-I_x(\ddot{\phi}\cos\theta - \dot{\phi}\dot{\theta}\sin\theta) - (I_z - I_y)\dot{\phi}\dot{\theta}\sin\theta = M_1 \quad (6.7.1f)$$

$$-I_y\ddot{\theta} - (I_x - I_z)\dot{\phi}^2\sin\theta\cos\theta = \frac{L}{2}A_3 \quad (6.7.1g)$$

$$I_z(\ddot{\phi}\sin\theta + \dot{\phi}\dot{\theta}\cos\theta) + (I_y - I_x)\dot{\phi}\dot{\theta}\cos\theta = M_3 - \frac{L}{2}A_2 \quad (6.7.1h)$$

Equations (6.7.1f), (6.7.1g), and (6.7.1h) may also be obtained directly from $d\vec{H}_C/dt = \vec{M}_C$ without either "consulting" or remembering Eqs. (6.4.9), (6.4.10), and (6.4.11). The expressions for $\vec{H}_C$ and $d\vec{H}_C/dt$ for doing so are as follows:

$$\vec{H}_C = \bar{\bar{I}}_C \bullet \vec{\omega} = I_z(\dot{\phi}\sin\theta)\hat{z} - I_y\dot{\theta}\hat{y} - I_x(\dot{\phi}\cos\theta)\hat{x} \quad (6.7.1i)$$

$$\frac{d}{dt}\vec{H}_C = I_z(\ddot{\phi}\sin\theta + \dot{\phi}\dot{\theta}\cos\theta)\hat{z} - I_y\ddot{\theta}\hat{y} - I_x(\ddot{\phi}\cos\theta - \dot{\phi}\dot{\theta}\sin\theta)\hat{x} + \vec{\omega}\times\vec{H}_C$$

$$= [(I_y - I_z)\dot{\phi}\dot{\theta}\sin\theta - I_x(\ddot{\phi}\cos\theta - \dot{\phi}\dot{\theta}\sin\theta)]\hat{x}$$

$$+ [(I_z - I_x)\dot{\phi}^2\sin\theta\cos\theta - I_y\ddot{\theta}]\hat{y}$$

$$+ [(I_y - I_x)\dot{\phi}\dot{\theta}\cos\theta + I_z(\ddot{\phi}\sin\theta + \dot{\phi}\dot{\theta}\cos\theta)]\hat{z} \quad (6.7.1j)$$

The expression for the absolute acceleration $\vec{a}_C = d^2\overrightarrow{OC}/dt^2$ of C is developed next, starting from the expression for $\overrightarrow{OC}$ as obtained by inspection of Fig. 6.20.

■ Absolute position vector for the center of mass C of the plate:

$$\vec{r}_C = \overrightarrow{OC} = \overrightarrow{OA} + \overrightarrow{AC} = R(\hat{x}\sin\theta + \hat{z}\cos\theta) + \frac{L}{2}\hat{x} \quad (6.7.1k)$$

■ Absolute velocity of C:

$$\vec{v}_C = \frac{d\vec{r}_C}{dt} = R\dot{\theta}(\hat{x}\cos\theta - \hat{z}\sin\theta) + \vec{\omega}\times\vec{r}_C = \left(R + \frac{L}{2}\sin\theta\right)\dot{\phi}\,\hat{y} + \frac{L}{2}\dot{\theta}\hat{z}$$

$$(6.7.1l)$$

■ Absolute acceleration of C:

$$\vec{a}_C = \frac{d\vec{v}_C}{dt} = \left[\left(R + \frac{L}{2}\sin\theta\right)\ddot{\phi} + \frac{L}{2}\dot{\phi}\dot{\theta}\cos\theta\right]\hat{y} + \frac{L}{2}\ddot{\theta}\hat{z} + \vec{\omega}\times\vec{v}_C$$

$$= -\left[\frac{L}{2}\dot{\theta}^2 + \left(R + \frac{L}{2}\sin\theta\right)\dot{\phi}^2\sin\theta\right]\hat{x} + \left[\left(R + \frac{L}{2}\sin\theta\right)\ddot{\phi} + L\dot{\phi}\dot{\theta}\cos\theta\right]\hat{y}$$

$$+ \left[\frac{L}{2}\ddot{\theta} - \left(R + \frac{L}{2}\sin\theta\right)\dot{\phi}^2\cos\theta\right]\hat{z} \quad (6.7.1m)$$

With $\vec{F}$ given by Eq. (6.7.1a) and $\vec{a}_C$ by Eq. (6.7.1m), the following scalar equations are then obtained from $\vec{F} = m\vec{a}_C$:

$$A_1 + mg\cos\theta = -m\left[\frac{L}{2}\dot\theta^2 + \left(R + \frac{L}{2}\sin\theta\right)\dot\phi^2\sin\theta\right] \quad (6.7.1n)$$

$$A_2 = m\left[\left(R + \frac{L}{2}\sin\theta\right)\ddot\phi + L\dot\phi\dot\theta\cos\theta\right] \quad (6.7.1o)$$

$$A_3 - mg\sin\theta = m\left[\frac{L}{2}\ddot\theta - \left(R + \frac{L}{2}\sin\theta\right)\dot\phi^2\cos\theta\right] \quad (6.7.1p)$$

The six unknowns θ, A_1, A_2, A_3, M_1, and M_3 can now be determined by solving the six equations (6.7.1f), (6.7.1g), (6.7.1h), (6.7.1n), (6.7.1o), and (6.7.1p). Equations (6.7.1g) and (6.7.1p) can be combined to yield

$$\left(\frac{L^2}{3} + \frac{b^2}{12}\right)\ddot\theta - \left[\frac{RL}{2} + \left(\frac{L^2}{3} - \frac{b^2}{12}\right)\sin\theta\right]\dot\phi^2\cos\theta + g\frac{L}{2}\sin\theta = 0 \quad (6.7.1q)$$

where use was made of the expressions for the moments of inertia I_x, I_y, and I_z shown in the equations of (6.7.1d). Equation (6.7.1q), whose only unknown is the angle θ when $\dot\phi$ is prescribed, is the differential equation of motion for the system.

Notice that the motion $\theta(t)$ does not depend on the lateral dimension a of the plate. Also notice that Eq. (6.7.1q) is the only equation that is needed if only the angle $\theta(t)$, and not reactions, are to be determined. In such a case the equations that involve the hinge reactions can be disregarded.

b1. When $\dot\phi = \text{constant} \triangleq \Omega$, Eq. (6.7.1q) admits the solution $\theta = \text{constant} \triangleq \theta_e$. For b small enough to be neglected (i.e., $b \ll 2L$), the equilibrium θ_e is obtained by solving the following equation, which is obtained from Eq. (6.7.1q):

$$\sin\theta_e - \frac{\Omega^2 L}{g}\left(\frac{R}{L} + \frac{2}{3}\sin\theta_e\right)\cos\theta_e = 0 \quad (6.7.1r)$$

The values of θ_e that satisfy Eq. (6.7.1r) may be obtained by plotting the left-hand side of that equation versus θ_e. Such a plot is shown in Fig. 6.21 for $\Omega^2 L/g = 1.5$ and $L = 3R$, and the solutions for θ_e are found to be

$$\theta_e \approx 52.2° \quad \text{and} \quad \theta_e \approx 194.2°$$

To investigate the motion about the equilibrium solution, one lets

$$\theta(t) \triangleq \theta_e + \theta_s(t)$$

For very small "perturbations" $\theta_s(t)$ about the equilibrium, linearization of Eq. (6.7.1q) in θ_s yields the following linear constant coefficient differential equation when $b \ll 2L$:

$$\ddot\theta_s + \frac{g}{L}\left[\frac{\Omega^2 L}{g}\left(\frac{3R}{2L}\sin\theta_e - \cos^2\theta_e + \sin^2\theta_e\right) + \frac{3}{2}\cos\theta_e\right]\theta_s = 0 \quad (6.7.1s)$$

Figure 6.21 Plot of $f \triangleq \sin\theta - (\Omega^2 L/g)(R/L + \frac{2}{3}\sin\theta)\cos\theta$ versus θ, for $\Omega^2 L/g = 1.5$ and $L = 3R$ for Example 6.7.1.

As first presented in Chapter 1, linearization about $\theta = \theta_e$ is done by expanding the nonlinear terms in a Taylor series about $\theta_s = 0$ and disregarding all the resulting polynomial nonlinearities in the series.

We then have for each equilibrium when $\Omega^2 L/g = 1.5$ and $L = 3R$:

For $\theta_e = 52.2°$:
$$\ddot{\theta}_s + 1.89\frac{g}{L}\theta_s = 0 \quad (6.7.1t)$$

For $\theta_e = 194.2°$:
$$\ddot{\theta}_s - 2.96\frac{g}{L}\theta_s = 0 \quad (6.7.1u)$$

These linearized differential equations disclose that the equilibrium solution $\theta_e = 52.2°$ is infinitesimally stable and that the one with $\theta_e = 194.2°$ is unstable. Equation (6.7.1q) should be integrated numerically with initial conditions near the equilibrium $\theta_e = 52.2°$ to check the stability of that equilibrium. If that is done, it can be seen that such an equilibrium is stable. The frequency of the small oscillations of the plate about its stable equilibrium $\theta_e = 52.2°$ is equal to $\sqrt{1.89g/L} \approx 1.37\sqrt{g/L}$. The period associated with this frequency is $2\pi/(1.37\sqrt{g/L}) \approx 4.59\sqrt{L/g}$. With $g = 9.81$ m/s^2 and $L = 1$ m, for example, the frequency of the small oscillations about $\theta_e \approx 52.2°$ is approximately 4.3 rad/s (for a period of $2\pi/4.3 \approx 1.46$ s).

b2. The numerical integration of Eq. (6.7.1q) when b is neglected, and the calculation of the components A_1, A_2, and A_3 of the reaction forces at the hinge as given by Eqs. (6.7.1n), (6.7.1o), and (6.7.1p), respectively, and for the components M_1 and M_3 of the hinge reaction moment given by Eqs. (6.7.1f) and (6.7.1h), respectively can be easily done with Simulink. Figure 6.22 shows a Simulink model for doing this.

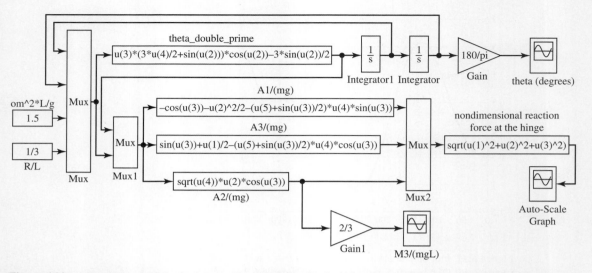

Figure 6.22 A Simulink model for integrating Eq. (6.7.1q) and calculating the hinge reactions for Example 6.7.1.

The plots of θ, and of the magnitude of the reaction force (normalized by mg) and of the reaction moment (normalized by mgL) at the hinge, versus the nondimensional time $\tau = t\sqrt{g/L}$ are shown in Fig. 6.23, for the case when $\Omega^2 L/g = 1.5$, $L = 3R$, and $b = 0$, when the motion is started with the initial conditions $\theta(0) = 30°$ and $\dot\theta(0) = 0$.

The following normalized components of the reactions are obtained from Eqs. (6.7.1n), (6.7.1o), (6.7.1g), (6.7.1f), and (6.7.1h):

$$\frac{A_1}{mg} = -\cos\theta - \frac{1}{2}\left(\frac{d\theta}{d\tau}\right)^2 - \frac{\Omega^2 L}{g}\left(\frac{R}{L} + \frac{1}{2}\sin\theta\right)\sin\theta$$

$$\frac{A_2}{mg} = \left(\Omega\sqrt{\frac{L}{g}}\right)\frac{d\theta}{d\tau}\cos\theta \qquad \frac{A_3}{mg} = \sin\theta + \frac{1}{2}\frac{d^2\theta}{d\tau^2} - \left(\frac{R}{L} + \frac{1}{2}\sin\theta\right)\frac{\Omega^2 L}{g}\cos\theta$$

$$\frac{M_1}{mgL} = 0 \qquad \frac{M_3}{mgL} = \frac{2}{3}\frac{A_2}{mg}$$

Such expressions are the ones used in the block diagram model shown in Fig. 6.22. It is suggested that you construct that model, modifying it to include an input for b/L, and investigate the effect of nonzero values of b/L on the motion and on the support reactions.

Notice that the response $\theta(\tau)$ in Fig. 6.23, which was started with the initial conditions $\theta(0) = 30°$ and $\dot\theta(0) = 0$, has a period of approximately $4.72\sqrt{L/g}$. This is slightly higher than the period of the very small oscillations about the equilibrium $\theta_e = 52.2°$, which was found to be $4.59\sqrt{L/g}$. This increase in the period with the amplitude of the motion is expected of a pendulum, and it was first presented in

SECTION 6.7 Solved Example Problems

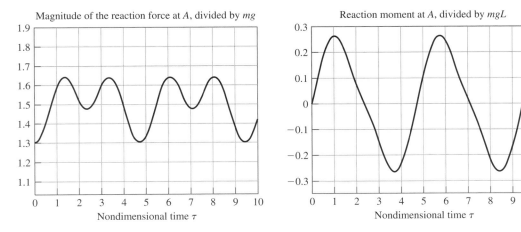

Figure 6.23 Angle θ and the normalized force and moment reactions at A (see Fig. 6.20) versus nondimensional time $\tau = \Omega\sqrt{g/L}$.

Chapter 1 (see Example 1.13.3, p. 43). The maximum value of the magnitude of the reaction force $A_1\hat{x}_1 + A_2\hat{x}_2 + A_3\hat{x}_3$ at the frictionless hinge is approximately $1.64mg$, and that of the reaction moment $M_1\hat{x}_1 + M_3\hat{x}_3$ is $0.26mgL$ (see Fig. 6.23).

EXAMPLE 6.7.2

A homogeneous rectangular plate of mass m and dimensions L, a, and b is connected to a horizontal shaft by a gimbal as shown in Fig. 6.24. The shaft is driven by a motor with a prescribed angular velocity $\Omega(t)$. Point O is the center of mass of the plate and is on the midline of the shaft. The gimbal allows the plate to rotate relative to the shaft, as indicated by the angle $\theta(t)$ in the figure. Friction in the hinges at A and B can be neglected. The

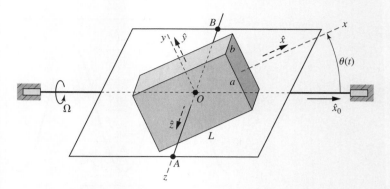

Figure 6.24 The plate and shaft system for Example 6.7.2.

orthogonal x, y, and z axes shown in the figure are principal axes of the rotating plate. The unit vectors $\hat{x}$ and $\hat{y}$ are parallel to those axes, and $\hat{z} \stackrel{\Delta}{=} \hat{x} \times \hat{y}$. The rotation of the plate relative to the shaft is about the z axis of the plate. The unit vector $\hat{x}_0$ is parallel to the shaft midline and is inertial (i.e., its direction does not change).

a. Obtain the differential equation for the angle θ, and verify that the system exhibits the equilibrium solutions $\theta = \text{constant} \stackrel{\Delta}{=} \theta_e$, but only $\theta_e = 0$ and $\theta_e = \pi/2$ correspond to physically different orientations of the plate relative to the shaft.
b. For $\Omega = \text{constant}$, analyze the stability of the small motions about the equilibrium solutions $\theta_e = 0$ and $\theta_e = \pi/2$, and show that for the equilibrium $\theta_e = 0$ to be stable the dimensions of the plate have to be such that $a > b$. Do the stability analysis by linearizing the differential equation obtained in part a.
c. Plot $\dot{\theta}$ versus θ and, from your plot, specify the region where the pair of initial conditions $\theta(0)$ and $\dot{\theta}(0)$ have to lie for the motion (even large motions) $\theta(t)$ to be oscillatory about the equilibrium $\theta_e = 0$.

■ **Solution**

a. The rotational motion of the plate is governed by the basic equation $d\vec{H}_O/dt = \vec{M}_O$, which yields Euler's three scalar equations. If we work with the principal axes of the plate, which are centered at point O, such equations are Eqs. (6.4.9), (6.4.10), and (6.4.11) (p. 359). The formulation is repeated here for this problem.

Inspection of Fig. 6.24 discloses that the absolute angular velocity $\vec{\omega}_{\text{plate}}$ of the plate is

$$\vec{\omega}_{\text{plate}} = \Omega \hat{z}_0 + \dot{\theta} \hat{z} = \Omega(\hat{x} \cos\theta - \hat{y} \sin\theta) + \dot{\theta}\hat{z} \stackrel{\Delta}{=} \omega_x \hat{x} + \omega_y \hat{y} + \omega_z \hat{z} \quad (6.7.2a)$$

Since the inertia dyadic of the plate is simply

$$\vec{\vec{I}}_{\text{plate}} = I_x \hat{x}\hat{x} + I_y \hat{y}\hat{y} + I_z \hat{z}\hat{z} \quad (6.7.2b)$$

where

$$I_x = \frac{m(a^2 + b^2)}{12} \qquad I_y = \frac{m(L^2 + b^2)}{12} \qquad I_z = \frac{m(L^2 + a^2)}{12} \quad (6.7.2c)$$

the absolute angular momentum $(\vec{H}_O)_{\text{plate}}$ of the rotating plate is then

$$(\vec{H}_O)_{\text{plate}} = \Omega(\hat{x}I_x \cos\theta - \hat{y}I_y \sin\theta) + I_z\dot{\theta}\hat{z} \tag{6.7.2d}$$

Therefore, as shown earlier, $d(\vec{H}_O)_{\text{plate}}/dt$ is obtained as follows:

$$\frac{d}{dt}(\vec{H}_O)_{\text{plate}} = I_x(\dot{\Omega}\cos\theta - \Omega\dot{\theta}\sin\theta)\hat{x} - I_y(\dot{\Omega}\sin\theta + \Omega\dot{\theta}\cos\theta)\hat{y}$$
$$+ I_z\ddot{\theta}\hat{z} + \vec{\omega}_{\text{plate}} \times \vec{H}_O$$
$$= [I_x(\dot{\Omega}\cos\theta - \Omega\dot{\theta}\sin\theta) + (I_y - I_z)\Omega\dot{\theta}\sin\theta]\hat{x}$$
$$- [I_y(\dot{\Omega}\sin\theta + \Omega\dot{\theta}\cos\theta) + (I_z - I_x)\Omega\dot{\theta}\cos\theta]\hat{y}$$
$$+ [I_z\ddot{\theta} + (I_x - I_y)\Omega^2(\sin\theta)\cos\theta]\hat{z} \tag{6.7.2e}$$

The plate is subjected to a moment $\vec{M}_O$ that is applied to it by the shaft. Such a moment is of the form

$$\vec{M}_O = M_x\hat{x} + M_y\hat{y} + M_z\hat{z} \tag{6.7.2f}$$

where the component M_z is only due to friction in the hinges at A and B (see Fig. 6.24). For frictionless hinges, $M_z = 0$. The $\hat{z}$ component of Eq. (6.7.2e) then yields the following equation:

$$I_z\ddot{\theta} + (I_x - I_y)\Omega^2(\sin\theta)\cos\theta = 0 \tag{6.7.2g}$$

Equation (6.7.2g), which only involves the unknown variable θ, is the differential equation of motion. For $\Omega = $ constant, the equation has constant coefficients, and this is the case considered here. The steps that led to Eq. (6.7.2g) could have been omitted, of course, and Eq. (6.7.2g) could have been obtained directly from Eq. (6.4.11) (p. 359), with ω_x, ω_y, and ω_z given by Eq. (6.7.2a) for the problem at hand. They were repeated here simply to avoid the impression that one needs to memorize (or even consult) Eqs. (6.4.9) to (6.4.11) to solve a problem. If you work in dynamics and do this type of work often, such memorization may occur naturally though, but the important point to keep in mind is that knowledge of the fundamental concepts, which goes together with the basic mathematical steps of the formulation of a dynamics problem, is what is most important.

Equation (6.7.2g) admits the equilibrium solution $\theta = $ constant $\triangleq \theta_e$, where either $\sin\theta_e = 0$ or $\cos\theta_e = 0$. Therefore, the equilibrium solutions corresponding to $\sin\theta_e = 0$ are $\theta_e = 0$ and $\theta_e = \pi$ (also 2π, 3π, etc., but they are equivalent to 0 and π), and those corresponding to $\cos\theta_e = 0$ are $\pi/2$ and $3\pi/2$. Of these four equilibrium solutions, only $\theta_e = 0$ and $\theta_e = \pi/2$ correspond to physically different orientations of the plate relative to the shaft.

b. To analyze the stability of the equilibrium of the system, let $\theta(t) \triangleq \theta_e + \theta_s(t)$ and proceed as indicated in Chapter 1. The following linearized differential equations are obtained for the perturbed motion about $\theta_e = 0$ and about $\theta_e = \pi/2$:

Linearization about $\theta_e = 0$: $\quad I_z\ddot{\theta}_s + (I_x - I_y)\Omega^2\theta_s = 0$

Linearization about $\theta_e = \pi/2$: $\quad I_z\ddot{\theta}_s - (I_x - I_y)\Omega^2\theta_s = 0$

These linearized differential equations disclose that when one equilibrium is infinitesimally stable, the other is unstable (because of the opposite signs of the θ_s terms in the differential equations). The equilibrium $\theta_e = 0$ is infinitesimally stable if $I_x > I_y$ and unstable if $I_x < I_y$. Other problems whose motion are governed by a differential equation similar to Eq. (6.7.2g) were analyzed in this book, and those analyses disclosed that the equilibrium $\theta_e = 0$ is stable when $I_x > I_y$. Thus, as per Eq. (6.7.2g) and the equations of (6.7.2c), the equilibrium $\theta_e = 0$ is stable if $a > L$ in Fig. 6.24 and unstable if $a < L$.

c. A plot of $d\theta/dt$ versus θ may be done by integrating Eq. (6.7.2g) numerically and then using the result of the numerical integration to generate the plot. However, there is no need to do the numerical integration at all since advantage may be taken of the fact that Eq. (6.7.2g) may be integrated once with respect to θ, as first presented in Chapter 1, to yield

$$\frac{1}{2}\dot\theta^2 - \frac{\Omega^2}{4}\frac{I_x - I_y}{I_z}(\cos^2\theta - \sin^2\theta) = \text{constant} \triangleq C$$

$$= \left[\frac{1}{2}\dot\theta^2 - \frac{\Omega^2}{4}\frac{I_x - I_y}{I_z}(\cos^2\theta - \sin^2\theta)\right]_{t=0} \quad \textbf{(6.7.2h)}$$

Figure 6.25 shows a plot of the angular velocity of the plate relative to the shaft, versus θ, for the case when $I_x > I_y$. The plot shows the curves that surround the stable equilibrium point $\theta_e = 0$ and that pass through the unstable equilibrium points $\theta_e = -\pi/2$ and $\pi/2$.

To avoid drawing a plot for a specific value of $\Omega^2(I_x - I_y)/I_z$, the normalized time $\tau = \Omega t\sqrt{(I_x - I_y)/I_z}$ was used and $d\theta/d\tau$, instead of $d\theta/dt$, was plotted

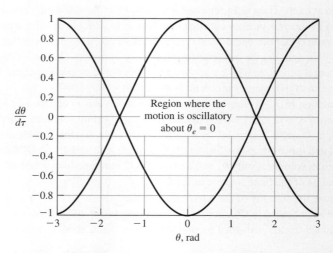

Figure 6.25 Plot of $d\theta/d\tau = \pm \cos\theta$ for Example 6.7.2, where $\tau = \Omega t\sqrt{(I_x - I_y)/I_z}$, for $I_x > I_z$.

versus θ. Such a plot[3] is valid for any positive value one wishes to choose for the quantity $\Omega\sqrt{(I_x - I_y)/I_z}$.

For an oscillatory motion about the equilibrium $\theta_e = 0$, the pair of values $\theta(0)$–$[d\theta/d\tau]_{\tau=0}$ must lie inside the region shown in Fig. 6.25. For the curves shown in Fig. 6.25 the value of the constant C in Eq. (6.7.2h) is obtained simply by setting $\dot\theta = 0$ and $\theta = \pi/2$ in that equation. That value is

$$C = \frac{\Omega^2}{4}\frac{I_x - I_y}{I_z}$$

Notice that this value of C yields the following relationship between $\dot\theta$ and θ for the curves shown in Fig. 6.25:

$$\frac{1}{2}\dot\theta^2 = \frac{\Omega^2}{4}\frac{I_x - I_y}{I_z}(\cos^2\theta - \sin^2\theta + 1) = \frac{\Omega^2}{2}\frac{I_x - I_y}{I_z}\cos^2\theta$$

or

$$\dot\theta = \pm\Omega\sqrt{\frac{I_x - I_y}{I_z}}|\cos\theta| \underbrace{=}_{\text{for the plate}} \pm\Omega\sqrt{\frac{a^2 - b^2}{a^2 + b^2}}\cos\theta \qquad (6.7.2\text{i})$$

As a specific example, for a plate with $a = 2L$ and $\pi/2 < \theta(0) < \pi/2$, this equation and inspection of Fig. 6.25 discloses that the motion is oscillatory as long as $|\dot\theta(0)| < \sqrt{\frac{3}{5}}|\Omega\cos\theta(0)|$.

6.8 Angular Orientation of a Rigid Body: Orientation Angles

It was seen that a rigid body rotating about a fixed point has three degrees of freedom, while a rigid body translating and rotating in space has six. Three of the six degrees of freedom of the body describe the translation of a point of the body (or of a rigid extension of it), which is normally chosen to be the center of mass C of the body (for which one has $\vec F = m\vec a_C$). The other three degrees of freedom describe the rotational motion of the body, for which one has $\vec M_C = d\vec H_C/dt$.

---

[3] The plot was generated with the following Simulink model with the function `[1,-1]*abs(cos(u))` typed in the MATLAB Function block, and the `output width` of the block set (after double clicking the block) to 2. In the `Solver` part of the `Simulation → Parameters` menu, the start time is -3 and the final time is 3. The time here is the same as the angle θ for the plot. The same plot may also be generated with the four MATLAB commands given here, which are separated by the semicolons.

```
theta=-3:0.01:3; f=abs(cos(theta)); plot(theta,f,'-',theta,-f,'-'); grid
```

The rotational motion of a body is described by its instantaneous orientation (also called the *attitude*) relative to some reference frame that is chosen to suit the particular application that is being investigated. For example, the convenient reference frame used for investigating the motion of an aircraft or a ship relative to the earth is a set of axes fixed to the earth. The orientation of an earth-orbiting satellite, for example, is normally referred to a set of earth-orbiting reference axes that consist of an axis that points to the center of the earth, a second axis that is perpendicular to the orbital plane of the center of mass of the satellite, and a third axis that is normal to the other two. As another example, the reference frame for investigating the trajectory of unwanted particles that may have entered a jet engine (and thus, are likely to cause erosion of the engine's rotating blades) might be a set of axes fixed to a rotating blade; in such a case, a set of coordinates chosen in that reference frame would be desirable to use so that the investigation of the particle's trajectories might disclose regions of the blade where erosion caused by those particles might occur.

> However, notice that when applying Newton's second law $\vec{F} = m\vec{a}_C$ and the moment equation (Euler's equation) $\vec{M}_C = d\vec{H}_C/dt$ to any problem, one ultimately needs to work with position vectors relative to an inertial point, with angles measured from nonrotating reference lines, and the acceleration of the center of mass C and the angular acceleration of the bodies must be *absolute*. In other words, after introducing convenient intermediate reference frames for a particular problem, one must also account for the motion of the reference frame relative to inertial space.

One widely used way to specify the orientation of a body relative to a desired reference frame was introduced by Euler. It consists of describing the orientation of a set of body-fixed axes, say x-y-z, relative to a chosen set of reference axes x_0-y_0-z_0, by first aligning x-y-z with x_0-y_0-z_0 and then performing three successive rotations θ_1, θ_2, θ_3 of the body axes in order to bring them into alignment with their actual orientation relative to the chosen reference frame. Such angles are called *orientation angles,* and there are 24 possible sets of them, 12 of which involve rotations about an updated orientation of the same axis (as originally introduced by Euler), as will be seen in the following. They are also often referred to as *Euler angles,* although the original Euler angles consist only of some of the 12 sets of orientation angles just mentioned. For some problems, a particular sequence of orientation angles is preferred over the others when it leads to equations that are simpler (but generally never simple!) to analyze. It is fair to say, however, that the use of a particular sequence over any other is often a matter of personal preference, and this can create some difficulties in comparing results of attitude dynamics problems done by different persons with different orientation angle sequences for the same problem. Figure 6.26a shows a possible orientation angle sequence θ_1, θ_2, and θ_3.

The sequence shown in Fig. 6.26 describes the orientation of a set of orthogonal body-fixed axes x-y-z (represented by the unit vectors $\hat{x}$, $\hat{y}$, and $\hat{z}$,

SECTION 6.8 Angular Orientation of a Rigid Body: Orientation Angles

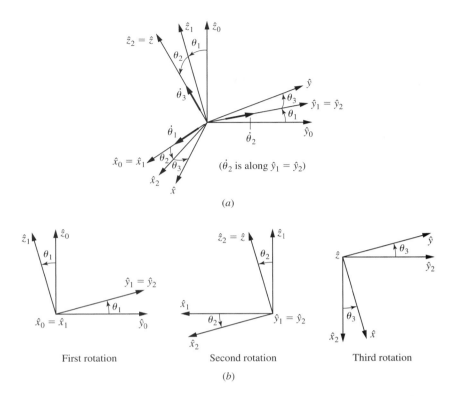

Figure 6.26 A three-axis orientation sequence to describe the orientation of a body.

respectively, that are parallel to those axes) with respect to the orthogonal reference axes x_0-y_0-z_0 (also represented by the unit vectors $\hat{x}_0$, $\hat{y}_0$, and $\hat{z}_0$, respectively). Other unit vectors that are parallel to the orientation of the axes after they are rotated are also shown. Referring to Fig. 6.26a, start by aligning x-y-z with x_0-y_0-z_0, i.e., start with $\hat{x} = \hat{x}_0$, $\hat{y} = \hat{y}_0$, and $\hat{z} = \hat{z}_0$. Then, perform the following successive rotations:

First rotation. A counterclockwise (counterclockwise simply by choice; it could be clockwise) rotation θ_1 about x_0 to bring the body axes to their first intermediate orientation x_1-y_1-z_1 (with $\hat{x}_1 = \hat{x}_0$). The angular velocity of the x_1-y_1-z_1 orthogonal axis system relative to the x_0-y_0-z_0 reference system, associated with this rotation, is simply $\dot{\theta}_1 \hat{x}_0$, and it is shown by a bold arrow in Fig. 6.26a.

Second rotation. A rotation θ_2 about the new orientation y_1 of the y axis to bring the body axes to their second intermediate orientation x_2-y_2-z_2 (with $\hat{y}_2 = \hat{y}_1$). The angular velocity of the x_2-y_2-z_2 system relative to the x_1-y_1-z_1 system, associated with this second rotation, is simply $\dot{\theta}_2 \hat{y}_1$, and it is also shown by a bold arrow in Fig. 6.26a.

Third and final rotation. A rotation θ_3 about the new and final orientation of the z axis to bring the body axes to their final orientation x-y-z (with $\hat{z} = \hat{z}_2$). The angular velocity of the body-fixed x-y-z axis system relative to the x_2-y_2-z_2 system, associated with this third rotation, is simply $\dot{\theta}_3 \hat{z}$, and it is also shown by a bold arrow in Fig. 6.26a.

The rotation sequence shown in Fig. 6.26 is widely used in aircraft and satellite dynamics, and in many applications in mechanical engineering and in structural dynamics. All three axes of the body were used in these rotations and, for this reason, they are known as *three-axis, body-fixed* angles. There are six possible rotation sequences of this type, with two different rotations performed after the first rotation.[4]

Note: Instead of using rotations about the updated orientations of the body-fixed axes, one could also use rotations that are always about the x_0, y_0, and z_0 axes. There are also $2 + 2 + 2 = 6$ possible combinations when all three axes are used, but such rotations are generally not used in practice. Such orientation angles are called *three-axis, space-fixed* angles.

Once an angle sequence is chosen to describe the orientation of the body, the next task is to obtain the expression for the absolute angular velocity $\vec{\omega}^B$ of the body. Having the expression for $\vec{\omega}^B$ one can then obtain the expression for the absolute angular momentum of the motion and, from there, the differential equations for investigating the rotational motion of the body.

The angular velocity $\vec{\omega}$ of the body, relative to the reference frame x_0-y_0-z_0, is simply the sum of the individual angular velocities $\dot{\theta}_1 \hat{x}_0$, $\dot{\theta}_2 \hat{y}_1$, and $\dot{\theta}_3 \hat{z}$, which are associated with each rotation. If the x_0-y_0-z_0 axes are inertial, $\vec{\omega} = \dot{\theta}_1 \hat{x}_0 + \dot{\theta}_2 \hat{y}_1 + \dot{\theta}_3 \hat{z}$ is the absolute angular velocity $\vec{\omega}^B$ of the body. If the x_0-y_0-z_0 directions are not inertial, the absolute angular velocity of the body is simply obtained by adding to $\vec{\omega}$ the absolute angular velocity of the x_0-y_0-z_0 frame.

The expression $\vec{\omega} = \dot{\theta}_1 \hat{x}_0 + \dot{\theta}_2 \hat{y}_1 + \dot{\theta}_3 \hat{z}$, obtained by inspection of Fig. 6.26, involves three nonorthogonal unit vectors. One now needs to express it in terms of an orthogonal triad, and such a transformation, first presented in Chapter 1, is simply obtained with the aid of the snapshots shown in Fig. 6.26b. In terms of the $\hat{x}$, $\hat{y}$, and $\hat{z}$ unit vectors that are parallel to the body-fixed x-y-z axes, the following expressions for $\hat{x}_0$ and $\hat{y}_1$ (which appear in the expression for $\vec{\omega}$) are written by inspection of Fig. 6.26b:

$$\hat{x}_0 = \hat{x}_1 = \hat{x}_2 \cos\theta_2 + \hat{z} \sin\theta_2 = (\hat{x} \cos\theta_3 - \hat{y} \sin\theta_3) \cos\theta_2 + \hat{z} \sin\theta_2$$
$$\hat{y}_1 = \hat{y}_2 = \hat{y} \cos\theta_3 + \hat{x} \sin\theta_3 \tag{6.8.1}$$

---

[4] For the sequence starting with a rotation about x_0, the second rotation could have been about z_1 (instead of y_1), and the third rotation about y. This accounts for the two possible sequences that start with θ_1 about x_0. Two other possible sequences would start with a rotation about y_0. Similarly, the other two would start with a rotation about z_0. This brings the total number of such rotation sequences to the six that were mentioned.

This then gives the following expression for the angular velocity $\vec{\omega}$ of the body, relative to the x_0-y_0-z_0 reference frame:

$$\vec{\omega} = \dot{\theta}_1 \hat{x}_0 + \dot{\theta}_2 \hat{y}_1 + \dot{\theta}_3 \hat{z}$$
$$= (\dot{\theta}_1 \cos\theta_2 \cos\theta_3 + \dot{\theta}_2 \sin\theta_3)\hat{x} + (\dot{\theta}_2 \cos\theta_3 - \dot{\theta}_1 \cos\theta_2 \sin\theta_3)\hat{y}$$
$$+ (\dot{\theta}_3 + \dot{\theta}_1 \sin\theta_2)\hat{z}$$
$$\triangleq \omega_x \hat{x} + \omega_y \hat{y} + \omega_z \hat{z} \qquad (6.8.2)$$

If the x_0-y_0-z_0 axes are nonrotating (i.e., inertial), these expressions for the angular velocity components ω_x, ω_y, and ω_z are the ones that go into Euler's equations (6.4.9) to (6.4.11), developed in Section 6.4.

Figure 6.26b can also be used for quickly developing expressions that relate any of the unit vectors in that figure. For example, the following relations involving the orientation of the unit vectors of a triad before and after a rotation, are readily obtained. They are written here in matrix form:

$$\begin{bmatrix} \hat{x}_0 \\ \hat{y}_0 \\ \hat{z}_0 \end{bmatrix} = \begin{bmatrix} 1 & 0 & 0 \\ 0 & \cos\theta_1 & -\sin\theta_1 \\ 0 & \sin\theta_1 & \cos\theta_1 \end{bmatrix} \begin{bmatrix} \hat{x}_1 \\ \hat{y}_1 \\ \hat{z}_1 \end{bmatrix} \triangleq [A_1] \begin{bmatrix} \hat{x}_1 \\ \hat{y}_1 \\ \hat{z}_1 \end{bmatrix} \qquad (6.8.3)$$

$$\begin{bmatrix} \hat{x}_1 \\ \hat{y}_1 \\ \hat{z}_1 \end{bmatrix} = \begin{bmatrix} \cos\theta_2 & 0 & \sin\theta_2 \\ 0 & 1 & 0 \\ -\sin\theta_2 & 0 & \cos\theta_2 \end{bmatrix} \begin{bmatrix} \hat{x}_2 \\ \hat{y}_2 \\ \hat{z}_2 \end{bmatrix} \triangleq [A_2] \begin{bmatrix} \hat{x}_2 \\ \hat{y}_2 \\ \hat{z}_2 \end{bmatrix} \qquad (6.8.4)$$

$$\begin{bmatrix} \hat{x}_2 \\ \hat{y}_2 \\ \hat{z}_2 \end{bmatrix} = \begin{bmatrix} \cos\theta_3 & -\sin\theta_3 & 0 \\ \sin\theta_3 & \cos\theta_3 & 0 \\ 0 & 0 & 1 \end{bmatrix} \begin{bmatrix} \hat{x} \\ \hat{y} \\ \hat{z} \end{bmatrix} \triangleq [A_3] \begin{bmatrix} \hat{x} \\ \hat{y} \\ \hat{z} \end{bmatrix} \qquad (6.8.5)$$

Each of these matrices, $[A_1]$, $[A_2]$, and $[A_3]$, is a simple *transformation matrix* that involves only one rotation. These matrices were first introduced in Chapter 1. Their determinant is always equal to 1, and their inverse is always equal to their transpose.

From Eqs. (6.8.3) to (6.8.5), the transformation from $\{\hat{x}, \hat{y}, \hat{z}\}$ to $\{\hat{x}_0, \hat{y}_0, \hat{z}_0\}$ (i.e., $\{\hat{x}_0, \hat{y}_0, \hat{z}_0\}$ written in terms of $\{\hat{x}, \hat{y}, \hat{z}\}$) is seen to be given as follows:

$$\begin{bmatrix} \hat{x}_0 \\ \hat{y}_0 \\ \hat{z}_0 \end{bmatrix} \triangleq [A] \begin{bmatrix} \hat{x} \\ \hat{y} \\ \hat{z} \end{bmatrix} \qquad (6.8.6)$$

where the transformation matrix $[A]$ is

$$[A] = [A_1][A_2][A_3]$$
$$= \begin{bmatrix} \cos\theta_2 \cos\theta_3 & -\cos\theta_2 \sin\theta_3 & \sin\theta_2 \\ \sin\theta_1 \sin\theta_2 \cos\theta_3 + \cos\theta_1 \sin\theta_3 & -\sin\theta_1 \sin\theta_2 \sin\theta_3 + \cos\theta_1 \cos\theta_3 & -\sin\theta_1 \cos\theta_2 \\ -\cos\theta_1 \sin\theta_2 \cos\theta_3 + \sin\theta_1 \sin\theta_3 & \cos\theta_1 \sin\theta_2 \sin\theta_3 + \sin\theta_1 \cos\theta_3 & \cos\theta_1 \cos\theta_2 \end{bmatrix}$$

$$(6.8.7)$$

EXAMPLE 6.8.1

For the rotation sequence shown in Fig. 6.26, with $\theta_1 = 20°, \theta_2 = 30°$, and $\theta_3 = 15°$, determine the smallest angle between the unit vectors $\hat{y}_0$ and $\hat{y}$, and also between $\hat{z}_1$ and $\hat{y}$.

■ Solution

As indicated in Chapter 1, the angle θ between two unit vectors $\hat{a}$ and $\hat{b}$ is simply obtained from the dot product $\hat{a} \bullet \hat{b} = |\hat{a}| |\hat{b}| \cos\theta = \cos\theta$. From Eq. (6.8.6), with $[A]$ given by Eq. (6.8.7), we then obtain

$$\hat{y}_0 \bullet \hat{y} = \cos\theta_1 \cos\theta_3 - \sin\theta_1 \sin\theta_2 \sin\theta_3 \approx 0.863 \stackrel{\Delta}{=} \cos\alpha$$

This gives $\alpha \approx 30.3°$ and $\alpha = 360 - 30.3 = 329.7°$. The smallest angle between $\hat{y}_0$ and $\hat{y}$ is then $\alpha = 30.3°$.

The angle (which we will call β) between $\hat{z}_1$ and $\hat{y}$ is obtained from the 3-2 element of the product $[A_2][A_3]$ of the $[A_2]$ and $[A_3]$ matrices given by Eqs. (6.8.4) and (6.8.5), respectively. You should find the following:

$$\hat{z}_1 \bullet \hat{y} = \sin\theta_2 \sin\theta_3 \approx 0.129 \stackrel{\Delta}{=} \cos\beta$$

This gives $\beta \approx 82.6°$ and $\beta = 360 - 82.6 = 277.4°$. The smallest angle between $\hat{z}_1$ and $\hat{y}$ is then $\beta = 82.6°$.

For some applications, such as the investigation of the motion of a top or a gyroscope, two-axis, body-fixed angles (the original Euler angles) are more "natural" to use. A rotation sequence, which we will call ϕ_1, ϕ_2, ϕ_3 here,[5] is illustrated in Fig. 6.27a. Figure 6.27b shows snapshots of the individual rotations.

The following sequence of rotations is performed in Fig. 6.27 after aligning $\hat{x}$ with $\hat{x}_0$, $\hat{y}$ with $\hat{y}_0$, and $\hat{z}$ with $\hat{z}_0$:

First rotation. A counterclockwise (again, counterclockwise simply by choice; it could be clockwise) rotation ϕ_1 about z_0 to bring the body axes to their first intermediate orientation x_1-y_1-z_1 (with $\hat{z}_1 = \hat{z}_0$). The angular velocity of the x_1-y_1-z_1 axis system relative to the x_0-y_0-z_0 reference system, associated with this rotation, is simply $\dot{\phi}_1 \hat{z}_0$, and it is shown by a bold arrow in Fig. 6.27a.

---

[5] If one is investigating the same problem using two different sequences of orientation angles, it is good practice to use different symbols for the angles of both sequences. Such a practice is advisable because the corresponding angles in two different sequences are different variables, and they will not be equal to each other. Using the same symbols for the angles of the two different sequences will cause confusion and can lead to erroneous conclusions about the motion.

SECTION 6.8 Angular Orientation of a Rigid Body: Orientation Angles

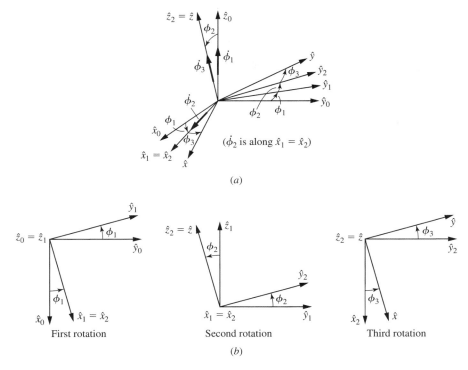

Figure 6.27 A two-axis orientation sequence to describe the orientation of a body.

Second rotation. A rotation ϕ_2 about the new orientation x_1 of the x axis to bring the body axes to their second intermediate orientation x_2-y_2-z_2 (with $\hat{x}_2 = \hat{x}_1$). The angular velocity of the x_2-y_2-z_2 system relative to the x_1-y_1-z_1 system, associated with this second rotation, is simply $\dot{\phi}_2 \hat{x}_1$, and it is also shown by a bold arrow in Fig. 6.27a.

Third and final rotation. A rotation ϕ_3 about the new and final orientation of the z axis, again, to bring the body axes to their final orientation x-y-z (with $\hat{z} = \hat{z}_2$). The angular velocity of the body-fixed x-y-z axis system relative to the x_2-y_2-z_2 system, associated with this third rotation, is simply $\dot{\phi}_3 \hat{z}$, and it is also shown by a bold arrow in Fig. 6.27a.

For two-axis orientation angles, the third and final rotation is performed about the updated position of the axes about which the first rotation was made. As for the case of three-axis, body-fixed orientation angles, there are also six possible rotation sequences for two-axis, body-fixed angles, with two different ones following the first rotation, as indicated earlier (which brings the total number of combinations to 12). In addition there are also 6 possible combinations for two-axis, space-fixed orientation angles (where the rotations are always about two of the x_0, y_0, and z_0 axes), thus bringing the total number of combinations to 24. Space-fixed orientation angles are generally not used in practice.

With the two-axis, space-fixed rotation sequence just illustrated, the angular velocity $\vec{\omega}$ of the body, relative to the reference frame x_0-y_0-z_0, is $\vec{\omega} = \dot{\phi}_1 \hat{z}_0 + \dot{\phi}_2 \hat{x}_1 + \dot{\phi}_3 \hat{z}$. Such an expression is converted to the triad $\{\hat{x}, \hat{y}, \hat{z}\}$ that is parallel to the body-fixed x-y-z axes with the following transformations that are obtained by inspection of the three snapshots in Fig. 6.27b.

$$\hat{z}_0 = \hat{z}_1 = \hat{z}\cos\phi_2 + \hat{y}_2 \sin\phi_2 = \hat{z}\cos\phi_2 + (\hat{y}\cos\phi_3 + \hat{x}\sin\phi_3)\sin\phi_2$$
$$\hat{x}_1 = \hat{x}_2 = \hat{x}\cos\phi_3 - \hat{y}\sin\phi_3 \tag{6.8.8}$$

This then gives the following expression for the angular velocity $\vec{\omega}$ of the body relative to the x_0-y_0-z_0 reference frame:

$$\begin{aligned}\vec{\omega} &= \dot{\phi}_1 \hat{z}_0 + \dot{\phi}_2 \hat{x}_1 + \dot{\phi}_3 \hat{z} \\ &= (\dot{\phi}_1 \sin\phi_2 \sin\phi_3 + \dot{\phi}_2 \cos\phi_3)\hat{x} + (\dot{\phi}_1 \sin\phi_2 \cos\phi_3 - \dot{\phi}_2 \sin\phi_3)\hat{y} \\ &\quad + (\dot{\phi}_3 + \dot{\phi}_1 \cos\phi_2)\hat{z} \\ &\triangleq \omega_x \hat{x} + \omega_y \hat{y} + \omega_z \hat{z}\end{aligned} \tag{6.8.9}$$

The individual transformations associated with each one of these rotations are given as

$$\begin{bmatrix}\hat{x}_0 \\ \hat{y}_0 \\ \hat{z}_0\end{bmatrix} = \begin{bmatrix}\cos\phi_1 & -\sin\phi_1 & 0 \\ \sin\phi_1 & \cos\phi_1 & 0 \\ 0 & 0 & 1\end{bmatrix} \begin{bmatrix}\hat{x}_1 \\ \hat{y}_1 \\ \hat{z}_1\end{bmatrix} \triangleq [B_1] \begin{bmatrix}\hat{x}_1 \\ \hat{y}_1 \\ \hat{z}_1\end{bmatrix} \tag{6.8.10}$$

$$\begin{bmatrix}\hat{x}_1 \\ \hat{y}_1 \\ \hat{z}_1\end{bmatrix} = \begin{bmatrix}1 & 0 & 0 \\ 0 & \cos\phi_2 & -\sin\phi_2 \\ 0 & \sin\phi_2 & \cos\phi_2\end{bmatrix} \begin{bmatrix}\hat{x}_2 \\ \hat{y}_2 \\ \hat{z}_2\end{bmatrix} \triangleq [B_2] \begin{bmatrix}\hat{x}_2 \\ \hat{y}_2 \\ \hat{z}_2\end{bmatrix} \tag{6.8.11}$$

$$\begin{bmatrix}\hat{x}_2 \\ \hat{y}_2 \\ \hat{z}_2\end{bmatrix} = \begin{bmatrix}\cos\phi_3 & -\sin\phi_3 & 0 \\ \sin\phi_3 & \cos\phi_3 & 0 \\ 0 & 0 & 1\end{bmatrix} \begin{bmatrix}\hat{x} \\ \hat{y} \\ \hat{z}\end{bmatrix} \triangleq [B_3] \begin{bmatrix}\hat{x} \\ \hat{y} \\ \hat{z}\end{bmatrix} \tag{6.8.12}$$

and this yields

$$\begin{bmatrix}\hat{x}_0 \\ \hat{y}_0 \\ \hat{z}_0\end{bmatrix} \triangleq [B] \begin{bmatrix}\hat{x} \\ \hat{y} \\ \hat{z}\end{bmatrix} \tag{6.8.13}$$

where

$$[B] = [B_1][B_2][B_3]$$
$$= \begin{bmatrix}\cos\phi_1 \cos\phi_3 - \sin\phi_1 \cos\phi_2 \sin\phi_3 & -\cos\phi_1 \sin\phi_3 - \sin\phi_1 \cos\phi_2 \cos\phi_3 & \sin\phi_1 \sin\phi_2 \\ \sin\phi_1 \cos\phi_3 + \cos\phi_1 \cos\phi_2 \sin\phi_3 & -\sin\phi_1 \sin\phi_3 + \cos\phi_1 \cos\phi_2 \cos\phi_3 & -\cos\phi_1 \sin\phi_2 \\ \sin\phi_2 \sin\phi_3 & \sin\phi_2 \cos\phi_3 & \cos\phi_2\end{bmatrix}$$
$$\tag{6.8.14}$$

Clearly, the transformation matrix $[B] = [B_1][B_2][B_3]$ obtained with the given rotation sequence is different from the $[A]$ matrix given by Eq. (6.8.7).

EXAMPLE 6.8.2

For the two-axis, body-fixed rotation sequence shown in Fig. 6.27, with $\phi_1 = 20°$, $\phi_2 = 30°$, and $\phi_3 = 15°$, determine the smallest angle between the unit vectors $\hat{y}_0$ and $\hat{y}$, and also between $\hat{z}_1$ and $\hat{y}$.

■ **Solution**

For this example we have for the angle between $\hat{y}_0$ and $\hat{y}$ (which we will call α_1),

$$\hat{y}_0 \bullet \hat{y} = \cos\phi_1 \cos\phi_2 \cos\phi_3 - \sin\phi_1 \sin\phi_3 \approx 0.697 \stackrel{\Delta}{=} \cos\alpha_1$$

This gives $\alpha_1 \approx 45.8°$ for the smallest angle between $\hat{y}_0$ and $\hat{y}$.

The angle (which we will call β_1) between $\hat{z}_1$ and $\hat{y}$ is also obtained from the 3-2 element of the product $[B_2][B_3]$, with the matrices $[B_2]$ and $[B_3]$ as given previously. You should find the following:

$$\hat{z}_1 \bullet \hat{y} = \sin\phi_2 \cos\phi_3 \approx 0.483 \stackrel{\Delta}{=} \cos\beta_1$$

This gives $\beta_1 \approx 61.1°$ for the smallest angle between $\hat{z}_1$ and $\hat{y}$.

EXAMPLE 6.8.3

Gravitational Moment on a Spacecraft

There are many cases where the external forces acting on a body depend on the orientation of the body. Such is the case for the aerodynamic and hydrodynamic forces acting, respectively, on aircraft and ships, for example. The gravitational force acting on an orbiting satellite is another example (see Problem 4.15, p. 263, which is based on Example 1.10.1, p. 23).

In this example, the expression for the gravitational moment $\vec{M}_C$ about the center of mass C of a rigid body of mass m in circular orbit around a planet is developed. The moment is certainly very small, but it cannot be neglected since it causes the orientation of orbiting bodies to slowly change with time. Such a moment depends on the spatial orientation of the body, and orientation angles are used for describing the rotational dynamics of the body relative to a convenient "orbiting reference frame."

The situation is illustrated in Fig. 6.28, where O is the center of the attracting planet. The planet is modeled as a homogeneous sphere of mass M and radius R, as done in Chapters 1 and 4. The scaling of the figure is highly exaggerated since only bodies with maximum dimension that is much smaller than $|\overrightarrow{OC}|$ are considered. This "restriction" is well justified in actual practical situations.

The convenient orbital reference frame previously mentioned is the set of orthogonal axes centered at C and parallel to the reference unit vectors $\hat{x}_0$ directed along the "local vertical" OC, $\hat{y}_0$ perpendicular to OC and in the orbital plane, and $\hat{z}_0 \stackrel{\Delta}{=} \hat{x}_0 \times \hat{y}_0$, which is perpendicular to the orbital plane. The radius of the circular orbit is $R = |\overrightarrow{OC}|$ and, as seen in Chapter 4, the orbital angular velocity is $\dot{\phi} \stackrel{\Delta}{=} \Omega = \sqrt{GM/R^3}$, where G is the universal gravitational constant. In terms of the mass M_E and radius R_E of the earth and of the acceleration g of gravity at the earth's surface, $GM_E = gR_E^2$.

Figure 6.28 shows an infinitesimal element of the body, with mass dm, and the infinitesimal gravitational force $d\vec{F}$ acting on it. The determination of the moment $\vec{M}_C$ is

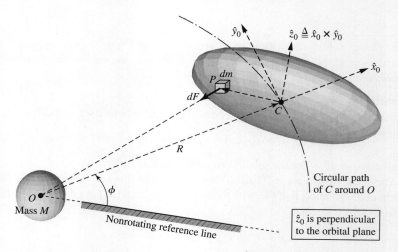

Figure 6.28 A rigid body subjected to the gravitational attraction of a planet, with scaling highly exaggerated for the sake of clarity.

done along the same steps followed in Example 1.10.1 (p. 23), with the exception that an integration that spans the entire volume of the rigid body needs to be done. The gravitational force $d\vec{F}$ acting on dm is calculated using Newton's law of universal gravitation (see Fig. 6.28):

$$d\vec{F} = -\frac{GM\,dm}{|\overrightarrow{OP}|^2}\frac{\overrightarrow{OP}}{|\overrightarrow{OP}|} \qquad (6.8.3a)$$

Therefore, the moment $\vec{M}_C$ is obtained by integrating $\overrightarrow{CP} \times d\vec{F}$ over the entire rigid body. With $\overrightarrow{OP} = \overrightarrow{OC} + \overrightarrow{CP} = R\hat{x}_0 + \overrightarrow{CP}$ (see Fig. 6.28), we then obtain

$$\vec{M}_C = \int_{body} \overrightarrow{CP} \times d\vec{F} = \int_{body} \overrightarrow{CP} \times \left(-\frac{GM\,dm}{|\overrightarrow{OP}|^2}\frac{\overrightarrow{OP}}{|\overrightarrow{OP}|}\right)$$

$$= -\int_{body} \frac{GM}{|\overrightarrow{OP}|^3}\overrightarrow{CP} \times (R\hat{x}_0 + \overrightarrow{CP})\,dm = -\int_{body} \frac{GMR}{|\overrightarrow{OP}|^3}(\overrightarrow{CP} \times \hat{x}_0)\,dm$$

$$(6.8.3b)$$

But (again, see Fig. 6.28),

$$|\overrightarrow{OP}|^2 = \overrightarrow{OP} \cdot \overrightarrow{OP} = (R\hat{x}_0 + \overrightarrow{CP}) \cdot (R\hat{x}_0 + \overrightarrow{CP}) = R^2 + 2R\hat{x}_0 \cdot \overrightarrow{CP} + |\overrightarrow{CP}|^2$$

$$= R^2\left[1 + 2\hat{x}_0 \cdot \frac{\overrightarrow{CP}}{R} + \underbrace{\left(\frac{|\overrightarrow{CP}|}{R}\right)^2}_{\text{negligible}}\right] \approx R^2\left(1 + 2\hat{x}_0 \cdot \frac{\overrightarrow{CP}}{R}\right) \qquad (6.8.3c)$$

where, for bodies whose maximum dimension is much smaller than R, the quadratic term $\left(|\overrightarrow{CP}|/R\right)^2$ is neglected when compared to $|\overrightarrow{CP}|/R$. In Eq. (6.8.3c), $|\overrightarrow{CP}|/R$ is certainly much smaller than 1, but if it is neglected, its small (but important) effect on the moment $\vec{M}_C$ is lost. Therefore, it must be kept.

By substituting in Eq. (6.8.3b) the expression for $|\overrightarrow{OP}|$ obtained from Eq. (6.8.3c), and expanding the term $|\overrightarrow{OP}|^{-3}$ in Taylor series for small $|\overrightarrow{CP}|/R$, the following approximate expression for $\vec{M}_C$ is obtained:

$$\vec{M}_C = -\int_{\text{body}} \frac{GM}{R^2}\left(1 - 3\hat{x}_0 \bullet \frac{|\overrightarrow{CP}|}{R}\right)(\overrightarrow{CP} \times \hat{x}_0)\, dm$$

$$= \frac{3GM}{R^3}\int_{\text{body}} (\hat{x}_0 \bullet \overrightarrow{CP})(\overrightarrow{CP} \times \hat{x}_0)\, dm \qquad (6.8.3\text{d})$$

In this equation, $\int_{\text{body}} \overrightarrow{CP}\, dm = 0$ because C is the center of mass of the body. Also, notice that terms that do not affect the integration in dm can be (and were) moved outside the integrand.

To proceed, the integral in Eq. (6.8.3d) needs to be evaluated. Since the integrand involves products of the components of the vector $\overrightarrow{CP}$, moments and products of inertia are expected to appear in the final result. To avoid products of inertia, and also moments of inertia that are time dependent, let us then proceed by using the central principal axes of the body. Let us denote such axes as x-y-z, and introduce a set of orientation variables to describe their orientation relative to x_0-y_0-z_0 orbital reference axes that are parallel to the $\hat{x}_0$-$\hat{y}_0$-$\hat{z}_0$ unit vector shown in Fig. 6.28. With $\hat{x}$-$\hat{y}$-$\hat{z}$ being unit vectors parallel to x-y-z, respectively, the orientation of the body may be described by first aligning $\hat{x}$-$\hat{y}$-$\hat{z}$ with $\hat{x}_0$-$\hat{y}_0$-$\hat{z}_0$ and then performing three successive rotations as shown earlier in this section. It was seen that the transformation relating $\hat{x}_0$-$\hat{y}_0$-$\hat{z}_0$ and $\hat{x}$-$\hat{y}$-$\hat{z}$ is of the form

$$\begin{bmatrix}\hat{x}_0\\\hat{y}_0\\\hat{z}_0\end{bmatrix} = \begin{bmatrix}A_{11} & A_{12} & A_{13}\\\ldots & \ldots & \ldots\\\ldots & \ldots & \ldots\end{bmatrix}\begin{bmatrix}\hat{x}\\\hat{y}\\\hat{z}\end{bmatrix} \triangleq [A]\begin{bmatrix}\hat{x}\\\hat{y}\\\hat{z}\end{bmatrix} \qquad (6.8.3\text{e})$$

The elements of matrix $[A]$ when three-axis and two-axis, body-fixed angles are used are given in Eqs. (6.8.7) (p. 397) and (6.8.14) (p. 400), respectively.

With $\hat{x}_0$ and $\overrightarrow{CP}$ expressed as

$$\hat{x}_0 = A_{11}\hat{x} + A_{12}\hat{y} + A_{13}\hat{z} \qquad (6.8.3\text{f})$$

$$\overrightarrow{CP} = x\hat{x} + y\hat{y} + z\hat{z} \qquad (6.8.3\text{g})$$

the integrand in Eq. (6.8.3f) becomes

$$(\hat{x}_0 \bullet \overrightarrow{CP})(\overrightarrow{CP} \times \hat{x}_0)$$
$$= (A_{11}x + A_{12}y + A_{13}z)[(yA_{13} - zA_{12})\hat{x} + (zA_{11} - xA_{13})\hat{y} + (xA_{12} - yA_{11})\hat{x}] \qquad (6.8.3\text{h})$$

and, when principal axes are used, this integrates to the following

$$\vec{M}_C = 3\frac{GM}{R^3}[(I_z - I_y)A_{12}A_{13}\hat{x} + (I_x - I_z)A_{11}A_{13}\hat{y} + (I_y - I_x)A_{11}A_{12}\hat{z}] \quad (6.8.3i)$$

where $I_x, I_y,$ and I_z are the three principal moments of inertia of the body, i.e.,

$$I_x = \int_{body}(y^2 + z^2)\,dm \qquad I_y = \int_{body}(x^2 + z^2)\,dm \qquad I_z = \int_{body}(x^2 + y^2)\,dm \quad (6.8.3j)$$

The expressions for $A_{11}, A_{12},$ and A_{13} for the three-axis, body-fixed angles are simpler than those for the two-axis, body-fixed angles, as disclosed by Eqs. (6.8.7) (p. 397) and (6.8.14) (p. 400), respectively. The three-axis, body-fixed angles presented in this section are the ones that appear widely in the technical literature dealing with investigations of rotational motions of satellites using Eq. (6.8.3i) for the gravitational moment. In terms of the angles $\theta_1, \theta_2,$ and θ_3 shown in Fig. 6.26, the expression for the moment $\vec{M}_C$ is then

$$\vec{M}_C = 3\frac{GM}{R^3}[-(I_z - I_y)(\sin\theta_2\cos\theta_2\sin\theta_3)\hat{x}$$
$$+ (I_x - I_z)(\sin\theta_2\cos\theta_2\cos\theta_3)\hat{y}$$
$$- (I_y - I_x)(\cos^2\theta_2\sin\theta_3\cos\theta_3)\hat{z}] \triangleq M_x\hat{x} + M_y\hat{y} + M_z\hat{z} \quad (6.8.3k)$$

> To analyze the rotational motion of a satellite, the expressions given for the moment components $M_x, M_y,$ and M_z are the ones that are used in Eqs. (6.4.9), (6.4.10), and (6.4.11), respectively, that were developed in Section 6.4. Such an analysis, which is an important problem in spacecraft dynamics, is left as a series of exercises (see Problems 6.21 to 6.24).

For a circular orbit, the term GM/R^3 in Eq. (6.8.3i) is constant and equal to the square of the orbital angular velocity Ω. As seen from the result given by Eq. (6.8.3i), the gravitational moment $\vec{M}_C$ acting on an orbiting body is proportional to the square of the orbital angular velocity $\Omega = \sqrt{GM/R^3}$, and it also depends on the principal inertia differences $I_x - I_y, I_x - I_z,$ and $I_y - I_z$. Since Ω is very small, the magnitude of $\vec{M}_C$ is also very small. Physically, such a moment is due to the change in the gravitational force of attraction with distance from the attracting planet. Such a moment is called the *gravity gradient moment*.

As an illustration of the orders of magnitude of the moment $\vec{M}_C$, consider a homogeneous thin bar of mass m and length L in a circular orbit 200 km above the surface of the earth. With the earth approximated as a sphere of radius $R_E = 6378$ km, and with $R = 6578$ km and $g = 9.81$ m/s^2, the orbital angular velocity in such a case is $\Omega = \sqrt{gR_E^2/R^3} \approx 0.0012$ rad/s. For motion in the orbital plane, for example, $\theta_1 = \theta_2 = 0$ in Fig. 6.26, and this yields

$$\vec{M}_C = -3\Omega^2\frac{mL^2}{12}(\sin\theta_3)(\cos\theta_3)\hat{z}$$

The maximum magnitude of $\vec{M}_C$ occurs for $\theta_3 = 45°$ in this case, and it is approximately equal to $1.75 \times 10^{-7}mL^2$ N, with m in kilograms and L in meters.

For the dumbbell of Example 1.10.1 (p. 23) and of Problem 4.15 (p. 263), with $I_x = 0$ and $I_y = I_z = mL^2/4$, one obtains $|\vec{M}_C|_{max} = 5.26 \times 10^{-7}\, mL^2$ N under the same conditions stated for the thin bar. Although such a moment is very small, it causes the orbiting dumbbell to oscillate about the local vertical with a long period of about 51 min for a circular orbit 200 km above the earth's surface (see Problem 4.15, p. 263, and also Problem 4.16).

A detailed analysis of the rotational dynamics of a spin-stabilized satellite, using the results developed in this section, is presented in Example 6.8.4. Such an analysis is another important problem in spacecraft dynamics. The entire analysis involves the formulation of the differential equations of motion, determination of the equilibrium solutions, linearization, and analysis of the stability of the equilibrium based on the linearized differential equations of motion.

Stability of a Spin-Stabilized Symmetric Satellite

EXAMPLE 6.8.4

Let the satellite illustrated in Fig. 6.28 (p. 402) be symmetric about its principal z axis so that $I_x = I_y$, as for a cylinder for example. The three-axis, body-fixed angles θ_1, θ_2, and θ_3 shown in Fig. 6.26 (p. 395) will be used for describing the rotational motion of the satellite relative to the orbital reference frame $\hat{x}_0$-$\hat{y}_0$-$\hat{z}_0$ seen in Fig. 6.28. The principal x, y, and z axes of the satellite are aligned, at all times, with the $\hat{x}$, $\hat{y}$, and $\hat{z}$ unit vectors, respectively, which are shown in Fig. 6.26. In the following, the differential equations for the rotational motion of a spin-stabilized satellite are developed. This is followed by the determination of the equilibrium solutions, and by the stability analysis.

Differential Equations of Motion

In terms of the angles θ_1, θ_2, and θ_3 shown in Fig. 6.26, the angular velocity of the satellite relative to the orbital reference frame $\hat{x}_0$-$\hat{y}_0$-$\hat{z}_0$ shown in Figs. 6.26 (p. 395) and 6.28 (p. 402) is given by Eq. (6.8.2) (p. 397). Since the absolute angular velocity of the $\{\hat{x}_0, \hat{y}_0, \hat{z}_0\}$ triad is $\dot{\phi}\hat{z}_0 \triangleq \Omega\hat{z}_0$ (see Fig. 6.28, p. 402), the absolute angular velocity of the satellite is then

$$\vec{\omega} = (\dot{\theta}_1 \cos\theta_2 \cos\theta_3 + \dot{\theta}_2 \sin\theta_3)\hat{x} + (\dot{\theta}_2 \cos\theta_3 - \dot{\theta}_1 \cos\theta_2 \sin\theta_3)\hat{y} + (\dot{\theta}_3 + \dot{\theta}_1 \sin\theta_2)\hat{z}$$
$$+ \Omega\hat{z}_0 \triangleq \omega_x \hat{x} + \omega_y \hat{y} + \omega_z \hat{z} \qquad (6.8.4a)$$

From either the three snapshots shown in Fig. 6.26 (p. 395) or from Eq. (6.8.6) (p. 397), the expression for $\hat{z}_0$ in terms of $\hat{x}$, $\hat{y}$, and $\hat{z}$ is [see Eq. (6.8.6), p. 397]:

$$\hat{z}_0 = (\sin\theta_1 \sin\theta_3 - \cos\theta_1 \sin\theta_2 \cos\theta_3)\hat{x}$$
$$+ (\sin\theta_1 \cos\theta_3 + \cos\theta_1 \sin\theta_2 \sin\theta_3)\hat{y} + (\cos\theta_1 \cos\theta_2)\hat{z} \qquad (6.8.4b)$$

This gives the following expressions for the components ω_x, ω_y, and ω_z of the absolute angular velocity $\vec{\omega}$ of the satellite:

$$\omega_x = (\dot{\theta}_1 \cos\theta_2 - \Omega \cos\theta_1 \sin\theta_2)\cos\theta_3 + (\dot{\theta}_2 + \Omega \sin\theta_1)\sin\theta_3 \qquad (6.8.4c)$$

$$\omega_y = (\dot{\theta}_2 + \Omega \sin\theta_1)\cos\theta_3 - (\dot{\theta}_1 \cos\theta_2 - \Omega \cos\theta_1 \sin\theta_2)\sin\theta_3 \qquad (6.8.4d)$$

$$\omega_z = \dot{\theta}_3 + \dot{\theta}_1 \sin\theta_2 + \Omega \cos\theta_1 \cos\theta_2 \qquad (6.8.4e)$$

These expressions are then used in Euler's differential equations, which are Eqs. (6.4.9) to (6.4.11), developed in Section 6.4. For $I_x = I_y$, and with $\vec{M}_C$ given by Eq. (6.8.3k) (p. 404), those equations yield the following differential equations for the rotational motion of the satellite

$$\dot{\omega}_x + k_1(\omega_y\omega_z + 3\Omega^2 \sin\theta_2 \cos\theta_2 \sin\theta_3) = 0 \quad (6.8.4\text{f})$$

$$\dot{\omega}_y - k_1(\omega_x\omega_z - 3\Omega^2 \sin\theta_2 \cos\theta_2 \cos\theta_3) = 0 \quad (6.8.4\text{g})$$

$$\dot{\omega}_z = 0 \quad (6.8.4\text{h})$$

where

$$k_1 \triangleq \frac{I_z - I_y}{I_x} = \frac{I_z}{I_x} - 1 \quad (6.8.4\text{i})$$

Notice that, using the definitions of moments of inertia given by the first three of Eqs. (6.3.6) (p. 340), it is seen that $k_1 = [\int_{\text{body}}(y^2 - z^2)\,dm]/[\int_{\text{body}}(y^2 + z^2)\,dm]$. Therefore, for any rigid body, the parameter k_1 is in the range $-1 \le k_1 \le 1$.

Equation (6.8.4h) immediately discloses that the $\hat{z}$ component of the absolute angular velocity of the satellite (i.e., its spin about its axis of symmetry) is constant, i.e.,

$$\omega_z = \dot{\theta}_3 + \dot{\theta}_1 \sin\theta_2 + \Omega \cos\theta_1 \cos\theta_2 = \text{constant} \quad (6.8.4\text{j})$$

Equilibrium Solutions

Equations (6.8.4f) and (6.8.4g) admit the equilibrium solutions $\theta_1 = \text{constant} \triangleq \theta_{1_e}$ and $\theta_2 = \text{constant} \triangleq \theta_{2_e}$, where $\theta_{1_e} = 0$ and $\theta_{1_e} = \pm\pi$, and $\theta_{2_e} = 0$ and $\theta_{2_e} = \pm\pi$. These equilibrium solutions represent the same physical situation, namely, a satellite with its symmetry axis perpendicular to the orbital plane and spinning relative to inertial space with a constant spin rate ω_z about that axis. The stability of such a state is analyzed next. The satellite is like a gyroscope in such a situation.

Stability Analysis of the Spin-Stabilized Satellite

The stability of each equilibrium is analyzed here by linearizing the differential equations of motion about the equilibrium solution $\theta_{1_e} = \theta_{2_e} = 0$, as just determined. Since $\theta_{1_e} = \theta_{2_e} = 0$, there is no need to introduce new symbols for the perturbed variables about the equilibrium. Thus, by letting $\theta_1(t)$ and $\theta_2(t)$ be small, the expressions for ω_x and ω_y in Eqs. (6.8.4c) and (6.8.4d) and for their first time derivatives become

$$\omega_x = (\dot{\theta}_1 - \Omega\theta_2)\cos\theta_3 + (\dot{\theta}_2 + \Omega\theta_1)\sin\theta_3 \quad (6.8.4\text{k})$$

$$\omega_y = (\dot{\theta}_2 + \Omega\theta_1)\cos\theta_3 - (\dot{\theta}_1 - \Omega\theta_2)\sin\theta_3 \quad (6.8.4\text{l})$$

$$\dot{\omega}_x = [\ddot{\theta}_1 - \Omega\dot{\theta}_2 + \dot{\theta}_3(\dot{\theta}_2 + \Omega\theta_1)]\cos\theta_3 + [\ddot{\theta}_2 + \Omega\dot{\theta}_1 - \dot{\theta}_3(\dot{\theta}_1 - \Omega\theta_2)]\sin\theta_3 \quad (6.8.4\text{m})$$

$$\dot{\omega}_y = [\ddot{\theta}_2 + \Omega\dot{\theta}_1 - \dot{\theta}_3(\dot{\theta}_1 - \Omega\theta_2)]\cos\theta_3 - [\ddot{\theta}_1 - \Omega\dot{\theta}_2 + \dot{\theta}_3(\dot{\theta}_2 + \Omega\theta_1)]\sin\theta_3 \quad (6.8.4\text{n})$$

Notice that no approximations should be made for θ_3 since that angle is continuously changing with time.

If the expressions for $\omega_x, \omega_y, \dot{\omega}_x$, and $\dot{\omega}_y$ given by Eqs. (6.8.4k) to (6.8.4n) are substituted into Eqs. (6.8.4f) and (6.8.4g), it can be verified that the following equations are obtained after linearization of the Ω^2 term due to the gravity gradient moment. The

SECTION 6.8 Angular Orientation of a Rigid Body: Orientation Angles

remaining lines of algebraic steps that lead to Eqs. (6.8.4o) and (6.8.4p) are left for you, the reader, to complete.

$$\ddot{\theta}_1 - \Omega\dot{\theta}_2 + [k_1 + \beta(1+k_1)]\Omega(\dot{\theta}_2 + \Omega\theta_1) = 0 \qquad (6.8.4o)$$

$$\ddot{\theta}_2 + \Omega\dot{\theta}_1 - [k_1 + \beta(1+k_1)]\Omega(\dot{\theta}_1 - \Omega\theta_2) + 3\Omega^2 k_1 \theta_2 = 0 \qquad (6.8.4p)$$

where β is a constant "spin parameter" defined as

$$\beta = \frac{\omega_z - \Omega}{\Omega} = \frac{\omega_z}{\Omega} - 1 \qquad (6.8.4q)$$

The general solutions to these equations are of the form $\theta_1 = Ae^{st}$ and $\theta_2 = Be^{st}$. As explained in Section 4.6.3 (p. 258), and also in Section 8.3, this yields the following algebraic equation (which is called the characteristic equation for this system) for the exponent s:

$$s^4 + a_2 \Omega^2 s^2 + a_0 \Omega^4 = 0 \qquad (6.8.4r)$$

where

$$a_2 = 1 + 3k_1 + k_1^2 + (1 + k_1)[2k_1 + (1 + k_1)\beta]\beta \qquad (6.8.4s)$$

$$a_0 = [k_1 + (1 + k_1)\beta][4k_1 + (1 + k_1)\beta] \qquad (6.8.4t)$$

As explained on p. 259 in Chapter 4, the conditions for the equilibrium to be infinitesimally stable are $a_2 > 0$, $a_0 > 0$, and $a_2^2 - 4a_0 > 0$. All of these three conditions must be satisfied for the equilibrium to be infinitesimally stable. The equilibrium is unstable if either $a_2 < 0$ or $a_0 < 0$ or $a_2^2 - 4a_0 < 0$. Numerical checks to verify stability by integrating the nonlinear differential equations of motion, i.e., Eqs. (6.8.4f) and (6.8.4g) with $\omega_z = (\beta + 1)\Omega$, are left as an exercise (see Problem 6.21).

Figure 6.29a shows the stability chart β versus k_1, for *any* spinning rigid symmetric satellite, disclosing the regions where the equilibrium is stable, and where it is unstable. Figure 6.29b illustrates, using a cylinder, how the parameter k_1 is affected by the shape of the body. For a homogeneous cylinder of radius a and length L, $k_1 = (3a^2 - L^2)/(3a^2 + L^2)$.

Looking at Fig. 6.29, it is seen that the equilibrium of a spinning satellite with $k_1 \approx 1$, for example, is unstable when $-\Omega < \omega_z < \Omega/2$. When $\omega_z = -\Omega$, the satellite is not spinning relative to inertial space. High spin rates, either clockwise (i.e., $\omega_z \gg 0$) or counterclockwise (i.e., $\omega_z \ll 0$), result in an infinitesimally stable equilibrium as indicated by the chart in Fig. 6.20. The satellite is like a fast spinning gyroscope in such cases.

The lines that separate the infinitesimally stable from the unstable regions in Fig. 6.29 are obtained from the equations $a_2 = 0$, $a_0 = 0$, and $a_2^2 - 4a_0 = 0$. The equation $a_2 = 0$ yields $\beta = (-k_1 \pm \sqrt{-1 - 3k_1})/(1 + k_1)$. The equation $a_0 = 0$ yields $\beta = -k_1/(1 + k_1)$ and $\beta = -4k_1/(1 + k_1)$. The equation $a_2^2 - 4a_0 = 0$ yields the following polynomial

$$\beta^4(1+k_1)^4 + 4k_1(1+k_1)^3\beta^3 + 2(1+k_1)^2(3k_1^2 + 3k_1 - 1)\beta^2$$
$$+ 4k_1(1+k_1)(k_1^2 + 3k_1 - 4) + (k_1^4 + 6k_1^3 - 5k_1^2 + 6k_1 + 1) = 0$$

whose real roots then yield additional lines in the β versus k_1 parameter space. The lines corresponding to $a_2 = 0$, $a_0 = 0$, and $a_2^2 - 4a_0 = 0$ are shown in Fig. 6.29 in order to

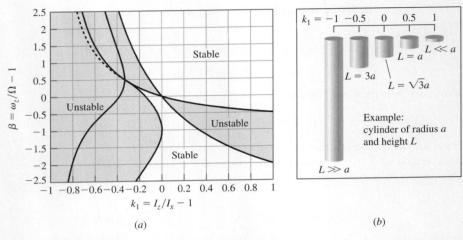

Figure 6.29 Stability chart for spinning satellites in circular orbit (dark region: unstable; white region: infinitesimally stable).

disclose how the stability chart was generated. To determine which region separated by those lines corresponds to an infinitesimally stable and to an unstable equilibrium, one only needs to test the signs of a_0, a_2, and $a_2^2 - 4a_0$ for any point on either "side" of those lines.

Except for the gray scaling coloring, which was done with a drawing software, the lines shown in Fig. 6.29 were generated with MATLAB, with the following commands. For help with any command that you may not understand, type `help command_name` in the MATLAB command window. For example, `help isreal` will show you what the `isreal` command does. The symbol ~ stands for the logical NOT operator. Also, see Section A.7 for a brief introduction to MATLAB.

```
k1=-0.99:0.01:1; len=length(k1); fig=figure; hold on; grid;
b1=(-k1+sqrt(-1-3*k1))./(1+k1); b2=(-k1-sqrt(-1-3*k1))./(1+k1);
b3=-k1./(1+k1); b4=4*b3; plot(k1,b1, k1,b2, k1,b3, k1,b4);
for kk=1:len,
    k=k1(kk), temp=1+k; c4=temp^4; c3=4*k*temp^3; c2=temp^2*(6*k^2+6*k-2);
    c1=4*k*temp*(k^2+3*k-4; c0=k^4+6*k^3-5k^2+6*k+1;
    r=roots([c4, c3, c2, c1, c0]);
    rr=[]; for kr=1:4, test=r(kr); if isreal(test), rr=[rr,test]; end; end;
    if ~isempty(rr), plot(k,rr); end;
end; axis([-1, 1, -2.5, 2.5])
```

6.9 Angular Velocity and the Orientation Angles: Notes

The rotational motion of a body is governed by the vector differential equation $d\vec{H}_C/dt = \vec{M}_C$, where $\vec{H}_C$ is related to the absolute angular velocity of the body by Eq. (6.3.7) (p. 340). When the body is able to move unconstrained in three-dimensional space, this gives the three scalar differential equations that govern the rotational motion of the body. When the moment $\vec{M}_C$ depends on the orientation of the body (as in the case of airplanes, ships, satellites, and artillery shells, for example), it is necessary to express the components of its absolute angular velocity in terms of the orientation angles and their time derivatives in order to be able to integrate the equations for determining the orientation of the body at any time. Assuming, for simplicity, that the orientation angles of the body are measured from a nonrotating reference frame x_0-y_0-z_0, the absolute angular velocity of the body is the one presented in Section 6.8 (and, thus, the components ω_x, ω_y, and ω_z are the ones that "go into" Euler's equations (6.4.9) to (6.4.11).

The time rate of change of the orientation angles θ_1, θ_2, and θ_3 may always be expressed in terms of the angular velocity components ω_x, ω_y, and ω_z. For the three-axis orientation angles θ_1, θ_2, and θ_3 illustrated in Fig. 6.26 (p. 395), for example, the following relations are obtained from the definitions of ω_x, ω_y, and ω_z introduced in Eq. (6.8.2) (p. 397):

$$\dot{\theta}_1 = \frac{\omega_x \cos\theta_3 - \omega_y \sin\theta_3}{\cos\theta_2}$$
$$\dot{\theta}_2 = \omega_x \sin\theta_3 + \omega_y \cos\theta_3 \qquad (6.9.1)$$
$$\dot{\theta}_3 = \omega_z - (\omega_x \cos\theta_3 - \omega_y \sin\theta_3)\tan\theta_2$$

Equations (6.9.1) disclose that the three-axis orientation angles of Fig. 6.26 have a "singularity" when $\cos\theta_2 = 0$, i.e., at $\theta_2 = \pm\pi/2$ rad (i.e., denominators become equal to zero). Actually, an inspection of Fig. 6.26 reveals that such a singularity occurs because the rotations θ_1 and θ_3 become indistinguishable from each other when $\theta_2 = \pm\pi/2$ (i.e., the rotations θ_1 and θ_3 are about the same axis in such a case).

For the two-axis orientation angles of Fig. 6.27, the expressions that relate $\dot{\phi}_1$, $\dot{\phi}_2$, and $\dot{\phi}_3$ to the new expressions of the components ω_x, ω_y, and ω_z defined by Eq. (6.8.9) (p. 400) are

$$\dot{\phi}_1 = \frac{\omega_x \sin\phi_3 + \omega_y \cos\phi_3}{\sin\phi_2}$$
$$\dot{\phi}_2 = \omega_x \cos\phi_3 - \omega_y \sin\phi_3 \qquad (6.9.2)$$
$$\dot{\phi}_3 = \omega_z - \frac{\omega_x \sin\phi_3 + \omega_y \cos\phi_3}{\tan\phi_2}$$

As indicated by Eqs. (6.9.2), the singularity for these two-axis orientation angles occurs when $\sin\phi_2 = 0$, i.e., when $\phi_2 = 0$ or $\phi_2 = \pi$ rad. By observing Fig. 6.27, the rotations ϕ_1 and ϕ_3 become indistinguishable from each other when either $\phi_2 = 0$ or $\phi_2 = \pi$ rad, and this is the cause of the singularity. Other orientation variables, such as *Euler parameters* can be used to avoid singularities. Interested readers can find details in more advanced books, such as Reference 13 in Appendix G.

6.10 Problems

6.1 Determine the central principal moments of inertia and the direction of the three principal axes of the bent bar of Example 6.3.1 (p. 344) (see Fig. 6.4). For this, take $L = 50$ cm, $R = 15$ cm, and the material density to be 0.63 g/cm, which is the approximate specific mass of a copper wire of 3 mm in diameter. Neglect the small diameter of the bar in your calculations.

6.2 Determine the central principal moments of inertia and the direction of the three principal axes of the triangular plate of Example 6.3.2 (p. 345) (see Fig. 6.5).

6.3 Two rectangular blocks are rigidly attached to each other as shown in Fig. 6.30, where $L = 1$ m, $a = 50$ cm, and $b = 40$ cm. The mass density of the larger block is $\rho_1 = 7700$ kg/m^3 and that of the smaller block is $\rho_2 = 2600$ kg/m^3. Determine, for the entire system:

a. The location of its center of mass
b. The central principal moments of inertia
c. The direction of the principal axes relative to the rectangular x-y-z coordinate system shown in the figure

Figure 6.30 Figure for Problem 6.3.

6.4 Based on the definitions of moments and products of inertia, which of the following are inertia matrices? Why?

$$\begin{bmatrix} 1 & -1 & \sqrt{3} \\ -1 & 2.5 & 0 \\ \sqrt{3} & 0 & 4 \end{bmatrix} \quad \begin{bmatrix} 3 & -2 & 3 \\ -2 & 9 & 2 \\ 3 & 2 & 7 \end{bmatrix} \quad \begin{bmatrix} 10 & 2 & -3 \\ 2 & 20 & -1 \\ -3 & -1 & 12 \end{bmatrix}$$

6.5 The inertia matrix of a body, referred to an orthogonal set of axes centered at the center of mass of the body, is

$$\begin{bmatrix} 5 & 0 & 3 \\ 0 & 7 & 0 \\ 3 & 0 & 4 \end{bmatrix} \text{ kg} \cdot \text{m}^2$$

What are the principal moments of inertia of the body?

6.6 A rigid body has the following moments and products of inertia about a set of orthogonal axes x-y-z that are centered at the center of mass C of the body and are fixed to the body (all quantities are in kg $\cdot$ m^2): $I_{xx} = 3$, $I_{yy} = 7$, $I_{zz} = 8$, $I_{xy} = 2$, $I_{xz} = 1$, $I_{yz} = -1$.

a. Using dyadics, determine the moments and products of inertia of the body about the axes x_s-y_s-z_s whose orientation with respect to the x-y-z axes is described by the rotation sequence shown in Fig. 6.31, with $\theta_1 = 30°$ and $\theta_2 = 45°$.

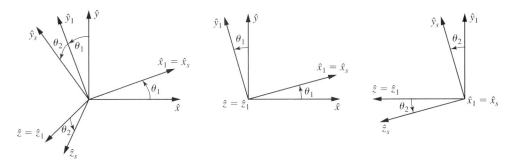

Figure 6.31 Figure for Problem 6.6.

b. Show that the solution to part a may also be obtained by matrix algebra as $[R]^T[I_C][R]$, where $[I_C]$ is the inertia matrix whose elements are as given previously and $[R]$ is the transformation matrix that relates the unit vectors $\hat{x}$-$\hat{y}$-$\hat{z}$ and $\hat{x}_s$-$\hat{y}_s$-$\hat{z}_s$ as

$$\begin{bmatrix} \hat{x} \\ \hat{y} \\ \hat{z} \end{bmatrix} = [R] \begin{bmatrix} \hat{x}_s \\ \hat{y}_s \\ \hat{z}_s \end{bmatrix}$$

Determine the nine elements of $[R]$.

6.7 A rigid body is translating and rotating in space with absolute angular velocity $\vec{\omega} = \omega_x \hat{x} + \omega_y \hat{y} + \omega_z \hat{z}$, where $\{\hat{x}, \hat{y}, \hat{z}\}$ are orthogonal unit vectors parallel to the x-y-z principal axes of the body. While rotating, the body is subjected to a moment $\vec{M}_C = M_x \hat{x} + M_y \hat{y} + M_z \hat{z}$ about its center of mass C. The components $\omega_x(t)$, $\omega_y(t)$, and $\omega_z(t)$ are measured by sensors, and the moment $\vec{M}_C$ is applied by actuators driven by a control system that guarantees that $M_x \omega_x + M_y \omega_y + M_z \omega_z = k \cos t$, where k is constant. Determine how the kinetic energy of the rotational motion of the body changes with time t for the cases when $k \neq 0$ and when $k = 0$. *Hint:* Use Euler's equations for principal axes.

6.8 A solid steel rotor (see Fig. 6.9, p. 355) of radius $R = 30$ cm and height $h = 30$ cm is given an initial absolute angular velocity $\vec{\omega} = 200\hat{z}$ rpm, where $\hat{z}$ is a unit vector parallel to the axial z axis shown in Fig. 6.9. The rotor is subjected to a moment $\vec{M}_C = k_1 t \hat{x} + k_2 \hat{y}$, where $k_1 = 4$ N · m/s and $k_2 = 2$ N · m, and $\hat{x}$ and $\hat{y} = \hat{z} \times \hat{x}$ are unit vectors parallel to the x and y axes shown in Fig. 6.9, which are fixed to the rotor. Determine the components $\omega_x(t)$, $\omega_y(t)$, and $\omega_z(t)$ of the absolute angular velocity $\vec{\omega} = \omega_x \hat{x} + \omega_y \hat{y} + \omega_z \hat{z}$ of the rotor. The density of steel is $\rho = 7700$ kg/m³, and the principal moments of inertia of a solid rotor are given in Example 6.3.7 (p. 354).

6.9 A solid steel rotor (see Fig. 6.9, p. 355) of radius $R = 30$ cm and height $h = 30$ cm is spinning about its axial z axis with a constant absolute angular velocity equal to 6000 rpm when a small disturbing moment of magnitude 0.1 N · m is applied to it in the direction perpendicular to the axial axis. *Estimate* the small change of orientation of the spin axis, in degrees, after 30 min have elapsed. The density of steel is $\rho = 7700$ kg/m³, and the principal moments of inertia of a solid rotor are given in Example 6.3.7 (p. 354).

6.10 A thin homogeneous rectangular plate of mass m and dimensions L and b is hinged to a rotating horizontal circular platform of radius $|\overrightarrow{OA}| = R$ as shown in

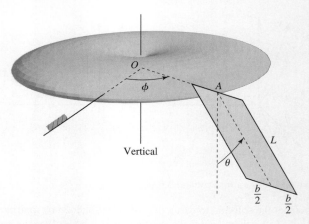

Figure 6.32 Figure for Problem 6.10.

Fig. 6.32. The center O of the platform is inertial, and the hinge at A only allows the plate to rotate about line OA relative to the platform. That rotation is tracked by the angle $\theta(t)$ shown in the figure. The absolute angular velocity of the platform is constant, i.e., $\dot{\phi} = \text{constant} \triangleq \Omega$.

a. Obtain the differential equation of motion for the system, all the equilibrium orientations $\theta = \text{constant} \triangleq \theta_e$ in terms of Ω, L, and b, and analyze their stability. Indicate whether they are stable, unstable, or asymptotically stable.

b. Determine the maximum and minimum values of $\theta(t)$ when $\theta(0) = 5°$ and $\dot{\theta}(0) = 0.08\sqrt{g/L}$, for $\Omega^2 L/g = 1$ and 2.

c. Verify the stability of each equilibrium solution for the system by sketching phase plane plots (i.e., $\dot{\theta}$ versus θ) for $\Omega^2 L/g = 1$ and 2.

6.11 For the system described in Problem 6.10, determine the value (or values) of θ that yield the maximum and minimum values of the component of the force reaction at A along the perpendicular to the plate when $\Omega^2 L/g = 1$ and when $\Omega^2 L/g = 3$. The motion is started with the initial conditions $\theta(0) = 30°$ and $\dot{\theta}(0) = 0$.

6.12 A thin homogeneous rectangular plate of mass M and dimensions a and B is welded to a shaft, in the plane of the plate, as shown in Fig. 6.33 (C is the center of mass of the plate). One of the sides of the plate is inclined by α degrees with the shaft, and the shaft is driven at a constant absolute angular velocity ω. It is desired to dynamically balance the system with four small masses (modeled as point masses), each with mass $m = M/48$, placed as shown in the figure. For $L = a/2$ and $b = 1.9L$, determine the angle β in order to achieve dynamic balancing.

6.13 A homogeneous steel rod PQ of mass m and length L is welded to a horizontal shaft as shown in Fig. 6.34, where $d_1 = |\overrightarrow{OA}|$ and $d_2 = |\overrightarrow{OB}|$. The bar is perpendicular to the shaft, but its center of mass C is at a distance e (called the *offset distance*) from the shaft midline. The shaft is driven at a constant absolute angular velocity Ω. For simplicity, neglect the mass of the shaft and approximate the bar as a thin line.

a. Determine the bearing reaction forces at A and B. Show that there is a component of those reaction forces that is proportional to $e\Omega^2$ and is

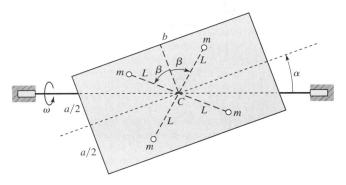

Figure 6.33 Figure for Problem 6.12.

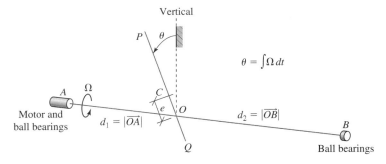

Figure 6.34 Figure for Problem 6.13.

perpendicular to the shaft, and rotates with the same angular velocity as the shaft.

b. What is the magnitude of the rotating component of the reaction force at A and B if $d_1 = d_2$, $m = 1$ kg, $\Omega = 600$ rpm, and $e = 6$ mm?

c. For $d_1 = d_2$, $m = 1$ kg, and $e = 6$ mm, what is the maximum value of Ω (in rpm) for the rotating component of the reaction force at A and B not to exceed 2 N?

6.14 The system shown in Fig. 6.35 consists of a thin homogeneous disc of radius R and mass m welded to a vertical shaft of mass m_{shaft} and length L. The plane of the disc is inclined by α degrees with the perpendicular to the shaft, and its center of mass is on the centerline of the shaft, with $|\vec{AC}| = L/2$. The shaft is rotating with a constant absolute angular velocity Ω and is supported by ball-and-socket joints at A and B.

a. Determine the components of the reaction forces at A and B that are perpendicular to the shaft. Express the answers in terms of the quantities m, R, L, α, and Ω.

b. For $m = 0.5$ kg, $R = 30$ cm, $L = 2$ m, and $\alpha = 30°$, determine the maximum value of Ω (in rpm) so that the rotating component of the reaction force at A indicated in part a does not exceed 2 N.

Figure 6.35 Figure for Problem 6.14. The bases at A and B are fixed.

Figure 6.36 Figure for Problem 6.15. The angle α is constant, and the base at A is fixed.

6.15 The system shown in Fig. 6.36 consists of a thin homogeneous disc of radius R and mass m welded to a bent shaft ABC, with lengths L_1 and L_2, respectively, and with C coinciding with the center of mass of the disc. Branch AB of the shaft is vertical, and the plane of the disc is perpendicular to branch BC of the shaft. The shaft is rotating with a constant absolute angular velocity Ω. Friction in the system, and the mass of the shaft, can be neglected.

  a. Determine the force and moment reactions at A.

  b. You should find a component of the reaction force and reaction moment that varies as Ω^2. Determine the maximum magnitude of those components in terms of m, R, L_1, L_2, and α.

  c. For $L_2 = 25$ cm and $\alpha = 30°$, what should be the maximum value of Ω so that the magnitude of the reaction force that is perpendicular to AB does not exceed mg?

6.16 The angular orientation of a body may be described by a set of three orientation angles, as presented in Section 6.8. As indicated in the note on p. 373 in Section 6.6.3, the differential equations of motion for the top analyzed in that section could have been formulated using the orientation angles for axes fixed to the rotating top. Such an orientation sequence is shown in Fig. 6.37, where the unit vectors $\hat{x}$, $\hat{y}$, and $\hat{z} \triangleq \hat{x} \times \hat{y}$ are parallel to principal x-y-z axes fixed to the rotating body.

The orientation of the system shown in Fig. 6.37 is described by first aligning the x-y-z body-fixed axes with the x_0-y_0-z_0 reference axes, as shown in Fig. 6.37a. The actual orientation of x-y-z relative to x_0-y_0-z_0 is then described by the rotation sequence ϕ-θ-ψ shown in Fig. 6.37b. This puts the system in its actual orientation shown in Fig. 6.37c. The moments of inertia of the system about the x-y-z axes that are centered at the fixed point O are I_x, $I_y = I_x$, and I_z. Using Euler's equations for principal axes, which are Eqs. (6.4.9) to (6.4.11) on p. 359:

  a. Obtain the differential equations of motion for the top in terms of the angles ϕ, θ, and ψ.

  b. Combine the appropriate equations obtained in part a and show that they yield Eqs. (6.6.3i) to (6.6.3k) that were formulated in Section 6.6.3, where I_n in those equations is the same as I_x in this problem.

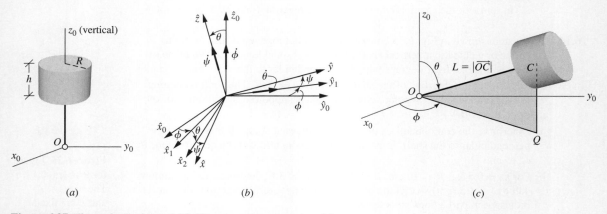

Figure 6.37 Figure for Problem 6.16. The axes x_0-y_0-z_0 are inertial.

6.17 For the rotation sequence shown in Fig. 6.26 (p. 395), $\theta_1 = 10°$, $\theta_2 = 30°$, and $\theta_3 = 20°$. Determine the angle the unit vector $\hat{y}$ makes with the unit vectors $\hat{x}_0$, $\hat{y}_0$, and $\hat{z}_0$.

6.18 For the rotation sequence shown in Fig. 6.26 (p. 395), $\theta_1 = 30°$, $\theta_2 = -30°$, and $\theta_3 = 45°$. Determine the direction cosines of the unit vectors $\hat{x}$, $\hat{y}$, and $\hat{z}$, relative to the unit vectors $\hat{x}_0$, $\hat{y}_0$, and $\hat{z}_0$.

6.19 For the rotation sequence shown in Fig. 6.27 (p. 399), $\phi_1 = 10°$, $\phi_2 = 30°$, and $\phi_3 = 20°$. Determine the angle the unit vector $\hat{y}$ makes with the unit vectors $\hat{x}_0$, $\hat{y}_0$, and $\hat{z}_0$.

6.20 For the rotation sequence shown in Fig. 6.27 (p. 399), $\phi_1 = 30°$, $\phi_2 = -30°$, and $\phi_3 = 45°$. Determine the direction cosines of the unit vectors $\hat{x}$, $\hat{y}$, and $\hat{z}$, relative to the unit vectors $\hat{x}_0$, $\hat{y}_0$, and $\hat{z}_0$.

6.21 Show that Eqs. (6.8.4f) and (6.8.4g) for a spin-stabilized symmetric satellite in circular orbit can be combined to yield the following two equations, which only involve θ_1 and θ_2:

$$\dot{\omega}_1 + \omega_2\dot{\theta}_3 + k_1\omega_z\omega_2 = 0$$

$$\dot{\omega}_2 - \omega_1\dot{\theta}_3 - k_1\omega_z\omega_1 + 3k_1\Omega^2 \sin\theta_2 \cos\theta_2 = 0$$

In these equations, $\omega_1 = \dot{\theta}_1 \cos\theta_2 - \Omega \cos\theta_1 \sin\theta_2$, $\omega_2 = \dot{\theta}_2 + \Omega \sin\theta_1$, $\dot{\theta}_3 = \omega_z - \dot{\theta}_1 \sin\theta_2 - \Omega \cos\theta_1 \cos\theta_2$, with $\omega_z = \Omega(\beta + 1)$ as given by Eq. (6.8.4q) in Example 6.8.4.

6.22 The center of mass C of a rigid body of central principal moments of inertia I_x, I_y, and I_z is in a circular orbit of radius R around the earth, as indicated in Fig. 6.28 (p. 402). The principal axes of the body, x-y-z, are aligned with the unit vectors $\hat{x}$-$\hat{y}$-$\hat{z}$ shown in Fig. 6.26 (p. 395).

a. What is the expression for the absolute angular velocity $\vec{\omega}$ of the body in terms of the three-axis, body-fixed orientation angles shown in Fig. 6.26 (p. 395), and of their time derivatives. Express the answer in the $\hat{x}$-$\hat{y}$-$\hat{z}$ triad.

b. Obtain the differential equations for the rotational motion of the body in terms of the three-axis, body-fixed orientation angles shown in Fig. 6.26 (p. 395). Approximate the earth as a solid homogeneous sphere of radius R_E, and treat its center as inertial.

c. Show that the differential equations admit the equilibrium solution $\theta_{1_e} = \theta_{2_e} = \theta_{3_e} = 0$, which physically means that the three principal axes of the body coincide with the orbital reference axes.

d. Linearize the differential equations obtained in part a about the equilibrium $\theta_{1_e} = \theta_{2_e} = \theta_{3_e} = 0$, and show that the linearized differential equations of motion are

$$\ddot{\theta}_1 + (k_1 - 1)\Omega\dot{\theta}_2 + k_1\Omega^2\theta_1 = 0$$

$$\ddot{\theta}_2 + (k_2 + 1)\Omega\dot{\theta}_1 - 4\Omega^2 k_2\theta_2 = 0$$

$$\ddot{\theta}_3 + 3k_3\Omega^2\theta_3 = 0$$

where

$$k_1 = \frac{I_z - I_y}{I_x} \qquad k_2 = \frac{I_x - I_z}{I_y} \qquad k_3 = \frac{I_y - I_x}{I_z}$$

e. Using the definitions of moments of inertia, show that
$k_3 = -(k_1 + k_2)/(1 + k_1 k_2)$ and that $-1 \leq k_1 \leq 1$ and $-1 \leq k_2 \leq 1$. The parameters k_1 and k_2 are widely used in satellite dynamics studies found in the more advanced technical literature.

Suggestion: For part b, use Euler's equations for principal axes, with $\vec{M}_C$ given by Eq. (6.8.3i) (p. 404). For part e, use the definition of moments of inertia.

6.23 The homogeneous rectangular plate shown in Fig. 6.6 (p. 347), with dimensions $L = 3b$ and $a = 1.5b$, is put in orbit around the earth. The center of mass of the plate is in a circular orbit of radius 300 km above the surface of the earth. Using the linearized differential equations given in Problem 6.22, determine, for each one of the following cases, whether the equilibrium $\theta_{1_e} = \theta_{2_e} = \theta_{3_e} = 0$ indicated in that problem is infinitesimally stable, asymptotically stable, or unstable. In real situations, one tries to put the body in orbit in an orientation that is as close as possible to a stable equilibrium orientation relative to the orbiting reference frame used in the book. Try to visualize each one of the given situations before you start solving the problem. Notice that each statement defines how the parameters k_1 and k_2 should be calculated.

a1. At equilibrium, the plate is oriented with its axis of minimum inertia aligned with the local vertical and its axis of maximum inertia perpendicular to the orbital plane.

a2. At equilibrium, the plate is oriented with its axis of minimum inertia aligned with the local vertical and its axis of maximum inertia tangent to the orbit (i.e., the same as the velocity of the center of mass of the plate).

b1. At equilibrium, the plate is oriented with its axis of minimum inertia tangent to the orbit and its axis of maximum inertia perpendicular to the orbital plane.

b2. At equilibrium, the plate is oriented with its axis of minimum inertia tangent to the orbit and its axis of maximum inertia aligned with the local vertical.

c1. At equilibrium, the plate is oriented with its axis of minimum inertia perpendicular to the orbital plane and its axis of maximum inertia tangent to the orbit.

c2. At equilibrium, the plate is oriented with its axis of minimum inertia perpendicular to the orbital plane and its axis of maximum inertia aligned with the local vertical.

From the results you obtain, specify in which orientation (or orientations, if that is the case) the plate should be put in orbit. Does the radius of the circular orbit affect the answer?

6.24 Using the linearized differential equations presented in Problem 6.22, analyze the stability of the equilibrium $\theta_{1_e} = \theta_{2_e} = \theta_{3_e} = 0$, and then generate a stability chart of k_2 versus k_1, showing the regions where the equilibrium is infinitesimally stable and those where it is unstable. *Hint:* You should get a characteristic equation of the form $(s^2 + b)(s^4 + a_2 s^2 + a_0) = 0$.

CHAPTER 7

Introduction to Analytical Dynamics

The use of Newton's second law $\vec{F} = d(m\vec{v})/dt = m\vec{a}$ for formulating the differential equations of motion of a single particle, and the use of its extension to the motion of the center of mass of a system of particles and continuum bodies, requires the inclusion of all forces applied to the individual particles (or bodies, for the case of multiple bodies) of the system. This includes even reaction forces that do not appear in the differential equations of motion, as they are eventually eliminated from the formulation by proper manipulation of the scalar equations that are obtained from $\vec{F} = m\vec{a}$. Many examples of this appear in the book, starting in Chapter 2. For instance, in Example 2.5.9 (p. 96), the normal reaction forces N_1 and N_2 appear in the free-body diagrams in Fig. 2.17 (p. 97), but they do not appear in the differential equations that are eventually obtained for the system. In Example 2.7.6 (p. 121), the reaction forces B_1 and B_2 appear in Eqs. (2.7.6a) and (2.7.6b) that are obtained from $\vec{F} = m\vec{a}$, but they do not appear in the differential equation that governs the motion of the system, which is Eq. (2.7.6d) (p. 124). Therefore, one natural question that comes to mind is, "Is there a better way to obtain the differential equations of motion of dynamical systems, by eliminating such forces right at the beginning of the formulation?" The answer is a definitive "yes," and the approach that makes that possible is a part of *analytical dynamics*. It is a powerful approach and surprisingly easy to use. Analytical dynamics is based on a few fundamental principles, starting with d'Alembert's principle, due to Jean Le Rond d'Alembert (1717–1783), and is based on work that was mostly developed by Joseph Louis Lagrange (1736–1813), by William Rowan Hamilton (1805–1865), and by Karl Gustav Jacobi (1804–1851). In addition to his far-reaching work on rigid body dynamics, Leonhard Euler (1707–1783) also did pioneering work on calculus of variations, which was contemporaneously advanced by Lagrange as applied to dynamics.

Analytical dynamics is a work-energy approach, and, as seen in this chapter, its fundamental quantities for obtaining the differential equations of motion of dynamical systems are the *virtual work* (defined in Section 7.2) done by the forces acting on the entire system and the *kinetic energy* of the motion of the entire system. Thus, formulation of expressions for acceleration are also avoided. Analytical dynamics views the system in its entirety, instead of looking at free-body diagrams of individual parts. Just like analytic geometry deals with geometry problems using calculus, *analytical dynamics* deals with dynamics of both simple and complex systems in a more mathematical manner than the vectorial approach.

7.1 Virtual Variations and Virtual Displacements

In basic calculus, the concepts of a function and the infinitesimal change and differential of a function are familiar. In analytical dynamics, the concept of *virtual variation* plays a significant role and *functionals* also appear. These concepts, which are from the *calculus of variations,* are introduced in Table 7.1, side by side with their analogies with calculus. Look at the three items in that table before you proceed any further.

Table 7.1 Analogies between calculus and calculus of variations

Calculus	Calculus of variations
An entity $y(x)$ is said to be a *function* of x when a value of x yields a value of $y(x)$. The *argument* x of a function $y(x)$ is the *independent variable*. For example, $y_1(x) = \sin(x)$ and $y_2(x) = \sqrt{1+x}$ are functions of x since a value of the independent variable x yields a value for both $y_1(x)$ and $y_2(x)$.	An entity $f(y(x))$ is called a *functional* when a function $y(x)$ of an independent variable x yields a value for f (i.e., a functional is a function of a function). For example, the entities $f_1 = \int_1^3 \sin(x)\,dx$, $f_2 = \int_2^5 \sqrt{1+x}\,dx$, $f_3 = \int_0^6 [(dy/dx)^2 - y^2 + x^2]\,dx$, where $y(x)$ is a function of x, are functionals. The values of these functionals depend, of course, on the functions $y(x)$ in the integrand. It turns out (as will be seen in Section 7.4.2) that functionals of the form $\int_{x_1}^{x_2} g(dy/dx, y, x)\,dx$, where y is an unknown function of the independent variable x, are the ones that appear in dynamics of point masses and rigid bodies.
If x denotes an independent variable, a change in x causes, of course, a change in a function $y(x)$. The quantity dx is called the *infinitesimal change of x*, and is equal to the difference between two infinitesimally close values of x. The corresponding infinitesimal change of a function $y(x)$ is denoted by dy and is called the *differential of the function*. For a function $y(x)$ of only one variable x, one has $dy = (dy/dx)\,dx$. The infinitesimal changes dx and dy are illustrated in Fig. 7.1a using the function $y(x) = x^2/4$.	An infinitesimal change of the argument $y(x)$ of the functional $f(y(x))$ to another function $y_{\text{new}}(x)$ is called the virtual variation of $y(x)$ and is denoted by δy (and pronounced "del y" instead of "delta y"). Notice that the change of a function $y(x)$ involves the change of x and, in contrast, the change of a functional involves the change of the argument $y(x)$ without changing the independent variable x. The virtual variation $\delta y = y_{\text{new}}(x) - y(x)$ of a function $y(x)$ that may appear in a functional $f(y(x))$ is illustrated in Fig. 7.1b.
The values of x for which $dy(x) = 0$ are called the values that make y stationary. When there are no restrictions on x, the points where $dy = 0$ are either maximum or minimum or inflection points of $y(x)$.	The functions $y(x)$ for which the variation $\delta f(y(x))$ of a functional f is zero are called the functions that make f stationary.

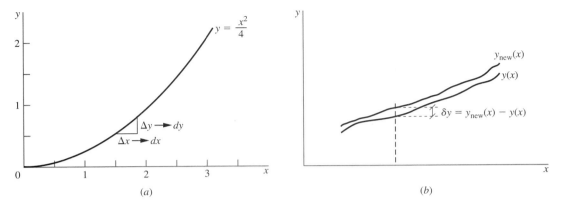

Figure 7.1 (*a*) Infinitesimal changes dx and dy of a function $y(x)$ and (*b*) virtual change δy of a function $y(x)$ that may appear in a functional $f(y(x))$.

Functionals As indicated in Table 7.1, a functional is a function of a function. It associates a number to a given function. For example, the integral $I = \int_{x_1}^{x_2} \sqrt{1 + (dy/dx)^2}\, dx$ is a functional whose meaning is the length of a curve $y(x)$ defined between the points $x = x_1$ and $x = x_2$. One might want to find the curve of minimum length, for example, that connects two given points $(x = x_1, y = y_1)$ and $(x = x_2, y = y_2)$ in space. Such a curve is the one that yields $\delta I = 0$, and it happens to be a straight line. The integral $\int_0^6 [(dy/dx)^2 - y^2 + x^2]\, dx$ is also a functional, which depends on $y(x)$ and on dy/dx; it evaluates to different values that depend on the function $y(x)$.

Total differential df and virtual variation δf Notice the difference between the infinitesimal change df of a function $f(q_1, q_2, \ldots, q_n; x)$ that depends on n functions $q_1(x), q_2(x), \ldots, q_n(x)$ (and maybe also depends explicitly on the independent variable x) and the virtual variation δf (in dynamics, the independent variable x is the time t). The total differential df is the actual infinitesimal change of f, and it is due to changes in $q_1, q_2, \ldots, q_n$ and also to the change in the independent variable x. As reviewed in Section 1.12 (p. 32):

$$df(q_1, q_2, \ldots, q_n; x) = \frac{\partial f}{\partial q_1} dq_1 + \frac{\partial f}{\partial q_2} dq_2 + \cdots + \frac{\partial f}{\partial q_n} dq_n + \frac{\partial f}{\partial x} dx \quad \textbf{(7.1.1)}$$

The coefficient of dq_1, for example, in Eq. (7.1.1) is the partial derivative of f with respect to q_1 (i.e., the derivative obtained when all the other arguments $q_2, \ldots, q_n$ and the independent variable x are held constant).

On the other hand, the virtual variation of $f(q_1, \ldots, q_n; x)$ is a variation that is obtained by holding the independent variable constant, i.e., "freezing" it, but changing the argument functions $q_1, \ldots, q_n$. The new symbol δ is used for such variations, and the virtual variation of f (which is due to virtual variations in $q_1, \ldots, q_n$) is then

$$\delta f(q_1, \ldots, q_n; x) = \frac{\partial f}{\partial q_1} \delta q_1 + \cdots + \frac{\partial f}{\partial q_n} \delta q_n \qquad (7.1.2)$$

In dynamics one encounters the kinetic energy T of the motion, which is a function that depends on n functions $q_1(t), q_2(t), \ldots, q_n(t)$, where t is time, and on the time derivatives $\dot{q}_1(t), \dot{q}_2(t), \ldots, \dot{q}_n(t)$, and maybe also depends explicitly on the independent variable t, i.e., $T(q_1, \ldots, q_n, \dot{q}_1, \ldots, \dot{q}_n; t)$. The total differential dT and the virtual variation δT of such a function is obtained as follows:

$$d(T(q_1, \ldots, q_n, \dot{q}_1, \ldots, \dot{q}_n; t)) = \frac{\partial T}{\partial q_1} dq_1 + \cdots + \frac{\partial T}{\partial q_n} dq_n + \frac{\partial T}{\partial \dot{q}_1} d\dot{q}_1$$

$$+ \cdots + \frac{\partial T}{\partial \dot{q}_n} d\dot{q}_n + \frac{\partial T}{\partial t} dt \qquad (7.1.3)$$

$$\delta(T(q_1, \ldots, q_n, \dot{q}_1, \ldots, \dot{q}_n; t)) = \frac{\partial T}{\partial q_1} \delta q_1 + \cdots + \frac{\partial T}{\partial q_n} \delta q_n + \frac{\partial T}{\partial \dot{q}_1} \delta \dot{q}_1$$

$$+ \cdots + \frac{\partial T}{\partial \dot{q}_n} \delta \dot{q}_n \qquad (7.1.4)$$

The variables $q_1, \ldots, q_n$ are the unknown variables that are chosen for describing the motion of any particular dynamical system under investigation. They may be the coordinates of a point or the angle a rotating line makes with a known reference line. The virtual variations of such variables are called *virtual displacements,* and the virtual variations of their time derivatives (i.e., $\delta \dot{q}_1, \ldots, \delta \dot{q}_n$) are called *virtual velocities.* Such variables become a known function of the independent variable (time) only when one integrates the differential equations that are obtained for them at the end of the *formulation phase* of a problem (i.e., the phase that consists in formulating the differential equations of motion for the system), after which the *analysis of the motion* phase starts. The virtual variations of such unknown variables are then arbitrary; i.e., they can be anything one desires, as long as they are infinitesimally small variations and do not violate any constraint relation that might exist between them (such as $q_1^2 + q_2^2 = $ constant, or any other relation). Generally one is able to choose a set of n independent (i.e., not restricted by any constraint relation) variables $q_1, \ldots, q_n$ to work with in a dynamics problem. The unknown variables $q_1, \ldots, q_n$ that are necessary for describing the motion of a system are called the *generalized coordinates* of the motion. Generally, generalized coordinates are independent of each other. However, there are cases when one is forced to work with constrained generalized coordinates, as in Example 7.9.2 (p. 475).

If a variable were known a priori, say $q_1 = \sin t$ (where, as before, t is the independent variable time), then one would have $dq_1 = (\cos t) \, dt$ and $\delta q_1 = (\cos t) \, \delta t = 0$ because $\delta t = 0$ by definition of a virtual variation. In other words, the virtual variation of any known (i.e., specified a priori) function of the independent variable is zero.

Relations between $d(\delta y)$ and $\delta(dy)$ and between $d(\delta y)/dt$ and $\delta(dy/dt)$

The virtual variations of y are defined as (see Fig. 7.1b):

$$\delta y \triangleq y_{\text{new}} - y \qquad (7.1.5)$$

In a similar manner, the virtual variations of dy and of dy/dt are defined as

$$\delta(dy) \triangleq dy_{\text{new}} - dy \qquad \delta\left(\frac{dy}{dt}\right) \triangleq \frac{dy_{\text{new}}}{dt} - \frac{dy}{dt} \qquad (7.1.6)$$

These definitions then yield the following results involving "a del of a d" and "a d of a del":

$$\delta(dy) \triangleq dy_{\text{new}} - dy = d(y_{\text{new}} - y) = d(\delta y) \qquad (7.1.7)$$

$$\delta\left(\frac{dy}{dt}\right) \triangleq \frac{d}{dt} y_{\text{new}} - \frac{d}{dt} y = \frac{d}{dt}(y_{\text{new}} - y) = \frac{d}{dt}(\delta y) \qquad (7.1.8)$$

> Conclusion: The operators d and δ are exchangeable.

The use of this conclusion will be crucial in Section 7.4.2. Such a conclusion might also have been inferred by saying that since the "d" and the "del" in Eqs. (7.1.1) and (7.1.2), respectively, are *linear operators,* they can be exchanged with each other (i.e., they *commute*).

7.2 Virtual Work, Generalized Forces, and Constraint Forces

> The material presented here, concerning generalized forces and potential energy, parallels that presented in Section 2.8. Therefore, reference will be made to that section when appropriate.

If $\vec{r}$ denotes the absolute position of a point P that is moving in space, i.e., the position vector of P relative to an inertial point O, the infinitesimal displacement of P relative to O is the vector $d\vec{r} \triangleq \vec{r}(t+dt) - \vec{r}(t)$ and the virtual displacement (vector) of P is the vector $\delta\vec{r}$. By definition of a virtual displacement, $\delta\vec{r} \triangleq \vec{r}_{\text{new}}(t) - \vec{r}(t)$ is an arbitrary infinitesimal change of the position of P that one makes at will, without the passage of the independent variable time. If a force $\vec{F}$ is acting at that point P, the virtual work done by that force, denoted by δW, is the scalar quantity defined as the dot product:

$$\delta W \triangleq \vec{F} \bullet \delta\vec{r} \qquad (7.2.1)$$

For a single particle P moving along a given surface in space, one of the forces acting on the particle is a reaction force $\vec{N}$ that is perpendicular to that surface at the point of contact with the particle. That reaction force is the one that is responsible for the motion of the particle to be constrained to that given surface, which is called the *surface of constraint*. The infinitesimal work $\vec{N} \bullet (d\vec{r})$ done by that constraint force is zero if the surface is not moving in space. However, $\vec{N} \bullet (d\vec{r})$ is not zero if the surface itself is given a known motion in any direction in space since the actual infinitesimal displacement $d\vec{r}$ of point P in this latter case is not necessarily perpendicular to the constraint force $\vec{N}$, which is still a force normal to the moving surface of constraint where the motion of particle P takes place. If, instead of working with the actual infinitesimal displacement $d\vec{r}$, one works with the virtual displacement $\delta\vec{r}$, the virtual work $\vec{N} \bullet (\delta\vec{r})$ done by the reaction $\vec{N}$ is equal to zero because as one freezes the independent variable t the virtual displacement $\delta\vec{r}$, which must not violate any constraint, must be along the surface where P moves on. In other words, $N \bullet (\delta\vec{r}) = 0$ because $\delta\vec{r}$ is normal to $\vec{N}$. By working with virtual displacements, one then eliminates that constraint force from the formulation.

In this example, there must also be an external force applied to the surface of constraint in order to make it move in the desired manner. Again, the infinitesimal work dW done by such a force is not zero in general, but the virtual work done by the same force is zero because time is frozen in any virtual displacement. In other words, that virtual displacement would be a virtual displacement of a known motion, which is equal to zero by definition.

In Newtonian dynamics, forces are classified into internal and external forces. Some internal forces (such as the normal reaction force due to contact between two bodies, the forces acting on any two particles whose distance between them is constant, and the tension in an inextensible bar) do not appear in the differential equations of motion of a body. However, they start showing up early in the formulation of those equations via Newton's second law $\vec{F} = m\vec{a}$.

In analytical dynamics, forces are classified into constraint and nonconstraint forces. Analytical dynamics considers the system in its entirety, without breaking it into parts and without dealing with individual free-body diagrams as done in Newtonian dynamics. Although the virtual work done by a specific constraint force in a free-body diagram may not be zero, the virtual work done by all constraint forces acting on the entire system (i.e., the sum of the virtual work done by every constraint force in the system) is equal to zero. The virtual work done by any force that needs to be applied to a point of the system to impose a *known* motion to that point is also equal to zero. This was illustrated in the example just discussed regarding a particle moving on a given surface. The double pendulum example shown in Fig. 7.2 also illustrates this point.

In Fig. 7.2, two mass particles at A and B, of masses m_1 and m_2, respectively, are connected by a rigid bar forming a double pendulum that moves on a vertical plane. The support point O' of the pendulum is given a known motion. Figure 7.2a shows the system, and two sets $\{\hat{r}_1, \hat{n}_1\}$ and $\{\hat{r}_2, \hat{n}_2\}$ of orthogonal unit vectors that will be used in our analysis. Figure 7.2b shows the internal constraint forces $\vec{N}_1$, $\vec{N}_2 = -\vec{N}_1$ (as per Newton's third law), $\vec{N}_3$, and $\vec{N}_4 = -\vec{N}_3$ (also as per Newton's third law). It also shows a force $\vec{F}_{O'}$ that is applied at point O' to make

SECTION 7.2 Virtual Work, Generalized Forces, and Constraint Forces

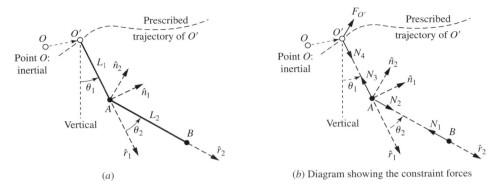

(a)

(b) Diagram showing the constraint forces

Figure 7.2 A double pendulum and free-body diagrams (also showing several unit vectors).

that point move in a desired manner. The virtual work done by all the forces shown in Fig. 7.2 is

$$\delta W = (\vec{F}_{O'} + \vec{N}_4) \bullet \underbrace{\delta(\overrightarrow{OO'})}_{=0}$$
$$+ \underbrace{(\vec{N}_3)}_{=-N_3 \hat{r}_1} \bullet \delta(\overrightarrow{OA}) + (\vec{N}_2) \bullet \delta(\overrightarrow{OA}) + (\vec{N}_1) \bullet \delta(\overrightarrow{OB})$$

In this equation, $\delta(\overrightarrow{OO'}) = 0$ by definition of virtual variation, i.e., because the motion $\overrightarrow{OO'}$ of O' is prescribed. Notice that the infinitesimal displacement of O', which is $d(\overrightarrow{OO'}) = [d(\overrightarrow{OO'})/dt] dt$ is not zero and, therefore, the infinitesimal work $(\vec{F}_{O'} + \vec{N}_4) \bullet d(\overrightarrow{OO'})$ is not zero.

Since (see Fig. 7.2) $\overrightarrow{OA} = \overrightarrow{OO'} + \overrightarrow{O'A}$, $\overrightarrow{O'A} = L_1 \hat{r}_1$, and $d(\overrightarrow{O'A})/dt = L_1 \dot{\theta}_1 \hat{n}_1$ [which discloses that $\delta(\overrightarrow{O'A}) = (L_1 \delta \theta_1) \hat{n}_1$], we then obtain

$$\delta(\overrightarrow{OA}) = \delta(\overrightarrow{OO'}) + \delta(\overrightarrow{O'A}) = (L_1 \delta \theta_1) \hat{n}_1$$

Also, since for the vectors $\vec{N}_2$ and $\vec{N}_1$, $\vec{N}_2 = -\vec{N}_1$ (see Fig. 7.2), the expression for δW then reduces to

$$\delta W = \underbrace{(-N_3 \hat{r}_1) \bullet [(L_1 \delta \theta_1) \hat{n}_1]}_{=0 \text{ because } \hat{r}_1 \bullet \hat{n}_1 = 0} + \vec{N}_1 \bullet [\delta(\overrightarrow{OB}) - \delta(\overrightarrow{OA})]$$

$$= \vec{N}_1 \bullet \delta \underbrace{(\overrightarrow{OB} - \overrightarrow{OA})}_{=\overrightarrow{AB} \text{ (see Fig. 7.2)}} = \vec{N}_1 \bullet \delta(\overrightarrow{AB})$$

$$= -N_1 \hat{r}_2 \bullet [L_2 (\delta \theta_1 + \delta \theta_2) \hat{n}_2] = 0 \qquad \text{since } \hat{r}_2 \bullet \hat{n}_2 = 0$$

These calculations and inspection of Fig. 7.2 also allow for the following explanations of why the virtual work of all the forces shown in that figure is equal to zero.

- For the pair of forces $\vec{N}_1$ and $\vec{N}_2 = -\vec{N}_1$, the virtual work is $\vec{N}_1 \bullet \delta(\overrightarrow{OB}) + \vec{N}_2 \bullet \delta(\overrightarrow{OA}) = \vec{N}_1 \bullet \delta(\overrightarrow{OB} - \overrightarrow{OA})$. But $\delta(\overrightarrow{OB} - \overrightarrow{OA})$ is the virtual displacement $\delta(\overrightarrow{AB})$ of point B relative to point A without violating the constraint $|\overrightarrow{AB}| = \text{constant} = L_2$. Thus, such a displacement takes place along a circle of radius L_2 centered at A. If the pendulum were moving in three-dimensional space, that displacement would be along a sphere of radius L_2 centered at A. That surface is the *surface of constraint* (which is reduced to a circle for the two-dimensional case) for the particle located at point B. The constraint forces $\vec{N}_1$ and $\vec{N}_2 = -\vec{N}_1$ are always perpendicular to that surface of constraint and, therefore, the virtual work $\vec{N}_1 \bullet \delta(\overrightarrow{AB})$ has to be equal to zero.

- The same explanation applies to the virtual work done by the forces $\vec{N}_3$ and $\vec{N}_4 = -\vec{N}_3$, which is $\vec{N}_3 \bullet \delta(\overrightarrow{O'A})$. If the motion were three dimensional, the surface of constraint in this case would be the sphere of radius L_1 centered at point O' (again, for the two-dimensional case, that surface of constraint reduces to a circle).

- The virtual work $\vec{F}_{O'} \bullet \delta(\overrightarrow{OO'})$ done by the force $\vec{F}_{O'}$ is equal to zero for a reason that is entirely independent of the trajectory of O'. It is zero because the motion of point O' is prescribed (i.e., it is a known desired motion) and, thus, $\delta(\overrightarrow{OO'}) = 0$. Therefore, $\vec{F}_{O'} \bullet \delta(\overrightarrow{OO'}) = 0$ no matter what the direction of $\vec{F}_{O'}$ is.

For a system composed of N particles (or, in general, a system composed of discrete particles and rigid bodies), the expression for the virtual work δW done by a set of forces $\vec{F}_1, \vec{F}_2, \ldots, \vec{F}_N$ acting on such a system is defined as the following sum of dot products:

$$\delta W \triangleq \vec{F}_1 \bullet \delta \vec{r}_1 + \vec{F}_2 \bullet \delta \vec{r}_2 + \cdots + \vec{F}_N \bullet \delta \vec{r}_N \quad (7.2.2)$$

The motion of such a system is described by a set of n variables $q_1(t)$, $q_2(t), \ldots, q_n(t)$, each of which must be single valued for any specific value of t. Frequently, such variables are either distances or angles, but generalized coordinates are not restricted to such choices. They may also be chosen to be something else, such as the sine of an angle (if that is more advantageous for a specific problem) or the amplitudes in a Fourier expansion of some variable. As mentioned earlier, they are called the *generalized coordinates* of the motion.

The absolute position vector $\vec{r}_i$ of any particle P_i is a function of the generalized coordinates $q_1, \ldots, q_n$ and may also be an explicit function of the time t, i.e.,

$$\vec{r}_i = \vec{r}_i(q_1, q_2, \ldots, q_n; t) \quad (7.2.3)$$

Therefore, after each dot product is performed in Eq. (7.2.2) for any particular problem, the expression that is obtained for the virtual work δW will be of the following form:

$$\delta W = \sum_{i=1}^{N} \vec{F}_i \bullet \delta \vec{r}_i = Q_{q_1}(q_1, q_2, \ldots, q_n, \dot{q}_1, \dot{q}_2, \ldots, \dot{q}_n; t)\, \delta q_1$$
$$+ Q_{q_2}(q_1, q_2, \ldots, q_n, \dot{q}_1, \dot{q}_2, \ldots, \dot{q}_n; t)\, \delta q_2 + \cdots$$
$$+ Q_{q_n}(q_1, q_2, \ldots, q_n, \dot{q}_1, \dot{q}_2, \ldots, \dot{q}_n; t)\, \delta q_n \quad (7.2.4)$$

which is of the same form as Eq. (2.8.2) (p. 152) without the term in dt and with virtual displacements $\delta q_1, \ldots, \delta q_n$ instead of the actual infinitesimal displacements $dq_1, \ldots, dq_n$.

The coefficients $Q_{q_1}, Q_{q_2}, \ldots, Q_{q_n}$ of each virtual displacement $\delta q_1, \delta q_2, \ldots, \delta q_n$ are called the *generalized forces* associated with the generalized coordinates $q_1, q_2, \ldots$. A generalized force Q_{q_k} has the unit of force if q_k is a linear displacement and the unit of a moment if q_k is an angular displacement. Generally, it is a relatively simple matter to obtain the generalized forces $Q_{q_1}, Q_{q_2}, \ldots, Q_{q_n}$, since all one has to do is to dot multiply all the forces by the respective virtual displacements of the point to which they are applied.

> The generalized forces are one of the two fundamental quantities that are needed to obtain the differential equations of motion of a dynamical system using the Lagrangian formulation of dynamics. As we will see shortly, the other fundamental quantity is the kinetic energy of the motion of the system.

In general, δW is not the "del" of (i.e., the virtual variation of) a function W. However, the expression for δW can always be separated into the sum of two parts: one part that *is* the del of a function U (to be determined) and is conventionally written with a minus sign as $-\delta(U)$, and another part that is not the del of anything but is simply whatever is left out of the expression for δW. Accordingly, one can then write[1]

$$\delta W = -\delta(U) + \underbrace{\text{whatever is left out of the expression of } \delta W}_{\overset{\Delta}{=}\,(\delta W)_{nc}} \quad (7.2.5)$$

Notice that $(\delta W)_{nc}$ consists of all the terms in δW that cannot be obtained from the "del of a function." The minus sign in the term $-\delta(U)$ in Eq. (7.2.5) is a convention that is adopted universally. With such a sign convention, the function U, which can only be a function of the generalized coordinates $q_1, \ldots, q_n$, and may also be an explicit function of time t (but never depend on the generalized velocities $\dot{q}_1, \ldots, \dot{q}_n$ because the expression for δW does not involve any $\delta \dot{q}$), is called the *potential energy* of the motion.

---

[1] $\delta(U)$, with the parentheses to make it clear that this is the del of a function, should be read as "del *of* U," where U is a function of $q_1, q_2, \ldots, q_n$ and may also be an explicit function of time t.

The remaining part of δW in Eq. (7.2.5), which is denoted by $(\delta W)_{nc}$ and is of the form

$$(\delta W)_{nc} \triangleq (Q_{q_1})_{nc} \delta q_1 + \cdots + (Q_{q_n})_{nc} \delta q_n$$

is the nonpotential part of δW. The coefficients of $\delta q_1, \ldots, \delta q_n$ in $(\delta W)_{nc}$, i.e., $(Q_{q_1})_{nc}, \ldots, (Q_{q_n})_{nc}$, are called the *generalized nonconservative forces* acting on the system.

The expression of $\delta(U)$ for any particular problem is of the form

$$\delta(U) = \frac{\partial U}{\partial q_1} \delta q_1 + \cdots + \frac{\partial U}{\partial q_n} \delta q_n$$

and the quantities $-\partial U/\partial q_1, \ldots, -\partial U/\partial q_n$ are called the *generalized conservative forces* acting on the system. Since the potential energy U is introduced through its partial derivatives $\partial U/\partial q_1, \ldots, \partial U/\partial q_n$, the expression for U is, then, obtained by integrations. Therefore, its expression will always contain an arbitrary constant of integration (which, actually, may also be an arbitrary function of time t). However, since only changes in potential energy are important, because motion is always accompanied by changes, the constant of integration is irrelevant. In this book, it will always be set to zero for convenience.

Several examples involving the development of expressions for potential energy were presented at the end of Section 2.8 (p. 155).[2] Additional examples are presented here.

EXAMPLE 7.2.1

The double pendulum shown in Fig. 7.3 consists of two rigid massless bars $O'A$ and AB of lengths L_1 and L_2, respectively. A particle of mass M_1 is placed at A and another of mass M_2 is placed at the end B of bar AB. Point O' represents a hinge that is forced to move (by an external force not shown in the figure) with a known motion represented by the dashed path in Fig. 7.3, and the motion of the system takes place in a vertical plane. The torsional spring connected to bars $O'A$ and AB is linear with stiffness k N · m/rad and is unstressed when $\theta_2 = \theta_u$ rad. Obtain the expression for the generalized forces acting on the system and for the potential energy of the motion.

■ **Solution**

The only forces and moments acting on this system for which the virtual work is not zero are the gravitational forces whose magnitudes are $M_1 g$ and $M_2 g$ (treating the gravitational field as uniform so that the gravitational forces are independent of distance), which are acting at points A and B, respectively, and the moment due to the torsional spring. Since they are only a function of position, their virtual work can be obtained from a potential

---

[2] Although in Chapter 2, the concept of virtual variation was not introduced at that level of presentation, the generation of the expression for U using Eq. (2.8.6) (p. 154) is fully consistent with the material presented here. The expression for δW corresponds to the one that is obtained by replacing d by δ in Eq. (2.8.6) and setting $\delta t = 0$ in the resulting equation.

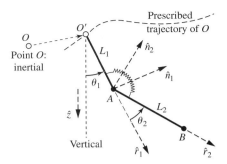

Figure 7.3 A double pendulum and several convenient unit vectors for Example 7.2.1.

energy U. From the presentation in Section 2.8, one could obtain the following expression for the potential energy simply by inspection of Fig. 7.3:

$$U = -M_1 g L_1 \cos\theta_1 - M_2 g[L_1 \cos\theta_1 + L_2 \cos(\theta_1 + \theta_2)] + \frac{k}{2}(\theta_2 - \theta_u)^2 \quad \text{(7.2.1a)}$$

from which the generalized forces Q_{θ_1}, associated with $\delta\theta_1$, and Q_{θ_2}, associated with $\delta\theta_2$, are

$$Q_{\theta_1} = -\frac{\partial U}{\partial \theta_1} = -(M_1 + M_2) g L_1 \sin\theta_1 - M_2 g L_2 \sin(\theta_1 + \theta_2) \quad \text{(7.2.1b)}$$

$$Q_{\theta_2} = -\frac{\partial U}{\partial \theta_2} = -M_2 g L_2 \sin(\theta_1 + \theta_2) - k(\theta_2 - \theta_u) \quad \text{(7.2.1c)}$$

If one does not feel either comfortable or very secure in writing such expressions of potential energy by inspection (which is actually not possible to do for examples that are either not familiar or, more likely, more complicated), one should always start by formulating the expression of the virtual work and, from there, obtain (if needed) the expression for potential energy. This is done as follows. Referring to Fig. 7.3, the following expression for δW is written, where the third term is the virtual work done by the spring force.

$$\delta W = M_1 g \hat{z} \cdot \delta \vec{r}_1 + M_2 g \hat{z} \cdot \delta \vec{r}_2 - k(\theta_2 - \theta_u)\, \delta\theta_2 \quad \text{(7.2.1d)}$$

But $\vec{r}_1 = \overrightarrow{OO'} + L_1 \hat{r}_1$ and $\vec{v}_1 = d\vec{r}_1/dt = d\overrightarrow{OO'}/dt + L_1 \dot\theta_1 \hat{n}_1$, and either one of these expressions gives $\delta \vec{r}_1 = (L_1 \delta\theta_1)\hat{n}_1$. Also, $\vec{r}_2 = \vec{r}_1 + L_2 \hat{r}_2$ and $\vec{v}_2 = d\vec{r}_2/dt = d\vec{r}_1/dt + L_2(\dot\theta_1 + \dot\theta_2)\hat{n}_2$, and either one of these expressions gives $\delta \vec{r}_2 = \delta \vec{r}_1 + L_2(\delta\theta_1 + \delta\theta_2)\hat{n}_2$. Therefore, the expression for δW yields

$$\delta W = (M_1 + M_2) g L_1 (\delta\theta_1) \underbrace{\hat{z} \cdot \hat{n}_1}_{=\,-\sin\theta_1} + M_2 g L_2 (\delta\theta_1 + \delta\theta_2) \underbrace{\hat{z} \cdot \hat{n}_2}_{=\,-\sin(\theta_1 + \theta_2)} - k(\theta_2 - \theta_u)\,\delta\theta_2$$

$$\stackrel{\Delta}{=} Q_{\theta_1} \delta\theta_1 + Q_{\theta_2} \delta\theta_2 \quad \text{(7.2.1e)}$$

As indicated in this equation with the $\stackrel{\Delta}{=}$ sign, the generalized forces Q_{θ_1} and Q_{θ_2} are, by definition, the coefficients of $\delta\theta_1$ and $\delta\theta_2$ in the expression for δW.

The search for the expression for the potential energy involves testing the expressions for the generalized forces, as done here. For this example we have

$$\frac{\partial Q_{\theta_1}}{\partial \theta_2} = -M_2 g L_2 \cos(\theta_1 + \theta_2) \tag{7.2.1f}$$

$$\frac{\partial Q_{\theta_2}}{\partial \theta_1} = -M_2 g L_2 \cos(\theta_1 + \theta_2) \tag{7.2.1g}$$

Since $\partial Q_{\theta_1}/\partial \theta_2 = \partial Q_{\theta_2}/\partial \theta_1$, the entire expression for δW is of the form $\delta W = -\delta(U) = -[(\partial U/\partial \theta_1)\, \delta \theta_1 + (\partial U/\partial \theta_2)\, \delta \theta_2]$, where

$$\frac{\partial U}{\partial \theta_1} = -Q_{\theta_1} = (M_1 + M_2) g L_1 \sin \theta_1 + M_2 g L_2 \sin(\theta_1 + \theta_2) \tag{7.2.1h}$$

$$\frac{\partial U}{\partial \theta_2} = -Q_{\theta_2} = M_2 g L_2 \sin(\theta_1 + \theta_2) + k(\theta_2 - \theta_u) \tag{7.2.1i}$$

To find U, start by integrating either $\partial U/\partial \theta_1$ with respect to θ_1 or $\partial U/\partial \theta_2$ with respect to θ_2. Integration of $\partial U/\partial \theta_2$ with respect to θ_2 yields, except for an arbitrary constant of integration that is set to zero for convenience,

$$U = -M_2 g L_2 \cos(\theta_1 + \theta_2) + \frac{k}{2}(\theta_2 - \theta_u)^2 + f(\theta_1) \tag{7.2.1j}$$

where $f(\theta_1)$ is to be determined so that $\partial U/\partial \theta_1$ matches the expression in Eq. (7.2.1h), i.e.,

$$\frac{\partial U}{\partial \theta_1} = M_2 g L_2 \sin(\theta_1 + \theta_2) + \frac{df}{d\theta_1} = (M_1 + M_2) g L_1 \sin \theta_1 + M_2 g L_2 \sin(\theta_1 + \theta_2) \tag{7.2.1k}$$

This then yields

$$\frac{df}{d\theta_1} = (M_1 + M_2) g L_1 \sin \theta_1$$

or

$$f(\theta_1) = -(M_1 + M_2) g L_1 \cos \theta_1$$

plus an arbitrary constant of integration that, for convenience, is again set to zero. With $f(\theta_1)$ as determined here, the expression for U given by Eq. (7.2.1j) is the same as the one in Eq. (7.2.1a) that was written by inspection of Fig. 7.3.

Note: Labeling the first approach to find U as an *inspection* process is actually a misnomer since there is only one way to obtain the expression for potential energy for any problem, i.e., by starting with δW and then proceeding to find which part of δW, if any, can be obtained from a $-\delta(U)$ term. For some forces, such as gravitational forces and spring forces, the expression that is obtained for U is either accompanied by a physical interpretation or by memorization, or both. The inspection process actually made use of such a physical interpretation and/or memorization. In the case of the potential energy due to a gravitational force that is associated with a constant acceleration g, for example, one might say that the physical interpretation of U is that it is equal to the energy mgh when the center of gravity of a body of mass m is at a distance h *above* a reference level that does not depend on the generalized coordinates. In the case of Fig. 7.3, $h = -L_1 \cos \theta_1$ and $h = -L_1 \cos \theta_1 - L_2 \cos(\theta_1 + \theta_2)$ for M_1 and M_2, respectively.

For a more elaborate version of this example, see Problem 7.7.

EXAMPLE 7.2.2

In terms of two generalized coordinates q_1 and q_2, the force acting on a particle P is given as

$$\vec{F} = \left(q_1 q_2^3 + q_1 + q_2^2\right)\hat{x}_1 + \left(\frac{3}{2}q_1^2 q_2^2 + q_2 + q_1^2\right)\hat{x}_2$$

while the absolute velocity of P, $\vec{v} = d\vec{r}/dt$, is

$$\vec{v} = (\dot{q}_1 + k \sin t)\hat{x}_1 + \dot{q}_2 \hat{x}_2 \text{ m/s}$$

where $k = 1$ m/s. Determine whether $\vec{F}$ is conservative or nonconservative, and obtain the expression for the potential energy of the motion of P.

■ **Solution**
From the expression for the velocity $\vec{v} = d\vec{r}/dt$ we can write

$$d\vec{r} = [dq_1 + (k \sin t)\, dt]\hat{x}_1 + (dq_2)\hat{x}_2 \quad (7.2.2a)$$

and, therefore, the virtual displacement $\delta \vec{r}$ is simply

$$\delta \vec{r} = (\delta q_1)\hat{x}_1 + (\delta q_2)\hat{x}_2 \quad (7.2.2b)$$

The virtual work done by $\vec{F}$ during a virtual displacement $\delta \vec{r}$ is then

$$\delta W = \vec{F} \bullet \delta \vec{r} = \left(q_1 q_2^3 + q_1 + q_2^2\right)\delta q_1 + \left(\frac{3}{2}q_1^2 q_2^2 + q_2 + q_1^2\right)\delta q_2$$
$$\stackrel{\Delta}{=} Q_{q_1} \delta q_1 + Q_{q_2} \delta q_2 \quad (7.2.2c)$$

The coefficients Q_{q_1} and Q_{q_2} of δq_1 and δq_2, respectively, in the expression for δW are the generalized forces associated with the generalized coordinates q_1 and q_2.

Since $\partial Q_{q_1}/\partial q_2 \neq \partial Q_{q_2}/\partial q_1$, δW is not the del of anything; i.e., there is no function W for which δW is of the form $\delta(W)$. Therefore, the force $\vec{F}$ is classified as nonconservative. The only parts $Q_{q_1}^*$ and $Q_{q_2}^*$ of Q_{q_1} and Q_{q_2}, respectively, for which $\partial Q_{q_1}^*/\partial q_2 = \partial Q_{q_2}^*/\partial q_1$ are $Q_{q_1}^* = q_1 q_2^3 + q_1$ and $Q_{q_2} = \frac{3}{2} q_1^2 q_2^2 + q_2$. The virtual work δW^* associated with these parts can be written in terms of a potential energy function $U(q_1, q_2)$ as

$$\delta W^* = \left(q_1 q_2^3 + q_1\right)\delta q_1 + \left(\tfrac{3}{2} q_1^2 q_2^2 + q_2\right) \delta q_2$$
$$= -\left(\frac{\partial U}{\partial q_1} \delta q_1 + \frac{\partial U}{\partial q_2} \delta q_2\right) \quad (7.2.2d)$$

The expression for the potential energy U is obtained by integrating the expressions for $\partial U/\partial q_1$ and $\partial U/\partial q_2$ as follows. For $\partial U/\partial q_1$, we have from Eq. (7.2.2d),

$$\frac{\partial U}{\partial q_1} = -q_1 q_2^3 - q_1 \qquad (7.2.2e)$$

and this gives, by integrating with respect to q_1,

$$U = -\frac{1}{2}q_1^2 q_2^3 - \frac{1}{2}q_1^2 + f(q_2) \qquad (7.2.2f)$$

where $f(q_2)$ is to be determined so that $\partial U/\partial q_2$ matches the expression in Eq. (7.2.2g), which follows.

For $\partial U/\partial q_2$, we have from Eq. (7.2.2d), and also from Eq. (7.2.2f),

$$\frac{\partial U}{\partial q_2} = -\frac{3}{2}q_1^2 q_2^2 - q_2 \underbrace{=}_{\text{from Eq. (7.2.2f)}} -\frac{3}{2}q_1^2 q_2^2 + \frac{df}{dq_2} \qquad (7.2.2g)$$

and since this gives $df/dq_2 = -q_2$, we then obtain for $f(q_2)$, with the constant of integration set to zero as done previously,

$$f = -\frac{1}{2}q_2^2 \qquad (7.2.2h)$$

The potential energy U is then, except for an arbitrary constant of integration,

$$U = -\frac{1}{2}\left(q_1^2 q_2^3 + q_1^2 + q_2^2\right) \qquad (7.2.2i)$$

EXAMPLE 7.2.3

A rigid body is subjected to a force $\vec{F}$ that is related to three generalized coordinates q_1, q_2, and q_3 (all of which are changing with time t, of course) as $\vec{F} = kq_1\hat{x}_1 + 3(\cos q_2)(\sin t)\hat{x}_2 + (q_1 \sin q_2)\hat{x}_3$ N, where $k = 1$ N/m. The force $\vec{F}$ is applied at a point P whose absolute position vector $\vec{r}$ is $\vec{r} = q_1\hat{x}_1 + (2\cos t)\hat{x}_2$ m. The unit vectors form an orthogonal triad, with $\hat{x}_3 = \hat{x}_1 \times \hat{x}_2$, and that triad rotates in space with absolute angular velocity $\vec{\omega} = (5\cos t)\hat{x}_1 + \dot{q}_2(3\hat{x}_2 + 2\hat{x}_3)$ rad/s.

a. Formulate the expression for the infinitesimal work and for the virtual work done by $\vec{F}$. What are the expressions for the generalized forces acting on the body?
b. Is $\vec{F}$ a conservative or nonconservative force? What is the expression for the potential energy of the motion?
c. If $\vec{F}$ is a nonconservative force, what are the expressions for the nonconservative generalized forces?

■ **Solution**

a. The absolute velocity $\vec{v}$ of point P is

$$\vec{v} = \frac{d}{dt}\vec{r} = \frac{d}{dt}[q_1\hat{x}_1 + (2\cos t)\hat{x}_2] = [\dot{q}_1\hat{x}_1 - (2\sin t)\hat{x}_2] + \vec{\omega} \times [q_1\hat{x}_1 + (2\cos t)\hat{x}_2]$$
$$= (\dot{q}_1 - 4\dot{q}_2 \cos t)\hat{x}_1 + (2q_1\dot{q}_2 - 2\sin t)\hat{x}_2 + (10\cos^2 t - 3q_1\dot{q}_2)\hat{x}_3 \qquad (7.2.3a)$$

SECTION 7.2 Virtual Work, Generalized Forces, and Constraint Forces

and, therefore, the infinitesimal displacement $d\vec{r} = \vec{v}\,dt$ is

$$d\vec{r} = [dq_1 - 4(\cos t)dq_2]\hat{x}_1 + [2q_1 dq_2 - 2(\sin t)\,dt]\hat{x}_2 + [10(\cos^2 t)\,dt - 3q_1 dq_2]\hat{x}_3$$

This then yields the following expression for the virtual displacement $\delta\vec{r}$:

$$\delta\vec{r} = [\delta q_1 - 4(\cos t)\delta q_2]\hat{x}_1 + (2q_1\delta q_2)\hat{x}_2 - (3q_1\delta q_2)\hat{x}_3 \quad (7.2.3\text{b})$$

The infinitesimal work dW done by the force $\vec{F}$ is then

$$dW \triangleq \vec{F}\bullet(d\vec{r}) = q_1\,dq_1 + \left[6q_1(\cos q_2)\sin t - 4q_1\cos t - 3q_1^2\sin q_2\right]dq_2$$
$$\quad + [10q_1(\sin q_2)\cos^2 t - 6(\cos q_2)\sin^2 t]\,dt$$
$$\triangleq Q_{q_1}\,dq_1 + Q_{q_2}\,dq_2 + a_0\,dt$$

while the virtual work δW done by $\vec{F}$ is

$$\delta W \triangleq \vec{F}\bullet(\delta\vec{r}) = q_1\,\delta q_1 + \left[6q_1(\cos q_2)\sin t - 4q_1\cos t - 3q_1^2\sin q_2\right]\delta q_2$$
$$\triangleq Q_{q_1}\,\delta q_1 + Q_{q_2}\,\delta q_2 \quad (7.2.3\text{c})$$

The coefficients of δq_1 and δq_2 seen in the expression for the virtual work δW (and also in the expression for dW) are the generalized forces Q_{q_1} and Q_{q_2} associated with the virtual displacements δq_1 and δq_2, respectively.

b. Inspection of Eq. (7.2.3c) discloses that

$$\frac{\partial Q_{q_1}}{\partial q_2} = 0 \qquad \frac{\partial Q_{q_2}}{\partial q_1} = 6(\cos q_2)\sin t - 4\cos t - 6q_1\sin q_2 \quad (7.2.3\text{d})$$

Since $\partial Q_{q_1}/\partial q_2 \neq \partial Q_{q_2}/\partial q_1$, the virtual work δW is not the del of anything. Therefore, the force $\vec{F}$ is nonconservative.

The potential energy U is obtained from a part of δW (if any) that allows one to write

$$\delta W \triangleq \underbrace{-\delta(U)}_{} + \text{ whatever is left out of the expression of }\delta W$$
$$= \frac{\partial U}{\partial q_1}\delta q_1 + \frac{\partial U}{\partial q_2}\delta q_2$$

Since $\partial Q_{q_1}/\partial q_2 = 0$, with $Q_{q_1} = q_1$, as seen from Eq. (7.2.3c), the term $Q_{q_1}\delta q_1 = q_1\delta q_1$ can be expressed as

$$q_1\delta q_1 = \delta\left(\frac{1}{2}q_1^2 + C\right) \quad (7.2.3\text{e})$$

where C is any desired constant (or even a specified function of time only), which will be set to zero since there is no need to keep it. Therefore, it is seen that a part of δW can be obtained from the potential energy

$$U = -\frac{1}{2}q_1^2 \quad (7.2.3\text{f})$$

c. The rest of the expression for δW is obtained from the following nonconservative generalized force, which is associated with q_2:

$$(Q_{q_2})_{nc} = 6q_1(\cos q_2)\sin t - 4q_1\cos t - 3q_1^2\sin q_2 \quad (7.2.3\text{g})$$

7.3 The Principles of d'Alembert and Virtual Work

Consider a system of N particles (which may also form any number of bodies of finite dimension) P_i of mass m_i ($i = 1, 2, \ldots, N$), subjected to a set of forces. For convenience, let us group the forces acting on particle P_i into two types: the resultant $\vec{F}_{i\_c}$ of the *constraint forces* if any and the resultant $\vec{F}_i$ of all other forces. Now, letting $\vec{r}_i$ denote the absolute position vector for the ith particle P_i, as shown in Fig. 7.4, and letting $\vec{v}_i = d\vec{r}_i/dt$ be the absolute velocity of P_i, one can then write according to Newton's second law:

$$\vec{F}_i + \vec{F}_{i\_c} = \frac{d}{dt}(m_i \vec{v}_i) \tag{7.3.1}$$

Notice that, in the most general situation, Eq. (7.3.1) represents $3N$ scalar equations if each particle P_i is able to move in three-dimensional space.

Although the virtual work of the force of constraint $\vec{F}_{i\_c}$ acting on particle P_i may be nonzero, the virtual work done by all forces of constraint acting together on the entire system is zero. This was illustrated in an example using Fig. 7.2 (p. 423). One can then take advantage of such a fact to eliminate the constraint forces right at the beginning of the formulation of the differential equations of motion for a particular system under investigation. Unless one wants to determine the constraint forces, this is better than carrying them throughout the formulation, which is what happens when free-body diagrams for each part of the system are used as required in the Newtonian approach. Their elimination is accomplished by dot multiplying both sides of Eq. (7.3.1) by the virtual displacement $\delta\vec{r}_i$ of particle P_i and then adding all the resulting equations as

$$\sum_{i=1}^{N} \vec{F}_i \bullet \delta\vec{r}_i + \sum_{i=1}^{N} \vec{F}_{i\_c} \bullet \delta\vec{r}_i = \sum_{i=1}^{N} \left[\frac{d}{dt}(m_i \vec{v}_i)\right] \bullet \delta\vec{r}_i$$

The second summation on the left-hand side of this equation is the virtual work done by all the constraint forces acting on the system and, therefore, is equal to zero. The first summation term is, by definition, the virtual work δW done by all the nonconstraint forces such as the externally applied forces that are not associated with a specified motion, friction forces, internal elastic forces due

Figure 7.4 Forces acting on a particle.

to springs, etc. Therefore, this equation reduces to

$$\delta W \triangleq \sum_{i=1}^{N} \vec{F}_i \bullet \delta \vec{r}_i = \sum_{i=1}^{N} \left[\frac{d}{dt}(m_i \vec{v}_i)\right] \bullet \delta \vec{r}_i \quad (7.3.2)$$

For continuous bodies, the mass m_i is replaced by an infinitesimal mass element $dm = \rho\, dV$ of the body, and the summation in Eq. (7.3.2) is replaced by an integral that should span the entire volume of the body. Here, ρ and dV are the volumetric material density and a volume element of the body, respectively.

Equation (7.3.2) is known as *Lagrange's form of d'Alembert's principle* (d'Alembert's principle was introduced on p. 227 in Chapter 4). For $d(m_i \vec{v}_i)/dt = 0$, that is, for problems for which $\vec{v}_i = $ constant $\neq 0$ or for those for which $\vec{v}_i = 0$ (i.e., statics problems), Eq. (7.3.2) reduces to $\delta W = 0$ and, in this case, it is known as the *principle of virtual work*.

As will be shown in the subsequent sections, Eq. (7.3.2) has a far-reaching consequence since it leads to *Hamilton's principle,* from which Lagrange's equation and the differential equations of motion of discrete systems can be readily obtained in a relatively simple manner.[3]

7.4 Hamilton's Principle and Lagrange's Equation

Let us look at Eq. (7.3.2) and the two ways of obtaining the virtual work δW done by the forces $\vec{F}_i$ during a virtual displacement $\delta \vec{r}_i$. By definition, δW is equal to the sum of the dot products of each $\vec{F}_i$ with the corresponding virtual displacements $\delta \vec{r}_i$. As already shown, this simply yields an expression of the form given by Eq. (7.2.4) (p. 425), from which the generalized forces, as defined by that same equation, are extracted. But making use of Newton's second law, Eq. (7.3.2) will also yield another expression for δW. As shown in the following, manipulation of the expression $\sum_{i=1}^{N}[d(m_i \vec{v}_i)/dt] \bullet \delta \vec{r}_i$, and a subsequent integration in time, yields *Hamilton's principle,* which is a fundamental principle in dynamics. Hamilton's principle, in turn, yields *Lagrange's equation* in terms of the n generalized coordinates $q_1, q_2, \ldots, q_n$ for any *discrete system,* which is a system of point masses and/or rigid bodies (i.e., systems whose motion are governed by ordinary differential equations).

7.4.1 Hamilton's Principle

To manipulate Eq. (7.3.2) one starts by making use of the general form of the expression for the absolute position vector $\vec{r}_i$ of any point P_i. That expression, which is Eq. (7.2.3) (p. 424), is repeated here for convenience.

$$\vec{r}_i = \vec{r}_i(q_1, q_2, \ldots, q_n; t)$$

---

[3] Another beauty, so to speak, of Hamilton's principle is that it also yields the partial differential equations of motion and a boundary condition equation for structural dynamics problems. The principle is widely used in dynamics and also in structural dynamics. The reader who is interested in such applications has much to gain by delving into such an approach.

The absolute velocity $\vec{v}_i = d\vec{r}_i/dt$ of a particle P_i is then of the following form:

$$\vec{v}_i = \frac{d}{dt}\vec{r}_i = \frac{\partial \vec{r}_i}{\partial q_1}\dot{q}_1 + \frac{\partial \vec{r}_i}{\partial q_2}\dot{q}_2 + \cdots + \frac{\partial \vec{r}_i}{\partial q_n}\dot{q}_n + \frac{\partial \vec{r}_i}{\partial t} = \sum_{k=1}^{n} \frac{\partial \vec{r}_i}{\partial q_k}\dot{q}_k + \frac{\partial \vec{r}_i}{\partial t} \quad (7.4.1)$$

The rest of the derivations in this section will soon involve multiple summations. The matrix notation, introduced next, will be used because it is less cumbersome to write and read.

- An underlined letter is used for denoting a column matrix, such as

$$\underline{f} = \begin{bmatrix} f_1 \\ f_2 \end{bmatrix}$$

  The transpose of such a column matrix is denoted by $\underline{f}^T$, i.e., $\underline{f}^T = [f_1 \ f_2]$.

- As per the notation previously introduced, the n generalized coordinates $q_1, q_2, \ldots, q_n$ are denoted by the elements of an $n \times 1$ column matrix $\underline{q}$ as

$$\underline{q} = \begin{bmatrix} q_1 \\ q_2 \\ \vdots \\ q_n \end{bmatrix}$$

  The transpose of $\underline{q}$ is $\underline{q}^T = [q_1 \ q_2 \ \cdots \ q_n]$. The explicit dependence of a function $\underline{f}$ on n generalized coordinates $q_1, \ldots, q_n$ (which implicitly depend on time t) and also on t is denoted by $\underline{f}(\underline{q}; t)$.

- The three scalar components of the absolute position vector $\vec{r}_i \triangleq x_i \hat{x} + y_i \hat{y} + z_i \hat{z}$ expressed in a triad $\{\hat{x}, \hat{y}, \hat{z} \triangleq \hat{x} \times \hat{y}\}$ are denoted by

$$\underline{r}_i = \begin{bmatrix} x_i \\ y_i \\ z_i \end{bmatrix}$$

  The transpose of $\underline{r}_i$ is $\underline{r}_i^T = [x_i \ y_i \ z_i]$, where $x_i = x_i(\underline{q}; t)$, $y_i = y_i(\underline{q}; t)$, and $z_i = z_i(\underline{q}; t)$.

- Similarly, the three scalar components of the absolute velocity vector $\vec{v}_i = d\vec{r}_i/dt \triangleq v_{ix}\hat{x} + v_{iy}\hat{y} + v_{iz}\hat{z}$ expressed in a triad $\{\hat{x}, \hat{y}, \hat{z} \triangleq \hat{x} \times \hat{y}\}$ are denoted by

$$\underline{v}_i = \begin{bmatrix} v_{ix} \\ v_{iy} \\ v_{iz} \end{bmatrix}$$

  The transpose of $\underline{v}_i$ is $\underline{v}_i^T = [v_{ix} \ v_{iy} \ v_{iz}]$.

  Notice that if the unit vector triad $\{\hat{x}, \hat{y}, \hat{z}\}$ is not rotating, $v_{ix} = \dot{x}_i$, $v_{iy} = \dot{y}_i$, and $v_{iz} = \dot{z}_i$, whose matrix representation is simply $d\underline{v}_i/dt = d\underline{r}_i/dt$. Otherwise, $\vec{v}_i = \dot{x}_i\hat{x} + \dot{y}_i\hat{y} + \dot{z}_i\hat{z} + \vec{\omega} \times \vec{r}_i$, where $\vec{\omega}$ is the absolute angular velocity of the $\{\hat{x}, \hat{y}, \hat{z}\}$ triad. With $\vec{\omega}$ expressed as

SECTION 7.4 Hamilton's Principle and Lagrange's Equation

$\vec{\omega} = \omega_x \hat{x} + \omega_y \hat{y} + \omega_z \hat{z}$, this would yield $\vec{v}_i = (\dot{x}_i + \omega_y z_i - \omega_z y_i)\hat{x} + (\dot{y}_i + \omega_z x_i - \omega_x z_i)\hat{y} + (\dot{z}_i + \omega_x y_i - \omega_y x_i)\hat{z}$, and its matrix representation has to be

$$\underline{v}_i = \frac{d}{dt}\underline{r}_i + \begin{bmatrix} 0 & -\omega_z & \omega_y \\ \omega_z & 0 & -\omega_x \\ -\omega_y & \omega_x & 0 \end{bmatrix} \underline{r}_i$$

In view of the differences in the representation of the derivative of a vector shown in this item, a nonrotating unit vector triad $\{\hat{x}, \hat{y}, \hat{z}\}$ will be used in order to simplify the derivations that will follow. Notice that this choice is made only for these derivations. For solving individual problems using the results obtained from the derivations, the use of rotating unit vectors is, in general, more convenient as it has already been demonstrated in the book.

- Using the compact matrix notation just described, the components of the linear momentum of a particle P_1 with mass m_i may be written as

$$\begin{bmatrix} m_i v_{ix} \\ m_i v_{iy} \\ m_i v_{iz} \end{bmatrix} = \begin{bmatrix} m_i & 0 & 0 \\ 0 & m_i & 0 \\ 0 & 0 & m_i \end{bmatrix} \begin{bmatrix} v_{ix} \\ v_{iy} \\ v_{iz} \end{bmatrix}$$

We are now ready to operate on Eq. (7.3.2), which is repeated here for convenience, using the compact matrix notation and also using $\vec{F}_i = d(m_i \vec{v}_i)/dt$, as per Newton's second law.

$$\delta W \triangleq \sum_{i=1}^{N} \vec{F}_i \bullet \delta \vec{r}_i = \sum_{i=1}^{N} \left[\frac{d}{dt}(m_i \vec{v}_i) \right] \bullet \delta \vec{r}_i$$

As shown here, the matrix representation of the first summation term in this equation (which is simply the definition of δW) is $\underline{Q}^T \delta \underline{q}$:

$$\delta W \triangleq \sum_{i=1}^{N} \vec{F}_i \bullet \delta \vec{r}_i = Q_{q_1} \delta q_1 + Q_{q_2} \delta q_2 + \cdots + Q_{q_n} \delta q_n$$
$$= [Q_{q_1} \; Q_{q_2} \; \cdots \; Q_{q_n}] \delta \underline{q} \triangleq \underline{Q}^T \delta \underline{q} \tag{7.4.2}$$

where

$$\underline{Q} \triangleq \begin{bmatrix} Q_{q_1} \\ Q_{q_2} \\ \vdots \\ Q_{q_n} \end{bmatrix}$$

is the generalized force column matrix.

The second summation term $\sum_{i=1}^{N}[d(m_i \vec{v}_i)/dt] \bullet \delta \vec{r}_i$ can be represented in matrix notation as

$$\sum_{i=1}^{N} \left[\frac{d}{dt}(m_i \vec{v}_i) \right] \bullet \delta \vec{r}_i = \left[\frac{d}{dt}(M \underline{v}) \right]^T \delta \underline{r} \tag{7.4.3}$$

with $\underline{r} = \underline{r}(\underline{q};t)$ and $\underline{v}(\underline{q},\underline{\dot{q}};t) = d\underline{r}/dt$, being the following $3N \times 1$ column matrices:

$$\underline{r} = \begin{bmatrix} x_1(\underline{q};t) \; y_1(\underline{q};t) \; z_1(\underline{q};t) \vdots \cdots \vdots x_N(\underline{q};t) \; y_N(\underline{q};t) \; z_N(\underline{q};t) \end{bmatrix}^T \quad (7.4.4)$$

$$\underline{v} = \frac{d\underline{r}}{dt} = \underbrace{\begin{bmatrix} \dfrac{\partial x_1}{\partial q_1} & \dfrac{\partial x_1}{\partial q_2} & \cdots & \dfrac{\partial x_1}{\partial q_n} \\ \dfrac{\partial y_1}{\partial q_1} & \dfrac{\partial y_1}{\partial q_2} & \cdots & \dfrac{\partial y_1}{\partial q_n} \\ & & \vdots & \\ \dfrac{\partial z_N}{\partial q_1} & \dfrac{\partial z_N}{\partial q_2} & \cdots & \dfrac{\partial z_N}{\partial q_n} \end{bmatrix}}_{\text{Call this matrix } \frac{\partial \underline{r}}{\partial \underline{q}}} \underline{\dot{q}} + \underbrace{\begin{bmatrix} \dfrac{\partial x_1}{\partial t} \\ \dfrac{\partial y_1}{\partial t} \\ \vdots \\ \dfrac{\partial z_N}{\partial t} \end{bmatrix}}_{\text{This is } \frac{\partial \underline{r}}{\partial t}} = \frac{\partial \underline{r}}{\partial \underline{q}} \underline{\dot{q}} + \frac{\partial \underline{r}}{\partial t} \quad (7.4.5)$$

and M being the $3N \times 3N$ symmetric *mass matrix* defined as

$$M = \begin{bmatrix} m_1 & 0 & 0 & \vdots & & & & & \\ 0 & m_1 & 0 & \vdots & & & & & \\ 0 & 0 & m_1 & \vdots & & & \text{zeros} & & \\ \cdots & \cdots & \cdots & & \ddots & & & & \\ & & & & & \ddots & & & \\ & & \text{zeros} & & & & \cdots & \cdots & \cdots \\ & & & & & \vdots & m_N & 0 & 0 \\ & & & & & \vdots & 0 & m_N & 0 \\ & & & & & \vdots & 0 & 0 & m_N \end{bmatrix}$$

Notice that Eq. (7.4.5) is the matrix representation of Eq. (7.4.1) for all particles $P_1, P_2, \ldots, P_N$.

Now work on Eq. (7.3.2), performing the transpose operation[4] and rearranging the intermediate results as follows:

$$\delta W = \left[\frac{d}{dt}(M\underline{v})\right]^T \delta\underline{r} = \left[\frac{d}{dt}(\underline{v}^T M)\right]\delta\underline{r} = \frac{d}{dt}(\underline{v}^T M \delta\underline{r}) - \underline{v}^T M \underbrace{\frac{d}{dt}(\delta\underline{r})}_{\text{same as } \delta\frac{d\underline{r}}{dt}}$$

$$= \frac{d}{dt}(\underline{v}^T M \delta\underline{r}) - \underbrace{\underline{v}^T M \delta \frac{d\underline{r}}{dt}}_{\text{same as } \underline{v}^T M \delta \underline{v}} = \frac{d}{dt}(\underline{v}^T M \delta\underline{r}) - \delta(T) \quad (7.4.6)$$

---

[4] Recall from matrix algebra that the transpose $(AB)^T$ of a product AB of two matrices A and B is equal to $B^T A^T$.

where
$$T = \frac{1}{2}\underline{v}^T M \underline{v} \qquad (7.4.7)$$
is the kinetic energy of the motion of the system.

By rearranging Eq. (7.4.6) and integrating in time the resulting equation between two arbitrary instants $t = t_1$ and $t = t_2$ and by noticing that

$$\int_{t=t_1}^{t=t_2} \frac{d}{dt}(\underline{v}^T M \delta \underline{r}) \, dt = \left(\underline{v}^T M \delta \underline{r}\right)\Big|_{t=t_1}^{t=t_2}$$

the following result is then obtained:

$$\boxed{\int_{t=t_1}^{t=t_2} [\delta W + \delta(T)] \, dt = \left(\underline{v}^T M \, \delta \underline{r}\right)\Big|_{t=t_1}^{t=t_2}} \qquad (7.4.8)$$

Equation (7.4.8) is called *Hamilton's extended principle,* and it is a fundamental principle in dynamics. The general form of the kinetic energy T for any dynamics problem is examined in Section 7.4.2, and the formulation is then continued in Section 7.4.3. The material presented in Section 7.4.3 will greatly reduce the mathematical operations that are needed in the search for an integral of the motion of a system. A number of solved examples using the theory presented in this section and in Sections 7.4.2 and 7.4.3 are included in Section 7.7.

Note: It is common to set the right-hand side of Eq. (7.4.8) to zero by considering variations for which $(\vec{r})_{t=t_1} = 0$ and $(\vec{r})_{t=t_2} = 0$, i.e., by considering variations for which $\vec{r}$ is known at two times t_1 and t_2. However, $(\vec{r})_{t=t_2}$ may not be equal to zero since, in dynamics problems, conditions at $t = t_1$ (i.e., initial conditions) for $\vec{r}$ and $\vec{v}$ are what is specified, thus making $\delta \vec{r}$ unknown at $t = t_2$ and, consequently, $(\delta \vec{r})_{t=t_2}$ different from zero before the motion becomes known. It turns out, though, that integration of $\delta(T)$ with respect to time yields another term that happens to cancel out the term in the right-hand side of Eq. (7.4.8). This is presented in Section 7.4.2, where Lagrange's equation is obtained from Eq. (7.4.8). Lagrange's equation is a fundamental equation in dynamics. Its use will be illustrated with a number of examples presented at the end of these derivations.

7.4.2 Lagrange's Equation (Discrete Systems) Obtained from Hamilton's Principle

Equation (7.4.8) yields an equation, called *Lagrange's equation,* that allows one to obtain the differential equations of motion of a system in terms of the generalized coordinates $q_1(t), \ldots, q_n(t)$. Lagrange's equation is obtained by making use of the general expression for δW in Eq. (7.4.8) as $\delta W = \underline{Q}^T \delta \underline{q}$ [see Eq. (7.4.2), p. 435], taking the variation $\delta(T)$ of the kinetic energy $T(q_1, \ldots, q_n, \dot{q}_1, \ldots, \dot{q}_n; t)$, which is

$$\delta(T) = \frac{\partial T}{\partial q_1}\delta q_1 + \cdots + \frac{\partial T}{\partial q_n}\delta q_n + \frac{\partial T}{\partial \dot{q}_1}\delta \dot{q}_1 + \cdots + \frac{\partial T}{\partial \dot{q}_n}\delta \dot{q}_n$$

$$= \left(\frac{\partial T}{\partial q_1} \cdots \frac{\partial T}{\partial q_n}\right)\delta \underline{q} + \left(\frac{\partial T}{\partial \dot{q}_1} \cdots \frac{\partial T}{\partial \dot{q}_n}\right)\delta \underline{\dot{q}} \triangleq \frac{\partial T}{\partial \underline{q}}\delta \underline{q} + \frac{\partial T}{\partial \underline{\dot{q}}}\delta \underline{\dot{q}} \quad (7.4.9)$$

and then proceeding as follows:

$$\int_{t=t_1}^{t=t_2} \left(\underline{Q}^T \delta \underline{q} + \frac{\partial T}{\partial \underline{q}}\delta \underline{q} + \frac{\partial T}{\partial \underline{\dot{q}}}\delta \underline{\dot{q}}\right) dt - \left(\underline{v}^T M \frac{\partial \underline{r}}{\partial \underline{q}}\delta \underline{q}\right)_{t=t_1}^{t=t_2} = 0 \quad (7.4.10)$$

Since

$$(\delta \underline{\dot{q}})\, dt = \left[\delta\left(\frac{d}{dt}\underline{q}\right)\right] dt = \left[\frac{d}{dt}(\delta \underline{q})\right] dt = d(\delta \underline{q}) \quad (7.4.11)$$

an integration by parts in Eq. (7.4.10) yields

$$\int_{t=t_1}^{t=t_2} \left(\underline{Q}^T + \frac{\partial T}{\partial \underline{q}} - \frac{d}{dt}\frac{\partial T}{\partial \underline{\dot{q}}}\right)\delta \underline{q}\, dt + \left[\left(\frac{\partial T}{\partial \underline{\dot{q}}} - \underline{v}^T M \frac{\partial \underline{r}}{\partial \underline{q}}\right)\delta \underline{q}\right]_{t=t_1}^{t=t_2} = 0 \quad (7.4.12)$$

Since $T = \frac{1}{2}\underline{v}^T M \underline{v}$, it follows that

$$\frac{\partial T}{\partial \underline{\dot{q}}} = \frac{1}{2}\underline{v}^T (M + M^T)\frac{\partial \underline{v}}{\partial \underline{\dot{q}}} \underset{\text{since } M^T = M}{=} \underline{v}^T M \frac{\partial \underline{v}}{\partial \underline{\dot{q}}} \quad (7.4.13)$$

and also, from Eq. (7.4.5), that[5]

$$\frac{\partial \underline{v}}{\partial \underline{\dot{q}}} = \frac{\partial \underline{r}}{\partial \underline{q}} \quad (7.4.14)$$

Therefore, as a consequence of Eqs. (7.4.13) and (7.4.14), the boundary terms in Eq. (7.4.12), i.e., the terms that are evaluated at $t = t_1$ and t_2, are identically

---

[5] More explicitly, Eq. (7.4.14) means

$$\begin{bmatrix} \frac{\partial v_{1x}}{\partial \dot{q}_1} & \frac{\partial v_{1x}}{\partial \dot{q}_2} & \cdots & \frac{\partial v_{1x}}{\partial \dot{q}_n} \\ \frac{\partial v_{1y}}{\partial \dot{q}_1} & \frac{\partial v_{1y}}{\partial \dot{q}_2} & \cdots & \frac{\partial v_{1y}}{\partial \dot{q}_n} \\ \frac{\partial v_{1z}}{\partial \dot{q}_1} & \frac{\partial v_{1z}}{\partial \dot{q}_2} & \cdots & \frac{\partial v_{1z}}{\partial \dot{q}_n} \\ \vdots & \vdots & & \vdots \\ \frac{\partial v_{Nx}}{\partial \dot{q}_1} & \frac{\partial v_{Nx}}{\partial \dot{q}_2} & \cdots & \frac{\partial v_{Nx}}{\partial \dot{q}_n} \\ \frac{\partial v_{Ny}}{\partial \dot{q}_1} & \frac{\partial v_{Ny}}{\partial \dot{q}_2} & \cdots & \frac{\partial v_{Ny}}{\partial \dot{q}_n} \\ \frac{\partial v_{Nz}}{\partial \dot{q}_1} & \frac{\partial v_{Nz}}{\partial \dot{q}_2} & \cdots & \frac{\partial v_{Nz}}{\partial \dot{q}_n} \end{bmatrix} = \begin{bmatrix} \frac{\partial r_{1x}}{\partial q_1} & \frac{\partial r_{1x}}{\partial q_2} & \cdots & \frac{\partial r_{1x}}{\partial q_n} \\ \frac{\partial r_{1y}}{\partial q_1} & \frac{\partial r_{1y}}{\partial q_2} & \cdots & \frac{\partial r_{1y}}{\partial q_n} \\ \frac{\partial r_{1z}}{\partial q_1} & \frac{\partial r_{1z}}{\partial q_2} & \cdots & \frac{\partial r_{1z}}{\partial q_n} \\ \vdots & \vdots & & \vdots \\ \frac{\partial r_{Nx}}{\partial q_1} & \frac{\partial r_{Nx}}{\partial q_2} & \cdots & \frac{\partial r_{Nx}}{\partial q_n} \\ \frac{\partial r_{Ny}}{\partial q_1} & \frac{\partial r_{Ny}}{\partial q_2} & \cdots & \frac{\partial r_{Ny}}{\partial q_n} \\ \frac{\partial r_{Nz}}{\partial q_1} & \frac{\partial r_{Nz}}{\partial q_2} & \cdots & \frac{\partial r_{Nz}}{\partial q_n} \end{bmatrix}$$

equal to zero, and Eq. (7.4.12) reduces to

$$\int_{t=t_1}^{t=t_2} \left[\underline{Q}^T + \frac{\partial T}{\partial \underline{q}} - \frac{d}{dt}\frac{\partial T}{\partial \underline{\dot{q}}} \right] \delta \underline{q}\, dt = 0 \qquad (7.4.15)$$

Equation (7.4.15) is satisfied only if the integrand is zero, i.e., if

$$\boxed{\begin{aligned}&\left(\frac{d}{dt}\frac{\partial T}{\partial \dot{q}_1} - \frac{\partial T}{\partial q_1} - Q_{q_1}\right)\delta q_1 + \left(\frac{d}{dt}\frac{\partial T}{\partial \dot{q}_2} - \frac{\partial T}{\partial q_2} - Q_{q_2}\right)\delta q_2 + \cdots \\ &+ \left(\frac{d}{dt}\frac{\partial T}{\partial \dot{q}_n} - \frac{\partial T}{\partial q_n} - Q_{q_n}\right)\delta q_n = 0\end{aligned}} \qquad (7.4.16)$$

Equation (7.4.16) is valid in the most general case, even if some of the variables $q_1, \ldots, q_n$ are related to each other. Such a case is dealt with in Section 7.8.

For the case when there are no constraints between any of the variables $q_1, \ldots, q_n$, the virtual displacements $\delta q_1, \ldots, \delta q_n$ are independent of each other. In this case, since each virtual displacement is arbitrary, each one of the bracketed terms in Eq. (7.4.16) must then be set to zero in order to satisfy that equation. This then yields the following equation that must be satisfied by each one of the n variables $q_1, \ldots, q_n$ when they are independent of each other:

$$\boxed{\frac{d}{dt}\frac{\partial T}{\partial \dot{q}_k} - \frac{\partial T}{\partial q_k} = Q_{q_k} \qquad \text{for } k = 1, 2, \ldots, n} \qquad (7.4.17)$$

Equation (7.4.17) is known as *Lagrange's equation*. Since the kinetic energy T is a quadratic function of the generalized velocities $\dot{q}_1, \dot{q}_2, \ldots, \dot{q}_n$ (as shown in Section 7.4.3), Lagrange's equation represents n second-order differential equations for the generalized coordinates $q_1, \ldots, q_n$. In general, the resulting differential equations are nonlinear and coupled.

> The fundamental quantities that are needed for formulating the second-order differential equations of motion of a dynamical system using Lagrange's equation are the kinetic energy T of the entire system and the generalized forces $Q_{q_1}, \ldots, Q_{q_n}$. The expressions for the generalized forces are simply the coefficients of $\delta q_1, \ldots, \delta q_n$ that appear in the expression for the virtual work δW. The general form of the expression for the kinetic energy for any dynamics problem is presented and discussed in Section 7.4.3.

Several examples using this formulation are presented in Section 7.7. No continuity is lost if you want to start looking at some of those examples now. If you decide to do so, return to this point of the theory and then proceed to Section 7.5, which shows how to check for *integrals of the motion* of Lagrange's equation. Such a check is actually relatively easy to do. It is also a very efficient and quick procedure to apply to specific problems.

7.4.3 General Form of the Kinetic Energy T

The velocity $\vec{v}$ is always an explicit linear function of the generalized velocities $\dot{q}_1, \ldots, \dot{q}_n$, and it may also explicitly depend (in general, in a nonlinear manner) on the generalized coordinates $q_1, \ldots, q_n$ and on the time t. The general form of the velocity $\vec{v}$ is given in matrix form in Eq. (7.4.5).

Since $T = \frac{1}{2}\underline{v}^T M \underline{v}$, use of Eq. (7.4.5) yields the following general result, which is valid for *all* dynamics problems:

$$T = \frac{1}{2}\underline{v}^T M \underline{v} = \frac{1}{2}\left[\frac{\partial \underline{r}}{\partial \underline{q}}\dot{\underline{q}} + \frac{\partial \underline{r}}{\partial t}\right]^T M \left[\frac{\partial \underline{r}}{\partial \underline{q}}\dot{\underline{q}} + \frac{\partial \underline{r}}{\partial t}\right]$$

$$= \underbrace{\frac{1}{2}\dot{\underline{q}}^T \left[\left(\frac{\partial \underline{r}}{\partial \underline{q}}\right)^T M \frac{\partial \underline{r}}{\partial \underline{q}}\right]\dot{\underline{q}}}_{\text{Call this part } T_2} + \underbrace{\left(\frac{\partial \underline{r}}{\partial t}\right)^T M \frac{\partial \underline{r}}{\partial \underline{q}}\dot{\underline{q}}}_{\text{Call this part } T_1} + \underbrace{\frac{1}{2}\left(\frac{\partial \underline{r}}{\partial t}\right)^T M \left(\frac{\partial \underline{r}}{\partial t}\right)}_{\text{Call this part } T_0} \quad (7.4.18)$$

or, summarizing,

$$\boxed{T = T_2 + T_1 + T_0} \quad (7.4.19)$$

As disclosed by Eq. (7.4.18), the general form of the kinetic energy is always the sum of three distinct parts:

- A part, denoted by $T_2 \triangleq \frac{1}{2}\dot{\underline{q}}^T A(\underline{q}; t)\dot{\underline{q}}$, where $A = (\partial \underline{r}/\partial \underline{q})^T M \, \partial \underline{r}/\partial \underline{q}$ is a symmetric matrix (because its transpose is equal to itself), that is a purely quadratic function of the generalized velocities $\dot{q}_1, \dot{q}_2, \ldots, \dot{q}_n$. In other words, T_2 is the part of T that is the sum of terms involving $\dot{q}_1^2, \dot{q}_1\dot{q}_2, \dot{q}_1^2$, $\dot{q}_1\dot{q}_3$, etc. For any specific problem, such a part is simply recognized by looking at the expression that is obtained for the kinetic energy T.
- A part, denoted by $T_1 \triangleq \underline{b}^T(\underline{q}, t)\dot{\underline{q}}$, where $\underline{b} = (\partial \underline{r}/\partial \underline{q})^T M \, \partial \underline{r}/\partial t$ is a column matrix, that is a linear function of the generalized velocities $\dot{q}_1, \dot{q}_2, \ldots, \dot{q}_n$. In other words, T_1 is of the form $b_1\dot{q}_1 + b_2\dot{q}_2 + \cdots + b_n\dot{q}_n$. Again, for any specific problem, such a part is simply recognized by looking at the expression that is obtained for the kinetic energy T.
- Finally, a part, denoted by T_0, which does not depend on the generalized velocities (but only on the generalized coordinates, and, maybe, on time explicitly). The term T_0 is also recognizable by inspection of any expression for T.

The classification of the different terms in the kinetic energy as T_2, T_1, and T_0 happens to be extremely convenient for obtaining the expression for the *Hamiltonian* of the motion. As will be seen, the Hamiltonian is one of the functions that is tested when one looks for possible integrals of the differential equations of motion of a system. In dynamics terminology, it is common to refer to those systems for which $T = T_2$ as *natural systems*.

7.5 Conservation Principles: Classical Integrals of Lagrange's Equation

7.5.1 What Is an Integral of Motion?

A second-order differential equation in a variable $q(t)$ has two integrals of motion. A *first* integral of motion is a relationship of the form $f_1(q, \dot{q}; t) = \text{constant} \triangleq C_1$; i.e., f_1 is a function of the variable q, of its first time derivative, and maybe is also an explicit function of the independent variable t. Thus, a first integral of the motion reduces the problem of finding $q(t)$ to the integration of a differential equation $f_1(q, \dot{q}; t) = C_1$ whose order is one less than the original second-order differential equation. For example, the differential equation $\ddot{q} + q = 0$ has the first integral of motion $\frac{1}{2}(\dot{q}^2 + q^2) = \text{constant}$.

Repeating the process for the equation $f_1(q, \dot{q}; t) = \text{constant}$, a *second* integral of motion is a relationship that involves derivatives up to one order smaller than the highest derivative in this latter differential equation. In this case, the second integral of motion is, then, a relationship of the form $f_2(q; t) = \text{constant} \triangleq C_2$ that only involves q and t. The solution to the original second-order differential equation is then the solution to this latter "algebraic" equation. The constants C_1 and C_2 are simply the functions f_1 and f_2, respectively, evaluated at the value of t when $q(t)$ and $\dot{q}(t)$ are given. The problem is said to be an *initial value problem* when initial values of q and $\dot{q}$, say $q(t_0)$ and $\dot{q}(t_0)$, are given. When the known conditions are split between two different values of t [as for example, when $q(t_0)$ and $\dot{q}(t_{\text{final}})$ are given], the problem is called a *two-point boundary value problem*. Classical dynamics problems are initial value problems. Problems in structural dynamics for example, are, in general, two-point boundary value problems.

Lagrange's equation, Newton's equation $\vec{F} = m\vec{a}$, and Euler's equation $\vec{M}_C = d\vec{H}_C/dt$, are equations that involve second derivatives, and no higher. The motion of such dynamical systems is governed by a set of n second-order differential equations, where n is called the *number of degrees of freedom* of the system. Therefore, the differential equations of motion for such systems exhibit $2n$ constants of the motion, n of them being *first integrals* and the other n being *second integrals* (which are integrals of motion of the first integrals).

7.5.2 Integrals of Motion Using Lagrange's Equation

Under certain conditions (listed subsequently in Cases 1 and 2 of this section), several integrals of motion can be obtained from Lagrange's equation, and the search for them turns out to be a relatively simple process in such cases. To begin the search, start by isolating all the terms in δW that can be obtained from a potential energy so that δW becomes

$$\delta W \triangleq -\delta(U) + (\delta W)_{nc} \tag{7.5.1}$$

When this is done, Eq. (7.4.17) can then be written as

$$\frac{d}{dt}\frac{\partial \mathcal{L}}{\partial \dot{q}_k} - \frac{\partial \mathcal{L}}{\partial q_k} = (Q_{q_k})_{nc} \qquad \text{for } k = 1, 2, \ldots, n \qquad (7.5.2)$$

where

$$\mathcal{L} \triangleq T - U = T_2 + T_1 + T_0 - U \qquad (7.5.3)$$

is called the *Lagrangian* of the motion.[6] There are two classical searches for integrals of motion.

CASE 1 A GENERALIZED MOMENTUM INTEGRAL OF MOTION

This is the simplest of the searches for an integral of motion. Equation (7.5.2) immediately discloses that $\partial \mathcal{L}/\partial \dot{q} = $ constant for any generalized coordinate q for which $\partial \mathcal{L}/\partial q = 0$ and also when the nonconservative generalized force associated with that generalized coordinate, i.e., $(Q_q)_{nc}$, is equal to zero. In summary,

$$\frac{\partial \mathcal{L}}{\partial \dot{q}} = \text{constant}$$

for any q for which

$$\frac{\partial \mathcal{L}}{\partial q} = 0 \quad \text{and} \quad (Q_q)_{nc} = 0$$

This integral of motion, if it exists for a particular problem, is called the *generalized momentum integral* because the quantity $p \triangleq \partial \mathcal{L}/\partial \dot{q}$ is either a linear momentum or an angular momentum. Any generalized coordinate q that does not appear in the Lagrangian (i.e., $\partial \mathcal{L}/\partial q = 0$) is called an *ignorable coordinate*.

CASE 2 AN ENERGY-LIKE INTEGRAL OF MOTION

To search for another integral of motion, let us see what happens when we postmultiply both sides of Eq. (7.5.2) by $\underline{\dot{q}}^T$. The idea here is to try to get an equation of the form d/dt of something = something else, and then see under what conditions the right-hand side of such an equation would be equal to zero. When this happens, such an equation can be immediately integrated to yield a constant of the motion.

Since $\mathcal{L} = \mathcal{L}(\underline{q}, \underline{\dot{q}}; t)$, the total time derivative of $\mathcal{L}$ is

$$\frac{d\mathcal{L}(\underline{q}, \underline{\dot{q}}; t)}{dt} = \frac{\partial \mathcal{L}}{\partial \underline{q}}\underline{\dot{q}} + \frac{\partial \mathcal{L}}{\partial \underline{\dot{q}}}\underline{\ddot{q}} + \frac{\partial \mathcal{L}}{\partial t} \qquad (7.5.4)$$

By replacing the term $\partial \mathcal{L}/\partial \underline{q}$ in this equation for its equivalent expression obtained from Eq. (7.5.2),

---

[6] Since the letter L is already used in this book for denoting distance, the symbol $\mathcal{L}$ is used for denoting the Lagrangian in order to avoid duplication of symbols (although no confusion should exist even if one wishes to use the same letter for denoting either quantity). The reader is, of course, free to use the letter L also for denoting a Lagrangian.

Eq. (7.5.4) can then be manipulated as follows:

$$\frac{d\mathcal{L}(\underline{q}, \underline{\dot{q}}; t)}{dt} = \left[\frac{d}{dt}\frac{\partial \mathcal{L}}{\partial \underline{\dot{q}}} - (\underline{Q}^T)_{nc}\right]\underline{\dot{q}} + \frac{\partial \mathcal{L}}{\partial \underline{\dot{q}}}\underline{\ddot{q}} + \frac{\partial \mathcal{L}}{\partial t}$$

$$\underbrace{=}_{\text{grouping the first and third terms}} \frac{d}{dt}\left(\frac{\partial \mathcal{L}}{\partial \underline{\dot{q}}}\underline{\dot{q}}\right) - (\underline{Q}^T)_{nc}\underline{\dot{q}} + \frac{\partial \mathcal{L}}{\partial t} \quad (7.5.5)$$

By defining the following function $\mathcal{H}$, which is called the *Hamiltonian* of the motion,[7]

$$\boxed{\mathcal{H} \triangleq \frac{\partial \mathcal{L}}{\partial \underline{\dot{q}}}\underline{\dot{q}} - \mathcal{L}} = \frac{\partial \mathcal{L}}{\partial \dot{q}_1}\dot{q}_1 + \frac{\partial \mathcal{L}}{\partial \dot{q}_2}\dot{q}_2 + \cdots$$

$$+ \frac{\partial \mathcal{L}}{\partial \dot{q}_n}\dot{q}_n - \mathcal{L} \quad (7.5.6)$$

from which it can be verified that $\partial \mathcal{H}/\partial t = -\partial \mathcal{L}/\partial t$, Eq. (7.5.5) yields

$$\boxed{\frac{d\mathcal{H}}{dt} = \underline{Q}_{nc}^T \underline{\dot{q}} + \frac{\partial \mathcal{H}}{\partial t}} = (Q_{q_1})_{nc}\dot{q}_1 + (Q_{q_2})_{nc}\dot{q}_2$$

$$+ \cdots + (Q_{q_n})_{nc}\dot{q}_n + \frac{\partial \mathcal{H}}{\partial t}$$
(7.5.7)

---

[7] To avoid, as much as possible, duplication of symbols, the symbol $\mathcal{H}$ is used in the book for denoting the Hamiltonian since the letter H is already used for denoting angular momentum. The reader is, of course, free to use either H or $\mathcal{H}$.

Thus, as indicated by this boxed equation, it is an easy task to take the time derivative of a Hamiltonian!

Equation (7.5.7) discloses that the Hamiltonian is constant when the Lagrangian $\mathcal{L}$ is not an *explicit* function of time (i.e., when $\partial \mathcal{L}/\partial t = 0$) and *also* when $\underline{Q}_{nc}^T \underline{\dot{q}} = 0$. Since the generalized coordinates are independent of each other, this last condition requires each one of the nonconservative generalized forces associated with each of the independent coordinates $q_1, \ldots, q_n$ to be equal to zero. The integral of the motion $\mathcal{H} = $ constant is called the *Jacobi integral*. In summary,

$$\boxed{\begin{array}{l} \mathcal{H} = \text{constant} \\ \text{when} \\ \dfrac{\partial \mathcal{L}}{\partial t} = 0 \quad \text{and} \\ (Q_{q_1})_{nc} = (Q_{q_2})_{nc} = \cdots = (Q_{q_n})_{nc} = 0 \end{array}}$$

The product

$$\underline{Q}_{nc}^T \underline{\dot{q}} = (Q_{q_1})_{nc}\dot{q}_1 + (Q_{q_2})_{nc}\dot{q}_2 + \cdots + (Q_{q_n})_{nc}\dot{q}_n$$

which is the sum of the product of a nonconservative generalized force (whose unit is either a force or a moment) by the corresponding generalized velocity, is the power into the system due to the generalized nonconservative forces.

The generalized momentum integral and the Jacobi integral, if they exist for a particular problem, are the two *classical* integrals of motion that are known for dynamical systems.

CASE 3 A NONCLASSICAL INTEGRAL OF MOTION

An obvious nonclassical integral of motion of Eq. (7.5.2) is immediately obtained when a generalized coordinate, say q, does not appear in the expression for the Lagrangian $\mathcal{L} = T - U$ (i.e., when $\partial \mathcal{L}/\partial q = 0$) and *also* when the generalized nonconservative force $(Q_q)_{nc}$ associated with q is the total time derivative of a function. This will most likely happen in the simpler case when $(Q_q)_{nc}$ is due to viscous damping and is of the form $(Q_q)_{nc} = -c\dot{q}$, where c is a constant viscous friction coefficient. In such a case, one obtains $\partial \mathcal{L}/\partial \dot{q} + cq = $ constant. There is no particular name for this integral of motion.

Other integrals of motion exist in very special cases that have been reported in the technical journals using methods that are beyond the scope of this book. Beyond that work, there are no known ways to find additional integrals of motion.

7.5.3 General Form of the Hamiltonian for Any Problem

The classification of the different terms in the kinetic energy as T_2, T_1, and T_0 turns out to be extremely convenient for obtaining the expression for the Hamiltonian of the motion for any problem. It is shown here that the Hamiltonian turns out to be simply $\mathcal{H} = T_2 + U - T_0$ and, thus, it can be obtained simply by using the expression for the potential energy U and the appropriate terms collected by inspection of the expression for the kinetic energy T. That is certainly much less "algebra" than computing it using its definition given by Eq. (7.5.6)!

To show that the Hamiltonian is simply $\mathcal{H} = T_2 + U - T_0$, let us make use of the general expression for the kinetic energy, and then generate the expression for $\mathcal{H}$. Since, for any specific problem, the expression for the kinetic energy is of the form given by Eq. (7.4.18), i.e.,

$$T = \frac{1}{2}\underline{\dot{q}}^T A(\underline{q};\, t)\, \underline{\dot{q}} + \underline{b}^T(\underline{q};\, t)\, \underline{\dot{q}} + T_0(\underline{q};\, t) \triangleq T_2 + T_1 + T_0$$

where A is always a symmetric matrix, the Hamiltonian $\mathcal{H}$ is then equal to

$$\mathcal{H} \triangleq \frac{\partial T}{\partial \underline{\dot{q}}}\underline{\dot{q}} - \mathcal{L} = \left[\frac{1}{2}\underline{\dot{q}}^T(A + A^T) + \underline{b}^T\right]\underline{\dot{q}} - (T - U)$$

$$= \underline{\dot{q}}^T A \underline{\dot{q}} + \underline{b}^T \underline{\dot{q}} - (T_2 + T_1 + T_0 - U) = (2T_2 + T_1) - (T_2 + T_1 + T_0 - U)$$

or, simply

$$\boxed{\mathcal{H} = T_2 + U - T_0} \qquad (7.5.8)$$

The result expressed by this equation shows that, once the expressions for the kinetic and potential energies are generated, no further mathematical manipulations are needed to obtain the Hamiltonian for any particular problem. To generate the expression for the Hamiltonian $\mathcal{H}$ for a dynamical system, one only needs to be able to recognize the different types of terms T_2, T_1, and T_0 that appear in the kinetic energy and then selectively extract the T_2 and T_0 terms in order to form the expression for the Hamiltonian $\mathcal{H}$ according to Eq. (7.5.8).

Notice that U and T_0 are of the same form $U(\underline{q};\, t)$ and $T_0(\underline{q};\, t)$. Thus, it is impossible to separate from $U - T_0$ what is U and what is T_0 when one is given an expression for a Hamiltonian $\mathcal{H}$ without being shown the expression for either T or U. In some cases, however, one might still be able to separate U from T_0 based on physical arguments. However, such a separation is not important at all. The function $U - T_0$ is called the *kinetic potential* of the motion.

As disclosed by Eq. (7.5.8), the Hamiltonian is an energy function and, in general, it is different from the total mechanical energy $E = T + U = T_2 + T_1 + T_0 + U$ of the motion. The Hamiltonian $\mathcal{H}$ is equal to the total mechanical energy

$E = T + U$ only when $T_1 = 0$ and $T_0 = 0$, i.e., when $T = T_2$. In such cases, the Jacobi integral, if it exists for a particular problem being analyzed, has the physical interpretation of being the total mechanical energy $T + U$ of the motion.

7.6 A Brief Summary

Here is a brief summary of the material that is sufficient for one to formulate the differential equations of motion using Lagrange's equation in terms of n independent generalized coordinates $q_1, q_2, \ldots, q_n$ and to test for the existence of classical first integrals of the motion.

1. Lagrange's equation is

$$\frac{d}{dt}\frac{\partial T}{\partial \dot{q}_k} - \frac{\partial T}{\partial q_k} = Q_{q_k} \qquad k = 1, \ldots, n \qquad (7.6.1)$$

where Q_{q_k} is the generalized force associated with the generalized coordinate q_k (which is simply the coefficient of δq_k in the expression for δW).

> The generalized forces $Q_{q_1}, \ldots, Q_{q_k}$ are determined from the expression for the virtual work δW done by all forces acting on the system. Such an expression is always of the form shown in Eq. (7.2.4) (p. 425), which may also be written as Eq. (7.2.5) (p. 425) by grouping terms that are derivable from a potential energy $U(q_1, q_2, \ldots, q_n; t)$.

Equation (7.6.1) yields n second-order differential equations for the n independent variables $q_1, \ldots, q_n$. The resulting equations are, in general, nonlinear and coupled to each other, but they can be integrated numerically with no major difficulty, especially when one uses Simulink, which does not involve computer programming. The differential equations of motion are linear in rare cases (for some simpler problems involving linear springs, for example).

2. The expression of the kinetic energy for any dynamics problem is always of the form $T = T_2 + T_1 + T_0$ (see Section 7.4.3, p. 440).

3. Equation (7.6.1) admits the following integrals of motion (i.e., functions involving the generalized coordinates $q_1, \ldots, q_n$, and their first time derivatives, that remain constant for all times):

    a. For any generalized coordinate that does not appear in the expression of the kinetic and potential energies, say q_ℓ, it follows that $\partial(T-U)/\partial \dot{q}_\ell$ = constant if the generalized nonconservative force associated with q_ℓ is equal to zero. Such an integral of motion has units of either linear or angular momentum, and, for this reason, it is called the *generalized momentum integral*. One also obtains $\partial(T-U)/\partial \dot{q}_\ell + cq_\ell$ = constant if the nonconservative force $(Q_{q_\ell})_{nc}$ is of the form $(Q_{q_\ell})_{nc} = -c\dot{q}_\ell$, with c being a constant viscous damping coefficient.

b. The Hamiltonian is constant if it does not depend explicitly on the time (i.e., when you inspect its expression and do not see time appearing by itself) and if the nonconservative generalized forces associated with the generalized coordinates $q_1, \ldots, q_n$ are zero. The Hamiltonian is defined by Eq. (7.5.6), but it is always equal to $H = T_2 + U - T_0$. Such an integral of motion is called the *Jacobi integral*.

7.7 Solved Example Problems

The differential equations of motion for the examples in this section were already formulated throughout the book by direct application of Newton's second law. They are considered here using the methods presented in this chapter so that you can see the difference in the two methodologies.

EXAMPLE 7.7.1 Differential Equations of Motion for Example 2.5.8

In this example, a body of mass m is thrown from an inertial point A with an initial absolute velocity of its center of mass P as shown in Fig. 7.5. In Example 2.5.8 (p. 95), air resistance is neglected and the only force acting on the body is the constant vertical gravitational force whose magnitude is mg. Here, in addition to the vertical force $-mg\hat{y}$, include the effect of air resistance modeled as $-c_1\vec{v} - c_2|\vec{v}|\vec{v}$. Obtain the differential equations of motion using Lagrange's equation, and check for the existence of any classical integrals of the motion. Also describe how to obtain the maximum height reached by the center of mass of the body.

Figure 7.5 Figure for Example 7.7.1.

■ **Solution**

Using the independent variables x and y as the generalized coordinates, the following is then obtained in terms of the unit vectors $\hat{x}$ and $\hat{y}$ shown in Fig. 7.5:

Absolute position vector for P: $\quad \vec{r} = x\hat{x} + y\hat{y}$

Absolute velocity of P: $\quad \vec{v} = \dfrac{d\vec{r}}{dt} = \dot{x}\hat{x} + \dot{y}\hat{y}$

Resultant force acting on P: $\quad \vec{F} = -mg\hat{y} - c_1(\dot{x}\hat{x} + \dot{y}\hat{y}) - c_2\sqrt{\dot{x}^2 + \dot{y}^2}(\dot{x}\hat{x} + \dot{y}\hat{y})$

The kinetic energy is then

$$T = \frac{1}{2}m\vec{v} \bullet \vec{v} = \frac{1}{2}m(\dot{x}^2 + \dot{y}^2) \qquad (7.7.1a)$$

which is of the form $T = T_2$, i.e., $T_1 = T_0 = 0$.

Inspection of the expression for $\vec{v}$ discloses that the virtual displacement of P is $\delta\vec{r} = (\delta x)\hat{x} + (\delta y)\hat{y}$. Therefore, the virtual work δW done by $\vec{F}$ and the generalized forces Q_x and Q_y are obtained as follows:

$$\delta W = \vec{F} \bullet (\delta\vec{r}) = -\left(c_1\dot{x} + c_2\dot{x}\sqrt{\dot{x}^2 + \dot{y}^2}\right)\delta x - \left(mg + c_1\dot{y} + c_2\dot{y}\sqrt{\dot{x}^2 + \dot{y}^2}\right)\delta y$$
$$\triangleq Q_x\,\delta x + Q_y\,\delta y \tag{7.7.1b}$$

Thus,

$$\frac{\partial T}{\partial \dot{x}} = m\dot{x} \qquad \frac{\partial T}{\partial \dot{y}} = m\dot{y} \qquad \frac{\partial T}{\partial x} = \frac{\partial T}{\partial y} = 0 \tag{7.7.1c}$$

and Lagrange's equation $d(\partial T/\partial \dot{q})/dt - \partial T/\partial q = Q_q$ yields, with $q = x$ and $q = y$, the following differential equations of motion:

$$m\ddot{x} = -c_1\dot{x} - c_2\dot{x}\sqrt{\dot{x}^2 + \dot{y}^2} \tag{7.7.1d}$$

$$m\ddot{y} = -mg - c_1\dot{y} - c_2\dot{y}\sqrt{\dot{x}^2 + \dot{y}^2} \tag{7.7.1e}$$

Integrals of Motion

There are no classical integrals of motion, unless $c_1 = c_2 = 0$. In such a case, there are two classical first integrals of motion, which are the generalized momentum $\partial T/\partial \dot{x} = m\dot{x}$ (thus, $\dot{x}$ = constant, which can also be integrated) and the Hamiltonian $\mathcal{H} = T_2 + U - T_0$, which is the same as $\mathcal{H} = T + U$ since $T = T_2$. The only part of δW that can be obtained as $-\delta(U)$ is the part $-mg\,\delta y$, which gives, except for an irrelevant constant of integration, the potential energy $U = mgy$. In such a case, $\mathcal{H}$ = constant yields $\mathcal{H} = \frac{1}{2}m(\dot{x}^2 + \dot{y}^2) + mgy$ = constant. The constant value of $\mathcal{H}$ is determined by the initial conditions of the motion as $\mathcal{H} = \frac{1}{2}m[\dot{x}^2(0) + \dot{y}^2(0)] + mgy(0)$. Notice that when $c_2 = 0$ and $c_1 \neq 0$, two nonclassical integrals of motion are obtained as $m\dot{x} + c_1 x$ = constant $\triangleq C_1$ and $m\dot{y} + c_1 y + mgt$ = constant $\triangleq C_2$.

When $c_2 = 0$, the differential equations of motion can be integrated analytically. In such a case, the maximum value of y can be obtained without the need for numerical integration. Numerical integration may also be used, of course.

When $c_2 \neq 0$, the differential equations have to be solved numerically because an analytical solution can no longer be obtained. The maximum value of y, for example, has to be determined from the numerical integration in such a case.

Differential Equations of Motion for Example 2.5.9 — EXAMPLE 7.7.2

In this example, as in Example 2.5.9 (p. 96), a block P_2 of mass m_2 slides on a face of a sliding block P_1, which is an inclined plane of mass m_1. P_2 is connected to P_1 by a linear spring of stiffness k N/m and unstressed length L_u m. Both blocks move on the same vertical plane. Friction in the system is viscous, with friction coefficients c_2 between P_2 and P_1, and c_1 between P_1 and the horizontal surface. The system, and two sets of unit vectors, are shown in Fig. 7.6, which is the same as Fig. 2.16 (p. 96), and is repeated here for convenience. Obtain the differential equations of motion using the Lagrangian formulation, and check for the existence of classical integrals of motion.

Figure 7.6 A block sliding on a sliding block for Example 7.7.2 (same as for Example 2.5.9).

■ **Solution**

To obtain the differential equations of motion using the Lagrangian formulation we need to develop expressions for the kinetic energy of the motion of the entire system and for the generalized forces acting on the system. To generate the expression for the kinetic energy we need to have the expression of the absolute velocity of each of the two blocks. We then have the following.

Absolute position vector $\vec{r}_1$ and absolute velocity $\vec{v}_1$ for block P_1:

$$\vec{r}_1 = x\hat{x}_1 \qquad \vec{v}_1 = \frac{d\vec{r}_1}{dt} = \dot{x}\hat{x}_1 \qquad (7.7.2a)$$

Absolute position vector $\vec{r}_2$, and absolute velocity $\vec{v}_2$ for block P_2:

$$\vec{r}_2 = x\hat{x}_1 + r\hat{y}_1 \qquad \vec{v}_2 = \frac{d\vec{r}_2}{dt} = \dot{x}\hat{x}_1 + \dot{r}\hat{y}_1 \qquad (7.7.2b)$$

The kinetic energy T is then

$$T = \frac{1}{2}m_1\vec{v}_1 \cdot \vec{v}_1 + \frac{1}{2}m_2\vec{v}_2 \cdot \vec{v}_2 = \frac{1}{2}m_1\dot{x}^2 + \frac{1}{2}m_2(\dot{x}^2 + \dot{r}^2 + 2\dot{x}\dot{r}\underbrace{\hat{x}_1 \cdot \hat{y}_1}_{=\cos\alpha}) \qquad (7.7.2c)$$

The generalized forces associated with the chosen generalized coordinates x and r are obtained from the expression for the virtual work δW of the forces acting on the system. The forces for which the virtual work is nonzero are

- The gravitational force $-m_2 g\hat{x}_2$ acting on P_2. The virtual work of the gravitational force $-m_1 g\hat{x}_2$ acting on block P_1 is zero because the virtual displacement of P_1 is $(\delta x)\hat{x}_1$, which is perpendicular to $\hat{x}_2$.
- The viscous friction force $-c_1\dot{x}\hat{x}_1$ acting on block P_1 due to contact with the horizontal surface.
- The viscous friction force proportional to $\dot{r}$ and acting parallel to the plane of contact between P_2 and P_1. Such a force does work only when one body moves relative to the other. The virtual displacement of P_2 relative to P_1 is $\delta\vec{r}_2 - \delta\vec{r}_1$, and since such a force is in the direction opposite to that relative displacement, its virtual work is equal to $-c_2\dot{r}\hat{y}_1 \cdot (\delta\vec{r}_2 - \delta\vec{r}_1)$.
- The force due to the spring that connects block P_2 to P_1. Such a force also does work only when one body moves relative to the other. Since the virtual displacement $\delta\vec{r}_2 - \delta\vec{r}_1$ of P_2 relative to P_1 stretches the spring (see Fig. 7.6), the spring force due

to such a displacement is equal to $-k(r - L_u)\hat{y}_1$. Therefore, the virtual work of the spring force is equal to $-k(r - L_u)\hat{y}_1 \bullet (\delta \vec{r}_2 - \delta \vec{r}_1)$.

The total virtual work δW is the sum of the virtual work of these forces just listed. Therefore, we have

$$\delta W = \underbrace{-m_2 g \ \hat{x}_2 \bullet (\delta \vec{r}_2)}_{= \hat{x}_2 \bullet [(\delta x)\hat{x}_1 + (\delta r)\hat{y}_1]} \underbrace{- c_1 \dot{x}\hat{x}_1 \bullet \delta \vec{r}_1}_{= (\delta x)\hat{x}_1} \underbrace{- c_2 \dot{r}\hat{y}_1 \bullet (\delta \vec{r}_2 - \delta \vec{r}_1)}_{= (\delta r)\hat{y}_1}$$

$$= -m_2 g(\delta r)\hat{x}_2 \bullet \hat{y}_1 = (m_2 g \sin \alpha)\delta r$$

$$- k(r - L_u)\hat{y}_1 \bullet (\delta \vec{r}_2 - \delta \vec{r}_1)$$

and the virtual work is

$$\delta W = [m_2 g \sin \alpha - c_2 \dot{r} - k(r - L_u)]\delta r - c_1 \dot{x} \ \delta x \stackrel{\Delta}{=} Q_r \ \delta r + Q_x \ \delta x \quad \text{(7.7.2d)}$$

The coefficients of δr and δx that appear in this expression are the generalized forces Q_r and Q_x associated with the generalized coordinates r and x, respectively.

Lagrange's equation $d(\partial T/\partial \dot{q})/dt - \partial T/\partial q = Q_q$ yields the following differential equations of motion for the system, with $q = x$ and $q = r$:

For $q = x$:

$$\frac{d}{dt}[(m_1 + m_2)\dot{x} + m_2 \dot{r} \cos \alpha] = Q_x = -c_1 \dot{x}$$

or

$$(m_1 + m_2)\dot{x} + m_2 \dot{r} \cos \alpha + c_1 x = \text{constant} \stackrel{\Delta}{=} C_1 \quad \text{(7.7.2e)}$$

For $q = r$:

$$\frac{d}{dt}[m_2(\dot{r} + \dot{x} \cos \alpha)] = Q_r = m_2 g \sin \alpha - c_2 \dot{r} - k(r - L_u)$$

or

$$m_2(\ddot{r} + \ddot{x} \cos \alpha) + c_2 \dot{r} + k(r - L_u) - m_2 g \sin \alpha = 0 \quad \text{(7.7.2f)}$$

Equations (7.7.2e) and (7.7.2f) are the differential equations of motion for the system, with Eq. (7.7.2f) being exactly the same as Eq. (2.5.9g) (p. 98). Equations (7.7.2e) and (7.7.2f) are equivalent to Eqs. (2.5.9g) and (2.5.9i) that were obtained in Chapter 2 via Newton's second law. To verify the equivalence, see Problem 7.3.

Integrals of Motion

The potential energy for this system is

$$U = -mgr \sin \alpha + \frac{k}{2}(r - L_u)^2 \quad \text{(7.7.2g)}$$

and, because $T = T_2$, as seen from Eq. (7.7.2c), the Hamiltonian is

$$\mathcal{H} = T_2 + U - T_0 = T + U \quad \text{(7.7.2h)}$$

The variable x is an ignorable coordinate because it does not appear explicitly in the Lagrangian $\mathcal{L} = T - U$, but the nonconservative generalized force is $Q_x = -c_1 \dot{x}$. Thus, the generalized momentum $\partial \mathcal{L}/\partial \dot{x}$ is constant only if $c_1 = 0$. However, since Q_x can be integrated with respect to time, the function $\partial \mathcal{L}/\partial \dot{x} + c_1 x = (m_1 + m_2)\dot{x} + m_2 \dot{r}\cos\alpha + c_1 x$ is constant, as already indicated by Eq. (7.7.2e).

The total time derivative of the Hamiltonian $\mathcal{H}$ is

$$\frac{d\mathcal{H}}{dt} = \underbrace{\frac{\partial \mathcal{H}}{\partial t}}_{=0} + (Q_x)_{nc}\,\dot{x} + (Q_r)_{nc}\,\dot{r} = -c_1\dot{x}^2 - c_2\dot{r}^2 \qquad (7.7.2i)$$

and, thus, the Hamiltonian is constant only if $c_1 = c_2 = 0$.

EXAMPLE 7.7.3 Differential Equations of Motion for Example 2.7.6

The system of Example 2.7.6 (p. 121) is shown in Fig. 7.7. At point B there is a particle with mass m. The following modifications are made here to make this example more general:

- Point A is given a known motion $\overrightarrow{OA} = x_A(t)\hat{x} + y_A(t)\hat{y}$ in any direction in the vertical plane.
- Link AB is a uniform bar of mass m_{bar} and length L. It is still subjected to the torque M_{bar} shown in the figure, which is applied by a motor.

Obtain all classical integrals of motion, if any, using the generalized coordinate θ shown in Fig. 7.7. Also, obtain the differential equation of motion for the system using Lagrange's equation.

Figure 7.7 A pendulum mounted on a moving base for Example 7.7.3.

■ Solution

The center of mass (C) of bar AB is in the middle of the bar. The absolute velocity $\vec{v}_C$ of C and the virtual variation $\delta(\overrightarrow{OC})$, are

$$\vec{v}_C = \frac{d}{dt}\overrightarrow{OC} = \frac{d}{dt}\overrightarrow{OA} + \frac{d}{dt}\left(\frac{L}{2}\hat{R}\right) = \dot{x}_A\hat{x} + \dot{y}_A\hat{y} + \frac{L}{2}\dot{\theta}\hat{\theta}$$

$$\delta(\overrightarrow{OC}) = \delta(\overrightarrow{OA}) + \frac{L}{2}(\delta\theta)\hat{\theta} \qquad (7.7.3a)$$

where $\delta(\overrightarrow{OA}) = 0$ because the motion of point A is prescribed.

SECTION 7.7 Solved Example Problems

The absolute velocity $\vec{v}_B$ of point B, and the virtual variation $\delta(\overrightarrow{OB})$, are

$$\vec{v}_B = \frac{d}{dt}\overrightarrow{OB} = \frac{d}{dt}\overrightarrow{OA} + \frac{d}{dt}(L\hat{R}) = \dot{x}_A\hat{x} + \dot{y}_A\hat{y} + L\dot{\theta}\hat{\theta}$$

$$\delta(\overrightarrow{OB}) = \delta(\overrightarrow{OA}) + L(\delta\theta)\hat{\theta} \tag{7.7.3b}$$

Therefore, the kinetic energy T of the motion is

$$T = \underbrace{\frac{1}{2}m_{\text{bar}}\vec{v}_C \bullet \vec{v}_C + \frac{1}{2}I_C\dot{\theta}^2}_{\text{This is the kinetic energy for the bar}} + \frac{1}{2}m\vec{v}_B \bullet \vec{v}_B$$

$$= \frac{1}{2}m_{\text{bar}}\left(\dot{x}_A^2 + \dot{y}_A^2 + \frac{L^2}{4}\dot{\theta}^2 + L\dot{\theta}\dot{x}_A \underbrace{\hat{x}\bullet\hat{\theta}}_{=\cos\theta} + L\dot{\theta}\dot{y}_A \underbrace{\hat{y}\bullet\hat{\theta}}_{=\sin\theta}\right) + \frac{1}{2}I_C\dot{\theta}^2$$

$$+ \frac{1}{2}m\left(\dot{x}_A^2 + \dot{y}_A^2 + L^2\dot{\theta}^2 + 2L\dot{\theta}\dot{x}_A \underbrace{\hat{x}\bullet\hat{\theta}}_{=\cos\theta} + 2L\dot{\theta}\dot{y}_A \underbrace{\hat{y}\bullet\hat{\theta}}_{=\sin\theta}\right) = T_2 + T_1 + T_0$$

$$\tag{7.7.3c}$$

where I_C is the moment of inertia of bar AB about an axis that passes through its center of mass C and is perpendicular to the plane of the motion, i.e.,

$$I_C = \frac{1}{12}m_{\text{bar}}L^2$$

The generalized forces associated with the chosen generalized coordinate θ is obtained from the expression for the virtual work δW of the forces acting on the system. The forces and moments for which the virtual work is nonzero are

- The gravitational force $-m_{\text{bar}}g\hat{y}$ acting at the center of mass of bar AB, and the constant gravitational force $-mg\hat{y}$ acting at point B. The virtual work of these forces is then $-m_{\text{bar}}g\hat{y} \bullet \delta(\overrightarrow{OC}) - mg\hat{y} \bullet \delta(\overrightarrow{OB}) = -g(m_{\text{bar}}L/2 + mL)(\sin\theta)\,\delta\theta$. The virtual work of these gravitational forces could also have been obtained more quickly as $-\delta(U)$, where $U = -m_{\text{bar}}g(L/2)\cos\theta - mgL\cos\theta$ (see Fig. 7.7). The quantities $(L/2)\cos\theta$ and $L\cos\theta$ are the vertical distances that the center of mass C and point B, respectively, are below point A in Fig. 7.7 (thus, the physical interpretation of the minus sign in U).
- The moment applied to bar AB by a motor. With $\hat{z} \triangleq \hat{R} \times \hat{\theta} = \hat{x} \times \hat{y}$ (see Fig. 7.7) and the absolute angular velocity of bar AB being $\dot{\theta}\hat{z}$, the infinitesimal work due to that moment is $(dW)_{\text{motor}} = (M_{\text{bar}}\hat{z}) \bullet (\dot{\theta}\hat{z}) = M_{\text{bar}}\,d\theta$. Therefore, the virtual work due to that moment is simply $(\delta W)_{\text{motor}} = M_{\text{bar}}\,\delta\theta$.

The total virtual work δW is the sum of the virtual work of these forces just listed. Therefore, we have

$$\delta W = \left[M_{\text{bar}} - \left(\frac{1}{2}m_{\text{bar}} + m\right)gL\sin\theta\right]\delta\theta \triangleq Q_\theta\,\delta\theta = -\delta(U) + M_{\text{bar}}\,\delta\theta \tag{7.7.3d}$$

where $U = -m_{\text{bar}}g(L/2)\cos\theta - mgL\cos\theta$. Notice that, since the moment M_{bar} is unspecified, it is not (and could not be) included in the expression for the potential energy U. Thus, it is regarded as a nonconservative moment. If M_{bar} were specified as a function that only depends on θ, $M_{\text{bar}}(\theta)$, for example, one would introduce the potential energy $U = -m_{\text{bar}}g(L/2)\cos\theta - mgL\cos\theta - \int M_{\text{bar}}\,d\theta$ and that moment would then be a conservative moment.

Tests for the Classical Integrals of the Motion

Generalized momentum integral. Since θ appears explicitly in the Lagrangian $\mathcal{L} = T - U$, it follows that $\partial \mathcal{L}/\partial \theta \neq 0$. Therefore, there is no momentum integral.

Testing the Hamiltonian. The Hamiltonian is $\mathcal{H} = T_2 + U - T_0$, where

$$T_0 = \frac{1}{2}(m_{\text{bar}} + m)\left(\dot{x}_A^2 + \dot{y}_A^2\right)$$

and

$$T_2 = \frac{1}{2}\left(I_C + m_{\text{bar}}\frac{L^2}{4} + mL^2\right)\dot{\theta}^2$$

are obtained by inspection of Eq. (7.7.3c). As per Eq. (7.5.7), the time derivative of $\mathcal{H}$ is

$$\frac{d\mathcal{H}}{dt} = (Q_\theta)_{nc}\dot{\theta} + \frac{\partial \mathcal{H}}{\partial t} = M_{\text{bar}}\dot{\theta} + \frac{\partial \mathcal{H}}{\partial t} = M_{\text{bar}}\dot{\theta} - \frac{1}{2}(m_{\text{bar}} + m)\frac{d}{dt}\left(\dot{x}_A^2 + \dot{y}_A^2\right)$$

Since the right-hand side of the expression for $d\mathcal{H}/dt$ is not zero when either (or both) $M_{\text{bar}} \neq 0$ or $\partial \mathcal{H}/\partial t = -\frac{1}{2}(m_{\text{bar}} + m)d(\dot{x}_A^2 + \dot{y}_A^2)/dt \neq 0$, the Hamiltonian is not constant.

When is the Hamiltonian $\mathcal{H}$ constant? That will be the case when $M_{\text{bar}} = 0$ at all times and $\dot{x}_A^2 + \dot{y}_A^2 = $ constant.

Differential Equation of Motion for the System

The differential equation of motion is readily obtained from Lagrange's equation $d(\partial T/\partial \dot{\theta})/dt - \partial T/\partial \theta = Q_\theta$.

We have

$$\frac{\partial T}{\partial \dot{\theta}} = \left(I_C + m_{\text{bar}}\frac{L^2}{4} + mL^2\right)\dot{\theta} + \left(\frac{1}{2}m_{\text{bar}} + m\right)L(\dot{x}_A\cos\theta + \dot{y}_A\sin\theta) \quad \text{(7.7.3e)}$$

$$\frac{\partial T}{\partial \theta} = \left(\frac{1}{2}m_{\text{bar}} + m\right)L(\dot{y}_A\cos\theta - \dot{x}_A\sin\theta)\dot{\theta} \quad \text{(7.7.3f)}$$

Therefore, Lagrange's equation yields the following differential equation of motion:

$$\left[m_{\text{bar}}\left(\frac{L^2}{4} + \frac{L^2}{12}\right) + mL^2\right]\ddot{\theta} + \left(\frac{1}{2}m_{\text{bar}} + m\right)gL\sin\theta$$

$$= M_{\text{bar}} - \left(\frac{1}{2}m_{\text{bar}} + m\right)L(\ddot{x}_A\cos\theta + \ddot{y}_A\sin\theta) \quad \text{(7.7.3g)}$$

When $m_{\text{bar}} = 0$ and $\ddot{y}_A(t) = 0$, which corresponds to the simpler case considered in Chapter 2, this differential equation becomes Eq. (2.7.6d) (p. 124), which was obtained by direct application of Newton's second law. The analysis of the motion is done in the manner presented in Chapter 2.

SECTION 7.7 Solved Example Problems

Differential Equations of Motion for Example 2.7.8

EXAMPLE 7.7.4

The system of Example 2.7.8 (p. 129) is shown in Fig. 7.8, where motion takes place in a vertical plane. The spring is linear, with stiffness k N/m and unstressed length L_u m, and there is no friction in the system. Obtain all classical integrals of the motion, if any, using the generalized coordinates r and θ shown in Fig. 7.8. Also, obtain the differential equations of motion for the system using Lagrange's equation.

■ **Solution**

Since the absolute velocity of point P is

$$\vec{v}_P = \frac{d}{dt}(\overrightarrow{OP}) = \frac{d}{dt}(r\hat{r}) = \dot{r}\hat{r} + r\dot{\theta}\hat{\theta} \qquad (7.7.4a)$$

the kinetic energy of the motion is

$$T = \frac{1}{2}m\vec{v}_P \bullet \vec{v}_P = \frac{1}{2}m(\dot{r}^2 + r^2\dot{\theta}^2) = T_2 \qquad (7.7.4b)$$

Inspection of Fig. 7.8 discloses that the potential energy U for this system is

$$U = -mgr\cos\theta + \frac{k}{2}(r - L_u)^2 \qquad (7.7.4c)$$

and since there are no other forces acting on the system that do virtual work, this then yields the following expression for the virtual work δW:

$$\delta W = -\delta(U) = -\left(\frac{\partial U}{\partial r}\delta r + \frac{\partial U}{\partial \theta}\delta\theta\right)$$

$$= [mg\cos\theta - k(r - L_u)]\delta r - (mgr\sin\theta)\delta\theta \stackrel{\Delta}{=} Q_r\,\delta r + Q_\theta\,\delta\theta \qquad (7.7.4d)$$

The expressions for the generalized forces Q_r and Q_θ are, respectively, the coefficients of δr and $\delta\theta$ in the expression for δW. Notice that writing the expression for potential energy by inspection (please see the note on p. 428), when it is possible to do so, provides the quickest way to obtain the conservative generalized forces acting on the system.[8]

The only classical integral of motion for the system being considered is the Hamiltonian $\mathcal{H} = T_2 + U - T_0$. Since $T = T_2$ by inspection of Eq. (7.7.4b), the Hamiltonian is the same as the total mechanical energy $T + U$. We then have

$$\mathcal{H} = \frac{1}{2}m(\dot{r}^2 + r^2\dot{\theta}^2) - mgr\cos\theta + \frac{k}{2}(r - L_u)^2 = \text{constant} \qquad (7.7.4e)$$

The quantities $\partial\mathcal{L}/\partial\dot{r}$ and $\partial\mathcal{L}/\partial\dot{\theta}$, where $\mathcal{L} = T - U$, are not constants because $\mathcal{L}$ depends explicitly on both r and θ.

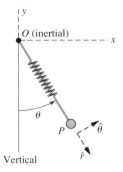

Figure 7.8 The spring-pendulum for Example 7.7.4 (same as for Example 2.7.8).

---

[8] The direct way to obtain the expression for δW is repeated here, in case of need.

$$\delta W = \left[\underbrace{mg(\hat{r}\cos\theta - \hat{\theta}\sin\theta)}_{\text{gravitational force}} \underbrace{- k(r - L_u)\hat{r}}_{\text{spring force}}\right] \bullet \underbrace{[(\delta r)\hat{r} + (r\,\delta\theta)\hat{\theta}]}_{\text{virtual displacement of } P}$$

$$= [mg\cos\theta - k(r - L_u)]\delta r - (mgr\sin\theta)\delta\theta \stackrel{\Delta}{=} Q_r\,\delta r + Q_\theta\,\delta\theta$$

The differential equations of motion are readily obtained from Lagrange's equation $d(\partial T/\partial \dot{q})/dt - \partial T/\partial q = Q_q$, with $q = r$ and $q = \theta$. Since

$$\frac{\partial T}{\partial \dot{r}} = m\dot{r} \qquad \frac{\partial T}{\partial \dot{\theta}} = mr^2\dot{\theta} \qquad \frac{\partial T}{\partial r} = mr\dot{\theta}^2 \qquad \frac{\partial T}{\partial \theta} = 0,$$

Lagrange's equation yields the following differential equations of motion for this system:

$$m(\ddot{r} - r\dot{\theta}^2) = mg\cos\theta - k(r - L_u) \qquad (7.7.4f)$$

$$m(r\ddot{\theta} + 2\dot{r}\dot{\theta}) = -mg\sin\theta \qquad (7.7.4g)$$

which are exactly of the same form as Eqs. (2.7.8f) and (2.7.8g) (p. 130). The analysis of motion was presented in Example 2.7.8.

EXAMPLE 7.7.5   Differential Equations of Motion for Example 2.7.9

In this example, a rigid body of mass $m = 0.5$ kg moves inside a horizontal tube of length $R = 2$ m, filled with oil and subjected to a moment M_{motor} applied by a motor as shown in Fig. 7.9. In Example 2.7.9 (p. 137) the body was a mass particle and the tube was approximated as massless. Here the dimensions of the body and the tube are taken into account. The vector $\vec{r}$ shown in Fig. 7.9 is the absolute position vector for the center of mass of the body that is moving inside the rotating tube. The tube is always in contact with a fixed horizontal table and is driven by a motor with a prescribed absolute angular velocity $\dot{\theta}\hat{z} \stackrel{\triangle}{=} \Omega(t)\hat{z}$ rad/s relative to inertial space. End O of the tube is inertial. Because of the viscosity of the oil, the body is subjected to a linear viscous friction force with friction coefficient $c = 1$ N/(m/s). For this, let $(I_O)_{\text{tube}}$ be the moment of inertia of the tube about an axis that passes through point O and that is perpendicular to the plane of the motion (which is the plane of the figure). Also, let $(I_C)_{\text{body}}$ be the moment of inertia of the body about a similar axis, but passing through the center of mass of the body. Obtain all the classical integrals of the motion, if any, and determine the time the center of mass of the body reaches the end of the tube if $\Omega = $ constant $= 5$ rad/s.

Figure 7.9 A constrained body in a rotating tube for Example 7.7.5 (same as for Example 2.7.9).

■ Solution

For this example, the expression for the kinetic energy is

$$T = \frac{m}{2}(\dot{r}^2 + r^2\dot{\theta}^2) + \frac{1}{2}(I_C)_{body}\dot{\theta}^2 + \frac{1}{2}(I_O)_{tube}\dot{\theta}^2$$

$$= \frac{m}{2}[\dot{r}^2 + r^2\Omega^2(t)] + \frac{1}{2}[(I_C)_{body} + (I_O)_{tube}]\Omega^2(t) = T_2 + T_0 \quad (7.7.5a)$$

which, as indicated in Eq. (7.7.5a), is of the form $T_2 + T_0$, where $T_2 = m\dot{r}^2/2$ and T_0 is the rest of the expression for T.

The only generalized force for which the virtual work is not equal to zero is the viscous friction force $-c\dot{r}\hat{r}$, and the virtual work of such a force is simply

$$\delta W = -c\dot{r}\hat{r} \bullet [(\delta r)\hat{r}] = -c\dot{r}\,\delta r \stackrel{\Delta}{=} Q_r\,\delta r \quad (7.7.5b)$$

If the moment M_{motor} shown in Fig. 7.9 were prescribed, instead of $\dot{\theta}$, its virtual work would be $M_{motor}\,\delta\theta$. However, since $\dot{\theta}$ is what is prescribed, $d\theta = \dot{\theta}\,dt$ and, thus, $\delta\theta = \dot{\theta}\,\delta t = 0$ as $\delta t = 0$ by definition of a virtual variation.

As for the classical integrals of motion (i.e., a generalized momentum equal to a constant, and the Hamiltonian also equal to a constant), there are none here! The generalized momentum $\partial T/\partial \dot{r} = m\dot{r}$ (which is a linear momentum) is not constant because $\partial T/\partial r \neq 0$. In addition, the generalized nonconservative force $-c\dot{r}$ is not zero either.

The Hamiltonian $\mathcal{H} = T_2 + U - T_0$, where $U = 0$, is not constant either. Its time derivative $d\mathcal{H}/dt$ is

$$\frac{d\mathcal{H}}{dt} = \frac{\partial \mathcal{H}}{\partial t} + \underbrace{(Q_r)_{nc}}_{=-c\dot{r}}\dot{r} = -[mr^2 + (I_C)_{body} + (I_O)_{tube}]\Omega\dot{\Omega} - c\dot{r}^2 \quad (7.7.5c)$$

The Hamiltonian would be constant only if $c = 0$ and $\Omega =$ constant. Notice that the Hamiltonian $\mathcal{H} = T_2 + U - T_0$ is not the same as the total mechanical energy $E \stackrel{\Delta}{=} T + U = T_2 + T_1 + T_0 + U$, where $T_1 = 0$ for this example. The two are the same only if $T = T_2$, i.e., if $T_1 = 0$ and $T_0 = 0$. In this example, the external motor must "pump" energy into the system in order to force the tube to rotate with a prescribed angular velocity $\dot{\theta} = \Omega(t)$, and, thus, the total mechanical energy E cannot be constant. Even if there were no damping in the system, the motor would still have to pump energy into the system to force the tube to rotate in the prescribed manner.

If the Hamiltonian were constant, the equation $\mathcal{H} =$ constant would be a first-order differential equation. Thus, in such a case, there would be no need to work with a second-order differential equation generated either by using Lagrange's equation or Newton's second law. However, even with $\Omega =$ constant, the Hamiltonian is not constant for this problem because the damping coefficient c is not equal to zero. We then need to generate the second-order differential equation of motion in order to solve the problem at hand.

We have

$$\frac{\partial T}{\partial \dot{r}} = m\dot{r} \qquad \frac{\partial T}{\partial r} = mr\dot{\theta}^2 = mr\Omega^2(t) \quad (7.7.5d)$$

and the following differential equation of motion is immediately obtained from Lagrange's equation:

$$m(\ddot{r} - \Omega^2 r) = -c\dot{r} \qquad (7.7.5e)$$

As presented in Example 2.7.9, the solution to this linear differential equation is

$$r(t) = A_1 e^{s_1 t} + A_2 e^{s_2 t} \qquad (7.7.5f)$$

For $\Omega = 5$ rad/s, $c = 1$ N/(m/s), and $m = 0.5$ kg, one obtains $A_1 \approx 0.6$ m, $A_2 \approx 0.4$ m, $s_1 \approx 4.1\,\text{s}^{-1}$, and $s_2 \approx -6.1\,\text{s}^{-1}$. For $r = 2$ m, Eq. (7.7.5f) yields $t \approx 0.29$ s.

EXAMPLE 7.7.6 Differential Equations of Motion for Example 2.7.13

In this example, as in Example 2.7.13 (p. 149), a point P of mass m slides on a circular frictionless track in a homogeneous ring of radius R and mass m_{ring}. The ring is forced, by a motor, to rotate with a given absolute angular velocity $\dot{\phi} \triangleq \Omega(t)$ about a vertical axis AA that passes through its center C as shown in Fig. 7.10. The center C of the ring is inertial (the unit vector triad $\{\hat{r}, \hat{\theta}, \hat{n} \triangleq \hat{r} \times \hat{\theta}\}$ shown in the figure will be used in the solution to the problem). With θ as the generalized coordinate, obtain the differential equation of motion for the system using Lagrange's equation and obtain the classical integrals of the motion (if any).

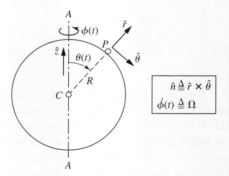

Figure 7.10 A particle moving on a rotating ring for Example 7.7.6.

■ **Solution**

Notice that this is a problem with a constraint, namely, $\dot{\phi} \triangleq \Omega(t)$, where $\Omega(t)$ is any given function of time. However, this is a simple constraint, in the sense that the dynamics of this system is governed by a single independent variable, which is chosen to be the angle $\theta(t)$.

Since the absolute velocity of point P is

$$\vec{v}_P = \frac{d}{dt}(R\hat{r}) = [\Omega(\hat{r}\cos\theta - \hat{\theta}\sin\theta) + \dot{\theta}\hat{n}] \times (R\hat{r}) = R\dot{\theta}\hat{\theta} + (R\sin\theta)\dot{\phi}\hat{n} \qquad (7.7.6a)$$

and the ring is rotating with absolute angular velocity $\dot\phi\hat z = \Omega\hat z$, the expression for the kinetic energy of the motion of this system is

$$T = \frac{mR^2}{2}(\dot\theta^2 + \dot\phi^2 \sin^2\theta) + \frac{1}{2}(I_{AA})_{\text{ring}}\dot\phi^2$$

$$= \underbrace{\frac{1}{2}mR^2\dot\theta^2}_{=T_2} + \underbrace{\frac{1}{2}[(I_{AA})_{\text{ring}} + mR^2\sin^2\theta]\Omega^2(t)}_{=T_0} \quad (7.7.6b)$$

where $(I_{AA})_{\text{ring}} = 2\int_0^\pi (R\sin\theta)^2[m_{\text{ring}}/(2\pi R)](R\,d\theta) = \frac{1}{2}m_{\text{ring}}R^2$ is the moment of inertia of the ring about axis AA.

As in Example 7.7.5, the virtual work of the moment applied by the motor is equal to zero. Notice that the actual infinitesimal work $dW = M_{\text{motor}}\,d\phi$ done by the motor is not zero, but that is not relevant for the approach taken in analytical dynamics. The only force for which the virtual work in this system with a frictionless track is not zero is the gravitational force $-mg\hat z$, and its virtual work can be obtained either as $\delta W = (-mg\hat z)\bullet[\delta(R\hat r)] = (-mg\hat z)\bullet[(R\,\delta\theta)\hat\theta]$ or from the potential energy $U = mg(R\cos\theta)$ as $\delta W = -\delta(U)$. This yields the following expression for the virtual work, from which the expression for the generalized force Q_θ is extracted:

$$\delta W = (mgR\sin\theta)\,\delta\theta \triangleq Q_\theta\,\delta\theta \quad (7.7.6c)$$

We then have

$$\frac{\partial T}{\partial \dot\theta} = mR^2\dot\theta \qquad \frac{\partial T}{\partial \theta} = mR^2\Omega^2(\sin\theta)\cos\theta \quad (7.7.6d)$$

Lagrange's equation $d(\partial T/\partial\dot\theta)/dt - \partial T/\partial\theta = Q_\theta$ yields the following differential equation of motion:

$$mR^2\ddot\theta - mR^2\Omega^2(\sin\theta)\cos\theta = Q_\theta = mgR\sin\theta \quad (7.7.6e)$$

As for the classical integrals of motion (i.e., a generalized momentum equal to a constant, and the Hamiltonian also equal to a constant), the generalized momentum $\partial T/\partial\dot\theta$ (which is an angular momentum) is not constant because $\partial T/\partial\theta \neq 0$.

The Hamiltonian

$$\mathcal{H} = T_2 + U - T_0 = \frac{1}{2}mR^2\dot\theta^2 + mgR\cos\theta - \frac{1}{2}[(I_{AA})_{\text{ring}} + mR^2\sin^2\theta]\Omega^2(t) \quad (7.7.6f)$$

is not constant either. Its time derivative $d\mathcal{H}/dt$ is

$$\frac{d\mathcal{H}}{dt} = \frac{\partial\mathcal{H}}{\partial t} + \underbrace{(Q_\theta)_{nc}}_{=0}\dot\theta = -[(I_{AA})_{\text{ring}} + mR^2\sin^2\theta]\Omega\dot\Omega \quad (7.7.6g)$$

The Hamiltonian is constant only if $\Omega = $ constant. Notice that, as in Example 7.7.5, the Hamiltonian $\mathcal{H} = T_2 + U - T_0$ is not the same as the total mechanical energy $E \triangleq T + U = T_2 + T_1 + T_0 + U$, where $T_1 = 0$ for this example. The two are the same only if $T = T_2$, i.e., if $T_1 = 0$ and $T_0 = 0$. In this example, the external motor must also pump energy into the system in order to force the ring to rotate with a prescribed angular velocity $\dot\phi = \Omega(t)$ and, thus, the total mechanical energy E cannot be constant.

A very convenient point to notice is that when $\Omega = $ constant, the equation $\mathcal{H} = T_2 + U - T_0 = $ constant is a first-order differential equation. Thus, there is no need to generate the second-order differential equation of motion because the equation $\mathcal{H} = $ constant is an integral of that second-order differential equation. Thus, all the work involving in generating that equation can be bypassed in such a case.

EXAMPLE 7.7.7 A Modified Form of Example 7.7.6

It is interesting and instructive to look at the same problem of Example 7.7.6 by considering a known moment $M_{\text{motor}} \hat{z}$ applied to the rotating ring, instead of specifying that the ring is to rotate with a prescribed angular velocity. In such a case, the ring will rotate with an unknown angular velocity $\dot{\phi}$, and the system has, then, two degrees of freedom: θ and ϕ. The differential equations of motion for this situation are also easily and readily obtained using the Lagrangian formulation, and the only modification to be made is in the expression for the virtual work δW, which now becomes

$$\delta W = (mgR\sin\theta)\,\delta\theta + M_{\text{motor}}\,\delta\phi \qquad (7.7.7\text{a})$$

The expression for the kinetic energy remains the same as Eq. (7.7.6b), but now $T = T_2$ (i.e., a purely quadratic form in the generalized velocities $\dot{\theta}$ and $\dot{\phi}$) since neither θ nor ϕ are known a priori. In order to avoid confusion, that expression is repeated here.

$$T = \frac{mR^2}{2}(\dot{\theta}^2 + \dot{\phi}^2 \sin^2\theta) + \frac{1}{2}(I_{AA})_{\text{ring}}\dot{\phi}^2 = T_2 \qquad (7.7.7\text{b})$$

Since the expression for the generalized force Q_θ also did not change, Eq. (7.7.6e) is one of the differential equations of motion, with Ω^2 in that equation being the unknown $\dot{\phi}^2$. Again to avoid confusion, that equation is repeated here.

$$mR^2\ddot{\theta} - mR^2\dot{\phi}^2(\sin\theta)\cos\theta = Q_\theta = mgR\sin\theta \qquad (7.7.7\text{c})$$

The other differential equation of motion is obtained from Lagrange's equation $d(\partial T/\partial\dot{\phi})/dt - \partial T/\partial\phi = Q_\phi$, where $Q_\phi = M_{\text{motor}}$, as seen by inspection of Eq. (7.7.7a). Since

$$\frac{\partial T}{\partial \dot{\phi}} = [(I_{AA})_{\text{ring}} + mR^2\sin^2\theta]\dot{\phi} \qquad \frac{\partial T}{\partial \phi} = 0 \qquad (7.7.7\text{d})$$

Lagrange's equation then yields the following second differential equation of motion:

$$\frac{d}{dt}\left\{[(I_{AA})_{\text{ring}} + mR^2\sin^2\theta]\dot{\phi}\right\} = M_{\text{motor}} \qquad (7.7.7\text{e})$$

For this form of the problem, M_{motor} is specified; i.e., it is a known function of time, and both Eqs. (7.7.7c) and (7.7.7e) have to be integrated simultaneously to obtain the solutions for $\theta(t)$ and $\phi(t)$. As for the classical integrals of the motion for this modified form of the problem, neither $\partial T/\partial\dot{\phi}$ is constant, of course (unless $M_{\text{motor}} = 0$, i.e., if the ring, with frictionless bearings, were not driven) nor $\partial T/\partial\dot{\theta}$ is constant. Since T is now

$T = T_2$, the Hamiltonian, which we shall denote by $\mathcal{H}_{\text{new}}$ in order not to confuse it with the $\mathcal{H}$ in Example 7.7.6, is

$$\mathcal{H}_{\text{new}} = T + U \qquad \text{(i.e., the total mechanical energy of the system)} \qquad (7.7.7f)$$

and its time derivative is

$$\frac{d\mathcal{H}_{\text{new}}}{dt} = \underbrace{\frac{\partial \mathcal{H}_{\text{new}}}{\partial t}}_{=0} + \dot{\theta}\underbrace{(Q_\theta)_{nc}}_{=0} + \dot{\phi}\underbrace{(Q_\phi)_{nc}}_{=M_{\text{motor}}} = M_{\text{motor}}\dot{\phi} \qquad (7.7.7g)$$

which is the power pumped by the motor into the system. Notice that $\partial \mathcal{H}_{\text{new}}/\partial t = 0$ because $\mathcal{H}_{\text{new}}$ is not an *explicit* function of time (i.e., the time t is not seen explicitly in the expression for $\mathcal{H}_{\text{new}}$).

Notice that when $\dot{\phi}$, instead of M_{motor}, is prescribed, one could modify the problem by ignoring the constraint $\dot{\phi} = \text{constant}$, applying the moment M_{motor} to the system, and then treating M_{motor} as an unknown. In such a case the virtual work associated with that moment would be $M_{\text{motor}}\,\delta\phi$, instead of zero as in the case when $\delta\phi = 0$ (i.e., when ϕ is prescribed). This modified approach yields a differential equation associated with $\delta\phi$ which is of the form $I\ddot{\phi} + \cdots = M_{\text{motor}}$. This may be done for any other problem and, if it is done, the kinetic energy will automatically be of the form $T = T_2$, and the Hamiltonian (i.e., $\mathcal{H}_{\text{new}}$) will automatically be equal to the total mechanical energy $T + U$. At the end of the formulation of the differential equations, the condition $\dot{\phi} = \text{constant}$ is then imposed in such a modified problem as being a solution to the differential equation $I\ddot{\phi} + \cdots = M_{\text{motor}}$. Thus, such an equation allows the motor torque M_{motor} to be determined. If the determination of the motor torque is of no interest, then there is no advantage in using such a modified approach.

EXAMPLE 7.7.8

Example 2.8.8 with Viscous Friction Included

In this example, as in Example 2.8.8 (p. 163), a small body P, modeled as a particle of mass m, is constrained to slide on a frictionless parabolic vertical track $y = ax^2$ m (see Fig. 7.11), where a is a given constant. The particle is also connected to a spring, and the other end of the spring is attached to an inertial point O. The spring is linear, with stiffness k N/m and unstressed length L_u m. Obtain the differential equation of motion for the system, including the effect of linear viscous friction with friction coefficient c N/(m/s), and check for classical integrals of motion.

Figure 7.11 A mass particle constrained to move on a parabolic track for Example 7.7.8.

■ **Solution**

Notice that this is also a problem with a constraint, namely, that particle P is constrained to move on the parabolic track $y = ax^2$. However, this is also a simple constraint that makes it possible to describe the motion using a single independent variable. Such types of constraints, and other types that require the introduction of additional variables, are discussed in Section 7.8 (p. 463).

In the absence of damping as done in Example 2.8.8 (i.e., for $c = 0$), the infinitesimal work dW for this problem happens to be an exact differential $d(W)$. This fact was then used in Chapter 2 to determine the maximum height reached by the particle for the undamped problem in Example 2.8.8 (p. 163) taking advantage of the basic (but limited in scope) equation $dW = d(T)$. That led to $T + U =$ constant, which is the integral of the motion that would have been obtained from Lagrange's equation. As mentioned in Chapter 2, use of the relation $dW = d(T)$ is actually very restricted for most problems. Such a restriction is overcome by the variational formulation presented in this chapter, which led to Hamilton's principle and to Lagrange's equation.

With the velocity of point P relative to the parabolic track $y = ax^2$ being equal to

$$\vec{v}_P = \dot{x}\hat{x} + \dot{y}\hat{y} = \dot{x}\hat{x} + 2ax\dot{x}\hat{y} \tag{7.7.8a}$$

the kinetic energy for the motion is

$$T = \frac{1}{2}m(\dot{x}^2 + \dot{y}^2) = \frac{1}{2}m\dot{x}^2(1 + 4a^2x^2) = T_2 \tag{7.7.8b}$$

The expression for the potential energy, which is obtained by inspection of Fig. 7.11, is

$$U = mgy + \frac{k}{2}\left[\sqrt{x^2 + (y-h)^2} - L_u\right]^2$$

$$= mgax^2 + \frac{k}{2}\left[\sqrt{x^2 + (ax^2-h)^2} - L_u\right]^2 \tag{7.7.8c}$$

The virtual work done by the gravitational force and by the spring force is equal to $-\delta(U)$. As for the viscous friction force, since the track is not rotating, the velocity of P relative to the track is also equal to $\vec{v}_P$. Thus, the viscous friction force is

$$\vec{F}_{\text{friction}} = -c\vec{v}_P = -c\dot{x}(\hat{x} + 2ax\hat{y}) \tag{7.7.8d}$$

With $\vec{v}_P$ as given previously, the virtual displacement of point P is then $(\delta x)\hat{x} + \delta y\hat{y} = (\hat{x} + 2ax\hat{y})\,\delta x$. Therefore, the virtual work done by the friction force is

$$(\delta W)_{nc} = -c\dot{x}(1 + 4a^2x^2)\,\delta x \triangleq (Q_x)_{nc}\,\delta x \tag{7.7.8e}$$

As indicated by this equation, the nonconservative generalized force $(Q_x)_{nc}$ is, by definition, the coefficient of δx in the expression for $(\delta W)_{nc}$.

The generalized momentum $\partial T/\partial \dot{x}$ is not constant whether c is zero or not, because x appears explicitly on T. Also, if $c \neq 0$, the Hamiltonian, which is the same as the total mechanical energy $T + U$ for this problem (because T is a purely quadratic form, T_2, in the generalized velocity $\dot{x}$), is not constant. Notice that the total time derivative of the

Hamiltonian is

$$\frac{d\mathcal{H}}{dt} = \underbrace{\frac{\partial \mathcal{H}}{\partial t}}_{=0} + (Q_x)_{nc}\dot{x} = -c\dot{x}^2(1+4a^2x^2) \qquad (7.7.8\text{f})$$

where $\partial \mathcal{H}/\partial t = 0$ because t does not appear explicitly in the expression for $\mathcal{H}$.

Lagrange's equation $d(\partial T/\partial \dot{x})/dt - \partial T/\partial x = Q_x$ yields the following differential equation of motion:

$$\underbrace{\frac{d}{dt}[m\dot{x}(1+4a^2x^2)]}_{\text{This is } \frac{\partial T}{\partial \dot{x}}} - \underbrace{4ma^2\dot{x}^2 x}_{\text{This is } \frac{\partial T}{\partial x}} = Q_x = -c\dot{x}(1+4a^2x^2) - \frac{\partial U}{\partial x}$$

$$(7.7.8\text{g})$$

or

$$(1+4a^2x^2)\ddot{x} + 4a^2x\dot{x}^2 + \frac{k}{m}\left[1 - \frac{L_u}{\sqrt{x^2+(ax^2-h)^2}}\right][2a(ax^2-h)+1]x$$

$$+ \frac{c}{m}\dot{x}(1+4a^2x^2) + 2agx = 0 \qquad (7.7.8\text{h})$$

Equation (7.7.8h) is the second-order differential equation that governs the motion of the system. Its numerical integration was presented in Example 2.8.9 (p. 166).

Notice that when $c = 0$, the integral of the motion $\mathcal{H}$ = constant is a first-order differential equation. To determine $x(t)$ numerically in this case, one may integrate either Eq. (7.7.8h), which is a second-order differential equation, or simply integrate the first-order differential equation $T + U$ = constant, which is more convenient since it only requires one integration.

EXAMPLE 7.7.9

Differential Equations of Motion for Example 2.7.10

In this example, as in Example 2.7.10 (p. 140), two particles P_1 and P_2 of masses m_1 and m_2, respectively, are connected by a massless inextensible string of length L that passes through a small hole on a fixed frictionless horizontal table as shown in Fig. 7.12. P_2 moves along the vertical. The system has only one degree of freedom because the string is inextensible. For $m_2 = 2m_1$, $L = 1$ m, $r(0) = 0.5$ m, $\dot{r}(0) = 1$ m/s, and $\dot{\theta}(0) = 2$ rad/s, determine the maximum and the minimum values of $r(t)$ by making use of the constants of motion of the system obtained with the Lagrangian formulation. Bypass the derivation of a second-order differential equation of motion.

■ **Solution**

Again, this is also a problem with a constraint, namely, that the length of the string connecting the two particles P_1 and P_2 is constant. However, just like the motion of a pendulum of constant length, this is also a simple constraint that makes it possible to describe the motion using a single independent variable, which is chosen to be the distance $r(t) = |\overrightarrow{OP_1}|$ shown in Fig. 7.12.

Figure 7.12 Two constrained particles for Example 7.7.9 (same as for Example 2.7.10).

The kinetic and potential energies of the system are readily obtained by inspection of Fig. 7.12 as

$$T = \frac{1}{2}m_1[\dot{r}^2 + (r\dot{\theta})^2] + \frac{1}{2}m_2\left[\frac{d}{dt}(L-r)\right]^2 = \frac{1}{2}(m_1+m_2)\dot{r}^2 + \frac{1}{2}m_1 r^2\dot{\theta}^2 = T_2$$
(7.7.9a)

$$U = -m_2 g(L-r) \quad \text{(7.7.9b)}$$

The entire expression for the kinetic energy is of the T_2 form. Since t does not appear explicitly in both T and U, the Hamiltonian $\mathcal{H} = T_2 + U - T_0$, which is $\mathcal{H} = T + U$ for this problem, is constant. Also, since $\partial \mathcal{L}/\partial \theta = 0$, where $\mathcal{L} = T - U$, the generalized momentum $\partial T/\partial \dot{\theta}$ is constant.

We then have (leaving out the irrelevant constant term $-m_2 g L$ in the expression for U):

$$\mathcal{H} = T_2 + U - T_0 = \frac{1}{2}[(m_1+m_2)\dot{r}^2 + m_1 r^2\dot{\theta}^2] + m_2 g r = \text{constant}$$

$$= \frac{1}{2}\left[(m_1+m_2)\dot{r}_0^2 + m_1 r_0^2\dot{\theta}_0^2\right] + m_2 g r_0 \quad \text{(7.7.9c)}$$

$$\frac{\partial T}{\partial \dot{\theta}} = m_1 r^2 \dot{\theta} = \text{constant} = m_1 r_0^2 \dot{\theta}_0 \quad \text{(7.7.9d)}$$

Equation (7.7.9d) can be readily solved for $\dot{\theta}$ in terms of r. When the result is substituted into Eq. (7.7.9c), the following first-order differential equation for r is obtained:

$$\frac{1}{2}(m_1+m_2)\dot{r}^2 + \frac{1}{2}m_1\frac{\left(r_0^2\dot{\theta}_0\right)^2}{r^2} + m_2 g r = \frac{1}{2}\left[(m_1+m_2)\dot{r}_0^2 + m_1 r_0^2\dot{\theta}_0^2\right] + m_2 g r_0 \quad \text{(7.7.9e)}$$

Equation (7.7.9e) is nonlinear, but it can be easily integrated numerically. That is the differential equation of motion for r. There is no need, of course, to generate the second-order differential equations of motion, either using Newton's second law or Lagrange's approach, because Eq. (7.7.9e) is a first integral of those second-order differential equation. For didactic reasons, however, those second-order differential equations are given in footnote 9.

The maximum and minimum values for r, say r_m, can also be obtained from Eq. (7.7.9e), without any numerical integration, just by setting $\dot{r} = 0$ and solving the resulting cubic polynomial equation for r_m. Only real and nonnegative roots should be accepted since they are the only ones that have physical meaning. It is suggested that you choose your own values for the initial conditions and determine the resulting maximum and minimum values of r. In addition, it is suggested that you also numerically integrate Eq. (7.7.9e) to verify your answers.

7.8 Constraints and Lagrange's Equation in the Presence of Constraints

Equation (7.4.17) (p. 439), is valid only if the n variables $q_1, \ldots, q_n$ are independent of each other. If some of the n variables are related to each other, then one has, in general, to work with Eq. (7.4.16) (p. 439), unless the constraint equations are very simple so that one can solve them to eliminate all the "extra" variables. To make this clear, consider the following examples.

- The motion of a bar of constant length L, such as the pendulum of Example 7.7.3 (p. 450). One certainly does not need to introduce an extra variable $r(t)$ with the constraint $r = $ constant $= L$ and then formulate the problem using two variables r and θ to obtain the differential equation for θ. The same was done for the two particles connected by an inextensible string of length L in Example 7.7.9 (p. 461).
- The motion of a particle P constrained to move on a parabolic track $y = ax^2$, as in Example 7.7.8 (p. 459). For such a constraint, the variable y is easily eliminated from the problem as done in that example.

7.8.1 Constraints

The simple constraints just exemplified are special cases of constraints of the form that follows, where some of the variables $q_1, q_2, \ldots, q_n$ may be related by

---

[9] The following differential equations are obtained from Lagrange's equation $d(\partial T/\partial \dot{q})/dt - \partial T/\partial q = Q_q$, with $q = r$ and $q = \theta$, where $Q_r = -\partial U/\partial r = -m_2 g$ and $Q_\theta = -\partial U/\partial \theta = 0$:

$$(m_1 + m_2)\ddot{r} - m_1 r \dot{\theta}^2 + m_2 g = 0$$

$$\frac{d}{dt}(m_1 r^2 \dot{\theta}) = 0 \quad \text{(thus } r^2 \dot{\theta} = \text{constant)}$$

For this problem, these equations happen to be exactly of the same form as Eqs. (2.7.10m) and (2.7.10k) that were obtained in Example 2.7.10, involving additional work that required free-body diagrams and consideration of internal forces.

m equations of constraint (where $m < n$).

$$a_1(q_1, q_2, \ldots, q_n; t) = 0$$
$$a_2(q_1, q_2, \ldots, q_n; t) = 0$$
$$\vdots$$
$$a_m(q_1, q_2, \ldots, q_n; t) = 0$$
(7.8.1)

Constraint equations of this form are *integral constraints* that, in dynamics, are called *holonomic constraints* (the word holonomic is of Greek origin). Time t may or may not appear explicitly in an equation of constraint. Each equation in (7.8.1) represents a *constraint surface* in the n-dimensional space $q_1, \ldots, q_n$, which also has a counterpart in a physical space of rectangular x-y-z coordinates. For example, for a particle moving on a sphere of radius $r = \text{constant} = R$, centered at the origin of such a coordinate system, the constraint surface $x^2 + y^2 + z^2 - R^2 = 0$ in polar coordinates is simply $r - R = 0$. The force that keeps the particle on the constraint surface is a *constraint force,* and that force is perpendicular to the constraint surface at the point of application of that constraint force. Therefore, it can also be said that the constraint force is parallel to the *gradient vector* of the constraint surface since, as known from analytic geometry, such a vector is perpendicular to the tangent to that surface (an illustration is given following the next paragraph).

A constraint surface is fixed in space if time t does not appear in a constraint equation, as in $a_1(q_1, q_2, \ldots, q_n) = 0$. Otherwise, the surface of constraint is moving if the constraint is of the form $a_1(q_1, q_2, \ldots, q_n; t) = 0$, i.e., if time appears explicitly in the equation of constraint. In either case, the virtual work of the constraint force is equal to zero by definition of a constraint force. Such a force is perpendicular to the constraint surface, and the virtual displacement of the point of application of such a force is tangent to that surface.

Figure 7.13 illustrates the notion of the *gradient vector,* using a function of two variables which, for simplicity, is taken to be a circle of radius L centered at the origin O' of a nonrotating (but translating) rectangular coordinate system x-y. An example of a system for which this constraint applies is a pendulum of constant length L, in planar motion, able to rotate about point O' which, in turn, is forced to move in the x-y plane with any desired known motion. The $\hat{x}$-$\hat{y}$ and $\hat{r}$-$\hat{n}$ unit vectors will be used later.

Point P of the pendulum shown in Fig. 7.13a is constrained to move, relative to point O', on the dashed circle shown. The constraint is holonomic in this case and is expressed either in terms of the x-y coordinates or the polar r-θ coordinates as follows:

$$a(x, y) = \sqrt{x^2 + y^2} - L = 0 \tag{7.8.2}$$

or

$$a(r) = r - L = 0 \tag{7.8.3}$$

SECTION 7.8 Constraints and Lagrange's Equation in the Presence of Constraints

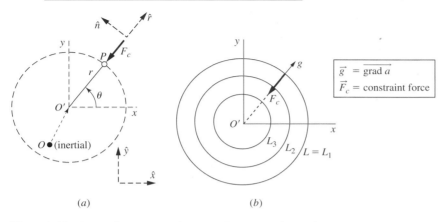

Figure 7.13 The gradient vector of a constraint and the force of constraint.

In terms of the x-y coordinates, the gradient vector of the curve $a(x, y) = \sqrt{x^2 + y^2} - L = 0$ is, from analytic geometry,

$$\overrightarrow{\text{grad }a} = \frac{\partial a}{\partial x}\hat{x} + \frac{\partial a}{\partial y}\hat{y} = \frac{x\hat{x} + y\hat{y}}{L} \quad (7.8.4)$$

while, using polar coordinates, and the simplest representation of the position vector of O relative to O', i.e., $\overrightarrow{OO'} = r\hat{r}$, the gradient vector is obtained as

$$\overrightarrow{\text{grad }a} = \frac{\partial a}{\partial r}\hat{r} = \hat{r} \quad (7.8.5)$$

Equations (7.8.4) and (7.8.5) are equivalent to each other, as can be easily verified with the coordinate transformations (see Fig. 7.13) $x = r\cos\theta$ and $y = r\sin\theta$ and the transformation of triads $\hat{x} = \hat{r}\cos\theta - \hat{n}\sin\theta$ and $\hat{y} = \hat{r}\sin\theta + \hat{n}\cos\theta$. Clearly, the gradient vector is normal to the constraint surface, and so is the force of constraint $\vec{F}_c$, which is shown in Fig. 7.13. Therefore, it is concluded that a constraint force is always proportional to the gradient of the constraint surface, and one can then write, with $\lambda = \lambda(t)$ being a proportionality factor (whose magnitude and sign are unknown),

$$\vec{F}_c = \lambda \overrightarrow{\text{grad }a} \quad (7.8.6)$$

A more convenient form of this expression is $\vec{F}_c = \lambda \overrightarrow{\text{grad }a}/|\overrightarrow{\text{grad }a}|$, where the vector $\overrightarrow{\text{grad }a}$ is normalized by its magnitude (in order to obtain a unit vector) so that $|\lambda|$ becomes the magnitude of the constraint force $\vec{F}_c$.

Returning now to the equations of (7.8.1), each one of those equations of a holonomic constraint represents a constraint surface in the space $q_1, \ldots, q_n$, which, in turn, represents an actual physical surface in three-dimensional space.

Figure 7.14 The spring-pendulum in three-dimensional space.

This is illustrated in Fig. 7.14 for the single-mass spring-pendulum system where the support point O is given a known motion $\overrightarrow{O'O}(t)$ and, thus, the virtual displacement $\delta(\overrightarrow{O'O})$ is equal to zero.

If the pendulum of Fig. 7.14 is constrained so that $r = \text{constant} \triangleq L$, point P is forced to move on a spherical constraint surface $a_r \triangleq r - L = 0$ centered at point O. Because of the motion of point O, the spherical constraint surface is moving in space. The gradient of that surface is $\overrightarrow{\text{grad } a_r} = \hat{r}$ (see $\hat{r}$ in Fig. 7.14), and the constraint force associated with that constraint surface is $\vec{F}_{c_1} = \lambda_1 \hat{r}$. If, in addition, the motion is also restricted so that $\phi = \text{constant} \triangleq \phi_0$, point P is also constrained to move on the plane that contains the vertical that passes through O and line OP. This second constraint surface is $a_\phi \triangleq \phi - \phi_0 = 0$ (which is also a moving surface if O is in motion). Its gradient is $\overrightarrow{\text{grad } a_\phi} = \hat{\phi}$ (see $\hat{\phi}$ in Fig. 7.14), and the constraint force associated with this latter constraint surface is $\vec{F}_{c_2} = \lambda_2 \hat{\phi}$. In the presence of the two constraints $a_r = r - L = 0$ and $a_\phi = \phi - \phi_0 = 0$, the motion of P is forced to be in the intersection of the two surfaces of constraint. For simple problems, such as the one involving a single mass particle (as just presented), it is seen that the λ multiplier method for finding the constraint forces can yield those forces without ambiguity. However, this is not the case for more complicated problems.

Analytical dynamics is not the best approach for finding constraint forces. In fact, the idea behind such an approach is to eliminate them right at the beginning of the formulation of any dynamics problem. In general, if one needs to determine constraint forces, it is easier, and more convenient, to go back to the individual free-body diagrams for each mass particle or each rigid body of a multibody system and reformulate the equations using the internal constraint forces in order to determine them. Such forces will clearly show up in those individual free-body diagrams and there should be no danger of misinterpreting them, which is likely to happen when one tries to find them using the λ multiplier method with Lagrange's equation without a free-body diagram.

SECTION 7.8 Constraints and Lagrange's Equation in the Presence of Constraints

Let us now go back to the main problem at hand, which is how to obtain the differential equations of motion of a system in the presence of constraint equations. To do so, one may start by solving the equations of (7.8.1) analytically for the m constrained variables, say, $q_1, q_2, \ldots, q_m$, in terms of the remaining variables $q_{m+1}, q_{m+2}, \ldots, q_n$, *if* that can be done. This, then, eliminates the constrained variables from the formulation, and one is left with the $n - m$ independent variables $q_{m+1}, q_{m+2}, \ldots, q_n$. But there is a big *if* here; i.e., the equations of constraint may be complicated enough to prevent one from analytically eliminating variables as indicated earlier. For examples of relatively simple illustrations where it is possible to analytically eliminate variables, see Example 7.7.5 (p. 454), for which the variable θ is prescribed as $\dot\theta = \Omega(t)$, and Example 7.7.6 (p. 456), for which the variable ϕ is prescribed as $\dot\phi = \Omega(t)$, where $\Omega(t)$ is a given function of time in both examples. The given constraint equations in both examples could be integrated to yield $\theta(t)$ and $\phi(t)$ for those examples. That is, both constraints were holonomic, despite being given in differential form, and easily solvable for the constrained variables. Also, another problem is that the constraint equations may also involve velocities in a manner that the equations cannot be integrated to yield constraint equations of the form given by the equations of (7.8.1). This is discussed next.

The most general form of constraint equations that may appear in a dynamics problem is of the form shown in the m equations of (7.8.7) that involve a total of n variables $q_1, q_2, \ldots, q_m, q_{m+1}, q_{m+2}, \ldots, q_n$, but only the $n - m$ variables $q_{m+1}, q_{m+2}, \ldots, q_n$ are independent of each other. The equations of (7.8.7) represent m first-order differential equations of constraint. The constraints are called *nonholonomic* if the equations of (7.8.7) cannot be integrated either directly or by introducing an integrating factor.[10]

$$B_{11}\dot q_1 + B_{12}\dot q_2 + \cdots + B_{1n}\dot q_n = a_1(q_1, q_2, \ldots, q_n; t)$$
$$B_{21}\dot q_1 + B_{22}\dot q_2 + \cdots + B_{2n}\dot q_n = a_2(q_1, q_2, \ldots, q_n; t)$$
$$\vdots$$
$$B_{m1}\dot q_1 + B_{m2}\dot q_2 + \cdots + B_{mn}\dot q_n = a_m(q_1, q_2, \ldots, q_n; t)$$

(7.8.7)

In general, the coefficients $B_{11}, B_{12}, \ldots, B_{mn}$ and the functions $a_1, a_2, \ldots, a_m$ depend on the variables $q_1, q_2, \ldots, q_n$, and also may explicitly depend on the independent variable t.

Holonomic constraints, i.e., integral constraints, are of the general form given by the equations of (7.8.1) and define a surface in space. Nonholonomic

---

[10] For example, the *expression* $2q_2\,dq_1 + q_1\,dq_2$ is not an exact differential since $\partial(2q_2)/\partial q_2 \ne \partial(q_1)/\partial q_1$. But the *equation* $2q_2\,dq_1 + q_1\,dq_2 = 0$, when multiplied by q_1, can be integrated because $\partial(2q_1 q_2)/\partial q_2 = \partial q_1^2/\partial q_1 = 2q_1$. The factor q_1 is the *integrating factor*, and the integrated form of constraint equations such as $2q_2\,dq_1 + q_1\,dq_2 = 0$ or $2q_2\dot q_1 + q_1\dot q_2 = 0$ is $q_1^2 q_2 = \text{constant} \stackrel{\Delta}{=} C$. Such a constraint, which at first glance seemed to be nonholonomic, is actually holonomic. This means that either variable q_1 or q_2 could be eliminated from a problem where two variables q_1 and q_2 were introduced, but constrained as $2q_2\,dq_1 + q_1\,dq_2$. In such a problem one could simply choose to work with q_1 and then eliminate q_2 by using the relation $q_2 = C/q_1^2$.

constraints involve expressions that cannot be integrated, without the knowledge of the motion, to be put in the form of the equations of (7.8.1). Thus, since the equation of a surface in the space $q_1, \ldots, q_n$ does not involve time derivatives of these variables, one cannot associate surfaces of constraint with nonholonomic constraints. The virtual work of the constraint forces (i.e., the sum of the virtual work of each constraint force) is still equal to zero, though, by definition of constraint forces.

7.8.2 Lagrange's Equation in the Presence of Constraints

The equations of constraint must be taken into consideration either in Hamilton's extended principle, i.e., Eq. (7.4.8) (p. 437), or in Eq. (7.4.16) (p. 439), which was obtained directly from that principle. A clever way to do this, which was devised by Lagrange, is to start by obtaining the virtual form of the constraint equations involving the virtual variations $\delta q_1, \ldots, \delta q_n$, as required by Eq. (7.4.16). The virtual form of the nonholonomic constraint equations of (7.8.7) is simply

$$B_{11}\,\delta q_1 + B_{12}\,\delta q_2 + \cdots + B_{1n}\,\delta q_n = 0$$
$$B_{21}\,\delta q_1 + B_{22}\,\delta q_2 + \cdots + B_{2n}\,\delta q_n = 0$$
$$\vdots$$
$$B_{m1}\,\delta q_1 + B_{m2}\,\delta q_2 + \cdots + B_{mn}\,\delta q_n = 0$$

(7.8.8)

Lagrange's method of dealing with constraints consists of introducing m new variables $\lambda_1, \lambda_2, \ldots, \lambda_m$ (i.e., one new variable for each equation of constraint) and then adding the sum

$$\lambda_1(B_{11}\,\delta q_1 + B_{12}\,\delta q_2 + \cdots + B_{1n}\,\delta q_n) + \cdots + \lambda_m(B_{m1}\,\delta q_1 + B_{m2}\,\delta q_2 + \cdots + B_{mn}\,\delta q_n)$$

to Eq. (7.4.16). This certainly does not change Eq. (7.4.16) because only zero is being added to it, but the cleverness of the method is in treating the quantities $\lambda_1, \lambda_2, \ldots, \lambda_m$ as m new independent variables that are chosen so that

$$\boxed{\frac{d}{dt}\frac{\partial T}{\partial \dot{q}_k} - \frac{\partial T}{\partial q_k} = Q_{q_k} + \lambda_1 B_{1k} + \lambda_2 B_{2k} + \cdots + \lambda_m B_{mk} \quad \text{for } k = 1, 2, \ldots, n}$$

(7.8.9)

The n second-order differential equations represented by Eq. (7.8.9), together with the m first-order differential equations of (7.8.7), form a set of $n+m$ equations in the $n+m$ variables $q_1, q_2, \ldots, q_n, \lambda_1, \lambda_2, \ldots, \lambda_m$. In general, such equations will have to be integrated numerically to obtain the solutions for $q_1, \ldots, q_n$, which are the variables of interest to a specific problem. The m auxiliary variables $\lambda_1, \lambda_2, \ldots, \lambda_m$ that are introduced to handle the constraint equations are called the *Lagrange multipliers*. They are the same λ multipliers that appear on p. 465, except that they are not associated with constraint surfaces when the constraints are nonholonomic.

SECTION 7.8 Constraints and Lagrange's Equation in the Presence of Constraints

This same procedure may also be used when the constraints are of the form given by the equations of (7.8.1). In such a case Eq. (7.8.9) becomes

$$\frac{d}{dt}\frac{\partial T}{\partial \dot{q}_k} - \frac{\partial T}{\partial q_k} = Q_{q_k} + \lambda_1 \frac{\partial a_1}{\partial q_k} + \lambda_2 \frac{\partial a_2}{\partial q_k} + \cdots + \lambda_m \frac{\partial a_m}{\partial q_k} \quad \text{for } k = 1, 2, \ldots, n$$

(7.8.10)

Example 7.9.3 in Section 7.9 illustrates a case where a reasonably complicated holonomic constraint is handled in this manner. The resulting equations consist of n second-order differential equations and m algebraic equations of constraint. In the dynamics and mathematics literature, such a system of equations is called a set of *differential-algebraic equations*. A strategy for solving such equations using Simulink is presented in Examples 7.9.1 (p. 470) and 7.9.3 (p. 477).

For holonomic constraints, each equation of constraint defines a constraint surface, and the Lagrange multipliers are proportional to a constraint force, which is normal to that surface. Such correspondence was presented in Section 7.8.1 (p. 463). However, either for multiparticle or multibody systems, the best and safer way to determine constraint forces, if one needs or desires to determine them, is to draw the free-body diagram for each component of the system and then reformulate the equations for those components using Newton's second law and Euler's equation. If one does not need to determine the constraint forces, Lagrange's multiplier method is a clever way to obtain the differential equations of motion of constrained systems.

Constraints can also appear in the form of inequalities. They occur when the motion of a system is limited to some region in space, such as the motion of a particle being limited to the region above the surface of the earth or above the surface defined by an inclined plane.

Generally, the analytical dynamics formulation is the most expeditious way to obtain the differential equations of motion of a dynamical system. Such an approach also has the advantage of providing a methodical search for integrals of the motion that is done in a relatively simple manner when one is able to work with a set of independent generalized coordinates. The analytical dynamics approach is also the simplest way for obtaining the differential equations of motion of constrained systems when one does not need to determine the constraint forces. However, if one needs to determine the forces of constraint in a constrained system, it would be actually better to use the Newtonian-Eulerian formulation since it unambiguously yields the constraint forces, both in magnitude and in direction, which would clearly appear in the free-body diagrams that one is required to draw in the vectorial formulation. A detailed example of such a situation was presented in Section 5.6 (p. 313), whose specific objective was to determine the internal forces on each element of a four-bar mechanism. In some problems, the best approach might be a combination of the analytical dynamics methods to obtain the differential equations and integrals of motion and of the Newtonian-Eulerian formulation to obtain the expressions involving the desired constraint forces in specific free-body diagrams that would be drawn showing those forces.

7.9 Solved Example Problems with Constraints

EXAMPLE 7.9.1

A rigid body of mass m is thrown from a point A so that the initial velocity of its center of mass C has magnitude v_0 and is directed at $\alpha = 30°$ with the horizontal as shown in Fig. 7.15. In addition to the gravitational force with constant magnitude mg, the body is also subjected to a force $\vec{F}$ that constrains the absolute velocity of C so that $\dot{y} = f_1(t)\dot{x} + f_2(t)$. The variables x and y are the coordinates of C in the rectangular nonrotating x-y coordinate system shown in Fig. 7.15, and $f_1(t)$ and $f_2(t)$ are given functions of time, with $v_0 \sin \alpha = f_1(0) v_0 \cos \alpha + f_2(0)$.

a. Obtain the differential equations that govern the motion of C.
b. For $v_0 = 5$ m/s and $x(0) = y(0) = 0$, plot $x(t)$ and $y(t)$ versus t and determine, from the plot, the maximum height reached by C.

Figure 7.15 Figure for Example 7.9.1.

■ Solution

a. Referring to Fig. 7.15, the absolute position vector $\vec{r}_C = \overrightarrow{AC}$ (assuming that point A is inertial) of the center of mass C may be expressed in terms of the nonrotating unit vectors $\hat{x}$ and $\hat{y}$ as

$$\vec{r}_C = x\hat{x} + y\hat{y} \qquad (7.9.1a)$$

The kinetic energy for the translational motion of C and the potential energy U are then

$$T = \frac{m}{2}(\dot{x}^2 + \dot{y}^2) \qquad (7.9.1b)$$

$$U = mgy \qquad (7.9.1c)$$

The constraint force $\vec{F}$ may be expressed as

$$\vec{F} = F_x \hat{x} + F_y \hat{y} \qquad (7.9.1d)$$

and its virtual work $(\delta W)_{\text{due to } \vec{F}} = \vec{F} \bullet \delta \vec{r}_C$ is equal to zero by definition of a constraint force. The virtual work δW done by all forces acting on the system is then

$$\delta W = -\delta(U) + (\delta W)_{\text{due to } \vec{F}} = -mg\, \delta y \triangleq Q_x\, \delta x + Q_y\, \delta y \qquad (7.9.1e)$$

where $Q_x = 0$.

SECTION 7.9 Solved Example Problems with Constraints

Unless $f_1(t)$ is constant, the given equation of constraint, $\dot{y} = f_1(t)\dot{x} + f_2(t)$, is nonholonomic because it cannot be converted into an integral form $f(x, y; t) = 0$. Because of the constraint imposed on the motion, the virtual displacements δx and δy are not independent of each other. Their relationship is obtained from the virtual form of the constraint equation. Since the differential form of the constraint equation $\dot{y} - f_1(t)\dot{x} - f_2(t) = 0$ is $dy - f_1(t)\, dx - f_2(t)\, dt = 0$, its virtual form is then

$$\delta y - f_1(t)\, \delta x = 0 \tag{7.9.1f}$$

We can now obtain the differential equation of motion either from Eq. (7.4.16) (p. 439), or using a Lagrange multiplier as indicated by Eq. (7.8.9) (p. 468) in Section 7.8. For completeness, both approaches, as well as direct application of Newton's second law, are shown here. All approaches yield the same results, which prove their validity.

Using Eq. (7.4.16)

For two variables x and y, Eq. (7.4.16) is

$$\left(\frac{d}{dt}\frac{\partial T}{\partial \dot{x}} - \frac{\partial T}{\partial x}\right)\delta x + \left(\frac{d}{dt}\frac{\partial T}{\partial \dot{y}} - \frac{\partial T}{\partial y}\right)\delta y - \delta W = 0 \tag{7.9.1g}$$

With $\delta y = f_1(t)\, \delta x$, as per Eq. (7.9.1f), this then yields

$$m\ddot{x} + (m\ddot{y} + mg)f_1(t) = 0 \tag{7.9.1h}$$

Equation (7.9.1h) and the constraint equation

$$\dot{y} = f_1(t)\dot{x} + f_2(t) \tag{7.9.1i}$$

are the differential equations of motion for the system. They can be combined to yield the following differential equation that only involves the variable x:

$$\left[1 + f_1^2(t)\right]\ddot{x} + \left[\dot{x}\dot{f}_1(t) + \dot{f}_2(t) + g\right]f_1(t) = 0 \tag{7.9.1j}$$

Using a Lagrange Multiplier

Only one Lagrange multiplier λ is introduced since there is only one equation of constraint. For the two variables x and y, Eq. (7.8.9) then implies

$$\frac{d}{dt}\frac{\partial T}{\partial \dot{x}} - \frac{\partial T}{\partial x} = Q_x + \lambda B_x \tag{7.9.1k}$$

$$\frac{d}{dt}\frac{\partial T}{\partial \dot{y}} - \frac{\partial T}{\partial y} = Q_y + \lambda B_y \tag{7.9.1l}$$

where $B_y = 1$ and $B_x = -f_1(t)$, as disclosed by the virtual form of the constraint given by Eq. (7.9.1f). Also, $Q_x = 0$ and $Q_y = -mg$, as disclosed by the expression for δW given by Eq. (7.9.1e). Equations (7.9.1k) and (7.9.1l) then yield

$$m\ddot{x} = -\lambda f_1(t) \tag{7.9.1m}$$

$$m\ddot{y} = -mg + \lambda \tag{7.9.1n}$$

or, by eliminating λ,

$$m\ddot{x} = -(m\ddot{y} + mg)f_1(t)$$

which is the same as Eq. (7.9.1h).

Using Newton's Second Law

Direct application of Newton's second law yields the following vector equation:

$$F_x\hat{x} + (F_y - mg)\hat{y} = m(\ddot{x}\hat{x} + \ddot{y}\hat{y}) \tag{7.9.1o}$$

This vector equation yields the following two scalar equations:

$$m\ddot{x} = F_x \tag{7.9.1p}$$

$$m\ddot{y} = F_y - mg \tag{7.9.1q}$$

Equations (7.9.1p) and (7.9.1q), together with the equation of constraint, Eq. (7.9.1i), form a set of three equations in four unknowns, namely, x, y, F_x, and F_y. Thus, a fourth equation is needed. Either F_x or F_y, or a combination of both, must be specified for the problem to be completely solvable. The problem statement specified that the only forces acting on the body are the gravitational force and the constraint force $\vec{F} = F_x\hat{x} + F_y\hat{y}$. Since the virtual work of the constraint force is zero, that *is* the fourth equation, i.e.,

$$F_x\,\delta x + F_y\,\delta y = F_x\,\delta x + F_y f_1(t)\,\delta x = 0 \tag{7.9.1r}$$

from which one obtains

$$F_x = -f_1(t)F_y \tag{7.9.1s}$$

It can be readily verified that Eqs. (7.9.1p), (7.9.1q), and (7.9.1s) yield Eq. (7.9.1j). Also, comparison of Eqs. (7.9.1m) and (7.9.1p), and Eqs. (7.9.1n) and (7.9.1q), discloses that the Lagrange multiplier λ is equal to the force F_y, and that $F_x = -\lambda f_1(t)$.

A Note of Caution

Resist the temptation to make shortcuts when dealing with nonholonomic constraints, such as substituting velocity (or velocities) given by the constraint equation (or equations) into the kinetic energy and then using Eq. (7.4.16) (p. 439), because this will yield wrong equations. You can verify this in this example by substituting $\dot{y} = f_1(t)\dot{x} + f_2(t)$ into Eq. (7.9.1b) and then applying Eq. (7.4.16) to the result. If you compare the resulting differential equation with that obtained here by three methods, including the direct application of Newton's second law, you will see that it is wrong. For nonholonomic constraints, such a shortcut is inconsistent with the formulation that yields Eq. (7.4.16). The formulation is based on the use of all the constrained and unconstrained variables that are introduced in the problem and the variational form of the equations of constraint. The shortcut works for holonomic constraints because, in such a case, constraint equations of the form $(\partial a_i/\partial q_1)\,\delta q_1 + \cdots + (\partial a_i/\partial q_n)\,\delta q_n = 0$, for $i = 1,\ldots, m$, imply the integrated equations of (7.8.1), while the equations of (7.8.8) for nonholonomic constraints cannot be expressed in the integrated form of the equations of (7.8.7). In other words, the equations of (7.8.7) imply the equations of (7.8.8), while the converse is not true.

---

b. Figure 7.16 shows a Simulink model for simultaneously integrating Eq. (7.9.1j) [which was obtained by hand manipulation of Eqs. (7.9.1h) and (7.9.1i)] and the

SECTION 7.9 Solved Example Problems with Constraints

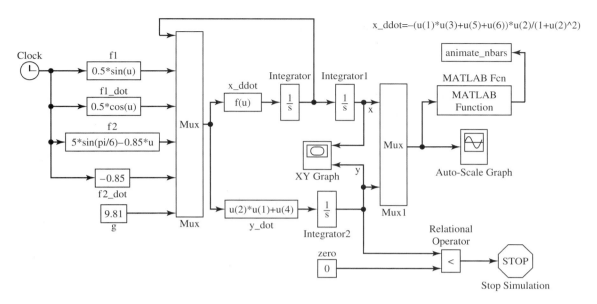

Figure 7.16 A Simulink model for integrating Eqs. (7.9.1i) and (7.9.1j).

constraint equation, which is Eq. (7.9.1i). Figure 7.17 shows the result of the numerical integration of the equations with $f_1 = 0.5 \sin t$ and $f_2 = v_0 \sin \alpha - 0.85t$, for $v_0 = 5$ m/s, $\alpha = 30°$, and the initial conditions $x(0) = y(0) = 0$. As disclosed by Fig. 7.15, the initial conditions for $\dot{x}$ and $\dot{y}$ are $\dot{x}(0) = v_0 \cos \alpha$ and $\dot{y}(0) = v_0 \sin \alpha$.

The maximum value for $y(t)$, obtained from Fig. 7.17, is $y_{max} \approx 4.2$ m.[11] The curve $y(t)$ versus $x(t)$ followed by the center of mass of the body is shown in Fig. 7.17b, with the arrow indicating how the curve is traced. The curve resembles the trajectory of a boomerang!

⟹ The Simulink model shown in Fig. 7.16 is the file `boomerang1.mdl` that is on the enclosed CD. To use it, either type `boomerang1` in the MATLAB command window or type `simulations_menu` and choose the `boomerang1` model from a pull-down menu that appears on the computer screen. Double click the `animate_nbars` block to change the axes limits of the animation figure, if desired. Look at the `Simulation → Parameters` menu in your computer to see the default settings for the simulation, and change the maximum step size of the integration scheme if the simulation runs too fast or too slow in your computer.

⟹ Figure 7.18 shows another Simulink model for integrating the differential equations for this same problem using a Lagrange multiplier and the strategy that follows. The differential equations are Eqs. (7.9.1m) and (7.9.1n), which are second-order

---

[11] Precise readings can be obtained from the figure generated by the autoscale graph block in the model shown in Fig. 7.16 by clicking on the zoom icon in that figure and then selecting, with the computer mouse, the area to zoom in.

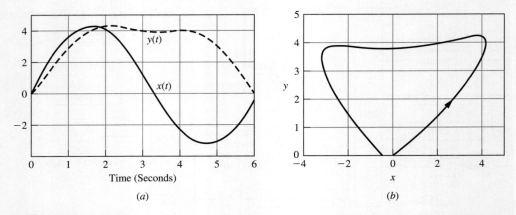

Figure 7.17 (a) Displacements $x(t)$ and $y(t)$ and (b) the trajectory $y(t)$ versus $x(t)$ for Example 7.9.1.

Figure 7.18 A Simulink model that uses the matrix equation (7.9.1t).

differential equations in x and y. With this model, the additional hand manipulations that were done previously to eliminate the Lagrange multiplier from the equations is avoided. The model is the file boomerang2.mdl that is also included on the enclosed CD. To use it, proceed in the same way as previously.

Since Eqs. (7.9.1m) and (7.9.1n) involve a third unknown, which is the Lagrange multiplier λ, a third equation is needed, which has to be the constraint equation. The following four-step strategy is used for solving those constrained differential-algebraic equations by introducing a Lagrange multiplier. The complete

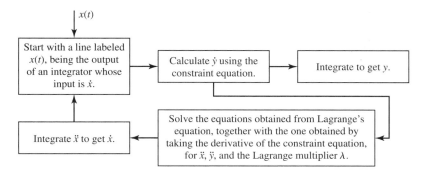

Figure 7.19 Flowchart for constructing a Simulink model to determine $x(t)$ and $y(t)$ for Example 7.9.1 using a Lagrange multiplier λ.

implementation of the strategy using Simulink is summarized in the flowchart shown in Fig. 7.19. The flowchart also answers the question, "How does one start to construct models such as the ones shown in Figs. 7.16 and 7.18?"

Step 1. Start by assuming that x and $\dot{x}$, which are the output of two integrators, are available.

Step 2. Calculate $\dot{y}$ using the constraint equation, which is Eq. (7.9.1i) (p. 471). Then, integrate $\dot{y}$ to get $y(t)$.

Step 3. Solve the three equations involving $\ddot{x}$, $\ddot{y}$, and λ for such variables. Such equations are written in matrix form as follows:

$$\begin{bmatrix} 1 & 0 & f_1 \\ 0 & 1 & -1 \\ -f_1 & 1 & 0 \end{bmatrix} \begin{bmatrix} \ddot{x} \\ \ddot{y} \\ \frac{\lambda}{m} \end{bmatrix} = \begin{bmatrix} 0 \\ -g \\ \dot{f}_1 \dot{x} + \dot{f}_2 \end{bmatrix} \qquad (7.9.1t)$$

The first two rows correspond to Eqs. (7.9.1m) and (7.9.1n), and the third row corresponds to the time derivative of the constraint equation. The solution is implemented in the `MATLAB Function` block labeled `x_ddot, y_ddot, lambda` in Fig. 7.18. The user-specified inputs to that block are f_1, $\dot{f}_1$, f_2, $\dot{f}_2$, and g (the acceleration of gravity), as required by Eq. (7.9.1t).

Step 4. Select the first output of the `MATLAB Function` block labeled `x_ddot, y_ddot, lambda` in Fig. 7.18, which is $\ddot{x}$, and then integrate it twice to generate $\dot{x}$ and x, which were required in step 1. The other two outputs, which are $\ddot{y}$ and λ/m, are ignored.

Constraint Equation for a Classical Example of a Nonholonomic Constraint

EXAMPLE 7.9.2

A classical example of a nonholonomic constraint is a vertical disk of radius R rolling without slipping on a horizontal surface, with the disk being able to turn about a vertical axis as shown in Fig. 7.20. The more complicated situation, and more realistic, of course, is when the disk also tilts so that it does not remain vertical at all times. The case shown in Fig. 7.20 illustrates the simpler nonholonomic constraint when the disk remains vertical.

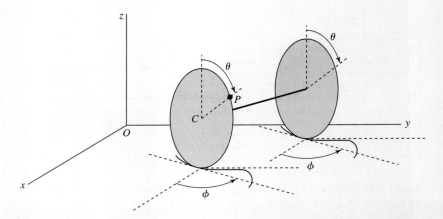

Figure 7.20 A vertical disk rolling, without slipping, on a horizontal surface.

The path shown in Fig. 7.20, followed by the point of contact of the disk with the x-y plane, may or may not be prescribed. The angle ϕ shown in the figure is the angle between the reference x axis and the tangent of that path at the point of contact. The angle θ is used for tracking the orientation of line CP relative to the reference z axis, where C is the center of the disk and P is a point fixed to its periphery. What are the constraint equations for this problem?

For rolling without slipping, the virtual displacement of point C must match that of the point of contact of the disk. Therefore, with x_C and y_C denoting the x and y coordinates of point C, the rolling with no slipping condition implies the following virtual equations of constraint (see Fig. 7.20):

$$\delta x_C = (R\,\delta\theta)\cos\phi \tag{7.9.2a}$$

$$\delta y_C = (R\,\delta\theta)\sin\phi \tag{7.9.2b}$$

These equations cannot be expressed in integral form, i.e., in a form that involves only the variables x_C, y_C, ϕ, and θ, but not their time derivatives. Therefore, the constraint Eqs. (7.9.2a) and (7.9.2b) are nonholonomic. Thus, one is forced to work with the four constrained variables x_C, y_C, ϕ, and θ. Equations (7.9.2a) and (7.9.2b) can be expressed in integral form only if $\phi = $ constant, i.e., if the disk rolls along a straight line. The case where $\phi = $ constant is a simpler classical case of rolling, and it was analyzed in detail in Section 5.4 (p. 299).

Notice that in terms of velocity, if the velocity of the center C of the disk is to be a vector that is in the plane of the disk and parallel to the tangent line shown in Fig. 7.20 (i.e., no lateral slip either), such a condition implies that

$$\dot{x}_C - R\dot\theta\cos\phi = 0 \tag{7.9.2c}$$

$$\dot{y}_C - R\dot\theta\sin\phi = 0 \tag{7.9.2d}$$

and these equations are of the form of the general equations of constraint [Eq. (7.8.7)] with their right-hand sides equal to zero. The differential equation (or equations) of motion for rolling problems may be obtained by the analytical dynamics approach. Notice that for

SECTION 7.9 Solved Example Problems with Constraints

problems with Coulomb friction, the force of friction depends on the constraint force. Generally, such problems are best handled by the Newtonian-Eulerian approach.

Dynamics of a Four-Bar Mechanism   EXAMPLE 7.9.3

The four-bar mechanism shown in Fig. 7.21 is driven by a motor that applies to bar AB a known moment $M_{\text{motor}}(t)$. The motion takes place in a horizontal plane and is started from rest [i.e., with $\dot{\theta}_2(0) = 0$] with a given initial value of θ_2. The mechanism consists of three moving thin homogeneous bars AB, BC, and CD of lengths L_2, L_3, and L_4 and masses m_2, m_3, and m_4, respectively. Obtain the differential equation of motion for this constrained system, and integrate it numerically. For $L_1 = 1$ m, $L_2 = 0.5$ m, $L_3 = 0.8$ m, $L_4 = 0.7$ m, $m_2 = 3.75$ kg, $m_3 = 3$ kg, $m_4 = 3$ kg, and $M_{\text{motor}}(t) = \pi/6 + 1.5\sin(kt)$ N·m (where $k = 2$ rad/s), plot θ_4 and θ_2 versus t for t in a reasonable range, determine the maximum and minimum values of $\theta_4(t)$, and show an animation of the motion. The motion is started with $\theta_2(0) = 30°$.

Figure 7.21 A four-bar mechanism subjected to a known moment M_{motor} applied to bar AB.

■ Solution

In terms of the unit vectors $\hat{x}$ and $\hat{y}$ shown in Fig. 7.21, the absolute velocity $\vec{v}$ of the center of mass C_3 of bar BC may be written as[12]:

$$\vec{v}_{C_3} = \frac{d}{dt}\overrightarrow{DC} + \frac{d}{dt}\overrightarrow{CC_3} = \frac{d}{dt}[L_4(\hat{x}\cos\theta_4 + \hat{y}\sin\theta_4)] + \frac{d}{dt}\left[\frac{L_3}{2}(-\hat{x}\cos\theta_3 - \hat{y}\sin\theta_3)\right]$$

$$= \left(\frac{L_3}{2}\dot{\theta}_3\sin\theta_3 - L_4\dot{\theta}_4\sin\theta_4\right)\hat{x} + \left(L_4\dot{\theta}_4\cos\theta_4 - \frac{L_3}{2}\dot{\theta}_3\cos\theta_3\right)\hat{y} \quad (7.9.3a)$$

The following expression is then obtained for the kinetic energy of the motion:

$$T = \frac{1}{2}I_A\dot{\theta}_2^2 + \frac{1}{2}I_D\dot{\theta}_4^2 + \frac{1}{2}I_3\dot{\theta}_3^2 + \frac{1}{2}m_3\vec{v}_{C_3}\cdot\vec{v}_{C_3} = \frac{1}{2}I_A\dot{\theta}_2^2 + \frac{1}{2}I_D\dot{\theta}_4^2 + \frac{1}{2}I_3\dot{\theta}_3^2$$

$$+ \frac{1}{2}m_3\left[\left(L_4\dot{\theta}_4\cos\theta_4 - \frac{L_3}{2}\dot{\theta}_3\cos\theta_3\right)^2 + \left(L_4\dot{\theta}_4\sin\theta_4 - \frac{L_3}{2}\dot{\theta}_3\sin\theta_3\right)^2\right]$$

$$= \frac{1}{2}I_A\dot{\theta}_2^2 + \frac{1}{2}I_D\dot{\theta}_4^2 + \frac{1}{2}I_3\dot{\theta}_3^2 + \frac{1}{2}m_3\left[L_4^2\dot{\theta}_4^2 + \frac{1}{4}L_3^2\dot{\theta}_3^2 - L_3L_4\dot{\theta}_3\dot{\theta}_4\cos(\theta_4 - \theta_3)\right]$$

$$(7.9.3b)$$

---
[12] The expression for $\vec{v}_{C_3}$ could also have been written as $\vec{v}_{C_3} = d\overrightarrow{AB}/dt + d\overrightarrow{BC_3}/dt$. It makes no difference which form is used.

In Eq. (7.9.3b), $I_A = m_2 L_2^2/3$ is the moment of inertia of bar AB about the vertical axis that passes through A, $I_D = m_4 L_4^2/3$ is the moment of inertia of bar CD about the vertical axis that passes through D, and $I_3 = m_3 L_3^2/12$ is the moment of inertia of bar BC about the vertical axis that passes through its center of mass. The use of these expressions for the moments of inertia imply that the thin bars are being approximated as lines AB, CD, and BC, respectively.

The angles θ_2, θ_3, and θ_4 are dependent on each other. As first presented in Chapter 3, the following constraint relations between them are obtained by inspection of Fig. 7.21:

$$\underbrace{L_2 \cos\theta_2 + L_3 \cos\theta_3 - L_4 \cos\theta_4 - L_1}_{\triangleq\, a_1(\theta_2, \theta_3, \theta_4)} = 0 \qquad (7.9.3c)$$

$$\underbrace{L_2 \sin\theta_2 + L_3 \sin\theta_3 - L_4 \sin\theta_4}_{\triangleq\, a_2(\theta_2, \theta_3, \theta_4)} = 0 \qquad (7.9.3d)$$

The virtual work δW done by all forces acting on the system is simply

$$\delta W = M_{\text{motor}}\, \delta\theta_2 \triangleq Q_{\theta_2}\, \delta\theta_2 \qquad (7.9.3e)$$

Generally, it is more convenient to use Lagrange multipliers to handle constraint equations. Since there are two of such equations, two multipliers $\lambda_1(t)$ and $\lambda_2(t)$ are needed. For the three variables θ_2, θ_3, and θ_4, Eq. (7.8.10) (p. 469) then implies

$$\frac{d}{dt}\frac{\partial T}{\partial \dot\theta_2} - \frac{\partial T}{\partial \theta_2} = Q_{\theta_2} + \lambda_1 \frac{\partial a_1}{\partial \theta_2} + \lambda_2 \frac{\partial a_2}{\partial \theta_2} \qquad (7.9.3f)$$

$$\frac{d}{dt}\frac{\partial T}{\partial \dot\theta_3} - \frac{\partial T}{\partial \theta_3} = Q_{\theta_3} + \lambda_1 \frac{\partial a_1}{\partial \theta_3} + \lambda_2 \frac{\partial a_2}{\partial \theta_3} \qquad (7.9.3g)$$

$$\frac{d}{dt}\frac{\partial T}{\partial \dot\theta_4} - \frac{\partial T}{\partial \theta_4} = Q_{\theta_4} + \lambda_1 \frac{\partial a_1}{\partial \theta_4} + \lambda_2 \frac{\partial a_2}{\partial \theta_4} \qquad (7.9.3h)$$

where $Q_{\theta_2} = M_{\text{motor}}$, $Q_{\theta_3} = Q_{\theta_4} = 0$, and the functions $a_1(\theta_2, \theta_3, \theta_4)$ and $a_2(\theta_2, \theta_3, \theta_4)$ are defined in Eqs. (7.9.3c) and (7.9.3d).

Equations (7.9.3f) to (7.9.3h) yield the following second-order differential equations:

$$\frac{1}{3} m_2 L_2^2 \ddot\theta_2 = M_{\text{motor}} - \lambda_1 L_2 \sin\theta_2 + \lambda_2 L_2 \cos\theta_2 \qquad (7.9.3i)$$

$$\frac{1}{3} m_3 L_3^2 \ddot\theta_3 - \frac{1}{2} m_3 L_3 L_4 \bigl[\ddot\theta_4 \cos(\theta_4 - \theta_3) - \dot\theta_4^2 \sin(\theta_4 - \theta_3) \bigr]$$
$$= -\lambda_1 L_3 \sin\theta_3 + \lambda_2 L_3 \cos\theta_3 \qquad (7.9.3j)$$

$$\left(\frac{1}{3} m_4 + m_3\right) L_4^2 \ddot\theta_4 - \frac{1}{2} m_3 L_3 L_4 \bigl[\ddot\theta_3 \cos(\theta_4 - \theta_3) + \dot\theta_3^2 \sin(\theta_4 - \theta_3) \bigr]$$
$$= \lambda_1 L_4 \sin\theta_4 - \lambda_2 L_4 \cos\theta_4 \qquad (7.9.3k)$$

Equations (7.9.3i) to (7.9.3k) involve five unknowns, which are θ_2, θ_3, θ_4, and the two Lagrange multipliers λ_1 and λ_2. They can be solved using the strategy presented in Example 7.9.1. That strategy involves augmenting Eqs. (7.9.3i) to (7.9.3k) with Eqs. (7.9.3l)

SECTION 7.9 Solved Example Problems with Constraints

and (7.9.3m) that follow, which are obtained by taking the second time derivative of the algebraic constraint equations. The five equations are combined into a single matrix equation, Eq. (7.9.3n), which is easily solved numerically by Simulink using a `MATLAB Function` block.

$$L_2\ddot{\theta}_2 \sin\theta_2 + L_3\ddot{\theta}_3 \sin\theta_3 - L_4\ddot{\theta}_4 \sin\theta_4 + L_2\dot{\theta}_2^2 \cos\theta_2 + L_3\dot{\theta}_3^2 \cos\theta_3 - L_4\dot{\theta}_4^2 \cos\theta_4 = 0 \tag{7.9.3l}$$

$$L_2\ddot{\theta}_2 \cos\theta_2 + L_3\ddot{\theta}_3 \cos\theta_3 - L_4\ddot{\theta}_4 \cos\theta_4 - L_2\dot{\theta}_2^2 \sin\theta_2 - L_3\dot{\theta}_3^2 \sin\theta_3 + L_4\dot{\theta}_4^2 \sin\theta_4 = 0 \tag{7.9.3m}$$

$$\begin{bmatrix} \frac{1}{3}m_2L_2^2 & 0 & 0 & L_2\sin\theta_2 & -L_2\cos\theta_2 \\ 0 & \frac{1}{3}m_3L_3^2 & -\frac{1}{2}m_3L_3L_4\cos(\theta_4-\theta_3) & L_3\sin\theta_3 & -L_3\cos\theta_3 \\ 0 & -\frac{1}{2}m_3L_3L_4\cos(\theta_4-\theta_3) & \left(\frac{1}{3}m_4+m_3\right)L_4^2 & -L_4\sin\theta_4 & L_4\cos\theta_4 \\ L_2\sin\theta_2 & L_3\sin\theta_3 & -L_4\sin\theta_4 & 0 & 0 \\ L_2\cos\theta_2 & L_3\cos\theta_3 & -L_4\cos\theta_4 & 0 & 0 \end{bmatrix}$$

$$\times \begin{bmatrix} \ddot{\theta}_2 \\ \ddot{\theta}_3 \\ \ddot{\theta}_4 \\ \lambda_1 \\ \lambda_2 \end{bmatrix} = \begin{bmatrix} M_{\text{motor}} \\ -\frac{1}{2}m_3L_3L_4\dot{\theta}_4^2 \sin(\theta_4-\theta_3) \\ \frac{1}{2}m_3L_3L_4\dot{\theta}_3^2 \sin(\theta_4-\theta_3) \\ L_4\dot{\theta}_4^2 \cos\theta_4 - L_3\dot{\theta}_3^2 \cos\theta_3 - L_2\dot{\theta}_2^2 \cos\theta_2 \\ -L_4\dot{\theta}_4^2 \sin\theta_4 + L_3\dot{\theta}_3^2 \sin\theta_3 + L_2\dot{\theta}_2^2 \sin\theta_2 \end{bmatrix} \tag{7.9.3n}$$

The Simulink implementation of this strategy is summarized in the flowchart shown in Fig. 7.22.

⟹ The Simulink model that implements the calculations outlined in Fig. 7.22 is shown in Fig. 7.23. The model is the file `fourbar_mechanism_dynamics.mdl` that is on the enclosed CD. To use it, either type

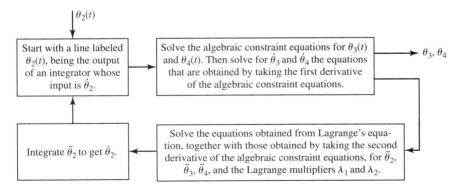

Figure 7.22 Flowchart for solving the differential-algebraic equations for the dynamics of a four-bar mechanism using Simulink.

Figure 7.23 A Simulink model implementing the solution strategy shown in Fig. 7.22.

`fourbar_mechanism_dynamics` in the MATLAB command window or type `simulations_menu` and choose the `fourbar_mechanism_dynamics` model from a pull-down menu that appears on the computer screen.

The block diagram model shown in Fig. 7.23 was constructed simply by modifying the model `fourbar_mechanism` shown in Fig. 3.9 (p. 193). The user inputs to the new model are the desired values for $L_1, L_2, L_3, L_4, m_2, m_3, m_4$, the `solution number` (see the second paragraph that follows), and the desired expression for the motor torque M_{motor}.

SECTION 7.9 Solved Example Problems with Constraints

The MATLAB Function block labeled `solution of Eq. (3.2.5)` in Fig. 7.23 is the same as the one with the same name in Fig. 3.13 (p. 198). By comparing Figs. 3.9 (p. 193) and 7.23, it is seen that one of the modifications made for constructing the `fourbar_mechanism_dynamics` model shown in Fig. 7.23 consists of the inclusion of two integrators for generating $\dot{\theta}_2$ and θ_2 using the solution for $\ddot{\theta}_2$ obtained from Eq. (7.9.3n). Conforming to Eq. (7.9.3n), that solution is the first of the five outputs of the MATLAB Function block labeled `theta2_ddot, ..., lambda from Eq. (7.9.3n)`. That output is selected at the function block named `theta2_ddot` and then connected to the input of the integrator labeled `Integrator1`, thus closing the loop in the model of Fig. 7.23 and in the flowchart of Fig. 7.22.

The input labeled `solution_number` is the same as the one described in Section 3.2.3 (p. 193). It is used for selecting the two possible initial orientations for bars BC and CD (seen in Fig. 7.21, p. 477) for the initial value chosen for θ_2. The desired initial values $\theta_2(0)$ and $\dot{\theta}_2(0)$ are specified after double clicking the two integrators in the `fourbar_mechanism_dynamics` model.

Figure 7.24 shows a solution for $\theta_4(t)$ and $\theta_2(t)$, starting with the configuration defined by `solution number 1`, with the data specified at the beginning of this example and with $\dot{\theta}_2(0) = 0$. The maximum and minimum values that are obtained for θ_4 are $\theta_{4\,\text{max}} = 180°$ and $\theta_{4\,\text{min}} \approx 84.8°$. The plots and the animation show that θ_4 oscillates between these two values, while θ_2 continually increases. This happens because the average value of the motor torque M_{motor} is nonzero, positive, and equal to $\pi/6$ N·m. Run the simulation for `solution number 2` to see what happens. Double click the `Simulation` → `Parameters` menu to see the default choices for the integration, including the values for the relative and absolute tolerances of the calculations. Change the relative tolerance to 10^{-3}, and also to 10^{-9}, to see if the results change. The acceptable values for the relative tolerance are those for which no changes are seen to occur in the solution.

 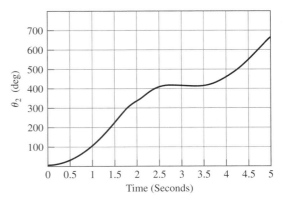

Figure 7.24 $\theta_4(t)$ and $\theta_2(t)$ generated by the Simulink model of Fig. 7.23.

7.10 Problems

7.1 Obtain the differential equations of motion for the double pendulum in Problem 2.43 (p. 176) and all classical integrals of motion, using Lagrange's formulation.

7.2 Show that the differential equations of motion you obtained for Problem 2.43 in Chapter 2 are equivalent to those that you obtained in Problem 7.1 using Lagrange's formulation. In other words, if they look different, you have to show that an appropriate combination of the equations obtained in one problem yields the equations obtained in the other problem.

7.3 Show that Eqs. (7.7.2e) (p. 449) and (7.7.2f), which were derived using Lagrange's equation, are equivalent to Eqs. (2.5.9g) (p. 98) and (2.5.9i), derived using Newton's second law. In other words, you have to show that an appropriate combination of the equations in one set yields the equations in the other set.

7.4 Obtain the differential equation of motion for Problem 2.26 (p. 172) using Lagrange's equation and all the classical integrals of motion.

7.5 Obtain the differential equations of motion for Problem 2.27 (p. 173) using Lagrange's equation and all the classical integrals of motion.

7.6 A small block of mass m is able to slide inside a frictionless groove that rotates about a vertical axis with a constant angular velocity $\dot{\theta} \triangleq \Omega$ as shown in Fig. 7.25. The block is connected by a linear spring to a point A at a constant distance h above a fixed point B on the axis of rotation, as shown in Fig. 7.25. Motion takes place in a horizontal plane.

Figure 7.25 Figure for Problem 2.6.

The spring stiffness is k N/m, and its unstressed length is $L_u = ah$ m, where a is given. The motion of the block is started with the initial conditions $r(0) = 0$ and $\dot{r}(0) = \Omega h$, where $r(t)$ is shown in Fig. 7.25. Making use of classical integrals of motion obtained with Lagrange's formulation, determine, for $m\Omega^2/k = 1$, and for $a = 2$ and $a = 1/2$, the maximum value of $r(t)/h$. Disregard the moment of inertia of the small block about its center of mass.

7.7 Consider the double pendulum shown in Fig. 7.26 where a linear torsional dashpot (i.e., a viscous torsional damper) with stiffness c N·m/(rad/s) is added in parallel to the torsional spring in that figure, and where the links $O'A$ and AB are modeled as thin homogeneous bars of masses m_1 and m_2 kg, respectively. The

Figure 7.26 Figure for Problem 7.7.

spring is linear with stiffness k N·m/rad, and it is unstressed when $\theta_2 = \theta_1$. The bars can only move in the plane of the figure. Point O' represents a hinge that is forced to move (by an external force not shown in the figure) with a known vertical displacement $e(t)$. Obtain the expressions for the potential energy of the motion, for the virtual work, and for the generalized conservative and nonconservative forces associated with θ_1 and θ_2. Obtain the differential equations of motion for the system using Lagrange's equation. Also, for $c \geq 0$, obtain all classical integrals of the motion. For what type of $e(t) \neq 0$ is the Hamiltonian a constant?

7.8 Consider the system of Example 2.8.8 (p. 163), where the vertical parabolic track is spun about the vertical y axis shown in Fig. 2.51 (p. 163) with a constant angular velocity $\Omega = 50$ rpm relative to inertial space. Neglect friction in the track. Making use of classical integrals of motion obtained with Lagrange's formulation, determine the maximum ($y_{\max}$) and the minimum ($y_{\min}$) values of the height y reached by the particle. The particle is released with $x(0) = -1.5$ m, and $\sqrt{\dot{x}^2(0) + \dot{y}^2(0)} = 6$ m/s. What are the corresponding values of the distance x? Use the data given in Example 2.8.8.

7.9 For the system in Problem 7.8, plot all the equilibrium solutions $x = $ constant $\triangleq x_e$ versus Ω, for $0 \leq \Omega \leq 50$ rpm. Use the following data: $m = 20$ g, $a = \frac{1}{3}$ m^{-1}, $k = 9$ N/m, $L_u = h = 1.2$ m. Also, use the expression for the integral of motion obtained in that problem to draw an accurate *phase plane plot* $\dot{x}$ versus x, with $\Omega = 20$ and 30 rpm. Superimpose all the plots drawn for the same value of Ω. Draw the plots using $\dot{x}(0) = 0$ and several values of $x(0)$ near $x = 0$ to obtain different curves. Determine, from any of your plots, the value of Ω when multiple equilibrium solutions start to occur. How does that value compare with the one that is obtained from the linearized differential equation about the equilibrium $x_e = 0$?

7.10 Verify the results of Problems 7.8 and 7.9 by numerical integration of the differential equation of motion of the system.

7.11 Consider the system shown in Fig. 2.41 (p. 140) described in Example 2.7.10. However, instead of particle P_2 being constrained to move along a vertical, let it move in three-dimensional space. Introduce two angles to describe the orientation of line OP_2, and then use Lagrange's formulation to obtain the differential equations of motion for the system, and all classical integrals of the motion. Neglect friction in the system.

7.12 Use Lagrange's formulation to obtain the differential equations of motion and all the classical integrals of motion for the two-body problem described in Example 4.3.2 (p. 232). Use the variables r and θ shown in Fig. 4.5 (p. 233).

7.13 The Geneva wheel mechanism shown in Fig. 3.15 (p. 199) is driven by a motor that applies to bar AB a known moment $M_{\text{motor}}(t)$. The motion takes place in a vertical plane. The mechanism consists of a homogeneous bar AB of constant length r_1 and mass m_1, with a pin at B that slides on a slotted bar CP of length L_2 and mass m_2. As shown in Fig. 3.15, the distance L between the fixed pins at A and C is constant. Neglect friction in the system. Using Lagrange's formulation, obtain the equations that govern the motion of that constrained system.

7.14 The driver in the slider crank mechanism shown in Fig. 3.20 (p. 209) is a homogeneous bar of length L_1 and mass m_1. Bar AB, of length L_2, is homogeneous and its mass is m_2. The mass of the block at B is m_B, and $D = 0$. A motor applies to bar OA a known moment $M_{\text{motor}}(t)$. The motion takes place in a horizontal plane. Neglect friction in the system. Using Lagrange's formulation, obtain the equations that govern the motion of that constrained system.

7.15 The scotch yoke mechanism shown in Fig. 3.28 (p. 218) consists of a thin homogeneous wheel of radius R and mass m_1, with a small pin P at its periphery, and a slotted body of mass m_2 that slides along a straight track. Using Lagrange's formulation, obtain the differential equation of motion for the system when the wheel is driven by a motor at O that applies a known moment $M_{\text{motor}}(t)$ to the system. Take point O to be inertial, and neglect friction in the system. For $M_{\text{motor}} = M \sin \tau$, where $\tau = \Omega t$ and M and Ω are constants, obtain a numerical solution for the angle θ shown in Fig. 3.28 and plot it versus the nondimensional time τ when $m_2 = m_1/5$ and $M/(m_1 \Omega^2 R^2) = 2$. Start the motion of the system from rest and with $\theta(0) = 0$. Motion takes place in a horizontal plane.

7.16 Consider the mechanism described in Problem 3.27 (p. 222) when it is driven by a motor, instead of θ_2 being a known function of time. The mechanism is shown in Fig. 3.34, and the motor applies a moment $M_{\text{motor}} = 3 \sin(\Omega t)$ N·m to bar AB, where $\Omega = 2$ rad/s. Bars AB, PC, and CD are very thin and homogeneous, and their masses are m_2, m_3, and m_4, respectively, while their lengths are L_2, L, and L_4. The quantity L_3 shown in the figure is the variable distance between B and C. Points A and D are inertial. Using Lagrange's formulation, obtain the equations that govern the constrained motion of the mechanism when motion is in a horizontal plane.

7.17 Using Lagrange's formulation, obtain the differential equations of motion and all the classical integrals of motion for the systems shown in Figs. D.2 and D.4 in Appendix D. Both of them are described in that appendix.

7.18 Use Lagrange's formulation to obtain the differential equations of motion and all the classical integrals of motion for the restricted three-body problem presented in Chapter 4 (i.e., for the motion of particle P_3 shown in Fig. 4.12, p. 250). Use the variables r, θ, and z, shown in Fig. 4.12a for describing the motion of P_3.

7.19 Obtain the differential equations of motion and all the classical integrals of motion for Example 5.3.12 (p. 291) using Lagrange's formulation.

7.20 Obtain the differential equations of motion and all the classical integrals of motion for Example 5.3.13 (p. 292) using Lagrange's formulation. Are there any nonclassical integrals of motion that can be easily obtained?

7.21 Obtain the differential equation of motion and all the classical integrals of motion for Problem 5.27 (p. 327) using Lagrange's formulation. For this, neglect friction in the system and consider the case when the moment M_{motor} is a given function of θ (in practice, such a function is generated with a feedback control system).

7.22 Obtain the differential equation of motion and all the classical integrals of motion for Examples 5.4.1 (p. 300) and 5.4.3 (p. 308) using Lagrange's formulation. The cylinder rolls without slipping in both cases.

7.23 Using Lagrange's formulation, obtain all the classical integrals of motion for Example 6.7.1 (p. 382) when $\dot{\phi} = \text{constant} \triangleq \Omega$. The principal moment of inertia of the rotating platform about the vertical axis is I_O. What is the second-order differential equation of motion for the system obtained from Lagrange's equation?

7.24 Obtain the differential equation of motion and all the classical integrals of motion for Example 6.7.2 (p. 389) using Lagrange's formulation.

7.25 By making use of classical integrals of motion obtained with Lagrange's formulation, solve part b of Problem 6.10 (p. 411).

7.26 Obtain all the classical integrals of motion for the system shown in Fig. 6.37 (p. 414), which is described in Problem 6.16. What are the differential equations of motion for the system?

7.27 Using Lagrange's formulation, obtain all the classical integrals of motion for the symmetric spin stabilized satellite in circular orbit that was considered in Example 6.8.4 (p. 405). Use those same orientation angles. What is the expression for the potential energy U associated with the virtual work δW of the gravitational moment $\vec{M}_C$? Is the Hamiltonian the same as the total mechanical energy $E = T + U$ for such a problem? *Hint:* The virtual work associated with $\vec{M}_C$ is obtained by generating the expression for $\vec{M}_C \bullet (\vec{\omega}\, dt)$ and then replacing "d" by "δ" in the resulting expression. The expressions for $\vec{M}_C$ and for the absolute angular velocity $\vec{\omega}$ were formulated in Examples 6.8.3 and 6.8.4.

7.28 Verify that $\left(\dot{q}^2 + cq\dot{q} + kq^2\right) e^{ct}$ is an integral of the motion governed by the differential equation $\ddot{q} + c\dot{q} + kq = 0$, where c and k are constants.

CHAPTER 8

Vibrations/Oscillations of Dynamical Systems

The subject matter of vibrations/oscillations of dynamical systems was already introduced in the book in a natural manner, starting in Chapter 1, including basic material needed for analyzing the motion of dynamical systems. Numerical integration of differential equations was also widely used throughout the book.

This short chapter complements the rest of the book with two topics presented in Sections 8.2 and 8.3. Two important problems, namely, the response of a vibration absorber device and the analysis of Foucault pendulum (described on p. 71 in Chapter 2), are presented in Sections 8.4 and 8.5. More advanced and specific material in the field of vibrations can be found in Reference 11 listed in Appendix G, and in many other books that deal specifically with the subject matter.

8.1 What Was Already Presented on Vibrations/Oscillations in Previous Chapters

The differential equations of motion of dynamical systems are second-order differential equations in terms of the motion coordinates and, with rare exceptions, are nonlinear. For motion governed by a differential equation of the form $\ddot{x} + f(x) = 0$, it was shown in Chapter 1 that a first integral of the motion, namely, $\dot{x}^2/2 + \int f(x)\,dx =$ constant, is readily obtained by use of the chain rule of differentiation. Such an integral of motion yields important information about the motion even when $x(t)$ itself is not known. The search for integrals of motion is better handled by the more advanced and more powerful analytical dynamics methodology presented in Chapter 7, although more advanced application of that material to specific problems is actually a relatively simple matter.

In order to be able to investigate the motion with analytical techniques, so that as much as possible can be learned about why the system behaves in the

SECTION 8.2 Response of a Linear System to Sinusoidal Forcing Functions

way it does, approximations need to be introduced in the nonlinear differential equations of motion so that techniques that have been already discovered can be applied. Such analytical techniques are applicable to simpler equations that are obtained by restricting oneself to the analysis of "small" motion about a particular solution of the nonlinear equations. A technique that is widely used involves *linearization* of the differential equations for such small motions about each *equilibrium* exhibited by them. Some of the more advanced techniques that deal with nonlinearities and are used for investigating nonlinear oscillations can be found in Reference 11 listed in Appendix G.

Linearization was introduced in Section 1.14 (p. 48), which begins with a presentation about Taylor series of a function of either one or several variables. Several single-degree-of-freedom examples were presented in that section. Other examples were presented in Sections 2.5 and 2.7, including a reasonably detailed analysis of a two-degree-of-freedom system (the spring-pendulum, in Example 2.7.8, p. 129). A reasonably detailed analysis of a more complicated three-degree-of-freedom system was presented in Section 4.6 (the restricted three-body problem, p. 249). Generalization of the techniques presented in those examples to systems with many degrees of freedom is presented in this chapter. Stability analysis based on linearized differential equations was introduced in Chapter 1 and used throughout the book. For more advanced studies of stability, see Reference 12 listed in Appendix G, which is a classic in the subject matter.

Before we deal with multiple-degree-of-freedom systems, the simpler analysis of the response of a linear system to a sinusoidal forcing function is presented in Section 8.2.

8.2 Response of a Linear System to Sinusoidal Forcing Functions

Consider the linear differential equation $m\ddot{x} + c\dot{x} + kx = f(t)$ with $m > 0$, $c \geq 0$, and $k > 0$. As seen in this book, such an equation describes, for example, the translational motion of a mass connected to a linear spring and to a linear dashpot, when the motion takes place along the vertical and a known force $f(t)$ is applied to the mass.[1] However, it is generally obtained after linearization of a nonlinear differential equation about a stable equilibrium of a system, such as the motion of a pendulum. As seen in Chapter 1, the equilibrium is asymptotically stable if $c > 0$. The case $c = 0$ occurs when there is no damping in the system. The function $f(t)$ is also referred to either as the *input* or the *excitation* applied to the system. The analysis of the response $x(t)$ to an input that is either $f(t) = F\sin(\Omega t)$ or $f(t) = F\cos(\Omega t)$ is considered in the following. Without loss of generality, one can consider either one of such inputs since $\sin(\Omega t) = \cos(\Omega t - \pi/2)$ and

---

[1] See Fig. 1.16 (p. 48) for such a system without a dashpot.

$\cos(\Omega t) = \sin(\Omega t + \pi/2)$. Let us use the following equation:

$$m\ddot{x} + c\dot{x} + kx = F\cos(\Omega t) \tag{8.2.1}$$

The solution to Eq. (8.2.1) is the sum of the homogeneous solution $x_h = Be^{st}$, as presented in Chapter 1 (p. 38), and of a particular solution $x_p(t)$. For the homogeneous solution, the exponent s is determined from $ms^2 + cs + k = 0$. Let us write the particular solution as $x_p(t) = A_1\cos(\Omega t) + A_2\sin(\Omega t)$, where A_1 and A_2 are constants. It satisfies the differential equation. It turns out to be more convenient to write the particular solution as either $x_p = A\cos(\Omega t + \alpha)$ or $x_p = A\sin(\Omega t + \beta)$ where $|A|$ is the maximum value of x_p [which one would have to calculate if $x_p(t) = A_1\cos(\Omega t) + A_2\sin(\Omega t)$ were used]. Thus, by expressing the general solution to Eq. (8.2.1) as

$$x = Be^{st} + A\cos(\Omega t + \alpha) \tag{8.2.2}$$

and by substituting such an expression in Eq. (8.2.1), the following equation is obtained:

$$\underbrace{(ms^2 + cs + k)}_{=0}Be^{st} + A[(k - m\Omega^2)\cos(\Omega t + \alpha) - c\Omega\sin(\Omega t + \alpha)]$$
$$= F\cos(\Omega t) \tag{8.2.3}$$

To collect sine and cosine terms that have the same arguments in Eq. (8.2.3), one may express $\cos(\Omega t)$ as

$$\cos(\Omega t) = \cos(\Omega t + \alpha - \alpha) = [\cos(\Omega t + \alpha)]\cos\alpha + [\sin(\Omega t + \alpha)]\sin\alpha \tag{8.2.4}$$

Then, collecting terms of $\cos(\Omega t + \alpha)$ and $\sin(\Omega t + \alpha)$ in both sides of the resulting equation, the following simple equations for the unknowns A and α are obtained:

$$A(k - m\Omega^2) = F\cos\alpha \tag{8.2.5}$$

$$-cA\Omega = F\sin\alpha \tag{8.2.6}$$

Since $\sin^2\alpha + \cos^2\alpha = 1$, the solution for A is readily obtained by squaring these equations and adding the results. This yields

$$A = \pm\frac{F}{\sqrt{(k - m\Omega^2)^2 + (c\Omega)^2}} \tag{8.2.7}$$

The absolute value of A is called the *amplitude* of the particular solution. Since the motion is oscillatory, there is no loss of generality by using only the $+$ sign to calculate A from Eq. (8.2.7). The angle α is readily obtained from Eqs. (8.2.5) and (8.2.6) as

$$\alpha = \arctan\frac{\Omega c}{k - m\Omega^2} \tag{8.2.8}$$

SECTION 8.2 Response of a Linear System to Sinusoidal Forcing Functions

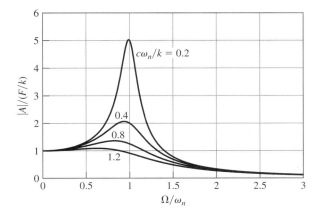

Figure 8.1 Nondimensional amplitude $|A|/(F/k)$ versus Ω/ω_n for several values of the nondimensional damping coefficient $c\omega_n/k$.

Figure 8.1 shows the plot of the nondimensional amplitude $|A|/(F/k)$ versus the nondimensional frequency Ω/ω_n for several values of the nondimensional damping coefficient $c\omega_n/k$, where $\omega_n = \sqrt{k/m}$ is the undamped natural frequency of the system.

The plot in Fig. 8.1 is based on the following equation, which is obtained simply by rearranging Eq. (8.2.7).

$$|A| = \frac{F/k}{\sqrt{\left(1 - \Omega^2/\omega_n^2\right)^2 + \left[(c\omega_n/k)(\Omega/\omega_n)\right]^2}} \quad (8.2.9)$$

The advantage of presenting the plot in terms of the nondimensional quantities previously indicated is that only one parameter is necessary to generate such plots, namely, $c\omega_n/k$, instead of having to specify values for F, k, m, and c as they appear in Eq. (8.2.7).

From these results, we obtain the following conclusions:

Conclusion 1. The particular solution is periodic and has the same frequency Ω of the periodic input to the system. Its amplitude depends on the excitation frequency Ω and the viscous damping coefficient c. For $c > 0$, its maximum amplitude occurs at a frequency smaller than the natural frequency ω_n of the system. The response amplitude tends to infinity as c tends to zero.

Conclusion 2. The amplitude of the periodic response given by the particular solution is proportional to the amplitude of the input excitation to the system.

Conclusion 3. The particular solution is in phase with the input (i.e., $\alpha = 0$) if $c = 0$ (i.e., no damping) but is out of phase if $c \neq 0$. Its phase

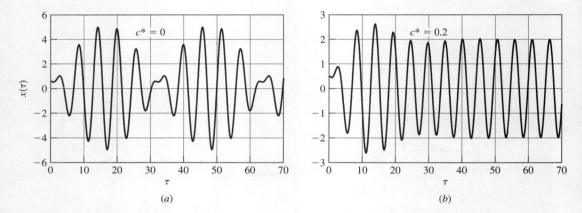

Figure 8.2 Solution of Eq. (8.2.10) with the initial conditions $x(0) = 0.5$ and $\dot{x}(0) = 0$, for (a) $c^* = 0$ and (b) $c^* = 0.2$.

angle is independent of the excitation amplitude F. Notice that when $c > 0$ the homogeneous solution tends to zero as $t \to \infty$ and the response tends to a steady periodic state that is the same as the particular solution determined earlier.

Figure 8.2 shows typical responses of Eq. (8.2.1) to a sinusoidal excitation with $F/k = 1$ and $\Omega = 1.2\omega_n$, where $\omega_n = \sqrt{k/m}$ as indicated earlier. They were obtained by numerical integration of the normalized form of Eq. (8.2.1), which is

$$\frac{d^2x}{d\tau^2} + c^* \frac{dx}{d\tau} + x = \frac{F}{k} \sin(\Omega^* \tau) \qquad (8.2.10)$$

where $\tau = \omega_n t$, $\Omega^* = \Omega/\omega_n$, and $c^* = c\omega_n/k$.

Notice how the presence of the homogeneous solution causes the response to be quite different from the steady-state solution, which is observed in the presence of damping after the transient homogeneous solution is dissipated. Also notice that the amplitude of the steady-state solution, which is seen in Fig. 8.2b after the homogeneous solution is dissipated, is the same as that obtained from Fig. 8.1.

The conclusions presented earlier apply only to linear systems. If Eq. (8.2.1) were obtained by linearization of a nonlinear differential equation about an equilibrium of the system, nonlinear terms that were disregarded in such a process would, of course, affect the motion. Such an effect would become more accentuated as the amplitude of the excitation is increased. However, even for very small excitation amplitudes, the nonlinearities can have a significant effect if the excitation frequency Ω is near values such as ω_n, $3\omega_n$, and $\omega_n/3$. Such cases are called *nonlinear external resonances*, and their analyses are done with methods that are known as *perturbation methods*. Since such analyses are beyond the scope of this book, the reader is referred to some of the more advanced references listed in Appendix G, such as Reference 11. It should also be noted that many nonlinear effects, such as the ones previously mentioned, can be essentially eliminated by increasing the damping in the system.

8.3 Multiple-Degree-of-Freedom Linear Systems

The general form of a set of n second-order linear homogeneous differential equations with constant coefficients, involving the variables $x_1(t), x_2(t), \ldots, x_n(t)$, is

$$M_{11}\ddot{x}_1 + \cdots + M_{1n}\ddot{x}_n + C_{11}\dot{x}_1 + \cdots + C_{1n}\dot{x}_n + K_{11}x_1 + \cdots + K_{1n}x_n = 0$$
$$M_{21}\ddot{x}_1 + \cdots + M_{2n}\ddot{x}_n + C_{21}\dot{x}_1 + \cdots + C_{2n}\dot{x}_n + K_{21}x_1 + \cdots + K_{2n}x_n = 0$$
$$\vdots$$
$$M_{n1}\ddot{x}_1 + \cdots + M_{nn}\ddot{x}_n + C_{n1}\dot{x}_1 + \cdots + C_{nn}\dot{x}_n + K_{n1}x_1 + \cdots + K_{nn}x_n = 0$$
$$(8.3.1)$$

Generally, such a set of linear equations is obtained only after linearizing, about an equilibrium solution, the nonlinear differential equations that govern the motion of a dynamical system.

For compactness, the set of equations of (8.3.1) is written in matrix form as

$$[M]\underline{\ddot{x}} + [C]\underline{\dot{x}} + [K]\underline{x} = \underline{0} \qquad (8.3.2)$$

where $\underline{x}$ and $\underline{0}$ are $n \times 1$ column matrices defined as $\underline{x} = [x_1 \; x_2 \; \cdots \; x_n]^T$, $\underline{0} = [0 \; 0 \; \cdots \; 0]^T$ (where the superscript T denotes *transpose*), while $[M]$, $[C]$, and $[K]$ are $n \times n$ matrices of the coefficients in the n equations of (8.3.1).

Since the exponential function is the only one for which its derivatives are proportional to the function itself, the general solution to Eq. (8.3.2) is $\underline{x} = \underline{b}_0 e^{st}$. In such a solution, s is constant and $\underline{b}_0 = [b_{01} \; b_{02} \; \cdots \; b_{0n}]^T$ is an $n \times 1$ column matrix of constants that are determined by the initial conditions of the motion. Substitution of the general solution into Eq. (8.3.2) yields

$$[[M]s^2 + [C]s + [K]]\underline{b}_0 e^{st} = \underline{0} \qquad (8.3.3)$$

If the matrix $[[M]s^2 + [C]s + K]$ has an inverse, the only solution to Eq. (8.3.3) is $\underline{b}_0 = \underline{0}$. Therefore, the only way to obtain a nonzero solution is for the determinant of the matrix $[[M]s^2 + [C]s + [K]]$ to be equal to zero, i.e.,

$$\det([M]s^2 + [C]s + [K]) = 0 \qquad (8.3.4)$$

This equation is a polynomial equation of degree $2n$ in s. In general, its solutions are complex numbers, which are of the form

$$s = \sigma + i\omega \qquad (8.3.5)$$

where $i = \sqrt{-1}$ and σ and ω are real. Such solutions occur in complex conjugate pairs $s = \sigma + i\omega$ and $s = \sigma - i\omega$. Thus, the general solution to the equations of (8.3.1) is a sum of n terms of the form $B_k e^{\sigma_k t} \cos(\omega_k t + \alpha_k)$, where the B_k's and the α_k's are constants that are determined by the initial conditions of the motion.

The equilibrium solution of the system, about which its nonlinear differential equations were linearized to yield the equations of (8.3.1), is unstable if the real part of *any* of the $2n$ eigenvalues $s_1, s_2, \ldots, s_{2n}$ is positive. The equilibrium is asymptotically stable if the real parts of *all* of them are negative. If there is no damping in the system, the equilibrium is stable [*infinitesimally stable* if the equations of (8.3.1) were obtained by linearization] if all the eigenvalues are purely imaginary. The eigenvalues of a matrix and the roots of a characteristic equation may be readily obtained with the MATLAB commands `eig` and `roots`, respectively.

8.4 Analysis of a Vibration Absorber Device

Figure 8.3 is a model of a system that consists of a block of mass m_1 mounted on an elastic foundation that is represented by several linear springs that are equivalent to a single spring of stiffness k_1 N/m and unstressed length L_{u_1} m. A system consisting of a smaller block of mass m_2 connected to a set of identical springs is attached to the larger block. Those springs are also approximated as linear. They are equivalent to a single spring of stiffness k_2 N/m and unstressed length L_{u_2}. Motion of the system takes place along the vertical and is tracked by the distances r_1 and r_2 shown in the figure. The larger block is subjected to a periodic force $F \sin(\Omega t)$ N, causing the system to vibrate. The objective is to investigate the motion of the system in order to see if it is possible to reduce, or even eliminate, the vibrations of the larger block by choosing appropriate values of m_2 and k_2 to accomplish such a task.

The differential equations of motion for the system may be obtained by drawing the free-body diagram for each one of the blocks shown in Fig. 8.3 and then using Newton's second law for each block. They may also be readily obtained by using Lagrange's equation, as presented in Chapter 7. The differential equations of motion are presented here; their derivation is left as an exercise for the reader.

$$m_1 \ddot{r}_1 + k_1(r_1 - L_{u_1}) + k_2(r_1 - r_2 + h + L_{u_2}) = F \sin(\Omega t) - m_1 g \quad (8.4.1)$$

$$m_2 \ddot{r}_2 + k_2(r_2 - r_1 - h - L_{u_2}) = -m_2 g \quad (8.4.2)$$

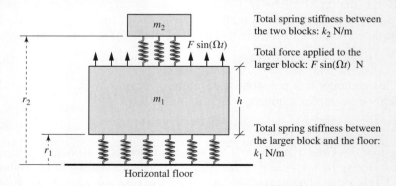

Figure 8.3 A vibration absorber.

SECTION 8.4 Analysis of a Vibration Absorber Device

To analyze the motion, first determine the equilibrium solution(s) to Eqs. (8.4.1) and (8.4.2) and then perturb the equilibrium. For the equilibrium solution, $r_1 = \text{constant} \triangleq r_{1e}$ and $r_2 = \text{constant} \triangleq r_{2e}$, where r_{1e} and r_{2e} must satisfy the following equations that are obtained from Eqs. (8.4.1) and (8.4.2):

$$k_1(r_{1e} - L_{u_1}) + k_2(r_{1e} - r_{2e} + h + L_{u_2}) = -m_1 g \quad (8.4.3)$$

$$k_2(r_{2e} - r_{1e} - h - L_{u_2}) = -m_2 g \quad (8.4.4)$$

These equations can be easily solved for r_{1e} and r_{2e}. Since the equations are linear, they have only one solution, instead of multiple solutions.

By letting, for convenience, $r_1 \triangleq r_{1e} + x_1(t)$ and $r_2 \triangleq r_{2e} + x_2(t)$, substitution of these expressions into Eqs. (8.4.1) and (8.4.2) yield the following differential equations for the "perturbations" $x_1(t)$ and $x_2(t)$:

$$m_1 \ddot{x}_1 + k_1 x_1 + k_2(x_1 - x_2) = F \sin(\Omega t) \quad (8.4.5)$$

$$m_2 \ddot{x}_2 + k_2(x_2 - x_1) = 0 \quad (8.4.6)$$

The solution to Eqs. (8.4.5) and (8.4.6) is the sum of the homogeneous solution (i.e., the one obtained with $F = 0$) and the particular solution. The homogeneous solution determines the stability of the equilibrium. Before anything else is done, it is important to look at the question of stability in order to ascertain that we are dealing with an equilibrium that is not unstable.

8.4.1 Homogeneous Solution and Stability of the Equilibrium

With $F = 0$, the solution to Eqs. (8.4.5) and (8.4.6) is of the form $x_1 = A_1 e^{st}$ and $x_2 = A_2 e^{st}$. As indicated in Section 8.3, the exponent s is obtained from

$$\det\left(\begin{bmatrix} m_1 s^2 + k_1 + k_2 & -k_2 \\ -k_2 & m_2 s^2 + k_2 \end{bmatrix}\right)$$

$$= m_1 m_2 \left[s^4 + \left(\frac{k_1 + k_2}{m_1} + \frac{k_2}{m_2}\right)s^2 + \frac{k_1}{m_1}\frac{k_2}{m_2}\right] = 0 \quad (8.4.7)$$

The roots of Eq. (8.4.7) are

$$s^2 = -\frac{1}{2}\left(\frac{k_1 + k_2}{m_1} + \frac{k_2}{m_2}\right) \pm \sqrt{\frac{1}{4}\left(\frac{k_1 + k_2}{m_1} + \frac{k_2}{m_2}\right)^2 - \frac{k_1}{m_1}\frac{k_2}{m_2}} \quad (8.4.8)$$

Since the radicand in Eq. (8.4.8) is a positive quantity[2] and its square root is greater than $\frac{1}{2}[(k_1 + k_2)/m_1 + k_2/m_2]$, both roots s^2 of Eq. (8.4.8) are negative. Therefore, the four quantities $s = \pm\sqrt{s^2}$ are purely imaginary, and the equilibrium is stable.

---

[2] This can be shown by simply expanding that radicand, which becomes $\frac{1}{4}[(k_1/m_1 - k_2/m_2)^2 + (k_2/m_1)^2 + 2(k_1/m_1 + k_2/m_2)k_2/m_1]$.

With $s = i\omega$, where $i = \sqrt{-1}$, the squares of the natural frequencies of the free undamped oscillation of the system are then

$$\left.\begin{array}{c}\omega_1^2\\ \omega_2^2\end{array}\right\} = \frac{1}{2}\left(\frac{k_1+k_2}{m_1} + \frac{k_2}{m_2}\right) \mp \sqrt{\frac{1}{4}\left(\frac{k_1+k_2}{m_1} + \frac{k_2}{m_2}\right)^2 - \frac{k_1}{m_1}\frac{k_2}{m_2}} \quad (8.4.9)$$

The free oscillation is the solution of Eqs. (8.4.5) and (8.4.6) and is of the form $a_1\cos(\omega_1 t + \alpha_1) + a_2\cos(\omega_2 t + \alpha_2)$, where a_1, a_2, α_1, and α_2 are determined by the initial conditions of the motion.

8.4.2 Particular Solution: Response to a Sinusoidal Force

Similar to what was presented in Section 8.2, the particular solutions to Eqs. (8.4.5) and (8.4.6) are also sinusoidal. Since there is no damping in the system, the response is in phase with the excitation $F\sin(\Omega t)$. Thus, by expressing $x_1(t)$ and $x_2(t)$ as

$$x_1(t) = B_1 \sin(\Omega t) \quad (8.4.10)$$

$$x_2(t) = B_2 \sin(\Omega t) \quad (8.4.11)$$

Eqs. (8.4.5) and (8.4.6) yield

$$\begin{bmatrix} k_1 + k_2 - m_1\Omega^2 & -k_2 \\ -k_2 & k_2 - m_2\Omega^2 \end{bmatrix} \begin{bmatrix} B_1 \\ B_2 \end{bmatrix} = \begin{bmatrix} F \\ 0 \end{bmatrix} \quad (8.4.12)$$

The solution to Eq. (8.4.12) is

$$B_1 = \frac{(k_2 - m_2\Omega^2)F}{D} = \left(\frac{k_2/m_2 - \Omega^2}{(\Omega^2 - \omega_1^2)(\Omega^2 - \omega_2^2)}\right)\frac{F}{m_1} \quad (8.4.13)$$

$$B_2 = \frac{k_2 F}{D} = \left(\frac{k_2/m_2}{(\Omega^2 - \omega_1^2)(\Omega^2 - \omega_2^2)}\right)\frac{F}{m_1} \quad (8.4.14)$$

where

$$D = m_1 m_2 \Omega^4 - [m_1 k_2 + m_2(k_1 + k_2)]\Omega^2 + k_1 k_2 = m_1 m_2 (\Omega^2 - \omega_1^2)(\Omega^2 - \omega_2^2) \quad (8.4.15)$$

is the determinant of the 2×2 matrix that appears in Eq. (8.4.12).

> Inspection of Eq. (8.4.13) immediately discloses that the forced response of the larger block (i.e., x_1) can be suppressed (i.e., $B_1 = 0$) by tuning the m_2-k_2 system so that $k_2/m_2 = \Omega^2$.

Also, notice that both B_1 and B_2 tend to zero as $\Omega \to \infty$ (i.e., very high frequency excitation) and tend to $k_2 F/(m_1 m_2 \omega_1^2 \omega_2^2) = F/k_1$ [since, as disclosed by Eq. (8.4.15), $m_1 m_2 \omega_1^2 \omega_2^2 = k_1 k_2$] as $\Omega \to 0$. Figure 8.4 shows a typical

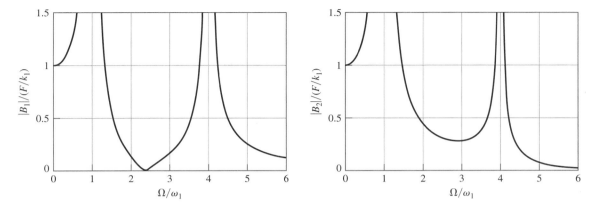

Figure 8.4 A typical amplitude-frequency response curve for the system of Fig. 8.3.

amplitude-frequency response curve for the system. The practical region of operation of the device is in a frequency range where $\sqrt{k_2/m_2}$ is near Ω, for which the response amplitude $|B_1|$ is a small acceptable value dictated by the user of the device.

In practice, there is always some amount of damping in the system. It turns out that the presence of damping has the undesirable effect of preventing the amplitude $|B_1|$ of $x_1(t)$ from becoming zero. This can be verified either analytically or by numerical integration of the differential equations after including a damping term on them, such as $c_1 \dot{x}_1 + c_2(\dot{x}_1 - \dot{x}_2)$ in the left-hand side of Eq. (8.4.5) and $c_2(\dot{x}_2 - \dot{x}_1)$ in the left-hand side of Eq. (8.4.6). For simplicity, one could take $c_1 = c_2 \triangleq c$.

8.5 Analysis of the Foucault Pendulum

8.5.1 Problem Description

In 1851, an important experiment was performed by the French physicist Jean-Bernard-Léon Foucault (1819–1868), in which he used a special pendulum he constructed for demonstrating the rotation of the earth about its own axis. Foucault was a physicist at the Paris observatory and received the French Legion of Honor medal for his numerous experimental contributions to mechanics, especially his remarkable experiment with his pendulum.

When the pendulum was given a small displacement in a vertical plane, it was observed that the oscillating pendulum rotated with respect to the earth with a period that was approximately equal to $24/\sin \lambda$ h, with λ being the latitude of the pendulum's location on the surface of the earth. The pendulum originally used by Foucault consisted of a 28-kg ball suspended by a 67-m wire. He repeated his experiment at the Panthéon in Paris. Today, Foucault's pendulum can be seen in many museums throughout the world.

The differential equations that govern the motion of the pendulum relative to the rotating earth are formulated in this section. The equations are then used for analyzing the effect of the spin of the earth on the rotation of the oscillating pendulum and to compare the results with the assertion made about the period of the rotation.

8.5.2 Formulation Using Rectangular Coordinates

Consider a pendulum consisting of a particle P of mass m and a wire (approximated as massless) of length L, and whose support is at a point Q of latitude λ near the surface of the earth as shown in Fig. 8.5. For simplicity, the vertical distance h between point O', which is on the surface of the earth, and the support at Q is taken to be $h = L$.

The scale of Fig. 8.5 is grossly exaggerated since it is assumed that the length L of the pendulum is much smaller than the radius R_E of the earth, with the earth being approximated as a homogeneous sphere of mass M_E. It is also assumed that the only forces acting on P are the tension in the wire QP, of magnitude N, and the gravitational force of attraction of the earth. The magnitude of the gravitational force acting on P is approximated as $GM_E m/R_E^2 = mg$, where $g \approx 9.81$ m/s^2 is the acceleration of gravity at the earth's surface. Also, since L is very small compared to R_E, the gravitational force acting on P is practically in the $-\hat{z}$ direction, where $\hat{z}$ is a unit vector directed "upward" along the local vertical line OO'. The rectangular coordinates x, y, and z shown in Fig. 8.5 will be used for analyzing the effect of Ω on the motion of the pendulum relative to the rotating earth.

Note: At this point, a natural question arises, namely, "Why work with rectangular coordinates instead of the 'more natural' independent variables θ and ϕ shown in Fig. 8.5?" The answer has to do with the complexity of the resulting differential equations of motion that are obtained in terms of either (θ, ϕ) or

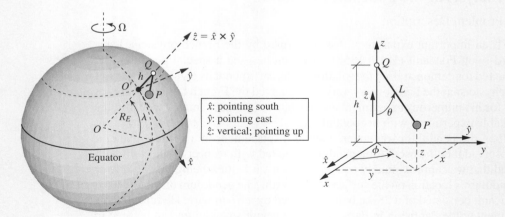

Figure 8.5 Foucault's pendulum (a pendulum for demonstrating the rotation of the earth).

SECTION 8.5 Analysis of the Foucault Pendulum

(x, y). To see this, one has to develop both types of equations. When this is done, it can then be seen that the equations developed with rectangular coordinates are simpler to use for determining the effect of the earth spin on the motion of the pendulum.

In terms of the earth-fixed unit vector triad $\{\hat{x}, \hat{y}, \hat{z} = \hat{x} \times \hat{y}\}$ shown in Fig. 8.5, the resultant force $\vec{F}$ acting on P is then approximately equal to

$$\vec{F} \approx -mg\hat{z} - N\frac{\overrightarrow{QP}}{|\overrightarrow{QP}|} = -mg\hat{z} - N\left(\frac{x\hat{x} + y\hat{y} + (z-L)\hat{z}}{L}\right) \quad (8.5.1)$$

where x, y, and z are the rectangular coordinates of P relative to earth-fixed x-y-z axes centered at point O'. Notice that, since the length L of the pendulum is constant, the variables x, y, and z are related as follows:

$$x^2 + y^2 + (z-L)^2 = L^2 \quad (8.5.2)$$

With λ being the latitude at location O' on the earth, the absolute angular velocity $\vec{\omega}$ of the $\{\hat{x}, \hat{y}, \hat{z}\}$ triad fixed to the earth as shown in Fig. 8.5 is approximated as

$$\vec{\omega} \approx \Omega(\hat{z} \sin \lambda - \hat{x} \cos \lambda) \quad (8.5.3)$$

where $\Omega = 2\pi$ rad/day $\approx 7.27 \times 10^{-5}$ rad/s. In Eq. (8.5.3), the rotation of the line from the sun to the center O of the earth, with angular velocity of 2π rad/year, is neglected since it is much smaller than the spin Ω of the earth about its own axis.

With the position vector from O' to P expressed as $\overrightarrow{O'P} = x\hat{x} + y\hat{y} + z\hat{z}$ and the center O of the earth now being approximated as inertial, the absolute position vector for P is then

$$\vec{r}_P = R_E \hat{z} + x\hat{x} + y\hat{y} + z\hat{z}$$

Therefore, the absolute velocity $\vec{v}_P$ and acceleration $\vec{a}_P$ of P are

$$\vec{v}_P = \frac{d\vec{r}_P}{dt} = \dot{x}\hat{x} + \dot{y}\hat{y} + \dot{z}\hat{z} + \vec{\omega} \times \vec{r}_P$$
$$= (\dot{x} - \Omega y \sin \lambda)\hat{x} + [\dot{y} + \Omega x \sin \lambda + \Omega(R_E + z) \cos \lambda]\hat{y} + (\dot{z} - \Omega y \cos \lambda)\hat{z}$$
$$(8.5.4)$$

$$\vec{a}_P = \frac{d\vec{v}_P}{dt} = \left(\frac{d\vec{v}_P}{dt}\right)_{\{\hat{x},\hat{y},\hat{z}\}} + \vec{\omega} \times \vec{v}_P$$
$$= [\ddot{x} - 2\Omega \dot{y} \sin \lambda - \Omega^2 x \sin^2 \lambda - \Omega^2 (R_E + z)(\sin \lambda) \cos \lambda]\hat{x}$$
$$+ (\ddot{y} + 2\Omega \dot{x} \sin \lambda + 2\Omega \dot{z} \cos \lambda - \Omega^2 y)\hat{y}$$
$$+ [\ddot{z} - 2\Omega \dot{y} \cos \lambda - \Omega^2 x (\sin \lambda) \cos \lambda - \Omega^2 (R_E + z) \cos^2 \lambda]\hat{z} \quad (8.5.5)$$

With $\vec{F}$ and $\vec{a}_P$ expressed as shown in Eqs. (8.5.1) and (8.5.5), the following scalar equations are obtained directly from $\vec{F} = m\vec{a}_P$:

$$m[\ddot{x} - 2\Omega\dot{y}\sin\lambda - \Omega^2 x \sin^2\lambda - \Omega^2(R_E + z)(\sin\lambda)\cos\lambda] = -\frac{x}{L}N \quad (8.5.6)$$

$$m(\ddot{y} + 2\Omega\dot{x}\sin\lambda + 2\Omega\dot{z}\cos\lambda - \Omega^2 y) = -\frac{y}{L}N \quad (8.5.7)$$

$$m[\ddot{z} - 2\Omega\dot{y}\cos\lambda - \Omega^2 x(\sin\lambda)\cos\lambda - \Omega^2(R_E + z)\cos^2\lambda] = -mg - \frac{z-L}{L}N \quad (8.5.8)$$

Equations (8.5.6) to (8.5.8), together with the constraint Eq. (8.5.2), govern the motion of the pendulum relative to the rotating earth. The effect of Ω on the small motion of the pendulum is now addressed in detail.

Since $\Omega = 2\pi$ rad/day $\approx 7.27 \times 10^{-5}$ rad/s is very small, approximations are now made in order to reduce the complexity of Eqs. (8.5.6) to (8.5.8), but still retaining the effect of Ω, of course. One simplification consists of neglecting the much smaller Ω^2 terms. When this is done, the equilibrium solutions $x = \text{constant} \triangleq x_e$, $y = \text{constant} \triangleq y_e$ and $z = \text{constant} \triangleq z_e$ to those equations are $x_e = 0$, $y_e = 0$, $z_e = 0$ and $x_e = 0$, $y_e = 0$, $z_e = L$, with $z_e = 0$ and $z_e = L$ obtained from the constraint equation, which is Eq. (8.5.2). The equilibrium for which $z_e = L$ corresponds to the upside down pendulum and is not of interest.

The next approximation consists of limiting the analysis to small motions about the equilibrium $x_e = 0$, $y_e = 0$, $z_e = 0$ and linearizing the equations for small x, y and z. By noticing that Eq. (8.5.2) is simply $x^2 + y^2 + z^2 - 2zL = 0$, its linearization yields $z \approx 0$, and Eq. (8.5.8) yields $N \approx m(g - 2\Omega\dot{y}\cos\lambda)$. With this expression for N, Eqs. (8.5.6) and (8.5.7) then immediately yield the following linearized differential equations involving the small independent variables $x(t)$ and $y(t)$, with Ω^2 terms neglected:

$$\ddot{x} - (2\Omega\sin\lambda)\dot{y} + \frac{g}{L}x = 0 \quad (8.5.9)$$

$$\ddot{y} + (2\Omega\sin\lambda)\dot{x} + \frac{g}{L}y = 0 \quad (8.5.10)$$

To see the effect Ω has on the motion, it helps to integrate Eqs. (8.5.9) and (8.5.10) numerically. A typical result of the integration, with $L = 10$ m and the initial conditions $x(0) = 0.01L$, $y(0) = 0$, and $\dot{x}(0) = \dot{y}(0) = 0$, is shown in Fig. 8.6. The integration was done using the block diagram model shown in Fig. 8.7, with $\Omega\sin\lambda = 0.0727$ in order to speed up the response.

$\Longrightarrow$ The model shown in Fig. 8.7 is the file `foucault.mdl` on the enclosed CD.

Figure 8.6 shows that the oscillating pendulum rotates slowly about the vertical line $O'Q$ seen in Fig. 8.5. The rotation is clockwise for a positive value of $\Omega\sin\lambda$ (i.e., in the northern hemisphere) and counterclockwise for negative values (i.e., in the southern hemisphere). The path followed by point P is indicated

SECTION 8.5 Analysis of the Foucault Pendulum

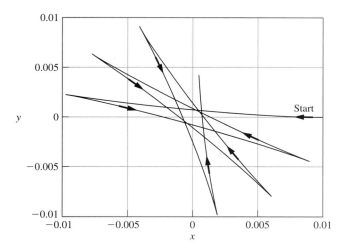

Figure 8.6 Typical result of the numerical integration of Eqs. (8.5.9) and (8.5.10).

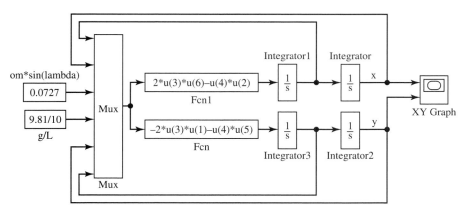

Figure 8.7 Numerical integration of Eqs. (8.5.9) and (8.5.10).

by the arrows that were added to the plot of Fig. 8.6. If you use the model shown in Fig. 8.7 with $\Omega = 7.27 \times 10^{-5}$, you will see that the rotation of the oscillating pendulum is indeed quite slow.

In view of the observed motion shown in Fig. 8.6, let us see what happens if the motion is expressed in terms of coordinates x_1 and y_1 that are rotated by a time-varying angle $\alpha(t)$ as shown in Fig. 8.8. According to that figure, x and y are expressed in terms of x_1 and y_1 as follows:

$$x(t) = x_1 \cos\alpha + y_1 \sin\alpha \quad (8.5.11)$$

$$y(t) = y_1 \cos\alpha - x_1 \sin\alpha \quad (8.5.12)$$

Figure 8.8 The x_1-y_1 axes rotated by $\alpha(t)$ relative to the x-y earth-fixed axes.

These equations then yield the following expressions for $\dot{x}$ and $\dot{y}$:

$$\dot{x}(t) = (\dot{x}_1 + \dot{\alpha} y_1) \cos \alpha + (\dot{y}_1 - \dot{\alpha} x_1) \sin \alpha \qquad (8.5.13)$$

$$\dot{y}(t) = (\dot{y}_1 - \dot{\alpha} x_1) \cos \alpha - (\dot{x}_1 + \dot{\alpha} y_1) \sin \alpha \qquad (8.5.14)$$

Now, it is not difficult to verify that, in terms of x_1 and y_1, Eqs. (8.5.9) and (8.5.10) become

$$\left[\ddot{x}_1 + \ddot{\alpha} y_1 + 2(\dot{\alpha} - \Omega \sin \lambda) \dot{y}_1 + \left(\frac{g}{L} - \dot{\alpha}^2 + 2\dot{\alpha} \Omega \sin \lambda \right) x_1 \right] \cos \alpha$$

$$+ \left[\ddot{y}_1 - \ddot{\alpha} x_1 + 2(\Omega \sin \lambda - \dot{\alpha}) \dot{x}_1 + \left(\frac{g}{L} - \dot{\alpha}^2 + 2\dot{\alpha} \Omega \sin \lambda \right) y_1 \right] \sin \alpha = 0 \qquad (8.5.15)$$

$$\left[\ddot{y}_1 - \ddot{\alpha} x_1 + 2(\Omega \sin \lambda - \dot{\alpha}) \dot{x}_1 + \left(\frac{g}{L} - \dot{\alpha}^2 + 2\dot{\alpha} \Omega \sin \lambda \right) y_1 \right] \cos \alpha$$

$$- \left[\ddot{x}_1 + \ddot{\alpha} y_1 + 2(\dot{\alpha} - \Omega \sin \lambda) \dot{y}_1 + \left(\frac{g}{L} - \dot{\alpha}^2 + 2\dot{\alpha} \Omega \sin \lambda \right) x_1 \right] \sin \alpha \qquad (8.5.16)$$

These two equations are satisfied only if

$$\ddot{x}_1 + \ddot{\alpha} y_1 + 2(\dot{\alpha} - \Omega \sin \lambda) \dot{y}_1 + \left(\frac{g}{L} - \dot{\alpha}^2 + 2\dot{\alpha} \Omega \sin \lambda \right) x_1 = 0 \qquad (8.5.17)$$

$$\ddot{y}_1 - \ddot{\alpha} x_1 + 2(\Omega \sin \lambda - \dot{\alpha}) \dot{x}_1 + \left(\frac{g}{L} - \dot{\alpha}^2 + 2\dot{\alpha} \Omega \sin \lambda \right) y_1 = 0 \qquad (8.5.18)$$

Notice that the angle $\alpha(t)$ is arbitrary. By observing Eqs. (8.5.17) and (8.5.18), it is seen that they can be decoupled by choosing α so that

$$\dot{\alpha} = \Omega \sin \lambda \qquad (8.5.19)$$

and this yields

$$\ddot{x}_1 + \left(\frac{g}{L} + \underbrace{\Omega^2 \sin \lambda}_{\text{negligible}} \right) x_1 = 0 \quad \text{and} \quad \ddot{y}_1 + \left(\frac{g}{L} + \underbrace{\Omega^2 \sin \lambda}_{\text{negligible}} \right) y_1 = 0 \qquad (8.5.20)$$

The equations of (8.5.20) describe two uncoupled oscillators, and they govern the motion of the pendulum as viewed by an observer fixed to the x_1-y_1 axes. As seen from Fig. 8.8, such axes are rotating with respect to the x-y earth-fixed axes. As indicated earlier, in the northern hemisphere, where $\sin \lambda > 0$, the rotation is clockwise. It is counterclockwise in the southern hemisphere, where $\sin \lambda < 0$. The period $T_{2\pi}$ of such a rotation is

$$T_{2\pi} = \frac{2\pi}{\dot{\alpha}} = \frac{2\pi}{\Omega \sin \lambda} = \frac{24}{\sin \lambda} \text{ h}$$

which is in complete agreement with the results originally observed and explained by Foucault. At the poles, the period $T_{2\pi}$ is a minimum and equal to 24 h.

SECTION 8.5 Analysis of the Foucault Pendulum

Summary For an observer fixed to the rotating earth, the oscillating pendulum wire rotates relative to the earth, about the vertical, with a period $T_{2\pi} = 2\pi/(\Omega \sin \lambda) = 24/\sin \lambda$ h. The rotation is clockwise if the pendulum is in the northern hemisphere and counterclockwise in the southern hemisphere. At the equator ($\lambda = 0$), no rotation due to the earth's spin occurs when the very small terms in Ω^2 are neglected, while at the poles the period $T_{2\pi}$ is minimum and equal to 24 h. At those locations, the oscillating pendulum wire rotates $360/4 = 90°$ every $24/4 = 6$ h. In Washington, D.C. (United States), Paris (France), and Rio de Janeiro (Brazil), a rotation of $90°$ takes approximately 9 h and 30 min, 8 h, and 15 h and 20 min, respectively.

APPENDIX A

Simulink Tutorial and a Brief Introduction to MATLAB

A.1 Introduction

Simulink is a software that is used for simulating dynamical systems and integrating differential equations, without the user having to write a computer program. It is the graphical extension of MATLAB, where one simply works with simulation blocks, such as blocks for integration and function generation, among many others. The user formulates the equations that govern the motion of the dynamical system under study and then simply prepares a Simulink block diagram model to do the calculations by connecting the necessary blocks in the appropriate manner. Such diagrams are much like the ones many readers have seen in books on dynamical systems and control systems. Simulink and MATLAB are widely used at many universities and companies worldwide. Simulink is easy to use, and not all of its features are needed for dynamics. Thus, this tutorial is relatively brief compared to the vast amount of information available in the Simulink manual.

Simulink and MATLAB run on personal computers (PCs) with Windows, UNIX engineering workstations (such as SUN and Silicon Graphics), and Macintosh computers. There is also a student version that is distributed by many university bookstores. These software are periodically upgraded by The MathWorks, Inc., and new simulation blocks and new features frequently appear in newer versions.

The files included in the `CdS_Dynamics` folder on the enclosed CD run on PCs with Windows, UNIX workstations, and Macintosh computers. That folder contains all the Simulink models (and MATLAB files that are "transparent" to the user but are needed for some of the models) that appear in this appendix and in other parts of the book. Copy the `CdS_Dynamics` folder to the hard disk of your computer, following the instructions in Section A.2.

> Your version of MATLAB/Simulink may display a warning message saying that the model from the enclosed CD was prepared with an earlier version of Simulink. If that happens, it may also display the name of the command you should "run" (i.e., by typing that command in the MATLAB command window) for updating the model. However, the model will be automatically updated when you save it. Saving the model is as simple as answering yes if the software asks you to save it or choosing File → Save with the mouse (i.e., clicking the word File in the model window with the mouse, and then selecting Save in the menu that opens up). The enclosed models run on Simulink 4 or higher and MATLAB 6.0 or higher (on a Macintosh, you need to have Simulink 5 or higher and MATLAB 6.5 or higher).

Some of the figures you see in this tutorial may look a little different from what you see on your computer screen, but this should be no cause for concern.

> The best way to use this tutorial is to use the computer to practice what is presented here while you read and study the material that follows.

To use Simulink, MATLAB has to be started. On a PC or Macintosh, this may be done either by double clicking a MATLAB icon or by selecting MATLAB in a menu. On a UNIX machine, this is generally done by typing matlab in a UNIX command window. A typical window that opens up when MATLAB starts on a PC is shown in Fig. A.1, where the part with the prompt >> is the MATLAB command window.

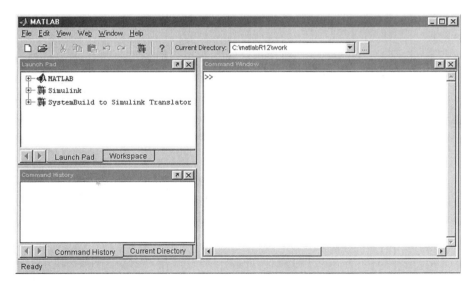

Figure A.1 A typical MATLAB window on a PC, showing the command window and two other windows (selected from the View menu in this illustration).

APPENDIX A Simulink Tutorial and a Brief Introduction to MATLAB

Figure A.2 A typical MATLAB command window on a PC.

Do not be concerned if the window that opened on your computer looks somewhat different, such as the one shown in Fig. A.2, and/or is not displaying the additional windows shown in Fig. A.1. Click on the `View` word, seen in Figs. A.1 and A.2, to observe the menu that opens up.

To start Simulink, type `simulink` in the MATLAB command window. This will open up either a "Simulink Library Browser" window, such as the one illustrated in Fig. A.3 or, in some computers, a "Simulink Library window" such as the one illustrated in Fig. A.4. Although the Library Browser window makes it easier to search for a specific block by typing its name in the `Find` part of that window (and then pressing the `return` key on the keyboard), it makes no difference which window appears on your computer screen.

Again do not be concerned if the windows on your computer screen do not look exactly like the ones shown here. At this time, briefly explore the contents of the window on your computer screen, and look at the menus that appear when you click the left button of the computer mouse on the words `File`, `Edit`, etc., that appear on them.

The icons labeled `Continuous`, `Sources`, `Sinks`, etc., that are seen in Figs. A.3 and A.4 represent Simulink library directories that contain a number of different blocks used for constructing Simulink models of dynamical systems. A typical contents of the `Continuous` library is seen in Fig. A.3. The `Integrator` block is the only one of those shown that is used in this book. Do not pay any attention to the *controls symbol* $\frac{1}{s}$ that appears on the integrator icon. A better symbol to use there might have been $\int$. Typical contents of the `Sources` and of the `Sinks` libraries are shown in Fig. A.5.

The `Clock` and `Constant` blocks from the `Sources` library are used in many of the models in this book. The digital `Display` and the `Stop Simulation` from the `Sinks` library are also widely used. These and other blocks used in the models on the enclosed CD are described in Section A.3. To look at the contents of other libraries, left click the mouse button in a desired

SECTION A.1 Introduction

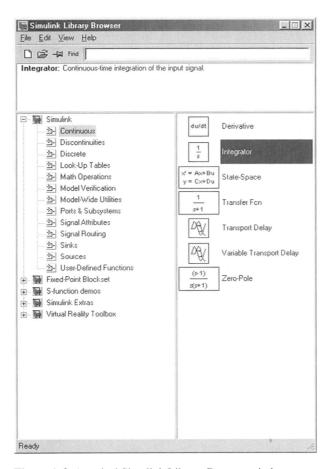

Figure A.3 A typical Simulink Library Browser window on a PC.

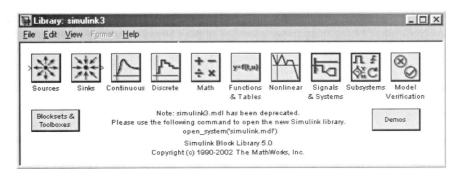

Figure A.4 A typical Simulink Library window.

Figure A.5 Typical contents of the Sources and of the Sinks libraries.

library, seen in Fig. A.3, or double click the left button in a library icon, seen in Fig. A.4. It is suggested that you do so before you proceed any further.

Icons in the Sources library (see Fig. A.5) provide input signals for models. They include a Clock whose output is the simulation time t and a Constant block whose output is a constant that is specified after double clicking the block. The output of the Clock, for example, may be used as input to the blocks named Fcn and MATLAB Fcn for generating a desired function of time. The Fcn and MATLAB Fcn blocks are located in the User Defined Functions library in Figs. A.3 and A.5, and in the Functions & Tables library in Fig. A.4. Those and other blocks are described in Section A.3.

The icons in the Sinks library include a Scope (for *oscilloscope*) block and an XY Graph block, both of which allow signals to be viewed as the simulation progresses, and a Display block that shows the digital value of its input signal. Many of the icons in the other libraries represent blocks that perform linear and nonlinear operations on their input signals (such as integration and sum and multiplication of several signals).

A Simulink block diagram model is constructed in a Simulink *model window*. Detailed explanations on the construction of a model are given in Section A.5. To create a model window on a PC, select New → Model in the File menu of any of the windows shown in Figs. A.1 to A.4. A model window labeled untitled, such as the one shown in Fig. A.6, will then appear on your computer screen. That window, and others, may be resized and moved on the computer screen the

Figure A.6 A typical Simulink model window.

same way you resize or move any other window on your computer. The model window also has several menus (`File`, `Edit`, etc.). Open them now to briefly get acquainted with what they do and to get accustomed to using them.[1]

Several of the Simulink blocks that are used the most in connection with the material in this book are described in Section A.3, in alphabetical order. As you see the block's icon in Section A.3 and read its description to the right of the block, you may want to try to find in which of the Simulink libraries that block is located in your version of Simulink. In the Library Browser window shown in Fig. A.3 you can type (in the little window to the right of the word `Find` in that figure) the name of the block you want to use (such as an `Integrator` or a `Mux`) and the browser will find it after the `Return` key on the keyboard is pressed.

A.2 How to Use the Enclosed Software

The easiest way to use the software on the enclosed CD is to follow the two steps below since this will allow you to easily access any of the Simulink models provided with the book. This is to be done only once.

Step 1. Copy the `CdS_Dynamics` folder from the enclosed CD to your computer disk. On PCs and Macintoshes, simply use the mouse to drag that folder to a directory on your computer hard disk, such as to `C:\CdS_Dynamics` on a PC or to `Macintosh HD:CdS_Dynamics` on a Macintosh. On a UNIX workstation, use the UNIX `cp` command (you may need to consult your UNIX administrator to do this) to copy it either to your home directory or a subdirectory of your choice.

---

[1] On a PC, by clicking on `View` with the mouse, for example, you may choose (on a menu that opens up) to either display or hide the "bar" (called the `Tool bar`) that shows several icons, and also the `Status bar`, which is the little part with the word `Ready` at the bottom of the model window.

Step 2. Append the pathname of the folder in step 1 to the MATLAB path and save it for future sessions. This can be done by simply clicking on `File → Set Path` in the window opened by MATLAB and continuing with the obvious choices that appear after that. Assuming that you copied the `CdS_Dynamics` folder to the directory indicated in step 1, this can also be done by typing the following in the MATLAB command window and then pressing the keyboard `Return` key:

On a PC. `addpath('C:\CdS_Dynamics')`
On a Macintosh. `addpath('Macintosh HD:CdS_Dynamics')`
On a UNIX machine. (If needed, consult your UNIX administrator.) The path might be added with the MATLAB command `addpath('/home/CdS_Dynamics')` if you copied the `CdS_Dynamics` folder to your home directory.

To save the new path for future sessions so that you do not have to do this step again in the future, click on `File → Set Path` in the MATLAB window, and then on `Save` in the window that opens up.

Note: MATLAB puts you in a default directory, as the one indicated in Fig. A.1, but it is best that you create one or more directories of your own for your work. PC and Macintosh users create new directories simply with a few mouse clicks. On a PC for example, one may start by double clicking on `My Computer`, and then on `(C:)` and on other desired subdirectories, and finally by single clicking on `File → New Folder` and renaming the new folder. In UNIX workstations, type `help mkdir` and `help cd` in the MATLAB command window and you will see the information for making directories and for changing to another directory. It is suggested that you also put such a folder (or folders) in your MATLAB path. When you do, changing to a new directory (folder) in MATLAB is simply done by selecting one from the `Current Directory` menu seen in Fig. A.1.

A.3 Description of Selected Simulink Blocks

Block icon	Block description
Abs	The output of the `Abs` block is simply the absolute value of the input to the block.
Auto-Scale Graph	The `Auto-Scale Graph` block is described on p. 511.

SECTION A.3   Description of Selected Simulink Blocks

Block icon	Block description
Clock	The output of the `Clock` block is the current simulation time. Connect it to an `Fcn` block (see description in this section) to generate a desired function of time [which may not be provided by the `Signal Generator` (see description in this section)] at the output of the `Fcn` block.
Constant	The `Constant` block generates a constant value which is specified by the user by double clicking the block with the mouse and entering the desired value in a little window that appears on the computer screen. The constant can also be a list of values ([0.1, 0.5, 2], for example[2]).
Dead Zone	As suggested by the plot that appears in the `Dead Zone` block's icon, the output $o(t)$ is zero if the input $f(t)$ is in a certain range (called the *dead zone*). The lower and upper limits (say, f_{lower} and f_{upper}) of the dead zone are specified by the user in a little window that appears by double clicking the icon with the mouse. The output of this block is equal to the following: $$o(t) = \begin{cases} 0 & \text{if } f_{\text{lower}} < f(t) < f_{\text{upper}} \\ f(t) - f_{\text{lower}} & \text{if } f(t) \leq f_{\text{lower}} \\ f(t) - f_{\text{upper}} & \text{if } f(t) \geq f_{\text{upper}} \end{cases}$$ For an example of the use of the `Dead Zone` block, see Fig. 2.13 (p. 94).
Display	The `Display` block displays the value(s) of its input signal(s). Multiple inputs are provided with a `Mux` block (see description in this section). You can choose the format of the displayed numbers in a little window that appears when you double clicked the block's icon with the mouse. For examples where this block is used, see the simulation of the mechanisms that are analyzed in Chapter 3. When needed, resize the block to see all its multiple displays if that is the case.
Fcn	The `Fcn` math function block can take several inputs [which are internally named $u(1)$, $u(2), \ldots, u(n)$ in the Simulink software] and generates a scalar output $f(u(1), u(2), \ldots)$ that is a function of the inputs specified by the user. If there is only a single input, that input may be referred to as either $u(1)$ or $u[1]$ (or simply u). Multiple inputs are provided to the `Fcn` block by connecting its input to the output of a `Mux` block (see description in this section). You can specify the desired function f by typing its expression in a little window that appears when you double click the block with the mouse. For example, the output $u_1 u_2^2 (\sin u_3) + |u_4| - \text{sign}(u_5)$ should be typed in that window either as `u(1)*u(2)^2*sin(u(3))+abs(u(4))-sgn(u(5))` or `u[1]*u[2]^2*sin[u[3]]+abs[u[4]]-sgn(u[5])`
Gain	The output of the `Gain` block is equal to its input signal multiplied by a desired constant "gain." The value of the desired constant gain is specified by the user in a little window that appears by double clicking the block with the mouse. The gain may also be a list of values, such as [0.1; 0.5; 2] or [0.1, 0.5, 2], to output a "vector line" (i.e., either a column or a row matrix whose dimension is equal to the number of elements of the list).

---

[2]The syntax for specifying a list of values starts and ends with a square bracket. The elements of the list are separated either by a comma (a space is also acceptable) or a semicolon. The list stands for a row matrix when a comma (or a space) is used and for a column matrix when a semicolon is used.

Block icon	Block description
Hit Crossing	The `Hit Crossing` block is used for detecting when the input crosses a desired value (called the *hit crossing offset*) in a desired direction, both of which are chosen after double clicking the block. The output of the block is equal to 1 when the hit crossing is detected, and equal to 0 otherwise. For an example of its use, see Fig. 2.13 (p. 94).
Integrator or f(t), g_0 → Integrator	The output of the `Integrator` block is equal to $g_0 + \int_{t_0}^{t} f(\tau)\,d\tau$, where $f(t)$ is the input signal and g_0 is the initial value of the output. The initial value g_0 (i.e., the *initial condition* for the integrator) is specified *internally* in a little window that appears when you double click the `Integrator` block with the mouse (the default value is 0.0). The initial condition may also be specified *externally,* in which case the `Integrator` icon is similar to the second icon shown. For such a case, double click the `Integrator` icon and choose `external`, for the "initial condition source," in the same window previously mentioned. An example of such a use is in Section 4.5. The initial condition may also be a list of values ([0.1, 0.5, 2], for example), to generate as many output signals as the number of elements in the list.[3] As mentioned earlier, do not pay any attention to the $\frac{1}{s}$ symbol that appears on the `Integrator` icon; a better symbol to use there might have been $\int$.
Logical Operator	The `Logical Operator` block performs the logical operations `AND`, `OR`, `NOT`, `NAND`, `NOR`, and `XOR` (i.e., "exclusively or") in its inputs. The desired operator and the number of inputs to the block are chosen in a little window that appears when you double click the block with the mouse. The output of the block is equal to 1 if the operation that is being performed is `TRUE` and equal to 0 if it is `FALSE`. For an example of its use, see Fig. 2.13 (p. 94).
MATLAB Function MATLAB Fcn	Unlike the `Fcn` block, the `MATLAB Fcn` block can generate a *vector output* (i.e., a list) $[f_1(u(1), u(2), \ldots, u(n)), f_2(u(1), u(2), \ldots, u(n)), \ldots, f_m(u(1), u(2), \ldots, u(n))]$, using n inputs $u(1), u(2), \ldots, u(n)$ specified by the user. Multiple inputs are provided to this block by connecting its input to the output of a `Mux` block. The expression for generating the m outputs $f_1, f_2, \ldots, f_m$ is typed, in a single line, in a little window that appears when you double click the block with the mouse. Be careful if you use spaces when entering the elements of a list because that might introduce a new undesired element. Therefore, it is advisable to avoid spaces (as in [sin(u) + 2, cos(u)], which creates the list [sin(u), 2, cos(u)] instead of the list [sin(u) + 2, cos(u)]). The *output dimension m* must also be specified after double clicking the block. For example, suppose a `Mux` at the input of the Matlab `Fcn` block has six inputs, of which the first four are supposed to be the four elements of a matrix[4] $A = [a_{11}, a_{12}; a_{21}, a_{22}]$ and the last two are supposed to be the elements of a column matrix (or "vector") $b = [b_1; b_2]$. If the inputs to the `Mux` (see the description in this section), starting with the top input, are $a_{11}, a_{12}, a_{21}, a_{22}, b_1$, and b_2, the MATLAB command `[u(1), u(2); u(3), u(4)]^(-1)*[u(5); u(6)]` causes the block to output the two elements of the 2×1 column matrix that is the result of the operation $A^{-1}b$. The output dimension in this case must be set to 2 (the default value is -1, which may be used only if the output dimension is the same as the input). An error message is displayed on the computer screen if the user specifies the wrong output dimension.

---

[3]The Integrator block may also be configured as a *reset integrator* to reset its output back to the initial condition, as commanded by a desired reset input signal. For this feature, see your Simulink manual.

[4]Square brackets are used for generating matrices. Row elements are separated either by a comma or by a space. Semicolons separate the columns.

SECTION A.3 Description of Selected Simulink Blocks

Block icon	Block description
Demux — or — Mux, or — Demux	Mux is for *multiplexer* and Demux is for *demultiplexer*. The Mux block combines several input lines into a single "vector output line" (the word *vector* here means a column matrix). The symbols $u(1)$, $u(2)$, etc., with $u(1)$ being the top input line of the icon, $u(2)$ the next input line, etc., must be used when you generate expressions using the signals from those input lines. This is an internal requirement of the software. The Demux block performs the inverse operation, i.e., its input is a vector line and its outputs are the individual components of the input line. The Mux block is very convenient for constructing compact Simulink models that might include the Fcn block (see its use in the example in this appendix). Double click these blocks to change their appearance, if your version of Simulink allows it (the display option in that menu are bar for the dark blocks on the left, and none for those on the right).
Product	The output of the Product block is the product of its inputs. The number of inputs to the block is specified in a little window that appears when you double click the block with the mouse.
Relational Operator	The Relational Operator block compares its two inputs and generates an output that is equal either to zero or one. You can choose the comparison operator by double clicking the block with the mouse. The choices are $=$, $\neq$, $<$, $\leq$, $\geq$, and $>$. For example, if $\geq$ is chosen, the output of the block is $+1$ if the input at the top input line of the block is greater than or equal to the input at the bottom input line, and 0 otherwise.
Scope / XY Graph / Auto-Scale Graph	As their names suggest, these are *oscilloscope* blocks for displaying the results of a simulation. The Scope block can take a *vector line* as input; in such a case, it will display several plots using the same axes. A vector line is generated as the output of either a Mux block, when several input lines are connected to that block, or a MATLAB Fcn block. The XY Graph block takes two inputs and plots the signal carried by the bottom input line to the block (the Y in the XY label) versus the signal carried by the top input line to the block (the X in the XY label). Therefore, X appears along the horizontal axis and Y appears along the vertical axis in the plot. In the models used in this book, the XY figures are drawn in a *figure window* (you can drag a copy of that scope from one of the models on the CD to new models that you create). To edit that window in those models (to include grids, etc), select Figure Toolbar after clicking the View button in that figure window. The *Auto-Scale Graph* is described below.

$\Longrightarrow$ The Auto-Scale Graph scope block is an alternative to the Scope block and is included on the enclosed CD by special permission and courtesy of The MathWorks, Inc. It is the file auto_scale_scope.mdl (and auto_scale_graph.mdl) on the enclosed CD.

To use the Auto-Scale Graph scope, type either auto_scale_scope or auto_scale_graph in the MATLAB command window,[5] and the scope

---

[5] It is assumed that you have set your MATLAB path as indicated in Section A.2. Otherwise, you will have to change directory to where the file auto_scale_scope.mdl is, type auto_scale_scope there to get the scope, and then go back to your working directory. Obviously, setting the path will eliminate all these very inconvenient steps.

icon will appear in a little window on the computer screen. You can then drag a copy of that icon to your Simulink model. It puts a figure window on the computer screen that can be manipulated (such as adding labels to the axes and changing axis limits) simply by using the icons that appear in that window. At the time of publication of this book, the regular scope did not have such a feature. One may also specify the color and type of the lines (e.g., full or dashed) in the plots displayed by the `Auto-Scale Graph` scope simply by entering the desired choices in a menu that appears when you double click that block's icon with the mouse. The plots may also be printed by selecting `Print` from the `File` menu of the activated figure window (a window will be "activated" when you place the mouse cursor on it and click the mouse). Usage of the `Auto-Scale Graph` scope is illustrated in the examples in this tutorial.

Block icon	Block description
Sign	The `Sign` block implements the sign (sgn) function. Its output is $+1$ if the input is positive, 0 if the input is zero, and -1 if the input is negative.
Signal Generator	The `Signal Generator` block generates sine waves, square waves, and a few other signals. The choices appear when you double click the block with the mouse.
Stop Simulation	The `Stop Simulation` block stops the simulation for any nonzero value of its input signal, and it is used in many models in this book.
Sum (or)	The output of the `Sum` block is the sum of its inputs. When you double click the block with the mouse, a little window appears where you can specify the signs for the operations with the input signals [for example, by specifying $+ + -$, the output of the `Sum` block will be $u(1) + u(2) - u(3)$, where $u(1)$, $u(2)$, etc., are the inputs to the block, starting from the top input shown on the rectangular icon]. The shape of the block may also be changed in that same window.
Switch	The `Switch` block passes to its output either its top input or its bottom input, depending on the value of its middle input (which is called the *control input*). The block passes the first input if the middle input is greater than or equal to the *threshold* value of the control input. Otherwise, it passes the third input. You can specify the threshold value in a little window that appears when you double click the block with the mouse. This block is used in Example 2.7.6 (see Fig. 2.31, p. 126). It could also be used, for example, to switch from one differential equation of motion to another, in problems whose different parts of a trajectory are governed by different differential equations.
Terminator	In rare situations, one may build a Simulink model and leave open an unimportant output of a block. Simulink will then issue a warning message indicating that open connection. Such a message can be eliminated by connecting a `Terminator` block to that open output. A warning message is not an error message since the simulation/animation runs properly even if you do not use a `Terminator` block.

Block icon	Block description
	A warning message for a model may also be disabled by clicking on the `Diagnostics` tab of the `Simulation Parameters` menu of the model, highlighting the selection `Unconnected block output` (in the `Configuration options` part of that menu), and then clicking on `None` under the `Action` choices to take.
simout To Workspace	The `To Workspace` block is used only if you want to send data from your Simulink simulation to the MATLAB workspace for manipulating the data in the MATLAB command window using MATLAB commands. It stores its output in a variable that you name by double clicking the block with the mouse and entering the desired name for that variable in the window that will appear on the computer screen. Choose `Array` in the `Save format` part of that window to save the data as a matrix. That matrix will be in the MATLAB workspace after the simulation stops. The number of columns in the matrix is determined internally by MATLAB. This block can take several inputs. The number of rows in the matrix that is generated is equal to the number of inputs to the block. The number of columns of the matrix is equal to the number of time steps that were used in the simulation after it stops. You may not want to send anything to the MATLAB workspace, but one reason you may want to do so is to use MATLAB commands to create superimposed plots of your liking with the data generated by the Simulink model after the simulation stops.
sin Trigonometric Function	The output of the `Trigonometric Function` block is a trigonometric function of its input u, such as $atan2(u1, u2)$ and others. Double click the block with the mouse to select the desired function. For $u1 = \sin\alpha$ and $u2 = \cos\alpha$, $atan2(u1, u2)$ uses the values of $\sin\alpha$ and $\cos\alpha$ to give the angle α in the correct quadrant, i.e., in the entire range $-\pi \leq \alpha \leq \pi$. This block, with the $atan2$ function, is extensively used in Chapter 3.

A.4 Special Simulink Blocks Prepared by the Author

When added to a Simulink model, the blocks shown in Fig. A.7 display on the computer screen a drawing that resembles a system of interconnected bars, undergoing motion governed by the equations that are formulated for that system. They are used in several models that are presented throughout this book and are a useful tool that helps with the investigation of motion. Assuming that you have set your MATLAB path as indicated in Section A.2, you can access these blocks by typing `animation_blocks` on the MATLAB command window. Also, you can access any Simulink model on the enclosed CD by typing the name of the model (without the `.mdl`) in the MATLAB command window. You can also access them by typing `simulations_menu` in the MATLAB command window and then using the left button of the computer mouse to select a model from a pull-down menu that appears on the computer screen. A list of all the files on the enclosed CD is given in Appendix E.

With the `animate_nbars` and `animate3d_nbars` blocks you can "construct" a system made of any number of interconnected straight bars (either of constant or variable length) and point masses and connect any of them. They

Figure A.7 Simulink animation blocks for visualizing the motion of a number of dynamical systems.

are described in Section A.5 (p. 520) and are used in many models that are developed in this book. The `two_body_animate` block is described and used in Chapter 4.

The `animate_nbars` block is for two-dimensional motion, and the `animate3d_nbars` block animates motion in three-dimensional space. Such a visualization capability is very helpful for better understanding the resulting motion governed by the differential equations of motion that are formulated for an object to model its dynamic behavior and to perform a number of numerical and visual experiments using such equations. By using this capability you will be able to change parameters and initial conditions in the differential equations of motion that you formulate and see the effect such changes have on the dynamic behavior of that object. In engineering practice, such "numerical experimentations" are used in design that involves dynamical systems and in research work.

A.5 How to Construct a Simulink Model

Let us construct a Simulink model to obtain a numerical solution to the differential equation of motion of a dynamical system. For this, let us use the spring-mass systems shown in Fig. A.8. The model is on the enclosed CD, but you should construct it following the instructions given here so that you learn how to use Simulink if you are not familiar with it.

The system in Fig. A.8a consists of a spring (which will be modeled as massless) of unstressed length L_u m, a block of mass m kg that is subjected to a known applied force $F(t)$, and a viscous damper (called a *dashpot*). Motion is on a horizontal plane. During the motion, the actual length $r(t)$ of the spring changes by $x(t)$ m from its unstressed length L_u. Let the reaction force applied by the spring be of the form $k_1 x + k_3 x^3$ (which is normally determined from a laboratory measurement), where k_1 is in N/m and k_3 in N/m^3. Also, let the reaction force applied by the dashpot be proportional to the velocity $\dot{x}$, with the

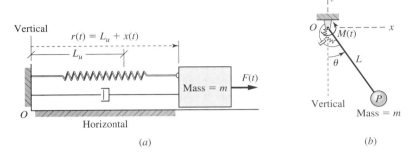

Figure A.8 Two examples of a spring-mass-dashpot system.

proportionality constant being c N/(m/s). As in the rest of this book, dots over a variable in a mathematical expression denote the time derivative of that variable. The differential equation that governs the motion of this system is

$$m\ddot{x} + c\dot{x} + k_1 x + k_3 x^3 = F(t) \tag{A.5.1}$$

The system in Fig. A.8b is a pendulum whose motion takes place on a vertical plane. It consists of a bar (which will be modeled as massless) of length L m and subjected to a known applied moment $M(t)$, a torsional spring that is unstressed when $\theta(t) = 0$, a block of mass m kg (which will be modeled as a point mass P), and a viscous torsional damper (called a *torsional dashpot*). The dashpot reaction moment is proportional to $\dot{\theta}$, with the proportionality constant being c N·m/(rad/s). The x-y coordinate system shown in the figure is not important now; its need will appear only in connection with an animation block that is presented on p. 520. Let the reaction moment applied by the torsional spring also be of the form $k_1\theta + k_3\theta^3$, where k_1 is in N·m/rad and k_3 in N·m/rad^3. With $g = 9.81$ m/s^2 being the acceleration of gravity, the differential equation that governs the motion of this system is

$$mL^2\ddot{\theta} + c\dot{\theta} + k_1\theta + k_3\theta^3 + mgL\sin\theta = M(t) \tag{A.5.2}$$

Since Eq. (A.5.1) can be viewed as a particular case of Eq. (A.5.2), simply by setting $L = 1$, leaving the g term out, and replacing $M(t)$ by $F(t)$, we only need to concentrate on Eq. (A.5.2). Notice that the equation is linear if $k_3 = 0$ and g is set to zero (in which case it can be solved analytically), and nonlinear otherwise. It is desired to plot $M(t)/(mL^2)$ and $\theta(t)$ in the same graph and to see how the system responds to a square wave $M(t)$, which is available from the `Signal Generator` block in Simulink.

To construct the Simulink model, start by copying some of the blocks in the Simulink libraries to your Simulink model window. If a model window called `untitled` does not exist yet, create one by selecting with the mouse New → Model in the `File` menu of any of the windows shown in Figs. A.1 to A.4. Since a second-order differential equation has to be integrated, two integrators are needed, one that integrates $\ddot{\theta}(t)$ to give us $\dot{\theta}(t)$, and the other one to integrate $\dot{\theta}(t)$ to give us $\theta(t)$. A scope will also be needed so that the result of the integration is displayed while the integration is in progress.

- Start constructing the Simulink model with an integrator whose output is the solution $\theta(t)$ of the differential equation. Look for the `Integrator` icon in the Simulink library (see either Fig. A.3 or A.4). To put a copy of the `Integrator` block in your model window, place the mouse pointer on the `Integrator` icon, press and hold the mouse button, drag the `Integrator` icon to the window named `untitled`, and then release the mouse button. To avoid having too many open windows, it is suggested that you close those library windows that you opened after you no longer need to copy blocks from them.

    Now, make a second copy of the `Integrator` block using the following instructions.

- To make a copy of a block in the same model window, you can select (with a single mouse click) the block icon that is now in your `untitled` model window. Press and hold the left button of the mouse (this also works for a single-button mouse), together with the `Control` key on the keyboard, drag the icon to where you want to put the copy, and then release the mouse button and the control key. If you do not press and hold the control key, the icon will be moved, instead of copied, to a new location in the window. Copies may also be made by using the right button (without pressing the `Control` key) or by using `copy` and `paste` from the `Edit` menu of the model window.

- Put in your Simulink model window a copy of the other blocks that are needed, such as a `Gain` block (for the coefficients that appear in the differential equation), a `Product` block (to generate the term θ^3), a `Scope` block, a `Signal Generator` block (to generate one of the functions provided by that block), and a `Mux` block (you will see why in a moment). Find them in the Simulink library browser, or by opening the proper library icons in the Simulink library window, whichever you prefer to use or is available in your version of the software. Let us also put in the model window a `Clock` block and a `To Workspace` block (just to demonstrate their use), and an `Auto-scale Graph` scope.

- As indicated earlier, to get the `Auto-scale Graph` scope, type either `auto_scale_scope` or `auto_scale_graph` in the MATLAB command window.[6] Then, put in your model a copy of the scope that is in a little window that appears on your computer screen. That little window can be closed after you have copied that scope to your model since it will no longer be needed. At this point, your Simulink model window may look like Fig. A.9.

---

[6] From now on, it will be assumed that you have set your MATLAB path as indicated in Section A.2.

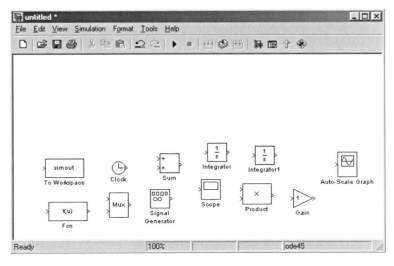

Figure A.9 Blocks that will be used to prepare a Simulink model to integrate Eq. (A.5.2).

Rearrange, in an organized manner, the blocks in the Simulink model window, and start to connect them to make a Simulink model that will solve the given differential equation. Start with the integrators. We already know that we need two of them because Eq. (A.5.2) is a second-order differential equation. We also know that $\theta(t)$ is the output of the integrator that integrates $\dot{\theta}(t)$. To obtain $\dot{\theta}(t)$ we need to integrate $\ddot{\theta}(t)$. Since we want to solve Eq. (A.5.2), we must obtain the expression for $\ddot{\theta}$ from that equation and input that expression, which is $\ddot{\theta} = M(t)/(mL^2) - [c/(mL^2)]\dot{\theta} - [k_1/(mL^2)]\theta - [k_3/(mL^2)]\theta^3 - (g/L)\sin\theta$, to an integrator. Therefore, to accomplish all this, connect the two integrators in series and input the right-hand side of the expression for $\ddot{\theta}$ to the integrator whose input is still unconnected, as indicated in the following. A Sum block with five inputs (double click the Sum block to be able to change the default list of signs to + − − − −) may be used to generate the sum of each one of the five terms that appear in the expression for $\ddot{\theta}$. After you connect the two integrators and reconfigure the Sum block, your model window may look like Fig. A.10.

To make the block diagram easier to read, the outputs of the two integrators shown in Fig. A.10 were labeled as theta and theta_dot (for $\dot{\theta}$), respectively, and the input of one of the integrators was labeled theta_ddot (for $\ddot{\theta}$). Also, some of the blocks shown in Fig. A.10 were resized for convenience. The instructions to do the steps that led to the arrangement shown in Fig. A.10 are listed here.

- To move a block from one location to another, click the block with the left button of the mouse, hold down the button, and drag the block to the desired location.

518 APPENDIX A Simulink Tutorial and a Brief Introduction to MATLAB

Figure A.10 The Simulink model window, after some of the blocks were rearranged.

- To resize a block, first "select" it by clicking it with the left button of the mouse. Then, with the mouse button released, move the mouse pointer to one of the corners of the block. When the pointer changes shape, hold down the mouse button and drag the mouse until you get the desired size for the block you are resizing. Single click the mouse in a blank space of the model window to "deselect" that block.
- To connect the output of a block to the input of another block, drag the mouse pointer from the desired output to the desired input (or vice versa), with the mouse button pressed. If you release the mouse button before you are too close to the input of the other block, that end of the line will be unconnected (that is how the line connected to the output of the integrator named Integrator was placed in Fig. A.10; it will be connected to the scope later).
- To insert text in a desired location of the Simulink model window, place the mouse pointer at that location, double click the mouse button, and start typing. When you finish typing, just single click the mouse elsewhere in the model window so that you can do something else with your model. Texts can be modified by using the mouse and the keyboard in the same way that is done in word processing software. The entire block of text can be moved to another desired location in the model window by simply dragging it with the mouse to that location.

One possible Simulink model for solving the problem at hand is obtained by connecting the remaining blocks as shown in Fig. A.11 (specific details follow).

Figure A.11 Simulink model (a fully functional block diagram) for solving Eq. (A.5.2).

The `animate_nbars` block (with its required inputs) is placed in the model so that one can actually see how the system moves as its differential equation of motion is being solved (instead of only generating plots of the solution) (specific details follow). The input to the scopes are connected to a `Mux` block [whose inputs are labeled `M(t)/(mL^2)` and `theta`] so that the plots for $M(t)/(mL^2)$ and $\theta(t)$ are superimposed on the same graph (as specified in the problem description). Notice that the three `Gain` blocks in the Simulink model shown in Fig. A.11 are "flipped" (i.e., rotated by 180°) so that the model is nicely organized. Also, their gains were changed to the values shown in the figure, and their labels were changed to `c/(mL^2)`, `k1/(mL^2)`, and `k3/(mL^2)` for easier identification.

- To flip a block, select that block with the mouse, and then select `Flip Block` in the `Format` menu of the model window. Experiment with some of the other choices in the `Format` menu, such as `Flip Name` and `Rotate Block`, to see what they do.
- To change the label of any block, single click the mouse at the block's label and insert or delete text as you would do in any word processor. If you highlight any part of the label (by dragging the mouse over that part with the left button pressed), pressing the `Delete` key on the keyboard will delete that part. Type the desired new label and, when finished, single click the mouse on a blank part of the model window so that you can continue your work.

520 APPENDIX A Simulink Tutorial and a Brief Introduction to MATLAB

- To connect the input of a block to an appropriate line, such as the connection between the input $u(1)$ of the Mux block (i.e., the first input line from the top of the Mux icon), and the output line of the Signal Generator block, place the mouse pointer on the input of that Mux block and drag the mouse to the appropriate line. By appropriately moving the mouse, you can make a 90° angle change of direction to create two orthogonal lines like the ones you see in Fig. A.11.
- If you want to move the location of the intersection of two orthogonal lines, place the mouse pointer at the intersection of the lines and drag the mouse. Experiment with it to familiarize yourself with these operations.

The animate_nbars block generates a drawing based on an x-y coordinate system centered at any desired point and with the $+y$ axis at 90° counterclockwise from the chosen direction for the $+x$ axis as shown in Fig. A.8. The input to the animate_nbars block must be a single multisignal line, carrying four signals for each bar to be drawn; such a line is provided at the output of either a Mux or a MATLAB Function block.

Figures consisting of any number n of bars bar_1, bar_2, ..., bar_n, can be "constructed" using the animate_nbars block. The inputs to the block must be the x and y coordinates of one end of a bar followed by the x and y coordinates of the other end of the same bar, followed by the same information for another bar, etc. To connect any number of bars at some point P, say, simply assign the same x-y coordinates of P to the desired end points of the bars that are to be connected. The first bar that is drawn on the computer screen is red, by default, and the others are blue. Double click the animate_nbars block to specify the axes limits on the drawing, the number of bars to be drawn, and the desired width of each bar. The settings of the animate_nbars block for the model shown in Fig. A.11 are shown in Fig. A.12.

Figure A.12 The settings of the animate_nbars menu for the model of Fig. A.11.

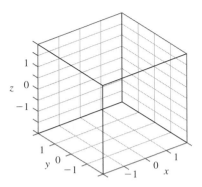

Figure A.13 The x-y-z axes in a three-dimensional animation.

> The `animate3d_nbars` block is for three-dimensional animation, and it requires six inputs for each bar: the x, y, and z coordinates of one end of a bar followed by the x, y, and z coordinates of the other end of the same bar, followed by the same information for another bar, etc. In such an animation, the x, y, and z axes are oriented as illustrated in Fig. A.13. For an example, see the model in Fig. 6.17 (p. 380).

The following are the inputs to the `Mux` block labeled `Mux1` that make up the single multisignal line that is the input of the `animate_nbars` block in Fig. A.11 for showing a single bar to imitate the pendulum of Fig. A.8. The x-y coordinate system used here is the one shown in Fig. A.8, which was chosen to be centered at point O.

Inputs 1 and 2. (Begin counting from the top of the `Mux`.) The constants $x = 0$ and $y = 0$ coordinates of point O (generated with the `Constant` block described in Section A.3)

Inputs 3 and 4. The xP/L $= \sin\theta$ and yP/L $= -\cos\theta$ nondimensional coordinates of point P of bar OP.

Save your model (see the following instructions) using a name of your choice. As mentioned earlier, it is best to work in a directory that you have created for your own work, which is *your working directory,* and not to save anything in the directory where MATLAB and Simulink are installed, for example. Also, it is best to save your work early on, and keep saving it as you make a number of changes, in case your computer crashes and you lose what you have done.

- The first time you save a model, use the mouse to choose `Save As` from the `File` menu that you see when the Simulink model window is activated. A window will then open up showing the directory you are in and the file default name that will be used when you save your model. Change the first part (the part before the file extension which, in Simulink is `mdl`) of the default name to an appropriate name of your liking and save the model. You also have, of course, the option of saving the model in any directory of your choice.

- The next time you save your model (after you make any changes you desire) use Save, instead of Save As. The model shown in Fig. A.11 was saved as model1_appendix_a, and it is the file model1_appendix_a.mdl on the enclosed CD. The name that is chosen automatically appears in the model window.

A Scope and an Auto-Scale Graph scope (either one is enough, of course) were put in the Simulink model shown in Fig. A.11 to illustrate the difference between their displays. Two To Workspace blocks were included in the model to demonstrate, in Section A.7, some MATLAB commands to generate plots involving variables in case you want Simulink to send them to the MATLAB workspace during the simulation. The default name for the output variable of a To Workspace block appears in the icon for that block when it was copied from its Simulink library. If you want to change that name in the copy of the block you put in your Simulink model, double click that copy of the block and enter the desired new name in the window that will appear on the screen. In the model shown in Fig. A.11, the names of the variables that are sent to the MATLAB workspace were chosen to be time_t and input_and_theta. Double click the To Workspace block and choose Array in the Save format option that appears in the menu for that block. When this is done, the variable it sends to the MATLAB workspace is a matrix. In the model1_appendix_a model shown in Fig. A.11, the variables time_t and input_and_theta are, respectively, $m \times 1$ and $m \times 2$ column matrices, where m is the number of time steps that was used by the software during the simulation.

- Double click the Signal Generator to choose one of the available waveforms (sine wave, square wave, etc.), its amplitude, and its frequency. Let us choose a square wave, with the parameter named amplitude set to -1 and frequency equal to 0.2 Hz (or cycles per second). This will generate a square wave that starts at $+1$, changes to -1 at the end of half a period,[7] when $t = \frac{1}{2} \times (1/0.2) = 2.5$ time units, and then starts to repeat itself.

- Double click the Auto-Scale Graph scope and see that the number of storage points is set to a high value for this simulation (the default set by the author is 5000, which is high enough for this example). The menu that appears is shown in Fig. A.14. The line color and type for the plots displayed by that auto_scale_scope may also be chosen in the menu

---

[7] If the parameter "amplitude" were set to $+1$ (which would really be the amplitude of the wave since *amplitude* is, by definition, a positive quantity), the square wave generated by the Signal Generator block would start at -1, and then change to $+1$ at the end of half a period. This "flipping" may or may not be changed by The MathWorks, Inc., in other versions of its software.

Figure A.14 The menu window associated with the `Auto-Scale Graph` scope.

shown in Fig. A.14. The default is `'k-/k--/r-.'`, etc., which means *black and solid line* (`k` is the symbol used for black) for the first plot, *black and dashed line* for the second plot, *red and dash-dot line* for the third plot, etc.[8] When the *sample time* Δt of the `Auto-Scale Graph` scope is set to -1 (the default), that scope inherits the actual value of the sampling time from the simulation. Change the sample time Δt to a small positive value if the plot consists of widely spaced connected points. Also, experiment using smaller values of the maximum step size in the `Simulation` $\rightarrow$ `Parameters` menu to get a plot whose quality is acceptable to you. The simulation runs more slowly as values are decreased.

■ Before a simulation is started, the desired initial conditions $\theta(0)$ and $\dot{\theta}(0)$ should be entered for the integrators. This is done by double clicking the integrators in the Simulink model and then entering the desired values in the appropriate place of a window that opens up. Since the pendulum is driven by an external moment, let us use the default values $\theta(0) = 0$ and $\dot{\theta}(0) = 0$ for both integrators in the model of Fig. A.11. Also, several simulation parameters need to be set in the `Simulation` $\rightarrow$ `Parameters` menu.

---

[8] Other available colors are `c` (cyan), `m` (magenta), `y` (yellow), `g` (green), `b` (blue), and `w` (white), and other line types are : (dotted), * (asterisk), o (circle), + (for using a plus sign to indicate a point).

APPENDIX A Simulink Tutorial and a Brief Introduction to MATLAB

Figure A.15 A typical Simulation → Parameters menu.

This is done by placing the mouse pointer on the word Simulation in the model window, pressing and holding the mouse button, selecting Simulation Parameters, and then releasing the mouse button. A typical menu that appears is shown in Fig. A.15.

As shown in Fig. A.15, there are several self-explanatory choices one may make in the Simulation → Parameters menu, including the choosing of an integration algorithm in the part named Solver in that window. Click the Solver tag, if it is not already showing up by default, to choose the integration stop time, the numerical algorithm for the integration, the maximum step size, etc. The default choice for the integration algorithm is probably set as a variable step ode45 algorithm, which is a Runge-Kutta algorithm that is a good choice for most problems. Set the Stop time to a desired value (say, 70, to stop the simulation at the end of 70 time units, or to a very large number such as 9999 so that you stop the simulation as you watch it). Set the Max step size either to auto (if that is not the default) or experiment by using a small number (say, 0.02), and close that window. After the simulation stops, it is wise to experiment with different values of the relative and/or absolute error tolerances of the simulation to see if the results are noticeably affected. Details about other possible choices in the Simulation → Parameters menu, such as the Output options (which is a choice that is not changed in any of the models in the book), may be found in your Simulink manual. Their default choices are used in this book.

■ Double click the Scope icon to make its display visible during the simulation. Notice that the Auto-Scale Graph scope described on

p. 511 opens up automatically when the simulation starts. We are now ready to start the simulation.

- To start the simulation, simply select `Start` (the word `Start` changes to `Stop` after the simulation starts) in the `Simulation` menu of the model window. To stop the simulation, either select `Stop` in the `Simulation` menu or choose a reasonable value for the `Stop time` in the `Simulation` → `Parameters` menu.

Figure A.16 shows the input $M(t)/(mL^2)$ and the solution $\theta(t)$ in the same plot, with the first two line types in the `Auto-Scale Graph` scope set to `'k--/k-'`. Figure A.16a is the `Scope` display that you will see after you double click that block to open it and click on the binoculars icon for automatically scaling the plot. The icons to the left of the binoculars are for zooming in on desired parts of a plot. Figure A.16b is the display of the `Auto-Scale Graph` scope, which opens automatically when the simulation is started, automatically adjusts the scaling of the axes as the simulation progresses, and also has a zooming feature. The figure may also be rotated, which is a desirable feature for viewing three-dimensional figures from different angles, and labels may be easily inserted as indicated in the following.

The display of the `Auto-Scale Graph` scope is in a *figure window* (although the figure number is not displayed), which is a window where annotations such as a title (the name of the simulation model automatically appears as the default for the title) and axes labels can be added to it. The texts `c/(mL^2)=0.005`, `k₁/(mL^2)=0.2`, etc., that appear in the figure window were inserted simply by clicking on the letter A next to the ↖ (the `Edit Plot` arrow), clicking the mouse at the place where they appear, and then typing `c/(mL^2)=0.005`, etc.

(a)

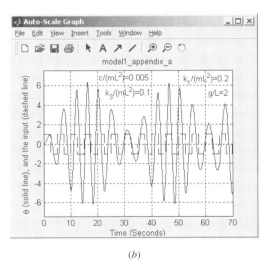

(b)

Figure A.16 Output of the simulation for $c/(mL^2) = 0.005$, $k_1/(mL^2) = 0.2$, $k_3/(mL^2) = 0.1$, and $g/L = 2$, as shown (a) by the `Scope`, and (b) by the `Auto-Scale Graph` scope.

526 APPENDIX A   Simulink Tutorial and a Brief Introduction to MATLAB

The subscripted symbol k_1 in $k_1/(mL^2) = 0.2$, is created by typing k_1. Typing k_1, with the backslash, would generate k_1. Try that! When you finish typing, click the mouse elsewhere in the figure window to leave the area where you typed. A block of text can be moved elsewhere by simply dragging it with the mouse. Greek letters are created by typing \alpha, \beta, etc.

To change the font type and size of the text you inserted, click the ↖ to highlight it, click the text to highlight it, and then look for a figure window edit menu that says Current Object Properties or similarly descriptive words. Explore the menu that opens up, and you should find the options for rotating text, changing fonts, etc. The ↖ acts like a toggle switch.

A number of changes in the figure window can also be made by opening up an edit menu for that window that says Axes Properties or similarly descriptive words. Such a menu also opens by highlighting the ↖ and double clicking any of the axes in the figure. Look at it to see the available choices.

Try running the simulation with different values of the parameters $c/(mL^2)$, $k_1/(mL^2)$, $k_3/(mL^2)$, and g/L in Eq. (A.5.2) and by changing the frequency of the square wave. Also, run the simulation with a sinusoidal input to see how the model responds to such an input.

There are several ways one can construct a Simulink block diagram model for solving the same problem. Figure A.17 shows a more compact model for solving Eq. (A.5.2), making liberal use of function blocks. An XY Graph Scope that draws figures in a *figure window* that can be edited (see its description in

Figure A.17 Another Simulink model for solving Eq. (A.5.2).

Section A.3) was also added. When needed, drag a copy of it from the models shown in this appendix to models you construct.

The "construction style" in Fig. A.17 is the one that is used in most of the models included on the enclosed CD. The expression for $\ddot{\theta}$ obtained from Eq. (A.5.2) is in the `Fcn` function block whose label has been changed to `theta_ddot` in Fig. A.17. That expression in that block is (double click the block to see it)

`u(1)-u(2)*u(7)-u(3)*u(6)-u(4)*u(3)^3-u(5)*sin(u(6))`

Also, the use of the `MATLAB Fcn` block is demonstrated in Fig. A.17 by using it to provide the four inputs that are needed by the `animate_nbars` block. Those inputs are typed in that block as (double click the block to see it)

`[0,0,sin(u),-cos(u)]`

and the output dimension of the block is set to 4 in the same window where the above expression is entered. One input for that model is the function $M(t)/(mL^2) = \sin^2(t)$ that is seen in the `Fcn` block whose label has been changed to `M(t)/(mL^2)`. To specify a different expression, double click that block and type the one you would like. The other inputs to the model are the four coefficients in Eq. (A.5.2), generated with `Constant` blocks described in Section A.3. The model shown in Fig. A.17 is the file `model2_appendix_a.mdl` on the enclosed CD. To use it, either type `model2_appendix_a` in the MATLAB command window or type `simulations_menu` to select it as described in the first paragraph of Section A.4.

Sometimes, one wants to group a collection of blocks in a Simulink model into a single block (which is called a *subsystem* in Simulink), especially when a model starts to become too big to be displayed in a clear manner on the computer screen. The creation of a subsystem is illustrated with the aid of the model shown in Fig. A.17. To do this, first select an area of the model containing the blocks to be grouped together by dragging the mouse, with its left button pressed, and then releasing the mouse button after the desired area is defined, as illustrated in Fig. A.18a. One may also select one by one the blocks and lines in that area by holding down the `Shift` key in the keyboard while clicking the blocks and lines to be selected with the mouse. Then, use the mouse to choose `Create Subsystem` in the `Edit` menu of the model window. The result is a model similar to the one shown in Fig. A.18b (after resizing and rearranging some of the blocks and lines to make the diagram look nicer!).

If you double click the subsystem, you will see its block arrangement; its input labels `In1`, `In2`, etc., and also any output label, may be changed to any other strings of your choice.

A.6 Invalid Loops in Simulink

There are several numerical methods for finding approximate solutions to a differential equation. They include Runge-Kutta methods and Adams corrector-predictor, and the integrators in Simulink implement a number of them. Say the

528 APPENDIX A Simulink Tutorial and a Brief Introduction to MATLAB

(a)

(b)

Figure A.18 (a) How to make a subsystem, and (b) the subsystem that was made.

variable to be solved for is $y(t)$. The numerical methodology involves calculating values for $y(t)$ at discrete values of t separated by small time steps Δt. The time steps may be fixed or variable for better error control. A value of y at time t is obtained by using previously calculated values of y at $t - \Delta_1 t, t - \Delta_2 t$, etc. Thus, in a block diagram, the output of an integrator at time t is used for calculating the new output at the next time step. Such loops are valid, of course.

There are blocks that simply generate an output at time t using values of their inputs at the same time t, such as function blocks and sum blocks. Thus, using the output of such blocks as inputs to the same blocks creates an invalid loop. In Simulink, such loops are called *algebraic loops,* and Simulink generates an error message if any of such loops are found in a block diagram model. If that happens when you construct a model, you need to look at your model more closely and make appropriate changes in order to eliminate such loops. Hopefully that will not happen to you, but if it does, be patient if your model is somewhat complicated and does not work the first time you make a change.

A.7 Brief Introduction to MATLAB

It is easy to do calculations using MATLAB using the self-explanatory mathematical operators +, -, *, / (for division), and ^ (for raising a variable to a power). Just like in mathematics, = is used for assigning a value to a variable. Enter the commands in the MATLAB command window, and then press the Return key. For example, typing a=2 creates the variable *a* with a value of 2 assigned to it. MATLAB displays a=2 after the Return key is pressed and released. Using a semicolon after an assignment suppresses the display. As in any mathematical package, MATLAB also has commands such as *sqrt,* for extracting square roots as in sqrt(23.2), exp [for exponentiation as in exp(3.2)], log (for the natural logarithm), log10 (for the logarithm in base 10), sin, cos, and asin (for arcsin). MATLAB also has predefined variables such as pi (for π), i (and j) for $\sqrt{-1}$, and inf for infinity.

For any MATLAB command, type help command_name, where command_name is the name of the command for which you want to get more information from MATLAB. The commands det, trace, and eig are for obtaining the determinant, the trace, and the eigenvalues of a matrix, respectively. Try help det, help trace, and help eig to see their descriptions. Also try help NaN and help format to see what they are.

The syntax for specifying a list of values starts and ends with a square bracket. The elements of the list are separated either by a comma (a space is also acceptable) or by a semicolon. The list stands for a row matrix when a comma (or a space) is used and for a column matrix when a semicolon is used. For each of the following command lines, input the commands shown and then press the Return key at the end of the line for MATLAB to process the commands. Explanations are provided following each command line.

    a=2, b=5, c=[2, 4, 8], d=[5; 10; 20]

The quantities *c* and *d* are a 1 × 3 row matrix and a 3 × 1 column matrix, respectively. The commas between the commands are for separating the commands. Semicolons can also be used. With a comma, MATLAB will display the result of the operations, while a semicolon inhibits the display (which is always stored in memory). No comma is necessary after the statement `d=[5; 10; 20]` since there is no subsequent command in that line. A semicolon after the assignments `a=2` and `d=[5; 10; 20]`, for example, would inhibit the display of the obvious result. The commands `max` and `min` give the maximum and minimum elements of either an *n* × 1 or a 1 × *n* list. Try `max(c)` and `min(c)`. You should get 8 and 2, respectively. Type `help max` and `help min` to get additional information.

The elements of the list *c* are $c(1) = 2$, $c(2) = 4$, and $c(3) = 8$. If you type `c(2)`, for example, in the MATLAB command window, MATLAB returns the value of that element in the list *a*. For a list $m = [2, 4, 8, 10]$, for example, if you type `m(2:4)`, MATLAB returns elements 2 to 4. Try it.

`d=a/b, e=1/a, f=c/a, g=a*b, h=a^3, k=a^(-1)`

This line illustrates the division, multiplication, and exponentiation operations.

`clear, a=[2,4,8], b=[1,2;5,8], c=a.^(-1), d=b.^(-1)`

The command `clear` erases variables and functions from memory. Type `help clear` for additional information. The operation `c=a.^(-1)` produces an array *c* whose elements are the inverse of each element of the array *a*. The operator `.^` is a special operator for arrays (see the footnotes in Example 1.13.3 for a typical use). Operators for arrays start with a . (as in `.^`, `.*`, and `./`). The operation `1./a` is the same as `a.^(-1)`, but the operation `1/a` does not make sense because *a* is not a scalar. To invert a square matrix such as matrix *b*, type either `b^(-1)` or `inv(b)` (type `help inv` for additional information).

`a=[1; 2], c=b^(-1), d=[5, 6, 7; 8, 9, 10]`

Type this command line and look at the results. Here, *a* is a 2 × 1 column matrix, *b* and *c* are 2 × 2 matrices, and *d* is a 2 × 3 matrix. The matrix *b* was defined earlier. MATLAB will tell you what variables are in memory if you type `who` in the command window (type `help who` for additional information). Check to see that *c* is the inverse of *b* by multiplying them either as $c*b$ or $b*c$. Also, try the command `b*a` and see if it returns the result you expect.

The command `size` gives a 1 × 2 row containing the number of rows and columns in a matrix. For example, `size(a)` displays [1 2] without the brackets. The command `length` gives the maximum element that is returned by `size`; `length(a)`, for example, displays 2. You might need to use these commands if you send to the MATLAB workspace a matrix generated with Simulink in order to determine how many elements were generated.

The elements of d are $d(1,1) = 5$, $d(1,2) = 6$, etc. If you type d(2,1), for example, MATLAB returns the value of that element of d, which is 8. The command d(:,2) returns $d(1,2)$ and $d(2,2)$, i.e., all the elements in column 2 of d. The command d(1,:) returns all the elements of row 1 of d. The value of any element of a matrix can be changed by assigning a new desired value to that element. Type, for example, the command d(2,2)=15. By typing d, MATLAB will display your new matrix d. To transpose a matrix, use a single quote, as in a' or d'. Try it.

Use MATLAB to obtain the solution to the following equation, with $\theta = \pi/6$ rad.

$$\begin{bmatrix} \cos\theta & \sin\theta \\ -\sin\theta & \cos\theta \end{bmatrix} x = \begin{bmatrix} 3 \\ 4 \end{bmatrix}$$

Did you get the following output from MATLAB?

```
ans =
    0.5981
    4.9641
```

Since this means

$$x = \begin{bmatrix} 0.5981 \\ 4.9641 \end{bmatrix}$$

try to write the command so that MATLAB returns

```
x =
    0.5981
    4.9641
```

A list or row matrix with equally spaced elements may be generated using colons, as in theta=0:0.1:5. The first and last values are the first and last elements of the list, and the middle one is the step size. The default step size is 1 if it is not specified, as in theta=0:5.

The rest of this appendix is for those readers who either prefer or need to generate plots with MATLAB, instead of using Simulink. It is also needed for superimposing XY plots (when X is not time t) as long as the XY Graph scope described on p. 511 in Section A.3 does not have such a capability. Plots are generated with the plot (for two-dimensional plots) and plot3 (for three-dimensional plots) commands. For additional information, type help(plot) and help(plot3). As an example, type the following commands to generate three arrays θ (theta in the command line), x, and y (the semicolon after each command is used for suppressing the long displays since each array has 1001 elements).

theta=0:0.01:10;x=40*sin(theta);y=theta.^2.*sin(theta);

The basic plot command plot(theta,x) plots x versus θ. To put a grid on the plot, type grid. Such a command acts like a toggle switch. Typing grid again removes the grid.

The command plot(theta, x, theta, y) plots x versus θ, and y versus θ.

The more sophisticated command

```
plot(theta, x, '-', theta, y, '- -' )
```
plots x versus θ with solid lines and y versus θ with dashed lines.

For even more sophistication, the command

```
plot(theta, x, 'g-', theta, y, 'r--' )
```
plots x versus θ with solid green lines and y versus θ with dashed red lines.[9]

Annotations, such as a title, axes labels, and texts, can be easily inserted in a figure window as indicated on p. 525. If one wishes to do so, they may also be inserted using MATLAB commands. For example, the command (with a string required to be enclosed in single quotes)

```
title('40sin\theta\_and\_\theta^2sin\theta versus \theta')
```

inserts the title $40\sin\theta\_and\_\theta^2\sin\theta$ versus θ. Notice the need for using _ for generating an underscore as in $40\sin\theta\_and$. The command

```
title('40sin\theta and \theta^2sin\theta versus \theta')
```

would insert the title $40\sin\theta$ and $\theta^2\sin\theta$ versus θ.

The commands `xlabel` and `ylabel` are for inserting labels on the x and y axes, respectively. Try the following commands where, as indicated, the strings must be enclosed in single quotes. Notice the options for specifying the font size and the font rotation (in degrees).

```
xlabel('\theta (in radians)');
ylabel('two functions of \theta,''fontsize'',14,''rotation'', 90');
```

Try the following command, which places text in a figure in a location chosen with the mouse after the `Return` key is pressed. Type `help gtext` in the MATLAB command window to see a more detailed explanation.

```
text1=gtext('c/(mL^2)=0.005');
```

■ By assigning a name to a text command, such as `text1=gtext('c/(mL^2)=0.005')`, instead of simply typing `gtext('c/(mL^2)=0.005')`, MATLAB creates the "object" called `text1` whose properties (such as the text location, text font, and text string) can be changed by using the MATLAB `set` command (for more details, see your MATLAB manual and/or type `help set` in the MATLAB command window).[10] For example, after the string `c/(mL^2)=0.005` is placed in the figure window with the mouse by using the `gtext` command, the string can be changed to `the value of c/(mL^2) is 0.005` with the MATLAB command

```
set(text1,'string','the value of c/(mL^2) is 0.005')
```

---

[9] Other available colors and line types are indicated in the footnote on p. 523.

[10] If you do not need to manipulate any of the properties of the text `c/(mL^2)=0.005` created with the command `gtext`, just type the command as `gtext('c/(mL^2)=0.005')`.

- To remove the `text1` string that you placed on your graph, you can type in the MATLAB command window either the command `reset(text1)` or the command

    `set(text1,'string', ' ')`

- To see the properties of the object `text1`, type `get(text1)` in the MATLAB command window. You will then see that the property called "string" (which is case insensitive and, thus, can be referred to by using lowercase, uppercase, or a mix of both letters) is set to `c/(mL^2)=0.005` after you use the command `text1=gtext('c/(mL^2)=0.005')`.

- To get a list of all the properties of an object (such as a figure) that can be changed, type in the MATLAB command window `get object`, with the word `object` replaced by the so-called handle of the object (such as the string `text1` created earlier). If interested, see your MATLAB manual for these more advanced optional features.

As another example of manipulating plots, consider the plots involving the variables `input_and_theta` and `time_t` that were sent to the MATLAB workspace by the Simulink models presented in this appendix. Recall that the variable `input_and_theta` in the simulation diagram shown in Fig. A.11 is a two-column matrix. The first column is the function $M(t)/(mL^2)$ generated by the `Signal Generator` block, and the second column is the solution $\theta(t)$ of Eq. (A.5.2). The following MATLAB commands (with comment lines for explanations) illustrate the creation and manipulation of plots of the elements of that matrix. Anything written after a `%` in a line is treated by MATLAB as a comment. For the commands `ginput` and `zoom`, type `help ginput` and `help zoom` to see what they do.

```
newfig=figure    % this generates a new figure window and brings it to the
                 % front on the screen
plot(time_t, input_and_theta(:,2))

% the preceding command is to plot only θ(t) versus t in the new figure
grid    % puts a grid on the plot; to remove the grid, type grid again
title('Example in Appendix A')
xlabel('time t (in seconds)');
ylabel('\theta(t)');
%
% try the following command: axis([0 10 0 1]);
% and you will see the axes limits in the plot change to:
% 0 to 10 for t; 0 to 1 for the vertical axis
% Also try the following commands:
% ginput(1)
% zoom
% to print the plot, use the mouse and select Print in the File menu.
```

As a final example, we illustrate how to superimpose several plots of $\dot{x}$ versus x, with $\dot{x}$ obtained from the following equation for different values of the constant C.

$$\frac{\dot{x}^2}{2} - \cos(x) = \text{constant} \triangleq C$$

The plot is similar to the one in Fig. 1.13 (p. 44). Try putting some arrows in the plot (see Fig. 1.13) using the ↗ that is seen in Fig. 1.13. Also try to label the y axis as $\dot{x}$, and change the font size. This can be done by separately inserting an x and a dot (·), and moving them in the figure window with the mouse. The plot created with the following commands is shown in Fig. A.19.

```
phase_plot=figure   % to generate a new figure window
hold on; % to "hold" the plots on the window, thus superimposing new plots
x=-2*pi:0.01:2*pi; C=-0.8, 0.62, 1, 1.205]; steps=length(C);
for k=1:steps,    % starting a loop; type help(for) for details
y=sqrt(2*( C(k)+cos(x) ));   plot(x, y, x, -y);
end
axis([-2*pi, 2*pi, -2.5, 2.5]);
gtext('C=-0.8'); gtext('C=0.62'); gtext('C=1'); gtext('C=1.205');
```

To add a line passing through the points $(-2\pi, 0)$ and $(2\pi, 0)$, use the command `line([-2*pi, 2*pi], [0, 0])`. To put a box around the figure, type `box`. The `box` command is a toggle switch.

Figure A.19 Phase plot $\dot{x}$ versus x created with the given commands.

APPENDIX B

Notes on Sequential Rotations, Angular Velocity, and Acceleration

B.1 Sequential Rotations and Angular Velocity

The orientation of a line in three-dimensional space can be described by a set of angles, as first illustrated by Fig. 1.1 on p. 2. Unless the rotations are about the same axis (a case which can be replaced by a single rotation about the same line), the final orientation of the line depends on the sequence of the rotations when the rotations are finite instead of infinitesimal. This is illustrated in Fig. B.1 where three orthogonal axes (labeled x_1, x_2, and x_3) and a line (in boldface) are shown.

The line is originally aligned with the x_3 axis. Imagine, now, three axes x'_1, x'_2, and x'_3 fixed to the line and initially aligned with x_1, x_2, and x_3, respectively. Two sequences of 90° successive rotations are then performed, both involving a clockwise rotation θ_1 about one of the primed axes and a counterclockwise[1] rotation θ_2 about the new orientation of a different primed axis. It is suggested that you use either a book or a pencil to reproduce the demonstration illustrated in Fig. B.1.

The top part on the right of Fig. B.1 shows what happens to the line when the sequence consists of a clockwise rotation $\theta_1 = 90°$ about $x'_1 = x_1$, followed by a counterclockwise rotation $\theta_2 = 90°$ about the updated position of the x'_2 axis. The line that started along the horizontal ends up vertical. However, if the sequence of rotations is now reversed, i.e., if it consists of a counterclockwise rotation $\theta_2 = 90°$ about x_2, followed by a clockwise rotation $\theta_1 = 90°$ about the updated position of the x'_1 axis, the line ends up pointing toward you.

---

[1] There is no special relevance about choosing the rotations to be clockwise or counterclockwise. These choices were made arbitrarily.

APPENDIX B Notes on Sequential Rotations, Angular Velocity, and Acceleration

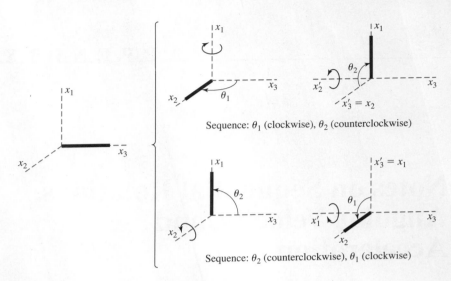

Figure B.1 Spatial orientation of a line using two sequential rotations.

This demonstration serves the purpose of showing that finite rotations cannot be represented as vectors because, as seen from Fig. B.1, the final direction of the line that underwent the rotation is not the vector sum of two rotations. If it were, the order of the rotations would be irrelevant as in the sum of two vectors.

Only infinitesimally small rotations (again, *not* finite rotations) can be represented as vectors because, as it turns out, the order of rotations for such a case is irrelevant. As you will see in the rest of this section, such a fact plays a decisive role in the determination of angular velocity when any line rotates in any manner in three-dimensional space.

To show the commutative property of infinitesimal rotations, consider a vector $\vec{r}$, of constant magnitude, undergoing a rotation $\Delta\theta_1$ about a line AA, followed by a rotation $\Delta\theta_2$ about line BB as shown in Fig. B.2. The unit vectors $\hat{a}$ and $\hat{b}$ are parallel to lines AA and BB, respectively. After the first rotation $\vec{r}$ becomes $\vec{r}_1$, and then it becomes $\vec{r}_2$ after the second rotation. The following expressions can be written by inspection of Fig. B.2:

$$\vec{r}_1 \approx \vec{r} + (\Delta\theta_1)\hat{a} \times \vec{r} \tag{B.1.1}$$

$$\vec{r}_2 \approx \vec{r}_1 + (\Delta\theta_2)\hat{b} \times \vec{r}_1 = \vec{r} + (\Delta\theta_1)\hat{a} \times \vec{r} + (\Delta\theta_2)\hat{b} \times [\vec{r} + (\Delta\theta_1)\hat{a} \times \vec{r}]$$

$$= \vec{r} + [(\Delta\theta_1)\hat{a} + (\Delta\theta_2)\hat{b}] \times \vec{r} + (\Delta\theta_1)(\Delta\theta_2)\hat{b} \times (\hat{a} \times \vec{r}) \tag{B.1.2}$$

When the rotations are infinitesimally small, the second-order infinitesimal quantity in the last term of Eq. (B.1.2) can be neglected. By denoting the infinitesimal rotations as $d\theta_1$ and $d\theta_2$, respectively, we then have in the limit as $\Delta t \to 0$,

$$\vec{r}_2 = \vec{r} + [(d\theta_1)\hat{a} + (d\theta_2)\hat{b}] \times \vec{r} \tag{B.1.3}$$

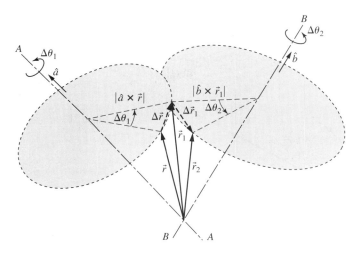

Figure B.2 Commutativity of infinitesimal rotations.

If the sequence of the rotations is reversed, so that the rotation $\Delta\theta_2$ about axis BB precedes the rotation $\Delta\theta_1$ about axis AA, the expression for the vector $\vec{r}_2$ in the limit as $\Delta t \to 0$ now becomes

$$\vec{r}_2 = \vec{r} + [(d\theta_2)\hat{b} + (d\theta_1)\hat{a}] \times \vec{r} \qquad \text{(B.1.4)}$$

> A comparison of Eqs. (B.1.3) and (B.1.4) reveals that the infinitesimal rotation $(d\theta_1)\hat{a}$ and $(d\theta_2)\hat{b}$ satisfy the commutative property $(d\theta_1)\hat{a} + (d\theta_2)\hat{b} = (d\theta_2)\hat{b} + (d\theta_1)\hat{a}$. Since angular velocity is calculated based on infinitesimal changes of angles, angular velocity is indeed a vector, and an angular velocity vector can be formulated even when rotation angles about several axes in space are used to describe the orientation of a line or of a body in space. The angular velocity (relative to the reference frame in which the rotations are referred to) associated with the infinitesimal rotation vector $(d\theta_1)\hat{a} + (d\theta_2)\hat{b}$ is simply $\vec{\omega} = \dot{\theta}_1\hat{a} + \dot{\theta}_2\hat{b}$.

B.2 A Note on the General Form of Acceleration Expressions

The problem of determining the absolute acceleration of a point, expressed in a rotating unit vector triad, was addressed in Sections 2.6 (p. 108) and 2.7 (p. 112) of Chapter 2. To summarize the general procedure, consider a particle P moving in space and a triad $\{x\} = \{\hat{x}_1, \hat{x}_2, \hat{x}_3\}$ that is rotating with angular velocity $\vec{\omega}$ relative to inertial space (i.e., relative to a nonrotating frame of reference). The chosen triad $\{x\}$ may be fixed, for example, to a rotating piece of machinery. In general, both the magnitude and the direction of $\vec{\omega}$ will be changing with time. Let us now go through the entire process of formulating the expression for the absolute velocity and absolute acceleration of P, expressing both vectors in the triad $\{x\}$ that one chooses to work with. The first thing to do is to start with the absolute

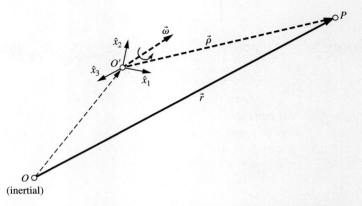

Figure B.3 Using an intermediate point when expressing the position vector of a point P.

position vector for P, i.e., with the vector $\vec{r} = \overrightarrow{OP}$ from an inertial point O to P as shown in Fig. B.3.

In general, for any particular problem under consideration, it is more convenient to express the vector $\vec{r}$ in terms of several "intermediate vectors" such as from O to a point O' whose motion is known, then from O' to some other point O_2 whose motion relative to O' is also known, etc. To account for such situations, let us then express the vector $\vec{r}$ as follows (see Fig. B.3):

$$\vec{r} = \overrightarrow{OO'} + \overrightarrow{O'P} \triangleq \overrightarrow{OO'} + \vec{\rho} \tag{B.2.1}$$

By expressing the position vector $\vec{\rho}$ of P relative to O' in terms of the chosen unit vectors $\{\hat{x}_1, \hat{x}_2, \hat{x}_3\}$, a general expression of the form $\vec{\rho} = \rho_{x_1}\hat{x}_1 + \rho_{x_2}\hat{x}_2 + \rho_{x_3}\hat{x}_3$ is obtained, as seen throughout this book. The absolute velocity of P is then obtained as

$$\vec{v} = \frac{d\vec{r}}{dt} = \frac{d\overrightarrow{OO'}}{dt} + \frac{d\vec{\rho}}{dt} = \vec{v}_{O'} + \left(\frac{d\vec{\rho}}{dt}\right)_{\{x\}} + \vec{\omega} \times \vec{\rho} \tag{B.2.2}$$

where $\vec{v}_{O'} \triangleq d\overrightarrow{OO'}/dt$ denotes the absolute velocity of point O' and $[d(\ )/dt]_{\{x\}}$ denotes the time derivative without taking into account the rotation of the unit vectors, such as $(d\vec{\rho}/dt)_{\{x\}} = \dot{\rho}_{x_1}\hat{x}_1 + \dot{\rho}_{x_2}\hat{x}_2 + \dot{\rho}_{x_3}\hat{x}_3$. The absolute velocity $\vec{v}_{O'}$ is known, and that is the reason the vector $\vec{r}$ was expressed as the sum of two parts. From Eq. (B.2.2), the expression for the absolute acceleration of P is then obtained as

$$\vec{a} = \frac{d\vec{v}}{dt} = \frac{d\vec{v}_{O'}}{dt} + \left[\left(\frac{d^2\vec{\rho}}{dt^2}\right)_{\{x\}} + \vec{\omega} \times \left(\frac{d\vec{\rho}}{dt}\right)_{\{x\}}\right]$$
$$+ \left[\frac{d}{dt}(\vec{\omega} \times \vec{\rho})_{\{x\}} + \vec{\omega} \times (\vec{\omega} \times \vec{\rho})\right]$$

$$= \vec{a}_{O'} + \left(\frac{d^2\vec{\rho}}{dt^2}\right)_{\{x\}} + 2\vec{\omega} \times \left(\frac{d\vec{\rho}}{dt}\right)_{\{x\}} + \left(\frac{d\vec{\omega}}{dt}\right)_{\{x\}} \times \vec{\rho} + \vec{\omega} \times (\vec{\omega} \times \vec{\rho})$$

(B.2.3)

where $\vec{a}_{O'} \triangleq d\vec{v}_{O'}/dt$ denotes the absolute acceleration of point O'.

Equation (B.2.3) discloses that when a position vector $\overrightarrow{OP}$ is "split," for convenience, into two parts as $\overrightarrow{OO'} + \overrightarrow{O'P}$, where the motion of O' is known, and the vector $\vec{\rho} = \overrightarrow{O'P}$ is expressed in a rotating unit vector triad $\{x\}$ whose absolute angular velocity is $\vec{\omega}$, the resulting expression that is obtained for the absolute acceleration contains five different types of terms.

The term $(d^2\vec{\rho}/dt^2)_{\{x\}}$ in Eq. (B.2.3) is interpreted as the acceleration of point P as seen by an observer fixed to the rotating triad $\{x\}$. The term $2\vec{\omega} \times (d\vec{\rho}/dt)_{\{x\}}$ is known as the *Coriolis acceleration* of P relative to O', while the term $\vec{\omega} \times (\vec{\omega} \times \vec{\rho})$ is known as the *centripetal acceleration* of P relative to O'. The negative of the centripetal acceleration is known as the *centrifugal acceleration*. The Coriolis acceleration is named after the French engineer Gaspard-Gustave Coriolis (1792–1843).

The equations that govern the motion of a particle P of mass m are simply obtained by equating the respective components of each side of the equation $\vec{F} = d(m\vec{v})/dt$, where $\vec{F}$ is the vector sum of all the *actual forces* (i.e., contact forces and field forces) acting on the particle, such as gravitational forces and reaction forces due to contact with another body (which includes friction). Such forces are obtained directly from a free-body diagram for the particle.

One may also write Newton's second law for a particle as $\vec{F} - m\,d\vec{v}/dt = \vec{F} - m\vec{a} = 0$, refer to the term $-m\vec{a}$ as an *inertial force,* and then proceed to include in the free-body diagram the five types of inertial forces as obtained from Eq. (B.2.3). Two such forces have received specific names. They are the *Coriolis force,* which is the term $-2m\vec{\omega} \times (d\vec{\rho}/dt)_{\{\hat{x},\hat{y},\hat{z}\}}$, and the *centrifugal force,* which is the term $-m\vec{\omega} \times (\vec{\omega} \times \vec{\rho})$ (the negative of the centrifugal force is known as the *centripetal force*). Clearly, such terms depend on the particular choice one makes for the $\{\hat{x}_1, \hat{x}_2, \hat{x}_3\}$ triad in which the acceleration is being expressed in and are nonzero only if the chosen triad is rotating. In this author's experience of teaching dynamics to both undergraduate and graduate students for many years, this approach is more confusing than helpful for most undergraduate students trying to understand the fundamentals of dynamics. In addition, it also requires that the person formulating the differential equations of motion of a system either memorize the exact form of the five terms in Eq. (B.2.3) or always consult a reference to use that "formula." To do so is unnecessary, and it is avoided in this book, as seen by the many examples that were presented. For each problem you have to solve, all you have to do to obtain the absolute acceleration $\vec{a}$ is to take the time derivative of the absolute velocity vector $\vec{v}$ (which, in turn, is obtained by taking the time derivative of the absolute position vector $\vec{r}$).

APPENDIX C

Properties of the Inertia Matrix of a Body

As seen in Chapter 6, the determination of the directions of the central principal axes of inertia of a body, and of the corresponding principal moments of inertia, involves solving an eigenvector–eigenvalue problem of matrix algebra. The solution to such a problem was presented in Chapter 6. The inertia matrix of a body is always real and symmetric, and this appendix presents important properties associated with such a matrix. The following quantities were used in Chapter 6.

- $[I_C]$. The inertia matrix of the body about a set of three orthogonal axes x, y, and z centered at the center of mass C of the body
- $(\hat{x}, \hat{y}, \hat{z} \triangleq \hat{x} \times \hat{y})$. Three orthogonal unit vectors parallel to the x-y-z axes
- $\vec{\omega} = \omega_x \hat{x} + \omega_y \hat{y} + \omega_z \hat{z}$. Absolute angular velocity of the body
- λ. Any of the three principal moments of inertia of the body

The determination of the direction of the principal axes and the principal moments of inertia leads to the matrix equation $[I_C]\underline{\omega} = \lambda \underline{\omega}$, i.e.,

$$\begin{bmatrix} I_{xx} - \lambda & I_{xy} & I_{xz} \\ I_{yx} & I_{yy} - \lambda & I_{yz} \\ I_{zx} & I_{zy} & I_{zz} - \lambda \end{bmatrix} \begin{bmatrix} \omega_x \\ \omega_y \\ \omega_z \end{bmatrix} = \begin{bmatrix} 0 \\ 0 \\ 0 \end{bmatrix} \quad \text{(C.1)}$$

This yields the following polynomial equation for λ:

$$\det\left(\begin{bmatrix} I_{xx} - \lambda & I_{xy} & I_{xz} \\ I_{yx} & I_{yy} - \lambda & I_{yz} \\ I_{zx} & I_{zy} & I_{zz} - \lambda \end{bmatrix}\right) \triangleq P(\lambda) \doteq -(\lambda^3 - a_2 \lambda^2 + a_1 \lambda - a_0) = 0$$

$$\text{(C.2)}$$

APPENDIX C Properties of the Inertia Matrix of a Body

where

$$a_2 = I_{xx} + I_{yy} + I_{zz} \tag{C.3}$$

$$a_1 = \begin{vmatrix} I_{xx} & I_{xy} \\ I_{yx} & I_{yy} \end{vmatrix} + \begin{vmatrix} I_{xx} & I_{xz} \\ I_{zx} & I_{zz} \end{vmatrix} + \begin{vmatrix} I_{yy} & I_{yz} \\ I_{zy} & I_{zz} \end{vmatrix} \tag{C.4}$$

$$a_0 = \det([I_C]) \tag{C.5}$$

Equation (C.2) admits three solutions $\lambda = \lambda_1$, $\lambda = \lambda_2$, and $\lambda = \lambda_3$.

The following properties of the inertia matrix $[I_C]$ are obtained from Eqs. (C.1) to (C.5).

- Since the polynomial $P(\lambda)$ given by Eq. (C.2) can also be written in terms of the principal moments of inertia λ_1, λ_2, and λ_3, which only depend on the mass distribution of the body, as

$$\begin{aligned} P(\lambda) &= -(\lambda^3 - a_2\lambda^2 + a_1\lambda - a_0) \\ &= -(\lambda - \lambda_1)(\lambda - \lambda_2)(\lambda - \lambda_3) \\ &= -[\lambda^3 - (\lambda_1 + \lambda_2 + \lambda_3)\lambda^2 + (\lambda_1\lambda_2 + \lambda_1\lambda_3 + \lambda_2\lambda_3) - \lambda_1\lambda_2\lambda_3] \end{aligned} \tag{C.6}$$

it follows that the three coefficients a_2, a_1, and a_0 do not depend on the orientation of the orthogonal axes $\{x, y, z\}$ that are centered at the center of mass C of the body. Therefore, we conclude that the trace (which is the sum of the diagonal elements of the inertia matrix), the determinant, and the sum of the principal 2 × 2 minor determinants [which are the determinants shown in Eq. (C.4)] of the inertia matrix of a body referred to any set of orthogonal axes $\{x, y, z\}$ fixed to the body are invariant upon rotation of such axes. Notice that one such conclusion has already been obtained in Section 6.3.1 [see Eq. (6.3.11), p. 341]. The existence of such invariants is useful, for example, for checking the results of the calculations involved in the determination of the principal moments of inertia.

- As mentioned in Chapter 6, the principal axes of inertia of a body corresponding to distinct eigenvalues of the inertia matrix are orthogonal to each other. To prove this statement, consider two distinct eigenvalues λ_1 and λ_2 of the inertia matrix $[I_C]$ and the corresponding eigenvectors $\underline{\omega}_1 = [\omega_{1x}, \omega_{1y}, \omega_{1z}]^T$ and $\underline{\omega}_2 = [\omega_{2x}, \omega_{2y}, \omega_{2z}]^T$, respectively. We can then write the following relations:

$$[I_C]\underline{\omega}_1 = \lambda_1 \underline{\omega}_1 \tag{C.7}$$

$$[I_C]\underline{\omega}_2 = \lambda_2 \underline{\omega}_2 \tag{C.8}$$

By premultiplying Eq. (C.7) by the transpose of $\underline{\omega}_2$, and Eq. (C.8) by the transpose of $\underline{\omega}_1$, the two results may be combined as

$$\underline{\omega}_2^T [I_C] \underline{\omega}_1 - \left(\underline{\omega}_1^T [I_C] \underline{\omega}_2 \right)^T = \lambda_1 \underline{\omega}_2^T \underline{\omega}_1 - \left(\lambda_2 \underline{\omega}_1^T \underline{\omega}_2 \right)^T$$

or
$$\underline{\omega}_2^T [I_C] \underline{\omega}_1 - \underline{\omega}_2^T [I_C]^T \underline{\omega}_1 = (\lambda_1 - \lambda_2) \underline{\omega}_2^T \underline{\omega}_1$$

Since the inertia matrix $[I_C]$ is symmetric, this equation yields

$$(\lambda_1 - \lambda_2) \underline{\omega}_2^T \underline{\omega}_1 = 0 \qquad (C.9)$$

This equation immediately discloses that $\underline{\omega}_2^T \underline{\omega}_1 = 0$ if $\lambda_1 \neq \lambda_2$. Since $\underline{\omega}_2^T \underline{\omega}_1$ is the same as the dot product of the vectors $\vec{\omega}_1 = \omega_{1x}\hat{x} + \omega_{1y}\hat{y} + \omega_{1z}\hat{z}$ and $\vec{\omega}_2 = \omega_{2x}\hat{x} + \omega_{2y}\hat{y} + \omega_{2z}\hat{z}$, Eq. (C.9) discloses that $\vec{\omega}_1$ and $\vec{\omega}_2$ are perpendicular to each other if $\lambda_1 \neq \lambda_2$. If, on the other hand, $\lambda_1 = \lambda_2$, Eq. (C.9) is identically satisfied regardless of the angle between $\vec{\omega}_1$ and $\vec{\omega}_2$.

■ The eigenvalues of the inertia matrix (or, more generally, of any real symmetric matrix), which are the solutions of the polynomial equation (C.2), are always real. To prove this statement, assume that Eq. (C.1) is satisfied by a complex eigenvalue λ, to which corresponds a complex eigenvector $\vec{\omega}$, which we represented in matrix form as $\underline{\omega}$. Since any complex solution to Eq. (C.1) must occur in complex conjugate pairs, we must have, with a * denoting the complex conjugate,

$$[I_C] \underline{\omega} = \lambda \underline{\omega} \qquad (C.10)$$

$$[I_C] \underline{\omega}^* = \lambda^* \underline{\omega}^* \qquad (C.11)$$

By premultiplying Eq. (C.10) by $\underline{\omega}^{*T}$ and Eq. (C.11) by $\underline{\omega}^T$, the result may be combined as

$$\underbrace{\underline{\omega}^{*T} [I_C] \underline{\omega} - (\underline{\omega}^T [I_C] \underline{\omega}^*)^T}_{=0} = \lambda \underline{\omega}^{*T} \underline{\omega} - (\lambda^* \underline{\omega}^T \underline{\omega}^*)^T$$

or

$$0 = (\lambda - \lambda^*) \underline{\omega}^{*T} \underline{\omega} \qquad (C.12)$$

The product $\underline{\omega}^{*T} \underline{\omega}$ in Eq. (C.12) is real and positive since the product of a complex number $a + \sqrt{-1}b$ by its complex conjugate $a - \sqrt{-1}b$ is equal to $a^2 + b^2$. Therefore, Eq. (C.12) is satisfied only if

$$\lambda^* = \lambda$$

which implies that the eigenvalue λ is real.

■ Another property of the inertia matrix, which helps one visualize the geometrical significance of its elements, is evidenced when we calculate the moment of inertia of the body about an arbitrary axis AA that passes through the origin of the chosen (x, y, z) coordinate system as shown in Fig. C.1.

Referring to Fig. C.1, the angles θ_x, θ_y, and θ_z are the angles line AA makes with the x, y, and z axes, respectively. Therefore, the following expressions can be written for the unit vector $\hat{a}$ along AA and for the

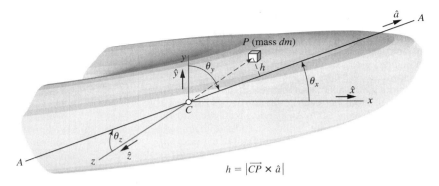

Figure C.1 Figure for calculating the moment of inertia I_{AA} of a body.

distance h from an arbitrary element of mass dm of the body (located at point P shown in the figure, with coordinates x_P, y_P, and z_P).

$$\hat{a} = \hat{x}\cos\theta_x + \hat{y}\cos\theta_y + \hat{z}\cos\theta_z \tag{C.13}$$

$$\begin{aligned}h &= |(x_P\hat{x} + y_P\hat{y} + z_P\hat{z}) \times \hat{a}| \\ &= \big[(y_P^2 + z_P^2)\cos^2\theta_x + (x_P^2 + z_P^2)\cos^2\theta_y + (x_P^2 + y_P^2)\cos^2\theta_z \\ &\quad - 2x_P y_P\cos\theta_x\cos\theta_y - 2x_P z_P\cos\theta_x\cos\theta_z \\ &\quad - 2y_P z_P\cos\theta_y\cos\theta_z\big]^{1/2}\end{aligned} \tag{C.14}$$

Making use of Eq. (C.14) and the definitions of moments and products of inertia I_{xx}, I_{xy}, etc., introduced in Chapter 6, the moment of inertia of the body about the AA axis is then obtained as follows:

$$\begin{aligned}I_{AA} &= \int_{\text{body}} h^2\, dm \\ &= I_{xx}\cos^2\theta_x + I_{yy}\cos^2\theta_y + I_{zz}\cos^2\theta_z + 2I_{xy}\cos\theta_x\cos\theta_y \\ &\quad + 2I_{xz}\cos\theta_x\cos\theta_z + 2I_{yz}\cos\theta_y\cos\theta_z\end{aligned} \tag{C.15}$$

It is interesting to notice that the following identities involving the direction cosines of line AA may be written if a vector $\overrightarrow{OQ} \triangleq x\hat{x} + y\hat{y} + z\hat{z}$ along that line is introduced in the formulation:

$$\cos\theta_x = \frac{x}{|\overrightarrow{OQ}|} \qquad \cos\theta_y = \frac{y}{|\overrightarrow{OQ}|} \qquad \cos\theta_z = \frac{z}{|\overrightarrow{OQ}|} \tag{C.16}$$

It is also interesting to note that Eq. (C.15) becomes

$$I_{xx}x^2 + I_{yy}y^2 + I_{zz}z^2 + 2I_{xy}xy + 2I_{xz}xz + 2I_{yz}yz = I_{AA}|\overrightarrow{OQ}|^2 \tag{C.17}$$

If the length $|\overrightarrow{OQ}|$ is chosen so that $|\overrightarrow{OQ}| = 1/\sqrt{I_{AA}}$, Eq. (C.17) is seen to represent an ellipsoid in the x-y-z space, centered at point C. Such an

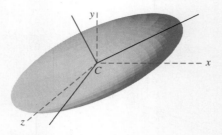

Figure C.2 The inertia ellipsoid of a body.

ellipsoid and the x-y-z axes are shown in Fig. C.2. It is called the *ellipsoid of inertia* of the body. Notice that the quantity $|\overrightarrow{OQ}| = 1/\sqrt{I_{AA}}$ is the distance from the origin C of the x-y-z axes (which is also the center of the ellipsoid of inertia) to any point on that ellipsoid.

The ellipsoid of inertia is the geometric representation of the inertia matrix of a body, and its principal axes, shown as three solid lines in Fig. C.2, are the principal axes of the body. The elongated axis shown in that figure corresponds to the principal axis of minimum inertia of the body.

In general, the inertia ellipsoid is not an ellipsoid of revolution. It is an ellipsoid of revolution only if two of the principal moments of inertia of the body are equal to each other. The ellipsoid of inertia of a pencil-shaped body is thin and elongated, while that of a body with three equal principal moments of inertia is a sphere. Such bodies are said to have spherical symmetry whether they are a sphere or a cube.

Two different bodies that have the same inertia ellipsoid (or, equivalently, the same inertia matrix) experience the same rotational motion when they are subjected to the same external moment. They will also experience the same translational motion if they have the same mass and are subjected to the same resultant force.

APPENDIX D

Suggested Computer Lab Assignments

Lab Assignment 1

Part 1

1. (10 points) A small projectile P is launched from the earth's surface with a velocity $\vec{v}_0$ that makes an angle α degrees with the horizontal and whose magnitude is $v_0 = |\vec{v}_0|$ m/s. Because of air resistance, the projectile is subjected to a *drag force;* the drag force is proportional to the square of the projectile's speed [with the proportionality coefficient being equal to c N/(m²/s²)] and is always directed opposite to the projectile's velocity vector $\vec{v}$. Use rectangular coordinates, and obtain the differential equations that govern the motion of the projectile by treating the earth's gravitational force of attraction as a constant, with the acceleration of gravity being equal to $g = 9.81$ m/s². Show the free-body diagram for P, properly labeling all quantities, and show all the steps of the formulation that lead to the differential equations, as done in the book. What are the initial conditions for your differential equations?

2. (40 points) A small block P of mass m is connected to a fixed pin O by a linear spring of stiffness k N/m and unstressed length L_u m. The block is constrained to move along a horizontal guide as shown in Fig. D.1. Friction between the block and the guide is of the linear viscous type, with the friction coefficient equal to c N/(m/s). The perpendicular distance from O to the guide is h m. The plane defined by the guide and the spring is vertical. Obtain the differential equation of motion of block P in terms of the variable $x(t)$ shown in the figure (for this, clearly show the free-body diagram for P, properly labeling all quantities, and show all the steps of the formulation that lead to the differential equation, as done in this book). For $c = 0$, $L_u = h$, $k/m = 0.5$, $x(0) = 5h$, and $\dot{x}(0) = Ah$ (with $A = 10$ s^{-1}), determine (analytically) the maximum and the minimum values of x/h.

Figure D.1 The system model for Lab Assignment 1.

Also, generate a plot of x_{max}/h and x_{min}/h versus $\dot{x}(0)/h$ for $c = 0$, $L_u = h$, $k/m = 0.5$, $x(0) = 5h$, and $\dot{x}(0) = Ah$, for $-10 \leq A \leq 10$.

Part 2 (50 points)

Integrate numerically the differential equation of motion for problem 2 in part 1 with $c/m = 0$ and $c/m = 0.3$, and plot x/h versus time, and $\dot{x}/h$ versus x/h, for both cases. What are the maximum and minimum values of x/h for both values of c/m? Compare the analytical answer obtained in part 1 (for $c = 0$) with that obtained from the numerical integration. If you use Simulink for the numerical integration, include the `animate_nbars` block in your diagram so that you can actually see how the block moves. Choose your own appropriate value for the final integration time.

Carefully observe the motion on the computer screen and the plots (including the $\dot{x}/h$ versus x/h plot), and include your observations in the comments section of your report. Your report should be complete enough to allow another engineer to reproduce your work.

Lab Assignment 2

> Virtually every dynamical system in nature is nonlinear; i.e., its motion is governed by one or more differential equations that are nonlinear. One characteristic of nonlinear systems is that they may exhibit more than one equilibrium.
>
> A system is said to be in *equilibrium* when the variable that is used to describe its motion remains constant for all times. For example, a pendulum (which everyone is familiar with) exhibits two physically distinct equilibria: one corresponding to the pendulum bar hanging down in the vertical position, and the other corresponding to the upside-down position. If the pendulum is displaced, even slightly, from the equilibrium position, a motion will occur. Consider now a small displacement from the equilibrium. What happens is that the motion will always stay in a small neighborhood of the equilibrium for the case of the hanging-down equilibrium; but that will never happen for the upside-down equilibrium. In technical language, the former equilibrium is said to be *stable,* while the latter is said to be *unstable*. For the stable equilibrium, the perturbed motion is an oscillation about that equilibrium, while for the unstable

equilibrium it is not (i.e., the motion that started in the neighborhood of the unstable equilibrium never stays in that neighborhood). When damping is added to the pendulum, the motion eventually dies out, getting farther and farther from the unstable equilibrium and tending, as time increases, to the stable equilibrium of the pendulum. The equilibrium that was classified as stable is now classified as *asymptotically stable*.

In this lab assignment you will play the role of an engineer investigating the motion of a mechanical device. The system is shown in Fig. D.2. It consists of a bar of length L m, small blocks pinned to the ends of the bar, a spring, a vertical track, and a horizontal track. The connecting bar is very light, and the effect of its mass on the motion of the system can be neglected.

The motion of the system is to be investigated using the model shown in Fig. D.2b. Block A (with mass m_1 kg) slides on the vertical track, and block B (with mass m_2 kg) slides on the horizontal track. Friction in the system is viscous, with friction coefficients c_1 N/(m/s) between A and the vertical track, and c_2 N/(m/s) between B and the horizontal track. Block B is attached to a horizontal linear spring of stiffness k N/m and unstretched length L_u m; the constant distance between the vertical track and end O of the spring is equal to $L \cos \theta_u + L_u$, where θ_u is the value of the angle θ when the spring is unstretched.

1. (50 points) Clearly show all the free-body diagrams, symbols, unit vectors, etc.
   a. Show that the differential equation of motion for the system is

$$\left(\cos^2 \theta + \frac{m_2}{m_1} \sin^2 \theta\right)\left(\frac{L}{g}\ddot{\theta}\right) + \left(\frac{m_2}{m_1} - 1\right)(\sin \theta)(\cos \theta)\left(\frac{L}{g}\dot{\theta}^2\right) + \cos \theta$$

$$+ (c_1^* \cos^2 \theta + c_2^* \sin^2 \theta)\left(\sqrt{\frac{L}{g}}\dot{\theta}\right) + \frac{kL}{m_1 g}(\cos \theta_u - \cos \theta) \sin \theta = 0$$

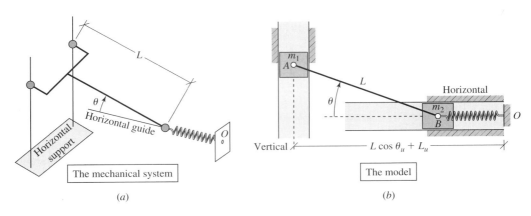

Figure D.2 (a) The system and (b) its model for Lab Assignment 2.

where $c_1^* = (c_1/m_1)\sqrt{L/g}$ and $c_2^* = (c_2/m_1)\sqrt{L/g}$ are nondimensional damping coefficients.

b. Show that by working with a nondimensional time τ, defined as $\tau = t\sqrt{g/L}$, the differential equation in item a becomes

$$\left(\cos^2\theta + \frac{m_2}{m_1}\sin^2\theta\right)\frac{d^2\theta}{d\tau^2} + \left(\frac{m_2}{m_1} - 1\right)(\sin\theta)(\cos\theta)\left(\frac{d\theta}{d\tau}\right)^2 + \cos\theta$$

$$+ (c_1^*\cos^2\theta + c_2^*\sin^2\theta)\frac{d\theta}{d\tau} + \frac{kL}{m_1 g}(\cos\theta_u - \cos\theta)\sin\theta = 0$$

c. Using the differential equation of motion for the system, obtain the equation that will allow you to determine the equilibrium values θ_e of the angle θ in terms of the nondimensional parameter $m_1 g/(kL)$, and θ_u.

2. (50 points) This is the numerical investigation part.
a. For $\theta_u = 50°$, $55°$, and $60°$, plot (on the same graph paper and properly identifying the curves) θ_e (in degrees) versus the nondimensional parameter $m_1 g/(kL)$ for θ_e in the range $-70° \leq \theta_e \leq 70°$. *Note:* If you do these plots by hand, they must be presented in a neat professional manner. For $\theta_u = 55°$, what are the values of θ_e for $m_1 g/(kL) = 0.15$?
b. Integrate numerically the nondimensional differential equation of motion for the system with $m_2/m_1 = 1$, $\theta_u = 55°$, and $m_1 g/(kL) = 0.15$. Do this for $c_1^* = c_2^* = 0$ and for $c_1^* = c_2^* = 0.2$. For each case, show plots of the trajectories $\theta' = d\theta/d\tau$ versus θ. If you use Simulink for the numerical integration, include the `animate_nbars` block in your block diagram so that you can actually see what the system is doing. In each plot mark with an × the location of each equilibrium point found in item a for $\theta_u = 55°$. For the undamped case, plot all curves on the same graph paper (this will allow you to better understand what goes on). Use the following sets of initial conditions to generate your plots:

$$[\theta(0) = 24°, \theta'(0) = 0] \quad [\theta(0) = 23°, \theta'(0) = 0],$$
$$[\theta(0) = 23°, \theta'(0) = 0.1] \quad [\theta(0) = 50°, \theta'(0) = 0]$$
$$[\theta(0) = -50°, \theta'(0) = 0].$$

Carefully observe the motion and the plots on the computer screen, and include your observations about the motions and about stability (indicating which equilibrium solutions you think are stable and which ones you think are unstable) in the comments section of your report. Put a little arrow on each θ' versus θ trajectory to indicate how it is actually traced (i.e., clockwise or counterclockwise) during the motion.

Lab Assignment 3

The system shown in Fig. D.3 consists of a bar of length $2L$, pinned at a fixed point O, a motor located at O, two equal lumped masses (each of mass m kg), and two linear springs (each of stiffness k N/m and unstressed length L_u. Each of the lumped masses is attached to the end of the bar, and the bar is approximated as massless. The springs are connected as shown, where points P and Q are fixed. The springs are unstretched when $\theta = 0$. The motor applies a moment M_{motor} to the bar, and there is viscous friction at the pin contact at O, with friction coefficient c N·m/(rad/s). The motion takes place in the horizontal plane.

1. (50 points) Clearly show all the free-body diagrams, symbols, unit vectors, etc.

    a. Obtain the differential equation of motion for the system, and then show that, in terms of a nondimensional time $\tau = t\sqrt{k/m}$ (where t is dimensional time), the differential equation is

    $$\frac{d^2\theta}{d\tau^2} + c^*\frac{d\theta}{d\tau} + \left(\sin\theta - \frac{h}{L}\cos\theta\right)$$
    $$\times \left[1 - \frac{1}{\sqrt{1 - 2(L/h)\sin\theta + 2(L/h)^2(1-\cos\theta)}}\right] = \frac{M_{\text{motor}}}{2kL^2}$$

    In this differential equation, c^* is a nondimensional viscous damping coefficient. What is the relation between c^* and c?

    Note: You can use either one of four ways to formulate the differential equation of motion: (1) directly from Newton's second law, (2) the moment equation (which is a consequence of Newton's second law), (3) the approach based on $dW = d(T)$ (which is also a consequence of Newton's second law), or (4) Lagrange's equation

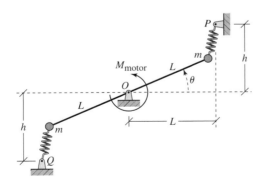

Figure D.3 The system model for Lab Assignment 3.

presented in Chapter 7. Each one of these approaches requires a different amount of effort on your part to arrive at the differential equation of motion.

b. With $M_{motor} = 0$, determine *all* the equilibrium values θ_e of the angle θ for $h/L = 2$.

c. One of the equilibrium values should be $\theta_e = 0$. Show that for very small angles θ (so that $\sin\theta \approx \theta$ and $\cos\theta \approx 1$, and products of small quantities, such as θ^2, are neglected), a Taylor series expansion about $\theta = 0$ of the nonlinear terms in the differential equation yields the following linearized differential equation of motion:

$$\frac{d^2\theta}{d\tau^2} + c^*\frac{d\theta}{d\tau} + \theta = \frac{M_{motor}}{2kL^2}$$

For $M_{motor} = 0$ and $c^* = 0$, what is the period of these small motions?

2. (50 points) Integrate numerically the nondimensional differential equation of motion for the system with $M_{motor} = 0$, $c = 0$, $h/L = 2$, and starting the system from rest. For $\theta(0)$ in the range $(0, 60]$ degrees, generate an *accurate* plot of the period of the oscillation versus the initial condition $\theta(0)$ in degrees (the period should be the vertical axis of your plot). Describe in your report how you gathered the data to generate such a plot, and how you plotted the data. How does your plot compare with your answer for the period for item 1c? Also, show a typical plot of $d\theta/d\tau$ and a typical plot of θ versus τ. If you use Simulink for the numerical integration, you can include the `animate_nbars` block in your diagram to help you visualize the motion.

Lab Assignment 4

An earth observation satellite is put in a circular orbit at an altitude of $h = 400$ km above the earth's surface. After a few orbits, one of the attitude control jets of the satellite, and the control system that controls the direction of the thrust produced by that jet, malfunctions. Onboard instruments determine that the jet is firing continuously, that it is applying a constant thrust to the satellite that is always pointing toward the center of the earth, and that there is no moment about the satellite's center of mass. The magnitude of that applied thrust is equal to mF_0 N, where m is the mass of the satellite and $F_0/g = 1/128$ (where $g = 9.81$ m/s^2).

Part 1

1. (25 points) Determine, analytically (and accurately), the maximum and the minimum altitude (i.e., distance from the earth's surface) of the satellite center of mass after the malfunction. Does the satellite crash on the earth's surface?

2. (25 points) For what range of the nondimensional parameter g/F_0 will the satellite crash on the earth's surface?

Note: Neglect the effect of the earth's atmosphere in your calculations, and treat the earth as a homogeneous sphere of radius $R_E = 6378$ km. Clearly show any free-body diagram you need to have, symbols, unit vectors, etc. Also, make sure that you clearly show all the steps that lead to the differential equations of motion, and all the steps of your analysis.

Part 2 (50 points)

Integrate numerically the differential equations that govern the motion of the center of mass of the satellite. Plot, for $F_0/g = 1/128$ and for two more values of F_0/g of your choice, with one (not both) of the values causing the satellite to crash on the earth's surface:

a. The trajectory of the satellite's center of mass (i.e., a plot of the position of the center of mass, using rectangular coordinates with the origin at the center of the earth);
b. The altitude of the center of mass of the satellite (in kilometers) versus time.

How much time elapsed (measured from the instant the malfunction occurred) until the satellite crashed? How much (in minutes) is the orbital period when the satellite does not crash? Include in your report your comments and conclusions, and an explanation of how the crash could have been avoided if the satellite designer had limited the value of F_0/g to be outside the range you determined in item 2 in part 1.

Lab Assignment 5

End B of a thin uniform bar AB of length L m and mass m kg is connected to a linear spring of unstressed length L_u m and stiffness k N/m. That end is free to slide along a frictionless horizontal floor, while the other end of the spring is pinned to a fixed point (see Fig. D.4). End A of the bar is free to slide along a frictionless vertical wall. The motion of the system takes place in a vertical plane,

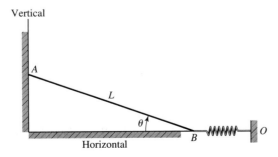

Figure D.4 The system model for Lab Assignment 5.

and the spring is unstressed when $\theta = \theta_u$ degrees. As seen in Fig. D.4, the floor and the vertical wall limit the angle θ to the range $0 \leq \theta \leq 90°$.

1. **(15 points)** Show that the differential equation of motion for the system is (see the note at the end of this assignment)

$$\frac{d^2\theta}{dt^2} + \frac{3g}{L}\left[\frac{1}{2}\cos\theta + \frac{kL}{mg}(\cos\theta_u - \cos\theta)\sin\theta\right] = 0$$

2. **(5 points)** For $\theta_u = 50°$ and $kL/(mg) = 5$, determine all the physically possible equilibrium values θ_e for the angle θ (this can be done using a hand-held calculator).

3. **(15 points)** By letting $\theta(t) = \theta_e + \theta_s(t)$, linearize the differential equation for small values of $\theta_s(t)$. Do not use any symbolic computation software to do this. Do it by hand, clearly showing all your steps, thus demonstrating that you have understood how to obtain a Taylor series and how to linearize a differential equation.

4. **(5 points)** Which of the equilibrium values determined in item 2 is stable (i.e., you are to determine for which equilibrium value θ_s is oscillatory)? For $\theta_u = 50°$ and $kL/(mg) = 5$, what is the frequency of the linearized θ_s oscillation about the stable equilibrium? Your answer will be in terms of g/L.

5. **(10 points)** For the initial conditions $\theta(0) = \theta_0$ and $\dot\theta(0) = 0$ obtain the solution for $\theta(t) = \theta_e + \theta_s(t)$ using the linearized differential equation of motion (you are going to use this in item 6).

6. Rewrite the given nonlinear differential equation in terms of the nondimensional time $\tau = t\sqrt{g/L}$ and, with $\theta_u = 50°$ and $kL/(mg) = 5$, numerically integrate the resulting equations (9 points for the integration) to address the following items. If you use Simulink to integrate the equation, include the `animate_nbars` block in your diagram to visualize the motion, if that is helpful to you.

    a. **(9 points)** With $\theta(0) = 43°$, plot $\theta(\tau)$ versus τ [meaning that $\theta(\tau)$ is to be the vertical axis of the plot] for $0 \leq \tau \leq 15$, and then superimpose, with dashed lines, the plot of your solution to the linearized equation for the given value of $\theta(0)$. Carefully observe the motion and/or your plot, and make notes for your report. Notice that if MATLAB is used for superimposing plots, the `hold` and `fplot` commands are convenient (in the MATLAB command window, type `help hold` and `help fplot` for a description of what they do).

    b. **(9 points)** Repeat the work in item a for $\theta(0) = 46°$. Again, carefully observe the motion and your plot, and make notes for your report.

    c. **(9 points)** Repeat the work in item b for $\theta(0) = 51.4°$ and $51.5°$. For this item, use $0 \leq \tau \leq \tau_{max}$, with a value for τ_{max} of your choice. Again, carefully observe the motion and/or your plot, and make notes for your report.

d. (14 points) Prepare a concise but meaningful report with your observations about what happened to the motion in items *a* to *c*. Specifically, you should comment on the following: How the nonlinear and the linearized motions compare in items *a* and *b* (and your reasons why they are what you saw); your explanation of what happened when you used $\theta(0) = 51.4°$; and your explanation of what happened when you used $\theta(0) = 51.5°$.

Note: As you have seen in the book, you can use several ways to formulate the differential equation(s) of motion for a rigid body. They include a combination of Newton's second law and the moment/angular-momentum equation (i.e., the Newton-Euler approach) and Lagrange's equation. Each one of these approaches requires a different amount of effort on your part to arrive at the differential equation(s) of motion. You may choose to use any of them. Clearly show all the free-body diagrams you may need, symbols and unit vectors you choose to work with, etc.

APPENDIX E

Simulink and MATLAB Files on the CD

The enclosed CD contains the files that are mentioned throughout the book. They are listed in this appendix, and may be used in any computer where MATLAB and Simulink are installed. They are portable from one computer to another.

> The easiest and best way to use the files on the enclosed CD is to copy the entire `CdS_Dynamics` folder to a directory of your choice in the hard disk of your computer, and then add that directory to the MATLAB path. The instructions to do this are given in Section A.2 (p. 507). Here, and throughout this book, it is assumed that you have implemented those instructions.

The `.mdl` files listed here are Simulink models, and the `.p` files are compiled MATLAB programs whose coding details are irrelevant to the studying of dynamics. The only user-callable `.p` files are `sgn.p` and `simulations_menu.p` (see Section E.2, p. 558). The `.m` files are *help files* (which are executed by typing `help` before the file name, such as `help animate_nbars`).

To use any of the Simulink models provided with this book, you can type the model name (the part before the `.mdl`) in the MATLAB command window, and the model will appear on your computer screen. However, the easiest way to use them is simply to type `simulations_menu` in the MATLAB command window. The window shown in Fig. E.1a will appear. Then, simply use the mouse to choose a simulation from the pull-down `Simulations_menu` (a portion of which is shown in Fig. E.1b), and that simulation model will open on your computer screen.

The software on the CD was tested with several versions of MATLAB and Simulink including the latest one available at the time of publication of the book. If you have a newer version of MATLAB/Simulink, it may display a warning message saying that the model you are using from the enclosed CD was prepared

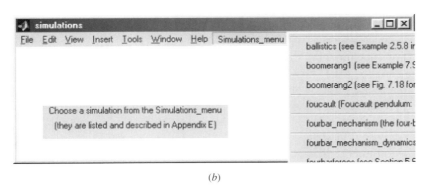

Figure E.1 (*a*) The window that opens when one types `simulations_menu`, and (*b*) part of the simulations menu that contains the list of all the Simulink models mentioned in the book.

with an earlier version of Simulink. That is not a problem at all, as the validity of the enclosed models are not affected. Your software may also display the name of the command you should run (i.e., by typing that command in the MATLAB command window) for updating the model to your later version. However, the model will be automatically updated by just saving it. Saving the model is as simple as answering yes if the software asks you to save it or choosing `File` → `Save` with the mouse (i.e., clicking the word `File` in the model window with the mouse, and then selecting `Save` in the menu that opens up).

E.1 Simulink Models

`animation_blocks.mdl`   Simulink model window with several animation blocks that can be used for animating simulations. The blocks are shown in Fig. A.7 (p. 514). Type `animation_blocks` in the MATLAB command window, and then use the mouse to drag a copy of the desired block to new simulations you decide to make. These blocks are used in many of the Simulink models in this book, which are on the enclosed CD. Double click the animation block in a simulation to choose values for several quantities described in the menu that opens up on the computer screen.

`auto_scale_graph.mdl`   Same as `auto_scale_scope.mdl`.

`auto_scale_scope.mdl`   (See Appendix A.) Simulink model for an Auto-Scale Graph scope that is described on p. 511 of Appendix A and included by special permission and courtesy of The MathWorks, Inc. Type either `auto_scale_scope` or `auto_scale_graph` in the MATLAB command window to get such a scope.

`ballistics.mdl`   (See Example 2.5.8, p. 95.) A simple model for ballistics problems.

`boomerang1.mdl`   (See Example 7.9.1, p. 470.) Simulink model for the x-y motion of the center of mass of a rigid body subjected to a constraint force so that $\dot{y} = f_1(t)\dot{x} + f_2(t)$, where $f_1(t)$ and $f_2(t)$ are given functions of time. The model is shown in Fig. 7.16 (p. 473).

`boomerang2.mdl`   (See Fig. 7.18, p. 474, and the explanation that follows that figure.) This is also a model for solving the problem in Example 7.9.1 (p. 470). It uses the solution strategy presented in the flowchart shown in Fig. 7.19.

`foucault.mdl`   (See Section 8.5.) Simulink model for the Foucault pendulum.

`fourbar_mechanism.mdl`   (See Section 3.2.3, p. 193, for a detailed explanation.) Simulink model of the four-bar mechanism. The model also displays the solution to Eqs. (3.2.1) and (3.2.2) that appear on p. 183 in Chapter 3.

`fourbar_mechanism_dynamics.mdl`   (See Example 7.9.3, p. 477.) Simulink model of the dynamics of the four-bar mechanism when a known torque is applied to one of the links of the mechanism. The model is shown in Fig. 7.23 (p. 480). It uses the solution strategy presented in the flowchart shown in Fig. 7.22.

`fourbarforces.mdl`   (See Section 5.6, p. 313.) Simulink model for determining the forces on each bar of the four-bar mechanism.

`geneva_wheel.mdl`   (See Section 3.3.1, p. 205.) Simulink model for the Geneva wheel mechanism.

`inc_plane_coulomb.mdl`   (See Example 2.5.7, p. 89.) Simulink model for the Coulomb friction case in Example 2.5.7.

`inc_plane_viscous.mdl`   (See Example 2.5.7, p. 89.) Simulink model for the viscous friction case in Example 2.5.7.

`inclined_plane.mdl`   (See Example 2.5.9, p. 96.) Simulink model for a system consisting of a block P_2 sliding on a sliding block P_1. Block P_2 is connected by a spring to block P_1.

`inclined_plane2.mdl`   (See Example 2.5.9.) The only difference between this model and the model `inclined_plane.mdl` is the use of the `MATLAB Function` block to generate the inputs to the `animate_nbars` block, thus resulting in a more compact model.

`model1_appendix_a.mdl`   (See Fig. A.11.) Simulink model for the example in the Simulink tutorial in Appendix A.

`model2_appendix_a.mdl`   (See Appendix A.) A more compact Simulink model for the example in the Simulink tutorial in Appendix A. The model is shown in Fig. A.17.

`parabola_track.mdl`   (See Example 2.8.9, p. 166.) Simulink model for a ball moving on a parabolic track.

`pend_moving_base.mdl`   (See Example 2.7.6, p. 121.) Simulink model for a pendulum mounted on a moving base in Example 2.7.6.

`restricted_3body.mdl`   (See Section 4.6, p. 249.) Simulink model for integrating Eqs. (4.6.13) to (4.6.15) for the restricted three-body problem of space mechanics.

`scotch_yoke.mdl`   (See Section 3.5, p. 218.) Simulink model for the Scotch yoke mechanism.

`sim_model_ex5_3_11.mdl`   (See Example 5.3.11, p. 286.) Simulink model for Example 5.3.11 (an L-bar pendulum mounted on a moving base).

`sim_model_ex5_3_13.mdl`   (See Example 5.3.13, p. 292.) Simulink model for Example 5.3.13.

`slider_crank.mdl`   (See Section 3.4.1, p. 214.) Simulink model for the slider crank mechanism.

`spring_pendulum.mdl`   (See Example 2.7.8, p. 129.) Simulink model for the spring pendulum in Example 2.7.8. The model demonstrates that nonlinearities can have a dominating effect on the motion of a dynamical system, in which case the motion modeled by linearized differential equations are completely different from the actual motion even for very small initial conditions.

`top_dynamics.mdl`   (See Section 6.6.3, p. 371.) Simulink model for a spinning top or gyroscope. The model is shown in Fig. 6.17 (p. 380). It displays a three-dimensional animation of the axial axis of the gyroscope.

`two_body_simul.mdl`   (See Section 4.5, p. 246.) Simulink model for the two-body problem of celestial mechanics (or for satellite problems).

`two_particles_table.mdl`   (See Example 2.7.10, p. 140.) Simulink model for two particles connected by a string that passes through a hole on a horizontal table. One particle moves on the table, while the other moves along the vertical.

E.2 MATLAB Files

`animate_nbars.p` (and `animate_nbars.m`) Type `help animate_nbars`, and detailed information on how to use the `animate_nbars` Simulink block shown in Fig. A.7 (p. 514) will appear on your computer screen. The information is the same as described on p. 520 of Appendix A. This function is called only by the `animate_nbars` block to animate a system of n bars in two-dimensional space.

`animate3d_nbars.p` (and `animate3d_nbars.m`) Type `help animate3d_nbars`, and detailed information on how to use the `animate3d_nbars` Simulink block shown in Fig. A.7 (p. 514) will appear on your computer screen. The information is the same as described on p. 521 of Appendix A. This function is called only by the `animate3d_nbars` block to animate any number of straight bars in three-dimensional space.

`sgn.p` (and `sgn.m`) This is a user-callable function with one argument, and it returns $+1$, -1, or 0, depending on whether the argument is a number that is positive, negative, or zero.

`sfunxyc.p` (and `sfunxyc.m`) A scope file. Included by special permission and courtesy of The MathWorks, Inc.

`sfunystc.p` (and `sfunystc.m`) A scope file. Included by special permission and courtesy of The MathWorks, Inc.

`simulations_menu.p` (and `simulations_menu.m`) This is a user-callable file that makes all the Simulink models on the enclosed CD easily accessible. Its use is described on p. 554 in this appendix.

`two_body_animate.p` (and `two_body_animate.m`) This function is called only by the `two_body_animate` block shown in Fig. A.7 (p. 514). It animates the two-body problem of space mechanics. A detailed analysis of the problem is presented in Section 4.4. Its Simulink implementation is presented in Section 4.5 (see Fig. 4.10, p. 246). When you type `help two_body_animate`, information will appear on your screen directing you to the instructions on p. 248 on how to use the `two_body_animate` Simulink block.

APPENDIX F

Answers to Selected Problems

Chapter 1
1.3 $\pm 14.8°$
1.4 0.093
1.7 $\vec{c} \approx 2.49\hat{x}_1 + 2.07\hat{x}_2 + 3.49\hat{x}_3$
1.17 $\int P(x)\,dx + k/\cos\theta + \text{constant}$
1.18 For $k = 2\Omega^2$, $\theta_{\max} \approx 39.6°$ and $\theta_{\min} = -\theta_{\max}$.
1.20 (a) Either $\ddot{x}_s + k\dot{x}_s/(1-x_e) + (\cos x_e)x_s + kx_s\dot{x}_s/(1-x_e)^2 + \cdots = 0$
or $(1-x_e-x_s)\ddot{x}_s + [(1-x_e)x_s - x_s^2]\cos x_e + k\dot{x}_s + \cdots = 0$
1.23 $x_{\min} \approx -0.48$, $x_{\max} \approx 0.36$
1.28 $x_{\max} \approx 0.174$

Chapter 2
2.3 (a) $\vec{v} \approx 2.297\hat{x} - 4.795\hat{y}$ m/s, and $\vec{a} \approx 27.769\hat{x} + 0.919\hat{y}$ m/s^2;
(b) $\vec{v} \approx 2.297\hat{x}$ m/s and $\vec{a} \approx 26.81\hat{x}$ m/s^2
2.5 (b) 0.866 km^{-1}
2.11 Constant gravitational force approximation: $y_{\max} \approx 157.6$ km after 189 s; using the inverse square law: $y_{\max} \approx 162$ km after 196.4 s
2.17 $\alpha \approx 17.7°$ and also 52.3°
2.19 Range ≈ 934.5 m when $c/m = 10^{-4}$ m^{-1}, and 600.6 m when $c/m = 10^{-3}$ m^{-1}
2.24 (a) 10 m; (b) $t \approx 4.2$ s
2.27 $r_e = 0.2$ m, $\omega \approx 3.5$ rad/s
2.33 $\theta \approx 48.2°$; P strikes the ground at $x \approx 1.125R$ to the right of point O.
2.34 $\theta \approx 43.3°$
2.39 For $c/\sqrt{mk} = 0.2$, $x_{\max} \approx 10.96h$ and $x_{\min} \approx -7.92h$.
2.49 $r_{\max} \approx 60.3$ cm, $r_{\min} \approx 11.2$ cm

Chapter 3
3.4 $\omega_{BC} \approx 0.14$ rad/s (clockwise), $\omega_{CD} \approx 6.94$ rad/s (counterclockwise)
3.8 $\omega_{BC} \approx 3.03$ rad/s (counterclockwise), $\omega_{CD} \approx 4.05$ rad/s (clockwise)
3.18 $\dot{\omega}_{BC} \approx 15.18$ rad/s^2 (counterclockwise), $\dot{\omega}_{CD} \approx 6.72$ rad/s^2 (counterclockwise)

APPENDIX F Answers to Selected Problems

3.22 $\dot{\omega}_{BC} \approx 39.6$ rad/s² (counterclockwise), $\dot{\omega}_{CD} \approx 48.05$ rad/s² (counterclockwise)
3.25 When $\theta_1 = 45°$: $\dot{\theta}_2 \approx 0.108$ rad/s, $\ddot{\theta}_2 \approx -0.35$ rad/s², $\dot{r}_2 \approx -0.88$ m/s, $\ddot{r}_2 \approx -0.728$ m/s²
3.28 $\dot{\theta}_3 = 100$ rpm, $\dot{\theta}_4 \approx 12.57$ rad/s

Chapter 4

4.2 $m \approx 4.362 M_E$, where M_E is the mass of the earth
4.6 $r_{max} \approx 6697$ km, $r_{min} \approx 6678$ km
4.11 (b) For $c_1/m = 0.0001$ s⁻¹ and $c_2/m = 5 \times 10^{-9}$ m⁻¹, $t_{final} \approx 36$ h. The satellite crashes on the earth.
4.12 For $C_D A/m = 5 \times 10^{-3}$ m²/kg, $t_{crash} \approx 8.17$ h.

Chapter 5

5.6 $I_C \approx 49.2$ kg·m²; period ≈ 2.9 s when the plate is hung from B.
5.8 44.9 cm $\leq x(0) \leq 73.2$ cm
5.11 The system does not oscillate when $c/(mL_1^2) = 4.5$ s⁻¹.
5.16 With $c_1^* = c_2^* = 0.1$, $[F_1/(m_2 g)]_{max} \approx 11.35$ and $[F_2/(m_2 g)]_{max} \approx 0.27$.
5.17 $|M_{ext}| \approx 1.886$ N·m (clockwise), $A_{r2} \approx -10.57$ N,
$A_{n2} \approx -4.715$ N $\left(\sqrt{A_{r2}^2 + A_{n2}^2} \approx 11.57 \text{ N}\right)$, $B_{r2} \approx 9.916$ N,
$B_{n2} \approx 4.715$ N $\left(\sqrt{B_{r2}^2 + B_{n2}^2} \approx 10.98 \text{ N}\right)$, $C_{r4} \approx -6.463$ N,
$C_{n4} \approx 1.653$ N $\left(\sqrt{C_{r2}^2 + C_{n2}^2} \approx 6.67 \text{ N}\right)$, $D_{r4} \approx 5.805$ N,
$D_{n4} \approx 0.827$ N $\left(\sqrt{D_{r2}^2 + D_{n2}^2} \approx 5.86 \text{ N}\right)$.
5.21 $|M_{ext}| \approx 15.7$ N·m (clockwise). The magnitude of reaction forces at A, B, and C are, approximately, 60.5, 57.7, and 19.5 N.
5.29 $\omega \approx 18.4$ rad/s
5.34 $\theta_e \approx -3.9°$, $\theta_e \approx 176.1°$, $\omega \approx 2.3$ rad/s
5.35 $\theta_e \approx -23.4°$, $\theta_e \approx 156.6°$, $|\vec{a}_C| = 2g/5$, $\omega \approx 6.9$ rad/s when $b = 35$ cm and ≈ 8.2 rad/s when $b = 25$ cm
5.38 $\omega \approx 2.5$ rad/s

Chapter 6

6.7 $T = k \sin t + $ constant
6.10 (b) $\theta_{max} \approx 8.1°$ when $\Omega^2 L/g = 1$, and $\approx 60.2°$ when $\Omega^2 L/g = 2$. For both cases, $\theta_{min} = -\theta_{max}$.
6.12 $56.5°$
6.14 (b) $\Omega_{max} \approx 28.6$ rad/s ≈ 273 rpm
6.17 $\pm 107.2°$, $\pm 26.4°$, and $\pm 70.6°$, respectively

Chapter 7

7.6 The Hamiltonian $\mathcal{H} = T_2 + U - T_0 = m(\dot{r}^2 - \Omega^2 r^2)/2 + k(\sqrt{r^2 + h^2} - L_u)^2/2$ is constant. For $a = 2$, $r_{max}/h \approx 7.09$; $r_{min} = -r_{max}$.
7.8 The Hamiltonian $\mathcal{H} = T_2 + U - T_0 = m(\dot{x}^2 + \dot{y}^2 - \Omega^2 x^2)/2 + mgy + k(\sqrt{x^2 + (y-h)^2} - L_u)^2/2$ is constant. For the data given in Example 2.8.8, $y_{max} \approx 92.5$ cm, $y_{min} = 0$, $x_{max} \approx 1.67$ m, $x_{min} = -x_{max}$.
7.9 $\Omega_{critical} = \sqrt{2ga}$ from the plots and from the linearized differential equation.

7.18 The Hamiltonian $\mathcal{H} = T_2 + U - T_0$ is the only classical integral of motion for this case, where

$$T_2 = \frac{1}{2}m_3(\dot{r}^2 + r^2\dot{\theta}^2 + \dot{z}^2)$$

$$T_0 = \frac{1}{2}m_3\left[r^2 + \left(\frac{m_2 L}{m_1 + m_2}\right)^2 - \frac{2m_2 Lr\cos\theta}{m_1 + m_2}\right]\Omega^2$$

$$U = -\frac{Gm_1 m_3}{\sqrt{r^2 + z^2}} - \frac{Gm_2 m_3}{\sqrt{L^2 + r^2 + z^2 - 2rL\cos\theta}}$$

The equation $\mathcal{H} = T_2 + U - T_0 =$ constant can be used as one of the differential equations of motion. Lagrange's equation yields Eqs. (4.6.13) to (4.6.15) shown in Section 4.6.

7.25 The Hamiltonian $\mathcal{H} = T_2 + U - T_0$ is constant. This gives

$$\left(\frac{mL^2}{8} + \frac{1}{2}I_x\right)\dot{\theta}^2 - mg\frac{L}{2}\cos\theta - \frac{\Omega^2}{2}(I_y\sin^2\theta + I_z\cos^2\theta)$$

$$-\frac{mL^2}{8}\Omega^2\sin^2\theta = \text{constant}$$

For values of $\theta_{\max}$ and $\theta_{\min}$, see the answer to Problem 6.10.

7.26 There are three classical integrals of motion: the Hamiltonian $\mathcal{H} = I_x(\dot{\theta}^2 + \dot{\phi}^2\sin^2\theta)/2 + I_z(\dot{\psi} + \dot{\phi}\cos\theta)^2/2 + mgL\cos\theta$, the generalized momentum $\partial T/\partial\dot{\psi} = I_z(\dot{\psi} + \dot{\phi}\cos\theta)$, and $\partial T/\partial\dot{\phi} = I_x\dot{\phi}\sin^2\theta + I_z(\dot{\psi}\cos\theta)\cos\theta$. Since there are three variables, these serve as the three differential equations of motion, and they are closer to the solutions θ, ϕ, and ψ than the corresponding second-order differential equations.

Appendix G

Some References for Advanced Studies

The following books are some of the suggestions for more advanced studies. Some of these books are *classics,* including those of Pars, Lanczos, and La Salle and Lefschetz.

1. Pars, P. (1979). *A Treatise on Analytical Dynamics.* Ox Bow Press, Woodbridge, Connecticut (first published in 1965 by Wiley).
2. Rosenberg, R. M. (1977). *Analytical Dynamics of Discrete Systems.* Plenum Press, New York.
3. Lanczos, C. (1966). *The Variational Principles of Mechanics.* University of Toronto Press, Toronto, Canada.
4. Goldstein, H. (1965). *Classical Mechanics.* Addison-Wesley, Reading, Massachusetts.
5. Arnold, R. N. and Maunder, L. (1961). *Gyrodynamics and its Engineering Applications.* Academic Press, New York.
6. Gray, A. (1959). *A Treatise on Gyrostatics and Rotational Motion: Theory and Applications.* Dover, New York.
7. Deimel, R. F. (1950). *Mechanics of the Gyroscope: the Dynamics of Rotation.* Dover, New York.
8. Perry, J. (1957). *Spinning Tops and Gyroscopic Motion: a Popular Exposition of Dynamics of Rotation.* Dover, New York.
9. Meirovitch, L. (1970). *Methods of Analytical Dynamics.* McGraw-Hill, New York.
10. Appell, P. (1924). *Traité de Mécanique Rationnelle,* Vol. 2. Gauthier-Villars, Paris, France.
11. Meirovitch, L. (2001). *Fundamentals of Vibrations.* McGraw-Hill, New York.
12. La Salle, J. and Lefschetz, S. (1961). *Stability by Liapunov's Direct Method with Applications.* Academic Press, New York.
13. Hughes, P. C. (1986). *Spacecraft Attitude Dynamics.* Wiley, New York.

INDEX

A

Absolute
  acceleration, 76
  angular momentum, 74, 228
  angular momentum (rigid body), 269, 336
  linear momentum, 74
  position vector, 14, 75
  velocity, 76
Acceleration of gravity, 16, 19
Algebraic loops, 529
Analytical dynamics, 155
Angular velocity, 110
Apocenter, 238
Apogee, 238
Apohelium, 238

B

Brahe (Tycho), 15

C

Cartesian coordinates, 3
Center
  of gravity, 21, 22
  of mass, 22, 226
  of percussion, 553
Central force problem, 108
Central principal axes of inertia, 351
Centrifugal acceleration, 539
Centripetal acceleration, 539
Chain rule of differentiation, 32
Characteristic equation, 38, 39

Conic section, 240
Conservation of angular momentum, 230
Conservation of linear momentum, 230, 232
Conservative forces, 426
Constraint
  forces, 421
  surface, 422, 424, 464
Constraints, 463
  holonomic, 464
  nonholonomic, 467
Contact forces, 27
Coriolis (Gaspard-Gustave), 539
Coriolis acceleration, 539
Coulomb (Charles Augustin), 31
Coulomb friction, 31
Couple, 27, 63, 268
Cross product, 9
Cylindrical coordinates, 4

D

d'Alembert (Jean Le Rond), 227, 417
d'Alembert's
  principle, 227
  principle: Lagrange's form, 433
Degree of freedom, 69
Differential
  equations, 35
  equations of motion, 28, 70
  of a function, 33
Differential-algebraic equations, 469, 474
Dot product, 8
Dry friction, 31
Dry friction coefficient, 31

Dumbbell, 23
Dyad, 334
Dyadic, 334
Dynamic balancing, 370

E

Eccentricity, 237
Eigenvalues of the inertia
    matrix, 352
Eigenvector of the inertia
    matrix, 352
Ellipsoid of inertia, 544
Elliptic orbit, 241
Equilibrium
    asymptotically stable, 54
    infinitesimally stable, 55
    stable, 53
    unstable, 54
Equilibrium solutions, 14, 38, 54
Equilibrium solutions
    (spring-pendulum), 261
Equivalent force systems, 20
Escape speed, 80, 81, 238
Escape velocity, 238
Euler (Leonhard), x, 417
Euler angles, 394
Euler's
    equation for nonprincipal axes, 359
    equation for principal axes, 359
    equations, 270
Exact differential, 33

F

Field forces, 27
Flight path angle, 244
Force of attraction: spherical
    body, 16
Foucault (Jean-Bernard-Léon), 71
Foucault pendulum, 71
Four-bar mechanism, 181,
    314, 477

Free-body diagram, 27
Friction, 31
Functional, 418

G

Generalized
    conservative forces, 426
    coordinates, 420, 424
    forces, 425
    momentum integral, 442, 445
    nonconservative forces, 426
Generalized forces, 421
Geneva wheel mechanism, 198
Gradient vector, 464
Gravitational force, 1, 19
Gravity gradient moment,
    23, 404
Gyroscope, 371

H

Hamilton (Sir William Rowan), x,
    155, 417
Hamiltonian, 443
    general form, 442
Hamilton's principle, 433
Holonomic constraints, 464
Homogeneous solution, 37
Hooke (Robert), 72
Hyperbolic orbit, 241

I

Ignorable coordinate, 442
Inertia
    dyadic (or tensor), 340
    ellipsoid, 544
    matrix, 340
    tensor (or dyadic), 340
Inertial
    point, 13, 14
    reference frame, 13

Infinitesimal
  quantity, 33
  work, 152
Initial value problem, 441
Instantaneous
  axis of rotation, 333
  center of rotation, 179
  center of zero velocity, 179
Integrals of motion, 151, 441

J

Jacobi (Karl Gustav), 155, 417
Jacobi integral, 443, 446

K

Kepler (Johannes), 15
Kepler's laws, 15
Kinematics, 68
Kinetic
  energy, 153
  energy (general form), 440
  potential, 444

L

Lagrange (Joseph Louis), x, 155, 417
Lagrange multipliers, 468
Lagrange's
  equation, 437
  equation with constraints, 468
  form of d'Alembert's principle, 433
Lagrangian, 442
Libration points, 256
Linear
  differential equations, 37
  vibrations, 39
  viscous damper, 41
  viscous friction, 31, 32

Linearization, 48
  of a differential equation, 54
Linearized (or *small*) oscillations, 264
Lyapunov (Aleksandr Mikhailovich), 55

M

Magnitude of a vector, 3
Mass moment of inertia, 271
Mass particle, 1
Matrix representation of a vector, 335
Measure numbers of a vector, 3
Mechanical energy, 154
Mechanism
  four-bar, 181
  Geneva wheel, 198
  scotch yoke, 218
  slider crank, 208
Moment of a vector, 12
Moment of inertia, 271, 341
Motion coordinates, 3, 69

N

Natural frequencies, 132
Natural systems, 440
Newton (Sir Isaac), x
Newton's law of universal gravitation, 15
Newton's second law, 70
Nonholonomic constraints, 467
Nonconservative forces, 426
Nondimensional damping coefficient, 37
Nondimensional time, 37
Nonlinear, 35
  oscillations, 39
  resonance, 62
  resonance phenomenon, 135
Nutation, 378

Index

O

Orbits
    elliptical, 241
    hyperbolic, 241
    parabolic, 241
Orientation angles, 394
    three-axis, body-fixed angles, 399
    three-axis, space-fixed angles, 396
    two-axis, body-fixed angles, 398
    two-axis, space-fixed angles, 399

P

Parabolic orbit, 241
Parallel axis theorem, 272, 343
Partial derivatives, 33
Particle, 1 (see also *Mass particle*)
Particular solution, 37
Pericenter, 238
Perigee, 238
Perihelium, 238
Perturbation about an equilibrium, 54
Perturbed variables, 261
Phase plane plot, 43
Polar coordinates, 4
Position vector, 2
Potential energy, 154
Precession, 374, 378
Principal
    axes of inertia, 342, 351
    moments of inertia, 342
Principle of virtual work, 433
Products of inertia, 341, 342
Projections of a unit vector, 5

R

Rectangular coordinates, 3
Reduced mass, 234
Reference frame, 13
Restricted circular three-body problem, 251
Restricted three-body problem, 249
Resultant force, 20
Right-hand rule, 9
Rigid body, 264
Rolling
    with slipping, 300
    without slipping, 300

S

Satellite
    dumbbell, 23
    gravity gradient moment, 401
    gravity gradient stabilization, 415
    spin stabilization, 405
Scalar components of a vector, 3
Scalar quantity, 1
Scotch yoke mechanism, 218
Screw driver rule, 10
Sign function, 32
Simplest representation of a vector, 5
Slider crank mechanism, 208
Speed, 2
Spherical body: force of attraction, 16
Spherical coordinates, 3
Spin stabilized satellite, 405
Spring
    extensional, 72
    hardening, 73
    linear, 73
    pendulum, 257
    potential energy, 157
    stiffness (or spring constant), 73
    torsional, 72
Stability test, 39
Standard atmosphere, 261
Static balancing, 371
Superposition principle, 37
Symmetric top, 371

T

Taylor (Brook), 48
Taylor series, 48

Tensor (second-order tensor), 334
Torque, 27
Trace of a matrix, 357
Transformation of triads, 5
Triads, 2
True anomaly, 240
Two-body problem, 232
Two-body problem: automated solution, 246
Two-point boundary value problem, 441

U

Unit dyadic, 334
Unit vector, 2

V

Vacant focus, 241
Vector, 1

Velocity, 2
Virtual
   displacement, 420
   variation, 419
   velocity, 420
   work, 421
Viscous
   damping, 41
   friction, 31, 41

W

Weight, 20
Weightlessness, 20